Surface Area

Right prism: $T = L + 2B$, where $L = hP$

Regular pyramid: $T = L + B$, where $L = \frac{1}{2}\ell P$

Right circular cylinder: $T = 2\pi rh + 2\pi r^2$

Right circular cone: $T = \pi r\ell + \pi r^2$

Sphere: $S = 4\pi r^2$

Volume

Right prism: $V = Bh$

Pyramid: $V = \frac{1}{3}Bh$

Right circular cylinder: $V = \pi r^2 h$

Right circular cone: $V = \frac{1}{3}\pi r^2 h$

Sphere: $V = \frac{4}{3}\pi r^3$

Trigonometry

$\sin\theta = \frac{\text{opposite}}{\text{hypotenuse}}$ $\cos\theta = \frac{\text{adjacent}}{\text{hypotenuse}}$ $\tan\theta = \frac{\text{opposite}}{\text{adjacent}}$

Selected Trigonometric Relationships

$\sin^2\theta + \cos^2\theta = 1$

$A_\triangle = \frac{1}{2}bc\sin\alpha$

$\frac{\sin\alpha}{a} = \frac{\sin\beta}{b} = \frac{\sin\gamma}{c}$

$c^2 = a^2 + b^2 - 2ab\cos\gamma$

π radians $= 180°$

Elementary Geometry for College Students

SECOND EDITION

Elementary Geometry for College Students

SECOND EDITION

DANIEL C. ALEXANDER
Parkland College

GERALYN M. KOEBERLEIN
Mahomet-Seymour High School

Houghton Mifflin Company

Boston New York

To my wife, Mary; and my children, Matthew, Phillip, and Sarah Alexander

To my husband, Richard; and family members Alice Koeberlein Sexton; Jamie and Morgan Sexton; and Laura Koeberlein

We appreciate your patience during the time that we spent revising the first edition of this text.

Senior Sponsoring Editor: Maureen O'Connor
Senior Associate Editor: Dawn Nuttall
Editorial Assistant: John Brister
Project Editor: Tamela Ambush
Editorial Assistant: Ryan Jones
Production/Design Coordinator: Jennifer Meyer Dare
Manufacturing Manager: Florence Cadran
Marketing Manager: Ros Kane

Cover design: Harold Burch Designs, NYC

Cover image: Atelier Kim Zwarts, Atelier Studios, Netherlands.

PHOTO CREDITS:

Chapter 1: Escher print: M.C. Escher's "Ascending and Descending," © 1998, Cordon Art B.V.—Baarn—Holland. All rights reserved; **Chapter 2:** vinyl siding: Michael Rizza/The Picture Cube; **Chapter 3:** roof truss: McLaughlin/The Image Works; **Chapter 4:** contemporary house: Stock Boston; **Chapter 5:** framed print: Corbis Bettmann; **Chapter 6:** engine: M. Antman/The Image Works; **Chapter 8:** building complex: Jim Sulley/The Image Works; **Chapter 10:** surveyor: Tony Stone Worldwide

Printed in the U.S.A.

Library of Congress Catalog Card Number: 98-71972
ISBN: 0-395-87055-0
3456789-QF-02-01-00-99

Contents

Preface

The second edition of *Elementary Geometry for College Students* was written for students who have not completed a course in geometry or for those who need to take a fresh look at geometry. The students should have some background in elementary algebra. The students will learn and apply the principles of geometry as well as recognize their relevance to the real world. In that many students will use this textbook to prepare for the further study of mathematics, we have completed a thorough and rich textbook.

Authors' Philosophy and Approach

The authors' philosophy toward the teaching of geometry has been the driving force in both the development and revision of this work. We believe the complete development of geometry begins with an idea, followed by examination and development of a theory, verification of the theory through deduction, and the application of resulting principles in the real world. Our approach to college geometry is largely visual, as it should be.

We present material in the same way that we present it in the classroom. We explain what we are doing, where we are going, and demonstrate relationships between topics. We use many paragraph proofs that are common at the college level. This method of proof can improve the student's writing style in that each paragraph must be ordered and justified.

This textbook, which parallels the goals of secondary level geometry textbooks, is heavily influenced by the standards recommended by both the National Council of Teachers of Mathematics (NCTM) and the American Mathematical Association of Two Year Colleges (AMATYC). For those interested in building a solid foundation in geometry, we believe that we have the most comprehensive textbook written for the college level. The content is both suitable for the student of geometry and for the future teacher of its topics. Furthermore, the first edition of this textbook was successfully used in geometry classes taught at the secondary level.

Goals of the Textbook

Specific outcomes for the student using this textbook are:

- Presentation of geometry essentials for the student who will use geometry in a vocation
- Preparation of the transfer student for further study of mathematics and related disciplines
- Exposing students to the step-by-step development of a logical mathematical system
- Providing discovery activities and exercises to maintain and advance student interest

New to the Second Edition

Content

- Separation of Chapter 3 of the first edition into two chapters to increase the amount of attention given to triangles and quadrilaterals
- Division of Section 6.5 of the first edition in order to separate the discussion of "locus of points" and "concurrence of lines"
- Division of Section 7.4 of the first edition into two sections, to make it more accessible
- Reordering Chapters 8 and 9 (solid geometry now before analytic geometry)
- Addition of appendices that focus upon the fundamentals of logic

- Movement of much of the algebra review into an appendix
- Rewriting selected pieces of the first edition for the purpose of clarification

Exercises

- An increase of approximately 50% in the number of applications in the textbook
- Expanded exercise sets, taking students from fundamental activities to a higher level of thinking

Features

- Increased attention to the inductive approach as evidenced by numerous **"Discover!"** activities
- Inclusion of new student-interest margin notes, **Geometry in Nature** and **Geometry in the Real World**
- Inclusion of new **tables that summarize properties** of geometric figures (such as triangles)
- Addition of a **glossary of terms** to the text
- Addition of an **Index of Applications**

Features Maintained for the Second Edition

Many of the popular features of the first edition were kept, including the following:

- Interesting **chapter openers** used to introduce the principal notion of the chapter
- A **Look Beyond** section completes each chapter, providing sketches that are interesting, sometimes historical, and always informative
- A comprehensive **chapter summary,** which reviews the chapter, previews the following chapter, and provides a list of the most important concepts of the chapter
- A **chapter review** that provides numerous review problems for the chapter
- Use of a second **color** to emphasize and highlight important features of the textbook
- **Challenge exercises** are indicated by a ➢ next to a problem number
- **Warnings,** placed in the left margin, are provided as needed
- References to **calculator** usage are made as needed

Supplements for the Instructor

Instructor Resource Manual with Solutions Manual The Instructor's Resource Manual provides suggestions for order and topics for the course, transparency masters, as well as suggestions for teaching each topic. There are also chapter tests and a quiz bank available as well as a complete Solutions Manual.

Computerized Test Generator The Computerized Test Generator is the electronic version of the quiz bank in the Instructor's Resource Manual. This user-friendly software permits an instructor to construct customized tests from the 500 items offered. **On-line testing** and **gradebook** functions are also provided. It is available for Windows for the IBM PC and compatible computers.

Supplements for the Student

Student Study Guide A Student Study Guide with suggestions for success in the study of geometry is available. This tool also provides a partial solutions manual for the student. The intent of the partial solutions manual is to provide guidance; it is not intended to complete the student's assignments.

Acknowledgments

There are many at Houghton Mifflin that we wish to thank. We are especially pleased that Maureen O'Connor, Senior Sponsoring Editor, recognized the quality of our first edition. We want to thank Dawn Nuttall, Senior Associate Editor, for her encouragement, guidance, and assistance in the completion of this textbook. Thanks must also go to Tamela Ambush, Project Editor, for her help and patience in taking this text through the production process. And thanks to Florence Powers, Senior Sales Representative, for her willingness to bring our book to the attention of Houghton Mifflin. Of course, there are many others at Houghton Mifflin who have contributed from behind the scenes and we thank you as well for your efforts.

Some who helped with the first edition and to whom we remain indebted are Theresa Grutz and Beth Dahlke. We would like to thank reviewers of the first edition, including: **Jane C. Beatie,** *University of*

South Carolina at Aiken; **Steven Blasberg,** *West Valley College;* **Patricia Clark,** *Indiana State University;* **George L. Holoway,** *Los Angeles Valley College;* **Tracy Hoy,** *College of Lake County;* **Josephine G. Lane,** *Eastern Kentucky University;* **James R. McKinney,** *Cal Poly at Pomona;* **Maurice Ngo,** *Chabot College;* **Ellen L. Rebold,** *Brookdale Community College;* and **Karen R. Swick,** *Palm Beach Atlantic College.*

We are most grateful to our second edition reviewers: **Paul Allen,** *University of Alabama;* **Barbara Brown,** *Anoka Ramsey Community College;* **Joyce Cutler,** *Framingham State College;* **Walter Czarnec,** *Framingham State College;* **Zoltan Fischer,** *Minneapolis Community and Technical College;* **Chris Graham,** *Mt. San Antonio Community College;* **Geoff Hagopian,** *College of the Desert;* **Edith Hays,** *Texas Woman's University;* **George L. Holloway,** *Los Angeles Valley College;* **John C. Longnecker,** *University of Northern Iowa;* **Nicholas Martin,** *Shepherd College;* **Jill McKenney,** *Lane Community College;* **James R. McKinney,** *Cal Poly at Pomona;* **Lauri Semarne; Joseph F. Stokes,** *Western Kentucky University;* **Steven L. Thomassin,** *Ventura College.*

Daniel C. Alexander
Geralyn M. Koeberlein

Foreword

The topics that comprise a basic course in plane geometry are found in Chapters 1–7. To enhance that material, the final chapters include:

Chapter 8: Solid Geometry
Chapter 9: Analytic Geometry
Chapter 10: Trigonometry

The order in which the chapters of the book can be studied is depicted in the following flow chart. It may be necessary to exclude parts of sections if a nonstandard sequence is chosen.

$$\begin{array}{c} \qquad\qquad\qquad\qquad\qquad\qquad\quad 8 \\ \qquad\qquad\qquad\qquad\qquad\qquad\quad \uparrow \\ 1 \to 2 \to 3 \to 4 \to 5 \to 6 \to 7 \to 9 \\ \qquad\qquad\qquad\qquad\qquad\qquad\quad \downarrow \\ \qquad\qquad\qquad\qquad\qquad\qquad\quad 10 \end{array}$$

For students who wish to review solving quadratic equations, Appendix A (Quadratic Equations) reviews factoring, the Square Roots method, and the Quadratic Formula.

For students who want a more complete background regarding the elements of logic, information can be found in Appendix B (Statements and Truth Tables) and Appendix C (Valid Arguments).

For a minimal course in geometry, some sections that can be treated as optional include the following:

Section 3.4 (Basic Constructions Justified)
Section 3.5 (Inequalities in a Triangle)
Section 5.5 (Segments Divided Proportionally)
Section 6.4 (Some Constructions and Inequalities for the Circle)
Section 7.5 (More Area Relationships in the Circle)

Daniel C. Alexander
Geralyn M. Koeberlein

Index of Applications

Chapter 1
Line and Angle Relationships

In geometry, figures can be drawn that create an illusion. For instance, our powers of reasoning suggest that something is wrong with the staircases in this Escher print. This chapter opens with a discussion of statements and the types of reasoning used in geometry. Section 1.2 focuses on the tools of geometry, such as the ruler and protractor. The remainder of the chapter begins the formal development of geometry by considering line and angle relationships. For any student who needs an algebra refresher or an introduction to logic, selected topics are conveniently found in the appendices. Other techniques from algebra are reviewed or developed in the textbook as needed.

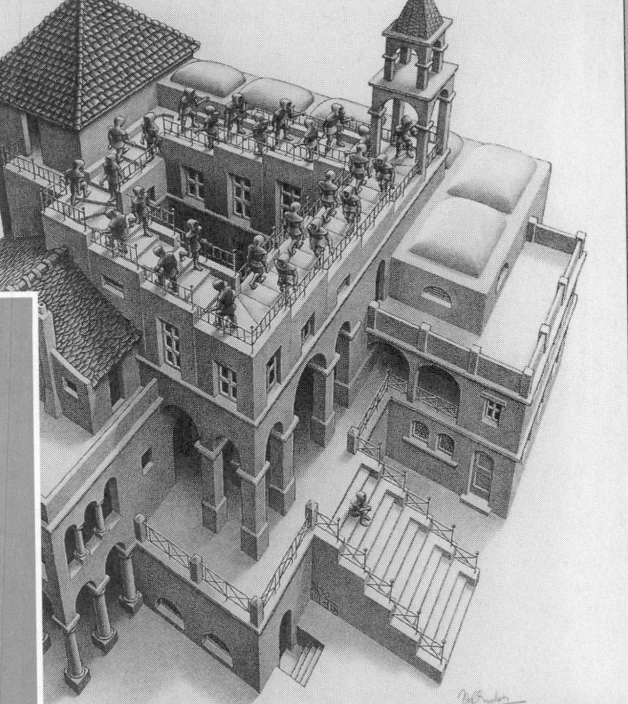

CHAPTER OUTLINE

1.1 Statements and Reasoning

A **statement** is a group of words and symbols that can be classified collectively as true or false. Some statements are classified as simple while others are classified as compound.

EXAMPLE 1

Classify each of the following as a true statement, a false statement, or neither.

1. $4 + 3 = 7$
2. An angle has two sides. (See Figure 1.1.)
3. Robert E. Lee played shortstop for the Yankees.
4. $7 < 3$ (read "7 is less than 3.")
5. Look out!

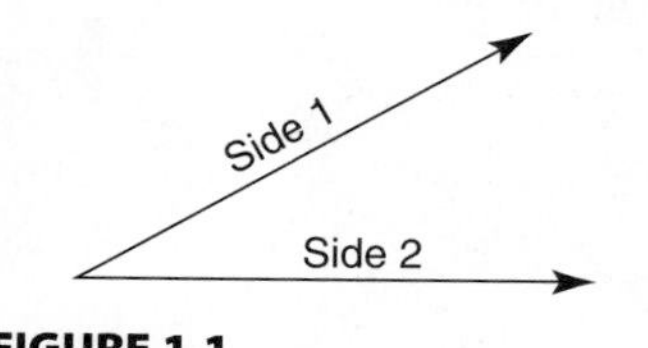

FIGURE 1.1

Solution 1 and 2 are true statements; 3 and 4 are false statements; 5 is not a statement.

Some statements contain one or more *variables*. As in algebra, a **variable** is a letter that represents a number. The claim "$x + 5 = 6$" is called an *open sentence* or *open statement* because it can be classified as true or false depending on the replacement value for the variable x. For instance, $x + 5 = 6$ is true if $x = 1$; for any other x, $x + 5 = 6$ is false. Some statements containing variables are classified as true statements because they are true for all replacements. Consider the Commutative Property of Addition, usually stated in the form $a + b = b + a$. In words, this property states that the same result is obtained when two numbers are added in either order; for instance, when $a = 4$ and $b = 7$, it follows that $4 + 7 = 7 + 4$.

Sometimes we form a statement by using other statements as "building blocks." In such cases, we may use letters such as P and Q to represent simple statements. For example, the letter P may refer to the statement "$4 + 3 = 7$" while the letter Q refers to "Babe Ruth was a U. S. President." The statement "$4 + 3 = 7$ *and* Babe Ruth was a U. S. President" has the form P *and* Q and is known as the **conjunction** of P and Q. The statement "$4 + 3 = 7$ *or* Babe Ruth was a U. S. President" has the form P *or* Q and is known as the **disjunction** of P and Q. A conjunction is true only when P and Q are *both* true. A disjunction is false only when P and Q are *both* false.

The **negation** of a given statement P makes a claim opposite that of the original statement. If the given statement is true, its negation is false, and vice versa. If P is a statement, we use $\sim P$ (read "not P") to indicate its negation.

EXAMPLE 2

Give the negation of each statement.

a) $4 + 3 = 7$ **b)** All fish can swim.

Solution **a)** $4 + 3 \neq 7$ ($\neq$ means "is not equal to.")

b) Some fish cannot swim. (To negate "All fish can swim," we say that at least one fish cannot swim.)

The statement "If P, then Q", which is known as a **conditional statement** (or **implication**), is classified as true or false as a whole. A statement of this form can be written in equivalent forms; for instance, the conditional statement "If an angle is a right angle, then it measures 90 degrees" is equivalent to the statement "All right angles measure 90 degrees."

EXAMPLE 3

Classify each conditional statement as true or false.

1. If an animal is a fish, then it can swim. (states, "All fish can swim.")
2. If two sides of a triangle are equal in length, then two angles of the triangle are equal in measure. (See Figure 1.2.)

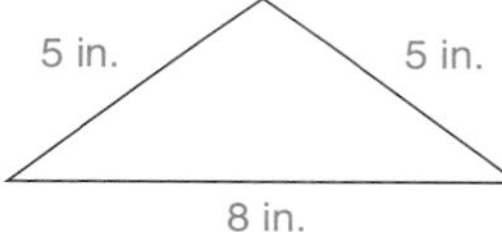

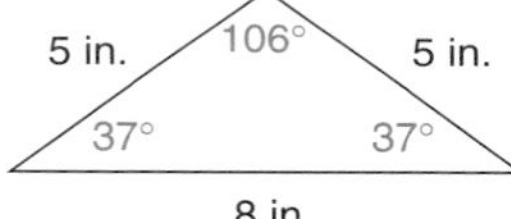

FIGURE 1.2

3. If Wendell studies, then he will receive an A on the test.

Solution Statements 1 and 2 are true. Statement 3 is false.

In the conditional statement "If P, then Q," P is the ***hypothesis*** and Q is the ***conclusion.*** In Example 3, statement 2, we have

Hypothesis: Two sides of a triangle are equal in length.

Conclusion: Two angles of the triangle are equal in measure.

For the true statement "If P, then Q," the hypothetical situation described in P implies the conclusion described in Q. This type of statement suggests some form of reasoning, so we turn our attention to this matter.

Reasoning

Learning geometry requires time, vocabulary development, attention to detail and order, supporting claims, and a lot of thinking. The following types of thinking or reasoning are used to develop mathematical principles.

1. Intuition	An inspiration leading to the statement of a theory
2. Induction	An organized effort to test the theory
3. Deduction	A formal argument that proves the tested theory

We are often inspired to think and say, "It occurs to me that . . . ." With **intuition,** a sudden insight allows one to make a statement without applying any formal reasoning. When intuition is used, we sometimes err by "jumping" to conclusions. In a cartoon, the character having the "bright idea" (using intuition) is shown with a light bulb next to her or his head.

EXAMPLE 4

Figure 1.3 is called a *regular pentagon* because its five sides have equal lengths and its angles have equal measures. What do you suspect is true of the lengths of the dashed parts of lines from B to E and from B to D?

Solution Using intuition, the lengths of the dashed parts of lines (known as *diagonals* of the pentagon) are the same.

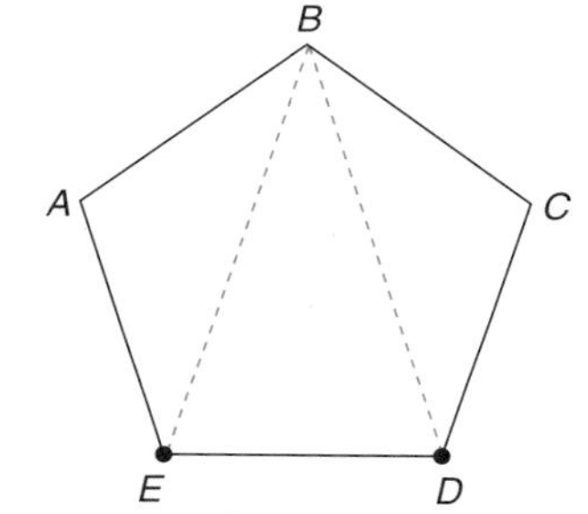

FIGURE 1.3

NOTE 1: A *ruler* can be used to verify that this claim is true. We will discuss measurement with the ruler in more detail in Section 1.2.

NOTE 2: Using methods found in Chapter 3, we could use deduction to prove that the two diagonals do indeed have the same length.

The role intuition plays in formulating mathematical thoughts is truly significant. But to have an idea is not enough! Testing a theory may lead to a revision or even total rejection of the theory. If a theory stands up to testing, it moves one step closer to becoming mathematical law.

We often use specific observations and experiments to draw a general conclusion. This type of reasoning is called **induction.** As you would expect, the observation/experimentation process is common in laboratory and clinical settings. Chemists, physicists, doctors, psychologists, weather forecasters, and many others use collected data as a basis for drawing conclusions . . . and so will we!

EXAMPLE 5

While in a grocery store, you examine several 8-oz cartons of yogurt. Although the flavors and brands differ, each carton is priced at 75 cents. What do you conclude?

Conclusion Every 8-oz carton of yogurt in the store costs 75 cents.

As you may already know (see Figure 1.2), a figure with three straight sides is called a *triangle.*

EXAMPLE 6

In a geometry class, you have been asked to measure the three interior angles of each triangle in Figure 1.4. You discover that triangles I, II, and IV have two angles (as marked) that have equal measures. What may you conclude?

Conclusion The triangles that have two sides of equal length also have two angles of equal measure.

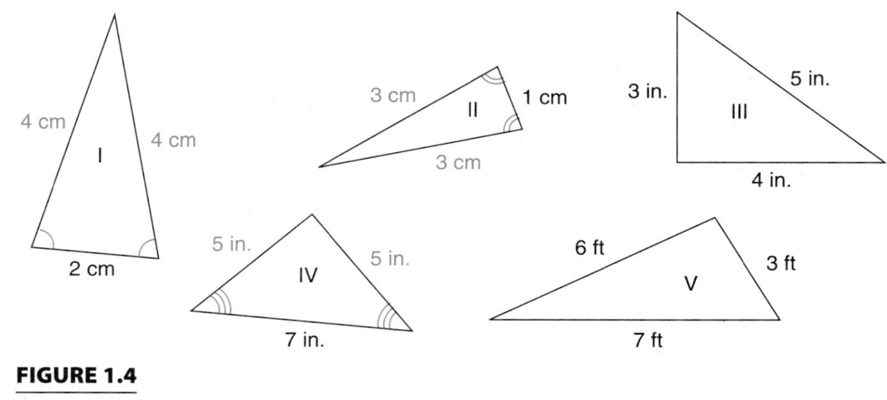

FIGURE 1.4

NOTE: A *protractor* can be used to support the conclusion found in Example 6. We will discuss the protractor in Section 1.2.

DEFINITION: **Deduction** is the type of reasoning in which the knowledge and acceptance of selected assumptions guarantees the truth of a particular conclusion.

A Valid Argument

In Example 7, we will illustrate the form of deductive reasoning used most frequently in the development of geometry. In this form, known as a **valid argument,** at least two statements are treated as facts; these assumptions are called the *premises* of the argument. Based upon the premises, a particular *conclusion* must follow. This form of deduction is called the Law of Detachment; for further information, see Appendices B and C.

EXAMPLE 7

If you accept the following statements 1 and 2 as true, what must you conclude?

1. If a student plays on the Rockville High School boys' varsity basketball team, then he is a talented athlete.
2. Todd plays on the Rockville High School boys' varsity basketball team.

Conclusion Todd is a talented athlete.

To make it easier to recognize this pattern for deductive reasoning, we use letters to represent statements in the following generalization.

LAW OF DETACHMENT

Let P and Q represent simple statements, and assume that statements 1 and 2 are true. Then a valid argument having conclusion C has the form

1. If P, then Q } premises
2. P

C. $\therefore Q$ } conclusion

NOTE: The symbol $\therefore$ means "therefore."

In the preceding form, the statement "If P, then Q" is often read "P implies Q." That is, when P is known to be true, Q must follow.

EXAMPLE 8

Is the following argument valid? Assume that premises 1 and 2 and true.

1. If it is raining, then Tim will stay in the house.
2. It is raining.

C. $\therefore$ Tim will stay in the house.

Conclusion The argument is valid because the form of the argument is

1. If P, then Q
2. P

C. $\therefore Q$

with P = "It is raining," and Q = "Tim will stay in the house."

EXAMPLE 9

Is the following argument valid? Assume that premises 1 and 2 are true.

1. If a man lives in London, then he lives in England.
2. William lives in England.

C. $\therefore$ William lives in London.

Conclusion The argument is not valid. Here, P = "A man lives in London," and Q = "A man lives in England." Thus the form of this argument is

1. If P, then Q
2. Q

C. $\therefore P$

But the Law of Detachment does not handle the question, "If Q, then what?" Even though statement Q is true, it does not enable us to draw a valid conclusion about P. Of course, if William lives in England, he *might* live in London; but he might instead live in Liverpool, Manchester, Coventry, or any of countless other places in England. Each of these possibilities is a **counterexample** disproving the validity of the argument. Remember that deductive reasoning is concerned with reaching conclusions that *must be true,* given the truth of the premises.

Warning

In the box, the argument on the left is valid and patterned after Example 8. The argument on the right is invalid; this form was given in Example 9.

VALID ARGUMENT	INVALID ARGUMENT
1. If P, then Q	1. If P, then Q
2. P	2. Q
C. $\therefore Q$	C. $\therefore P$

We will use deductive reasoning throughout our work in geometry. Suppose that you know these two facts:

1. If an angle is a right angle, then it measures 90°.
2. Angle A is a right angle.

Then you may conclude

C. Angle A measures 90°.

1.1 Exercises

In Exercises 1 and 2, which sentences are statements? If a sentence is a statement, classify it as true or false.

1. a) Where do you live?

b) $4 + 7 \neq 5$.

c) Washington was the first U.S. president.

d) $x + 3 = 7$ when $x = 5$.

2. a) Chicago is located in the state of Illinois.

b) Get out of here!

c) $x < 6$ (read as "x is less than 6") when $x = 10$.

d) Babe Ruth is remembered as a great football player.

In Exercises 3 and 4, give the negation of each statement.

3. a) Christopher Columbus crossed the Atlantic Ocean.

b) All jokes are funny.

4. a) No one likes me.

b) Angle 1 is a right angle.

In Exercises 5 to 10, classify each statement as simple, conditional, a conjunction, or a disjunction.

5. If Alice plays, the volleyball team will win.

6. Alice played and the team won.

7. The first-place trophy is beautiful.

8. An integer is odd or it is even.

9. Matthew is playing shortstop.

10. You will be in trouble if you don't change your ways.

In Exercises 11 to 18, state the hypothesis and the conclusion of each statement.

11. If you go to the game, then you will have a great time.

12. If two chords of a circle have equal lengths, then the arcs of the chords are congruent.

13. If the diagonals of a parallelogram are perpendicular, then the parallelogram is a rhombus.

14. If $\frac{a}{b} = \frac{c}{d}$ where $b \neq 0$ and $d \neq 0$, then $a \cdot d = b \cdot c$.

15. Corresponding angles are congruent if two parallel lines are cut by a transversal.
16. Vertical angles are congruent when two lines intersect.
17. All squares are rectangles.
18. Base angles of an isosceles triangle are congruent.

In Exercises 19 to 24, classify each statement as true or false.

19. If a number is divisible by 6, then it is divisible by 3.
20. Rain is wet and snow is cold.
21. Rain is wet or snow is cold.
22. If Jim lives in Idaho, then he lives in Boise.
23. Triangles are round or circles are square.
24. Triangles are square or circles are round.

In Exercises 25 to 32, name the type of reasoning (if any) used.

25. While participating in an Easter egg hunt, Sarah notices that each of the seven eggs she has found are numbered. Sarah concludes that all eggs used for the hunt are numbered.
26. You walk into your geometry class, look at the teacher, and conclude that you will have a quiz today.
27. Albert knows the rule "If a number is added to each side of an equation, then the new equation has the same solution set as the given equation." Given the equation $x - 5 = 7$, Albert concludes that $x = 12$.
28. You believe that "Anyone who plays major league baseball is a talented athlete." Knowing that Duane Gibson has just been called up to the major leagues, you conclude that Duane Gibson is a talented athlete.
29. As a handcuffed man is brought into the police station, you glance at him and say to your friend, "That fellow looks guilty to me."
30. While judging a science fair project, Mr. Cange finds that each of the first 5 projects is outstanding and concludes that all 10 will be outstanding.
31. You know the rule "If a person lives in the Santa Rosa Junior College district, then he or she will receive a tuition break at Santa Rosa." Candace tells you that she has received a tuition break. You conclude that she resides in the Santa Rosa Junior College district.
32. As Mrs. Gibson enters the doctor's waiting room, she concludes that it will be a long wait.

In Exercises 33 to 36, use intuition to state a conclusion.

33. You are told that the opposite angles formed when two lines cross are **vertical angles.** In the figure, angles 1 and 2 are vertical angles. Conclusion?

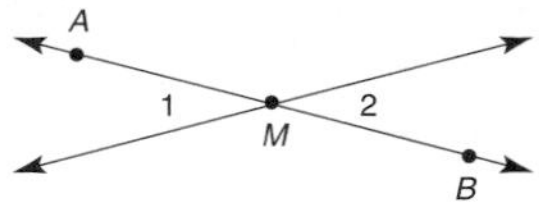

Exercises 33, 34

34. In the figure, point M is called the **midpoint** of line segment AB. Conclusion?
35. The two triangles shown are **similar** to each other. Conclusion?

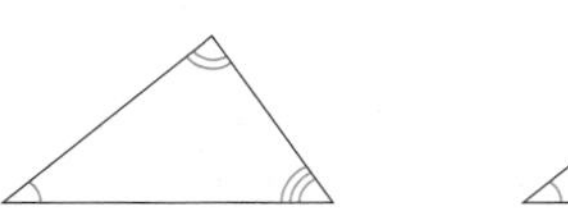
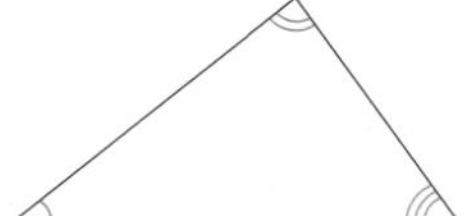

36. Observe (but do not measure) the following angles. Conclusion?

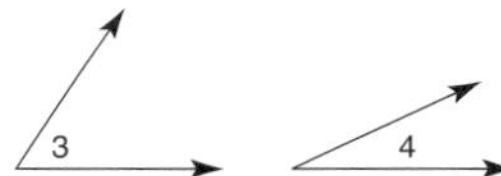

In Exercises 37 to 40, use induction to state a conclusion.

37. Several movies directed by Lawrence Garrison have won Academy Awards, while many others have received nominations. His latest work, *A Prisoner of Society,* is to be released next week. Conclusion?
38. On Monday, Matt says to you, "Andy hit his little sister at school today." On Tuesday, Matt informs you, "Andy threw his math book into the wastebasket during class." On Wednesday, Matt tells you, "Because Andy was throwing peas in the school cafeteria, he was sent to the principal's office." Conclusion?
39. While searching for a classroom, Tom stopped at an instructor's office to ask directions. On the office bookshelves are books titled *Intermediate Algebra, Calculus, Modern Geometry, Linear Algebra,* and *Differential Equations.* Conclusion?
40. At a friend's house, you see several food items, including apples, pears, grapes, oranges, and bananas. Conclusion?

In Exercises 41 to 50, use deduction to state a conclusion, if possible.

41. If the sum of the measures of two angles is 90°, then these angles are called "complementary." Angle 1 measures 27° and angle 2 measures 63°. Conclusion?
42. If a person attends college, then he or she will be a suc-

cess in life. Kathy Jones attends Dade County Community College. Conclusion?

43. All mathematics teachers have a strange sense of humor. Alex is a mathematics teacher. Conclusion?

44. All mathematics teachers have a strange sense of humor. Alex has a strange sense of humor. Conclusion?

45. If Stewart Powers is elected president, then every family will have an automobile. Every family has an automobile. Conclusion?

46. If Tabby is meowing, then she is hungry. Tabby is hungry. Conclusion?

47. If a person is involved in politics, then that person will be in the public eye. June Jesse has been elected to the Missouri state senate. Conclusion?

48. If a student is enrolled in a literature course, then he or she will work very hard. Bram Spiegel digs ditches by hand 6 days a week. Conclusion?

49. If a person is rich and famous, then he or she is happy. Marilyn is wealthy as well as well-known. Conclusion?

50. If you study hard and hire a tutor, then you will make an A in this course. You make an A in this course. Conclusion?

1.2 Informal Geometry and Measurement

FIGURE 1.5

In geometry, the terms *point, line,* and *plane* are described but not defined. Other concepts that are accepted intuitively, but never defined, include the *straightness* of a line, the *flatness* of a plane, the notion that a point on a line lies *between* two other points on the line, and the notion that a point lies in the *interior* or *exterior* of an angle. Some of the terms found in this section are formally defined in later sections of Chapter 1. The following are descriptions of some of the undefined terms.

A **point,** which is represented by a dot, has location but not size; that is, a point has no dimensions. An uppercase italic letter is used to name a point. Figure 1.5 shows points *A, B,* and *C.* ("Point" may be abbreviated "pt." for convenience.)

The second undefined term is **line.** Lines have a quality of "straightness" that is not defined but assumed. Given several points on a line, these points form a straight path. Whereas a point has no dimensions, a line is one-dimensional; that is, the distance between any two points on a given line can be measured. Line *AB,* represented symbolically by $\overleftrightarrow{AB}$, extends infinitely far in opposite directions, as suggested by the arrows on the line. A line may also be represented by a single lowercase letter. Figures 1.6(a) and (b) show the lines *AB* and *m.* When a lowercase letter is used to name a line, the line symbol is omitted.

Note the position of point *X* on $\overleftrightarrow{AB}$ in Figure 1.6(c). When three points such as *A, X,* and *B* are on the same line, they are said to be **collinear.** In the order shown, which is symbolized *A-X-B,* point *X* is said to be *between A* and *B.*

When no drawing is provided, the notation *A-B-C* means that these points are collinear, with *B between A* and *C.* When a drawing is provided, we assume that all points in the drawing that appear to be collinear are collinear, *unless otherwise stated.* Figure 1.6(d) shows that *A, B,* and *C* are collinear, with *B* between *A* and *C.*

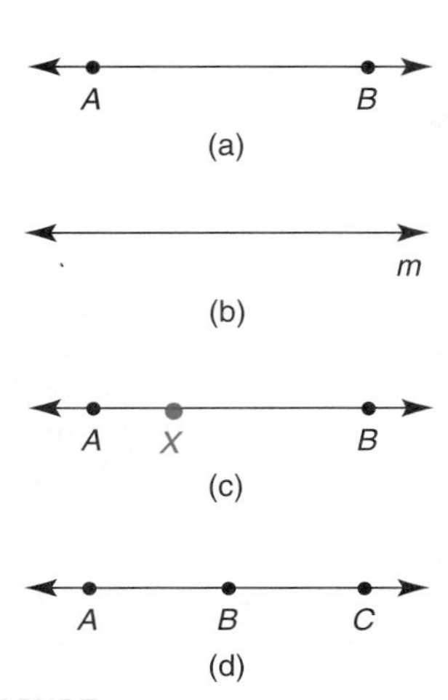

FIGURE 1.6

At this time, we will also informally introduce some terms that will be formally defined later. You have probably encountered the terms *angle*, *triangle*, and *rectangle* many times. An example of each is shown in Figure 1.7.

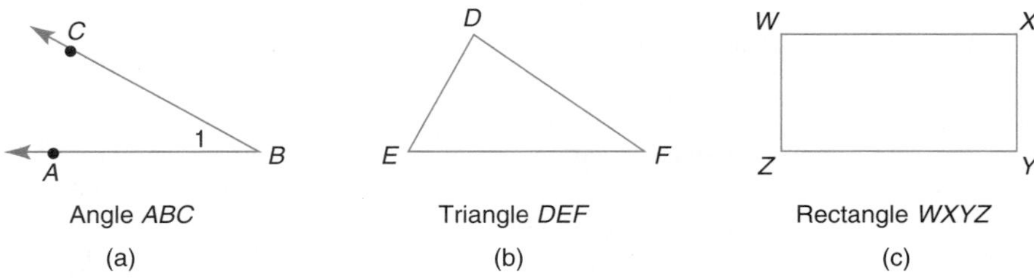

FIGURE 1.7

Using symbols and abbreviations, we refer to Figures 1.7(a), (b), and (c) as $\angle ABC$, $\triangle DEF$, and rect. *WXYZ*, respectively. Some caution must be used when naming figures; for instance, it is incorrect to describe the angle in Figure 1.7(a) as $\angle ACB$ because that order implies a path from point *A* to point *C* to point *B* . . . a different angle! In $\angle ABC$, the point *B* at which the sides meet is called the **vertex** of the angle. Because there is no confusion regarding the angle described, $\angle ABC$ is also known as $\angle B$ (using only the vertex) or as $\angle 1$. The points *D*, *E*, and *F* at which the sides of $\triangle DEF$ meet are called the *vertices* (plural of vertex) of the triangle. Similarly, *W*, *X*, *Y*, and *Z* are the vertices of the rectangle.

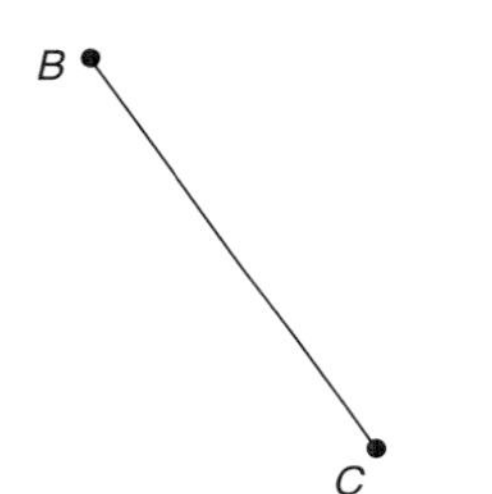

FIGURE 1.8

A **line segment** is part of a line. It consists of two distinct points on the line and all points between them. (See Figure 1.8.) Using symbols, the line segment is indicated by $\overline{BC}$; note that $\overline{BC}$ is a set of points but is not a number. We use *BC* (omitting the segment symbol) to indicate the *length* of this line segment; thus, *BC* is a number. The sides of a triangle or rectangle are line segments. The vertices of a rectangle are named in an order that identifies its line segment sides.

EXAMPLE 1

Can the rectangle in Figure 1.7(c) be named a) *XYZW*? b) *WYXZ*?

Solution

a) Yes, since the points taken in this order trace the figure.
b) No; for example, $\overline{WY}$ is not a side of the rectangle.

Measuring Line Segments

The instrument used to measure a line segment is a scaled straightedge such as a *ruler*, a *yardstick*, or a *meter stick*. Generally, we place the "0-point" of the ruler at one end of the line segment and find the numerical length as the number at the other end. Line segment *RS* ($\overline{RS}$ in symbols) in Figure 1.9 measures 5 centimeters. Because we express the length of $\overline{RS}$ by *RS* (with no bar), we write $RS = 5$ cm.

Because manufactured measuring devices such as the ruler, yardstick, or meter stick may lack perfection or be misread, there is a margin of error each time one is used. In Figure 1.9, for instance, *RS* may actually measure 5.02 cm (and that could be rounded from 5.023 cm, etc.).

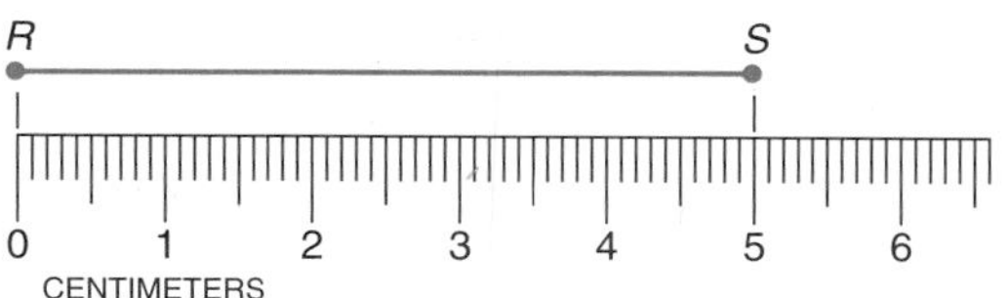

FIGURE 1.9

In Example 2, a ruler (not drawn to scale) is shown in Figure 1.10. In the drawing, the distance between consecutive marks on the ruler corresponds to one inch.

EXAMPLE 2

In rectangle $ABCD$ of Figure 1.10, the line segments $\overline{AC}$ and $\overline{BD}$ shown are the diagonals of the rectangle. How do the lengths of the diagonals compare?

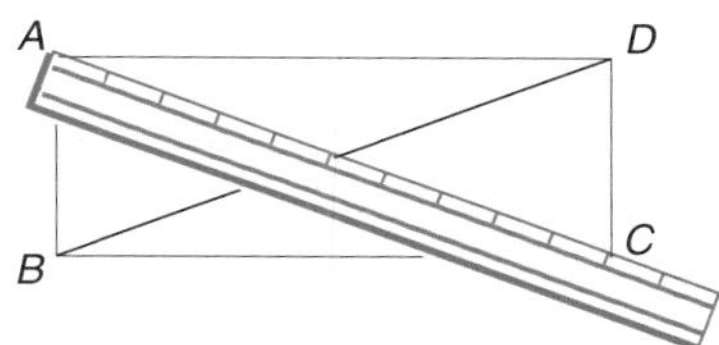

FIGURE 1.10

Solution As intuition suggests, the lengths of the diagonals are the same. As shown, $AC = 10''$ and $BD = 10''$.

NOTE: $10''$ means 10 inches while $10'$ means 10 feet.

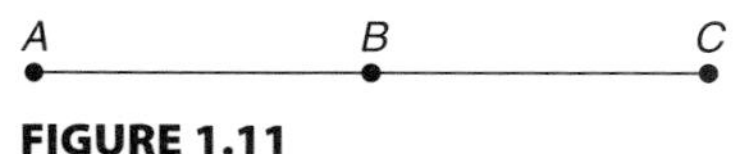

FIGURE 1.11

In Figure 1.11, point B lies **between** A and C on $\overline{AC}$. If $AB = BC$, then B is the **midpoint** of $\overline{AC}$. When $AB = BC$, the geometric figures $\overline{AB}$ and $\overline{BC}$ are said to be **congruent.** Numerical lengths may be equal, but the actual line segments (geometric figures) are congruent. The symbol for congruence is $\cong$; thus $\overline{AB} \cong \overline{BC}$ if B is the midpoint of $\overline{AC}$. Example 3 emphasizes the relationship between $\overline{AB}$, $\overline{BC}$, and $\overline{AC}$ when B lies between A and C.

EXAMPLE 3

In Figure 1.12, the lengths of $\overline{AB}$ and $\overline{BC}$ are $AB = 4$ and $BC = 8$. What is AC, the length of $\overline{AC}$?

FIGURE 1.12

Solution As intuition suggests, the length of $\overline{AC}$ equals $AB + BC$. Thus $AC = 4 + 8 = 12$.

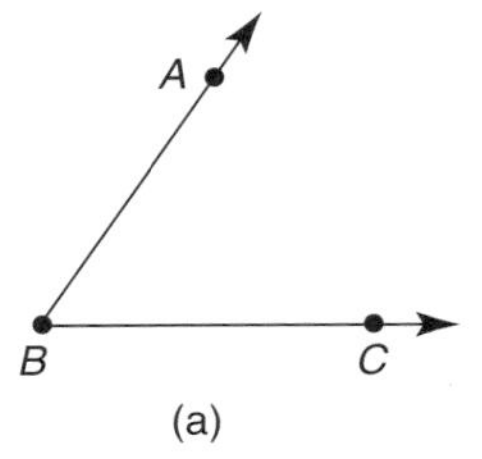

(a)

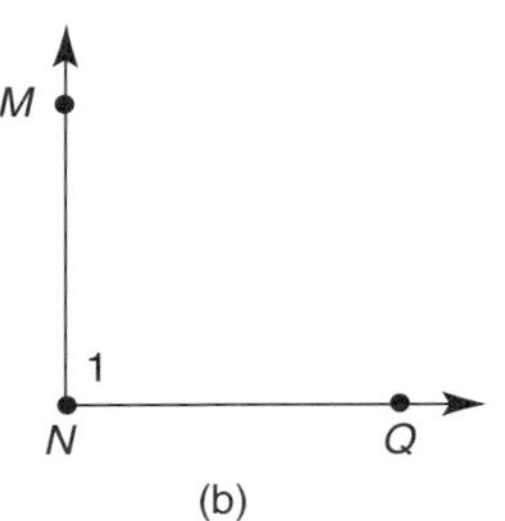

(b)

FIGURE 1.13

Measuring Angles

Although we formally define an angle in Section 1.4, we consider it intuitively at this time.

An angle's measure does not depend upon the length of its sides, but upon the amount of opening between its sides. In Figure 1.13, the arrows on the angles' sides suggest that the sides extend indefinitely.

You cannot measure an angle with a ruler! The instrument shown in Figure 1.14 (and used in the measurement of angles) is a **protractor.** For example, you would express the measure of $\angle RST$ by writing $m\angle RST = 50°$. When a lowercase m is used before the angle symbol $\angle$, it means the measure of the angle (in degrees). Measuring the angles in Figure 1.13 with a protractor, we would find that $m\angle B = 55°$ and $m\angle 1 = 90°$. Even if the degree symbol is missing, the measure is understood to be in degrees; thus, $m\angle 1 = 90$.

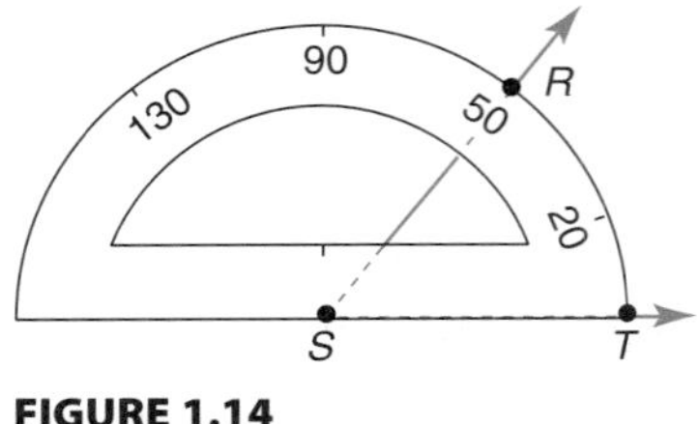

FIGURE 1.14

In practice, the protractor will show that the measure of an angle must be greater than 0° but less than or equal to 180°. To measure an angle with a protractor, the steps are:

1. Place the notch of the protractor at the point where the sides of the angle meet (the vertex). See point S in Figure 1.15.
2. Place the edge of the protractor along a side of the angle so that the scale reads "0." See point T in Figure 1.15.
3. Read the angle size by reading the degree measure that corresponds to the second side of the angle. CAUTION: Many protractors show dual scales. See point R in Figure 1.15.

EXAMPLE 4

For Figure 1.15, find the measure of $\angle RST$.

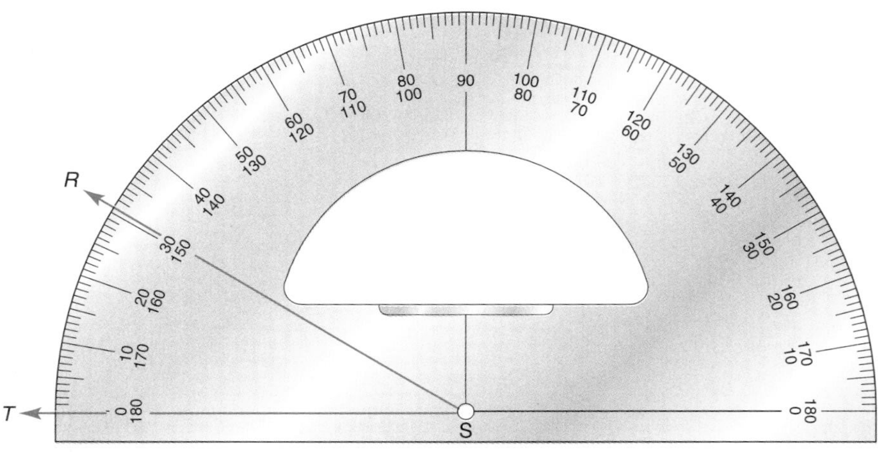

FIGURE 1.15

Solution Using the protractor, we find that the measure of angle RST is 31°. (In symbols, $m\angle RST = 31°$ or $m\angle RST = 31$.)

As with a ruler, measurement with a protractor will not be perfect.

The lines of a notebook are *parallel.* Informally, **parallel** lines won't cross over each other even if they are extended indefinitely. In Figure 1.16(a), we say that lines ℓ and m are parallel; notice here the use of a lowercase letter to name a line. We may say that line segments are parallel if they are parts of parallel lines; thus, $\overline{RS}$ is parallel to $\overline{MN}$ in Figure 1.16(b).

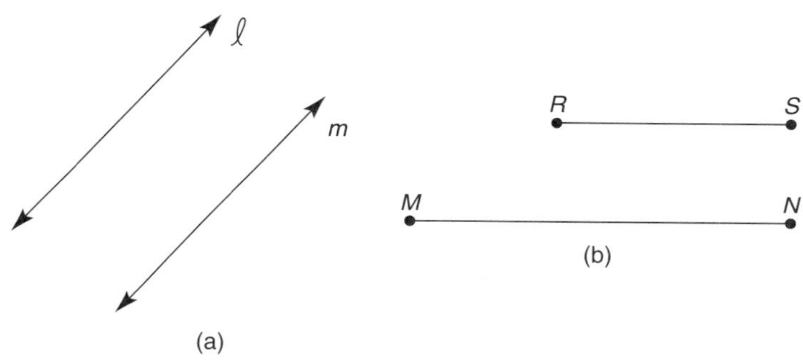

FIGURE 1.16

EXAMPLE 5

In Figure 1.17 the sides of angles ABC and DEF are parallel ($\overline{AB}$ to $\overline{DE}$ and $\overline{BC}$ to $\overline{EF}$). Use a protractor to decide whether these angles have equal measures.

FIGURE 1.17

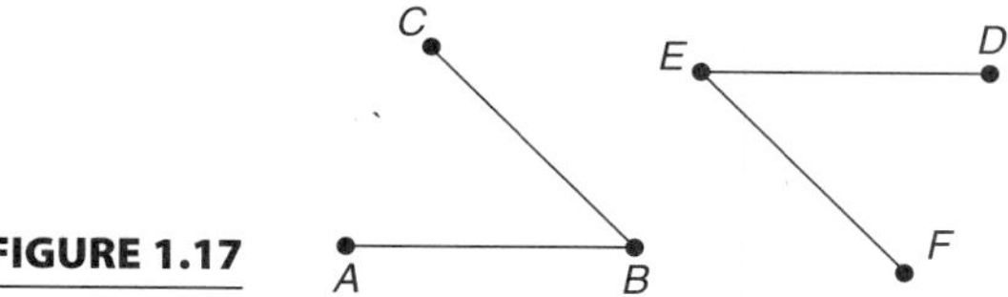

Solution The angles have equal measures. Both measure 44°.

Two angles with equal measures are said to be *congruent.* In Figure 1.17, $\angle ABC \cong \angle DEF$.

In Figure 1.18, angle ABD has been separated into smaller angles ABC and CBD; if the two smaller angles are congruent (have equal measures), then angle ABD has been *bisected.* In general, the word **bisect** means that a figure has been separated into two parts of equal measure. The largest angle measure that we consider in this text is 180°. Any angle having a 180° measure is called a **straight angle,** an angle whose sides are in opposite directions. See straight angle RST in Figure 1.19(a) on the next page. When a straight angle is bisected, as shown in Figure 1.19(b), the two angles formed are **right angles** (each measures 90°).

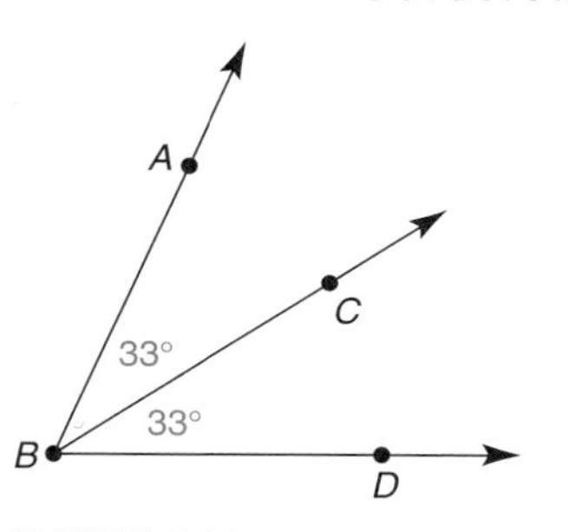

FIGURE 1.18

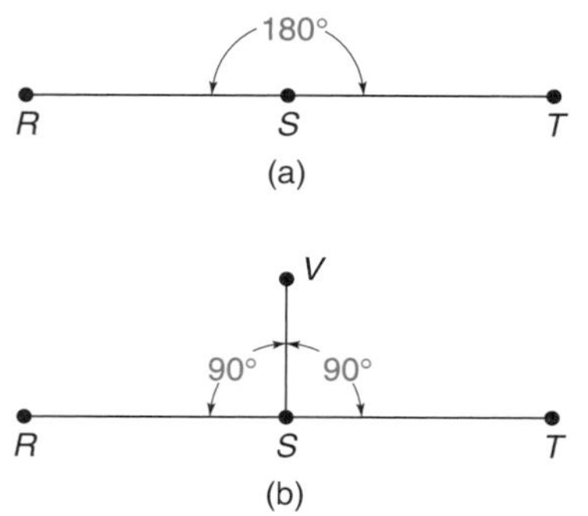

FIGURE 1.19

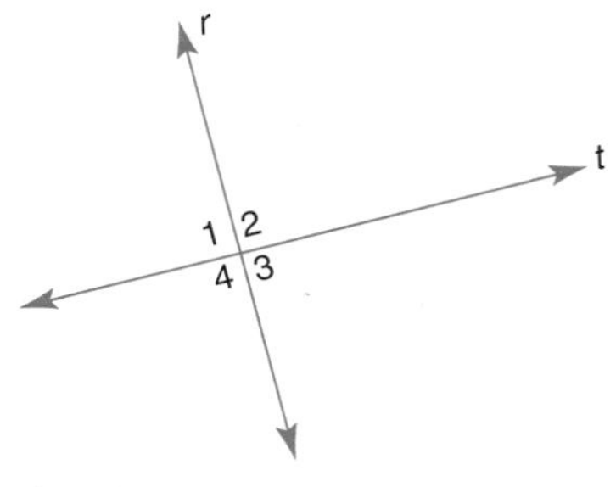

FIGURE 1.20

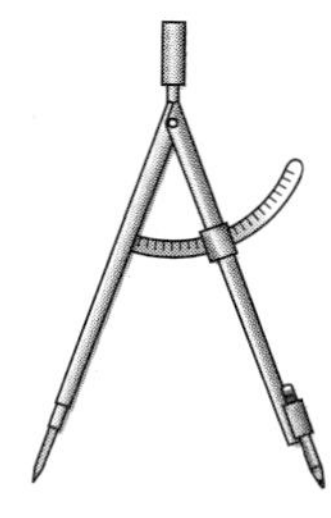

FIGURE 1.21

When two lines have a point in common, they are said to **intersect.** When two lines intersect and form right angles, they are said to be **perpendicular.**

EXAMPLE 6

In Figure 1.20, lines r and t are perpendicular. What is the measure of each of the angles formed?

Solution

Each of the marked angles (numbered 1, 2, 3, and 4) is a right angle and measures 90°.

Constructions

Another tool used in geometry is the **compass.** This instrument, shown in Figure 1.21, is used to construct circles and parts of circles. The compass and circle are discussed in the following paragraphs.

The ancient Greeks insisted that only two tools (a compass and a straightedge) be used for geometric **constructions,** which were idealized drawings assuming perfection in the use of these tools. The compass was used to create "perfect" circles and for marking off segments of "equal" length. The straightedge could be used to pass a line through two designated points.

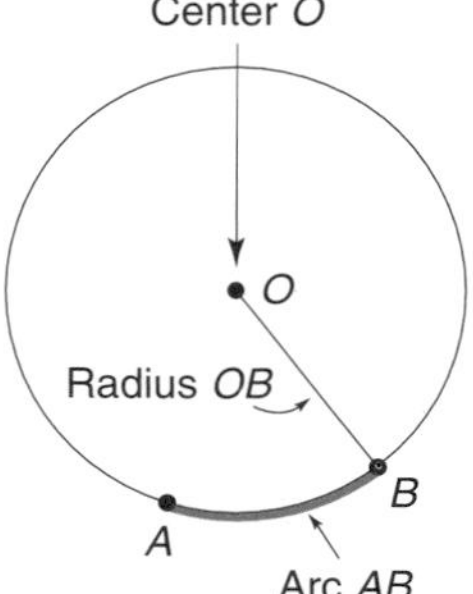

FIGURE 1.22

A **circle** is the set of all points in a plane that are at a given distance from a particular point (known as the "center" of the circle). The part of a circle between any two of its points is known as an **arc.** Any line segment joining the center to a point on the circle is a **radius** (plural, "radii") of the circle. See Figure 1.22.

Construction 1, which follows, is quite basic and depends only upon using arcs of the same radius length to construct line segments of the same length. Construction 2 is more difficult to perform and explain, so we will delay its explanation to a later chapter (see Section 3.4).

Construction 1:
To construct a segment congruent to a given segment.

Given: $\overline{AB}$ in Figure 1.23(a)

Construct: $\overline{CD}$ on line m so that $\overline{CD} \cong \overline{AB}$ (or $CD = AB$)

Construction: With your compass open to the length of $\overline{AB}$, place the stationary point of the compass at C and mark off a length equal to AB, as shown in Figure 1.23(b).

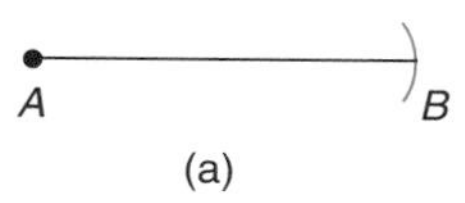

FIGURE 1.23

The following construction is shown step-by-step. Intuition suggests that point M in Figure 1.24(c) is the midpoint of $\overline{AB}$.

Construction 2:
To construct the midpoint M of a given line segment AB.

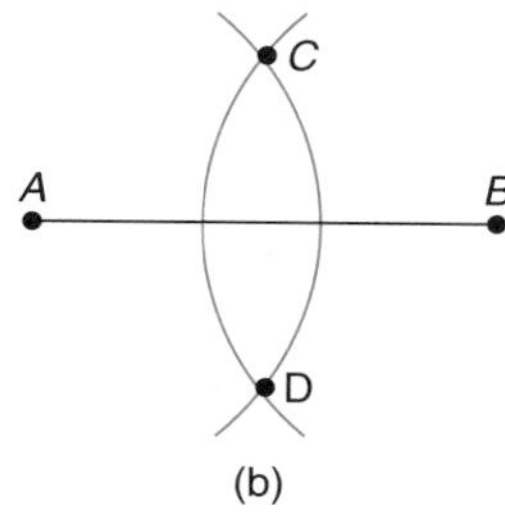

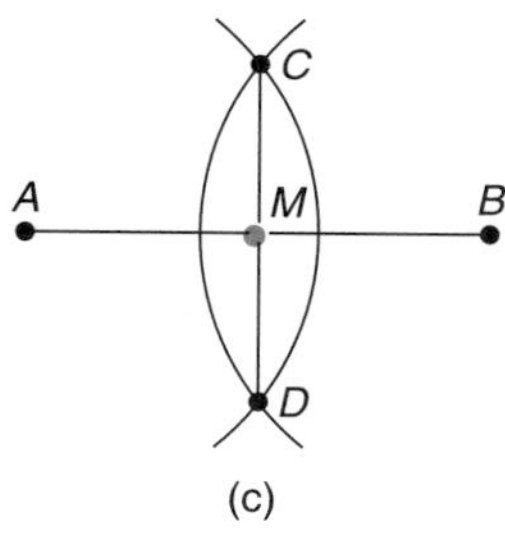

FIGURE 1.24

Given: $\overline{AB}$ in Figure 1.24(a)

Construct: M so that $AM = MB$ (M is the midpoint of $\overline{AB}$.)

Construction: Open your compass to a length greater than one-half of $\overline{AB}$. Using A as the center of the arc, mark off an arc that extends both above and below segment AB. With B as the center and keeping the same length of radius, mark off an arc that extends above and below $\overline{AB}$ so that two points (C and D) are determined where the arcs cross; see Figure 1.24(b). Now draw $\overline{CD}$. In Figure 1.24(c), the point where $\overline{CD}$ crosses $\overline{AB}$ is the midpoint M.

EXAMPLE 7

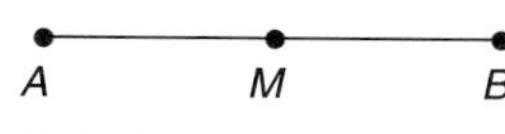

FIGURE 1.25

In Figure 1.25, M is the midpoint of $\overline{AB}$.

a) Find AM if $AB = 15$.
b) Find AB if $AM = 4.3$.
c) Find AB if $AM = 2x + 1$.

Solution

a) AM is one half of AB, so $AM = 7\frac{1}{2}$.
b) AB is twice AM, so $AB = 2(4.3)$ or $AB = 8.6$.
c) AB is twice AM, so $AB = 2(2x + 1)$ or $AB = 4x + 2$.

The technique from algebra used in Example 8 and also needed for Exercises 45 and 46 of this section depends upon the following properties of addition and subtraction.

If $a = b$ and $c = d$, then $a + c = b + d$.

Words: Equals added to equals provide equal sums.

Illustration: Since $0.5 = \frac{5}{10}$ and $0.2 = \frac{2}{10}$, it follows that $0.5 + 0.2 = \frac{5}{10} + \frac{2}{10}$.

If $a = b$ and $c = d$, then $a - c = b - d$.

Words: Equals subtracted from equals provide equal differences.

Illustration: Since $0.5 = \frac{5}{10}$ and $0.2 = \frac{2}{10}$, it follows that $0.5 - 0.2 = \frac{5}{10} - \frac{2}{10}$.

EXAMPLE 8

In Figure 1.26, point B lies on $\overline{AC}$ between A and C. If $AC = 10$ and AB is 2 units longer than BC, find the length x of $\overline{AB}$ and the length y of $\overline{BC}$.

Solution

Because $AB + BC = AC$, we have $x + y = 10$.
Because $AB - BC = 2$, we have $x - y = 2$.
Adding these equations,

$$\begin{array}{rl} x + y &= 10 \\ x - y &= 2 \\ \hline 2x &= 12 \end{array} \quad \text{so } x = 6.$$

If $x = 6$, then $x + y = 10$ becomes $6 + y = 10$ and $y = 4$.
Thus $AB = 6$ and $BC = 4$.

FIGURE 1.26

1.2 Exercises

1. If line segment AB and line segment CD are drawn to scale, what does intuition tell you about the lengths of these segments?

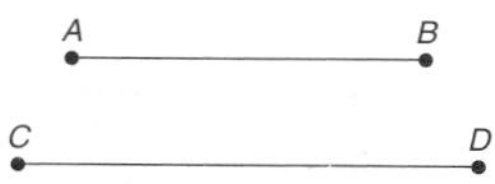

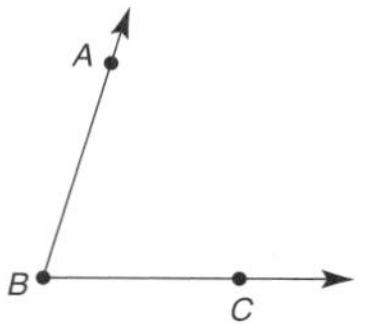

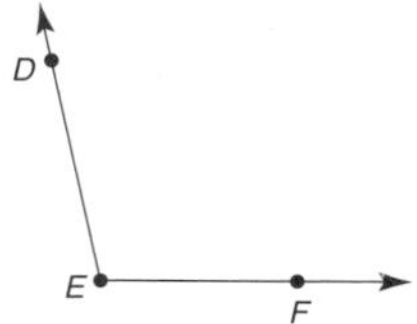

Exercise 2

2. If angles ABC and DEF were measured with a protractor, what does intuition tell you about the degree measures of these angles?

3. How many endpoints does a line segment have? How many midpoints does a line segment have?

4. Do the points A, B, and C appear to be collinear?

Exercises 4–6

5. How many lines can be drawn to contain both points A and B? How many lines can be drawn to contain points A, B, and C?

6. Consider noncollinear points A, B, and C. If each line must contain two of the points, what is the total number of lines that are determined by these points?

7. Name all the angles in the figure.

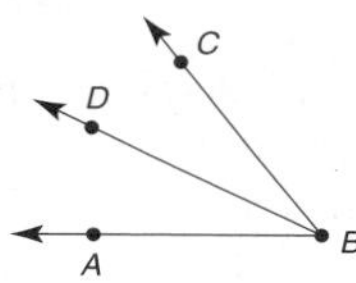

8. Which of the following measures can an angle have? 23°, 90°, 200°, 110.5°, −15°

9. Must two different points be collinear? Must three or more points be collinear? Can three or more points be collinear?

10. Which symbol(s) correctly expresses the order in which the points A, B, and X lie on the given line, A-X-B or A-B-X?

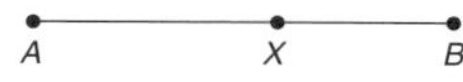

11. Which symbols correctly name the angle shown? $\angle ABC$, $\angle ACB$, $\angle CBA$

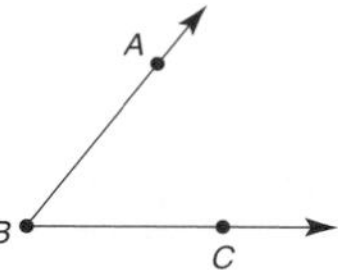

12. A triangle is named $\triangle ABC$. Can it also be named $\triangle ACB$? Can it be named $\triangle BAC$?

13. Consider rectangle $MNPQ$. Can it also be named rectangle $PQMN$? Can it be named rectangle $MNQP$?

14. When two lines cross (intersect), they share exactly one point in common. In the drawing, what is the point of intersection? How do the measures of $\angle 1$ and $\angle 2$ compare?

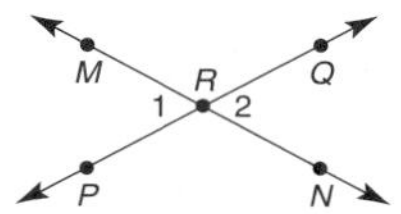

15. Judging from the ruler shown (not to scale) estimate the measure of each line segment.

a) AB **b)** CD

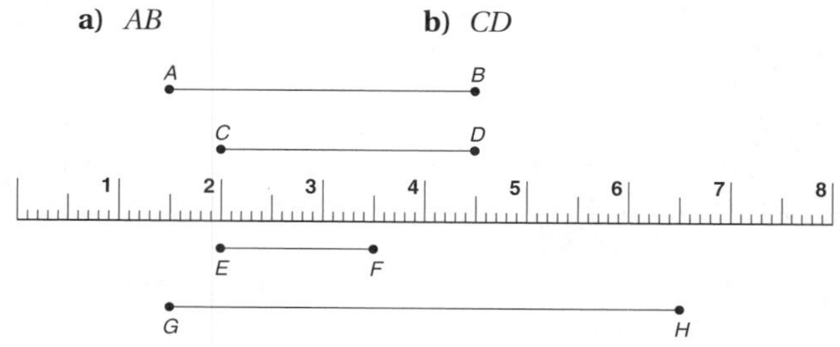

Exercises 15, 16

16. Judging from the ruler, estimate the measure of each line segment.

a) EF **b)** GH

17. Judging from the protractor provided, estimate the measure of each angle to the nearest multiple of 5° (e.g. 20°, 25°, 30°, etc.).

a) $m\angle 1$ **b)** $m\angle 2$

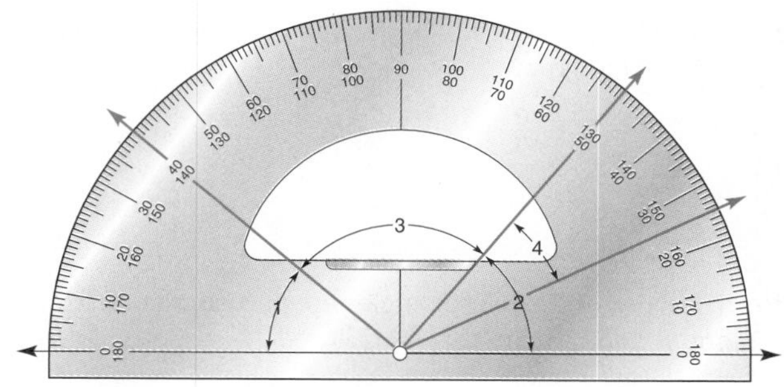

Exercises 17, 18

18. Judging from the protractor, estimate the measure of each angle to the nearest multiple of 5° (e.g. 20°, 25°, 30°, etc.).

a) $m\angle 3$ **b)** $m\angle 4$ (bisects $\angle 2$)

19. Consider the square at the right, $RSTV$. It has 4 right angles and 4 sides of the same length. How are sides $\overline{RS}$ and $\overline{ST}$ related? How are sides $\overline{RS}$ and $\overline{VT}$ related?

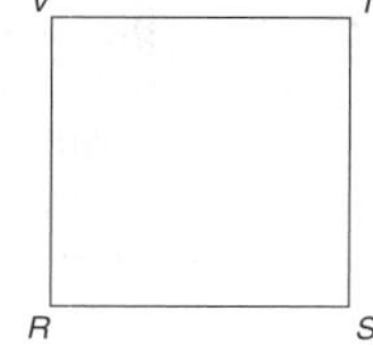

20. Square $RSTV$ has diagonals $\overline{RT}$ and $\overline{SV}$ (not shown). If the diagonals are drawn, how will their lengths compare? Do the diagonals of a square appear to be perpendicular?

21. Use a compass to draw a circle. Draw a radius, a line segment that connects the center to a point on the circle. Measure the length of the radius. Draw other radii and find their lengths. How do the lengths of the radii compare?

22. Use a compass to draw a circle of radius 1 inch. Draw a chord, a line segment that joins two points on the circle. Draw other chords and measure their lengths. What is the largest possible length of a chord in this circle?

23. The sides of the pair of angles are parallel. Are $\angle 1$ and $\angle 2$ congruent?

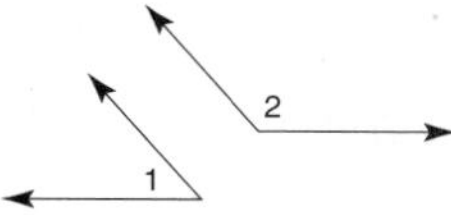

24. The sides of the pair of angles are parallel. Are $\angle 3$ and $\angle 4$ congruent?

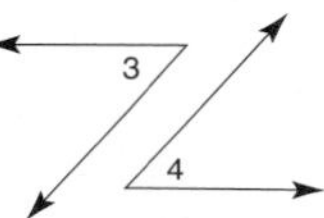

25. The sides of the pair of angles are perpendicular. Are $\angle 5$ and $\angle 6$ congruent?

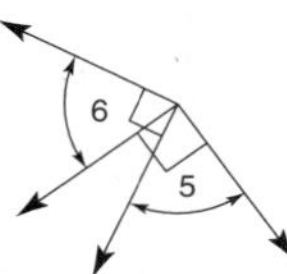

26. The sides of the pair of angles are perpendicular. Are $\angle 7$ and $\angle 8$ congruent?

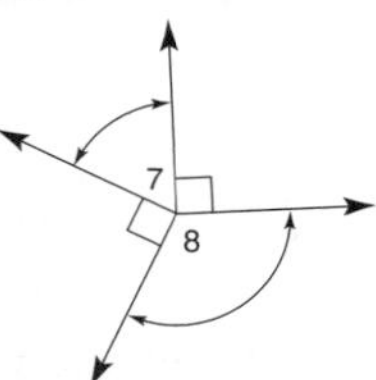

27. On a piece of paper, use your compass to construct a triangle that has two sides of the same length. Cut the triangle out of the paper and fold the triangle in half so that the congruent sides coincide (one lies over the other). What seems to be true of two angles of that triangle?

28. On a piece of paper, use your protractor to draw a triangle that has two angles of the same measure. Cut the triangle out of the paper and fold the triangle in half so that the angles of equal measure coincide (one lies over the other). What seems to be true of two of the sides of that triangle?

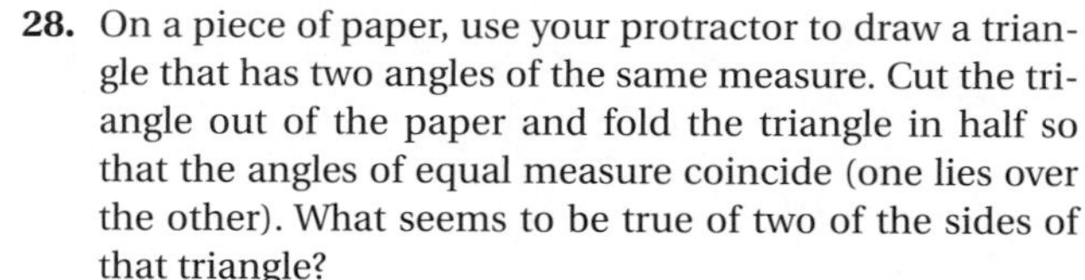

29. A trapezoid is a four-sided figure that contains one pair of parallel sides. Which sides of the trapezoid $MNPQ$ appear to be parallel?

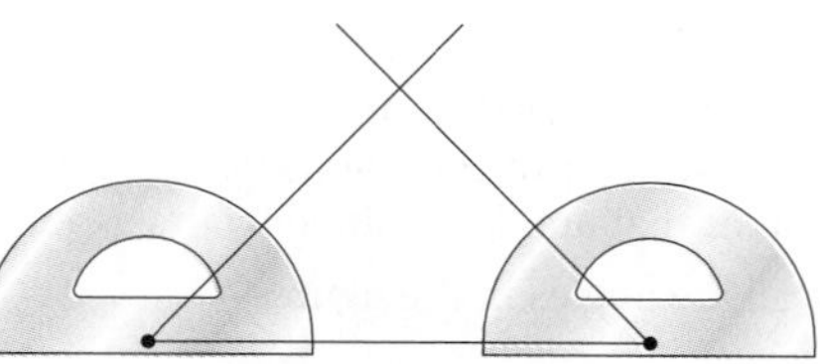

30. In the rectangle shown, what is true of the lengths of each pair of opposite sides?

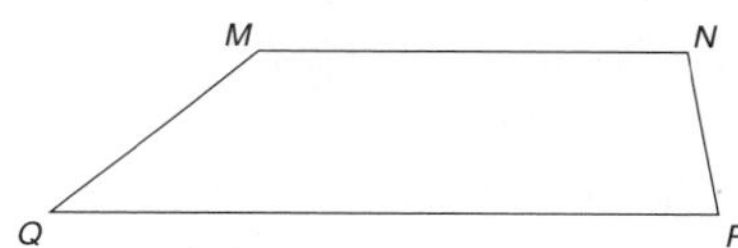

31. A line segment is bisected if its two parts have the same length. Which line segment, $\overline{AB}$ or $\overline{CD}$, is bisected at point X?

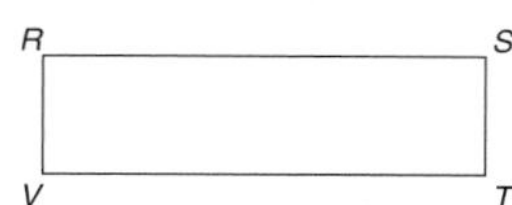

32. An angle is bisected if its two parts have the same measure. Use three letters to name the angle that is bisected.

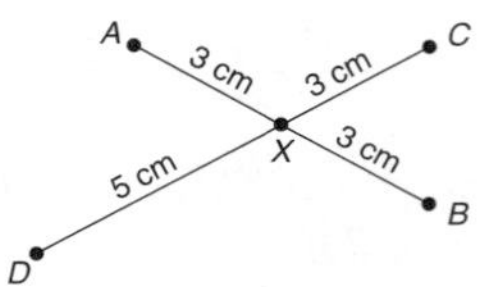

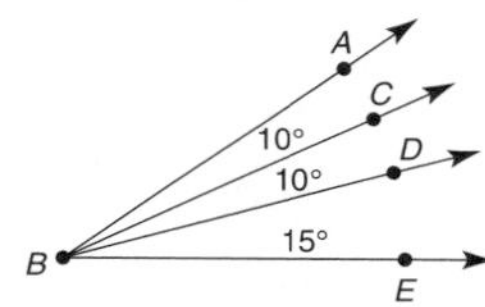

In Exercises 33 to 36, where A-B-C o[...]C, it follows that AB + BC = AC.

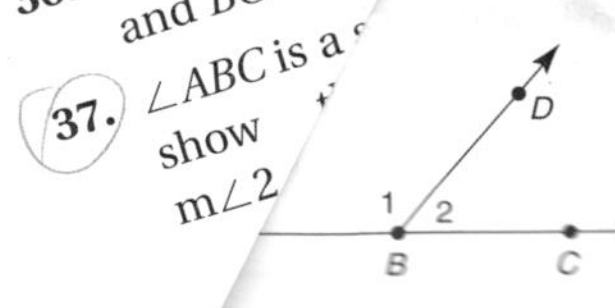

Exercises 33–36

33. Find AC if AB = 9 and BC [...]
34. Find AB if AC = 25 and [...]d AC = 21.
35. Find x if AB = x, BC [...]e length of $\overline{AC}$) if AB = x
36. Find an expressi[...] and BC = y.

37. ∠ABC is a [...] show [...]sing your protractor, you can [...] m∠2 [...]∠2 = 180°. Find m∠1 if

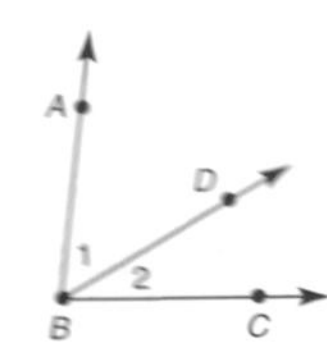

[...]∠1 = 2x and m∠2 = x.
[...]xercise 37.)

[...] 42, m∠1 + m∠2 = m∠ABC.

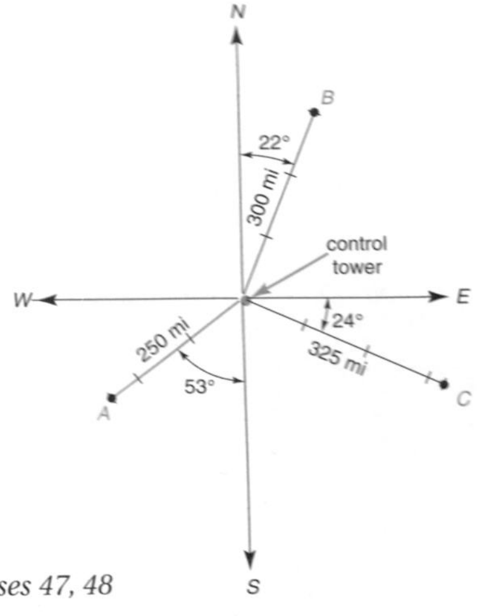

Exercises 39–42

39. Find m∠ABC if m∠1 = 32° and m∠2 = 39°.
40. Find m∠1 if m∠ABC = 68° and m∠1 = m∠2.
41. Find x if m∠1 = x, m∠2 = 2x + 3. and m∠ABC = 72°.
42. Find an expression for m∠ABC if m∠1 = x and m∠2 = y.
43. A compass was used to mark off three congruent segments, $\overline{AB}$, $\overline{BC}$, and $\overline{CD}$. Thus, $\overline{AD}$ has been trisected at points B and C. If AD = 32.7, how long is $\overline{AB}$?

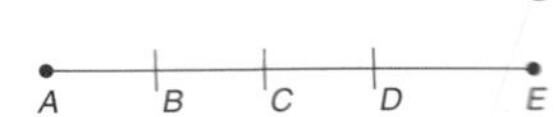

44. Use your compass and straightedge to bisect $\overline{EF}$.

45. ➢ In the figure, m∠1 = x and m∠2 = y. If x − y = 24°, find x and y.
(**HINT:** m∠1 + m∠2 = 180°.)

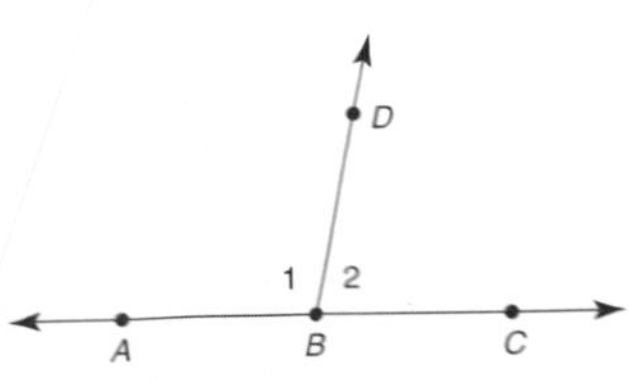

46. ➢ In the drawing, m∠1 = x and m∠2 = y. If m∠RSV = 67° and x − y = 17°, find x and y.
(**HINT:** m∠1 + m∠2 = m∠RSV.)

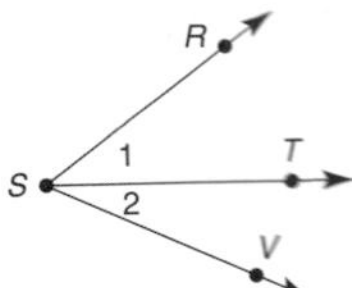

For Exercises 47 and 48, use the following information. Relative to its point of departure or some other point of reference, the angle that is used to locate the position of a ship or airplane is called its bearing. The bearing may also be used to describe the direction in which the airplane or ship is moving. By using an angle between 0° and 90°, a bearing is measured from the North-South line toward the East or West. In the diagram, airplane A (which is 250 miles from Chicago's O'Hare airport's control tower) has a bearing of S 53° W.

A
D
1
2
B
C

Exercises 47, 48

47. Relative to the control tower, find the bearing of airplane B.
48. Relative to the control tower, find the bearing of airplane C.

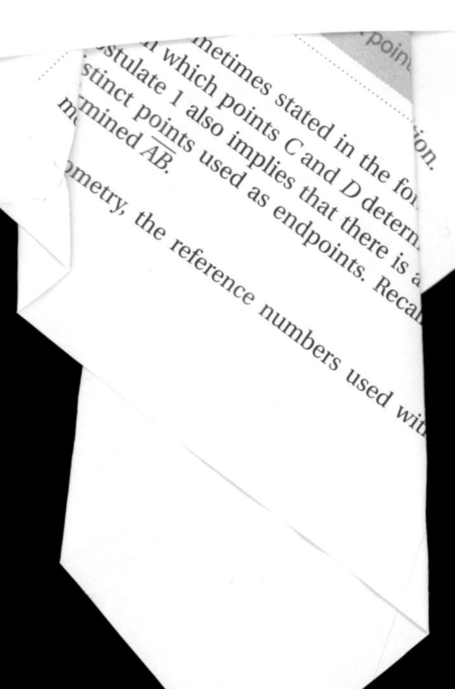

AB. This
mportant).
on of
replace the

1.3 Early Definitions and Postulates

A Mathematical System

Like algebra, the branch of mathe... a **mathematical system.** Each syste... properties. In the formal study of geom... terms. Building upon this foundation, additional... Once the terminology is sufficiently developed, cert... tics) of the system become apparent. These properti... **postulates** of the system; more generally, such st... **tions.** Having developed a vocabulary and acc... principles will follow logically using deductive... be proved and are called **theorems.** The foll... nents of a mathematical system (sometime... system).

FOUR PARTS OF ...

1. Undefined terms
2. Defined terms
3. Axioms or postulates
4. Theorems

Characteri...

Terms suc... cause they d... mined. Terr... *is a good* ... using w... this w...

... *ang*... ent) is ... statements.

"If ...

FIGURE 1.30

As we saw in Section 1.2, there is a relationship between the lengths of the line segments determined in Figure 1.30. This relationship is stated in the third postulate.

> **POSTULATE 3: (Segment-Addition Postulate)**
> If X is a point of $\overline{AB}$ and A-X-B, then $AX + XB = AB$.

EXAMPLE 2

In Figure 1.30, find AB if

a) $AX = 7.32$ and $XB = 6.19$. **b)** $AX = 2x + 3$ and $XB = 3x - 7$.

Solution

a) $AB = 7.32 + 6.19$, so $AB = 13.51$.
b) $AB = (2x + 3) + (3x - 7)$, so $AB = 5x - 4$.

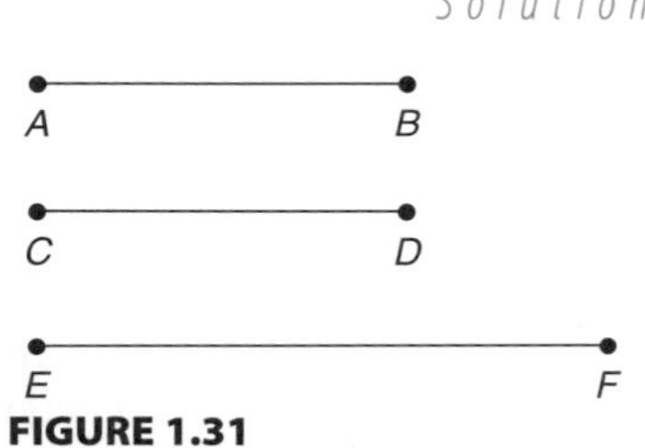

FIGURE 1.31

> **DEFINITION: Congruent (≅) segments** are two segments that have the same length.

In general, geometric figures that can be made to coincide (fit perfectly one on top of the other) are said to be **congruent.** The symbol ≅ is a combination of the symbol ~, which means that the figures have the same shape, and =, which means that the corresponding parts of the figure have the same measure. In Figure 1.31, $\overline{AB} \cong \overline{CD}$, but $\overline{AB} \ncong \overline{EF}$. Does it follow that $\overline{CD} \cong \overline{EF}$?

In Figure 1.32, if A, M, and B are collinear and $\overline{AM} \cong \overline{MB}$, then M is the **midpoint** of $\overline{AB}$. Equivalently, M is the midpoint of $\overline{AB}$ if $AM = MB$. Also, if $\overline{AM} \cong \overline{MB}$, then $\overline{CD}$ is described as a **bisector** of $\overline{AB}$. Under what condition would $\overline{AB}$ be a bisector of $\overline{CD}$?

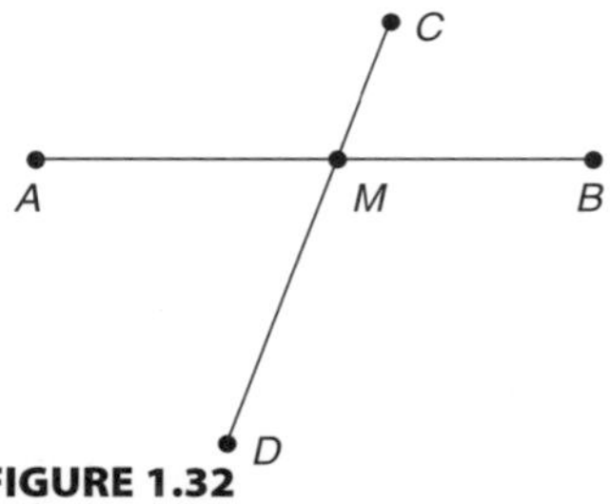

FIGURE 1.32

EXAMPLE 3

Given: M is the midpoint of $\overline{EF}$ (not shown). $EM = 3x + 9$ and $MF = x + 17$
Find: x and EM

Solution

Since M is the midpoint of $\overline{EF}$, $EM = MF$. Then

$$\begin{aligned} 3x + 9 &= x + 17 \\ 2x + 9 &= 17 \\ 2x &= 8 \\ x &= 4 \end{aligned}$$

By substitution, $EM = 3(4) + 9 = 12 + 9 = 21$.

In geometry, the word **union** is used to describe the joining or combining of two figures or sets of points.

DEFINITION: **Ray** AB, denoted by $\overrightarrow{AB}$, is the union of $\overline{AB}$ and all points X on $\overleftrightarrow{AB}$ such that B is between A and X.

In Figure 1.33, $\overleftrightarrow{AB}$, $\overrightarrow{AB}$, and $\overrightarrow{BA}$ are shown; note that $\overrightarrow{AB}$ and $\overrightarrow{BA}$ are not the same ray.

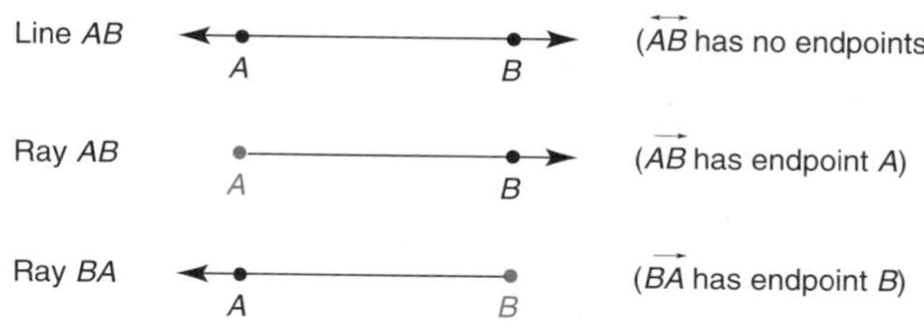

FIGURE 1.33

Opposite rays are two rays that share a common endpoint; the union of opposite rays is a straight line. In Figure 1.35(a), $\overrightarrow{BA}$ and $\overrightarrow{BC}$ are opposite rays.

The **intersection** of two geometric figures is the set of points that the two figures share in common. In everyday life, the intersection of Bradley Avenue and Neil Street is the part of the roadway that the two roads have in common (Figure 1.34).

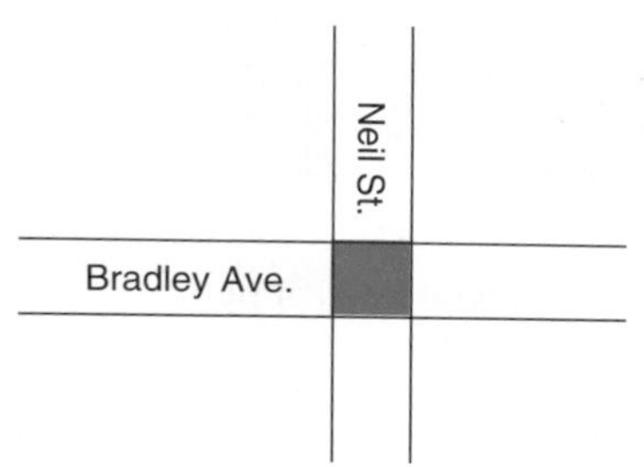

FIGURE 1.34

POSTULATE 4:
If two lines intersect, they intersect at a point.

In Figure 1.35(b), lines ℓ and m intersect at point P.

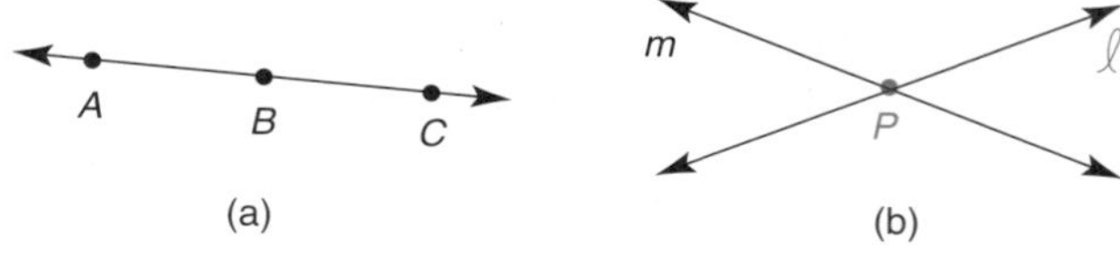

FIGURE 1.35

DEFINITION: **Parallel lines** are lines that lie in the same plane but do not intersect.

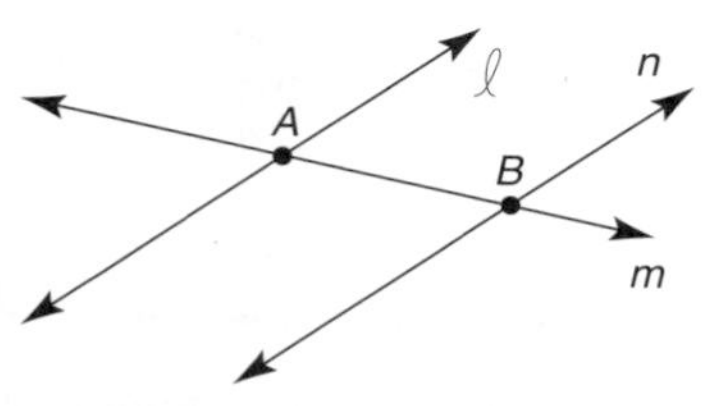

FIGURE 1.36

In Figure 1.36, ℓ and n are parallel; in symbols, $\ell \parallel n$. However, ℓ and m are not parallel; in symbols, $\ell \nparallel m$.

EXAMPLE 4

In Figure 1.36, what is the intersection of

a) line ℓ and m? **b)** line ℓ and line n?

Solution

a) Point A **b)** Parallel lines do not intersect.

Another undefined term in geometry is **plane.** A plane is two-dimensional; that is, it has infinite length and infinite width, but no thickness. Except for its limited size, a flat surface such as the top of a table could be used as an example of a plane. An uppercase letter can be used to name a plane. Because a plane (like a line) is infinite, we can only show a portion of the plane or planes, as in Figure 1.37.

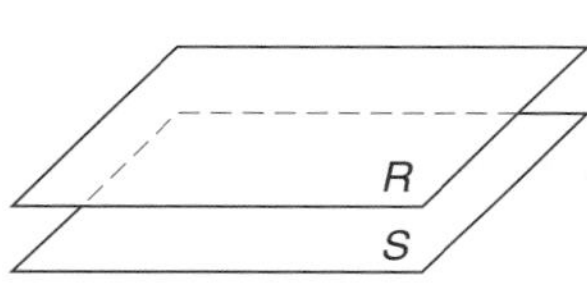

Planes *R* and *S*

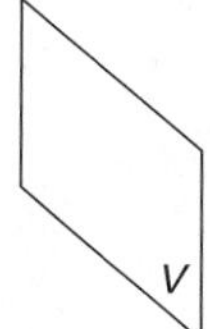

Planes *T* and *V*

FIGURE 1.37

Because a plane is two-dimensional, it consists of an infinite number of points and contains an infinite number of lines. Two distinct points may determine (or "fix") a line; likewise, exactly three noncollinear points determine a plane. Just as collinear points lie on the same line, **coplanar points** lie in the same plane. In Figure 1.38, points *B, C, D,* and *E* are coplanar, while *A, B, C,* and *D* are noncoplanar.

FIGURE 1.38

In this book, points shown in figures are assumed to be coplanar unless otherwise stated. For instance, points *A, B, C, D,* and *E* are coplanar in Figure 1.39(a) as are points *F, G, H, J,* and *K* in Figure 1.39(b).

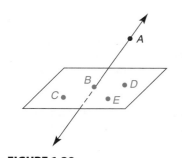

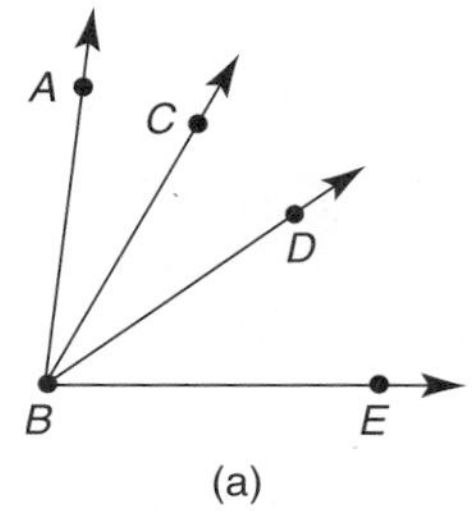

(a)

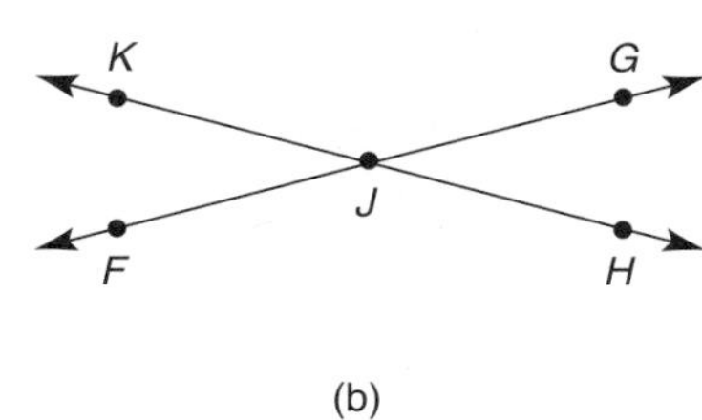

(b)

FIGURE 1.39

POSTULATE 5:
Through three noncollinear points, there is exactly one plane.

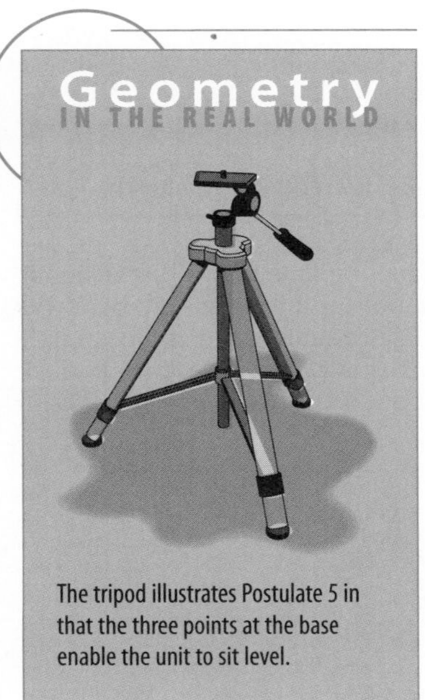

The tripod illustrates Postulate 5 in that the three points at the base enable the unit to sit level.

Based upon Postulate 5, we can see why a three-legged table sits evenly, but a four-legged table would "wobble" if the legs were of unequal length.

Space is the set of all possible points. It is three-dimensional, having qualities of length, width, and depth. When two planes intersect in space, their intersection is a line. An opened greeting card suggests this relationship, as does Figure 1.40(a). This notion gives rise to our next postulate.

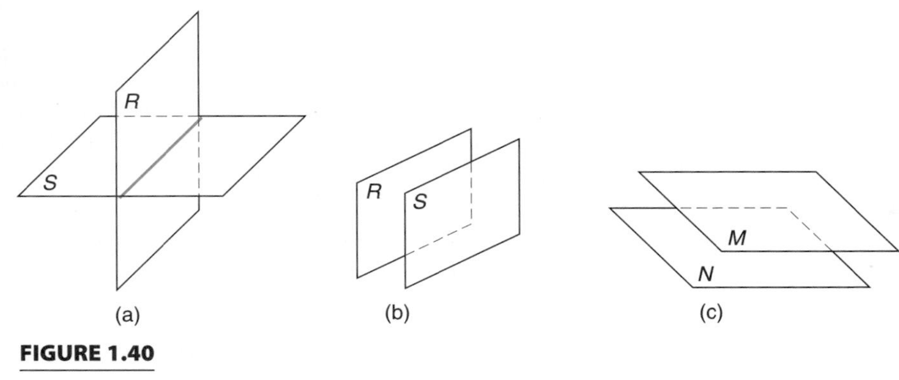

FIGURE 1.40

POSTULATE 6:
If two distinct planes intersect, then their intersection is a line.

The intersection of two infinite planes is infinite because it is a line. [See Figure 1.40(a).] If two planes do not intersect, then they are **parallel.** The parallel **vertical** planes in Figure 1.40(b) may remind you of the opposite walls of your classroom. The parallel **horizontal** planes in Figure 1.40(c) suggest the relationship between ceiling and floor.

Imagine a plane and two points of that plane, points *A* and *B*. Now think of the line containing the two points and its relationship to the plane. Perhaps your conclusion can be summed up as follows:

POSTULATE 7:
Given two distinct points in a plane, the line containing these points also lies in the plane.

Because the uniqueness of the midpoint of a line segment can be justified, we call the following statement a theorem.

THEOREM 1.3.1: The midpoint of a line segment is unique.

FIGURE 1.41

If M is the midpoint of $\overline{AB}$ in Figure 1.41, then no other point can separate $\overline{AB}$ into two congruent parts. The proof of this theorem is based upon the Ruler Postulate. M is *the* point that is located $\frac{1}{2}(AB)$ units from A (and from B).

The numbering system used to identify Theorem 1.3.1 need not be memorized. However, this theorem number may be used in a later reference. The numbering system works as follows:

1	3	1
CHAPTER where found	SECTION where found	ORDER found in section

A summary of the theorems presented in this textbook can be found at the end of the book.

1.3 Exercises

In Exercises 1 to 20, use the drawings as needed to answer the following questions.

1. Name three points that appear to be

a) collinear. **b)** noncollinear.

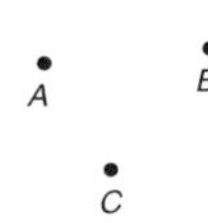

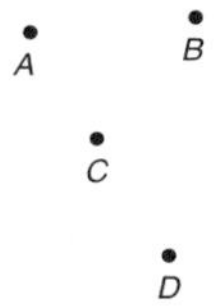

Exercises 1, 2

2. How many lines can be drawn through

a) point A?
b) points A and B?
c) points A, B, and C?
d) points A, B, and D?

3. Give the meaning of $\overleftrightarrow{CD}$, $\overline{CD}$, CD, and $\overrightarrow{CD}$.

4. Explain the difference, if any, between

a) $\overleftrightarrow{CD}$ and $\overleftrightarrow{DC}$.
b) $\overline{CD}$ and $\overline{DC}$.
c) CD and DC.
d) $\overrightarrow{CD}$ and $\overrightarrow{DC}$.

5. Name two lines that appear to be

a) parallel. **b)** nonparallel.

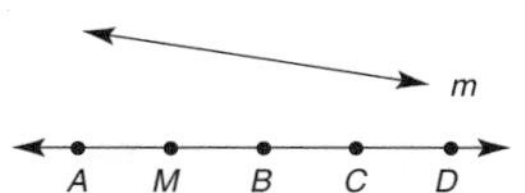

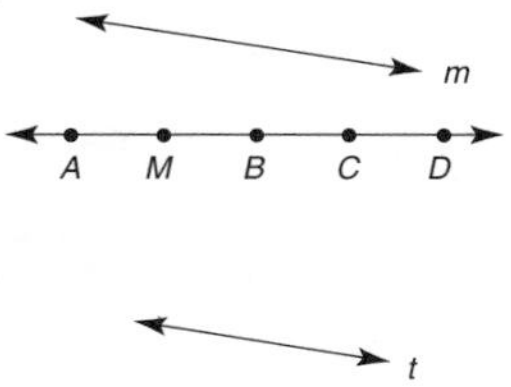

Exercises 5–9

6. Classify as true or false:

a) $AB + BC = AD$
b) $AD - CD = AB$
c) $AD - CD = AC$
d) $AB + BC + CD = AD$
e) $AB = BC$

7. *Given:* M is the midpoint of $\overline{AB}$
$AM = 2x + 1$ and $MB = 3x - 2$
Find: x and AM

8. *Given:* M is the midpoint of $\overline{AB}$
$AM = 2(x + 1)$ and $MB = 3(x - 2)$
Find: x and AB

9. *Given:* $AM = 2x + 1$, $MB = 3x + 2$, and $AB = 6x - 4$
Find: x and AB

10. Can a segment bisect a line? a segment? Can a line bisect a segment? a line?

11. In the figure, name

a) two opposite rays.
b) two rays that are not opposite.

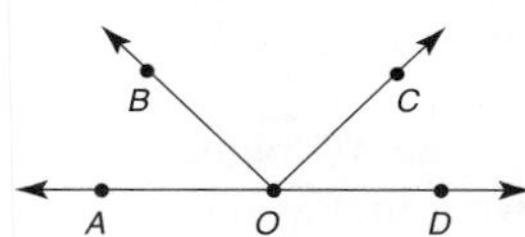

12. Suppose that (a) point C lies in plane X and (b) point D lies in plane X. What may you conclude regarding $\overleftrightarrow{CD}$?

13. Make a sketch of

a) two intersecting lines that are perpendicular.
b) two intersecting lines that are *not* perpendicular.
c) two parallel lines.

14. Make a sketch of

a) two intersecting planes.
b) two parallel planes.
c) two parallel planes intersected by a third plane that is not parallel to the first or second plane.

15. Suppose that (a) planes M and N intersect, (b) point A lies in both planes M and N, and (c) point B lies in both planes M and N. What may you conclude regarding $\overleftrightarrow{AB}$?

16. Suppose that (a) points A, B, and C are collinear and (b) $AB > AC$. Which point can you conclude *cannot* lie between the other two?

17. Suppose that points A, R, and V are collinear. If $AR = 7$ and $RV = 5$, then which point cannot possibly lie between the other two?

18. Points A, B, C, and D are coplanar; B, C, and D are collinear; point E is not in plane M. How many planes contain

a) points A, B and C?
b) points B, C, and D?
c) points A, B, C, and D?
d) points A, B, C, and E?

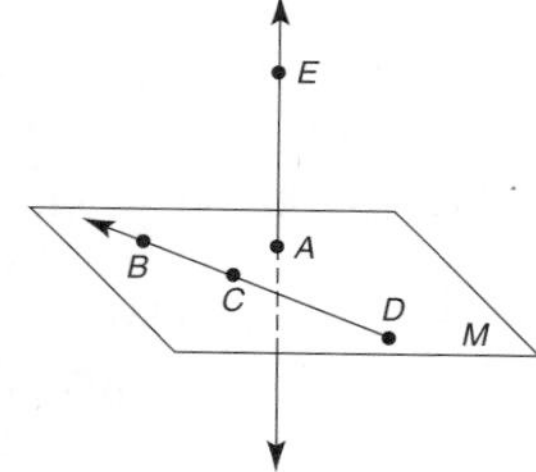

19. Using the number line provided, name the point that

a) is the midpoint of $\overline{AE}$.
b) is the endpoint of a segment of length 4, if the other endpoint is point G.
c) has a distance from B equal to $3(AC)$.

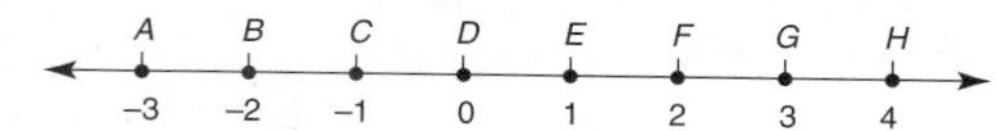

Exercises 19, 20

20. Given that B is the midpoint of $\overline{AC}$ and C is the midpoint of $\overline{BD}$, what may you conclude about the lengths of

a) $\overline{AB}$ and $\overline{CD}$?
b) $\overline{AC}$ and $\overline{BD}$?
c) $\overline{AC}$ and $\overline{CD}$?

In Exercises 21 to 24, use only a compass and a straightedge to complete each construction.

21. *Given:* $\overline{AB}$ and $\overline{CD}$ $(AB > CD)$
Construct: $\overline{MN}$ on line ℓ so that $MN = AB + CD$

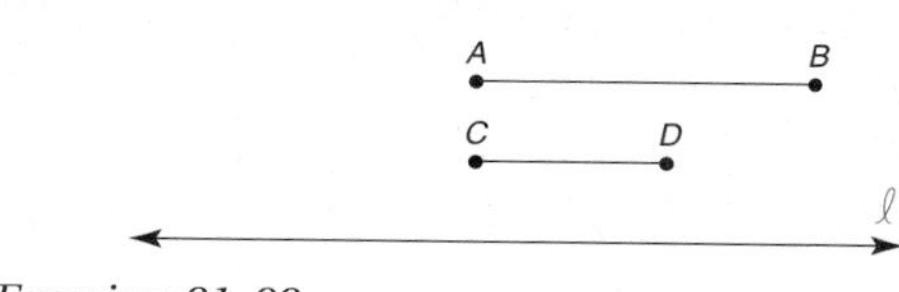

Exercises 21, 22

22. *Given:* $\overline{AB}$ and $\overline{CD}$ $(AB > CD)$
Construct: $\overline{EF}$ so that $EF = AB - CD$

23. *Given:* $\overline{AB}$ as shown in the figure.
Construct: $\overline{PQ}$ on line n so that $PQ = 3(AB)$

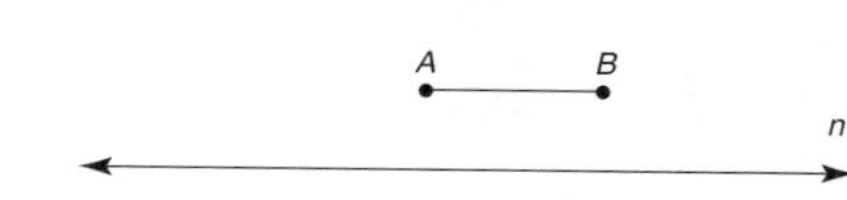

Exercises 23, 24

24. *Given:* $\overline{AB}$ as shown in the figure.
Construct: $\overline{TV}$ on line n so that $TV = \frac{1}{2}(AB)$

25. Can you use the construction for the midpoint of a segment to divide a line segment into

a) three congruent parts?
b) four congruent parts?
c) six congruent parts?
d) eight congruent parts?

26. Generalize your findings in Exercise 25.

27. Consider noncollinear points A, B, C, and D. Using two points at a time (such as A and B), how many lines are determined by these points?

28. Consider noncoplanar points A, B, C, and D. Using three points at a time (such as A, B and C), how many planes are determined by these points?

29. Line ℓ is parallel to plane P (that is, it will not intersect P even if extended). Line m intersects line ℓ. What may you conclude about m and P?

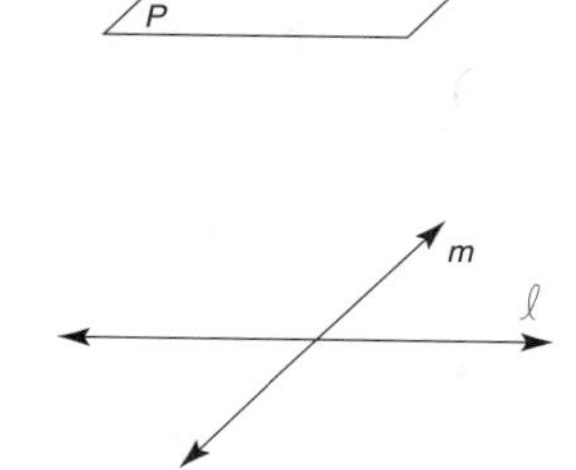

30. $\overleftrightarrow{AB}$ and $\overleftrightarrow{EF}$ are said to be **skew** lines because they neither intersect nor are parallel. How many planes are determined by

a) parallel lines AB and DC?
b) intersecting lines AB and BC?
c) skew lines AB and EF?
d) lines AB, BC, and DC?
e) points A, B, and F?
f) points A, C, and H?
g) points A, C, F, and H?

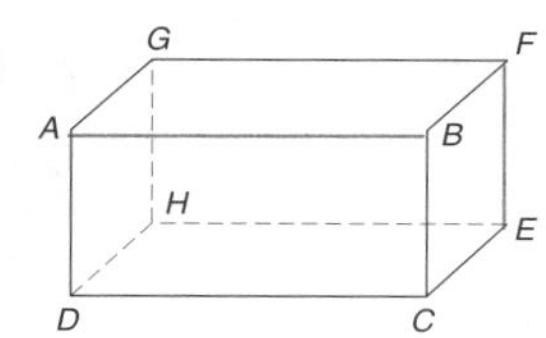

1.4 Angles and Their Relationships

This section introduces you to the language of angles. Recall from Section 1.3 that the word *union* means that two objects are joined.

DEFINITION: An **angle** is the union of two rays that share a common endpoint.

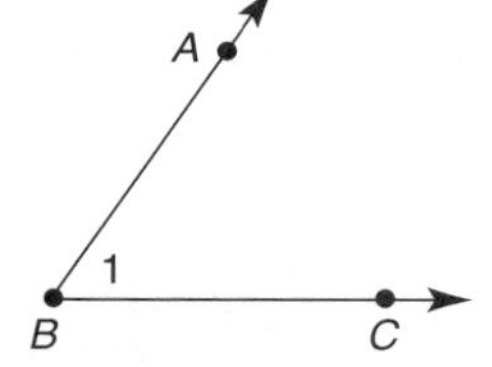

FIGURE 1.42

In Figure 1.42, the angle is symbolized by $\angle ABC$ or $\angle CBA$. The rays BA and BC are known as the **sides** of the angle. B, the common endpoint of these rays, is known as the **vertex** of the angle. When three letters are used to name an angle, the vertex is always named in the middle. In many instances, a single letter or numeral is used to name the angle. The angle in Figure 1.42 may be described as $\angle B$ (the vertex) or as $\angle 1$.

POSTULATE 8: (Protractor Postulate)
The measure of an angle is a unique positive number.

NOTE: In Chapters 1 to 9, the measures of angles are between 0° and 180°, including 180°. Angles with negative measures or measures greater than 180° are discussed in Chapter 10.

Types of Angles

An angle whose measure is less than 90° is an **acute angle.** If the angle's measure is exactly 90°, the angle is a **right angle.** If the angle's measure is between 90° and 180°, the angle is **obtuse.** Finally, an angle whose measure is exactly 180° is a **straight angle;** alternatively, a straight angle is one whose sides form opposite rays (a straight line). See Table 1.2 on the next page.

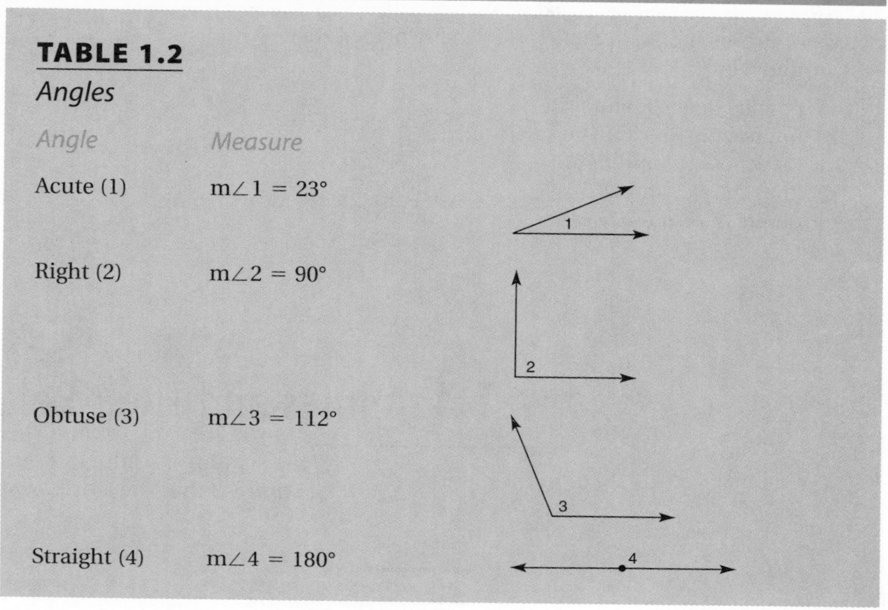

TABLE 1.2
Angles

Angle	Measure
Acute (1)	$m\angle 1 = 23°$
Right (2)	$m\angle 2 = 90°$
Obtuse (3)	$m\angle 3 = 112°$
Straight (4)	$m\angle 4 = 180°$

DISCOVER!

An index card can be used to categorize the type of angle that is displayed. In each sketch, an index card is placed over an angle. A dashed ray indicates that a side is hidden. What type of angle is shown in each figure? (Notice the placement of the card in each figure.)

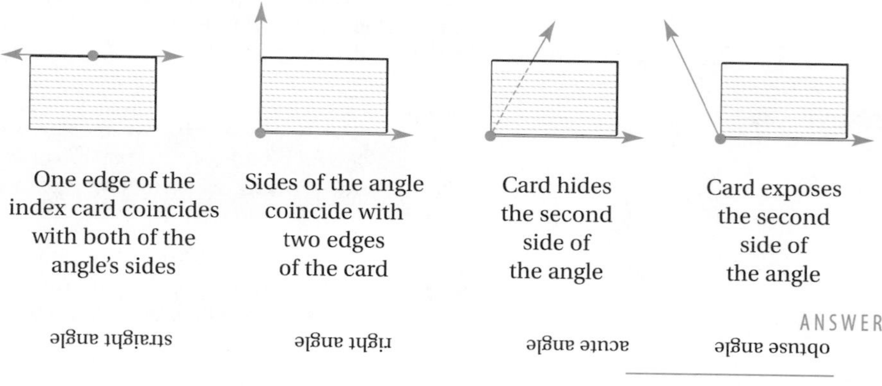

One edge of the index card coincides with both of the angle's sides
Sides of the angle coincide with two edges of the card
Card hides the second side of the angle
Card exposes the second side of the angle
ANSWER
straight angle
right angle
acute angle
obtuse angle

In Figure 1.43(a), $\angle ABC$ contains the noncollinear points A, B, and C. These three points, in turn, determine a plane. The plane containing $\angle ABC$ is separated into three subsets by the angle:

Points like D are the *interior* of $\angle ABC$.

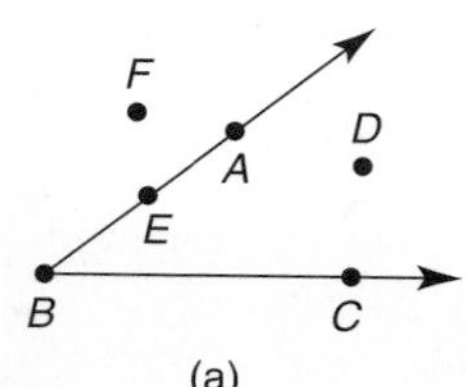

(a)

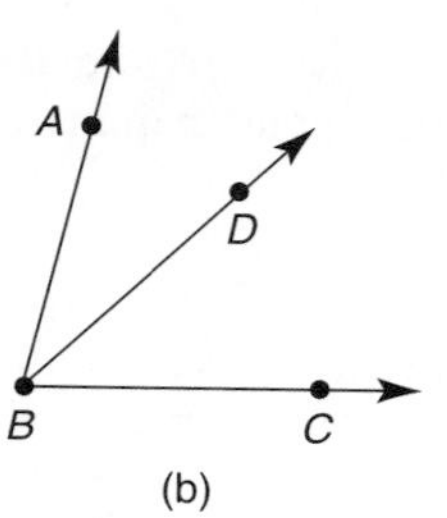

(b)

FIGURE 1.43

Points like E are said to be *on* $\angle ABC$.

Points like F are in the *exterior* of $\angle ABC$.

With this description, it is possible to state the counterpart of the Segment-Addition Postulate!

POSTULATE 9: (Angle-Addition Postulate)
If a point D lies in the interior of an angle ABC, then
$m\angle ABD + m\angle DBC = m\angle ABC$.

Figure 1.43(b) illustrates the Angle-Addition Postulate. Given a different figure, the theorem could allow a statement such as $m\angle MNP + m\angle PNQ = m\angle MNQ$.

EXAMPLE 1

Use Figure 1.43(b) to find $m\angle ABC$ if:

a) $m\angle ABD = 27°$ and $m\angle DBC = 42°$
b) $m\angle ABD = x°$ and $m\angle DBC = (2x - 3)°$

Solution

a) Using the Angle-Addition Postulate,
$m\angle ABC = m\angle ABD + m\angle DBC$. That is, $m\angle ABC = 27° + 42° = 69°$.
b) $m\angle ABC = m\angle ABD + m\angle DBC = x° + (2x - 3)° = (3x - 3)°$

Classifying Pairs of Angles

In Figure 1.43(b), $\angle ABD$ and $\angle DBC$ are also said to be adjacent. Two angles are **adjacent** if they share a common side and a common vertex but have no interior points in common. In Figure 1.43(b), $\angle ABC$ and $\angle ABD$ are not adjacent because they have interior points in common.

DEFINITION: **Congruent angles ($\cong \angle$s)** are two angles with the same measure.

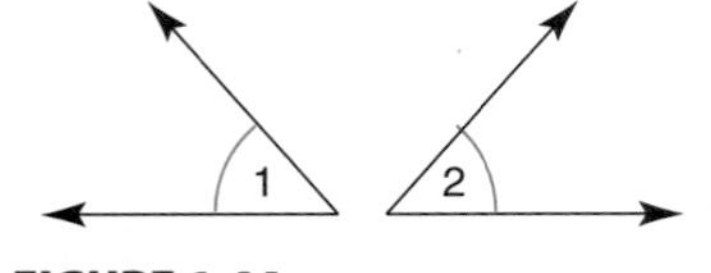

FIGURE 1.44

Congruent angles must coincide when one is placed over the other. (Do not consider that the sides appear to have different lengths; remember that rays are infinite in length!) In symbols, $\angle 1 \cong \angle 2$ if $m\angle 1 = m\angle 2$. In Figure 1.44, similar markings indicate that $\angle 1 \cong \angle 2$.

EXAMPLE 2

Given: $\angle 1 \cong \angle 2$
$m\angle 1 = 2x + 15$
$m\angle 2 = 3x - 2$

Find: x

Solution $\angle 1 \cong \angle 2$ means $m\angle 1 = m\angle 2$. Therefore

$$\begin{aligned} 2x + 15 &= 3x - 2 \\ 17 &= x \qquad \text{or} \qquad x = 17 \end{aligned}$$

(Notice that $m\angle 1 = 2(17) + 15 = 49°$ and $m\angle 2 = 3(17) - 2 = 49°$.)

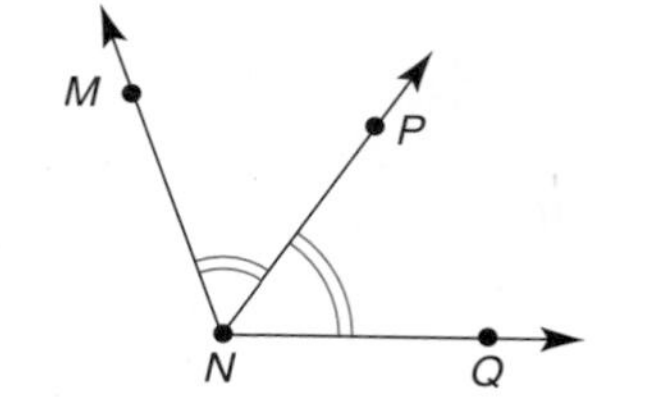

FIGURE 1.45

When P is located in the interior of $\angle MNQ$ so that $\angle MNP \cong \angle PNQ$, $\overrightarrow{NP}$ is said to **bisect** $\angle MNQ$. Equivalently, $\overrightarrow{NP}$ is the **bisector** of $\angle MNQ$. Thus, the angle is bisected when $m\angle MNP = m\angle PNQ$ (see Figure 1.45).

Many angle relationships involve a pair of angles. For instance, two angles whose measures add up to 90° are **complementary,** and each angle is the **complement** of the other. Similarly, two angles whose measures add up to 180° are known as **supplementary,** and each angle is the **supplement** of the other.

EXAMPLE 3

Given: $\angle P$ and $\angle Q$ are complementary so that

$$m\angle P = \frac{x}{2} \qquad \text{and} \qquad m\angle Q = \frac{x}{3}$$

Find: x, $m\angle P$, and $m\angle Q$

Solution

$$\begin{aligned} m\angle P + m\angle Q &= 90 \\ \frac{x}{2} + \frac{x}{3} &= 90 \end{aligned}$$

Multiplying by 6 (the least common denominator, or LCD, for 2 and 3),

$$\begin{aligned} 6 \cdot \frac{x}{2} + 6 \cdot \frac{x}{3} &= 6 \cdot 90 \\ 3x + 2x &= 540 \\ 5x &= 540 \\ x &= 108 \end{aligned}$$

$$m\angle P = \frac{x}{2} = \frac{108}{2} = 54°$$

$$m\angle Q = \frac{x}{3} = \frac{108}{3} = 36°$$

NOTE: $m\angle P = 54°$ and $m\angle Q = 36°$, so their sum is exactly 90°.

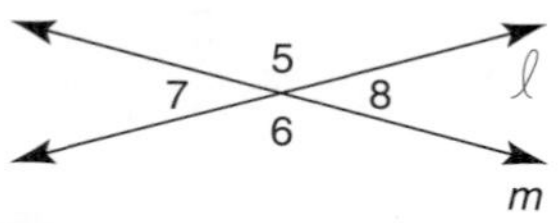

FIGURE 1.46

When two straight lines intersect, the pairs of nonadjacent angles formed are each known as **vertical angles.** In Figure 1.46, $\angle 5$ and $\angle 6$ are vertical angles, as are $\angle 7$ and $\angle 8$. In addition, $\angle 5$ and $\angle 7$ can be described as adjacent and supplementary angles, as can $\angle 5$ and $\angle 8$. If $m\angle 7 = 30°$, then what is $m\angle 5$? What is $m\angle 8$? It is true in general that vertical angles are congruent, and we will prove this in Example 4.

Recall the Addition and Subtraction Properties of Equality: If $a = b$ and $c = d$, then $a \pm c = b \pm d$. These principles can be used in solving a system of equations such as the following one:

$$\begin{aligned} x + y &= 5 \\ 2x - y &= 7 \\ \hline 3x \quad &= 12 \qquad \text{left and right sides are added} \\ x &= 4 \end{aligned}$$

Now we can substitute 4 for x in either of our original equations, to solve for y:

$$\begin{aligned} x + y &= 5 \\ 4 + y &= 5 \qquad \text{by substitution} \\ y &= 1 \end{aligned}$$

If $x = 4$ and $y = 1$, then $x + y = 5$ *and* $2x - y = 7$.

When each term in an equation is multiplied by a number, the solutions of the equation are not changed. For instance, the equations $2x - 3 = 7$ and $6x - 9 = 21$ (each term multiplied by 3) both have the solution $x = 5$. Likewise, the values of x and y that make the equation $4x + y = 180$ true also make the equation $16x + 4y = 720$ (each term multiplied by 4) true. We use this method in Example 4.

EXAMPLE 4

Given: In Figure 1.46 on page 32, ℓ and m intersect so that

$$\begin{aligned} m\angle 5 &= 2x + 2y \\ m\angle 8 &= 2x - y \\ m\angle 6 &= 4x - 2y \end{aligned}$$

Find: x and y

Solution

$\angle 5$ and $\angle 8$ are supplementary (adjacent and together form a straight angle). Therefore $m\angle 5 + m\angle 8 = 180$. $\angle 5$ and $\angle 6$ are congruent (vertical). Therefore $m\angle 5 = m\angle 6$. Consequently, we have

$$\begin{aligned} (2x + 2y) + (2x - y) &= 180 \qquad (\text{supplementary } \angle\text{s 5 and 8}) \\ 2x + 2y &= 4x - 2y \qquad (\cong \angle\text{s 5 and 6}) \end{aligned}$$

Simplifying,

$$\begin{aligned} 4x + y &= 180 \\ 2x - 4y &= 0 \end{aligned}$$

Using the Multiplication Property of Equality, we multiply the first equation by 4, so the equivalent system is

$$\begin{aligned} 16x + 4y &= 720 \\ 2x - 4y &= 0 \\ \hline 18x \quad &= 720 \qquad \text{adding left, right sides} \\ x &= 40 \end{aligned}$$

From the equation $4x + y = 180$, it follows that

$$\begin{aligned} 4(40) + y &= 180 \\ 160 + y &= 180 \\ y &= 20 \end{aligned}$$

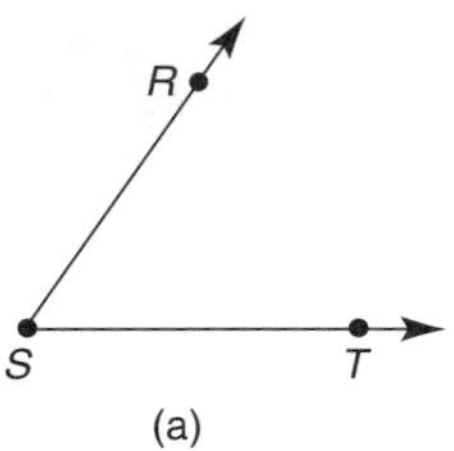

(a)

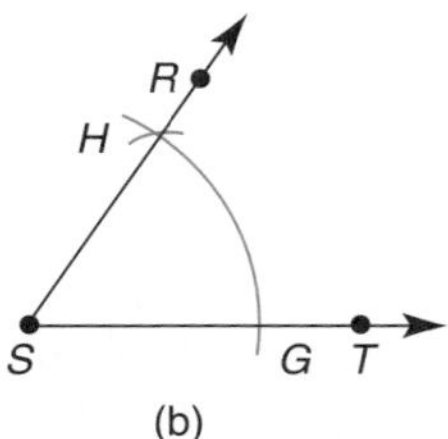

(b)

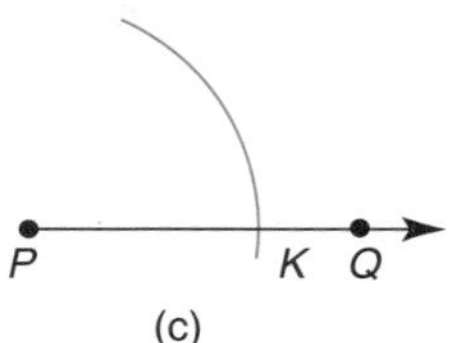

(c)

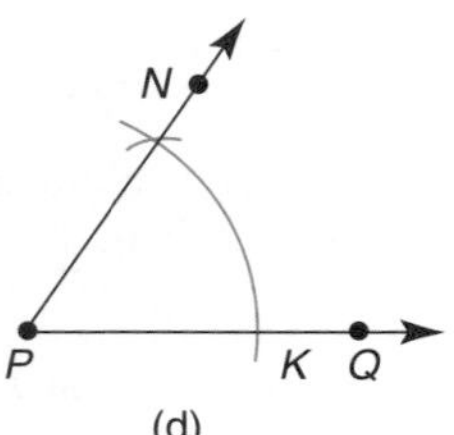

(d)

FIGURE 1.47

Summarizing, $x = 40$ and $y = 20$.

NOTE: $m\angle 5 = 120°$, $m\angle 8 = 60°$, and $m\angle 6 = 120°$.

Constructions with Angles

In Section 1.2, we considered Constructions 1 and 2 with line segments. Now consider two constructions that involve angle concepts. It will become clear why these methods are valid in Section 3.4. However, intuition suggests that the techniques are appropriate.

Construction 3:
To construct an angle congruent to a given angle.

Given: $\angle RST$ in Figure 1.47(a)

Construct: With $\overrightarrow{PQ}$ as one side, $\angle NPQ \cong \angle RST$

Construction: With a compass, mark an arc to intersect both sides of $\angle RST$ (at points G and H, respectively). [See Figure 1.47(b).] Without changing the radius, mark an arc to intersect $\overrightarrow{PQ}$ at K and the "would-be" second side of $\angle NPQ$. [See Figure 1.47(c).]

Now mark an arc to measure the distance from G to H. Using the same radius, mark an arc with K as center to intersect the would-be second side of the desired angle. Now draw the ray from P through the point of intersection of the two arcs. [See Figure 1.47(d).]

The resulting angle is the one desired, as we will prove in Section 3.4, Example 1.

Just as a line segment can be bisected, so can an angle. This takes us to a fourth construction method.

Construction 4:
To construct the angle bisector of a given angle.

Given: $\angle PRT$ in Figure 1.48(a).

Construct: $\overrightarrow{RS}$ so that $\angle PRS \cong \angle SRT$

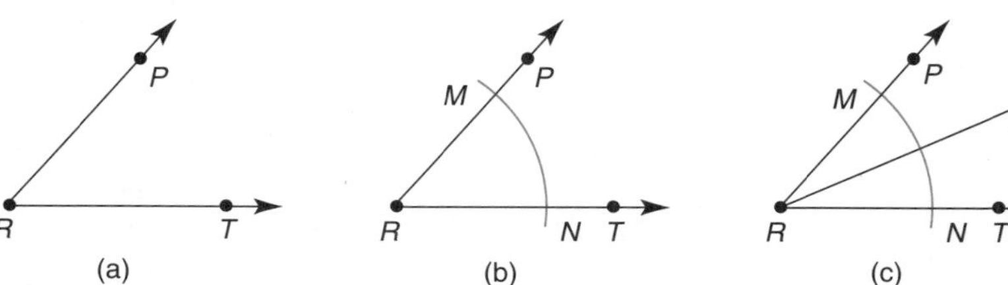

FIGURE 1.48

Construction: Using a compass, mark an arc to intersect the sides of $\angle PRT$ at points M and N. [See Figure 1.48(b).] Now, with M and N as centers, mark off two arcs with equal radii to intersect at point S in the interior of $\angle PRT$, as shown. Now draw ray RS, the desired angle bisector. [See Figure 1.48(c).]

Reasoning from the definition of an angle bisector, the Angle-Addition Postulate, and the Protractor Postulate, we can justify the following theorem.

THEOREM 1.4.1: There is one and only one angle bisector for a given angle.

This theorem is often stated, "The bisector of an angle is unique." This statement is proved in Example 5 of Section 2.2.

1.4 Exercises

Use drawings as needed to answer each of the following questions.

1. Must two rays with a common endpoint be coplanar? Must three rays with a common endpoint be coplanar?
2. Suppose that $\overrightarrow{AB}$, $\overrightarrow{AC}$, $\overrightarrow{AD}$, $\overrightarrow{AE}$, and $\overrightarrow{AF}$ are coplanar.

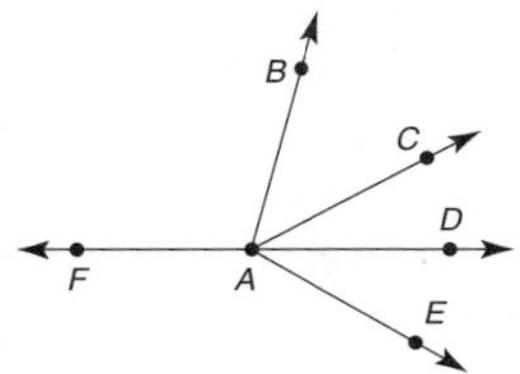

Exercises 2–5

Classify the following as true or false:

a) $m\angle BAC + m\angle CAD = m\angle BAD$
b) $\angle BAC \cong \angle CAD$
c) $m\angle BAE - m\angle DAE = m\angle BAC$
d) $\angle BAC$ and $\angle DAE$ are adjacent
e) $m\angle BAC + m\angle CAD + m\angle DAE = m\angle BAE$

3. Without using a protractor, name the type of angle represented by:

a) $\angle BAE$ b) $\angle FAD$ c) $\angle BAC$ d) $\angle FAE$

4. What, if anything, is wrong with the claim $m\angle FAB + m\angle BAE = m\angle FAE$?
5. $\angle FAC$ and $\angle CAD$ are adjacent and $\overrightarrow{AF}$ and $\overrightarrow{AD}$ are opposite rays. What may you conclude about $\angle FAC$ and $\angle CAD$?

Exercises 6–11

6. *Given:* $m\angle RST = 39°$
$m\angle TSV = 23°$
Find: $m\angle RSV$
7. *Given:* $m\angle RSV = 59°$
$m\angle TSV = 17°$
Find: $m\angle RST$
8. *Given:* $m\angle RST = 2x + 9$
$m\angle TSV = 3x - 2$
$m\angle RSV = 67°$
Find: x
9. *Given:* $m\angle RST = 2x - 10$
$m\angle TSV = x + 6$
$m\angle RSV = 4(x - 6)$
Find: x and $m\angle RSV$
10. *Given:* $m\angle RST = 5(x + 1) - 3$
$m\angle TSV = 4(x - 2) + 3$
$m\angle RSV = 4(2x + 3) - 7$
Find: x and $m\angle RSV$
11. *Given:* $\overrightarrow{ST}$ bisects $\angle RSV$
$m\angle RST = x + y$
$m\angle TSV = 2x - 2y$
$m\angle RSV = 64°$
Find: x and y

12. *Given:* $\overrightarrow{ST}$ bisects $\angle RSV$
$m\angle RST = 2x + 3y$
$m\angle TSV = 3x - y + 2$
$m\angle RSV = 80°$
Find: x and y

13. *Given:* $\overleftrightarrow{AB}$ and $\overleftrightarrow{AC}$ in plane P as shown
$\overleftrightarrow{AD}$ intersects P at point A
$\angle CAB \cong \angle DAC$
$\angle DAC \cong \angle DAB$
What may you conclude?

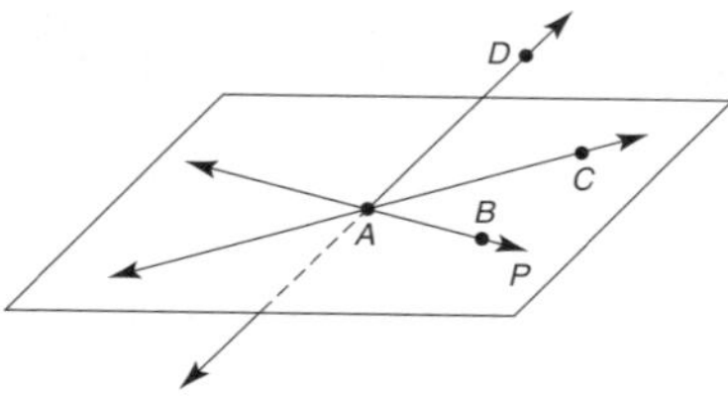

14. Two angles are complementary. One angle is 12° larger than the other. Using two variables x and y, find the size of each angle by solving a system of equations.

15. Two angles are supplementary. One angle is 24° more than twice the other. Using two variables x and y, find the measure of each angle.

16. For two complementary angles, find an expression for the measure of the second angle if the measure of the first is:

a) $x°$
b) $(3x - 12)°$
c) $(2x + 5y)°$

17. Suppose that two angles are supplementary. Find expressions for the supplements, using the expressions provided in Exercise 16, parts (a) to (c).

18. On the protractor shown, $\overrightarrow{NP}$ bisects $\angle MNQ$. Find x.

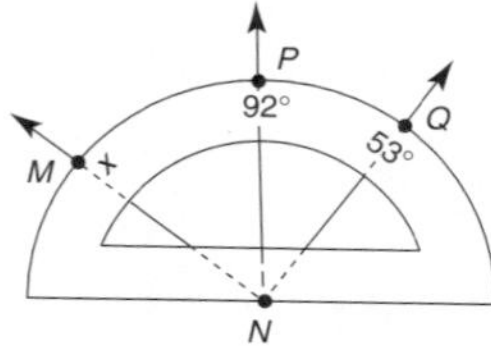

Exercises 18, 19

19. On the protractor shown, $\angle MNP$ and $\angle PNQ$ are complementary. Find x.

20. Classify as true or false:

a) If points P and Q lie in the interior of $\angle ABC$, then $\overline{PQ}$ lies in the interior of $\angle ABC$.

b) If points P and Q lie in the interior of $\angle ABC$, then $\overleftrightarrow{PQ}$ lies in the interior of $\angle ABC$.

c) If points P and Q lie in the interior of $\angle ABC$, then $\overrightarrow{PQ}$ lies in the interior of $\angle ABC$.

In Exercises 21 to 26, use only a compass and a straightedge to perform the indicated constructions.

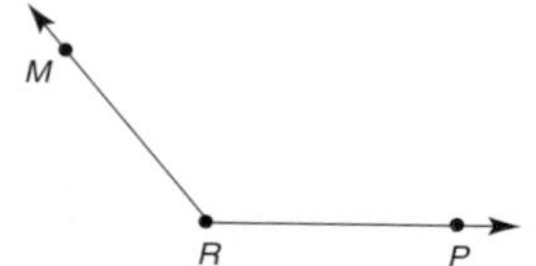

Exercises 21–23

21. *Given:* Obtuse $\angle MRP$
Construct: With $\overrightarrow{OA}$ as one side, an angle $\cong \angle MRP$

22. *Given:* Obtuse $\angle MRP$
Construct: $\overrightarrow{RS}$, the angle bisector of $\angle MRP$

23. *Given:* Obtuse $\angle MRP$
Construct: Rays $\overrightarrow{RS}$, $\overrightarrow{RT}$, and $\overrightarrow{RU}$ so that $\angle MRP$ is divided into four $\cong$ angles

24. *Given:* Straight $\angle DEF$
Construct: A right angle with vertex at E
(**HINT:** Use Construction 4.)

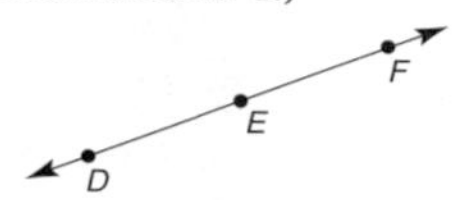

25. Draw a triangle with three acute angles. Construct angle bisectors for each of the three angles. Based on the appearance of your construction, what seems to be true?

26. *Given:* Acute $\angle 1$ and $\overline{AB}$
Construct: Triangle ABC with $\angle A \cong \angle 1$, $\angle B \cong \angle 1$, and base $\overline{AB}$

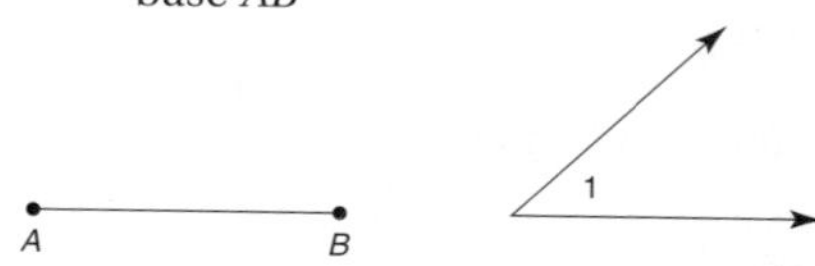

27. What seems to be true of two of the sides in the triangle you constructed in Exercise 26?

28. *Given:* Straight $\angle ABC$ and $\overrightarrow{BD}$
Construct: Bisectors of $\angle ABD$ and $\angle DBC$

What type of angle is formed by the bisectors of the two angles?

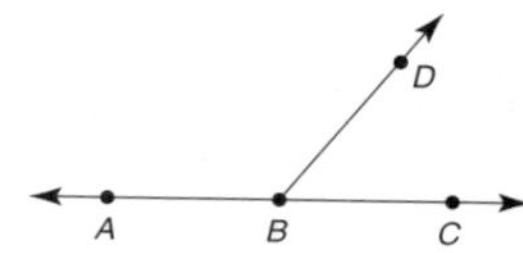

1.5 Introduction to Geometric Proof

This section introduces some guidelines for proving geometric properties. Several examples are offered to help you develop your own proofs. In the beginning, the form of proof will be a two-column proof, with Statements in the left column and Reasons in the right column. But where do the statements and reasons come from?

To deal with this question, you must ask "What" is it that is known (Given) and "Why" should the conclusion (Prove) follow from this information? Understanding the why may mean dealing with several related conclusions and thus several intermediate whys. In correctly piecing together a proof, you will usually scratch out several conclusions and reorder these. Of course, each conclusion must be justified by citing the Given (hypothesis), a previously stated definition or postulate, or a previously proven theorem.

Selected properties from algebra are often included in our list of "reasons" used to justify statements. For instance, we use the Addition Property of Equality to justify adding the same number to each side of an equation. Other reasons found in a proof may include the Subtraction Property of Equality, the Multiplication Property of Equality, and the Division Property of Equality. Substitution and the Transitive Property of Equality are often needed. The Transitive Property of Equality states, "If $a = b$ and $b = c$, then $a = c$."

DISCOVER!

In the diagram, the wooden trim pieces are mitered (cut at an angle) to be equal and to form a right angle when placed together. Use the properties of algebra to explain why the measures of $\angle 1$ and $\angle 2$ are each 45°. What you have done is an informal "proof."

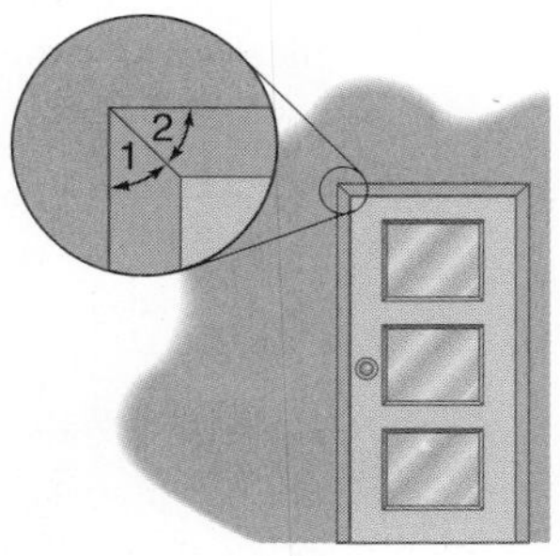

ANSWER

$m\angle 1 + m\angle 2 = 90°$. Since $m\angle 1 = m\angle 2$, we see that $m\angle 1 + m\angle 1 = 90°$. Thus $2 \cdot m\angle 1 = 90$ and, dividing by 2, we see that $m\angle 1 = 45°$. Then $m\angle 2 = 45°$ also.

This activity suggests that a more formal form of geometric proof exists.

The typical format for a geometric proof is as follows:

Given: ________ [Drawing]
Prove: ________

Consider this problem:

FIGURE 1.49

Given: A-P-B on $\overline{AB}$ (Figure 1.49)
Prove: $AP = AB - PB$

First consider the Drawing (Figure 1.49), and relate it to any additional information described by the Given. Then consider the Prove statement. Do you understand the claim, and does it seem reasonable? If it seems reasonable, the intermediate claims must be ordered and supported to form the contents of the proof. Since a proof must begin with the Given and conclude with the Prove, the proof of the preceding problem has this form:

PROOF

Statements	Reasons
1. A-P-B on $\overline{AB}$	1. Given
2. ?	2. ?
.	.
.	.
.	.
?. $AP = AB - PB$	?. ?

To construct the proof, you must glean from the Drawing and the Given that

$$AP + PB = AB$$

You can then deduce (through subtraction) that $AP = AB - PB$. Thus the complete proof problem will have this appearance.

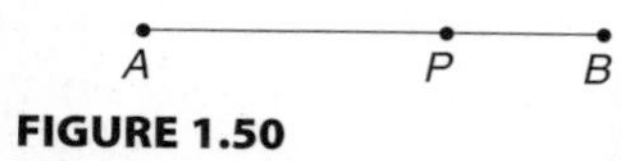
FIGURE 1.50

Given: A-P-B on $\overline{AB}$ (Figure 1.50)
Prove: $AP = AB - PB$

PROOF

Statements	Reasons
1. A-P-B on $\overline{AB}$	1. Given
2. $AP + PB = AB$	2. Segment-Addition Postulate
3. $AP = AB - PB$	3. Subtraction Property of Equality

Sample Proofs

Now consider this problem:

Given: $MN > PQ$
Prove: $MP > NQ$

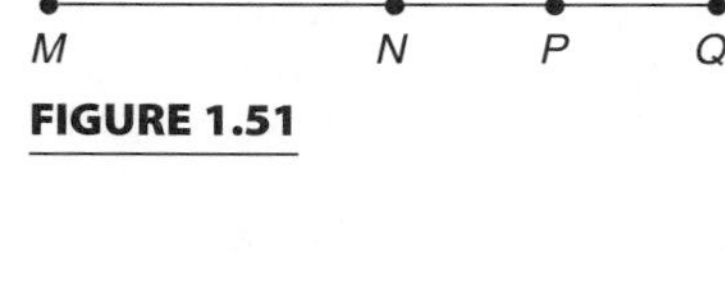

FIGURE 1.51

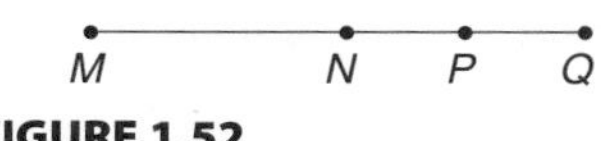

FIGURE 1.52

To understand the situation, first study the Drawing (Figure 1.51) and the related Given. Then read the Prove with reference to the drawing. Constructing the proof requires you to begin with the Given and end with the Prove. What may be confusing here is that the Given involves MN and PQ, while the Prove involves MP and NQ. However, this is easily remedied through the addition of NP to each side of the inequality $MN > PQ$; see step 2 in the following proof.

Given: $MN > PQ$ (Figure 1.52)
Prove: $MP > NQ$

PROOF

Statements	Reasons
1. $MN > PQ$	**1.** Given
2. $MN + NP > NP + PQ$	**2.** Addition Property of Inequality
3. But $MN + NP = MP$ and $NP + PQ = NQ$	**3.** Segment-Addition Postulate
4. $MP > NQ$	**4.** Substitution

NOTE: The final reason may come as a surprise! However, the Substitution Axiom of Equality allows you to replace a quantity with its equal in *any* statement—including an inequality!

Now that you have a better idea of the development of proofs, consider an example that involves an angle bisector.

EXAMPLE 1

Study this proof, noting the order of the statements and reasons.

Given: $\overrightarrow{ST}$ bisects $\angle RSU$
$\overrightarrow{SV}$ bisects $\angle USW$ (Figure 1.53)
Prove: $m\angle RST + m\angle VSW = m\angle TSV$

FIGURE 1.53

PROOF

Statements	Reasons
1. $\overrightarrow{ST}$ bisects $\angle RSU$	**1.** Given
2. $m\angle RST = m\angle TSU$	**2.** If an angle is bisected, then the measures of the resulting angles are equal
3. $\overrightarrow{SV}$ bisects $\angle USW$	**3.** Same as #1
4. $m\angle VSW = m\angle USV$	**4.** Same as #2
5. $m\angle RST + m\angle VSW = m\angle TSU + m\angle USV$	**5.** Addition Property of Equality (use the equations from statements 2 and 4)
6. $m\angle TSU + m\angle USV = m\angle TSV$	**6.** Angle-Addition Postulate
7. $m\angle RST + m\angle VSW = m\angle TSV$	**7.** Substitution

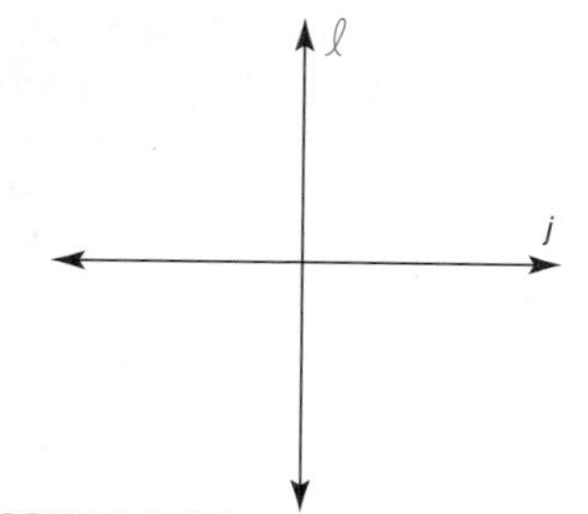

FIGURE 1.54

In the proof of Example 1, the Given information was split between statements 1 and 3; this was done for clarity, not out of necessity. In the proof, reason 2 was the definition of angle bisector. This should remind you of the importance of the role of definitions in a mathematical system.

Informally, a **vertical** line is one that extends up and down, like a flagpole. On the other hand, a line that extends left to right is **horizontal.** In Figure 1.54, ℓ is vertical and j is horizontal. Where lines ℓ and j intersect, they appear to form angles of equal measure.

DEFINITION: **Perpendicular lines** are two lines that meet to form congruent adjacent angles.

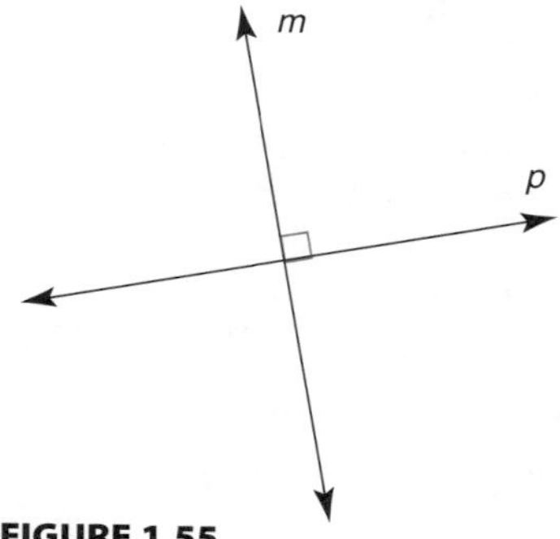

FIGURE 1.55

Perpendicular lines do not have to be vertical and horizontal. In Figure 1.55, the slanted lines m and p are perpendicular ($m \perp p$). As we have seen, a small square is often placed in the opening of an angle formed by perpendicular lines to signify that the lines are perpendicular.

The purpose of Example 2, which follows, is to establish the relationship between perpendicular lines and right angles. Example 2 is not a formal proof because it does not contain a statement of a theorem to be proven. Study this proof, noting the order of the statements and reasons. The numbers in parentheses to the left of the statements reference the earlier statement(s) upon which the new statement is based.

EXAMPLE 2

Given: $\overleftrightarrow{AB} \perp \overleftrightarrow{CD}$, intersecting at E (Figure 1.56)

Prove: $\angle AEC$ is a right angle

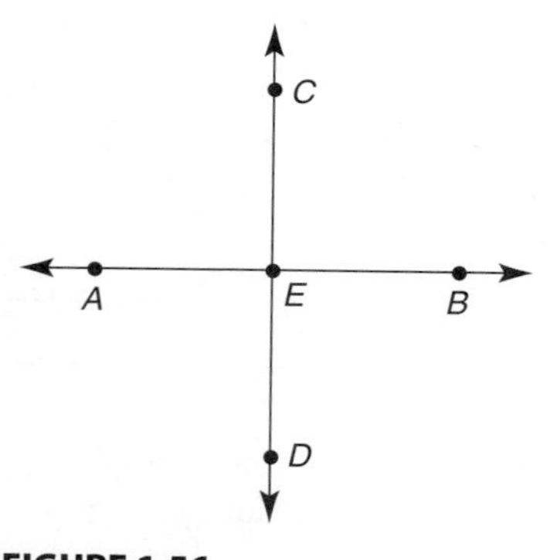

FIGURE 1.56

PROOF

	Statements	Reasons
	1. $\overleftrightarrow{AB} \perp \overleftrightarrow{CD}$, intersecting at E	1. Given
(1)	2. $\angle AEC \cong \angle CEB$	2. Perpendicular lines meet to form congruent adjacent angles (Definition)
(2)	3. $m\angle AEC = m\angle CEB$	3. If two angles are congruent, their measures are equal
	4. $\angle AEB$ is a straight angle and $m\angle AEB = 180°$	4. Measure of a straight angle equals 180°
	5. $m\angle AEC + m\angle CEB = m\angle AEB$	5. Angle-Addition Postulate
(4), (5)	6. $m\angle AEC + m\angle CEB = 180°$	6. Substitution
(3), (6)	7. $m\angle AEC + m\angle AEC = 180°$ or $2 \cdot m\angle AEC = 180°$	7. Substitution
(7)	8. $m\angle AEC = 90°$	8. Division Property of Equality
(8)	9. $\angle AEC$ is a right angle	9. If the measure of an angle is 90°, then the angle is a right angle

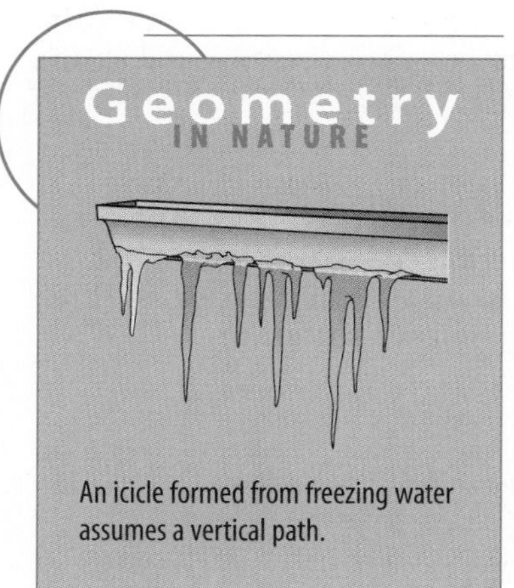

An icicle formed from freezing water assumes a vertical path.

Congruence of angles (or of line segments) is closely tied to equality of angle measures (or segment measures) by the definition of congruence. The following list gives some properties of the congruence of angles:

Reflexive: $\angle 1 \cong \angle 1$; an angle is congruent to itself.

Symmetric: If $\angle 1 \cong \angle 2$, then $\angle 2 \cong \angle 1$.

Transitive: If $\angle 1 \cong \angle 2$ and $\angle 2 \cong \angle 3$, then $\angle 1 \cong \angle 3$.

In later chapters, we will see that congruence of triangles and similarity of triangles also have reflexive, symmetric, and transitive properties.

Returning to the formulation of a proof, the final example in this section is based on the fact that vertical angles are congruent when two lines intersect. Because there are two pairs of congruent angles, the Prove could be stated:

Prove: $\angle 1 \cong \angle 3$ and $\angle 2 \cong \angle 4$

Such a conclusion is a conjunction and would be proved if both congruences were established. For simplicity, the Prove of Example 3 is stated:

Prove: $\angle 2 \cong \angle 4$

Study this proof, noting the order of the statements and reasons.

EXAMPLE 3

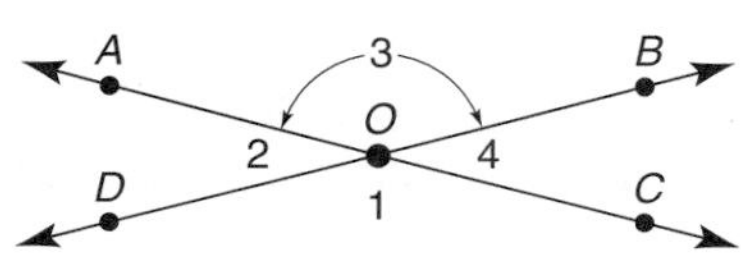

FIGURE 1.57

Given: $\overleftrightarrow{AC}$ intersects $\overleftrightarrow{BD}$ at O (Figure 1.57)

Prove: $\angle 2 \cong \angle 4$

PROOF

Statements	Reasons
1. $\overleftrightarrow{AC}$ intersects $\overleftrightarrow{BD}$ at O	**1.** Given
2. $\angle$s AOC and DOB are straight $\angle$s, with $m\angle AOC = 180$ and $m\angle DOB = 180$	**2.** The measure of a straight angle is 180°
3. $m\angle AOC = m\angle DOB$	**3.** Substitution
4. $m\angle 1 + m\angle 4 = m\angle DOB$ and $m\angle 1 + m\angle 2 = m\angle AOC$	**4.** Angle-Addition Postulate
5. $m\angle 1 + m\angle 4 = m\angle 1 + m\angle 2$	**5.** Substitution
6. $m\angle 4 = m\angle 2$	**6.** Subtraction Property of Equality
7. $\angle 4 \cong \angle 2$	**7.** If two angles are equal in measure, the angles are congruent
8. $\angle 2 \cong \angle 4$	**8.** Symmetric Axiom of Congruence of Angles

In the preceding proof, the degree symbol (°) has been omitted from the statements, and this will continue to be done in future proofs. Moreover, there is no need to reorder the congruent angles from statement 7 to statement 8 since congruence of angles is symmetric; in the later work, statement 7 will be written to match the Prove even if the previous line does not have the right order. The same type of thinking applies to proving lines perpendicular or parallel: the order is simply not important!

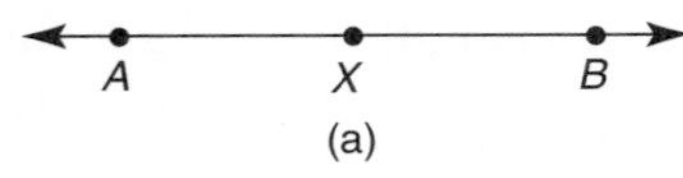

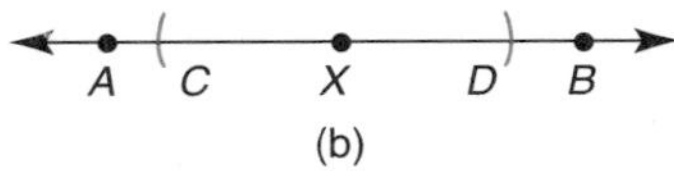

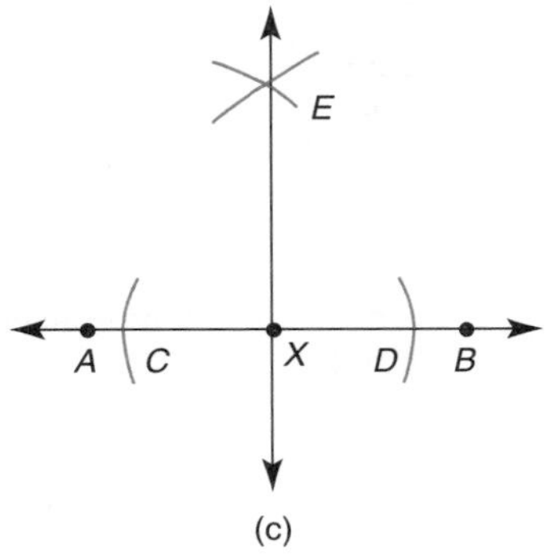

FIGURE 1.58

Constructions Leading to Perpendicular Lines

Construction 2 in Section 1.2 not only determined the midpoint of $\overline{AB}$ but also that of the **perpendicular bisector** of $\overline{AB}$. In many instances, we need the perpendicular line at a point other than the midpoint of a segment.

Construction 5:
To construct the line perpendicular to a given line at a specified point on the given line.

Given: $\overleftrightarrow{AB}$ with point X in Figure 1.58(a)

Construct: A line $\overleftrightarrow{EX}$, so that $\overleftrightarrow{EX} \perp \overleftrightarrow{AB}$

Construction: Using X as the center, mark off arcs of equal radii on each side of X to intersect $\overleftrightarrow{AB}$ at C and D. [See Figure 1.58(b).] Now, using C and D as centers, mark off arcs of equal radii with a length greater than XD so that these arcs intersect either above or below $\overleftrightarrow{AB}$.

Calling the point of intersection E, draw $\overleftrightarrow{EX}$, which is the desired perpendicular line. [See Figure 1.58(c).]

The theorem Construction 5 is based on is a consequence of the Protractor Postulate, and we state it without proof.

THEOREM 1.5.1: There is exactly one line perpendicular to a given line at any point on the line.

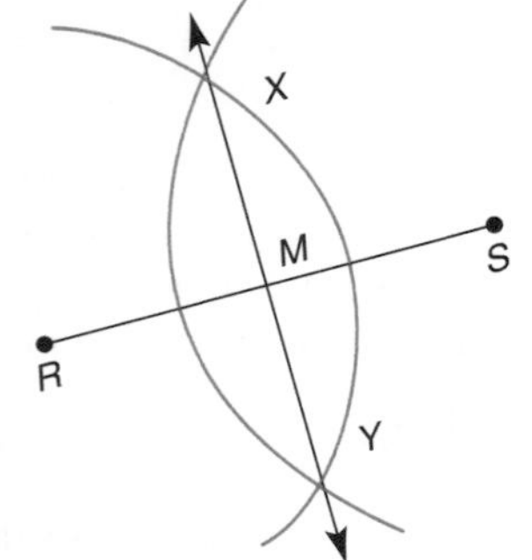

FIGURE 1.59

Construction 2, which was used to locate the midpoint of a line segment in Section 1.2, is also the method for constructing the perpendicular bisector of a line segment. In Figure 1.59, $\overleftrightarrow{XY}$ is the perpendicular bisector of $\overline{RS}$. The following theorem can be proved by methods developed later in this book.

THEOREM 1.5.2: The perpendicular bisector of a line segment is unique.

1.5 Exercises

In Exercises 1 and 2, supply reasons.

1. *Given:* ∠1 ≅ ∠3
Prove: ∠*MOP* ≅ ∠*NOQ*

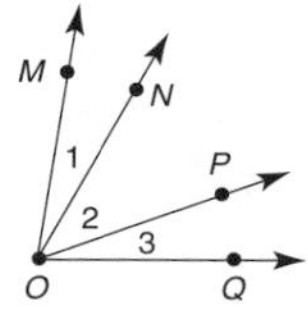

PROOF

Statements	Reasons
1. ∠1 ≅ ∠3	**1.** ?
2. m∠1 = m∠3	**2.** ?
3. m∠1 + m∠2 = m∠*MOP* and m∠2 + m∠3 = m∠*NOQ*	**3.** ?
4. m∠1 + m∠2 = m∠2 + m∠3	**4.** ?
5. m∠*MOP* = m∠*NOQ*	**5.** ?
6. ∠*MOP* ≅ ∠*NOQ*	**6.** ?

2. *Given:* $\overleftrightarrow{AB}$ intersects $\overleftrightarrow{CD}$ at *O* so that ∠1 is a right ∠ (Use the figure at the top of the next column.)
Prove: ∠2 and ∠3 are complementary

PROOF

Statements	Reasons
1. $\overleftrightarrow{AB}$ intersects $\overleftrightarrow{CD}$ at *O*	**1.** ?
2. ∠*AOB* is a straight ∠, so m∠*AOB* = 180	**2.** ?
3. m∠1 + m∠*COB* = m∠*AOB*	**3.** ?
4. m∠1 + m∠*COB* = 180	**4.** ?
5. ∠1 is a right angle	**5.** ?
6. m∠1 = 90	**6.** ?
7. 90 + m∠*COB* = 180	**7.** ?
8. m∠*COB* = 90	**8.** ?
9. m∠2 + m∠3 = m∠*COB*	**9.** ?
10. m∠2 + m∠3 = 90	**10.** ?
11. ∠2 and ∠3 are complementary	**11.** ?

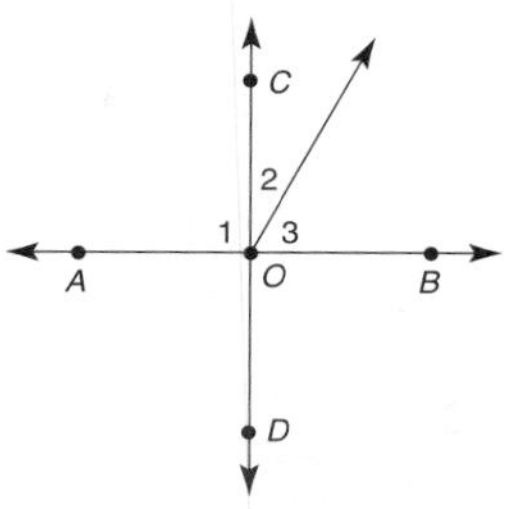

Exercise 2

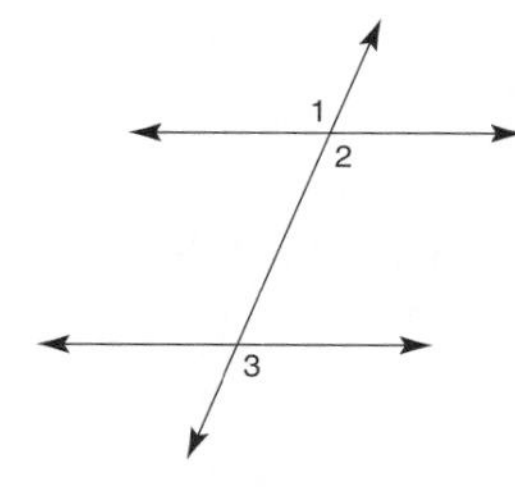

Exercise 3

In Exercises 3 and 4, supply statements.

3. *Given:* ∠1 ≅ ∠2 and ∠2 ≅ ∠3
Prove: ∠1 ≅ ∠3

PROOF

Statements	Reasons
1. ?	**1.** Given
2. ?	**2.** Transitive Property of Congruence

4. *Given:* m∠*AOB* = m∠1
m∠*BOC* = m∠1
Prove: $\overrightarrow{OB}$ bisects ∠*AOC*

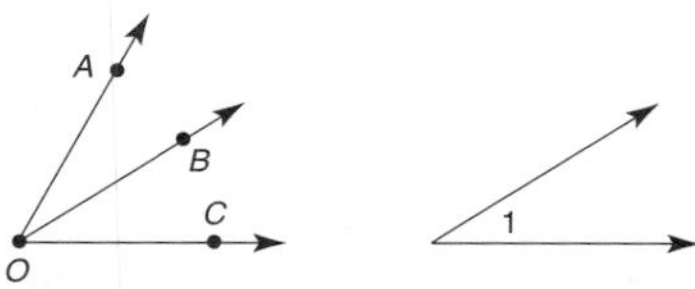

PROOF

Statements	Reasons
1. ?	**1.** Given
2. ?	**2.** Substitution
3. ?	**3.** Angles with equal measures are congruent
4. ?	**4.** If a ray divides an angle into two congruent angles, then the ray bisects the angle

In Exercises 5 to 9, use a compass and a straightedge to complete the constructions.

5. *Given:* Point N on line s
Construct: Line m through N so that $m \perp s$

6. *Given:* $\overrightarrow{OA}$
Construct: Right angle BOA
(**HINT:** Use a straightedge to extend $\overrightarrow{OA}$ to the left.)

7. *Given:* Line ℓ containing point A
Construct: A 45° angle with vertex at A

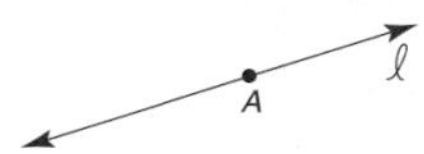

8. *Given:* $\overline{AB}$
Construct: The perpendicular bisector of $\overline{AB}$

9. *Given:* Triangle ABC
Construct: The perpendicular bisectors of sides $\overline{AB}$, $\overline{AC}$, and $\overline{BC}$

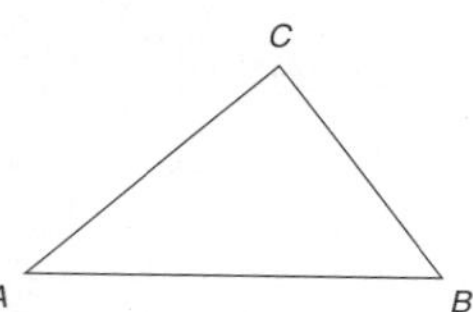

10. Draw a conclusion based on the results of Exercise 9.

In Exercises 11 and 12, provide the missing statements and reasons.

11. *Given:* $\angle$s 1 and 3 are complementary
$\angle$s 2 and 3 are complementary
Prove: $\angle 1 \cong \angle 2$

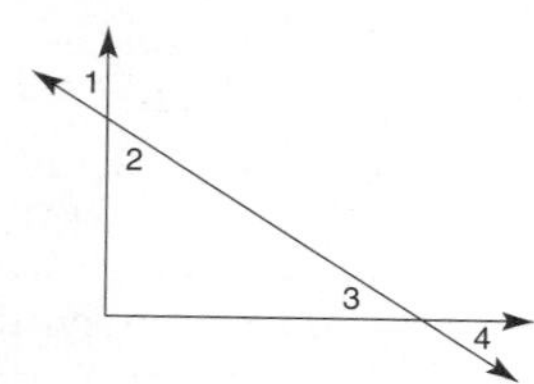

PROOF

Statements	Reasons
1. $\angle$s 1 and 3 are complementary; $\angle$s 2 and 3 are complementary	**1.** ?
2. $m\angle 1 + m\angle 3 = 90$; $m\angle 2 + m\angle 3 = 90$	**2.** The sum of the measures of complementary $\angle$s is 90
(2) **3.** $m\angle 1 + m\angle 3 = m\angle 2 + m\angle 3$	**3.** ?
4. ?	**4.** Subtraction Property of Equality
(4) **5.** ?	**5.** If two $\angle$s are = in measure, they are $\cong$

12. *Given:* $\angle 1 \cong \angle 2$; $\angle 3 \cong \angle 4$
$\angle$s 2 and 3 are complementary
Prove: $\angle$s 1 and 4 are complementary

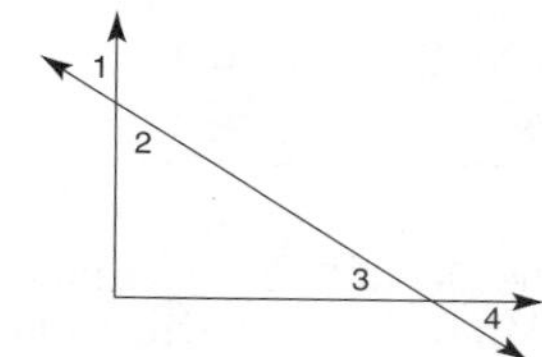

PROOF

Statements	Reasons
1. $\angle 1 \cong \angle 2$ and $\angle 3 \cong \angle 4$	**1.** ?
2. ? and ?	**2.** If two $\angle$s are $\cong$, then their measures are equal
3. $\angle$s 2 and 3 are complementary	**3.** ?
(3) **4.** ?	**4.** The sum of the measures of complementary $\angle$s is 90
(2), (4) **5.** $m\angle 1 + m\angle 4 = 90$	**5.** ?
6. ?	**6.** If the sum of the measures of two angles is 90, then the angles are complementary

13. Does the relation "is perpendicular to" have a reflexive property (consider line ℓ)? a symmetric property (consider lines ℓ and m)? a transitive property (consider lines ℓ, m, and n)?

14. Does the relation "is greater than" have a reflexive property (consider real number a)? a symmetric property (consider real numbers a and b)? a transitive property (consider real numbers a, b, and c)?

15. Does the relation "is a brother of" have a reflexive property (consider one male)? a symmetric property (consider two males)? a transitive property (consider three males)?

16. Does the relation "is in love with" have a reflexive property (consider one person)? a symmetric property (consider two people)? a transitive property (consider three people)?

17. By this time, the text has used numerous symbols and abbreviations. In this exercise, provide the *word* represented or abbreviated by each of the following:

a) $\perp$
b) $\angle$s
c) supp.
d) rt.
e) m$\angle 1$
f) adj.
g) comp.
h) $\overrightarrow{AB}$
i) $\cong$
j) vert.

18. If there were no understood restriction to lines in a plane in Theorem 1.5.1, the theorem would be false. Explain why the following statement is false: "In space, there is exactly one line perpendicular to a given line at any point on the line."

19. If there were no understood restriction to lines in a plane in Theorem 1.5.2, the theorem would be false. Explain why the following statement is false: "In space, the perpendicular bisector of a line segment is unique."

20. ➢ In the accompanying proof, provide the missing reasons.

Given: $\angle 1$ and $\angle 2$ are complementary
$\angle 1$ is acute
Prove: $\angle 2$ is also acute

PROOF

	Statements	Reasons
	1. $\angle 1$ and $\angle 2$ are complementary	**1.** ?
(1)	**2.** $m\angle 1 + m\angle 2 = 90$	**2.** ?
	3. $\angle 1$ is acute	**3.** ?
(3)	**4.** Where $m\angle 1 = x$, $0 < x < 90$	**4.** ?
(2)	**5.** $x + m\angle 2 = 90$	**5.** ?
(5)	**6.** $m\angle 2 = 90 - x$	**6.** ?
(4)	**7.** $-x < 0 < 90 - x$	**7.** ?
(7)	**8.** $90 - x < 90 < 180 - x$	**8.** ?
(7, 8)	**9.** $0 < 90 - x < 90$	**9.** ?
(6, 9)	**10.** $0 < m\angle 2 < 90$	**10.** ?
(10)	**11.** $\angle 2$ is acute	**11.** ?

1.6 The Formal Proof of a Theorem

Recall from Section 1.3 that statements that follow logically from known undefined terms, definitions, and postulates are called theorems. In other words, a theorem is a statement that can be proven. The formal proof of a theorem has several parts. To begin to understand how these parts are related, you need to consider carefully the terms "hypothesis" and "conclusion." The hypothesis of a statement describes the given situation (Given), while the conclusion describes what you need to establish (Prove). When a statement has the form "If H, then C," the hypothesis is the H statement and the conclusion is the C statement. Some theorems must be reworded to fit into "If . . . , then . . ." form so that the hypothesis and conclusion are easy to recognize.

EXAMPLE 1

Give the hypothesis H and conclusion C for each of these statements:

a) If two lines intersect, then the vertical angles formed are congruent.

b) All right angles are congruent.
c) Parallel lines do not intersect.
d) Lines are perpendicular when they meet to form congruent adjacent angles.

Solution

a) As is — H: Two lines intersect.
C: The vertical angles formed are congruent.

b) Reworded — If two angles are right angles, then these angles are congruent.
H: Two angles are right angles.
C: The angles are congruent.

c) Reworded — If two lines are parallel, then these lines do not intersect.
H: Two lines are parallel.
C: The lines do not intersect.

d) Reordered — When (if) two lines meet to form congruent adjacent angles, these lines are perpendicular.
H: Two lines meet to form congruent adjacent angles.
C: The lines are perpendicular.

Why do we need to distinguish between the hypothesis and the conclusion? For a theorem, the hypothesis determines the Drawing and the Given, providing a description of the Drawing's known characteristics. The conclusion determines what you wish to establish (the Prove) concerning the Drawing.

The Written Parts of a Formal Proof

The five necessary parts of a formal proof are listed in the accompanying box in the order in which they should be developed.

ESSENTIAL PARTS OF THE FORMAL PROOF OF A THEOREM

1. *Statement:* States the theorem to be proved.
2. *Drawing:* Represents the hypothesis of the theorem.
3. *Given:* Describes the Drawing according to the information found in the hypothesis of the theorem.
4. *Prove:* Describes the Drawing according to the claim made in the conclusion of the theorem.
5. *Proof:* Orders a list of claims (Statements) and justifications (Reasons), beginning with the Given and ending with the Prove; there must be a logical flow in this Proof.

The most difficult part of a formal proof is the thinking process that must take place between parts 4 and 5. This game plan or analysis involves deducing and ordering conclusions based on the given situation. One must be somewhat like a lawyer—selecting the claims that help prove the case, while discarding those that are superfluous. In the process of ordering the statements, it may be beneficial to think in reverse order, like so:

The Prove statement would be true if what else were true?

The final proof must be arranged in an order that allows one to reason from an earlier statement to a later claim by using deduction (perhaps several times).

H: hypothesis	⟵	statement of proof
P: principle	⟵	reason of proof
∴ C: conclusion	⟵	next statement in proof

Consider the following theorem whose proof is given in Example 2.

THEOREM 1.6.1: If two lines are perpendicular, then they meet to form right angles.

EXAMPLE 2

Write the parts of the formal proof of Theorem 1.6.1.

Solution

1. State the theorem.

   If two lines are perpendicular, then they meet to form right angles.

2. The hypothesis is H: Two lines are perpendicular.
   Make a Drawing to fit this description. (See Figure 1.60.)
3. Write the Given statement, using the Drawing and based on the hypothesis H: Two lines are $\perp$.
   Given: $\overleftrightarrow{AB} \perp \overleftrightarrow{CD}$ intersecting at E

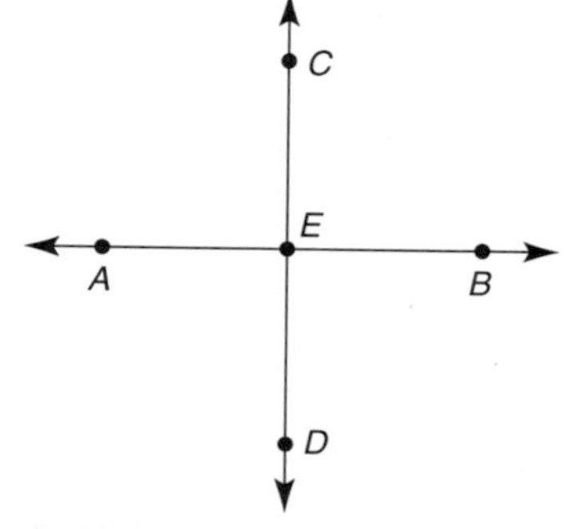

FIGURE 1.60

4. Write the Prove statement, using the Drawing and based on the conclusion C: They meet to form right angles.
   Prove: $\angle AEC$ is a right angle.
5. Construct the Proof. This proof is found in Example 2, Section 1.5. A formal proof would differ from Example 2 only by being preceded by the statement of the theorem!

Converse of a Statement

The **converse** of the statement "If *P*, then *Q*" is "If *Q*, then *P*." That is, the converse of a given statement interchanges its hypothesis and conclusion. Consider the following:

Statement: If a person lives in London, then that person lives in England.

Converse: If a person lives in England, then that person lives in London.

In this case, the given statement is true while its converse is false. For more information on converse statements, see Appendix B. Sometimes the converse of a true statement is also true. In fact, Example 3 presents the formal proof of a theorem that is the converse of Theorem 1.6.1.

■ **Warning**

You should not make a drawing that embeds qualities beyond those described in the hypothesis; neither should your drawing indicate fewer qualities than the hypothesis prescribes! ■

Once a theorem has been proved, it may be cited thereafter as a reason in future proofs. Thus, any theorem found in this section can be used for justification in later sections.

The proof that follows is nearly complete! It is difficult to provide a complete formal proof that explains the "how to" and simultaneously presents the final polished form. Thus, Example 2 aims at the how to, while Example 3 illustrates the polished form. What you do not see in Example 3 are the thought and scratch paper needed to piece this puzzle together.

The proof of a theorem is not unique! From the start, students' Drawings need not match, although the same relationships should be indicated. Certainly different letters are likely to be chosen in illustrating the hypothesis.

THEOREM 1.6.2: If two lines meet to form a right angle, then these lines are perpendicular.

EXAMPLE 3

Give a formal proof for Theorem 1.6.2.

If two lines meet to form a right angle, then these lines are perpendicular.

Given: $\overleftrightarrow{AB}$ and $\overleftrightarrow{CD}$ intersect at E so that $\angle AEC$ is a right angle (Figure 1.61)

Prove: $\overleftrightarrow{AB} \perp \overleftrightarrow{CD}$

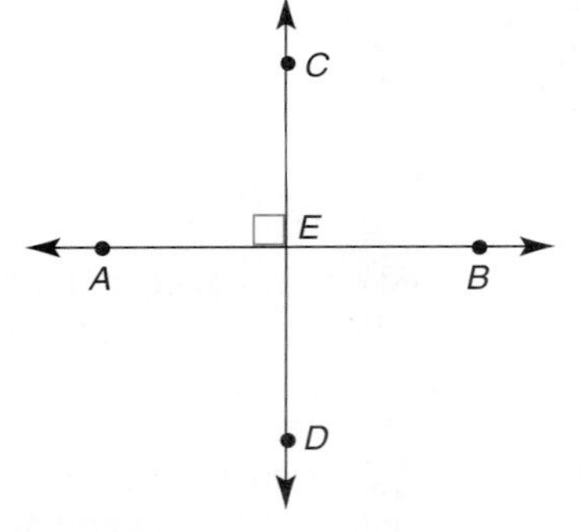

FIGURE 1.61

PROOF

	Statements	Reasons
	1. $\overleftrightarrow{AB}$ and $\overleftrightarrow{CD}$ intersect so that $\angle AEC$ is a right angle	**1.** Given
	2. $m\angle AEC = 90$	**2.** If an $\angle$ is a right $\angle$, its measure is 90
	3. $\angle AEB$ is a straight $\angle$, so $m\angle AEB = 180$	**3.** If an $\angle$ is a straight $\angle$, its measure is 180
	4. $m\angle AEC + m\angle CEB = m\angle AEB$	**4.** Angle-Addition Postulate
(2), (3), (4)	**5.** $90 + m\angle CEB = 180$	**5.** Substitution
(5)	**6.** $m\angle CEB = 90$	**6.** Subtraction Property of Equality
(2), (6)	**7.** $m\angle AEC = m\angle CEB$	**7.** Substitution
	8. $\angle AEC \cong \angle CEB$	**8.** If two $\angle$s have = measures, the $\angle$s are $\cong$
	9. $\overleftrightarrow{AB} \perp \overleftrightarrow{CD}$	**9.** If two lines form $\cong$ adjacent $\angle$s, these lines are $\perp$

Several other theorems are now stated, the proofs of which are left as exercises. This list contains theorems that are quite useful when cited as reasons in later proofs. Proof of Theorem 1.6.8 is provided.

THEOREM 1.6.3: If two angles are complementary to the same angle (or to congruent angles), then these angles are congruent.

THEOREM 1.6.4: If two angles are supplementary to the same angle (or to congruent angles), then these angles are congruent.

THEOREM 1.6.5: If two lines intersect, then the vertical angles formed are congruent.

THEOREM 1.6.6: Any two right angles are congruent.

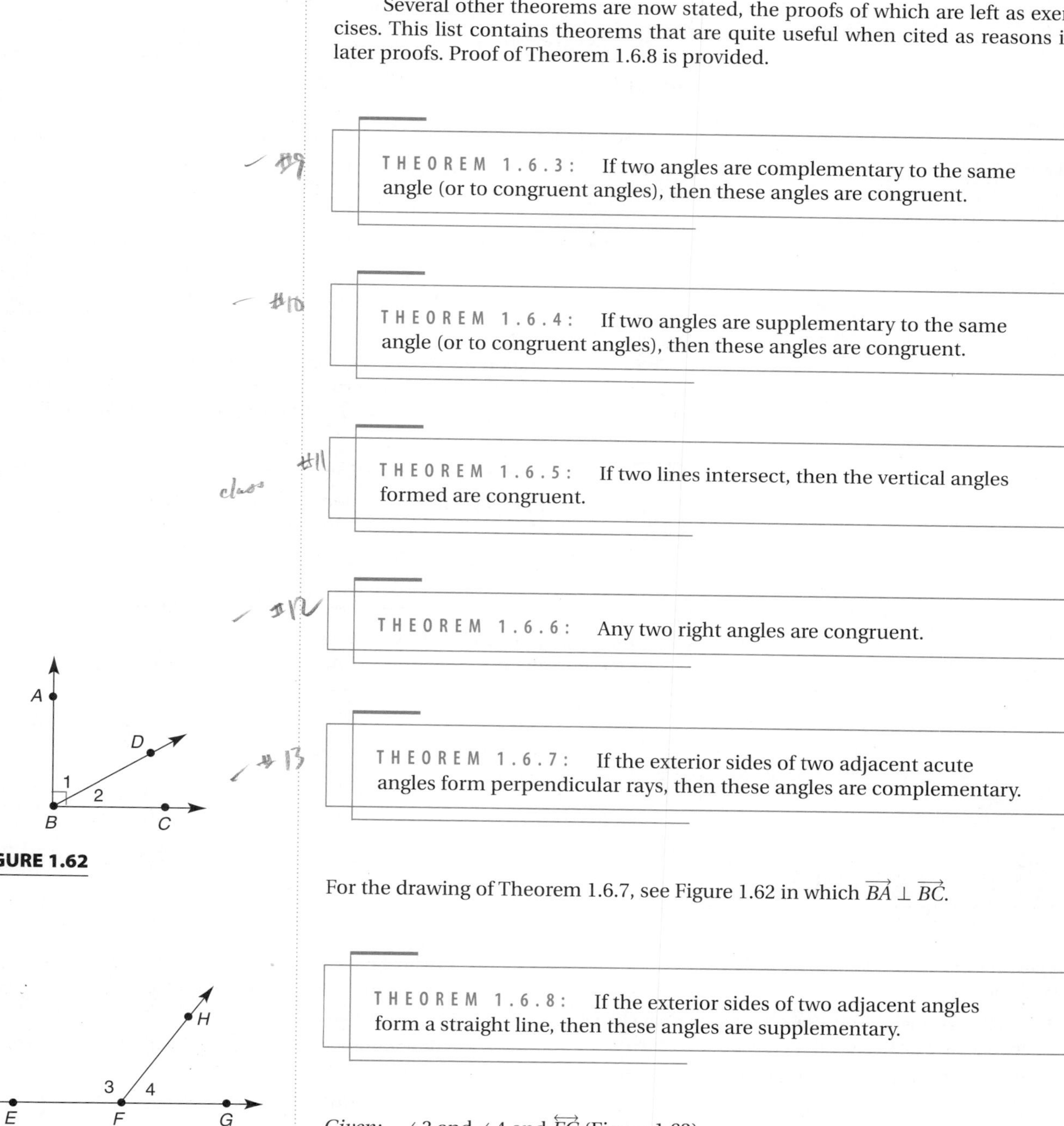

FIGURE 1.62

THEOREM 1.6.7: If the exterior sides of two adjacent acute angles form perpendicular rays, then these angles are complementary.

For the drawing of Theorem 1.6.7, see Figure 1.62 in which $\overrightarrow{BA} \perp \overrightarrow{BC}$.

THEOREM 1.6.8: If the exterior sides of two adjacent angles form a straight line, then these angles are supplementary.

FIGURE 1.63

Given: $\angle 3$ and $\angle 4$ and $\overleftrightarrow{EG}$ (Figure 1.63)

Prove: $\angle 3$ and $\angle 4$ are supplementary

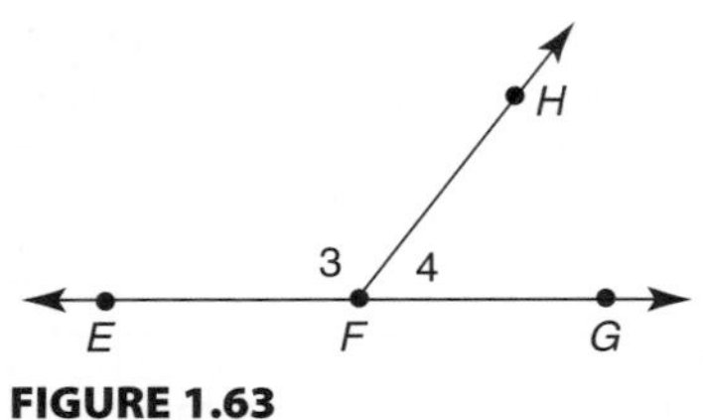

FIGURE 1.63

PROOF

Statements	Reasons
1. ∠3 and ∠4 and $\overleftrightarrow{EG}$	1. Given
2. m∠3 + m∠4 = m∠*EFG*	2. Angle-Addition Postulate
3. ∠*EFG* is a straight angle	3. If the sides of an ∠ are opposite rays, it is a straight ∠
4. m∠*EFG* = 180	4. The measure of a straight ∠ is 180
5. m∠3 + m∠4 = 180	5. Substitution
6. ∠3 and ∠4 are supplementary	6. If the sum of the measures of two ∠s is 180, the ∠s are supplementary

The final two theorems in this section are stated for convenience. When cited as "reasons," they will make later proofs easier to complete. We suggest that the student make drawings to illustrate Theorem 1.6.9 and Theorem 1.6.10.

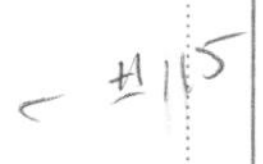

THEOREM 1.6.9: If two segments are congruent, then their midpoints separate these segments into four congruent segments.

THEOREM 1.6.10: If two angles are congruent, then their bisectors separate these angles into four congruent angles.

1.6 Exercises

In Exercises 1 to 6, state the hypothesis H and conclusion C for each statement.

1. If a line segment is bisected, then each of the equal segments has half the length of the original segment.
2. If two sides of a triangle are congruent, then the triangle is isosceles.
3. All squares are quadrilaterals.
4. Every regular polygon has congruent interior angles.
5. Two angles are congruent if each is a right angle.
6. The lengths of corresponding sides of similar polygons are proportional.
7. Name, in order, the five parts of the formal proof of a theorem.
8. Which part (hypothesis or conclusion) of a theorem determines the

   a) Drawing? **b)** Given? **c)** Prove?

In Exercises 9 to 17, complete the formal proof of each theorem.

9. If two angles are complementary to the same angle, then these angles are congruent.

Given: $\angle 1$ is comp. to $\angle 3$
$\angle 2$ is comp. to $\angle 3$
Prove: $\angle 1 \cong \angle 2$

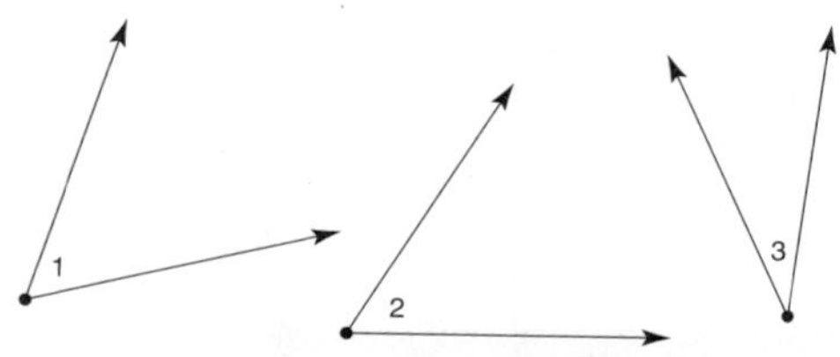

PROOF

Statements	Reasons
1. $\angle 1$ is comp. to $\angle 3$ $\angle 2$ is comp. to $\angle 3$	**1.** ?
2. $m\angle 1 + m\angle 3 = 90$ $m\angle 2 + m\angle 3 = 90$	**2.** ?
3. $m\angle 1 + m\angle 3 = m\angle 2 + m\angle 3$	**3.** ?
4. $m\angle 1 = m\angle 2$	**4.** ?
5. $\angle 1 \cong \angle 2$	**5.** ?

10. If two angles are supplementary to the same angle, then these angles are congruent.

Given: $\angle 1$ is supp. to $\angle 2$
$\angle 3$ is supp. to $\angle 2$
Prove: $\angle 1 \cong \angle 3$
(**HINT:** See Exercise 9 for help.)

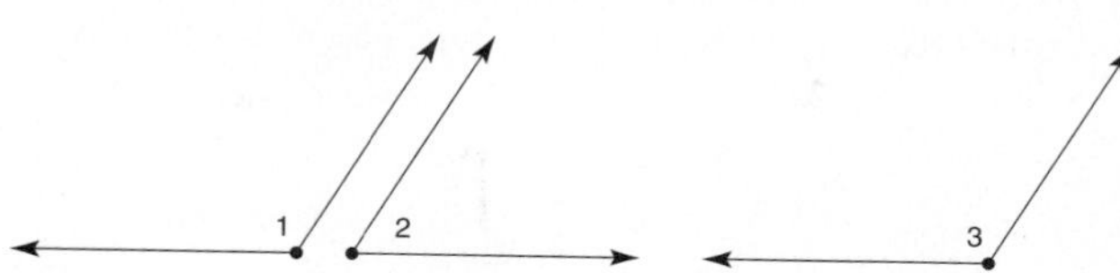

11. If two lines intersect, the vertical angles formed are congruent.
(**HINT:** See Example 3 of Section 1.5.)

12. Any two right angles are congruent.

13. If the exterior sides of two adjacent acute angles form perpendicular rays, then these angles are complementary.

Given: $\overleftrightarrow{BA} \perp \overrightarrow{BC}$
Prove: $\angle 1$ is comp. to $\angle 2$

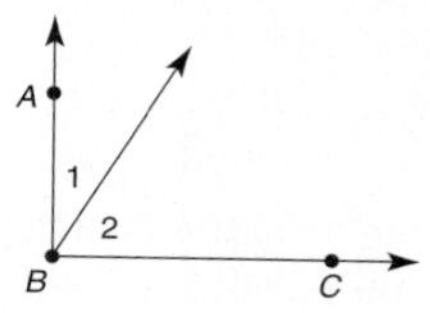

PROOF

Statements	Reasons
1. $\overleftrightarrow{BA} \perp \overrightarrow{BC}$	**1.** ?
2. ?	**2.** If two rays are $\perp$, then they meet to form a rt. $\angle$
3. $m\angle ABC = 90$	**3.** ?
4. $m\angle ABC = m\angle 1 + m\angle 2$	**4.** ?
5. $m\angle 1 + m\angle 2 = 90$	**5.** Substitution
6. ?	**6.** If the sum of the measures of two angles is 90, then the angles are complementary

14. If two line segments are congruent, then their midpoints separate these segments into four congruent segments.

Given: $\overline{AB} \cong \overline{DC}$
M is the midpoint of $\overline{AB}$
N is the midpoint of $\overline{DC}$
Prove: $\overline{AM} \cong \overline{MB} \cong \overline{DN} \cong \overline{NC}$

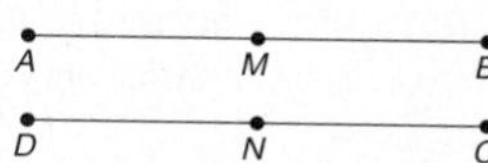

15. If two angles are congruent, then their bisectors separate these angles into four congruent angles.

Given: $\angle ABC \cong \angle EFG$
$\overrightarrow{BD}$ bisects $\angle ABC$
$\overrightarrow{FH}$ bisects $\angle EFG$
Prove: $\angle 1 \cong \angle 2 \cong \angle 3 \cong \angle 4$

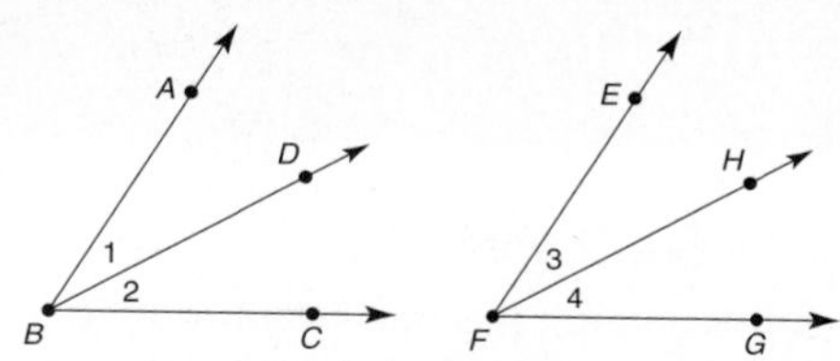

16. The bisectors of two adjacent supplementary angles form a right angle. (See the figure at the right.)

Given: $\angle ABC$ is supp. to $\angle CBD$
$\overrightarrow{BE}$ bisects $\angle ABC$
$\overrightarrow{BF}$ bisects $\angle CBD$

Prove: $\angle EBF$ is a right angle

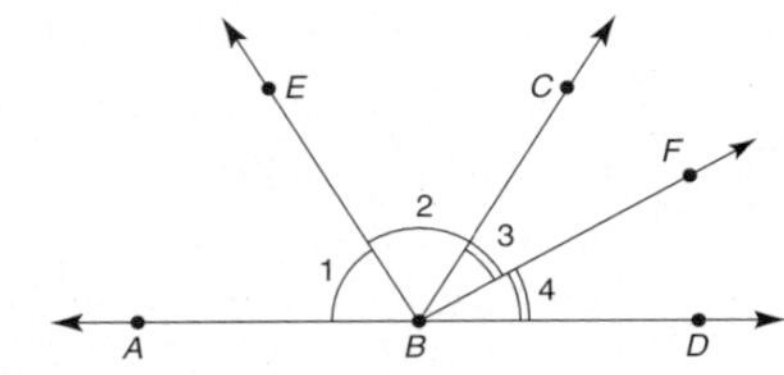

Exercise 16

17. The supplement of an acute angle is an obtuse angle. (**HINT:** Use Exercise 20 of Section 1.5 as a guide.)

A Look Beyond: Historical Sketch of Euclid

Names often associated with the early development of Greek mathematics, beginning in approximately 600 B.C., include Thales, Pythagoras, Archimedes, Appolonius, Diophantus, Eratosthenes, and Heron. However, the name most often associated with traditional geometry is that of Euclid, who lived around 300 B.C.

Euclid, himself a Greek, was asked to head the mathematics department at the University of Alexandria (in Egypt), which was the center of Greek learning. It is believed that Euclid told Ptolemy (the local ruler) that "There is no royal road to geometry," in response to Ptolemy's request for a quick and easy knowledge of the subject.

Euclid's best known work is the *Elements,* a systematic treatment of geometry with some algebra and number theory. That work, which consists of 13 volumes, has dominated the study of geometry for more than 2000 years. Most secondary-level geometry courses, even today, are based on Euclid's *Elements* and in particular on these volumes:

Book I: Triangles and congruence, parallels, quadrilaterals, the Pythagorean theorem, and area relationships

Book III: Circles, chords, secants, tangents, and angle measurement

Book IV: Constructions and regular polygons

Book VI: Similar triangles, proportions, and the Angle Bisector theorem

Book XI: Lines and planes in space, and parallelepipeds

One of Euclid's theorems was a forerunner of the theorem of trigonometry known as the Law of Cosines. Although it is difficult to understand now, it will make sense to you later. As stated by Euclid, "In an obtuse-angled triangle, the square of the side opposite the obtuse angle equals the sum of squares of the other two sides and the product of one side and the projection of the other upon it."

While it is believed that Euclid was a great teacher, he is also recognized as a great mathematician and as the first author of an elaborate textbook. In Chapter 2, Euclid's Parallel Postulate is central to the study of plane geometry.

Summary

A Look Back at Chapter 1

Our goal in this chapter has been to introduce geometry. We discussed the types of reasoning that are used to develop geometric relationships. The use of the tools of measurement (ruler and protractor) was described. We encountered the four elements of a mathematical system: undefined terms, definitions, postulates, and theorems. The undefined terms were needed to lay the foundation for defining new terms. The postulates were needed to lay the foundation for the theorems we proved here and for the theorems that lie ahead. Constructions presented in this chapter included the bisector of an angle and the perpendicular to a line at a point on the line.

■ *A Look Ahead to Chapter 2*

In the next chapter, the theorems we will prove are based on a postulate known as the Parallel Postulate. A new method of proof, called indirect proof, will be introduced and used in later chapters. While many of the theorems in Chapter 2 deal with parallel lines, several theorems in the chapter deal with the angles of a polygon.

■ *Important Terms and Concepts of Chapter 1*

1.1 Statement
Variable
Conjunction
Disjunction
Negation
Implication (Conditional)
Hypothesis
Conclusion
Intuition
Induction
Deduction
Argument (Valid and Invalid)
Law of Detachment
Counterexample

1.2 Point
Line
Collinear points
Vertex
Line segment
Betweenness of Points
Midpoint
Congruent
Protractor
Parallel Lines
Bisect
Straight Angle
Right Angle
Intersect
Perpendicular
Compass
Constructions
Circle
Arc
Radius

1.3 Mathematical System
Axiom or Postulate
Assumption
Theorem
Ruler Postulate
Segment-Addition Postulate
Congruent Segments
Midpoint of a Line Segment
Bisector of a Line Segment
Union
Ray
Opposite Rays
Intersection of two geometric figures
Parallel Lines
Plane
Coplanar Points
Space
Parallel, Vertical, Horizontal Planes

1.4 Angle
Sides of an Angle
Vertex of an Angle
Protractor Postulate
Acute, Right, Obtuse, and Straight Angles
Angle-Addition Postulate
Adjacent Angles
Congruent Angles
Bisector of an Angle
Complementary and Supplementary Angles
Vertical Angles

1.5 Vertical Lines and Horizontal Lines
Perpendicular Lines
Reflexive, Symmetric, and Transitive Properties of Congruence
Perpendicular Bisector of a Line Segment

1.6 Converse
Formal Proof of a Theorem

■ **A Look Beyond:** *Historical Sketch of Euclid*

Review Exercises

1. Name the four components of a mathematical system.
2. Name three types of reasoning.
3. Name the four characteristics of a good definition.

In Review Exercises 4 to 6, name the type of reasoning illustrated.

4. While watching the pitcher warm up, Phillip thinks, "I'll be able to hit against him."
5. Laura is away at camp. On the first day, her mother brings her additional clothing. On the second day, her mother brings her another pair of shoes. On the third day, her mother brings her cookies. Laura concludes that her mother misses her.
6. Sarah knows the rule, "A number (not 0) divided by itself equals 1." The teacher asks Sarah, "What is 5 divided by 5?" Sarah says, "The answer is 1."

In Review Exercises 7 and 8, state the hypothesis and conclusion for each statement.

7. If the diagonals of a trapezoid are equal in length, then the trapezoid is isosceles.
8. The diagonals of a parallelogram are congruent if the parallelogram is a rectangle.

In Review Exercises 9 to 11, draw a valid conclusion where possible.

9. 1. If a person has a good job, then that person has a college degree.
   2. Billy Fuller has a college degree.

   C. $\therefore$?

10. 1. If a person has a good job, then that person has a college degree.
    2. Jody Smithers has a good job.

    C. $\therefore$?

11. 1. If the measure of an angle is 90°, then that angle is a right angle.
    2. Angle A is not a right angle.

    C. $\therefore$?

12. A, B, and C are three points on a line. $AC = 8$, $BC = 4$, and $AB = 12$. Which point must be between the other two points?

13. Use three letters to name the angle shown. Also use one letter to name the same angle. Decide whether the angle is less than 90°, equal to 90°, or greater than 90°.

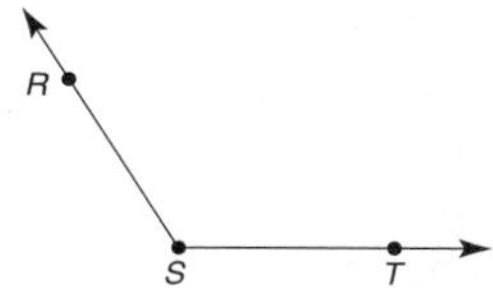

14. Figure $MNPQ$ is a rhombus. Draw diagonals $\overline{MP}$ and $\overline{QN}$ of the rhombus. How do $\overline{MP}$ and $\overline{QN}$ appear to be related?

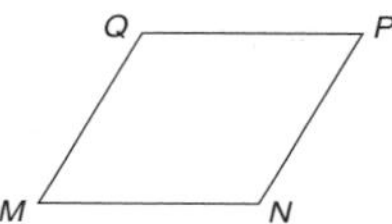

In Review Exercises 15 to 17, sketch and label the figures described.

15. Points A, B, C, and D are coplanar. A, B, and C are the only three of these points that are collinear.
16. Line ℓ intersects plane X at point P.
17. Plane M contains intersecting lines j and k.
18. *Given:* $\overrightarrow{BD}$ bisects $\angle ABC$
    $m\angle ABD = 2x + 15$
    $m\angle DBC = 3x - 2$
    Find: $m\angle ABC$
19. *Given:* $m\angle ABD = 2x + 5$
    $m\angle DBC = 3x - 4$
    $m\angle ABC = 86°$
    Find: $m\angle DBC$

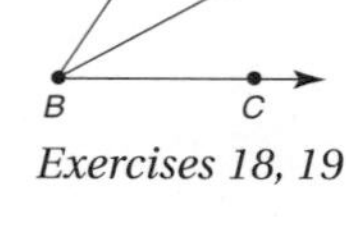

Exercises 18, 19

20. *Given:* $AM = 3x - 1$
    $MB = 4x - 5$
    M is the midpoint of $\overline{AB}$
    Find: AB
21. *Given:* $AM = 4x - 4$
    $MB = 5x + 2$
    $AB = 25$
    Find: MB

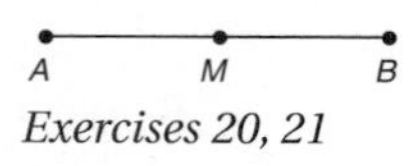

Exercises 20, 21

22. *Given:* D is the midpoint of $\overline{AC}$
$\overline{AC} \cong \overline{BC}$
$CD = 2x + 5$
$BC = x + 28$
Find: AC

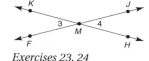

23. *Given:* $m\angle 3 = 7x - 21$
$m\angle 4 = 3x + 7$
Find: $m\angle FMH$

24. *Given:* $m\angle FMH = 4x + 1$
$m\angle 4 = x + 4$
Find: $m\angle 4$

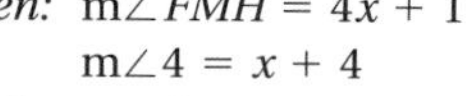
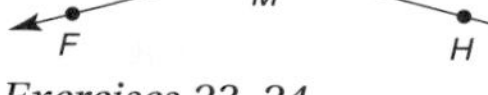

Exercises 23, 24

25. *Given:* $\angle EFG$ is a right angle
$m\angle HFG = 2x - 6$
$m\angle EFH = 3 \cdot m\angle HFG$
Find: $m\angle EFH$

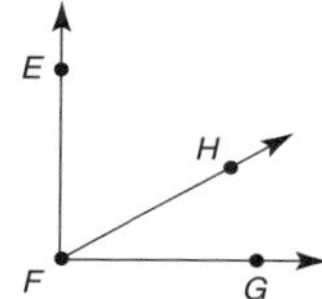

26. Two angles are supplementary. One angle is 40° more than four times the other. Find the measures of the two angles.

27. **a)** Write an expression for the perimeter of the triangle shown in the figure.
b) If the perimeter is 32 centimeters, find the value of x.
c) Find the length of each side of the triangle.

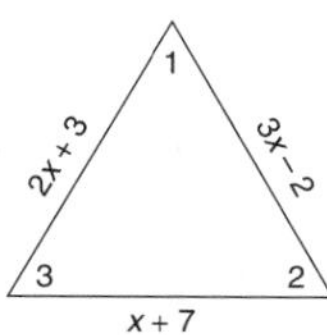

28. The sum of the measures of all three angles of the triangle in Review Exercise 27 is 180°. If the sum of the measures of angles 1 and 2 is more than 130°, what can you conclude about the measure of angle 3?

29. Susan wants to have a 4-ft board with some pegs on it. She wants to leave 6 in. on each end and 4 in. between each peg. How many pegs will fit on the board? [**HINT:** If n represents the number of pegs, then $(n - 1)$ represents the number of equal spaces.]

State whether the statements in Review Exercises 30 to 34 are always true (A), sometimes true (S), or never true (N).

30. If $AM = MB$, then A, M, and B are collinear.

31. If two angles are congruent, then they are right angles.

32. The bisectors of vertical angles are opposite rays.

33. Complementary angles are congruent.

34. The supplement of an obtuse angle is another obtuse angle.

35. Fill in the missing statements or reasons.

Given: $\angle 1 \cong \angle P$
$\angle 4 \cong \angle P$
$\overrightarrow{VP}$ bisects $\angle RVO$
Prove: $\angle TVP \cong \angle MVP$

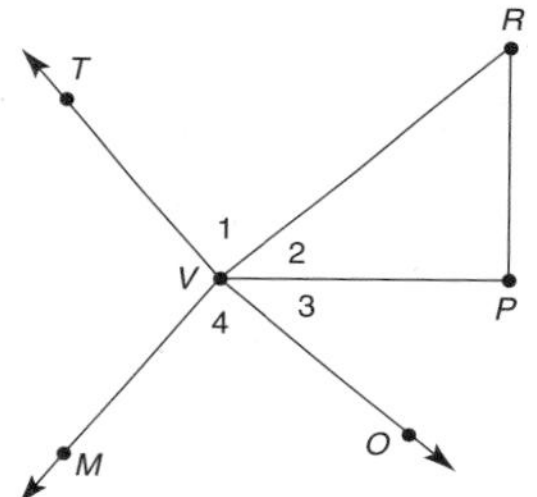

PROOF

	Statements	Reasons
	1. $\angle 1 \cong \angle P$	1. Given
	2. ?	2. Given
(1), (2)	3. ?	3. Transitive Prop. of $\cong$
(3)	4. $m\angle 1 = m\angle 4$	4. ?
	5. $\overrightarrow{VP}$ bisects $\angle RVO$	5. ?
	6. ?	6. If a ray bisects an $\angle$, it forms two $\angle$s of equal measure
(4), (6)	7. ?	7. Addition Prop. of Equality
	8. $m\angle 1 + \angle 2 = m\angle TVP$; $m\angle 4 + m\angle 3 = m\angle MVP$	8. ?
(7), (8)	9. $m\angle TVP = m\angle MVP$	9. ?
	10. ?	10. If two $\angle$s are = in measure, then they are $\cong$

Write two-column proofs for Review Exercises 36 to 43.

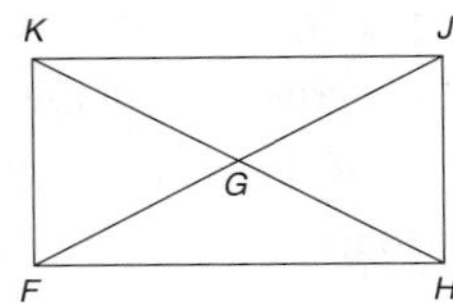

Exercises 36–38

36. *Given:* $\overline{KF} \perp \overline{FH}$
$\angle JHF$ is a right $\angle$
Prove: $\angle KFH \cong \angle JHF$

37. *Given:* $\overline{KH} \cong \overline{FJ}$
G is the midpoint of both $\overline{KH}$ and $\overline{FJ}$
Prove: $\overline{KG} \cong \overline{GJ}$

38. *Given:* $\overline{KF} \perp \overline{FH}$
Prove: $\angle KFJ$ is comp. to $\angle JFH$

39. *Given:* $\angle 1$ is comp. to $\angle M$
$\angle 2$ is comp. to $\angle M$
Prove: $\angle 1 \cong \angle 2$

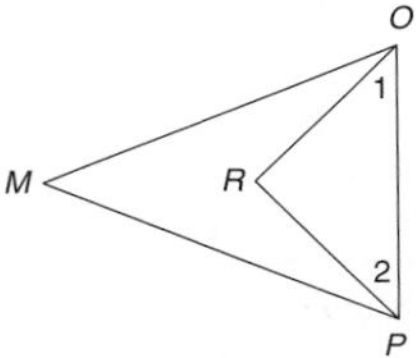

Exercises 39, 40

40. *Given:* $\angle MOP \cong \angle MPO$
$\overrightarrow{OR}$ bisects $\angle MOP$
$\overrightarrow{PR}$ bisects $\angle MPO$
Prove: $\angle 1 \cong \angle 2$

For Exercise 41, see the figure at the top of the next column.

41. *Given:* $\angle 4 \cong \angle 6$
Prove: $\angle 5 \cong \angle 6$

42. *Given:* Figure as shown
Prove: $\angle 4$ is supp. to $\angle 2$

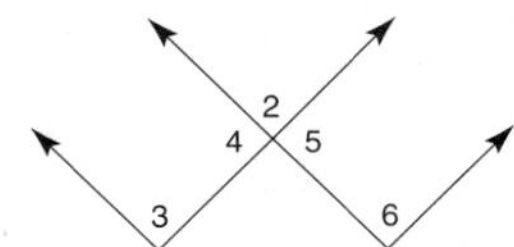

Exercises 41–43

43. *Given:* $\angle 3$ is supp. to $\angle 5$
$\angle 4$ is supp. to $\angle 6$
Prove: $\angle 3 \cong \angle 6$

44. *Given:* $\overline{VP}$
Construct: $\overline{VW}$ such that
$VW = 4 \cdot VP$

45. Construct a 135° angle.

46. *Given:* Triangle PQR
Construct: The three angle bisectors
What did you discover about the three angle bisectors of this triangle?

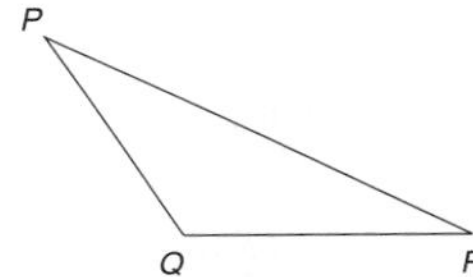

47. *Given:* $\overline{AB}$, $\overline{BC}$, and $\angle B$ as shown in Exercise 48
Construct: Triangle ABC

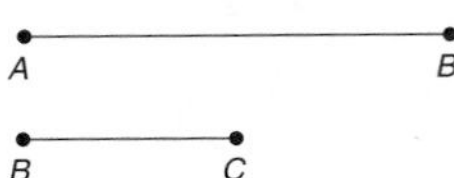

48. *Given:* $m\angle B = 50°$
Construct: An angle whose measure is 20°

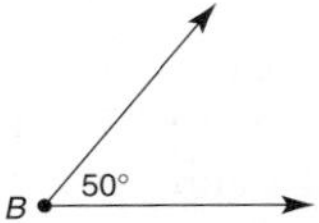

Chapter 2
Parallel Lines

The vinyl pieces of siding near the roofline of a house are parallel. Angles that are marked similarly appear to be congruent. As we shall see, this is not by chance! An experienced construction worker cuts the vinyl pieces so that the marked angles are congruent. In this manner, the vinyl strips of siding run parallel to each other.

CHAPTER OUTLINE

2.1 The Parallel Postulate and Special Angles

Perpendicular Lines

By definition, two lines (or segments or rays) are **perpendicular** if they meet to form congruent adjacent angles. Using this definition, we proved the theorem stating that "perpendicular lines meet to form right angles." We can also say that two rays or line segments are perpendicular if they are parts of perpendicular lines. We now consider a method for constructing a line perpendicular to a given line.

Construction 6:
To construct the line that is perpendicular to a given line from a point not on the given line.

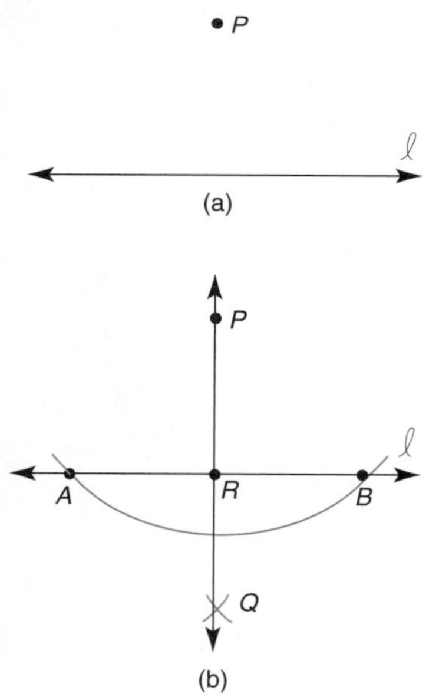

FIGURE 2.1

Given: In Figure 2.1(a), line ℓ and point P not on ℓ

Construct: $\overleftrightarrow{PQ} \perp \ell$

Construction: With P as the center, open the compass to a length large enough to intersect ℓ in two points A and B. [See Figure 2.1(b).]
With A and B as centers, mark off arcs of equal radii (using the same compass opening) to intersect at a point Q, as shown.
Draw $\overleftrightarrow{PQ}$ to complete the desired line.

In this construction, $\angle PRA$ and $\angle PRB$ are right angles. The arcs drawn from A and B to intersect below line ℓ provide greater accuracy than would arcs intersecting above ℓ.

Construction 6 suggests a uniqueness relationship that can be proved.

THEOREM 2.1.1: From a point not on a given line, there is exactly one line perpendicular to the given line.

The term "perpendicular" is extended to include line-plane and plane-plane relationships. The drawings in Figure 2.2 indicate two perpendicular lines, a line perpendicular to a plane, and two perpendicular planes.

Parallel Lines

Just as the word "perpendicular" can relate lines and planes, the word "parallel" can be used to describe possible relationships among lines and planes. However, parallel lines must lie in the same plane, as the following definition emphasizes.

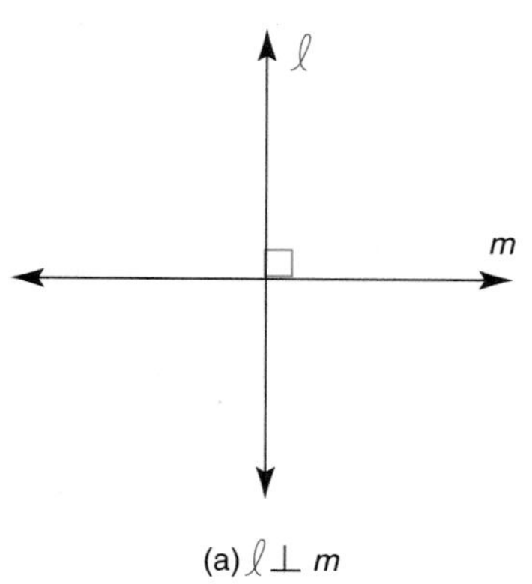

(a) $\ell \perp m$

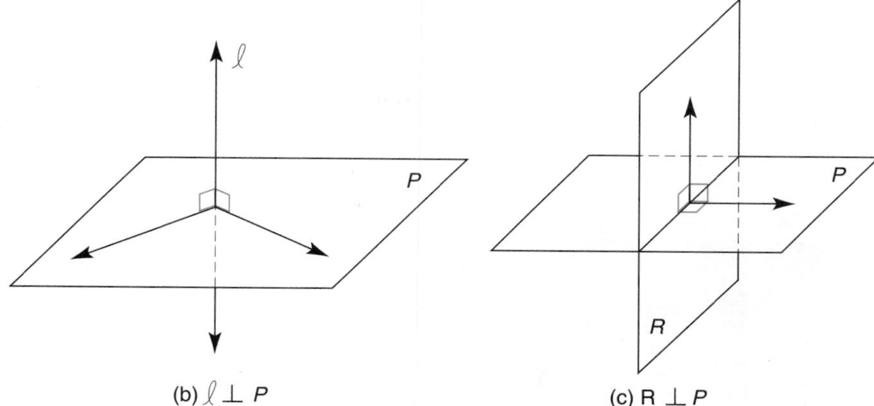

(b) $\ell \perp P$ (c) $R \perp P$

FIGURE 2.2

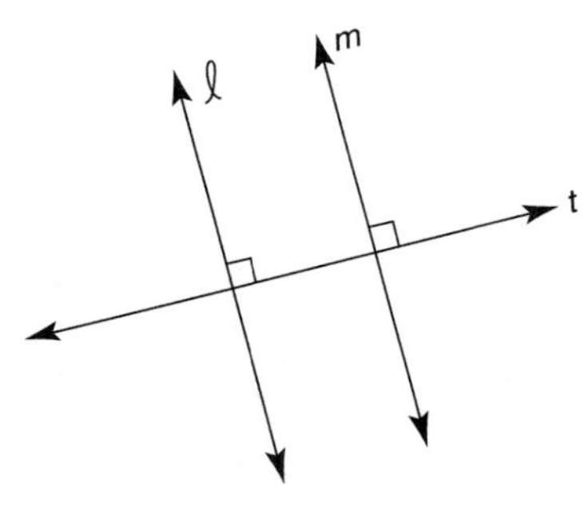

DEFINITION: **Parallel lines** are lines in the same plane that do not intersect.

DISCOVER!

In the sketch at the left, lines ℓ and m lie in the same plane with line t and are perpendicular to line t. How are the lines ℓ and m related to each other?

ANSWER

These lines are said to be *parallel*. They will not intersect.

More generally, two lines in a plane, a line and a plane, or two planes are parallel if they do not intersect (see Figure 2.3). Segments or rays are parallel if

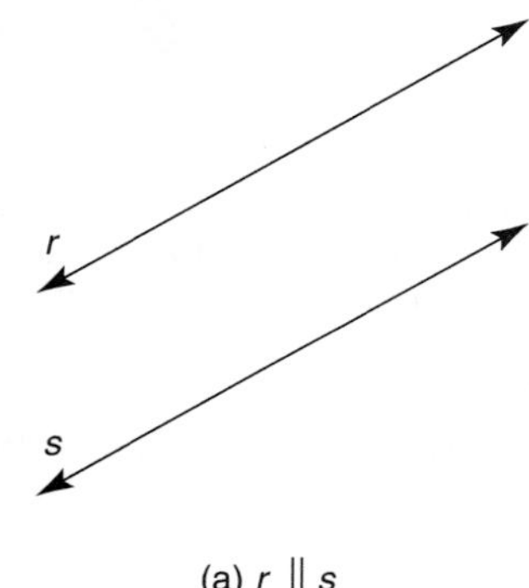

(a) $r \parallel s$

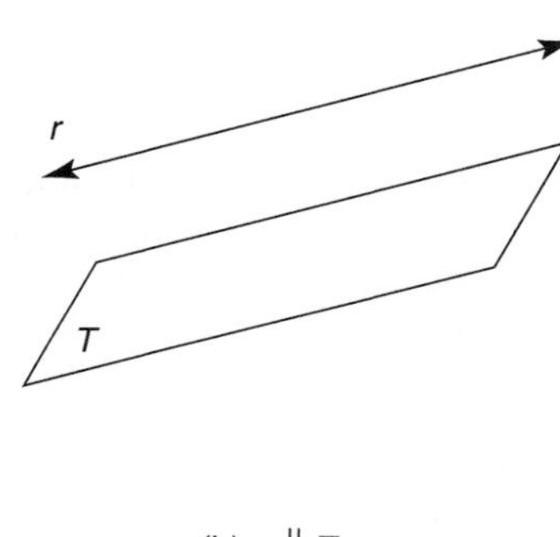

(b) $r \parallel T$

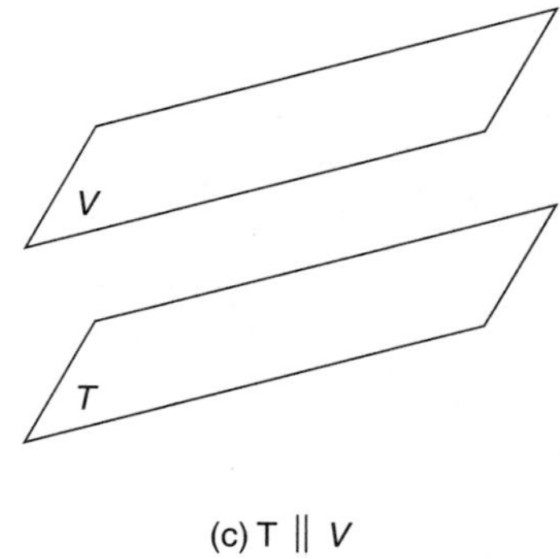

(c) $T \parallel V$

FIGURE 2.3

they are parts of two parallel lines. Figure 2.3 illustrates possible applications of the word "parallel." In Figure 2.4, two parallel planes M and N are intersected by a third plane G. How must lines of intersection a and b be related?

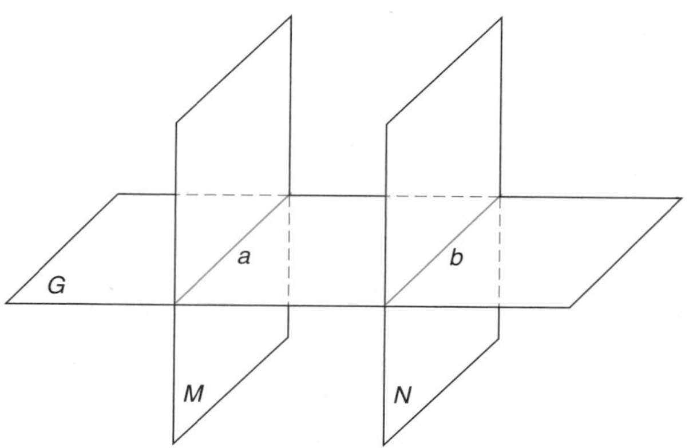

FIGURE 2.4

Geometry IN THE REAL WORLD

The rungs of a ladder are parallel line segments.

Euclidean Geometry

The type of geometry found in this textbook is known as Euclidean geometry. In this geometry, a plane is a flat, two-dimensional surface in which the line segment joining any two points of the plane lies entirely within the plane. While the postulate that follows characterizes Euclidean geometry, the Look Beyond section of this chapter discusses alternative geometries. Postulate 10, the Euclidean Parallel Postulate, is easy to accept because of the way we perceive a plane.

POSTULATE 10: (Parallel Postulate)
Through a point not on a line, exactly one line is parallel to the given line.

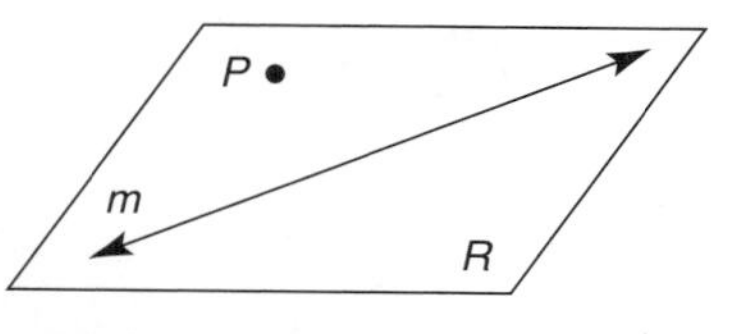

FIGURE 2.5

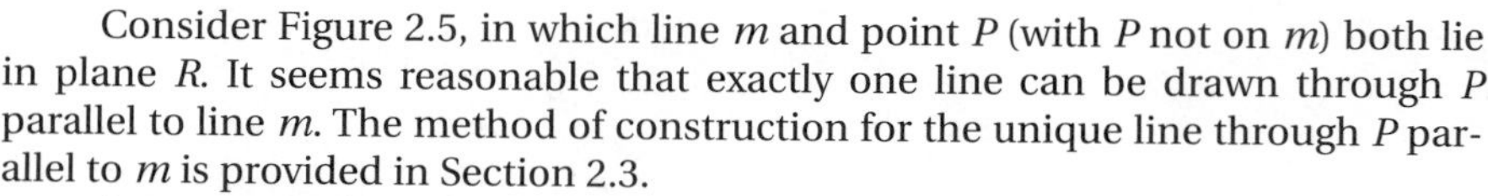

Consider Figure 2.5, in which line m and point P (with P not on m) both lie in plane R. It seems reasonable that exactly one line can be drawn through P parallel to line m. The method of construction for the unique line through P parallel to m is provided in Section 2.3.

A **transversal** is a line that intersects two (or more) other lines at distinct points; all of the lines lie in the same plane. In Figure 2.6, t is a transversal for lines r and s. Angles that are formed between r and s are **interior angles;** those outside r and s are **exterior angles.**

Interior angles: $\angle 3, \angle 4, \angle 5, \angle 6$

Exterior angles: $\angle 1, \angle 2, \angle 7, \angle 8$

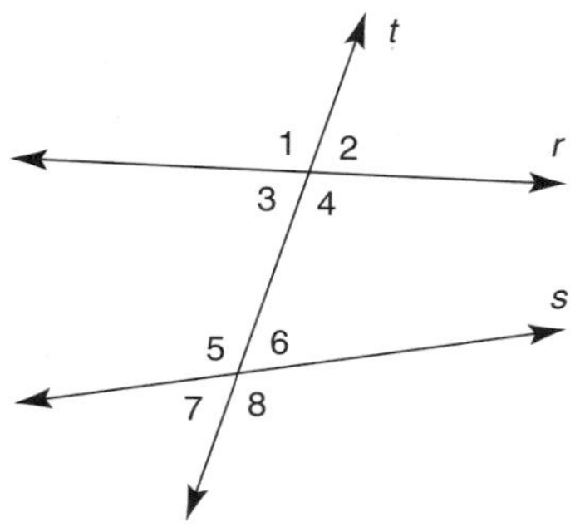

FIGURE 2.6

Two angles that lie in the same relative positions when two lines are cut by a transversal are **corresponding angles.** In Figure 2.6, $\angle 1$ and $\angle 5$ are corresponding angles; each angle is above the line and to the left of the transversal that helps form the angle.

Corresponding angles: (must be in pairs)	$\angle 1$ and $\angle 5$	above left
	$\angle 3$ and $\angle 7$	below left
	$\angle 2$ and $\angle 6$	above right
	$\angle 4$ and $\angle 8$	below right

Two interior angles that have different vertices and lie on opposite sides of the transversal are **alternate interior angles.** Two exterior angles that have different vertices and lie on opposite sides of the transversal are **alternate exterior angles.** Both types of alternate angles must occur in pairs; in Figure 2.6, these pairs of angles are numbered as follows:

Alternate interior angles:	$\angle 3$ and $\angle 6$
	$\angle 4$ and $\angle 5$
Alternate exterior angles:	$\angle 1$ and $\angle 8$
	$\angle 2$ and $\angle 7$

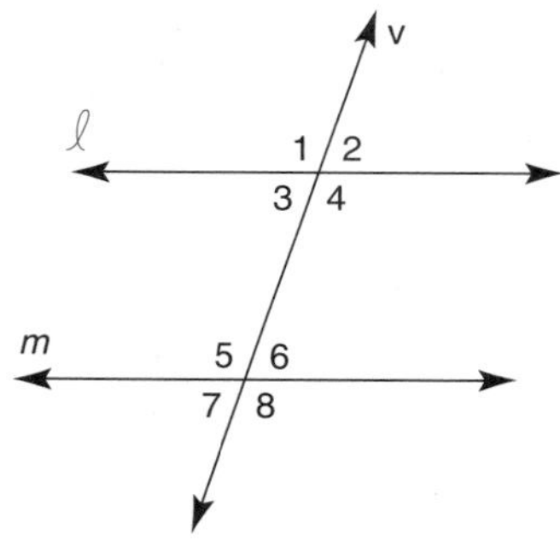

FIGURE 2.7

Parallel Lines and Congruent Angles

Now suppose that parallel lines ℓ and m in Figure 2.7 are cut by transversal v. If a protractor were used to measure $\angle 1$ and $\angle 5$, these corresponding angles would have equal measures; that is, they are congruent. Similarly, any other pair of corresponding angles will be congruent as long as $\ell \parallel m$.

POSTULATE 11:
If two parallel lines are cut by a transversal, then the corresponding angles are congruent.

EXAMPLE 1

In Figure 2.7, $\ell \parallel m$ and $m\angle 1 = 117$. Find:

a) $m\angle 2$
b) $m\angle 5$
c) $m\angle 4$
d) $m\angle 8$

Solution

a) $m\angle 2 = 63$ supplementary to $\angle 1$
b) $m\angle 5 = 117$ corresponding to $\angle 1$
c) $m\angle 4 = 117$ vertical $\angle$ to $\angle 1$
d) $m\angle 8 = 117$ corresponding to $\angle 4$ (found in part (c))

Several theorems follow from Postulate 11; for some theorems, formal proofs are provided. Study the proofs and be able to state all the theorems. Later, you can cite the theorems as reasons in subsequent proofs.

THEOREM 2.1.2: If two parallel lines are cut by a transversal, then the alternate interior angles are congruent.

Given: $a \parallel b$ in Figure 2.8; transversal k

Prove: $\angle 3 \cong \angle 6$

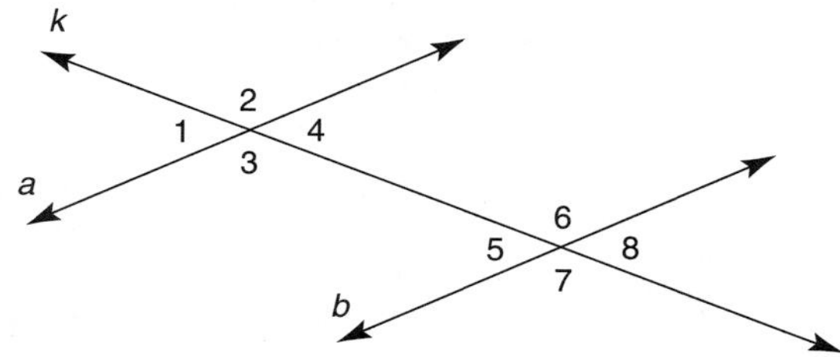

FIGURE 2.8

PROOF

Statements	Reasons
1. $a \parallel b$; transversal k	**1.** Given
2. $\angle 2 \cong \angle 6$	**2.** If two $\parallel$ lines are cut by a transversal, corresponding $\angle$s are $\cong$
3. $\angle 3 \cong \angle 2$	**3.** If two lines intersect, vertical $\angle$s formed are $\cong$
4. $\angle 3 \cong \angle 6$	**4.** Transitive (of $\cong$)

It is easy to prove that $\angle 4$ and $\angle 5$ are congruent because they are supplements to $\angle 3$ and $\angle 6$. Another theorem that is similar to Theorem 2.1.2 follows, but the proof is left as Exercise 22.

THEOREM 2.1.3: If two parallel lines are cut by a transversal, then the alternate exterior angles are congruent.

Parallel Lines and Supplementary Angles

Whenever two parallel lines are cut by a transversal, an interesting relationship exists between the two interior angles on the same side of the transversal. The following proof establishes that these interior angles are supplementary; a similar claim can be made for the pair of exterior angles on the same side of the transversal.

> **THEOREM 2.1.4:** If two parallel lines are cut by a transversal, then the interior angles on the same side of the transversal are supplementary.

Given: In Figure 2.9, $\overleftrightarrow{TV} \parallel \overleftrightarrow{WY}$ with transversal $\overleftrightarrow{RS}$

Prove: $\angle 1$ and $\angle 3$ are supplementary

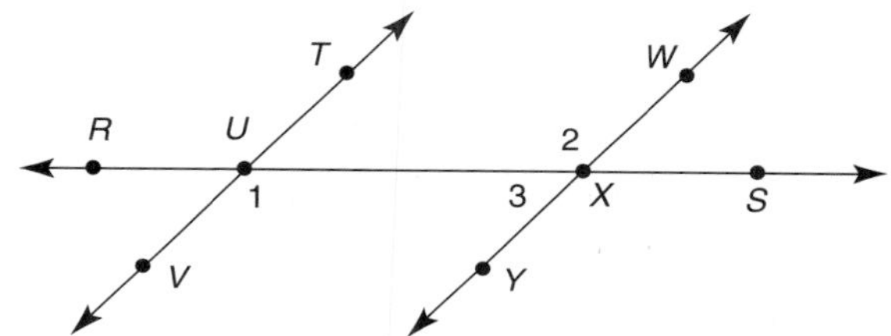

FIGURE 2.9

PROOF

Statements	Reasons
1. $\overleftrightarrow{TV} \parallel \overleftrightarrow{WY}$; transversal $\overleftrightarrow{RS}$	**1.** Given
2. $\angle 1 \cong \angle 2$	**2.** If two $\parallel$ lines are cut by a transversal, alternate interior $\angle$s are $\cong$
3. $m\angle 1 = m\angle 2$	**3.** If two $\angle$s are $\cong$, their measures are $=$
4. $\angle WXY$ is a straight $\angle$, so $m\angle WXY = 180$	**4.** If an $\angle$ is a straight $\angle$, its measure is 180
5. $m\angle 2 + m\angle 3 = m\angle WXY$	**5.** Angle-Addition Postulate
6. $m\angle 2 + m\angle 3 = 180$	**6.** Substitution
7. $m\angle 1 + m\angle 3 = 180$	**7.** Substitution
8. $\angle 1$ and $\angle 3$ are supplementary	**8.** If the sum of measures of two $\angle$s is 180, the $\angle$s are supplementary

The proof of the following theorem is left as an exercise.

> **THEOREM 2.1.5:** If two parallel lines are cut by a transversal, then the exterior angles on the same side of the transversal are supplementary.

The remaining examples in this section illustrate methods from algebra and deal with the angles formed when two parallel lines are cut by a transversal.

EXAMPLE 2

Given: $\overleftrightarrow{TV} \parallel \overleftrightarrow{WY}$ with transversal $\overleftrightarrow{RS}$

$$m\angle RUV = (x + 4)(x - 3)$$
$$m\angle WXS = x^2 - 3$$

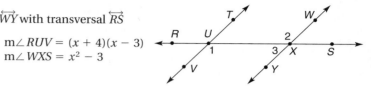

Find: x

Solution

$\angle RUV$ and $\angle WXS$ are alternate exterior angles, so they are congruent. Then $m\angle RUV = m\angle WXS$. Therefore

$$\begin{aligned}(x + 4)(x - 3) &= x^2 - 3\\ x^2 + x - 12 &= x^2 - 3\\ x - 12 &= -3\\ x &= 9\end{aligned}$$

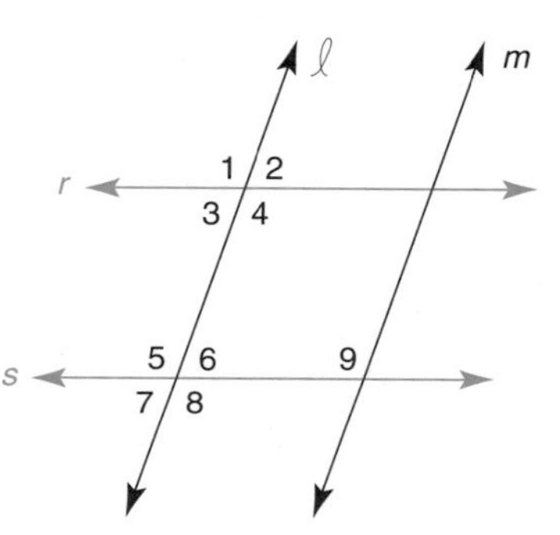

FIGURE 2.10

For Example 2, notice that both angles measure 78° when $x = 9$.

In Figure 2.10, lines r and s are parallel. However, ℓ and m are not necessarily parallel. Which angle, $\angle 1$ or $\angle 9$, must be congruent to $\angle 5$? To answer the question, it might be helpful to use a second color to draw in the lines that are known to be parallel. Now decide which transversal to use, ℓ or m. In this case, ℓ is the transversal for r and s that forms corresponding angles 5 and 1. Therefore, $\angle 1$ must be congruent to $\angle 5$ if $r \parallel s$.

EXAMPLE 3

Given: In Figure 2.10, $r \parallel s$ and transversal ℓ

$$m\angle 3 = 4x + y$$
$$m\angle 5 = 6x + 5y$$
$$m\angle 6 = 5x - 2y$$

Find: x and y

Solution

$\angle 3$ and $\angle 6$ are congruent alternate interior angles; also, $\angle 3$ and $\angle 5$ are supplementary angles according to Theorem 2.1.4. These facts lead to the following system of equations:

$$\begin{aligned}4x + y &= 5x - 2y\\ (4x + y) + (6x + 5y) &= 180\end{aligned}$$

These equations can be simplified to

$$\begin{aligned}x - 3y &= 0\\ 10x + 6y &= 180\end{aligned}$$

After we divide each term of the second equation by 2, the system becomes

$$\begin{aligned}x - 3y &= 0\\ 5x + 3y &= 90\end{aligned}$$

Addition leads to the equation $6x = 90$, so $x = 15$. Substituting 15 for x into the equation $x - 3y = 0$, we have

$$\begin{aligned} 15 - 3y &= 0 \\ 3y &= 15 \\ y &= 5 \end{aligned}$$

Our solution, $x = 15$ and $y = 5$, yields the following angle measures:

$$\begin{aligned} m\angle 3 &= 65 \\ m\angle 5 &= 115 \\ m\angle 6 &= 65 \end{aligned}$$

NOTE: The equation $x - 3y = 0$ could be multiplied by 2 to obtain $2x - 6y = 0$. Then the equations $2x - 6y = 0$ and $10x + 6y = 180$ could be added.

Notice that the angle measures determined in Example 3 are consistent with Figure 2.10 and the required relationships for the angles named. For instance, $m\angle 3 + m\angle 5 = 180$, and we see that interior angles on the same side of the transversal are indeed supplementary.

2.1 Exercises

Use drawings, as needed, to answer each question.

1. Does the relation "is parallel to" have a

a) reflexive property? (consider a line m)
b) symmetric property? (consider lines m and n in a plane)
c) transitive property? (consider coplanar lines m, n and q)

2. In a plane, $\ell \perp m$ and $t \perp m$. By appearance, how are ℓ and t related?

3. Suppose that $r \parallel s$. Each interior angle on the right side of the transversal t has been bisected. Using intuition, what appears to be true of $\angle 9$ formed by the bisectors?

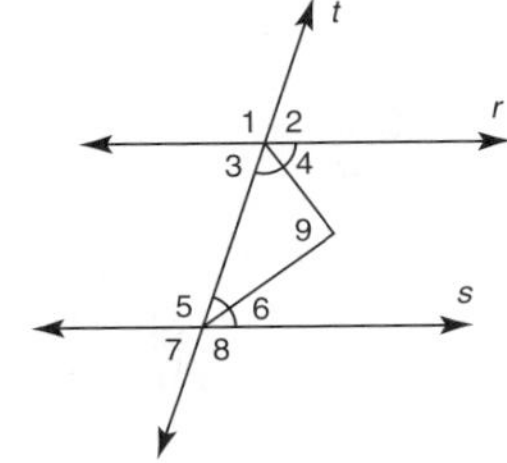

Exercise 3

4. Make a sketch to represent two planes that are

a) parallel.
b) perpendicular.

5. Suppose that r is parallel to s and $m\angle 2 = 87$. Find:

a) $m\angle 3$
b) $m\angle 6$ (formed by t and s)
c) $m\angle 1$
d) $m\angle 7$

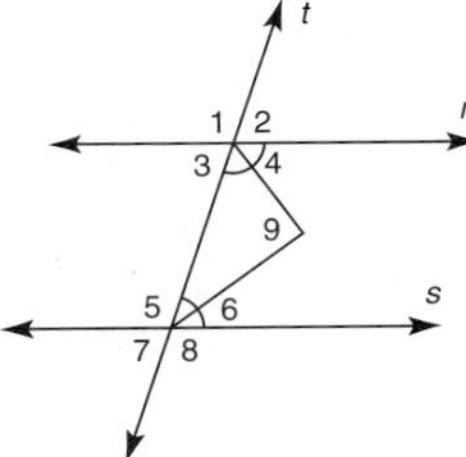

Exercise 5

6. In Euclidean geometry, how many lines can be drawn through a point P not on a line ℓ that are

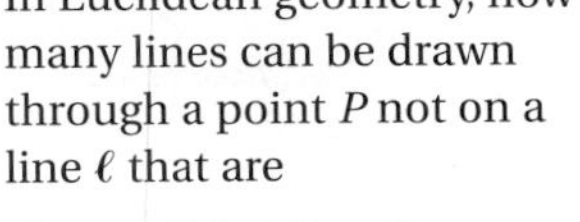

a) parallel to line ℓ?
b) perpendicular to line ℓ?

7. Lines r and s are cut by transversal t. Which angle

a) corresponds to $\angle 1$?
b) is the alternate interior $\angle$ for $\angle 4$ (formed by t and r)?
c) is the alternate exterior $\angle$ for $\angle 1$?
d) is the other interior angle on the same side of transversal t as $\angle 3$?

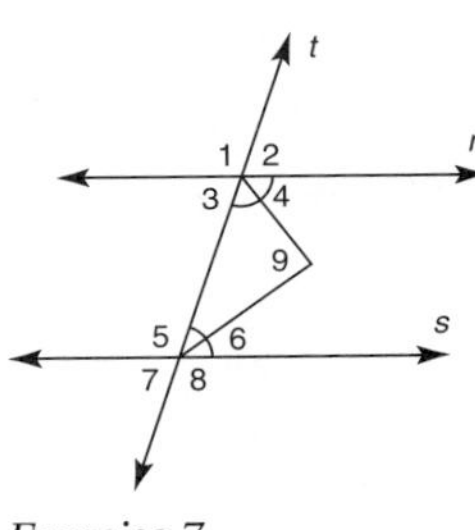

Exercise 7

8. $\overline{AD} \parallel \overline{BC}$, $\overline{AB} \parallel \overline{DC}$, and m$\angle A = 92°$. Find:

a) m$\angle B$
b) m$\angle C$
c) m$\angle D$

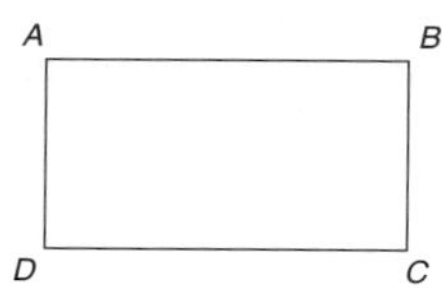

9. $\ell \parallel m$, with transversal t, and $\overrightarrow{OQ}$ bisects $\angle MON$. If m$\angle 1 = 112°$, find the following:

a) m$\angle 2$
b) m$\angle 4$
c) m$\angle 5$
d) m$\angle MOQ$

10. *Given:* $\ell \parallel m$
transversal t
m$\angle 1 = 4x + 2$
m$\angle 6 = 4x - 2$
Find: x and m$\angle 5$

Exercises 9, 10

11. *Given:* $m \parallel n$
transversal k
m$\angle 3 = x^2 - 3x$
m$\angle 6 = (x + 4)(x - 5)$
Find: x and m$\angle 4$

12. *Given:* $m \parallel n$
transversal k
m$\angle 1 = 5x + y$
m$\angle 2 = 3x + y$
m$\angle 8 = 3x + 5y$
Find: x, y, and m$\angle 8$

Exercises 11–13

13. *Given:* $m \parallel n$
transversal k
m$\angle 3 = 6x + y$
m$\angle 5 = 8x + 2y$
m$\angle 6 = 4x + 7y$
Find: x, y, and m$\angle 7$

14. In the three-dimensional figure, $\overline{CA} \perp \overline{AB}$ and $\overline{BE} \perp \overline{AB}$. Are $\overleftrightarrow{CA}$ and $\overleftrightarrow{BE}$ parallel to each other? (Compare to Exercise 2.)

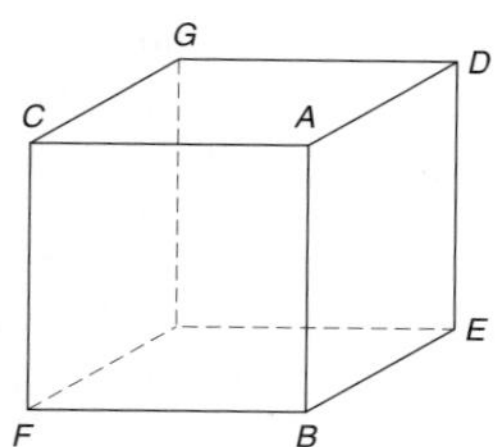

15. *Given:* $\ell \parallel m$ and $\angle 3 \cong \angle 4$
Prove: $\angle 1 \cong \angle 4$

PROOF

Statements	Reasons
1. $\ell \parallel m$	1. ?
2. $\angle 1 \cong \angle 2$	2. ?
3. $\angle 2 \cong \angle 3$	3. ?
4. ?	4. Given
5. ?	5. Transitive of $\cong$

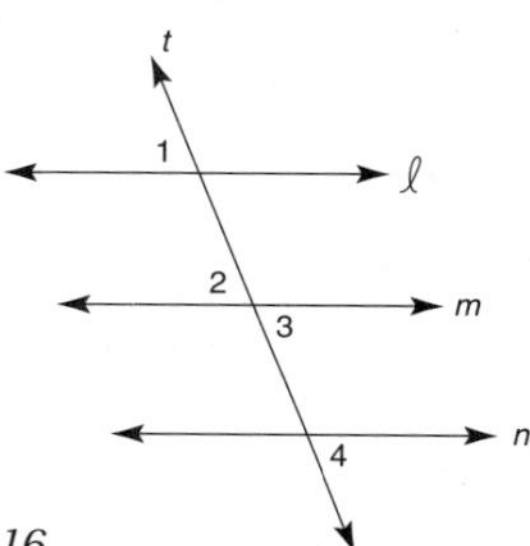

Exercises 15, 16

16. *Given:* $\ell \parallel m$ and $m \parallel n$
Prove: $\angle 1 \cong \angle 4$

PROOF

Statements	Reasons
1. $\ell \parallel m$	1. ?
2. $\angle 1 \cong \angle 2$	2. ?
3. $\angle 2 \cong \angle 3$	3. ?
4. ?	4. Given
5. $\angle 3 \cong \angle 4$	5. ?
6. ?	6. ?

17. *Given:* $\overleftrightarrow{CE} \parallel \overleftrightarrow{DF}$
transversal $\overleftrightarrow{AB}$
$\overrightarrow{CX}$ bisects $\angle ACE$
$\overrightarrow{DE}$ bisects $\angle CDF$
Prove: $\angle 1 \cong \angle 3$

18. *Given:* $\overleftrightarrow{CE} \parallel \overleftrightarrow{DF}$
transversal $\overleftrightarrow{AB}$
$\overrightarrow{DE}$ bisects $\angle CDF$
Prove: $\angle 3 \cong \angle 6$

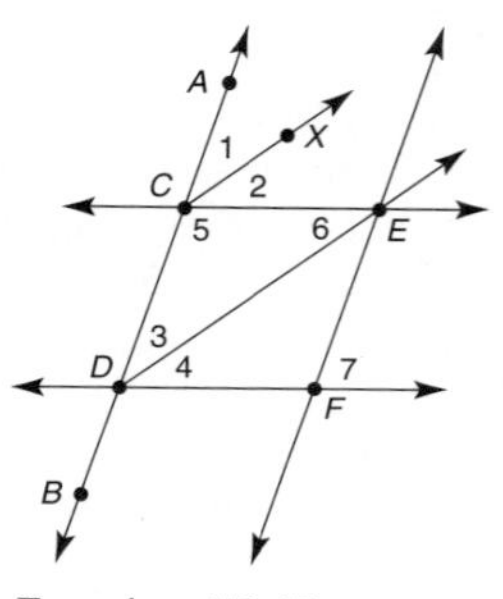

Exercises 17, 18

19. *Given:* $r \parallel s$
transversal t
$\angle 1$ is a right $\angle$
Prove: $\angle 2$ is a right $\angle$

20. *Given:* $r \parallel s$, $r \perp t$
Prove: $s \perp t$

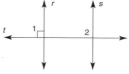

Exercises 19, 20

21. In triangle ABC, line t is drawn through vertex A so that $t \parallel \overline{BC}$.

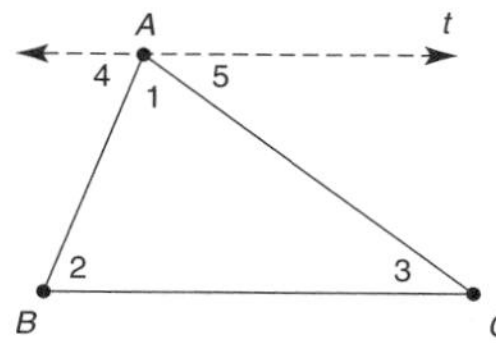

a) Which pairs of $\angle$s are $\cong$?
b) What is the sum of $m\angle 1$, $m\angle 4$, and $m\angle 5$?
c) What is the sum of measures of the $\angle$s of $\triangle ABC$?

In Exercises 22 to 24, write a formal proof of each theorem.

22. If two parallel lines are cut by a transversal, then the alternate exterior angles are congruent.

23. If two parallel lines are cut by a transversal, then the exterior angles on the same side of the transversal are supplementary.

24. If a transversal is perpendicular to one of two parallel lines, then it is also perpendicular to the other line.

25. Suppose that two lines are cut by a transversal in such a way that corresponding angles are not congruent. Can those two lines be parallel?

26. *Given:* Line ℓ and point P not on ℓ
Construct: $\overleftrightarrow{PQ} \perp \ell$

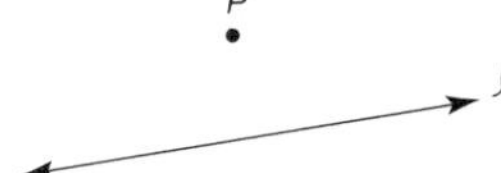

27. *Given:* Triangle ABC with three acute angles
Construct: $\overline{BD} \perp \overline{AC}$

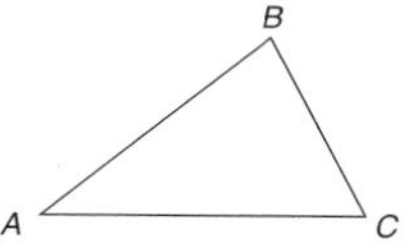

28. *Given:* Triangle MNQ with obtuse $\angle MNQ$
Construct: $\overline{NE} \perp \overline{MQ}$

29. *Given:* Triangle MNQ with obtuse $\angle MNQ$
Construct: $\overline{MR} \perp \overline{NQ}$
(**HINT:** Extend $\overline{NQ}$.)

Exercises 28, 29

30. *Given:* A line m and a point T not on m

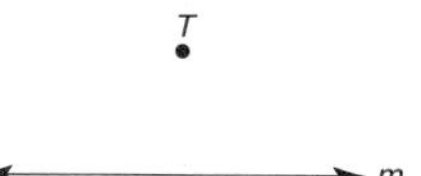

Suppose that you do the following:
i) Construct a perpendicular line r from T to m.
ii) Construct a line s perpendicular to line r at point T.
What relationship holds between s and m?

2.2 Indirect Proof

Let $P \to Q$ represent the statement "If P, then Q." The following statements are related to this conditional statement (Note: Recall that $\sim P$ represents the negation of P.).

Conditional (or Implication)	$P \to Q$	If P, then Q.
Converse of Conditional	$Q \to P$	If Q, then P.
Inverse of Conditional	$\sim P \to \sim Q$	If not P, then not Q.
Contrapositive of Conditional	$\sim Q \to \sim P$	If not Q, then not P.

For example, consider the following conditional statement.

If Tom lives in San Diego, then he lives in California.

This true statement has these related statements:

Converse: If Tom lives in California, then he lives in San Diego. (false)

Inverse: If Tom does not live in San Diego, then he does not live in California. (false)

Contrapositive: If Tom does not live in California, then he does not live in San Diego. (true)

In general, the conditional statement and its contrapositive are either both true or both false! In advanced courses, a statement of the form "If P, then Q" may be established by proving that its contrapositive is true, if that seems the easier proof to complete. Similarly, the converse and the inverse are also either both true or both false. See Appendix B for more information about the conditional and related statements.

EXAMPLE 1

For the conditional statement that follows, give the converse, the inverse, and the contrapositive. Then classify each as true or false.

If two angles are vertical angles, then they are congruent angles.

Solution

Converse: If two angles are congruent angles, then they are vertical angles. (false)

Inverse: If two angles are not vertical angles, then they are not congruent angles. (false)

Contrapositive: If two angles are not congruent angles, then they are not vertical angles. (true)

The Law of Negative Inference

Consider the following circumstances, and accept each premise as true:

1. If Matt cleans his room, then he will go to the movie. ($P \to Q$)
2. Matt does not get to go to the movie. ($\sim Q$)

What may you conclude? You should have deduced that Matt did not clean his room; if he had, he would have gone to the movie. This "backdoor" reasoning is based on the fact that the truth of $P \to Q$ implies the truth of $\sim Q \to \sim P$.

LAW OF NEGATIVE INFERENCE

$$\begin{array}{r} P \to Q \\ \sim Q \\ \hline \therefore \sim P \end{array}$$

Like the Law of Detachment from Section 1.1, the Law of Negative Inference is a form of deduction. While the Law of Detachment characterizes the method of "direct proof" encountered in preceding sections, the Law of Negative Inference is the backbone of the method of proof known as **indirect proof.**

Indirect Proof

You will need to know when to use the indirect method of proof. Often the theorem to be proved has the form $P \rightarrow Q$, in which Q contains a negation and denies some claim. For instance, an indirect proof might be best if Q reads in one of these ways:

c is *not* equal to d

ℓ is *not* perpendicular to m

However, we will see in Example 4 of this section that the indirect method can be used to prove that line ℓ is parallel to line m. Another instance in which indirect proof is used is for proving existence and uniqueness theorems; see Example 5.

The method of indirect proof is illustrated in Example 2. All indirect proofs in this book are given in paragraph form (as are many of the direct proofs).

In any paragraph proof, each statement must still be justified. Because of the need to order your statements properly, this type of proof may have a positive impact on the essays you write for your other classes.

EXAMPLE 2

Given: In Figure 2.11, $\overrightarrow{BA}$ is *not* perpendicular to $\overrightarrow{BD}$

Prove: $\angle 1$ and $\angle 2$ are *not* complementary

Proof: Suppose that $\angle 1$ and $\angle 2$ are complementary. Then $m\angle 1 + m\angle 2 = 90$ because the sum of the measures of two complementary $\angle$s is 90. We also know that $m\angle 1 + m\angle 2 = m\angle ABD$, by the Angle-Addition Postulate. In turn, $m\angle ABD = 90$ by substitution. Then $\angle ABD$ is a right angle. In turn, $\overrightarrow{BA} \perp \overrightarrow{BD}$. But this contradicts the given hypothesis; therefore the supposition must be false, and it follows that $\angle 1$ and $\angle 2$ are not complementary.

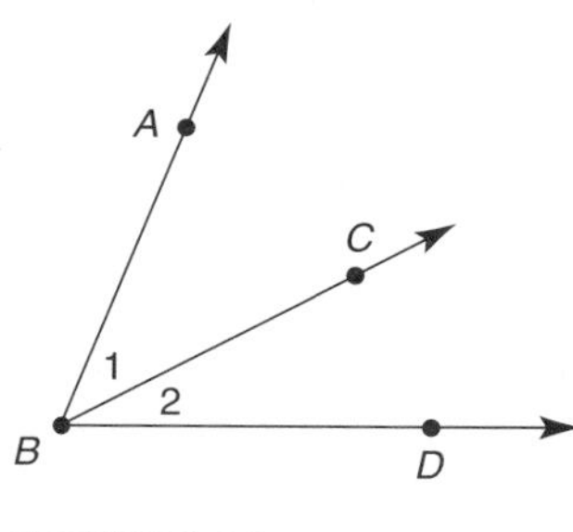

FIGURE 2.11

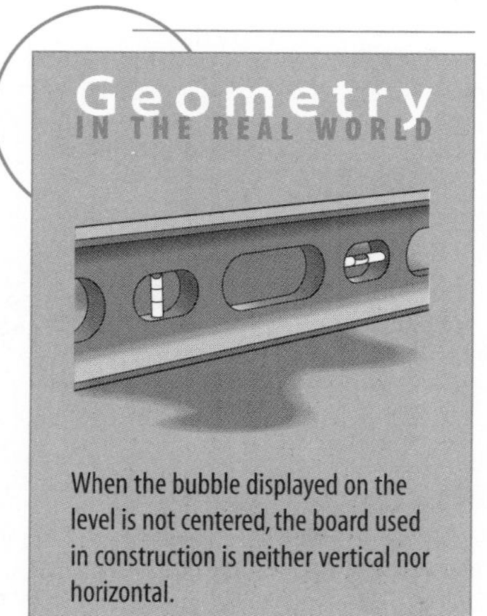

When the bubble displayed on the level is not centered, the board used in construction is neither vertical nor horizontal.

In Example 2 and in all indirect proofs, the first statement takes the form

"Suppose that . . ." or *"Assume that . . ."*

By its very nature, this statement cannot be supported even though every other statement in the proof can be justified; thus, when a contradiction is reached, the finger of blame points to the supposition. At this stage of the proof, we may say that the claim involving $\sim Q$ has failed and is false; thus, our only recourse is to conclude that Q is true. Following is an outline of this technique.

METHOD OF INDIRECT PROOF

To prove the statement $P \to Q$ or to complete the proof problem of the form

Given: P

Prove: Q

where Q is often a negation, use the following steps:

1. Suppose that $\sim Q$ is true.
2. Reason from the supposition until you reach a contradiction.
3. Note that the supposition claiming that $\sim Q$ is true must be false, and that Q must therefore be true.

Step 3 completes the proof.

The contradiction that is discovered in an indirect proof often has the form $\sim P$. Thus, the assumed statement $\sim Q$ has forced the conclusion $\sim P$, asserting that $\sim Q \to \sim P$ is true. Then the desired theorem $P \to Q$ (the contrapositive of $\sim Q \to \sim P$) is also true.

EXAMPLE 3

Complete a formal proof of the following theorem:

If two lines are cut by a transversal so that corresponding angles are not congruent, then the two lines are not parallel.

Given: In Figure 2.12, ℓ and m are cut by transversal t

$\angle 1 \not\cong \angle 5$

Prove: $\ell \nparallel m$

Proof: Assume that $\ell \parallel m$. When these lines are cut by transversal t, the corresponding angles (including $\angle 1$ and $\angle 5$) are congruent. But $\angle 1 \not\cong \angle 5$ by hypothesis. Thus the assumed statement, which claims that $\ell \parallel m$, must be false. It follows that $\ell \nparallel m$.

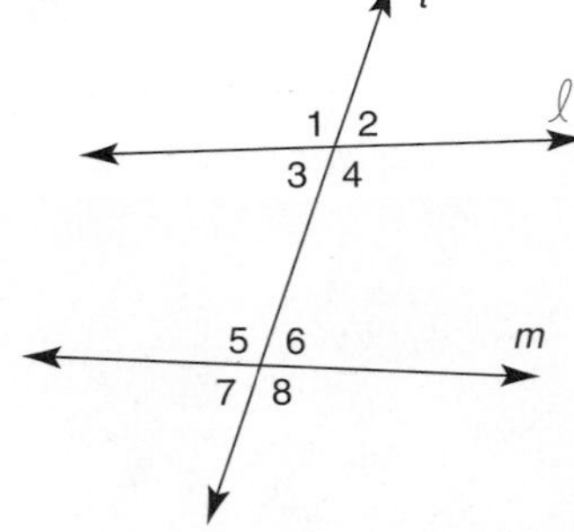

FIGURE 2.12

The versatility of the indirect proof is shown in the final examples of this section. The indirect proofs preceding Example 4 all contain a negation in the conclusion (Prove); the proofs in the final illustrations use the indirect method to achieve a positive conclusion.

EXAMPLE 4

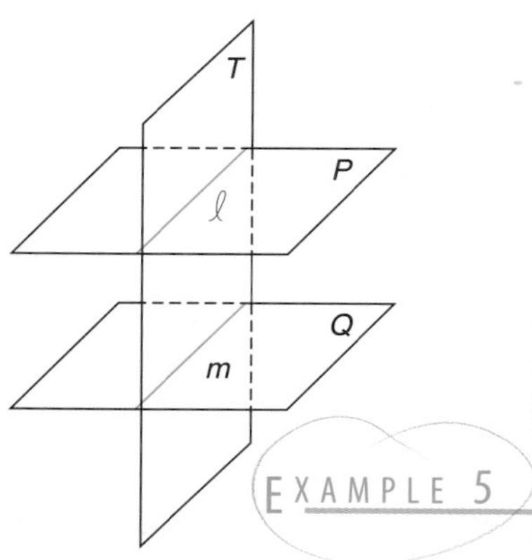

FIGURE 2.13

Given: In Figure 2.13, parallel planes P and Q are intersected by plane T in lines ℓ and m

Prove: $\ell \parallel m$

Proof: Assume that ℓ is not parallel to m. Then ℓ and m intersect at some point A. But if so, point A must be on both planes P and Q, which means that planes P and Q intersect; but P and Q are parallel by hypothesis. Therefore, the assumption that ℓ and m are not parallel must be false, and it follows that $\ell \parallel m$.

Indirect proofs are also used to establish uniqueness theorems, as Example 5 illustrates.

EXAMPLE 5

Prove the statement, "The angle bisector of an angle is unique."

Given: In Figure 2.14(a), $\overrightarrow{BD}$ bisects $\angle ABC$

Prove: $\overrightarrow{BD}$ is the only angle bisector for $\angle ABC$

Proof: $\overrightarrow{BD}$ bisects $\angle ABC$, so $m\angle ABD = \frac{1}{2}m\angle ABC$. Suppose that $\overrightarrow{BE}$ [as shown in Figure 2.14(b)] is also a bisector of $\angle ABC$ and that $m\angle ABE = \frac{1}{2}m\angle ABC$.

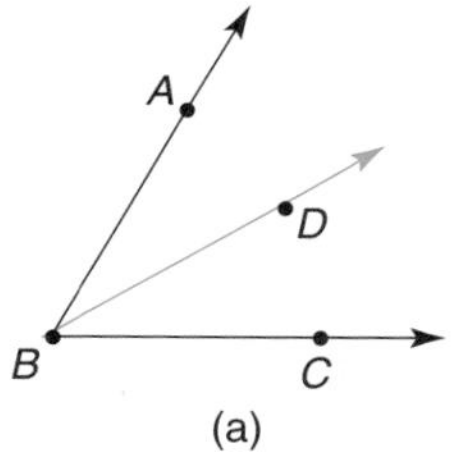

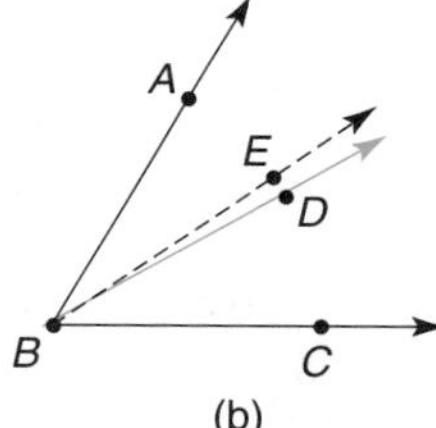

FIGURE 2.14

By the Angle-Addition Postulate, $m\angle ABD = m\angle ABE + m\angle EBD$. By substitution, $\frac{1}{2}m\angle ABC = \frac{1}{2}m\angle ABC + m\angle EBD$; but then $m\angle EBD = 0$ by subtraction. An angle with a measure of 0 contradicts the Protractor Postulate, which states that the measure of an angle is a unique positive number. Therefore the assumed statement must be false, and it follows that the angle bisector of an angle is unique.

2.2 Exercises

In Exercises 1 to 4, write the converse, the inverse, and the contrapositive of each statement. When possible, classify the statement as true or false.

1. If Juan wins the state lottery, then he will be rich.
2. If $x > 2$, then $x \neq 0$.

3. Two angles are complementary if the sum of their measures is 90°.

4. In a plane, if two lines are not perpendicular to the same line, then these lines are not parallel.

In Exercises 5 to 8, draw a conclusion where possible.

5. **a)** If two triangles are congruent, then the triangles are similar.
   b) Triangles *ABC* and *DEF* are not congruent.
   c) ?

6. **a)** If two triangles are congruent, then the triangles are similar.
   b) Triangles *ABC* and *DEF* are not similar.
   c) ?

7. **a)** If $x > 3$ and $x \neq 7$, then $x = 5$.
   b) $x > 3$ and $x \neq 7$
   c) ?

8. **a)** If $x > 3$ and $x \neq 7$, then $x = 5$.
   b) $x \neq 5$
   c) ?

9. Which of the following statements would you prove by the indirect method?
   a) In triangle *ABC*, if $m\angle A > m\angle B$, then $AC \neq BC$.
   b) If alternate exterior $\angle 1 \not\cong$ alternate exterior $\angle 8$, then ℓ is not parallel to m.
   c) If $(x + 2) \cdot (x - 3) = 0$, then $x = -2$ or $x = 3$.
   d) If two sides of a triangle are congruent, then the two angles opposite these sides are also congruent.
   e) The perpendicular bisector of a line segment is unique.

10. For each statement in Exercise 9 that can be proved by the indirect method, give the first statement in each proof.

11. A periscope uses an indirect method of observation. This instrument allows one to see what would otherwise be obstructed. Mirrors are located (see $\overline{AB}$ and $\overline{CD}$ in the drawing) so that an image is reflected twice. How are $\overline{AB}$ and $\overline{CD}$ related to each other?

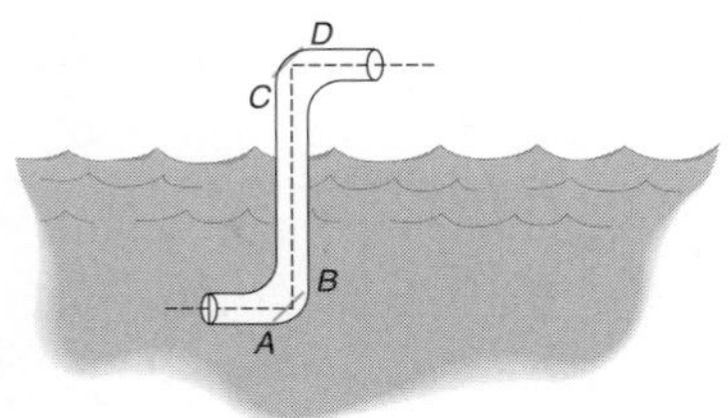

12. Some stores use an indirect method of observation. The purpose may be for safety (to avoid collisions) or to foil the attempts of would-be shoplifters. In this situation, a mirror (see $\overline{EF}$ in the drawing) is placed at the intersection of two aisles as shown. An observer at point *P* can then see any movement along the indicated aisle. In the sketch, what is the measure of $\angle GEF$?

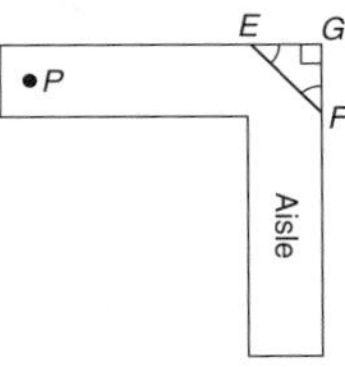

In Exercises 13 to 24, give the indirect proof for each problem or statement.

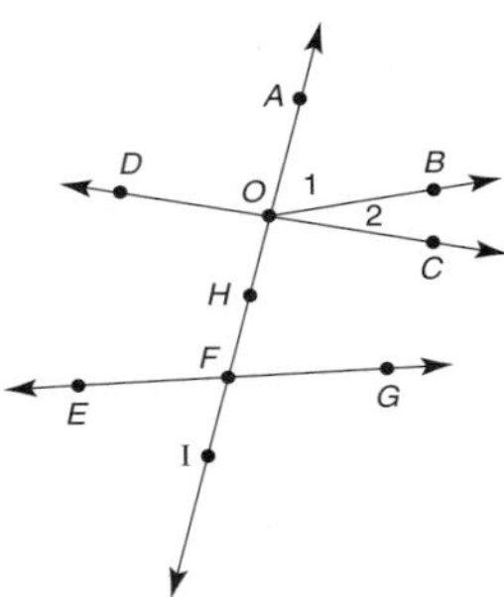

Exercises 13–16

13. *Given:* $\angle AOD \not\cong \angle AFE$
    Prove: $\overleftrightarrow{DC} \not\parallel \overleftrightarrow{EG}$

14. *Given:* $\angle 1 \not\cong \angle 2$
    Prove: $\overrightarrow{OB}$ does not bisect $\angle AOC$

15. *Given:* $m\angle AOD > m\angle AOC$
    Prove: $\overleftrightarrow{AI}$ is not perpendicular to $\overleftrightarrow{DC}$

16. *Given:* $AO > HF$
    $OH = FI$
    Prove: *H* is not the midpoint of $\overline{AI}$

17. If two angles are not congruent, then these angles are not vertical angles.

18. If $x^2 \neq 25$, then $x \neq 5$.

19. If alternate interior angles are not congruent when two lines are cut by a transversal, then the lines are not parallel.

20. If *a* and *b* are positive numbers, then $\sqrt{a^2 + b^2} \neq a + b$.

21. The midpoint of a line segment is unique.

22. There is exactly one line perpendicular to a given line at a point on the line.

23. ➢ In a plane, if two lines are parallel to a third line, then the two lines are parallel to each other.

24. ➢ In a plane, if two lines are intersected by a transversal so that the corresponding angles are congruent, then the lines are parallel.

2.3 Proving Lines Parallel

In Section 2.1, several methods for proving angles congruent or supplementary were developed by using parallel lines. Here is a quick review of the relevant postulate and theorems. Each has the hypothesis, "If two parallel lines are cut by a transversal, . . ."

POSTULATE 11: If two parallel lines are cut by a transversal, then the corresponding angles are congruent.

THEOREM 2.1.2: If two parallel lines are cut by a transversal, then the alternate interior angles are congruent.

THEOREM 2.1.3: If two parallel lines are cut by a transversal, then the alternate exterior angles are congruent.

THEOREM 2.1.4: If two parallel lines are cut by a transversal, then the interior angles on the same side of the transversal are supplementary.

THEOREM 2.1.5: If two parallel lines are cut by a transversal, then the exterior angles on the same side of the transversal are supplementary.

Suppose that we now wish to prove that two lines are parallel rather than to establish an angle relationship (as the previous statements do). At present, the only method we have of proving lines parallel is based on the definition of parallel lines. Establishing the conditions of the definition (that coplanar lines do *not* intersect) is virtually impossible! Thus we begin to develop methods for proving that lines in a plane are parallel by proving Theorem 2.3.1 by the indirect method. Counterparts of Theorems 2.1.2–2.1.5, namely Theorems 2.3.2–2.3.5, are proved directly but depend upon Theorem 2.3.1.

THEOREM 2.3.1: If two lines are cut by a transversal so that the corresponding angles are congruent, then these lines are parallel.

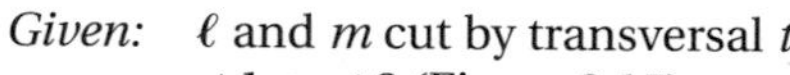

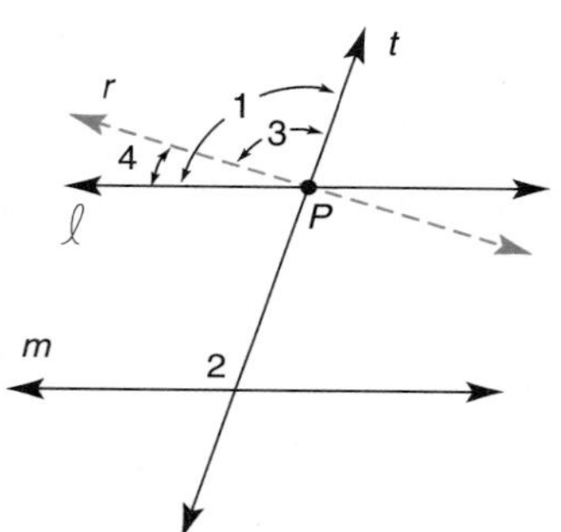

FIGURE 2.15

Given: ℓ and m cut by transversal t
$\angle 1 \cong \angle 2$ (Figure 2.15)

Prove: $\ell \parallel m$

Proof: Suppose that $\ell \nparallel m$. Then a line r can be drawn through point P so that it is parallel to m; this follows from the Parallel Postulate. If $r \parallel m$, then $\angle 3 \cong \angle 2$ since these angles correspond. But $\angle 1 \cong \angle 2$ by hypothesis. Now $\angle 3 \cong \angle 1$ by the Transitive Property of Congruence; therefore $m\angle 3 = m\angle 1$. But $m\angle 3 + m\angle 4 = m\angle 1$. (See Figure 2.15). Substituting, $m\angle 1 + m\angle 4 = m\angle 1$; and by subtraction, $m\angle 4 = 0$. This contradicts the Protractor Postulate, which states that the measure of any angle must be a positive number. Consequently, r and ℓ must coincide, and it follows that $\ell \parallel m$.

Once proved, Theorem 2.3.1 also opens the doors to a host of other methods for proving that lines are parallel. In all such claims, the lines are assumed to be coplanar. Each claim is the converse of its counterpart in Section 2.1.

THEOREM 2.3.2: If two lines are cut by a transversal so that the alternate interior angles are congruent, then these lines are parallel.

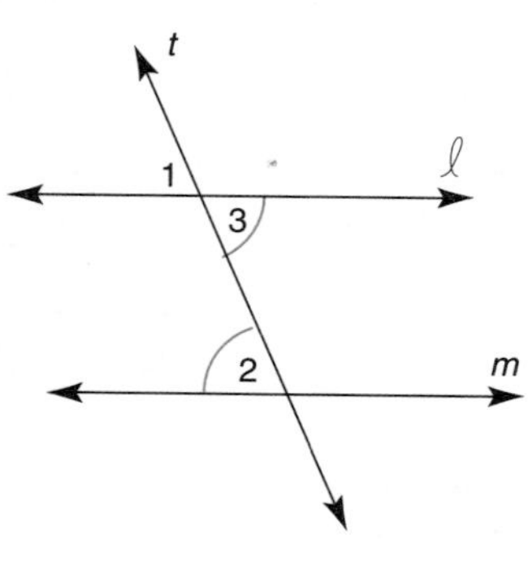

FIGURE 2.16

Given: Lines ℓ and m and transversal t
$\angle 2 \cong \angle 3$ (Figure 2.16)

Prove: $\ell \parallel m$

Plan for the proof: Show that $\angle 1 \cong \angle 2$ (corresponding angles). Then apply Theorem 2.3.1, in which $\cong$ corresponding $\angle$s imply parallel lines.

DISCOVER!

When a staircase is designed, "stringers" are cut for each side of the stairs as shown. How are angles 1 and 3 related? How are angles 1 and 2 related?

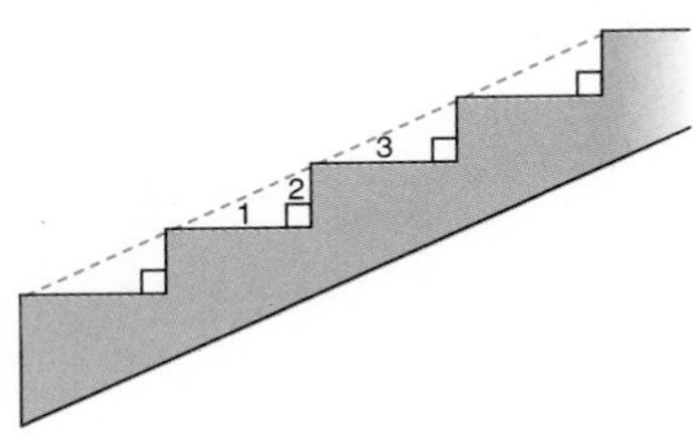

ANSWERS

(a) Congruent (b) Complementary

PROOF

Statements	Reasons
1. ℓ and m; trans. t; $\angle 2 \cong \angle 3$	1. Given
2. $\angle 1 \cong \angle 3$	2. If two lines intersect, vertical $\angle$s are $\cong$
3. $\angle 1 \cong \angle 2$	3. Transitive Property of Congruence
4. $\ell \parallel m$	4. If two lines are cut by a transversal so that corr. $\angle$s are $\cong$, then these lines are parallel.

The following theorem is proven in a manner much like the proof of Theorem 2.3.2. The proof is left as an exercise.

THEOREM 2.3.3: If two lines are cut by a transversal so that the alternate exterior angles are congruent, then these lines are parallel.

In a more involved drawing, it may be difficult to decide which lines are parallel because of congruent angles. Consider Figure 2.17. Suppose that $\angle 1 \cong \angle 3$. Which lines must be parallel? The resulting confusion (since it appears that a may be parallel to b *and* c may be parallel to d) can be overcome by asking, "Which lines help form $\angle 1$ and $\angle 3$?" In this case, $\angle 1$ and $\angle 3$ are formed by lines a and b with c as the transversal. Thus, $a \parallel b$.

EXAMPLE 1

In Figure 2.17, which lines must be parallel if $\angle 3 \cong \angle 8$?

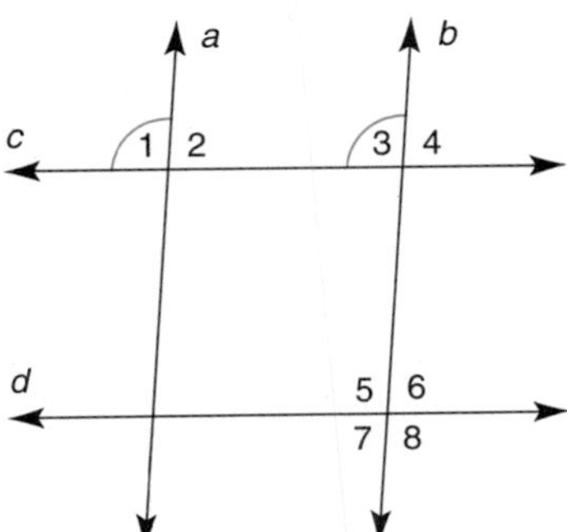

FIGURE 2.17

Solution $\angle 3$ and $\angle 8$ are the alternate exterior angles formed when lines c and d are cut by transversal b. Thus, $c \parallel d$.

Theorems 2.3.4 and 2.3.5 enable us to prove that lines are parallel when certain pairs of angles are supplementary.

THEOREM 2.3.4: If two lines are cut by a transversal so that the interior angles on the same side of the transversal are supplementary, then these lines are parallel.

EXAMPLE 2

Prove Theorem 2.3.4. (See Figure 2.18.)

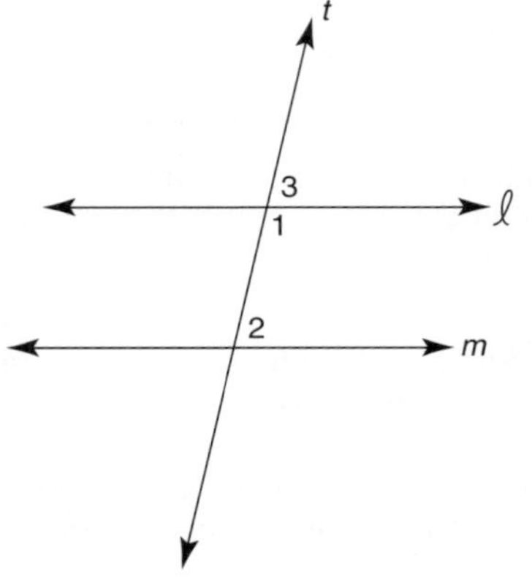

FIGURE 2.18

Given: Lines ℓ and m; transversal t;
$\angle 1$ is supplementary to $\angle 2$

Prove: $\ell \parallel m$

PROOF

Statements	Reasons
1. ℓ and m; trans. t; $\angle 1$ is supp. to $\angle 2$	**1.** Given
2. $\angle 1$ is supp. to $\angle 3$	**2.** If the exterior sides of two adjacent $\angle$s form a straight line, these $\angle$s are supplementary
3. $\angle 2 \cong \angle 3$	**3.** If two $\angle$s are supp. to the same $\angle$, they are $\cong$.
4. $\ell \parallel m$	**4.** If two lines are cut by a transversal so that corr. $\angle$s are $\cong$, then these lines are parallel

The proof of Theorem 2.3.5 is similar to that of Theorem 2.3.4. The proof is left as an exercise.

THEOREM 2.3.5: If two lines are cut by a transversal so that the exterior angles on the same side of the transversal are supplementary, then these lines are parallel.

EXAMPLE 3

In Figure 2.19, which line segments must be parallel if $\angle B$ and $\angle C$ are supplementary?

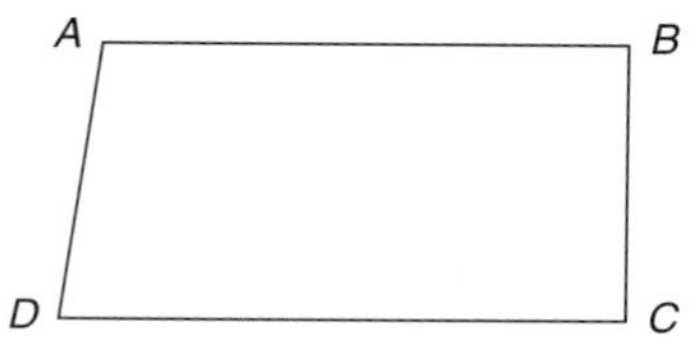

FIGURE 2.19

Solution

Again, the solution lies in the question, "Which line segments form $\angle B$ and $\angle C$?" With $\overline{BC}$ as a transversal, $\angle B$ and $\angle C$ are formed by $\overline{AB}$ and $\overline{DC}$. It follows that $\overline{AB} \parallel \overline{DC}$, since $\angle$s B and C are supplementary.

We include two final theorems that provide additional means of proving that lines are parallel. The proof of Theorem 2.3.6 requires an auxiliary line (a transversal). Proof of Theorem 2.3.7 is found in Example 4.

THEOREM 2.3.6: If two lines are each parallel to a third line, then these lines are parallel to each other.

THEOREM 2.3.7: If two coplanar lines are each perpendicular to a third line, then these lines are parallel to each other.

EXAMPLE 4

Complete the proof.

Given: $\overleftrightarrow{AC} \perp \overleftrightarrow{BE}$ and $\overleftrightarrow{DF} \perp \overleftrightarrow{BE}$ (Figure 2.20).

Prove: $\overleftrightarrow{AC} \parallel \overleftrightarrow{DF}$

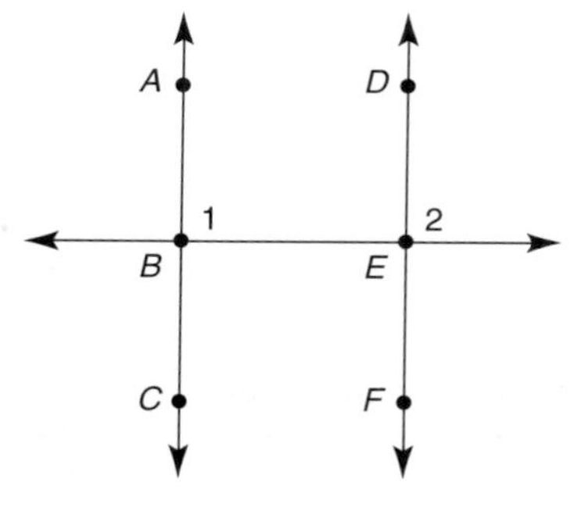

FIGURE 2.20

PROOF

Statements	Reasons
1. $\overleftrightarrow{AC} \perp \overleftrightarrow{BE}$ and $\overleftrightarrow{DF} \perp \overleftrightarrow{BE}$	**1.** Given
2. $\angle$s 1 and 2 are rt. $\angle$s	**2.** If two lines are perpendicular, they meet to form right $\angle$s
3. $\angle 1 \cong \angle 2$	**3.** All right angles are $\cong$
4. $\overleftrightarrow{AC} \parallel \overleftrightarrow{DF}$	**4.** If two lines are cut by a transversal so that corr. $\angle$s are $\cong$, then these lines are parallel

EXAMPLE 5

Complete the proof.

Given: $\overline{RV}$ and $\overline{ST}$ intersect at point X (See Figure 2.21 on the next page.)
$\angle 1$ and $\angle 3$ are complementary
$\angle 2$ and $\angle 4$ are also complementary

Prove: $\overline{RS} \parallel \overline{TV}$

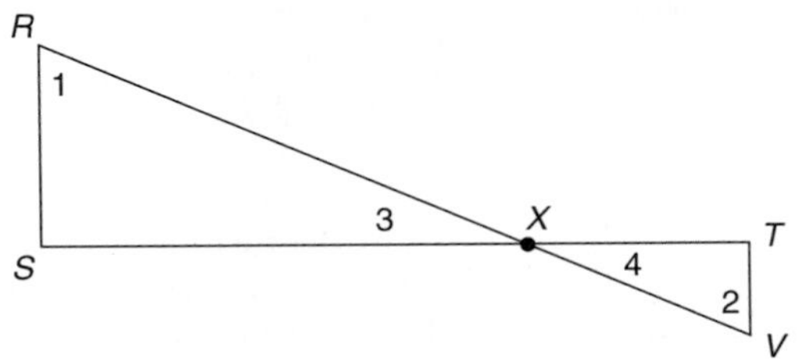

FIGURE 2.21

PROOF

Statements	Reasons
1. $\overline{RV}$ and $\overline{ST}$ intersect at point X $\angle 1$ and $\angle 3$ are complementary $\angle 2$ and $\angle 4$ are complementary	1. Given
2. $\angle 3 \cong \angle 4$	2. If two lines intersect, vertical $\angle$s are $\cong$
3. $\angle 1 \cong \angle 2$	3. Complements of $\cong$ $\angle$s are $\cong$
4. $\overline{RS} \parallel \overline{TV}$	4. If two lines are cut by a transversal so that alternate interior $\angle$s are $\cong$, these lines are parallel

Construction 7 depends upon Theorem 2.3.1, which is restated below.

> **THEOREM 2.3.1:** If two lines are cut by a transversal so that corresponding angles are congruent, then these lines are parallel.

Construction 7:
To construct the line parallel to a given line from a point not on that line.

Given: $\overleftrightarrow{AB}$ and point P not on $\overleftrightarrow{AB}$, as in Figure 2.22(a)

Construct: The line through point P parallel to $\overleftrightarrow{AB}$

Construction: Draw a line (to become a transversal) through point P and some point on $\overleftrightarrow{AB}$. For convenience, we choose point A and draw $\overleftrightarrow{AP}$ as in Figure 2.22(b). Using P as the vertex, construct the angle that corresponds to $\angle PAB$ so that this angle is congruent to $\angle PAB$. It may be necessary to extend $\overleftrightarrow{AP}$ upward to accomplish this. The line $\overleftrightarrow{PX}$ shown in Figure 2.22(c) is the desired line parallel to $\overleftrightarrow{AB}$.

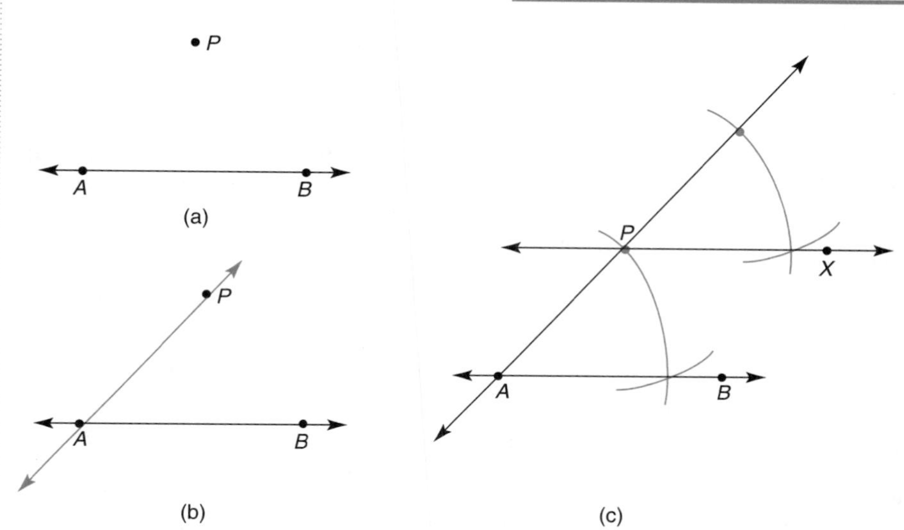

FIGURE 2.22

2.3 Exercises

In Exercises 1 to 10, name the lines (if any) that must be parallel under the given conditions.

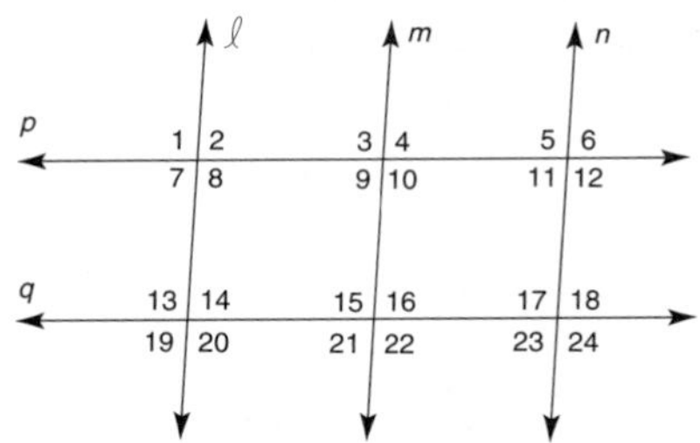

Exercises 1–10

1. $\angle 1 \cong \angle 20$
2. $\angle 3 \cong \angle 10$
3. $\angle 9 \cong \angle 14$
4. $\angle 7 \cong \angle 11$
5. $\ell \perp p$ and $n \perp p$
6. $\ell \parallel m$ and $m \parallel n$
7. $\ell \perp p$ and $m \perp q$
8. $\angle 8$ and $\angle 9$ are supplementary
9. $m\angle 8 = 110$, $p \parallel q$, and $m\angle 18 = 70$
10. The bisectors of $\angle 9$ and $\angle 21$ are parallel.

In Exercises 11 and 12, complete each proof by filling in the missing statements and reasons.

11. *Given:* $\angle 1$ and $\angle 2$ are complementary
$\angle 3$ and $\angle 1$ are complementary
Prove: $\overline{BC} \parallel \overline{DE}$

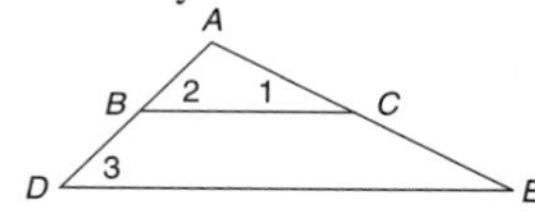

PROOF

Statements	Reasons
1. $\angle$s 1 and 2 are comp.; $\angle$s 3 and 1 are comp.	1. ?
2. $\angle 2 \cong \angle 3$	2. ?
3. ?	3. If two lines are cut by a transversal so that corr. $\angle$s are $\cong$, the lines are $\parallel$

12. *Given:* $\ell \parallel m$
$\angle 3 \cong \angle 4$
Prove: $\ell \parallel n$

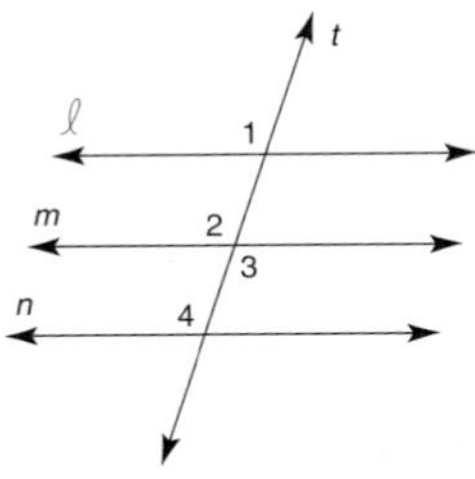

PROOF

Statements	Reasons
1. $\ell \parallel m$	**1.** ?
2. $\angle 1 \cong \angle 2$	**2.** ?
3. $\angle 2 \cong \angle 3$	**3.** If two lines intersect, the vertical $\angle$s formed are $\cong$
4. ?	**4.** Given
5. $\angle 1 \cong \angle 4$	**5.** Transitive Prop. of $\cong$
6. ?	**6.** ?

In Exercises 13 to 16, complete the proof.

13. *Given:* $\overline{AD} \perp \overline{DC}$
$\overline{BC} \perp \overline{DC}$
Prove: $\overline{AD} \parallel \overline{BC}$

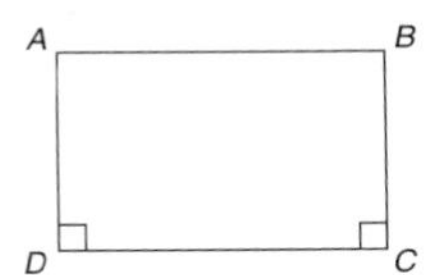

14. *Given:* $m\angle 2 + m\angle 3 = 90$
$\overrightarrow{BE}$ bisects $\angle ABC$
$\overrightarrow{CE}$ bisects $\angle BCD$
Prove: $\ell \parallel n$

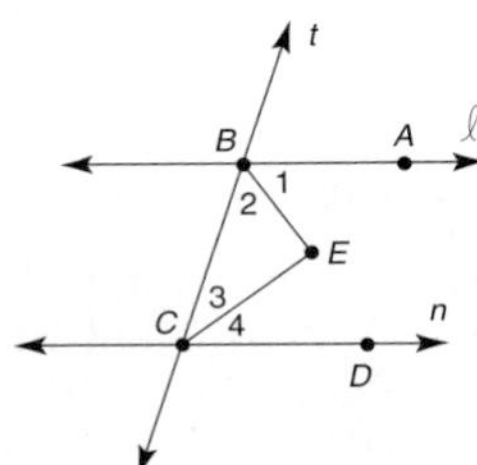

15. *Given:* $\overrightarrow{DE}$ bisects $\angle CDA$
$\angle 3 \cong \angle 1$
Prove: $\overline{ED} \parallel \overline{AB}$

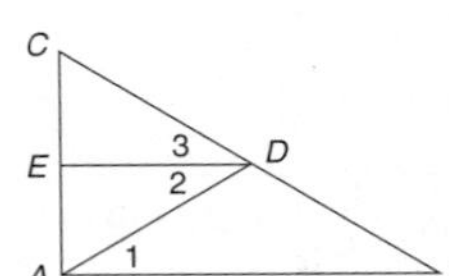

16. *Given:* $\overline{XY} \parallel \overline{WZ}$
$\angle 1 \cong \angle 2$
Prove: $\overline{MN} \parallel \overline{XY}$

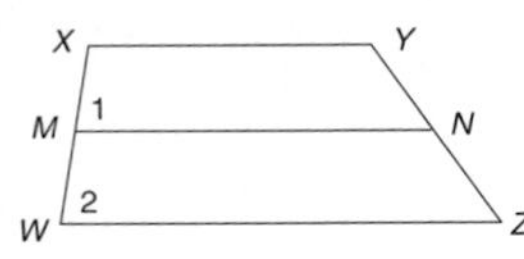

In Exercises 17 to 20, determine the value of x so that line ℓ will be parallel to line m.

17. $m\angle 6 = x^2 - 9$
$m\angle 2 = x(x - 1)$

18. $m\angle 4 = 2x^2 - 3x + 6$
$m\angle 5 = 2x(x - 1) - 2$

19. $m\angle 3 = (x + 1)(x + 4)$
$m\angle 5 = 16(x + 3) - (x^2 - 2)$

20. $m\angle 2 = (x^2 - 1)(x + 1)$
$m\angle 8 = 185 - x^2(x + 1)$

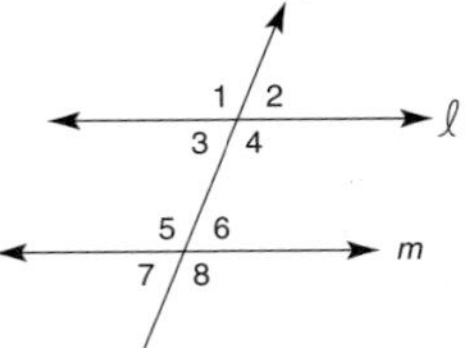

Exercises 17–20

In Exercises 21 to 23, give a formal proof for each theorem.

21. If two lines are cut by a transversal so that the alternate exterior angles are congruent, then these lines are parallel.

22. If two lines are cut by a transversal so that the exterior angles on the same side of the transversal are supplementary, then these lines are parallel.

23. If two lines are parallel to the same line, then these lines are parallel to each other. (Assume three coplanar lines.)

24. Explain why the statement in Exercise 23 remains true even if the three lines are not coplanar.

25. Given that point P does *not* lie on line ℓ, construct the line through point P that is parallel to line ℓ.

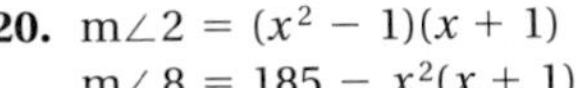

26. Given that point Q does *not* lie on $\overline{AB}$, construct the line through point Q that is parallel to $\overline{AB}$.

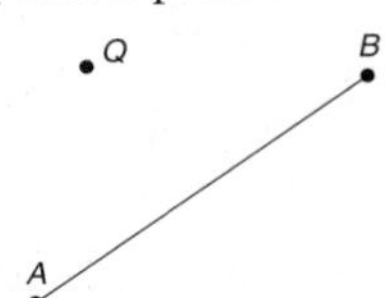

27. A carpenter drops a plumb line from point A to $\overline{BC}$. Assuming that $\overline{BC}$ is horizontal, the point D at which the plumb line intersects $\overline{BC}$ will determine the vertical line segment $\overline{AD}$. Use a construction to locate point D.

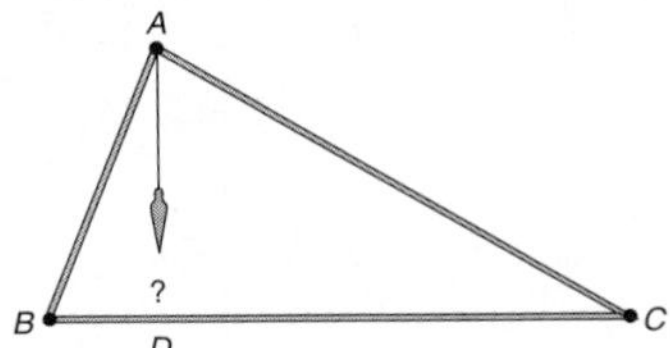

2.4 The Angles of a Triangle

Recall that in geometry, the word *union* means that figures are joined or combined.

> **DEFINITION:** A **triangle** is the union of three line segments that are determined by three noncollinear points.

Symbolized by $\triangle$, the triangle is a figure you have encountered many times. In the triangle in Figure 2.23, known as $\triangle ABC$, each point A, B, and C is a **vertex** of the triangle; collectively, these three points are the **vertices** of the triangle. $\overline{AB}$, $\overline{BC}$, and $\overline{AC}$ are the **sides** of the triangle. Point D is in the **interior** of the triangle; point E is on the triangle; and point F is in the **exterior** of the triangle.

Triangles may be categorized by the lengths of their sides. Table 2.1 presents the type of triangle, the relationship among its sides, and a drawing in which congruent parts are indicated.

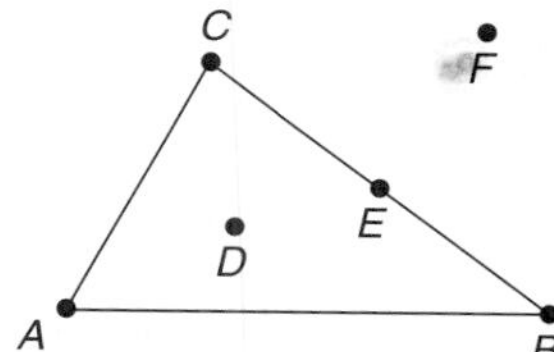

FIGURE 2.23

TABLE 2.1

Triangles Classified by Congruent Sides

Type		Number of Congruent Sides
Scalene		None
Isosceles		At least two congruent sides
Equilateral		All three sides congruent

Triangles may also be classified according to their angles (see Table 2.2).

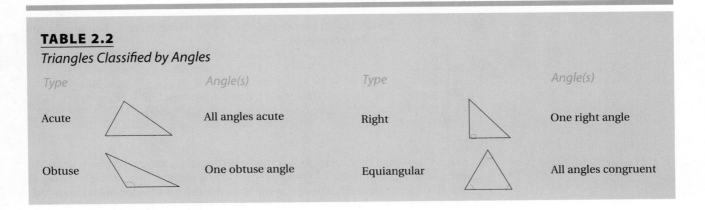

TABLE 2.2
Triangles Classified by Angles

Type	Angle(s)	Type	Angle(s)
Acute	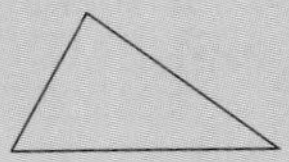 All angles acute	Right	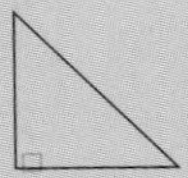 One right angle
Obtuse	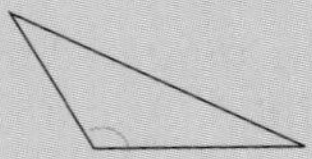 One obtuse angle	Equiangular	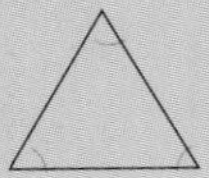 All angles congruent

EXAMPLE 1

In $\triangle HJK$ (not shown), $HJ = 4$, $JK = 4$, and $m\angle J = 90°$. Describe completely the type of triangle represented.

Solution

$\triangle HJK$ is a right isosceles $\triangle$, or $\triangle HJK$ is an isosceles right triangle.

DISCOVER!

From a paper triangle, cut the angles from the "corners." Now place the angles together at the same vertex as shown. What is the sum of the measures of the three angles?

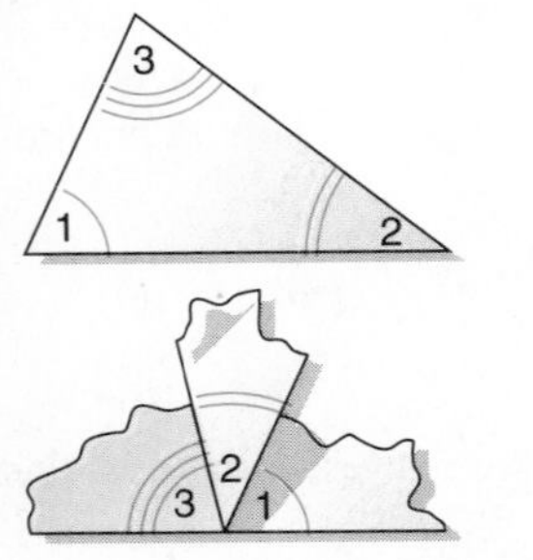

ANSWER
180°

In an earlier exercise, it was suggested that the sum of the measures of the three interior angles of a triangle is 180°. This is now stated as a theorem and proved through the use of an **auxiliary** (or helping) **line.** When an auxiliary line is added to the drawing for a proof, a justification must be given for the existence of that line. Justifications include statements such as

There is exactly one line through two distinct points.

An angle has exactly one bisector.

There is only one line perpendicular to another line at a point on that line.

When an auxiliary line is introduced into a proof, the original drawing is sometimes redrawn for the sake of clarity. This has been done in the proof of the following important theorem.

THEOREM 2.4.1: In a triangle, the sum of the measures of the interior angles is 180°.

Given: $\triangle ABC$ in Figure 2.24(a)

Prove: $m\angle A + m\angle B + m\angle C = 180°$

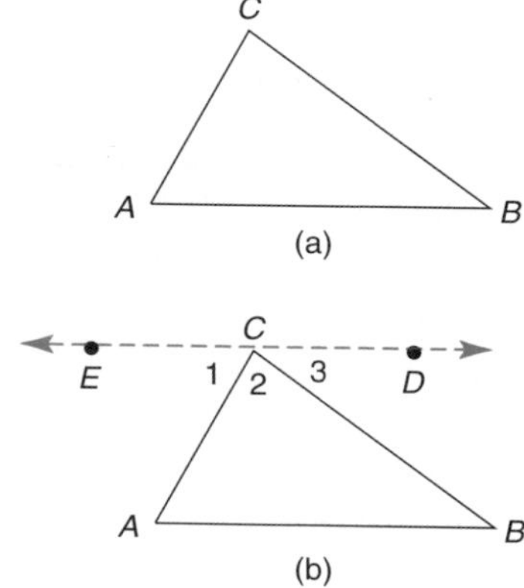

FIGURE 2.24

PROOF

	Statements	Reasons
	1. Draw $\overleftrightarrow{ED}$ through C so that $\overleftrightarrow{ED} \parallel \overleftrightarrow{AB}$; see Figure 2.24(b)	**1.** Through a point outside a line, exactly one line is parallel to the given line
	2. $m\angle 1 + m\angle 2 = m\angle ECB$	**2.** Angle-Addition Postulate
	3. $m\angle ECB + m\angle 3 = m\angle ECD$	**3.** Angle-Addition Postulate
(2), (3)	**4.** $m\angle 1 + m\angle 2 + m\angle 3 = m\angle ECD$	**4.** Substitution
	5. $\angle ECD$ is a straight $\angle$, so $m\angle ECD = 180°$	**5.** If an $\angle$ is a straight $\angle$, it measures 180°
(4), (5)	**6.** $m\angle 1 + m\angle 2 + m\angle 3 = 180°$	**6.** Substitution
	7. $\angle 1 \cong \angle A$ and $\angle 3 \cong \angle B$	**7.** If two $\parallel$ lines are cut by a transversal, alternate interior $\angle$s are $\cong$
	8. $m\angle 1 = m\angle A$ and $m\angle 3 = m\angle B$	**8.** If two $\angle$s are $\cong$, they have $=$ measures
(6), (8)	**9.** $m\angle A + m\angle B + m\angle ACB = 180°$	**9.** Substitution

Statements 7 and 8 are related so closely that we cannot have one without the other. At times, we will use these notions of the equality and congruence of angles interchangeably within a proof, without stating both.

EXAMPLE 2

In $\triangle RST$ (not shown), $m\angle R = 45°$ and $m\angle S = 64°$. Find $m\angle T$.

Solution

In $\triangle RST$, $m\angle R + m\angle S + m\angle T = 180°$, so $45° + 64° + m\angle T = 180°$. Thus $109° + m\angle T = 180°$ and $m\angle T = 71°$.

A theorem that follows directly from a previous theorem is known as a **corollary** of that theorem. Corollaries, like theorems, must be proved before they can be used. These proofs are often brief, but they depend on the related theorem. Here are some corollaries of Theorem 2.4.1. We suggest that the student make a drawing to illustrate each corollary.

COROLLARY 2.4.2: Each angle of an equiangular triangle measures 60°.

COROLLARY 2.4.3: The acute angles of a right triangle are complementary.

EXAMPLE 3

Given: $\angle M$ is a right $\angle$ in $\triangle NMQ$ (not shown)
$m\angle N = 57°$

Find: $m\angle Q$

Solution

Since the acute $\angle$s of a right triangle are complementary,

$$m\angle N + m\angle Q = 90°$$
$$\therefore 57° + m\angle Q = 90°$$
$$m\angle Q = 33°$$

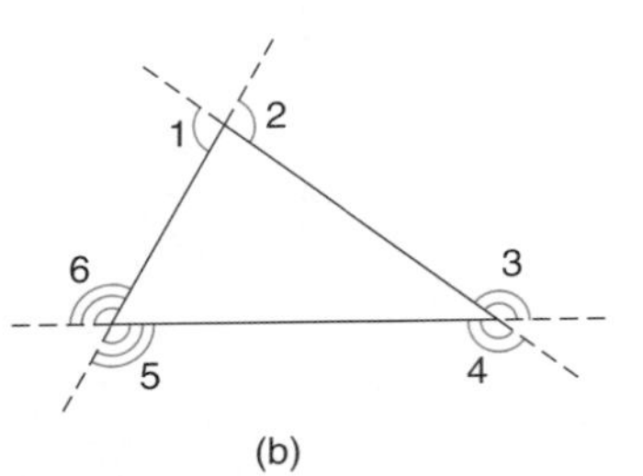

FIGURE 2.25

COROLLARY 2.4.4: If two angles of one triangle are congruent to two angles of another triangle, then the third angles are also congruent.

When the sides of a triangle are extended, each angle that is formed by a side and an extension of the adjacent side is an **exterior angle** of the triangle. In Figure 2.25(a), $\angle ACD$ is an exterior angle of $\triangle ABC$; for a triangle, there are a total of six exterior angles—two at each vertex. (See Figure 2.25(b).)

COROLLARY 2.4.5: The measure of an exterior angle of a triangle equals the sum of the measures of the two nonadjacent interior angles.

EXAMPLE 4

Given: In Figure 2.26,

$$m\angle 1 = x^2 + 2x$$
$$m\angle S = x^2 - 2x$$
$$m\angle T = 3x + 10$$

Find: x

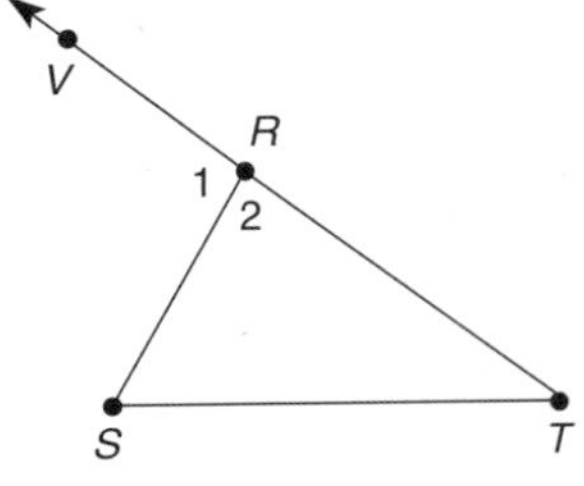

FIGURE 2.26

Solution By Corollary 2.4.5,

$$\begin{aligned} m\angle 1 &= m\angle S + m\angle T \\ x^2 + 2x &= (x^2 - 2x) + (3x + 10) \\ 2x &= x + 10 \\ x &= 10 \end{aligned}$$

Check: $m\angle 1 = 120°$, $m\angle S = 80°$, and $m\angle T = 40°$; so $120 = 80 + 40$, which satisfies the conditions of Corollary 2.4.5.

2.4 Exercises

In Exercises 1 and 2, make drawings as needed.

1. Suppose that for $\triangle ABC$ and $\triangle MNQ$, you know that $\angle A \cong \angle M$ and $\angle B \cong \angle N$. Explain why $\angle C \cong \angle Q$.
2. Suppose that T is a point on side $\overline{PQ}$ of $\triangle PQR$. Also, $\overrightarrow{RT}$ bisects $\angle PRQ$, and $\angle P \cong \angle Q$. If $\angle 1$ and $\angle 2$ are the angles formed when $\overrightarrow{RT}$ intersects $\overline{PQ}$, explain why $\angle 1 \cong \angle 2$.

In Exercises 3 to 5, $j \parallel k$ and $\triangle ABC$.

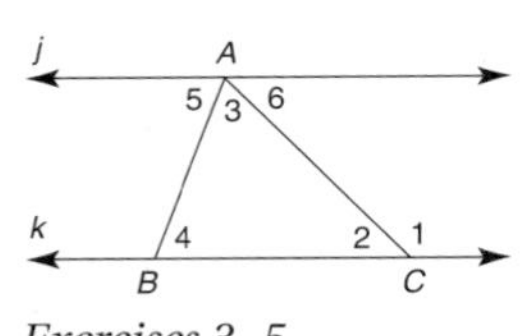

Exercises 3–5

3. *Given:* $m\angle 3 = 50°$, $m\angle 4 = 72°$
   Find: $m\angle 1$, $m\angle 2$, and $m\angle 5$
4. *Given:* $m\angle 3 = 55°$, $m\angle 2 = 74°$
   Find: $m\angle 1$, $m\angle 4$, and $m\angle 5$
5. *Given:* $m\angle 1 = 122.3°$, $m\angle 5 = 41.5°$
   Find: $m\angle 2$, $m\angle 3$, and $m\angle 4$
6. *Given:* $\overline{MN} \perp \overline{NQ}$ and $\angle$s as shown
   Find: x, y, and z

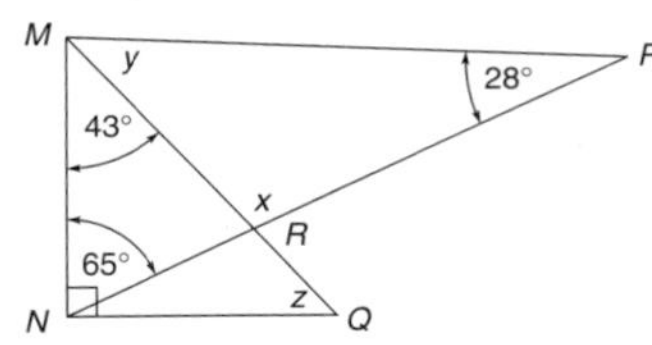

7. *Given:* $\overline{AB} \parallel \overline{DC}$
   $\overrightarrow{DB}$ bisects $\angle ADC$
   $m\angle A = 110°$
   Find: $m\angle 3$
8. *Given:* $\overline{AB} \parallel \overline{DC}$
   $\overrightarrow{DB}$ bisects $\angle ADC$
   $m\angle 1 = 36°$
   Find: $m\angle A$

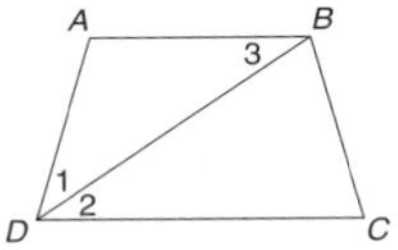

Exercises 7, 8

9. *Given:* $\triangle ABC$ with B-D-E-C
   $m\angle 3 = m\angle 4 = 30°$
   $m\angle 1 = m\angle 2 = 70°$
   Find: $m\angle B$

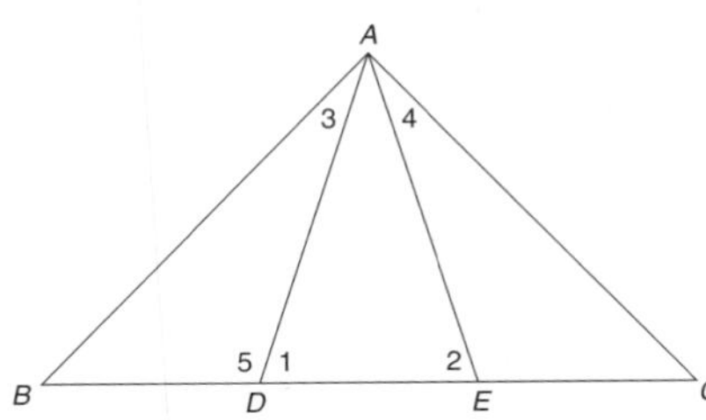

Exercises 9, 10

10. *Given:* $\triangle ABC$ with B-D-E-C
    $m\angle 1 = 2x$
    $m\angle 3 = x$
    Find: $m\angle B$ in terms of x
11. Consider any triangle and one exterior angle at each vertex. What is the sum of the measures of the three exterior angles of the triangle?

12. *Given:* Right $\triangle ABC$ with right $\angle C$
$m\angle 1 = 7x + 4$
$m\angle 2 = 5x + 2$
Find: x

A 1 C 2 3 B

Exercises 12, 13

13. *Given:* $m\angle 1 = x$
$m\angle 2 = y$
$m\angle 3 = 3x$
Find: x and y

14. *Given:* $m\angle 1 = 8(x + 2)$
$m\angle 3 = 5x - 3$
$m\angle 5 = 5(x + 1) - 2$
Find: x

15. *Given:* $m\angle 1 = x$
$m\angle 2 = 4y$
$m\angle 3 = 2y$
$m\angle 4 = 2x - y - 40$
Find: x, y, $m\angle 5$

5 1 2 3 4

Exercises 14, 15

16. *Given:* Equiangular $\triangle RST$
$\overrightarrow{RV}$ bisects $\angle SRT$
Prove: $\triangle RVS$ is a right $\triangle$

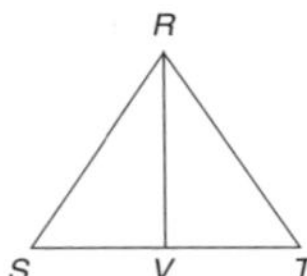

17. *Given:* $\overline{MN}$ and $\overline{PQ}$ intersect at K
$\angle M \cong \angle Q$
Prove: $\angle P \cong \angle N$

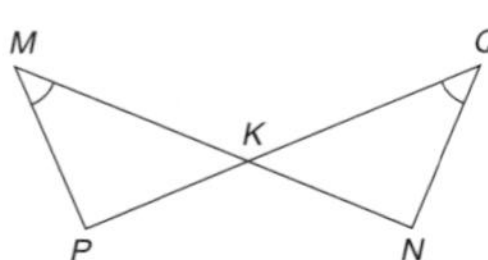

18. The sum of the measures of two angles of a triangle equals the measure of the third (largest) angle. What type of triangle is described?

19. Draw, if possible, an
   a) isosceles obtuse triangle.
   b) equilateral right triangle.

20. Draw, if possible, a
   a) right scalene triangle.
   b) triangle having both a right angle and an obtuse angle.

21. Along a straight shoreline, two houses are located at points H and M. The houses are 5000 feet apart. A small island lies in view of both houses, with angles as indicated. Find $m\angle I$.

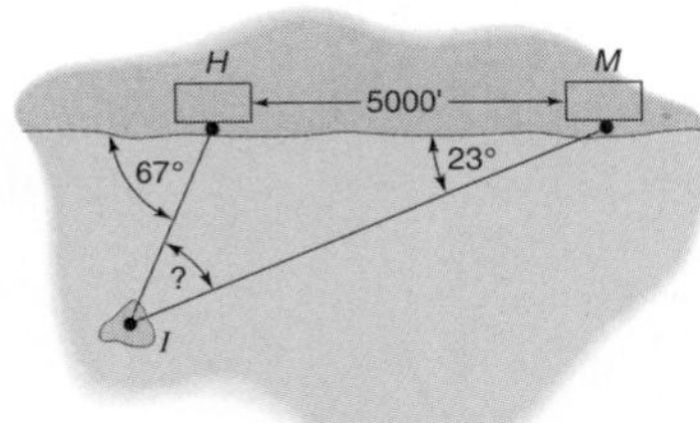

22. An airplane has leveled off (is flying horizontally) at an altitude of 12,000 feet. Its pilot can see each of two small towns at points R and T in front of the plane. With angle measures as indicated, find $m\angle R$.

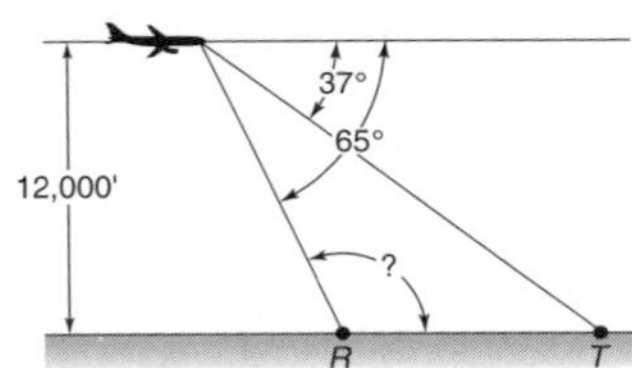

23. On a map, three Los Angeles suburbs are located at points N (Newport Beach), P (Pomona), and B (Burbank). With angle measures as indicated, determine $m\angle N$ and $m\angle P$.

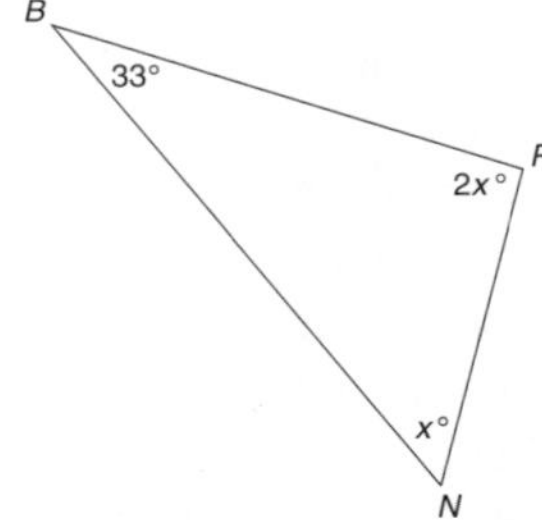

24. The roofline of a house shows the shape of right triangle ABC with $m\angle C = 90°$. If the measure of $\angle CAB$ is 24° larger than the measure of $\angle CBA$, then how large is each angle?

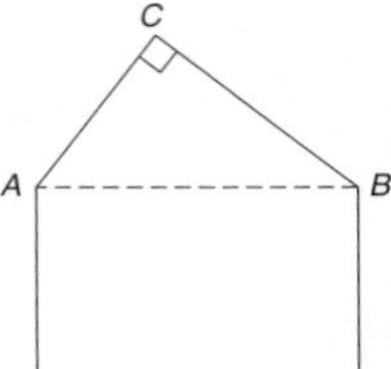

25. The triangular symbol on the "PLAY" button of a VCR has congruent angles at M and N. If $m\angle P = 30°$, what are the measures of angle M and angle N?

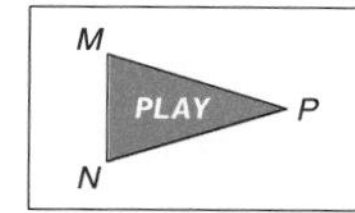

26. A polygon with four sides is called a *quadrilateral.* Consider the figure and the dashed auxiliary line. What is the sum of the measures of the four interior angles of this (or any other) quadrilateral?

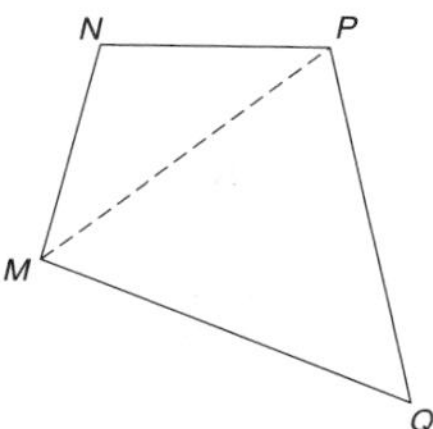

27. Explain why the following statement is true.

Each interior angle of an equiangular triangle measures 60°.

28. Explain why the following statement is true.

The acute angles of a right triangle are complementary.

In Exercises 29 to 31, write a formal proof for each corollary.

29. The measure of an exterior angle of a triangle equals the sum of the measures of the two nonadjacent interior angles.

30. If two angles of one triangle are congruent to two angles of another triangle, then the third angles are also congruent.

31. Use an indirect proof to establish the following theorem: A triangle cannot have more than one right angle.

32. *Given:* $\overleftrightarrow{AB}$, $\overleftrightarrow{DE}$, and $\overleftrightarrow{CF}$
$\overleftrightarrow{AB} \parallel \overleftrightarrow{DE}$
$\overrightarrow{CG}$ bisects $\angle BCF$
$\overrightarrow{FG}$ bisects $\angle CFE$
Prove: $\angle G$ is a right angle

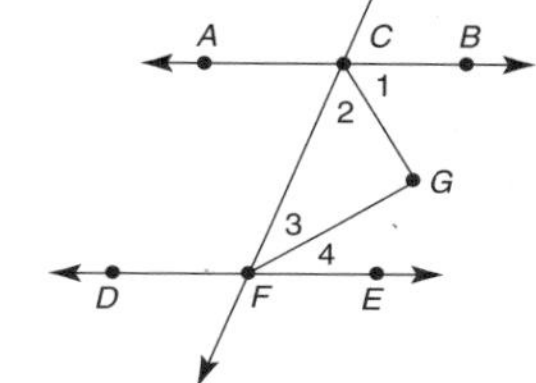

33. *Given:* $\overrightarrow{NQ}$ bisects $\angle MNP$
$\overrightarrow{PQ}$ bisects $\angle MPR$
$m\angle Q = 42$
Find: $m\angle M$

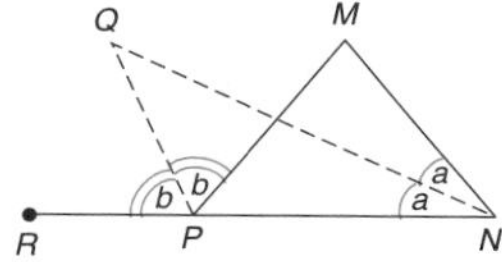

2.5 Convex Polygons

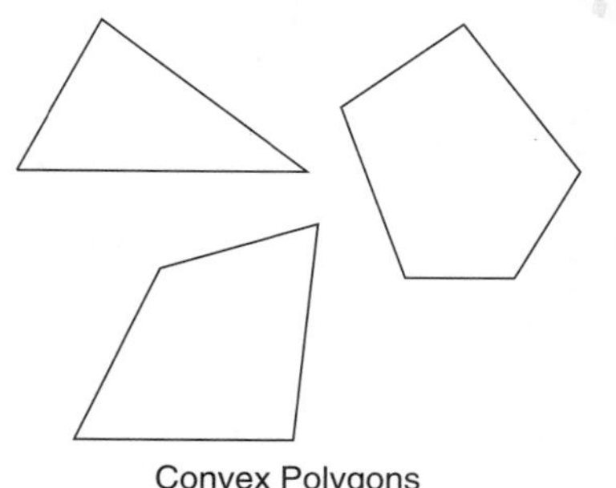

Convex Polygons

FIGURE 2.27

DEFINITION: A **polygon** is a closed plane figure whose sides are line segments that intersect only at the endpoints.

The polygons we generally consider in this textbook are **convex;** the angle measures of convex polygons are between 0° and 180°. Convex polygons are shown in Figure 2.27; those in Figure 2.28 on the next page are **concave.** A line segment joining two points of a concave polygon can contain points in the exterior of the polygon. Figure 2.29 shows some figures that aren't polygons at all!

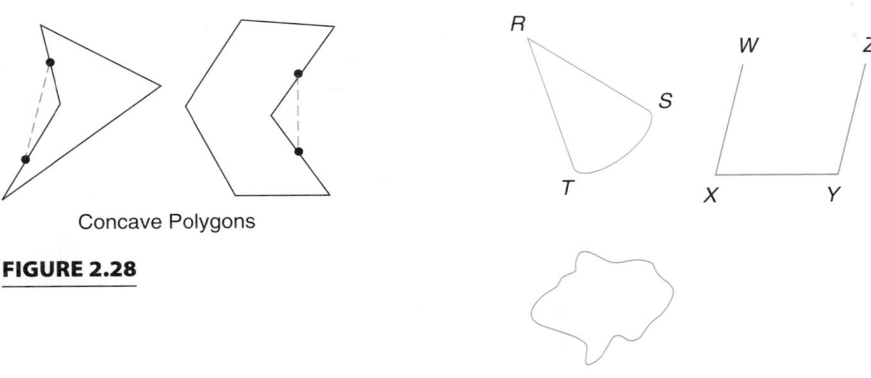

FIGURE 2.28

FIGURE 2.29

Table 2.3 shows some special names for polygons with fixed numbers of sides.

TABLE 2.3

Polygon	Number of Sides	Polygon	Number of Sides
Triangle	3	Heptagon	7
Quadrilateral	4	Octagon	8
Pentagon	5	Nonagon	9
Hexagon	6	Decagon	10

Diagonals of a Polygon

A **diagonal** of a polygon is a line segment that joins two nonconsecutive vertices.

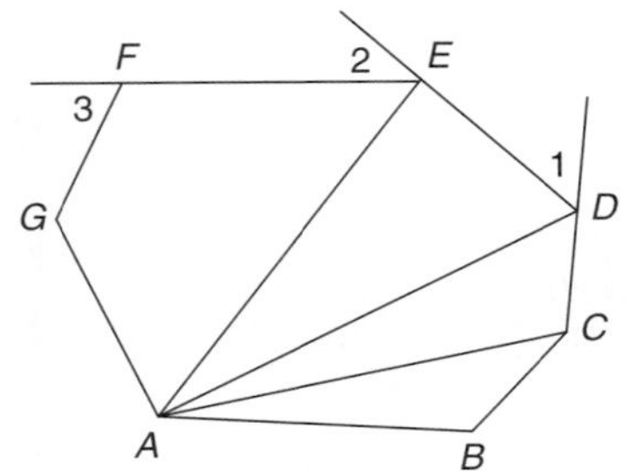

FIGURE 2.30

Figure 2.30 shows heptagon *ABCDEFG* for which $\angle GAB$, $\angle B$, and $\angle BCD$ are some of the interior angles and $\angle 1$, $\angle 2$, and $\angle 3$ are some of the exterior angles. $\overline{AB}$, $\overline{BC}$, and $\overline{CD}$ are some of the sides of the heptagon, since these join consecutive vertices. $\overline{AC}$, $\overline{AD}$, and $\overline{AE}$ are among the many diagonals of the polygon, since each joins nonconsecutive vertices of *ABCDEFG*.

Table 2.4 illustrates polygons having different numbers of sides and the corresponding numbers of diagonals.

When the number of sides is small, we can list all diagonals by name. In pentagon *ABCDE*, we see diagonals $\overline{AC}$, $\overline{AD}$, $\overline{BD}$, $\overline{BE}$, and $\overline{CE}$, a total of five. As the number of sides increases, it is more difficult to count all the diagonals. In such a case, the formula of Theorem 2.5.1 is most convenient to use. Although this theorem is given without proof, Exercise 33 of this section provides some insight for the proof.

TABLE 2.4

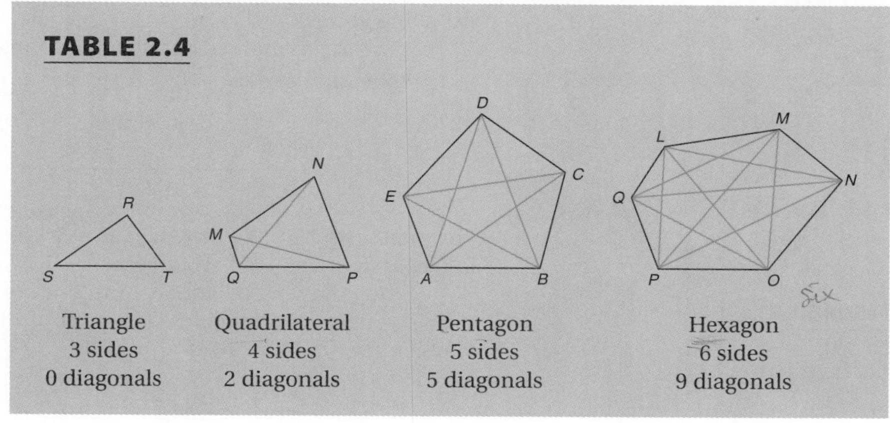

THEOREM 2.5.1: The total number of diagonals D in a polygon of n sides is given by the formula $D = \frac{n(n-3)}{2}$.

This theorem reminds us that a triangle has no diagonals, since $D = \frac{3(3-3)}{2} = 0$.

EXAMPLE 1

Use the formula of Theorem 2.5.1 to find the number of diagonals for any pentagon.

Solution To use the formula of Theorem 2.5.1, we note that $n = 5$. Then $D = \frac{5(5-3)}{2} = \frac{5(2)}{2} = 5$.

Sum of the Interior Angles of a Polygon

The following theorem provides the formula for the sum of the interior angles of any polygon.

THEOREM 2.5.2: The sum S of the measures of the interior angles of a polygon with n sides is given by $S = (n-2) \cdot 180°$. Note that $n > 2$ for any polygon.

Let us consider an informal proof of Theorem 2.5.2 for the special case of a pentagon. The proof would change for a polygon of a different number of sides, but only by the number of triangles into which the polygon can be separated. Although Theorem 2.5.2 is true for concave polygons, we consider the proof only for the case of the convex polygon.

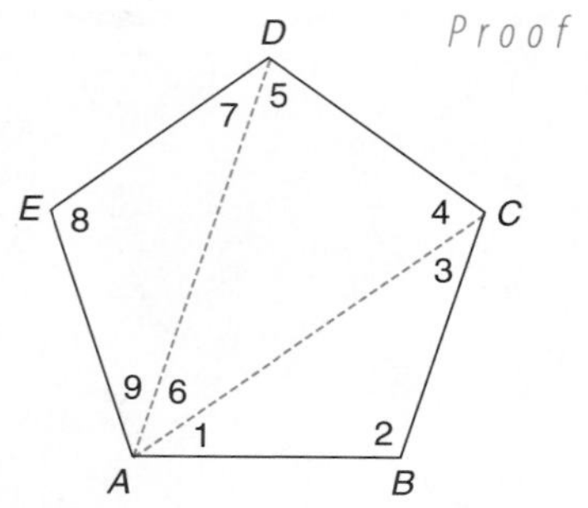

FIGURE 2.31

Proof Consider the pentagon $ABCDE$ in Figure 2.31 with auxiliary segments (diagonals from one vertex) as shown.

With angles marked as shown in triangles ABC, ACD, and ADE,

$$\begin{array}{rl} m\angle 1 + \qquad m\angle 2 + m\angle 3 &= 180° \\ m\angle 6 + m\angle 5 \qquad + m\angle 4 &= 180° \\ m\angle 8 + m\angle 9 + m\angle 7 \qquad &= 180° \\ \hline m\angle E + m\angle A + m\angle D + m\angle B + m\angle C &= 540° \end{array}$$

adding

For pentagon $ABCDE$, in which $n = 5$, the sum of the measures of the interior angles is $(5 - 2) \cdot 180°$, which equals $540°$.

When drawing diagonals from one vertex of a polygon of n sides, we can always form $(n - 2)$ triangles. The sum of the measures of the interior angles always equals $(n - 2) \cdot 180°$.

EXAMPLE 2

Find the sum of the measures of the interior angles of a hexagon. Then find the measure of each interior angle of an equiangular hexagon.

Solution For the hexagon, $n = 6$, so the sum of the measures of the interior angles is $(6 - 2) \cdot 180°$ or $4(180°)$ or $720°$.

In an equiangular hexagon, each of the six interior angles measures $\frac{720°}{6}$ or $120°$.

EXAMPLE 3

Find the number of sides in a polygon whose sum of interior angles is $2160°$.

Solution Here $S = 2160$ in the formula of Theorem 2.5.2. Since $(n - 2) \cdot 180 = 2160$, we have $180n - 360 = 2160$.

Then
$$180n = 2520$$
$$n = 14.$$

The polygon has 14 sides.

Regular Polygons

Figure 2.32 shows polygons that are, respectively: (a) **equilateral,** (b) **equiangular,** and (c) **regular** (both sides and angles are congruent). Note the parts that are marked congruent.

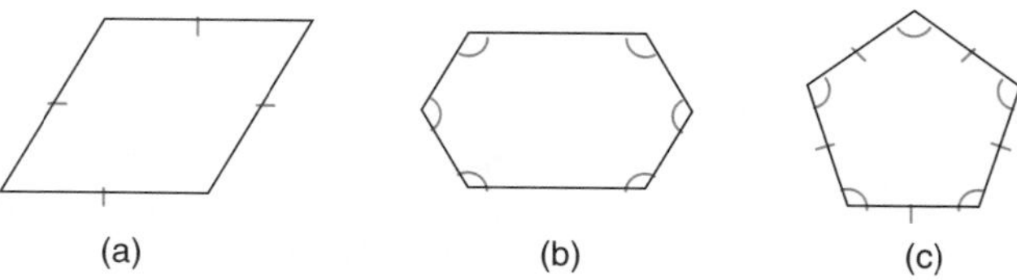

FIGURE 2.32

DEFINITION: A **regular polygon** is a polygon that is both equilateral and equiangular.

The polygon in Figure 2.32(c) is a *regular pentagon.* Other examples of regular polygons include the equilateral triangle and the square.

Based upon the formula $S = (n - 2) \cdot 180°$ from Theorem 2.5.2, there is also a formula for the measure of each interior angle of a regular polygon. It applies to equiangular polygons as well.

COROLLARY 2.5.3: The measure I of each interior angle of a regular polygon of n sides is

$$I = \frac{(n - 2) \cdot 180°}{n}.$$

EXAMPLE 4

Find the measure of each interior angle of a ceramic floor tile in the shape of an equiangular octagon (Figure 2.33).

Solution

For an octagon, $n = 8$.

Then
$$I = \frac{(8 - 2) \cdot 180}{8}$$
$$I = \frac{6 \cdot 180}{8}$$
$$I = \frac{1080}{8}, \text{ so } I = 135°.$$

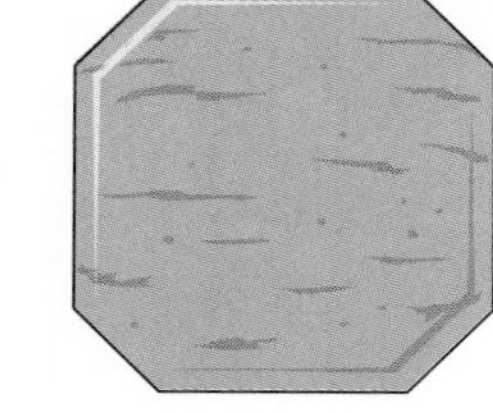

FIGURE 2.33

Each interior angle of the tile measures 135°.

EXAMPLE 5

Each interior angle of a certain regular polygon has a measure of 144°. Find its number of sides, and identify the type of polygon it is.

Solution

Let n be the number of sides the polygon has. All n of the interior angles are equal in measure.

The measure of each interior angle is given by

$$I = \frac{(n - 2) \cdot 180}{n}$$

and

$$I = 144$$

Then
$$\frac{(n-2)\cdot 180}{n} = 144$$
$$(n-2)\cdot 180 = 144n$$
$$180n - 360 = 144n$$
$$36n = 360$$
$$n = 10$$

The polygon is a regular decagon.

A second corollary to Theorem 2.5.2 concerns the sum of the interior angles of any quadrilateral. For the proof, we simply let $n = 4$ in the formula $S = (n - 2)\cdot 180°$.

COROLLARY 2.5.4: The sum of the four interior angles of a quadrilateral is 360°.

Based upon Corollary 2.5.4, it is clearly the case that each interior angle of a square or rectangle measures 90°.

The following interesting corollary to Theorem 2.5.2 can be established through algebra.

COROLLARY 2.5.5: The sum of the measures of the exterior angles of a polygon, one at each vertex, is 360°.

We now consider an algebraic proof for Corollary 2.5.5.

Proof

A polygon of n sides has n interior angles and n exterior angles, if one is considered at each vertex. As shown in Figure 2.34, these interior and exterior angles may be grouped into pairs of supplementary angles. Since there are n pairs of angles, the sum of the measures of all pairs is $180 \cdot n$ degrees.

In turn, the sum of the measures of the interior angles is $(n - 2)\cdot 180°$.

In words, we have

$$\text{Sum of Measures of Interior Angles} + \text{Sum of Measures of Exterior Angles} = \text{Sum of Measures of All Supplementary Pairs}$$

Let S represent the sum of the measures of the exterior angles.

$$(n-2)\cdot 180 + S = 180n$$
$$180n - 360 + S = 180n$$
$$-360 + S = 0$$
$$\therefore S = 360$$

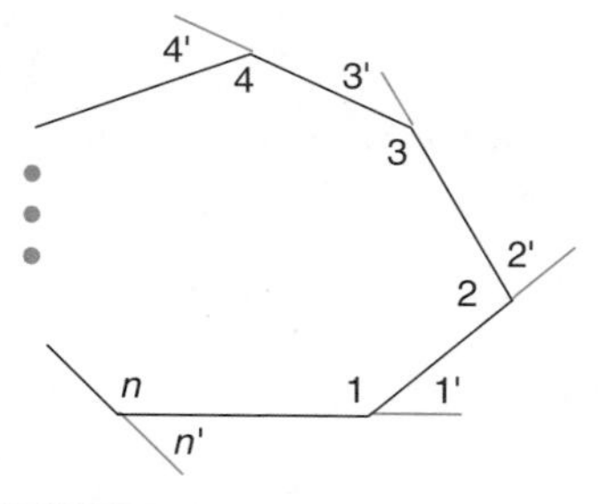

FIGURE 2.34

EXAMPLE 6

Use Corollary 2.5.5 to find the number of sides of a regular polygon if each interior angle measures 144°. (Notice that we are repeating Example 5.)

Solution

If each interior angle measures 144°, then each exterior angle measures 36° (they are supplementary, since exterior sides of these adjacent angles form a straight line).

Now each of the n exterior angles has the measure

$$\frac{360°}{n}$$

In this case, $\frac{360}{n} = 36$, and it follows that $36n = 360$, so $n = 10$. The polygon (a decagon) has 10 sides.

The next corollary follows from Corollary 2.5.5. The claim made in Corollary 2.5.6 was illustrated in Example 6.

COROLLARY 2.5.6: The measure E of each exterior angle of a regular polygon of n sides is $E = \frac{360°}{n}$.

Polygrams

A **polygram** is the star-shaped figure that results when the sides of certain polygons are extended. The polygon must be convex with five or more sides. When the polygon is regular, the resulting polygram is also regular—that is, the interior acute angles are congruent and all sides are congruent. The names of polygrams come from the names of the polygons whose sides were extended. Figure 2.35 shows a pentagram, a hexagram, and an octagram. With congruent angles and sides, these figures are **regular polygrams.**

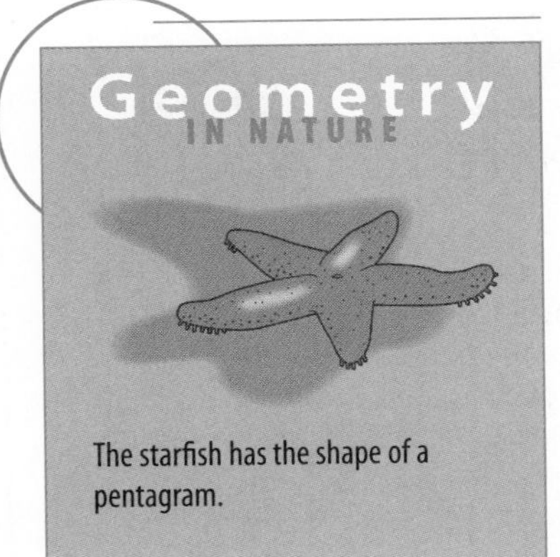

The starfish has the shape of a pentagram.

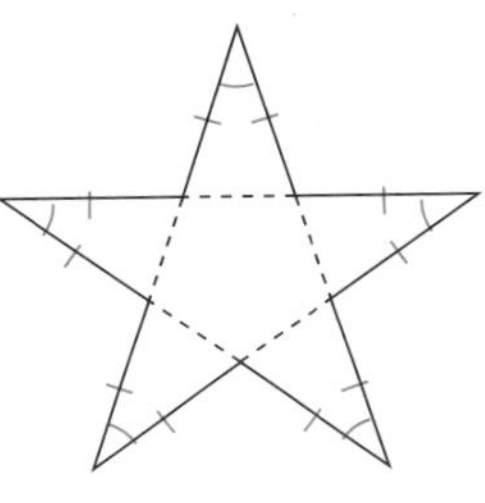

Pentagram

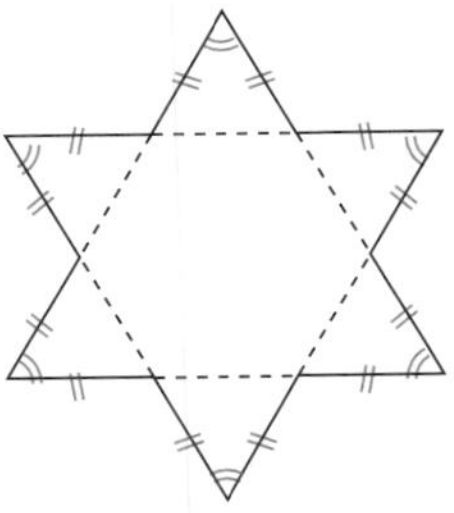

Hexagram

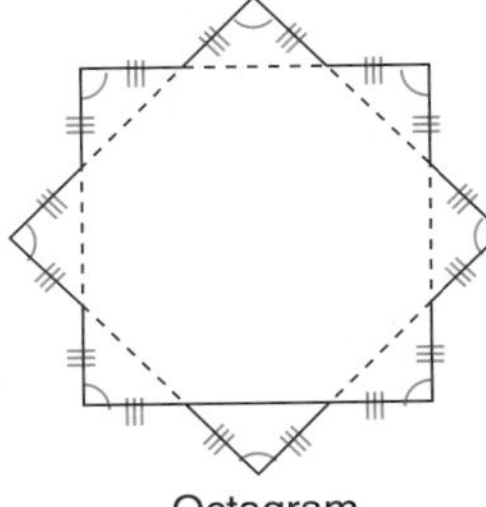

Octagram

FIGURE 2.35

2.5 Exercises

1. As the number of sides of a regular polygon increases, does each interior angle increase or decrease in measure?

2. As the number of sides of a regular polygon increases, does each exterior angle increase or decrease in measure?

3. *Given:* $\overline{AB} \parallel \overline{DC}$, $\overline{AD} \parallel \overline{BC}$, $\overline{AE} \parallel \overline{FC}$, with angle measures as indicated.
   Find: x, y, and z

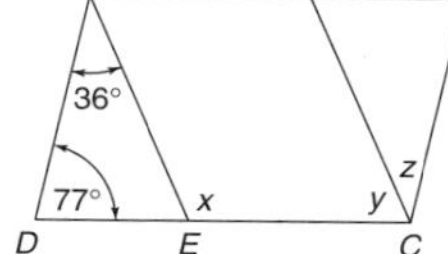

4. In pentagon $ABCDE$ with angle measures as indicated, find the measure of interior angle EDC.

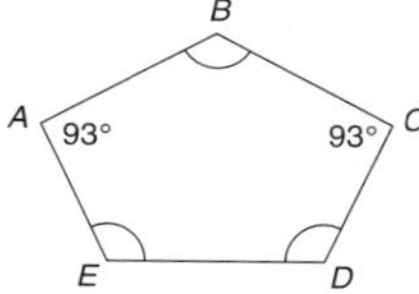

5. Find the total number of diagonals for a polygon of n sides if:
   a) $n = 5$ b) $n = 10$

6. Find the total number of diagonals for a polygon of n sides if:
   a) $n = 6$ b) $n = 8$

7. Find the sum of measures of the interior angles of a polygon of n sides if:
   a) $n = 5$ b) $n = 10$

8. Find the sum of measures of the interior angles of a polygon of n sides if:
   a) $n = 6$ b) $n = 8$

9. Find the measure of each interior angle of a regular polygon of n sides if:
   a) $n = 4$ b) $n = 12$

10. Find the measure of each interior angle of a regular polygon of n sides if:
    a) $n = 6$ b) $n = 10$

11. Find the measure of each exterior angle of a regular polygon of n sides if:
    a) $n = 4$ b) $n = 12$

12. Find the measure of each exterior angle of a regular polygon of n sides if:
    a) $n = 6$ b) $n = 10$

13. Find the number of sides that a polygon has if the sum of measures of its interior angles is:
    a) 900° b) 1260°

14. Find the number of sides that a polygon has if the sum of measures of its interior angles is:
    a) 1980° b) 2340°

15. Find the number of sides that a regular polygon has if the measure of each interior angle is:
    a) 108° b) 144°

16. Find the number of sides that a regular polygon has if the measure of each interior angle is:
    a) 150° b) 168°

17. Find the number of sides in a regular polygon whose exterior angles each measure:
    a) 24° b) 18°

18. Find the number of sides in a regular polygon whose exterior angles each measure:
    a) 45° b) 9°

19. What is the measure of each interior angle of a stop sign?

20. Lug bolts are equally spaced about the wheel to form the equal angles shown in the figure. What is the measure of each of the equal acute angles?

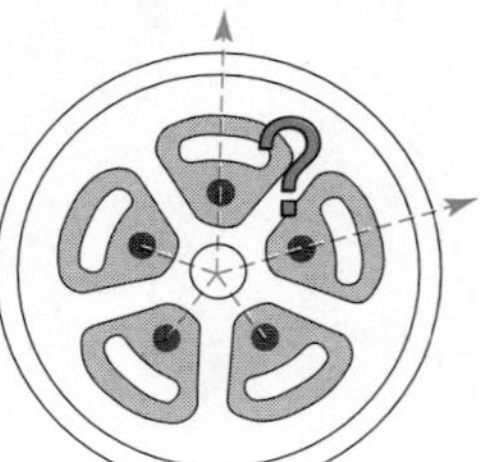

21. *Given:* Quadrilateral $RSTQ$ with exterior $\angle$s at R and T
    Prove: $m\angle 1 + m\angle 2 = m\angle 3 + m\angle 4$

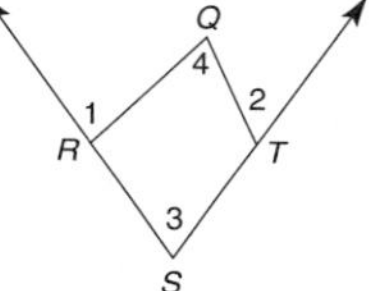

22. *Given:* Regular hexagon $ABCDEF$ with diagonal $\overline{AC}$ and exterior $\angle 1$
Prove: $m\angle 2 + m\angle 3 = m\angle 1$

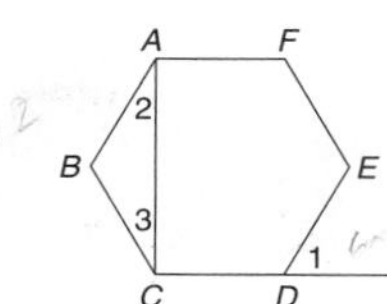

23. *Given:* Quadrilateral $RSTV$ with diagonals $\overline{RT}$ and $\overline{SV}$ intersecting at W
Prove: $m\angle 1 + m\angle 2 = m\angle 3 + m\angle 4$

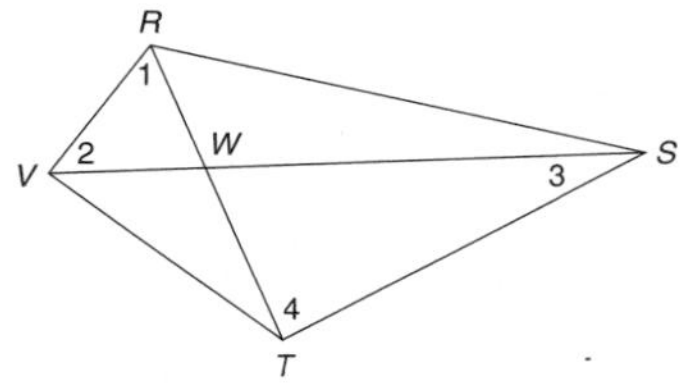

24. *Given:* Quadrilateral $ABCD$ with $\overline{BA} \perp \overline{AD}$ and $\overline{BC} \perp \overline{DC}$
Prove: $\angle$s B and D are supplementary

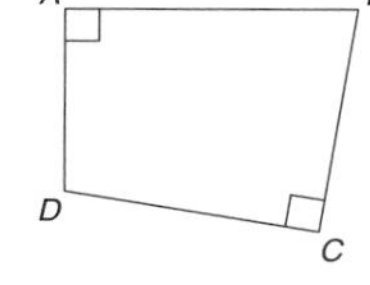

25. A father wishes to make a home plate for his son to use in practicing baseball. Find the size of each of the equal angles if the home plate is modeled on the one in (a) and (b).

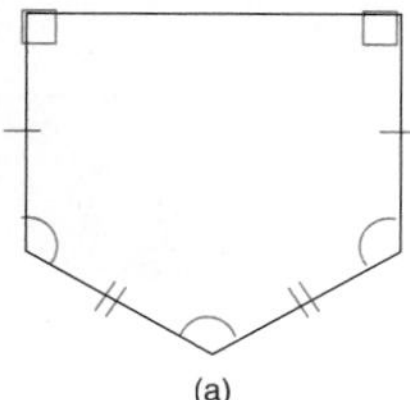

(a)

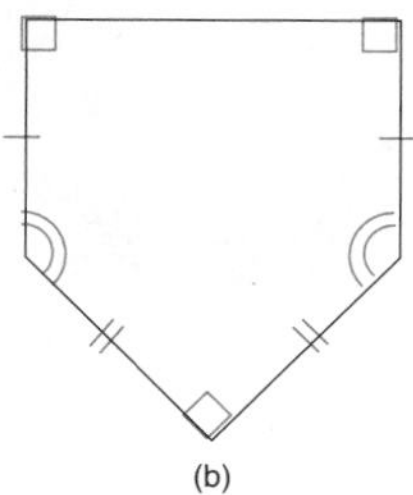

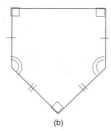

(b)

26. The adjacent interior and exterior angles of a certain polygon are supplementary, as indicated in the drawing. Assume that you know that the measure of each interior angle of a regular polygon is $\frac{(n-2)180}{n}$.

a) Express the measure of each exterior angle as the supplement of the interior angle.

b) Simplify the expression in part (a) to show that each exterior angle has a measure of $\frac{360}{n}$.

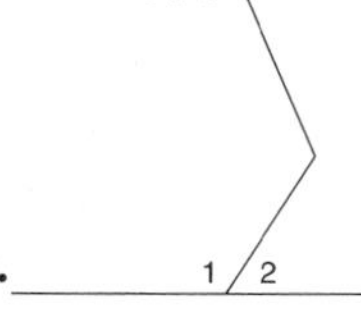

27. Find the measure of each acute interior angle of a regular pentagram.

28. Find the measure of each acute interior angle of a regular octagram.

29. Consider any regular polygon; find and join (in order) the midpoints of the sides. What does intuition tell you about the resulting polygon?

30. Consider a regular hexagon $RSTUVW$. What does intuition tell you about $\triangle RTV$, the result of drawing diagonals $\overline{RT}$, $\overline{TV}$, and $\overline{VR}$?

31. The face of a clock has the shape of a regular polygon with 12 sides. What is the measure of the angle formed by two consecutive sides?

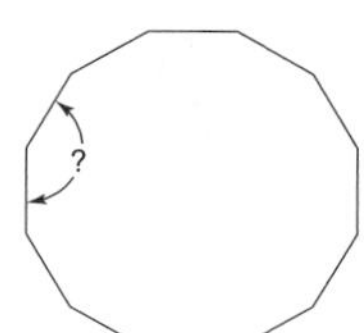

32. The top surface of a picnic table is in the shape of a regular hexagon. What is the measure of the angle formed by two consecutive sides?

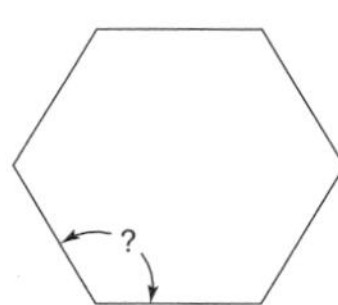

33. ➢ Consider a polygon of n sides determined by the n noncollinear vertices A, B, C, D, and so on.

a) Choose any vertex of the polygon. To how many of the remaining vertices of the polygon can the selected vertex be joined to form a diagonal?

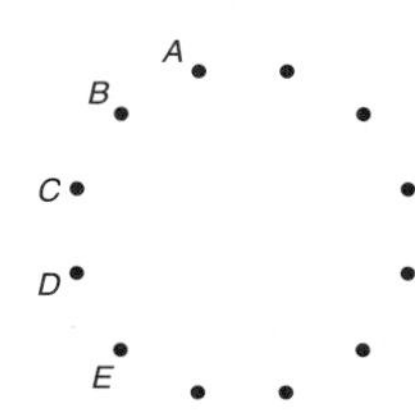

b) Considering that each of the n vertices in (a) can be joined to any one of the remaining $(n - 3)$ vertices to form diagonals, the product $n(n - 3)$ appears to represent the total number of diagonals possible. However, this number includes duplications, such as $\overline{AC}$ and $\overline{CA}$. What expression actually represents D, the total number of diagonals in a polygon of n sides?

34. For the concave quadrilateral $ABCD$, explain why the sum of the interior angles is 360°. (**HINT:** Draw $\overline{BC}$.)

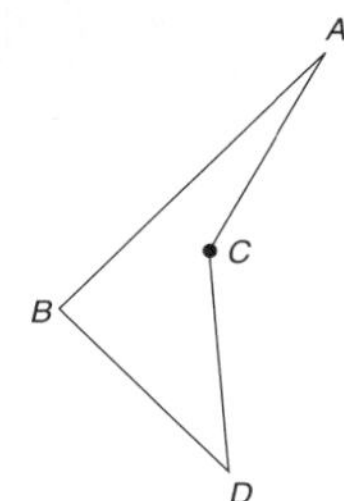

A Look Beyond: Non-Euclidean Geometries

The geometry we present in this book is often described as Euclidean geometry. A non-Euclidean geometry is a geometry characterized by the existence of at least one contradiction of a Euclidean geometry postulate. To appreciate this subject, you need to realize the importance of the word "plane" in the Parallel Postulate. Thus the Parallel Postulate is now restated.

PARALLEL POSTULATE:
In a plane, through a point not on a line, exactly one line is parallel to the given line.

The Parallel Postulate characterizes a course in plane geometry; it corresponds to the theory that "the earth is flat." On a small scale (most applications aren't global), the theory works well and serves the needs of carpenters, designers, and most engineers.

To begin the move to a different geometry, consider the surface of a sphere (like the earth). (See Figure 2.36.) By definition, a **sphere** is the set of all points in space that are at a fixed distance from a given point. If a line segment on the surface of the sphere is extended to form a line, it becomes a great circle (like the equator of the earth). Each line in this geometry, known as "spherical geometry," is the intersection of a plane containing the center of the sphere with the sphere.

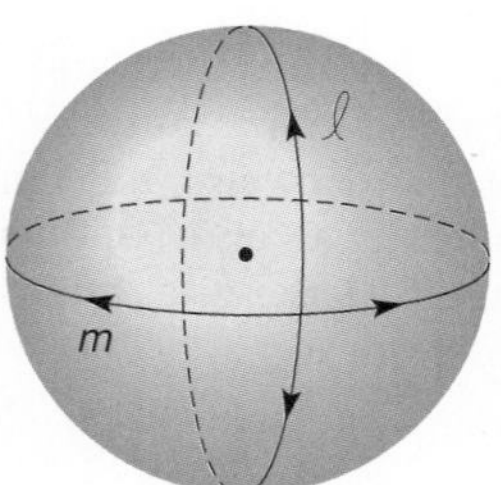

(a) ℓ and *m* are lines in spherical geometry

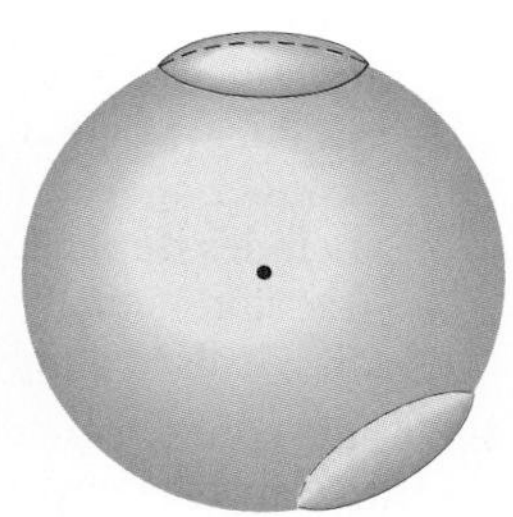

(b) These circles are *not* lines in spherical geometry

FIGURE 2.36

Spherical geometry (or elliptic geometry) is actually a model of Riemannian geometry, named in honor of Georg F. B. Riemann (1826–1866), the German mathematician responsible for the next postulate. The Reimannian Postulate is not numbered in this book, because it does not characterize Euclidean geometry.

RIEMANNIAN POSTULATE:
Through a point not on a line, there are no lines parallel to the given line.

To understand the Reimannian Postulate, consider a sphere (Figure 2.37) containing line ℓ and point *P* not on ℓ. Any line drawn through point *P* must intersect ℓ in two points. To see this develop, follow the frames in Figure 2.38, which depict an attempt to draw a line parallel to ℓ through point *P*.

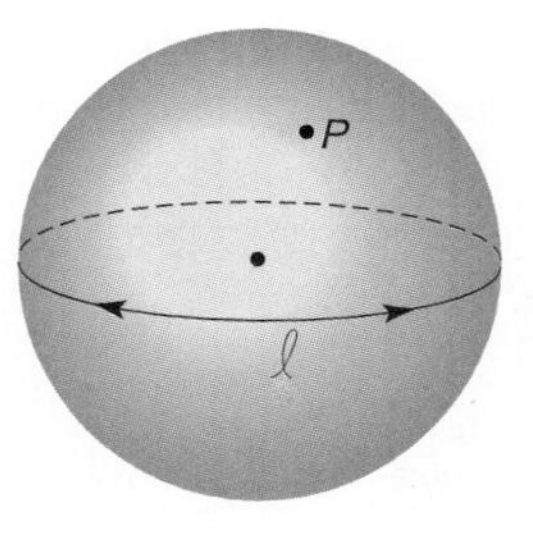

(a)

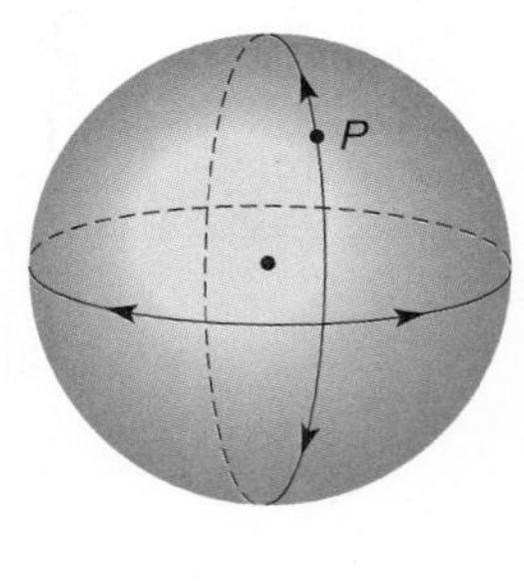

(b)

FIGURE 2.37

Consider the natural extension to Riemannian geometry of the claim that the shortest distance between two points is a straight line. For the sake of efficiency and common sense, a person traveling from New York City to London will follow the path of a line as it is known in spherical geometry. As you might guess, this is used to chart international flights between cities. In Euclidean geometry, the claim suggests that a person tunnel under the earth's surface from one city to the other.

A second type of non-Euclidean geometry is attributed to the works of a German, Karl F. Gauss (1777–1855); a Russian, Nikolai Lobachevski (1793–1856);

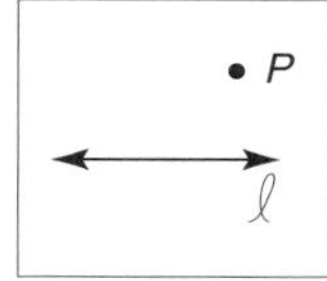

(a) Small part of surface of the sphere

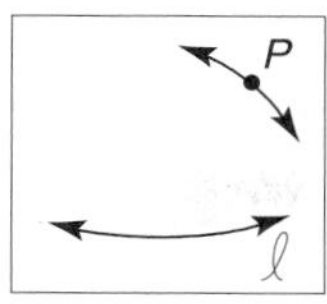

(b) Line through P "parallel" to ℓ on larger part of surface

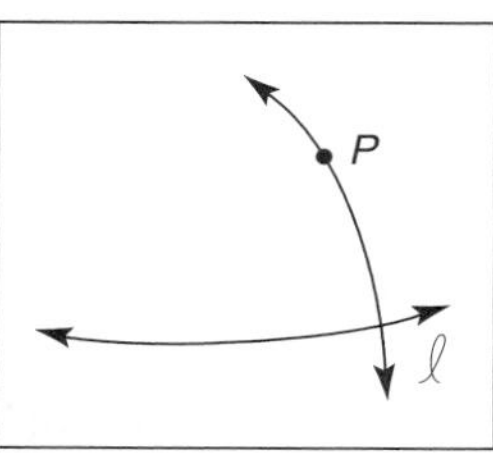

(c) Line through P shown to intersect ℓ on larger portion of surface

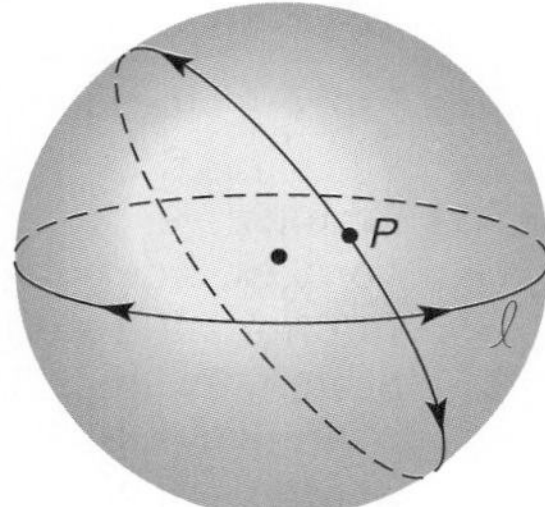

(d) All of line ℓ and the line through P shown on entire sphere

FIGURE 2.38

and a Hungarian, Johann Bolyai (1802–1862). The postulate for this system of non-Euclidean geometry is as follows.

LOBACHEVSKIAN POSTULATE:
Through a point not on line, there are infinitely many lines parallel to the given line.

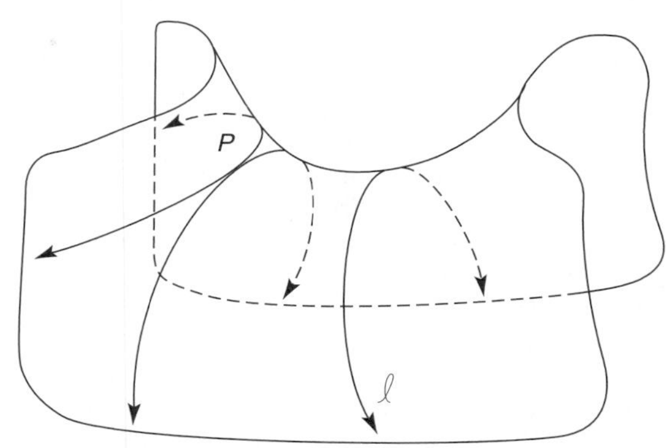

FIGURE 2.39

This form of non-Euclidean geometry is termed "hyperbolic geometry." Rather than use the plane or sphere as the surface for study, mathematicians use a saddle-like surface known as a **hyperbolic paraboloid.** (See Figure 2.39.) A line ℓ is the intersection of a plane with this surface. Clearly, more than one plane can intersect this surface to form a line containing P that does not intersect ℓ. In fact, an infinite number of planes intersect the surface in an infinite number of lines parallel to ℓ and containing P. Table 2.5 compares the three types of geometry.

TABLE 2.5
Comparison of Types of Geometry

Postulate	Model	Line	Number of Lines through P Parallel to ℓ
Parallel (Euclidean)	Plane geometry	Intersection of two planes	One
Riemannian	Spherical geometry	Intersection of plane with sphere (plane contains center of sphere)	None
Lobachevskian	Hyperbolic geometry	Intersection of plane with hyperbolic paraboloid	Infinitely many

Summary

A Look Back at Chapter 2

The goal of this chapter has been to prove several theorems based on the postulate, "If two parallel lines are cut by a transversal, then the corresponding angles are congruent." The method of indirect proof was introduced as a basis for proving lines parallel if the corresponding angles are congruent. Several methods of proving lines parallel were then demonstrated by the direct method. The Parallel Postulate was used to prove that the sum of the measures of the interior angles of a triangle is 180°. Several corollaries followed naturally from this theorem. A sum formula was then developed for the interior angles of any polygon.

A Look Ahead to Chapter 3

In the next chapter, the concept of congruence will be extended to triangles, and several methods of proving triangles congruent will be developed. Several theorems dealing with the inequalities of a triangle will also be proved.

Important Terms and Concepts of Chapter 2

2.1 Perpendicular Lines, Perpendicular Planes
Parallel Lines, Parallel Planes
Parallel Postulate
Transversal
Interior Angles, Exterior Angles
Corresponding Angles, Alternate Interior Angles, Alternate Exterior Angles

2.2 Conditional, Converse, Inverse, Contrapositive
Law of Negative Inference
Indirect Proof

2.3 Proving Lines Parallel

2.4 Triangle
Vertices, Sides of a Triangle
Interior and Exterior of a Triangle
Scalene Triangle, Isosceles Triangle, Equilateral Triangle
Acute Triangle, Obtuse Triangle, Right Triangle, Equiangular Triangle
Auxiliary Line
Corollary
Exterior Angle of a Triangle

2.5 Convex Polygons (Triangle, Quadrilateral, Pentagon, Hexagon, Heptagon, Octagon, Nonagon, Decagon)
Concave Polygon
Diagonals of a Polygon
Regular Polygon, Equilateral Polygon, Equiangular Polygon
Polygram

A Look Beyond: Non-Euclidean Geometries

Review Exercises

1. If $m\angle 1 = m\angle 2$, which lines are parallel?

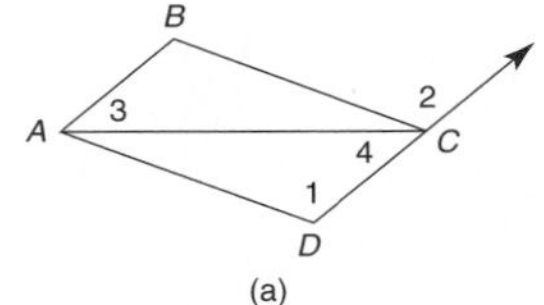

(a)

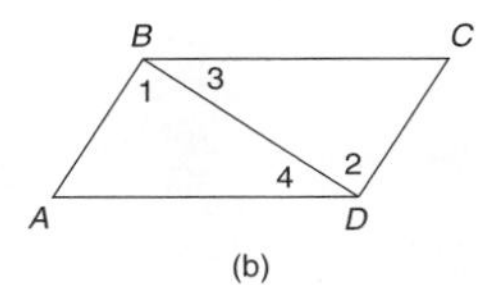

(b)

2. *Given:* m∠13 = 70°
Find: m∠3

3. *Given:* m∠9 = 2*x* + 17
m∠11 = 5*x* − 94
Find: *x*

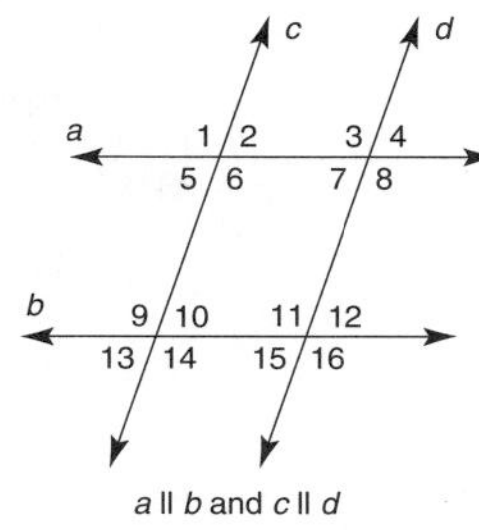

Exercises 2–3

4. *Given:* m∠*B* = 75°, m∠*DCE* = 50°
Find: m∠*D* and m∠*DEF*

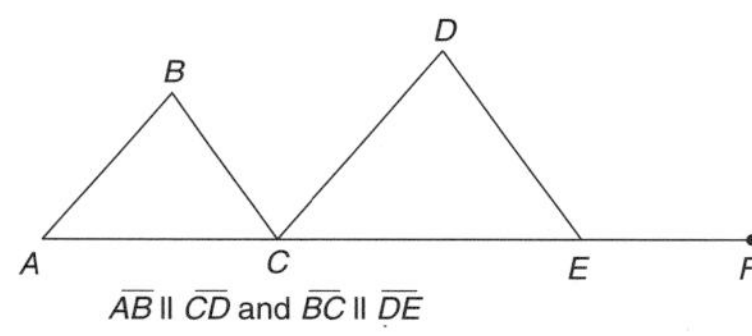

Exercises 4–5

5. *Given:* m∠*DCA* = 130°
m∠*BAC* = 2*x* + *y*
m∠*BCE* = 150°
m∠*DEC* = 2*x* − *y*
Find: *x* and *y*

6. *Given:* In the drawing for Review Exercises 6 to 11,
$\overline{AC} \parallel \overline{DF}$
$\overline{AE} \parallel \overline{BF}$
m∠*AEF* = 3*y*
m∠*BFE* = *x* + 45
m∠*FBC* = 2*x* + 15
Find: *x* and *y*

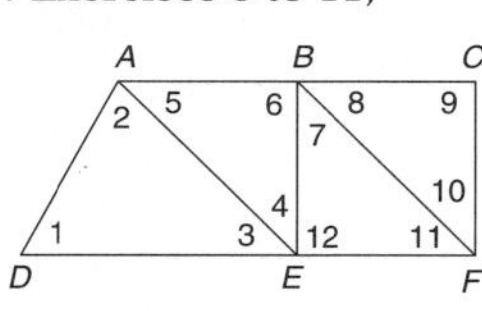

Exercises 6–11

Use the given information to name the segments that must be parallel. If there are no such segments, write "none." Assume A-B-C and D-E-F. (Use the drawing from Exercise 6.)

7. ∠3 ≅ ∠11
8. ∠4 ≅ ∠5
9. ∠7 ≅ ∠10
10. ∠6 ≅ ∠9
11. ∠8 ≅ ∠5 ≅ ∠3

For Review Exercises 12 to 15, find the values of x and y.

12.

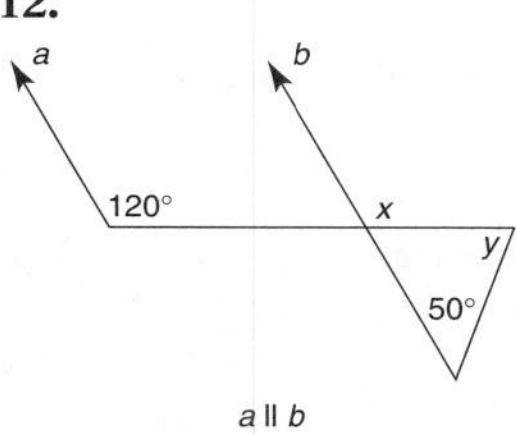

13.

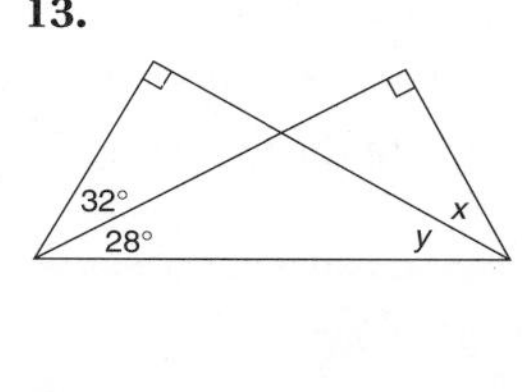

14.

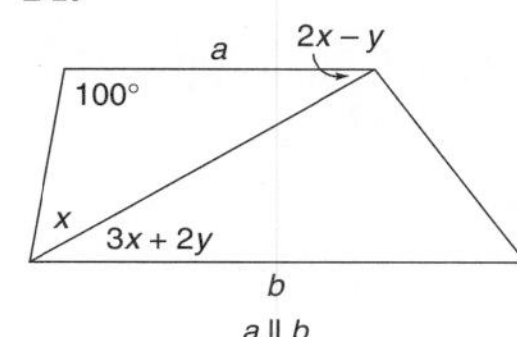

15.

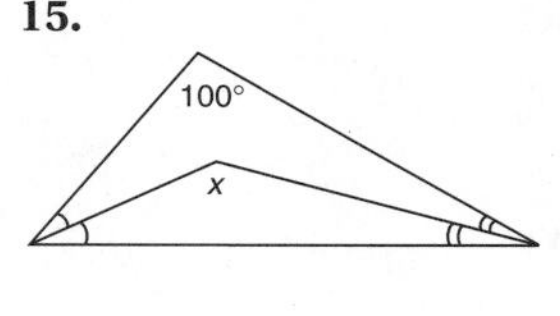

16. *Given:* m∠1 = $x^2 - 12$
m∠4 = *x*(*x* − 2)
Find: *x* so that $\overline{AB} \parallel \overrightarrow{CD}$

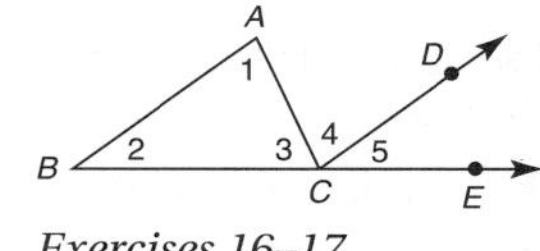

Exercises 16–17

17. *Given:* $\overline{AB} \parallel \overrightarrow{CD}$
m∠2 = $x^2 - 3x + 4$
m∠1 = $17x - x^2 - 5$
m∠*ACE* = 111
Find: m∠3, m∠4, m∠5

18. *Given:* $\overline{DC} \parallel \overline{AB}$
∠*A* ≅ ∠*C*
m∠*A* = 3*x* + *y*
m∠*D* = 5*x* + 10
m∠*C* = 5*y* + 20
Find: m∠*B*

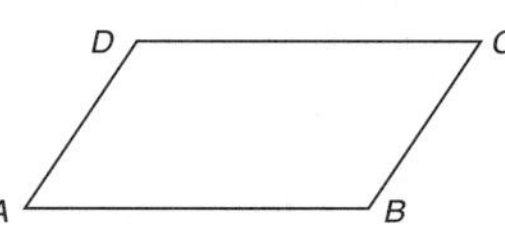

For Review Exercises 19 to 24, decide whether the statements are always true (A), sometimes true (S), or never true (N).

19. An isosceles triangle is a right triangle.
20. An equilateral triangle is a right triangle.
21. A scalene triangle is an isosceles triangle.

22. An obtuse triangle is an isosceles triangle.

23. A right triangle has two congruent angles.

24. A right triangle has two complementary angles.

25. Complete the following table for regular polygons.

Number of sides	8	12	20				
Measure of each exterior ∠				24	36		
Measure of each interior ∠						157.5	178
Number of diagonals							

For Review Exercises 26 to 29, sketch, if possible, the polygon described.

26. A quadrilateral that is equiangular but not equilateral

27. A quadrilateral that is equilateral but not equiangular

28. A triangle that is equilateral but not equiangular

29. A hexagon that is equilateral but not equiangular

For Review Exercises 30 and 31, write the converse, inverse, and contrapositive of each statement.

30. If two angles are right angles, then the angles are congruent.

31. If it is not raining, then I am happy.

32. Which statement—the converse, the inverse, or the contrapositive—always has the same truth or falsity as a given implication?

33. *Given:* $\overline{AB} \parallel \overline{CF}$
$\angle 2 \cong \angle 3$
Prove: $\angle 1 \cong \angle 3$

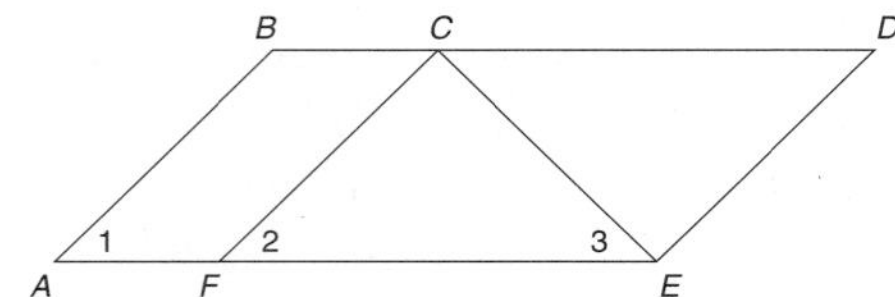

34. *Given:* $\angle 1$ is complementary to $\angle 2$
$\angle 2$ is complementary to $\angle 3$
Prove: $\overline{BD} \parallel \overline{AE}$

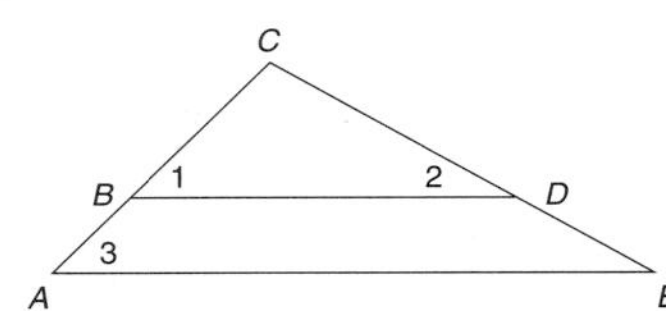

35. *Given:* $\overline{BE} \perp \overline{DA}$
$\overline{CD} \perp \overline{DA}$
Prove: $\angle 1 \cong \angle 2$

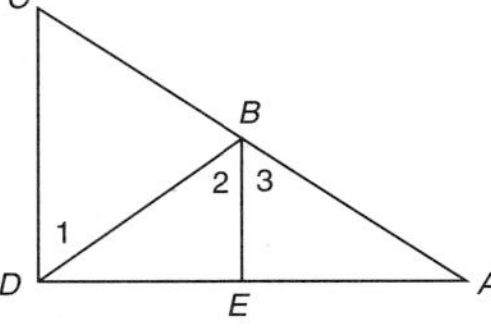

36. *Given:* $\angle A \cong \angle C$
$\overline{DC} \parallel \overrightarrow{AB}$
Prove: $\overline{DA} \parallel \overline{CB}$

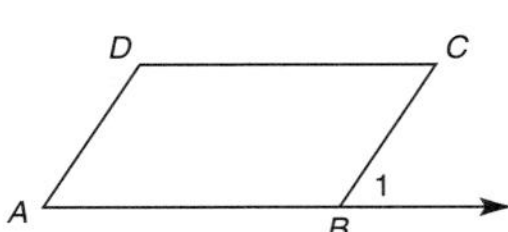

For Exercises 37 to 40, give the first statement for an indirect proof.

37. If $x^2 + 7x + 12 \neq 0$, then $x \neq -3$.

38. If two angles of a triangle are not congruent, then the sides opposite those angles are not congruent.

39. *Given:* $m \nparallel n$
Prove: $\angle 1 \ncong \angle 2$

40. *Given:* $\angle 1 \ncong \angle 3$
Prove: $m \nparallel n$

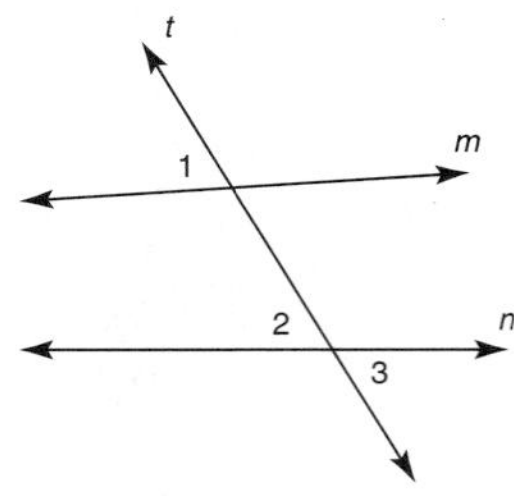

Exercises 39–40

41. Construct the line through C parallel to $\overline{AB}$.

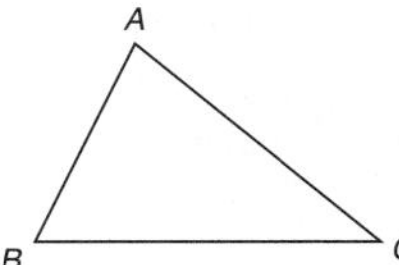

42. Construct an equilateral triangle ABC with side $\overline{AB}$.

A B

Chapter 3
Triangles

In much of this chapter, we will be using the notion of congruent triangles. A pair of triangles are congruent if one fits perfectly over the other. In roof trusses such as the ones shown, we invariably see congruent triangles.

In Chapter 4, we will study quadrilaterals (figures with four sides). Many of the properties of quadrilaterals depend upon the properties of triangles developed in Chapter 3.

CHAPTER OUTLINE

3.1 Congruent Triangles

Two triangles are **congruent** if one coincides (fits perfectly) with the other. In Figure 3.1, we say that $\triangle ABC \cong \triangle DEF$ if these congruences hold:

$$\angle A \cong \angle D \qquad \overline{AB} \cong \overline{DE}$$
$$\angle B \cong \angle E \qquad \overline{BC} \cong \overline{EF}$$
$$\angle C \cong \angle F \qquad \overline{AC} \cong \overline{DF}$$

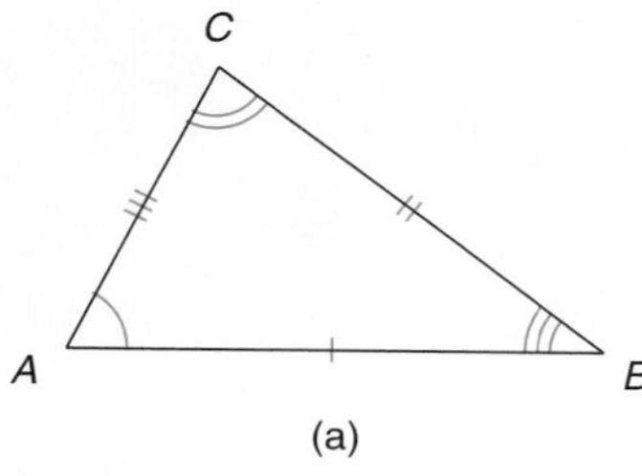

(a)

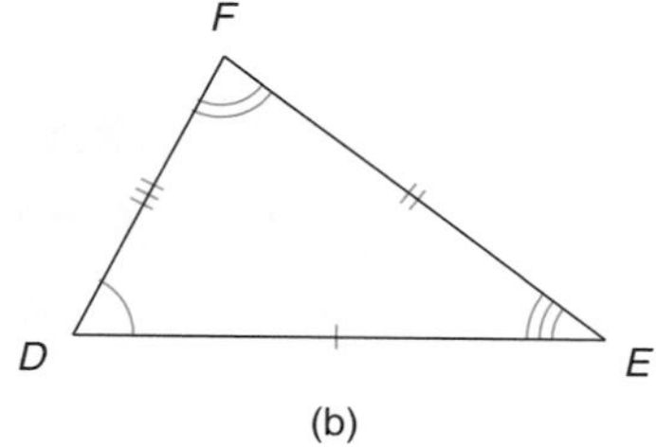

(b)

FIGURE 3.1

From the indicated congruences, we also say that vertex A corresponds to vertex D, as does B to E and C to F. In symbols, this correspondence is represented by

$$A \leftrightarrow D \qquad B \leftrightarrow E \qquad C \leftrightarrow F$$

The claim $\triangle MNQ \cong \triangle RST$ orders corresponding vertices, so we can conclude from this statement that

$$M \leftrightarrow R, \qquad N \leftrightarrow S, \qquad \text{and} \qquad Q \leftrightarrow T$$

This correspondence of vertices implies the congruence of corresponding parts such as $\angle M \cong \angle R$ and $\overline{NQ} \cong \overline{ST}$.

Conversely, if the correspondence of vertices of two congruent triangles is $M \leftrightarrow R$, $N \leftrightarrow S$, and $Q \leftrightarrow T$, we can write $\triangle MNQ \cong \triangle RST$, $\triangle NQM \cong \triangle STR$, and so on.

DISCOVER!

Holding two sheets of construction paper together, use scissors to cut out two triangles. How do the triangles compare?

ANSWER

The triangles are congruent.

DEFINITION: Two triangles are **congruent** when the six parts of the first triangle are congruent to the six corresponding parts of the second triangle.

As always, any definition is reversible! If two triangles are known to be congruent, we may conclude that the corresponding parts are congruent. Moreover, if the six pairs of parts are known to be congruent, then so are the triangles! From the congruent parts indicated in Figure 3.2, we may conclude that $\triangle MNQ \cong \triangle RST$.

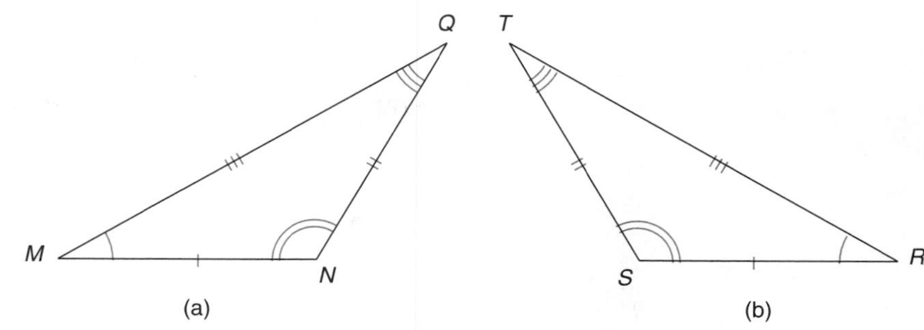

FIGURE 3.2

Some of the properties of congruent triangles that are useful in later proofs and explanations are:

> **1.** $\triangle ABC \cong \triangle ABC$ (Reflexive Property of Congruence)
>
> **2.** If $\triangle ABC \cong \triangle DEF$, then $\triangle DEF \cong \triangle ABC$. (Symmetric Property of Congruence)
>
> **3.** If $\triangle ABC \cong \triangle DEF$ and $\triangle DEF \cong \triangle GHI$, then $\triangle ABC \cong \triangle GHI$. (Transitive Property of Congruence)

It would be difficult to establish that triangles were congruent if six pairs of congruent parts had to be verified first. Fortunately, it is possible to prove triangles congruent with fewer than six pairs of congruences. To suggest a first method, consider the construction in Example 1.

EXAMPLE 1

Construct a triangle whose sides have the lengths of the segments provided in Figure 3.3(a).

Solution Choose $\overline{AB}$ as the first side of the triangle and mark its length as shown in Figure 3.3(b).

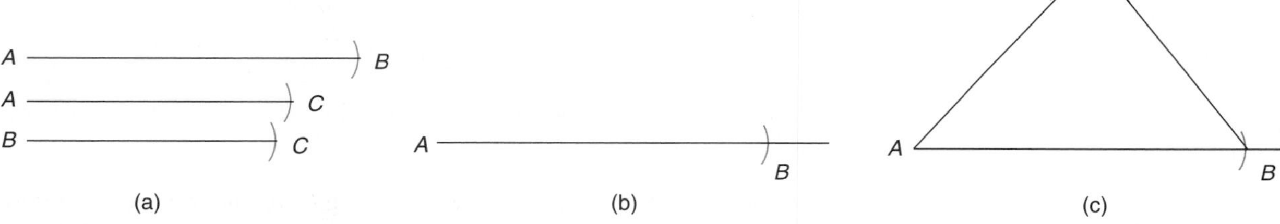

FIGURE 3.3

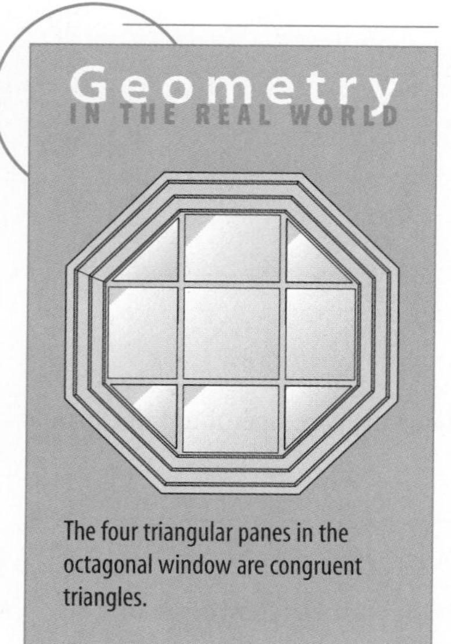

The four triangular panes in the octagonal window are congruent triangles.

Using the left endpoint A, mark off an arc of length equal to that of $\overline{AC}$. Now mark off an arc the length of $\overline{BC}$ from the right endpoint B so that these arcs intersect at C, the third vertex of the triangle. Joining point C to A and then to B completes the desired triangle. [See Figure 3.3(c).]

Consider Example 1 once more. If a "different" triangle had been constructed by using $\overline{AC}$ as the first side, it would be congruent to the one shown. It might be necessary to flip or rotate it to have corresponding vertices match, but that is perfectly acceptable! The point of Example 1 is that it does provide a method for establishing the congruence of triangles, using only three pairs of parts. If corresponding angles are measured in the previous triangle or in any other triangle constructed with the same lengths for sides, these pairs of angles will also be congruent!

SSS (Method for Proving Triangles Congruent)

POSTULATE 12:
If the three sides of one triangle are congruent to the three sides of a second triangle, then the triangles are congruent (SSS).

The designation SSS will be cited as a reason in the proof that follows. The three S letters refer to the three pairs of congruent sides.

EXAMPLE 2

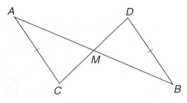

Given: $\overline{AB}$ and $\overline{CD}$ bisect each other at M
$\overline{AC} \cong \overline{DB}$
(See Figure 3.4.)

Prove: $\triangle AMC \cong \triangle BMD$

FIGURE 3.4

PROOF

Statements	Reasons
1. $\overline{AB}$ and $\overline{CD}$ bisect each other at M	**1.** Given
2. $\overline{AM} \cong \overline{MB}$ $\overline{CM} \cong \overline{MD}$	**2.** If a segment is bisected, the segments formed are $\cong$
3. $\overline{AC} \cong \overline{DB}$	**3.** Given
4. $\triangle AMC \cong \triangle BMD$	**4.** SSS

NOTE: In steps 2 and 3, the three pairs of sides were shown to be congruent; thus, SSS is cited to justify that $\triangle AMC \cong \triangle BMD$.

The two sides that form an angle of a triangle are said to **include that angle** of the triangle. In $\triangle TUV$ in Figure 3.5(a), sides $\overline{TU}$ and $\overline{TV}$ form $\angle T$; therefore, $\overline{TU}$ and $\overline{TV}$ include $\angle T$. In turn, $\angle T$ is said to be the included angle for $\overline{TU}$ and

$\overline{TV}$. Similarly, any two angles of a triangle must have a common side, and these two angles are said to **include that side.** In $\triangle TUV$, $\angle U$ and $\angle T$ share the common side $\overline{UT}$; therefore, $\angle U$ and $\angle T$ include the side $\overline{UT}$. $\overline{UT}$ is the side included by $\angle U$ and $\angle T$.

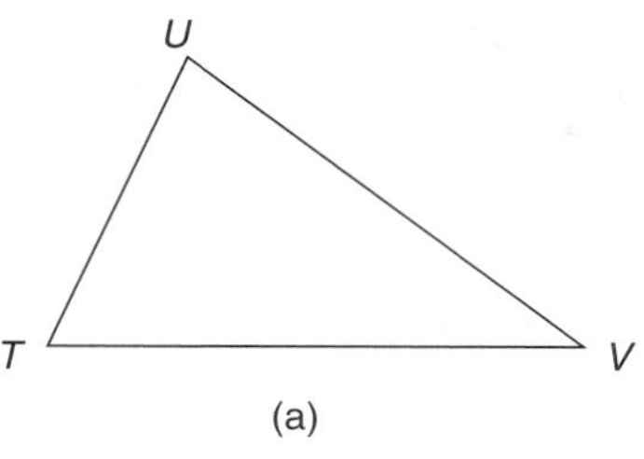

(a)

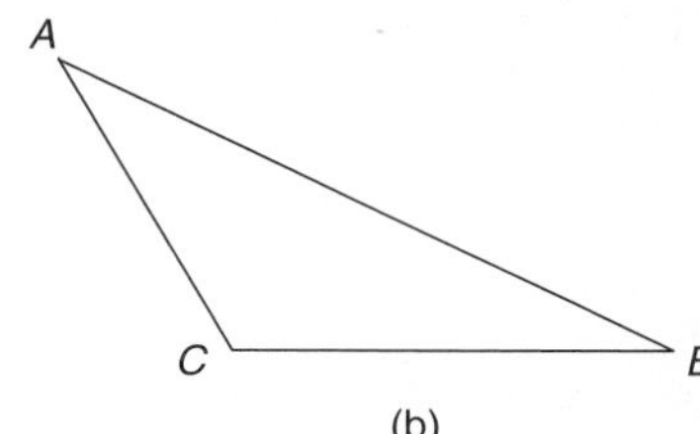

(b)

FIGURE 3.5

EXAMPLE 3

In $\triangle ABC$ of Figure 3.5(b):

a) Which angle is included by $\overline{AC}$ and $\overline{CB}$?
b) Which sides include $\angle B$?
c) What is the included side for $\angle A$ and $\angle B$?
d) Which angles include $\overline{CB}$?

Solution

a) $\angle C$ (since it is formed by $\overline{AC}$ and $\overline{CB}$)
b) $\overline{AB}$ and $\overline{BC}$ (since these form $\angle B$)
c) $\overline{AB}$ (since it is the common side for $\angle A$ and $\angle B$)
d) $\angle C$ and $\angle B$ (since $\overline{CB}$ is a side of each angle)

SAS (Method for Proving Triangles Congruent)

A second way of establishing that two triangles are congruent involves showing that two sides and the included angle of one triangle are congruent to two sides and the included angle of a second triangle. If two people each draw a triangle so that two of the sides measure 2 cm and 3 cm and the included angle measures 54°, then those triangles are congruent. (See Figure 3.6.)

2
54°
3
(a)

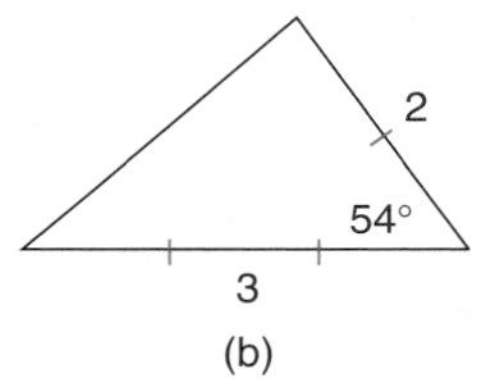

(b)

FIGURE 3.6

> **POSTULATE 13:**
> If two sides and the included angle of one triangle are congruent to two sides and the included angle of a second triangle, then the triangles are congruent (SAS).

The order of the letters SAS in Postulate 13 helps us remember that the two sides that are named have the angle "between" them. That is, in each triangle, the two sides form the angle.

In Example 4, which follows, the two triangles to be proven congruent share a common side; the statement $\overline{PN} \cong \overline{PN}$ is justified by the Reflexive Property of Congruence, sometimes referred to as Identity. However stated, this justification applies when triangles (or perhaps other polygons) have a part in common. In Example 4, notice the use of SAS as the final reason.

EXAMPLE 4

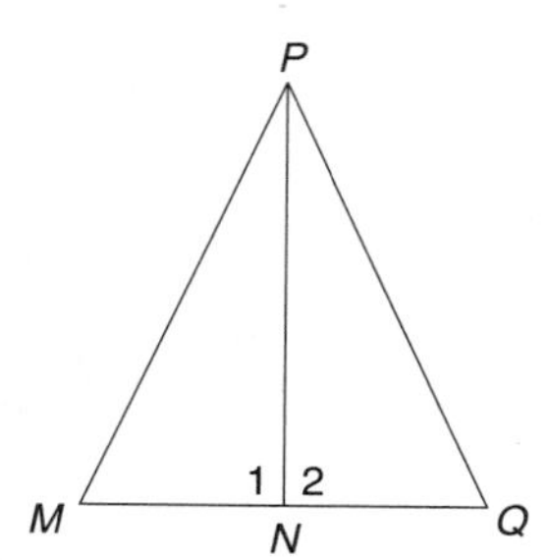

FIGURE 3.7

Given: $\overline{PN} \perp \overline{MQ}$
$\overline{MN} \cong \overline{NQ}$
(See Figure 3.7.)

Prove: $\triangle PNM \cong \triangle PNQ$

PROOF

Statements	Reasons
1. $\overline{PN} \perp \overline{MQ}$	**1.** Given
2. $\angle 1 \cong \angle 2$	**2.** If two lines are $\perp$, they meet to form $\cong$ adjacent $\angle$s
3. $\overline{MN} \cong \overline{NQ}$	**3.** Given
4. $\overline{PN} \cong \overline{PN}$	**4.** Identity (or Reflexive)
5. $\triangle PNM \cong \triangle PNQ$	**5.** SAS

NOTE: In $\triangle PNM$, $\overline{MN}$ (step 3) and $\overline{PN}$ (step 4) include $\angle 1$; thus, SAS is used to verify that $\triangle PNM \cong \triangle PNQ$.

ASA (Method for Proving Triangles Congruent)

The next method for proving triangles congruent requires a combination of two angles and the included side. If two people each draw a triangle for which two of the angles measure 33° and 47° and the included side measures 5 centimeters, then those triangles are congruent. See Figure 3.8.

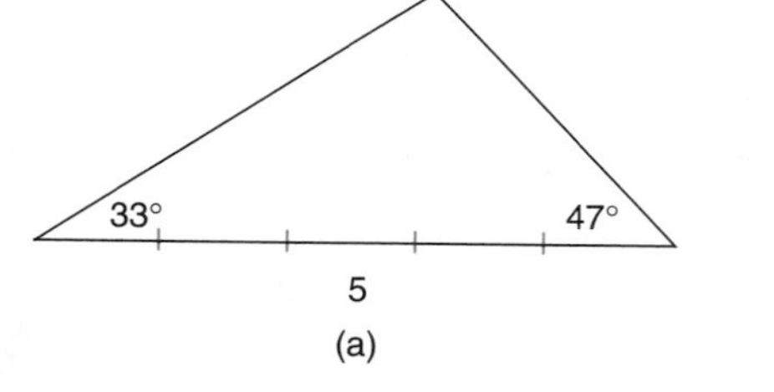

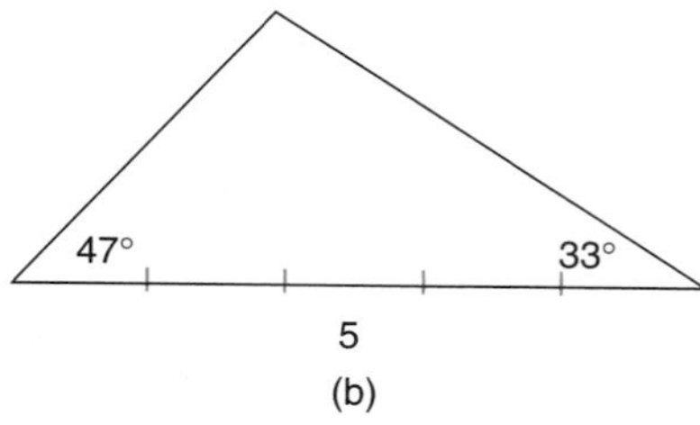

FIGURE 3.8

POSTULATE 14:
If two angles and the included side of one triangle are congruent to two angles and the included side of a second triangle, then the triangles are congruent (ASA).

While this method is written compactly as ASA, you must be careful as you write these abbreviations! For example, ASA refers to two angles and the in-

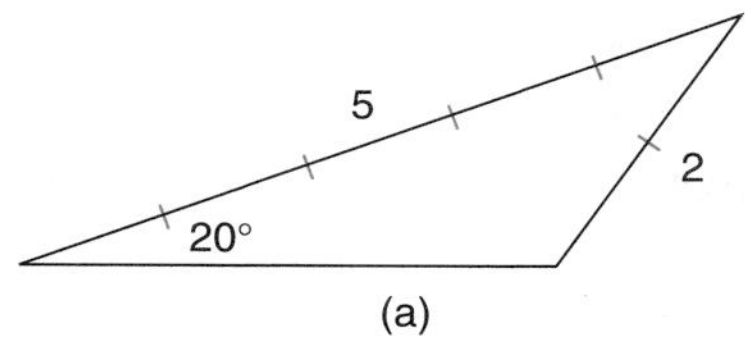

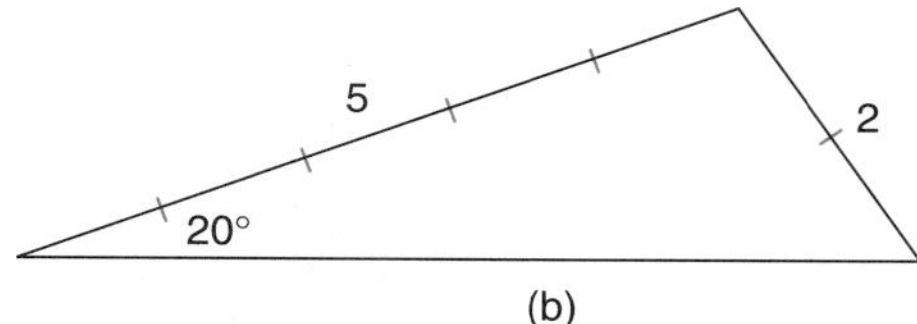

FIGURE 3.9

cluded side while SAS refers to two sides and the included angle. To use either postulate, the specific conditions described in it must be satisfied.

While SSS, SAS, and ASA are all valid methods of proving triangles congruent, SSA is *not* a method and *cannot* be used. In Figure 3.9, the two triangles are marked to show SSA, and yet the two triangles are *not* congruent.

Another combination that cannot be used to prove triangles congruent is AAA. See Figure 3.10. Three congruent pairs of angles in two triangles do not guarantee congruent pairs of sides!

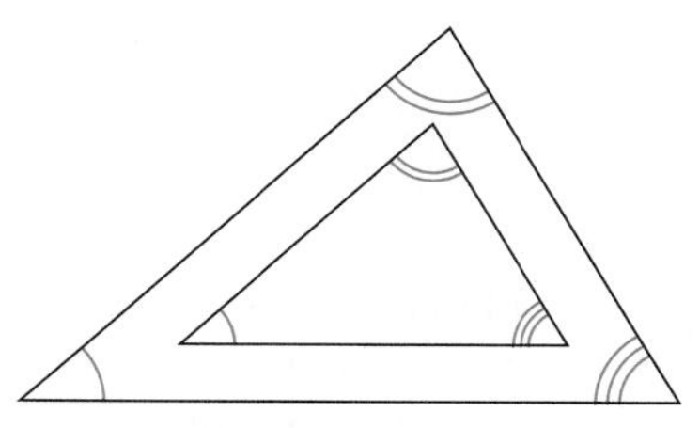

FIGURE 3.10

In Example 5, the triangles to be proven congruent overlap (see Figure 3.11). For that reason, the triangles have been redrawn separately in Figure 3.12. Notice the parts marked congruent as established in the proof. Note that Identity (or Reflexive) can also be used to say that an angle is congruent to itself.

EXAMPLE 5

Given: $\overline{AC} \cong \overline{DC}$
$\angle 1 \cong \angle 2$
(See Figure 3.11.)

Prove: $\triangle ACE \cong \triangle DCB$

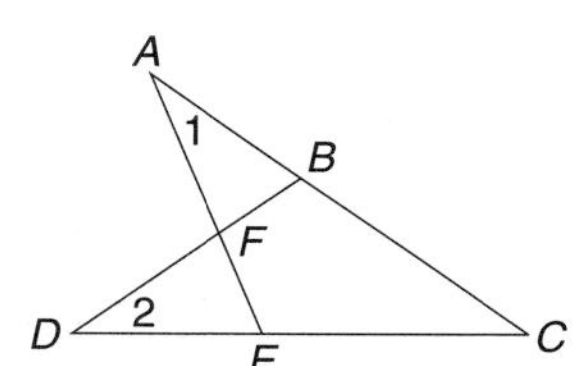

FIGURE 3.11

PROOF

Statements	Reasons
1. $\overline{AC} \cong \overline{DC}$ (See Figure 3.12.)	**1.** Given
2. $\angle 1 \cong \angle 2$	**2.** Given
3. $\angle C \cong \angle C$	**3.** Identity
4. $\triangle ACE \cong \triangle DCB$	**4.** ASA

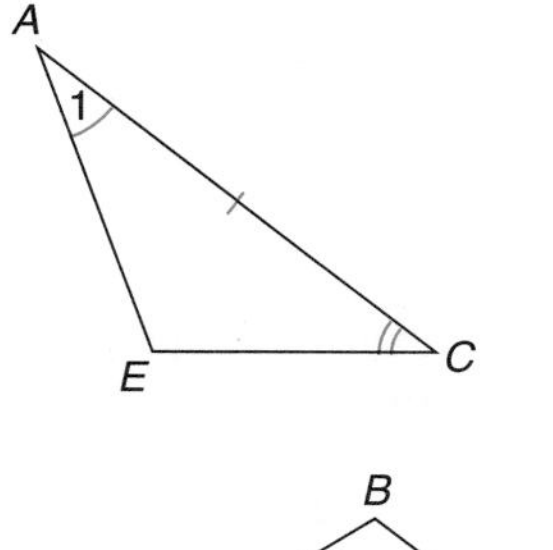

FIGURE 3.12

Next we consider a theorem (proven by the ASA postulate) that is convenient as a reason in many proofs.

AAS (Method for Proving Triangles Congruent)

THEOREM 3.1.1: If two angles and the nonincluded side of one triangle are congruent to two angles and the nonincluded side of a second triangle, then the triangles are congruent (AAS).

Given: $\angle T \cong \angle K$, $\angle S \cong \angle J$, and $\overline{SR} \cong \overline{HJ}$ (See Figure 3.13 on the next page.)

Prove: $\triangle TSR \cong \triangle KJH$

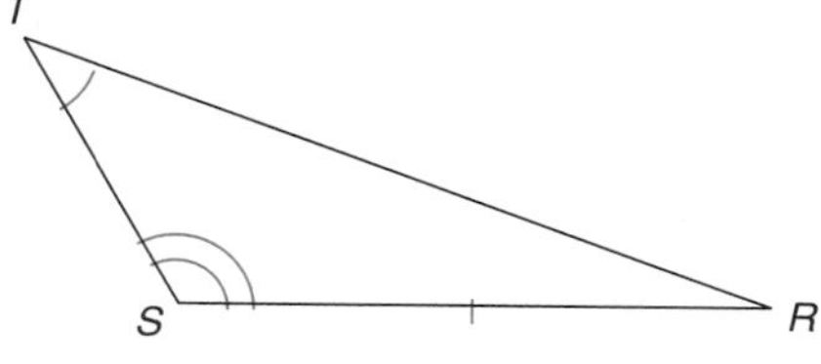

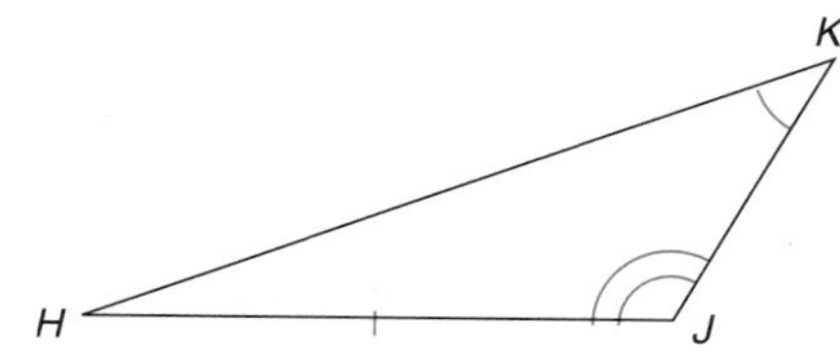

FIGURE 3.13

■ *Warning*

Do not use AAA or SSA, since they are simply not valid for proving triangles congruent; with AAA the triangles have the same shape but are not necessarily congruent. ■

PROOF

Statements	Reasons
1. $\angle T \cong \angle K$ $\angle S \cong \angle J$	**1.** Given
2. $\angle R \cong \angle H$	**2.** If two $\angle$s of one $\triangle$ are $\cong$ to two $\angle$s of another $\triangle$, then the third $\angle$s are also congruent
3. $\overline{SR} \cong \overline{HJ}$	**3.** Given
4. $\triangle TSR \cong \triangle KJH$	**4.** ASA

In summary, you may use SSS, SAS, ASA, or AAS to prove that triangles are congruent. Identity or Reflexive may be used to state a self-congruence when a side or angle is common to two triangles.

3.1 Exercises

In Exercises 1 to 6, use drawings as needed to answer each question.

1. Name a common angle and a common side for $\triangle ABC$ and $\triangle ABD$. If $\overline{BC} \cong \overline{BD}$, can you conclude that $\triangle ABC$ and $\triangle ABD$ are congruent? Can SSA be used as a reason for proving triangles congruent?

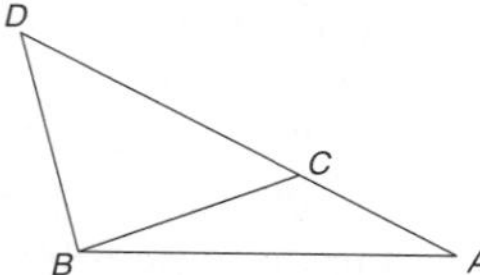

2. With corresponding angles indicated, the triangles are congruent. Find values for a, b, and x.

3. In a right triangle, the sides that form the right angle are the **legs;** the longest side (opposite the right angle) is the **hypotenuse.** When two right triangles have congruent pairs of legs, some textbooks say that the right triangles are congruent by the reason LL. In our work, LL is just a special case of one of the postulates in this section. Which postulate is that?

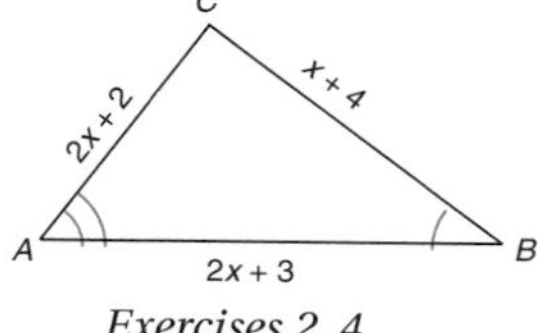

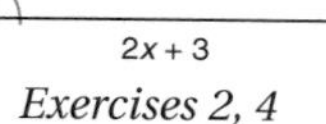
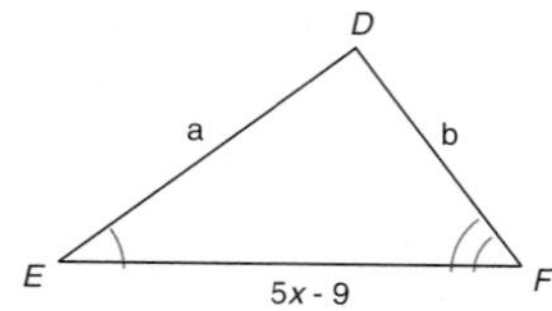

Exercises 2, 4

4. In the figure for Exercise 2, write a statement that the triangles are congruent, with due attention to the order of corresponding vertices.

5. In $\triangle ABC$, the midpoints of the sides are joined. What does intuition tell you about the relationship between $\triangle AED$ and $\triangle FDE$? (We will prove this relationship later.)

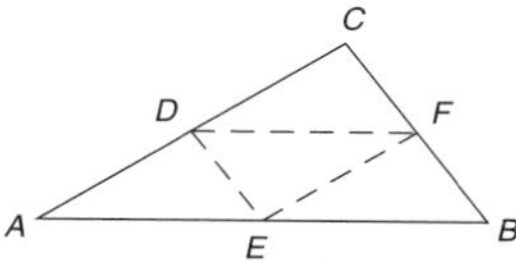

6. Suppose that you wish to prove that $\triangle RST \cong \triangle SRV$. Using the reason Identity, name one pair of corresponding parts that are congruent.

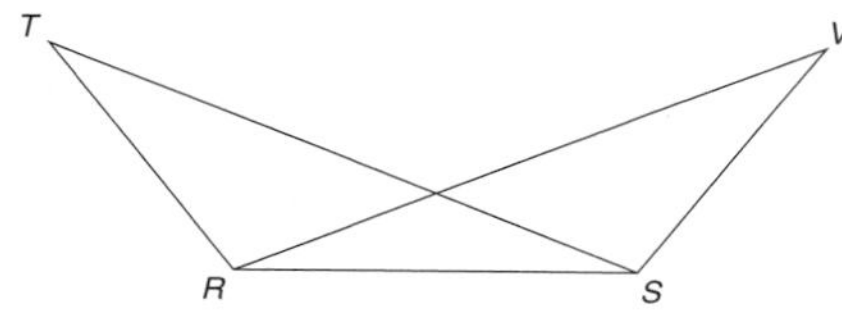

In Exercises 7 to 10, congruent parts are indicated by like dashes (sides) or arcs (angles). State which method (SSS, SAS, ASA, or AAS) would be used to prove the two triangles congruent.

7.

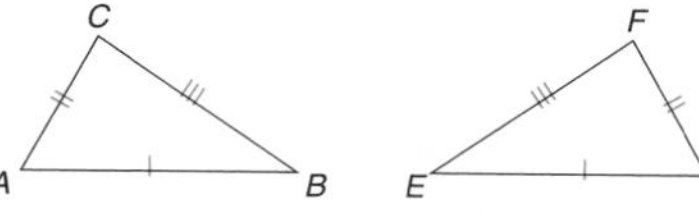

8.

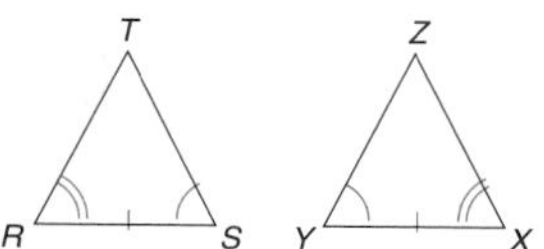

9.

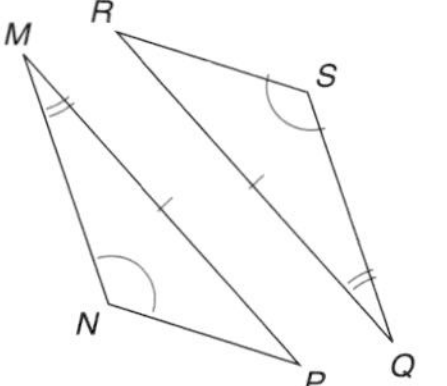

10.

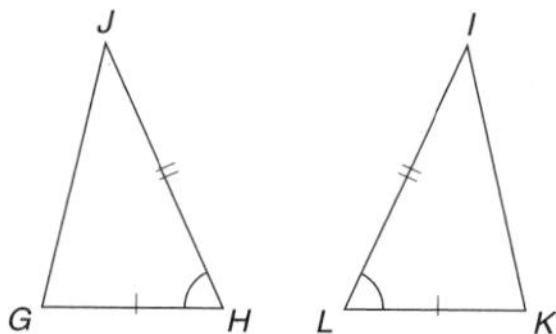

In Exercises 11 to 16, use the given information to state the reason why $\triangle ABC \cong \triangle DBC$. Use marks like those used in Exercises 7–10.

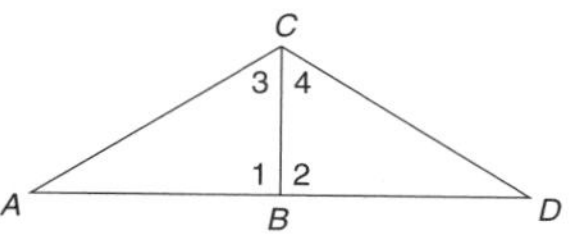

Exercises 11–16

11. $\angle A \cong \angle D$, $\overline{AB} \cong \overline{BD}$, and $\angle 1 \cong \angle 2$

12. $\angle A \cong \angle D$, $\overline{AC} \cong \overline{CD}$, and B is the midpoint of $\overline{AD}$

13. $\angle A \cong \angle D$, $\overline{AC} \cong \overline{CD}$, and $\overrightarrow{CB}$ bisects $\angle ACD$

14. $\angle A \cong \angle D$, $\overline{AC} \cong \overline{CD}$, and $\overline{AB} \cong \overline{BD}$

15. $\overline{AC} \cong \overline{CD}$, $\overline{AB} \cong \overline{BD}$, and $\overline{CB} \cong \overline{CB}$ (by Identity)

16. $\angle 1$ and $\angle 2$ are right $\angle$s, $\overline{AB} \cong \overline{BD}$, and $\angle A \cong \angle D$

In Exercises 17 and 18, the triangles to be proven congruent have been redrawn separately.

a) *Name an additional pair of parts that are congruent by Identity.*
b) *Considering the congruent parts, state the reason why the triangles must be congruent.*

17. $\triangle ABC \cong \triangle AED$

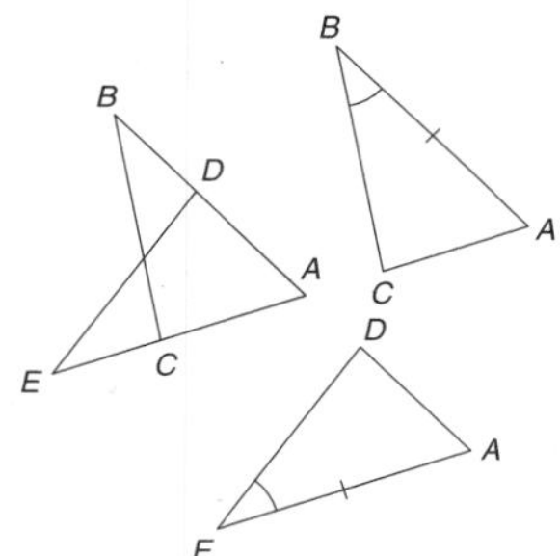

18. $\triangle MNP \cong \triangle MQP$

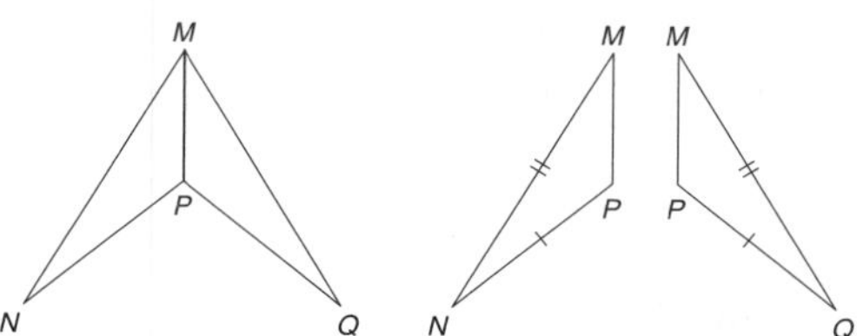

In Exercises 19 to 22, prove that the triangles named are congruent. Considering the congruent pairs marked, name the additional pair of parts that must be congruent to use the method named.

19. SAS

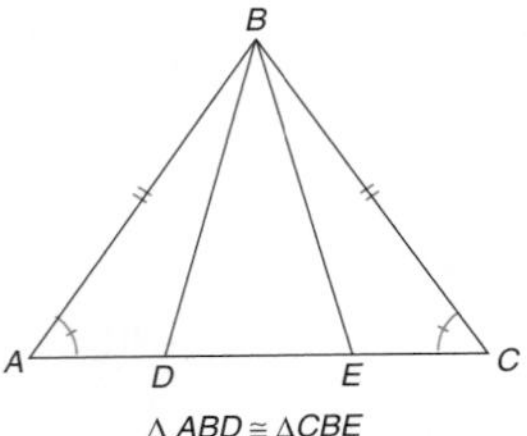

$\triangle ABD \cong \triangle CBE$

20. ASA

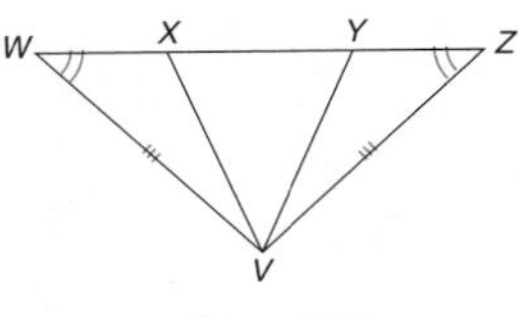

$\triangle WVY \cong \triangle ZVX$

21. SSS

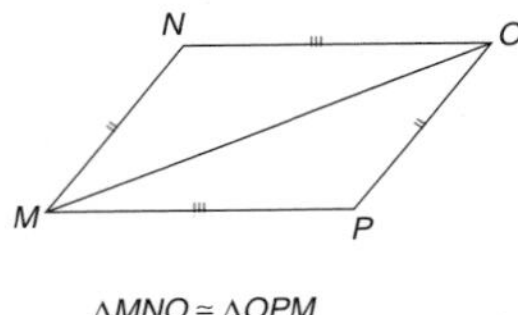

$\triangle MNO \cong \triangle OPM$

22. AAS

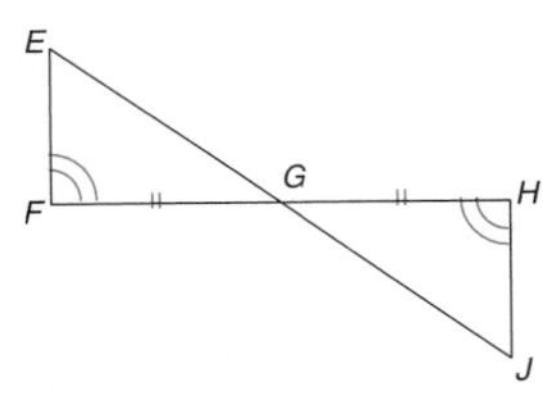

$\triangle EFG \cong \triangle JHG$

In Exercises 23 and 24, complete each proof.

23. *Given:* $\overline{AB} \cong \overline{CD}$ and $\overline{AD} \cong \overline{CB}$
Prove: $\triangle ABC \cong \triangle CDA$

PROOF

Statements	Reasons
1. $\overline{AB} \cong \overline{CD}$ and $\overline{AD} \cong \overline{CB}$	**1.** ?
2. ?	**2.** Identity
3. $\triangle ABC \cong \triangle CDA$	**3.** ?

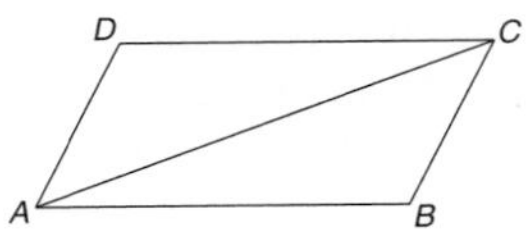

Exercises 23, 24

24. *Given:* $\overline{DC} \parallel \overline{AB}$ and $\overline{AD} \parallel \overline{BC}$
Prove: $\triangle ABC \cong \triangle CDA$

PROOF

Statements	Reasons
1. $\overline{DC} \parallel \overline{AB}$	**1.** ?
2. $\angle DCA \cong \angle BAC$	**2.** ?
3. ?	**3.** Given
4. ?	**4.** If two $\parallel$ lines are cut by a transversal, alt. int. $\angle$s are $\cong$
5. $\overline{AC} \cong \overline{AC}$	**5.** ?
6. ?	**6.** ASA

In Exercises 25 to 30, use SSS, SAS, ASA, or AAS to prove that the triangles are congruent.

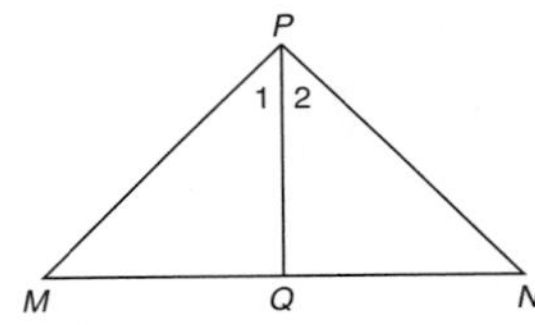

Exercises 25, 26

25. *Given:* $\overrightarrow{PQ}$ bisects $\angle MPN$
$\overline{MP} \cong \overline{NP}$
Prove: $\triangle MQP \cong \triangle NQP$

26. *Given:* $\overline{PQ} \perp \overline{MN}$ and $\angle 1 \cong \angle 2$
Prove: $\triangle MQP \cong \triangle NQP$

27. *Given:* $\overline{AB} \perp \overline{BC}$ and $\overline{AB} \perp \overline{BD}$
$\overline{BC} \cong \overline{BD}$
Prove: $\triangle ABC \cong \triangle ABD$

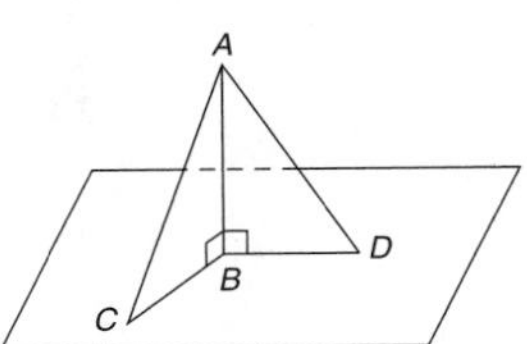

28. *Given:* $\overline{PN}$ bisects $\overline{MQ}$
$\angle M$ and $\angle Q$ are right angles
Prove: $\triangle PQR \cong \triangle NMR$

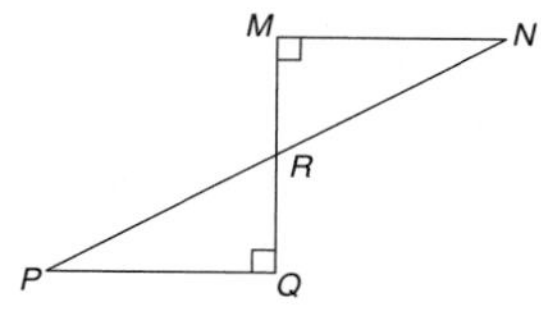

29. *Given:* $\angle VRS \cong \angle TSR$ and $\overline{RV} \cong \overline{TS}$
Prove: $\triangle RST \cong \triangle SRV$

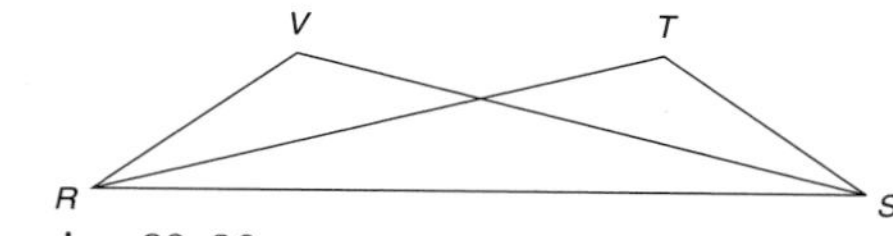

Exercises 29, 30

30. *Given:* $\overline{VS} \cong \overline{TR}$ and $\angle TRS \cong \angle VSR$
Prove: $\triangle RST \cong \triangle SRV$

In Exercises 31 to 34, the methods to be used are SSS, SAS, ASA, and AAS.

31. Given that $\triangle RST \cong \triangle RVU$, does it follow that $\triangle RSU$ is also congruent to $\triangle RVT$? Name the method, if any, used in arriving at this conclusion.

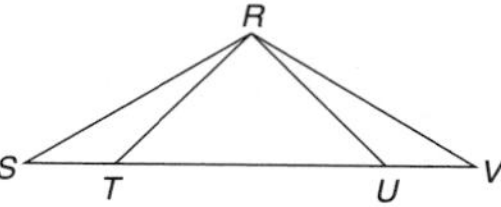

Exercises 31, 32

32. Given that $\angle S \cong \angle V$ and $\overline{ST} \cong \overline{UV}$, does it follow that $\triangle RST \cong \triangle RVU$? Which method, if any, did you use?

33. Given that $\angle A \cong \angle E$ and $\angle B \cong \angle D$, does it follow that $\triangle ABC \cong \triangle DEC$? If so, cite the method used in arriving at this conclusion.

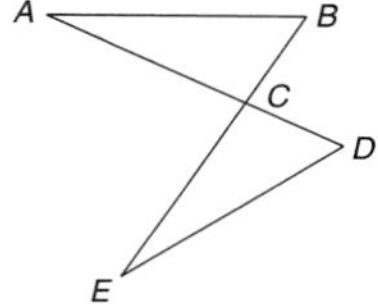

Exercises 33, 34

34. Given that $\angle A \cong \angle E$ and $\overline{BC} \cong \overline{DC}$, does it follow that $\triangle ABC \cong \triangle DEC$? Cite the method, if any, used in reaching this conclusion.

35. In quadrilateral $ABCD$, $\overline{AC}$ and $\overline{BD}$ are perpendicular bisectors of each other. Name *all* triangles that are congruent to:

a) $\triangle ABE$ **b)** $\triangle ABC$ **c)** $\triangle ABD$

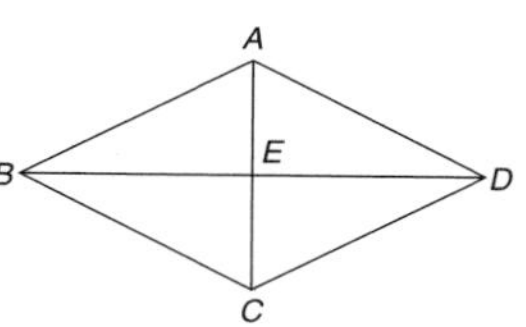

36. In $\triangle ABC$ and $\triangle DEF$, you know that $\angle A \cong \angle D$, $\angle C \cong \angle F$, and $\overline{AB} \cong \overline{DE}$. Before concluding that the triangles are congruent by ASA, you need to show that $\angle B \cong \angle E$. State the postulate or theorem that allows you to confirm this statement ($\angle B \cong \angle E$).

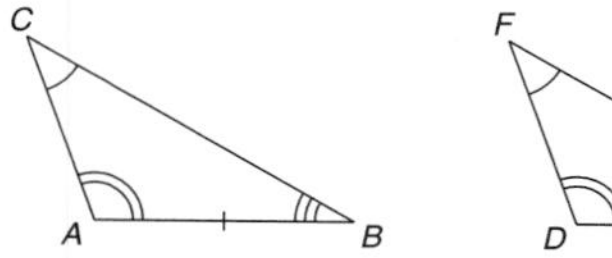

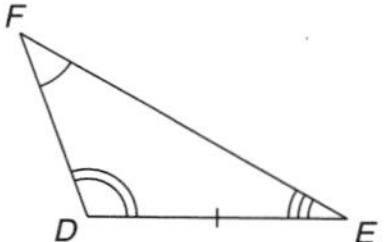

In Exercises 37 and 38, complete each proof.

37. *Given:* Plane M
C is the midpoint of $\overline{EB}$
$\overline{AD} \perp \overline{BE}$ and $\overline{AB} \parallel \overline{ED}$
Prove: $\triangle ABC \cong \triangle DEC$

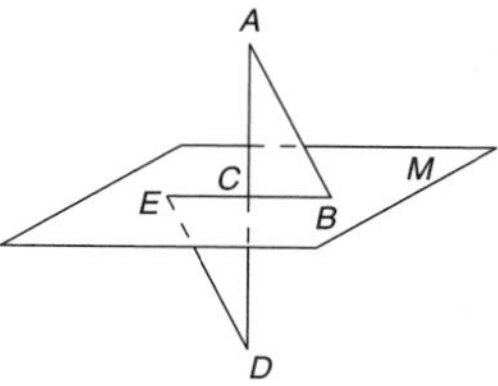

38. *Given:* $\overline{SP} \cong \overline{SQ}$ and $\overline{ST} \cong \overline{SV}$
Prove: $\triangle SPV \cong \triangle SQT$ and $\triangle TPQ \cong \triangle VQP$

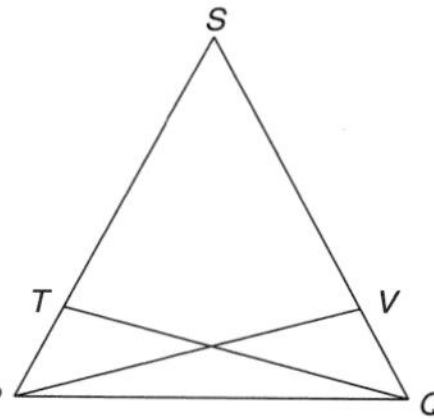

3.2 Corresponding Parts of Congruent Triangles

Recall that the definition of congruent triangles states that *all* six parts (three sides and three angles) of one triangle are congruent respectively to the six corresponding parts of the second triangle. If we have proved that $\triangle ABC \cong \triangle DEF$ by SAS (the congruent parts are marked in Figure 3.14), then we can draw conclusions such as $\angle C \cong \angle F$ and $\overline{AC} \cong \overline{DF}$. The following reason is often cited for drawing such conclusions and is based on the definition of congruent triangles.

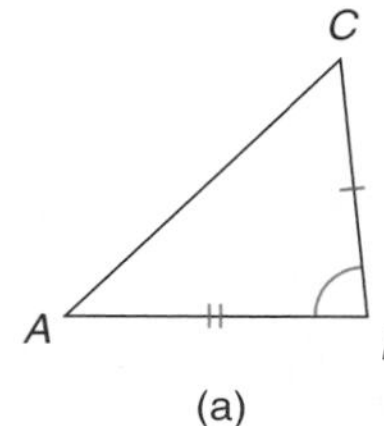

(a)

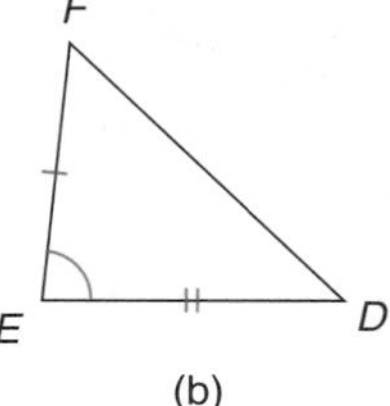

(b)

FIGURE 3.14

CPCTC: Corresponding parts of congruent triangles are congruent.

For triangles that have been proven congruent, CPCTC may be used to establish that either two line segments (corresponding sides) or two angles (corresponding angles) are congruent.

EXAMPLE 1

Given: $\overrightarrow{WZ}$ bisects $\angle TWV$
$\overline{WT} \cong \overline{WV}$
(See Figure 3.15.)

Prove: $\overline{TZ} \cong \overline{VZ}$

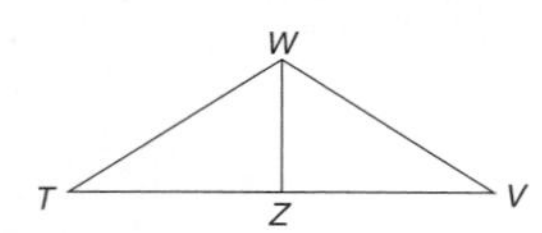

FIGURE 3.15

PROOF

Statements	Reasons
1. $\overrightarrow{WZ}$ bisects $\angle TWV$	**1.** Given
2. $\angle TWZ \cong \angle VWZ$	**2.** The bisector of an angle separates it into two $\cong$ $\angle$s
3. $\overline{WT} \cong \overline{WV}$	**3.** Given
4. $\overline{WZ} \cong \overline{WZ}$	**4.** Identity
5. $\triangle TWZ \cong \triangle VWZ$	**5.** SAS
6. $\overline{TZ} \cong \overline{VZ}$	**6.** CPCTC

In Example 1, we could just as easily have used CPCTC to prove that two angles are congruent. If we had been asked to prove that $\angle T \cong \angle V$, then the final statement would have read

6. $\angle T \cong \angle V$	**6.** CPCTC

We can take the proof a step further by proving triangles congruent and then using CPCTC to reach another conclusion, such as parallel or perpendicular lines. In Example 1, suppose we had been asked to prove that $\overline{WZ}$ bisects $\overline{TV}$. Then steps 1–6 on page 112 would remain as is and a seventh step would read

7. $\overline{WZ}$ bisects $\overline{TV}$	**7.** If a line segment is divided into two $\cong$ parts, then it has been bisected

In our study of triangles, we will establish three types of conclusions:

1. *Proving triangles congruent,* like $\triangle TWZ \cong \triangle VWZ$
2. *Proving corresponding parts of congruent triangles congruent,* like $\overline{TZ} \cong \overline{VZ}$ (Notice that two $\triangle$s have to be proved $\cong$ before CPCTC can be used.)
3. *Establishing a further relationship,* like $\overline{WZ}$ bisects $\overline{TV}$ (Notice that we must establish that two $\triangle$s are $\cong$ and also apply CPCTC before this goal can be reached.)

While little is stated in this book about a "plan for proof," every geometry student and teacher must have a plan before a proof can be completed.

EXAMPLE 2

Given: $\overline{ZW} \cong \overline{YX}$
$\overline{ZY} \cong \overline{WX}$
(See Figure 3.16.)

Prove: $\overline{ZY} \parallel \overline{WX}$

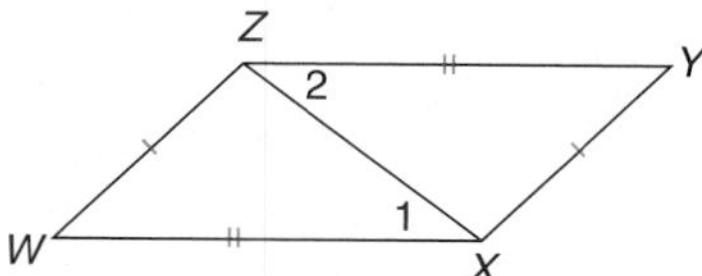

FIGURE 3.16

Plan for Proof: By showing that $\triangle ZWX \cong \triangle XYZ$, we can show that $\angle 1 \cong \angle 2$ by CPCTC. Then $\angle$s 1 and 2 are congruent alternate interior angles for $\overline{ZY}$ and $\overline{WX}$, which must be parallel.

PROOF

Statements	Reasons
1. $\overline{ZW} \cong \overline{YX}$; $\overline{ZY} \cong \overline{WX}$	**1.** Given
2. $\overline{ZX} \cong \overline{ZX}$	**2.** Identity
3. $\triangle ZWX \cong \triangle XYZ$	**3.** SSS
4. $\angle 1 \cong \angle 2$	**4.** CPCTC
5. $\overline{ZY} \parallel \overline{WX}$	**5.** If two lines are cut by a transversal so that the alt. int. $\angle$s are $\cong$, these lines are $\parallel$

Suggestions for Proving Triangles Congruent

Because each proof depends upon establishing congruent triangles, we offer the following suggestions.

FIGURE 3.17

Suggestions for a proof that involves congruent triangles:

1. Mark the figures systematically, using:
   - **a)** a *square* in the opening of a right angle;
   - **b)** the same number of *dashes* on congruent sides; and
   - **c)** the same number of *arcs* on congruent angles.
2. Trace the triangles to be proven congruent in different colors.
3. If the triangles overlap, draw them separately.

NOTE: See Figure 3.17 for reference.

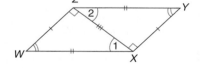

Right Triangles

In a right triangle, the side opposite the right angle is the **hypotenuse** of the triangle, while the sides of the right angle are the **legs** of the triangle. These parts of a right triangle are illustrated in Figure 3.18.

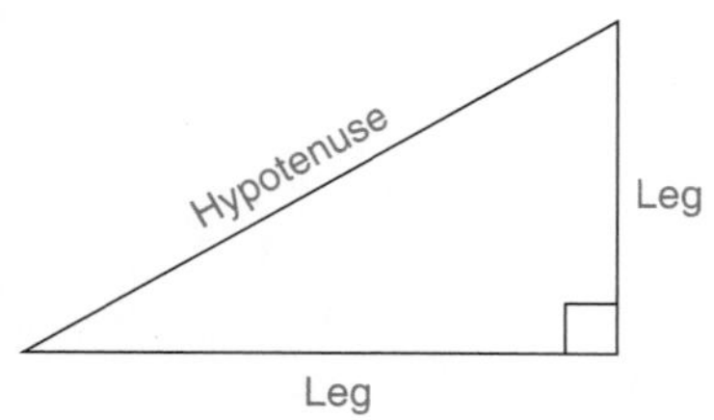

FIGURE 3.18

In addition to the methods discussed earlier for proving triangles congruent, we also have the HL method, which applies exclusively to right triangles. In HL, H refers to hypotenuse and L refers to leg. The proof of this method will be delayed until Section 5.3.

HL (Method for Proving Triangles Congruent)

THEOREM 3.2.1: If the hypotenuse and a leg of one right triangle are congruent to the hypotenuse and a leg of a second right triangle, then the triangles are congruent (HL).

The relationship described in Theorem 3.2.1 is illustrated in Figure 3.19. In Example 3, the construction leads to a unique right triangle.

EXAMPLE 3

Given: $\overline{AB}$ and $\overline{CA}$ in Figure 3.20(a)

Construct: The right triangle with hypotenuse of length equal to AB and one leg of length equal to CA.

Solution

First we construct $\overleftrightarrow{CQ}$ perpendicular to $\overleftrightarrow{EF}$ at point C [See Figure 3.20(b).] Second, mark off the length of $\overline{CA}$ on $\overleftrightarrow{CQ}$, as shown in Figure 3.20(c).

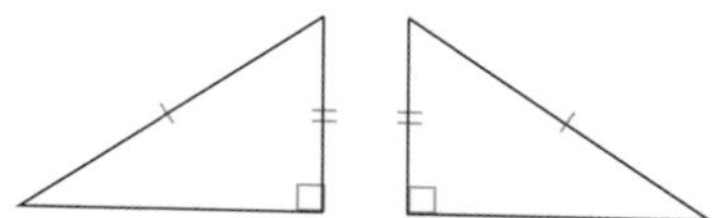

FIGURE 3.19

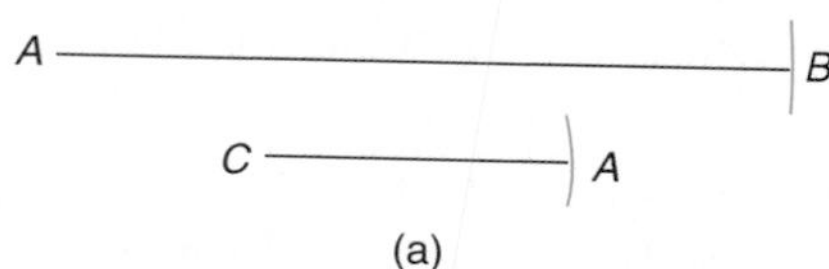

(a)

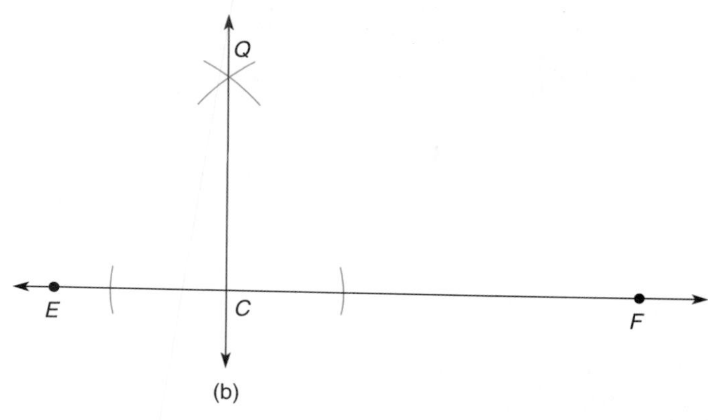

(b)

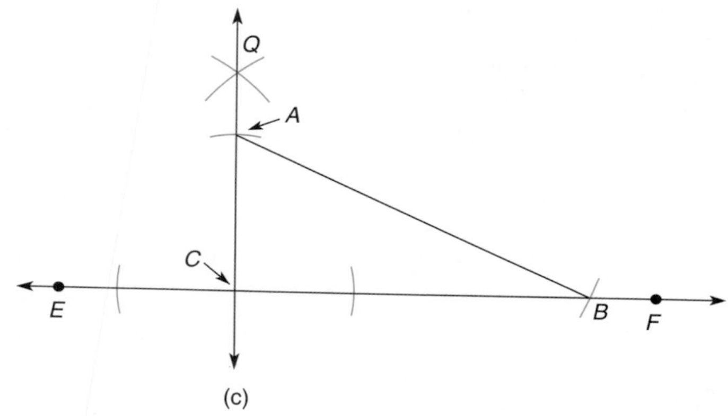

(c)

FIGURE 3.20

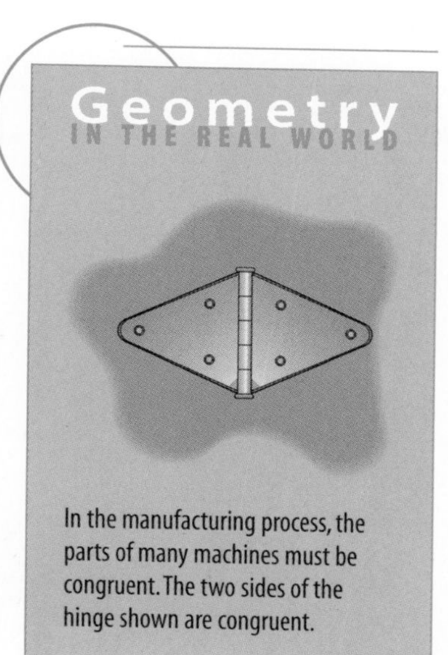

Geometry
IN THE REAL WORLD
In the manufacturing process, the parts of many machines must be congruent. The two sides of the hinge shown are congruent.

Finally, with point A as center, mark off a length equal to that of $\overline{AB}$, as shown in Figure 3.20(c). $\triangle ABC$ is the desired right $\triangle$.

EXAMPLE 4

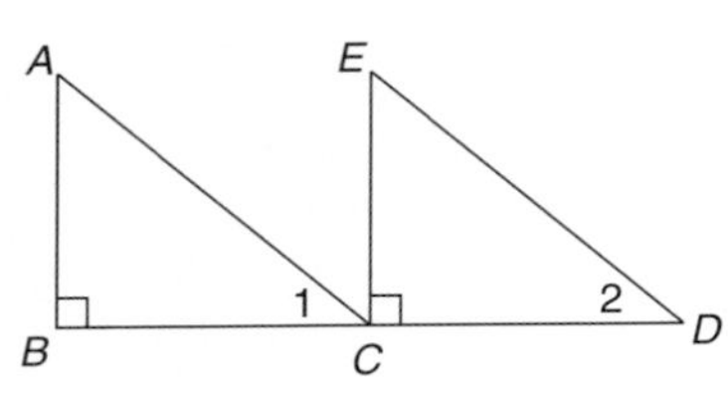

FIGURE 3.21

Cite the reason why the right triangles $\triangle ABC$ and $\triangle ECD$ in Figure 3.21 are congruent if:

a) $\overline{AB} \cong \overline{EC}$ and $\overline{AC} \cong \overline{ED}$
b) $\angle A \cong \angle E$ and C is the midpoint of $\overline{BD}$
c) $\overline{BC} \cong \overline{CD}$ and $\angle 1 \cong \angle 2$
d) $\overline{AB} \cong \overline{EC}$ and $\overline{EC}$ bisects $\overline{BD}$

Solution

a) HL **b)** AAS **c)** ASA **d)** SAS

3.2 Exercises

In Exercises 1 to 8, plan and write the two-column proof for each problem.

1. *Given:* $\angle 1$ and $\angle 2$ are right $\angle$s
$\overline{CA} \cong \overline{DA}$
Prove: $\triangle ABC \cong \triangle ABD$

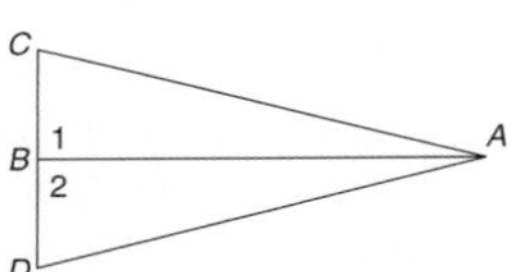

Exercises 1, 2

2. *Given:* $\angle 1$ and $\angle 2$ are right $\angle$s
$\overrightarrow{AB}$ bisects $\angle CAD$
Prove: $\triangle ABC \cong \triangle ABD$

3. *Given:* P is the midpoint of both $\overline{MR}$ and $\overline{NQ}$
Prove: $\triangle MNP \cong \triangle RQP$

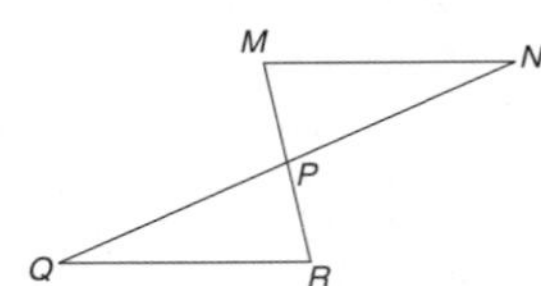

Exercises 3, 4

4. *Given:* $\overline{MN} \parallel \overline{QR}$ and $\overline{MN} \cong \overline{QR}$
Prove: $\triangle MNP \cong \triangle RQP$

5. *Given:* $\angle R$ and $\angle V$ are right $\angle$s
$\angle 1 \cong \angle 2$
Prove: $\triangle RST \cong \triangle VST$

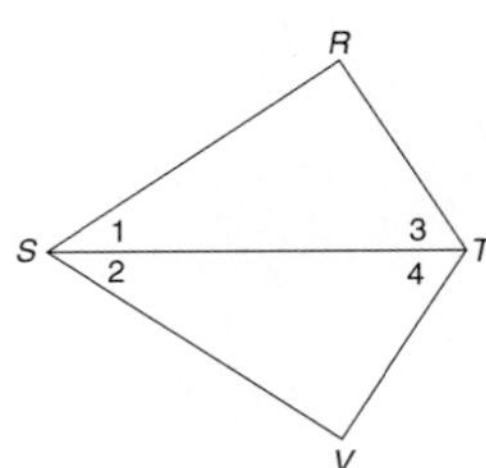

Exercises 5–8

6. *Given:* $\angle 1 \cong \angle 2$ and $\angle 3 \cong \angle 4$
Prove: $\triangle RST \cong \triangle VST$

7. *Given:* $\overline{SR} \cong \overline{SV}$ and $\overline{RT} \cong \overline{VT}$
Prove: $\triangle RST \cong \triangle VST$

8. *Given:* $\angle R$ and $\angle V$ are right $\angle$s
$\overline{RT} \cong \overline{VT}$
Prove: $\triangle RST \cong \triangle VST$

9. *Given:* $\overline{UW} \parallel \overline{XZ}$, $\overline{VY} \perp \overline{UW}$, and $\overline{VY} \perp \overline{XZ}$
$m\angle 1 = m\angle 4 = 42°$
Find: $m\angle 2$, $m\angle 3$, $m\angle 5$, and $m\angle 6$

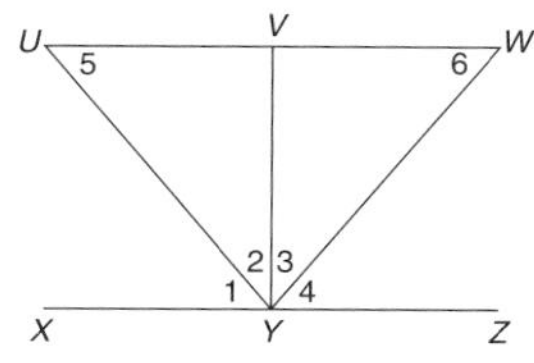

Exercises 9, 10

10. *Given:* $\overline{UW} \parallel \overline{XZ}$, $\overline{VY} \perp \overline{UW}$, and $\overline{VY} \perp \overline{XZ}$
$m\angle 1 = m\angle 4 = 4x + 3$
$m\angle 2 = 6x - 3$
Find: $m\angle 1$, $m\angle 2$, $m\angle 3$, $m\angle 4$, $m\angle 5$, and $m\angle 6$

In Exercises 11 and 12, complete each proof.

11. *Given:* $\overline{HJ} \perp \overline{KL}$ and $\overline{HK} \cong \overline{HL}$
Prove: $\overline{KJ} \cong \overline{JL}$

PROOF

Statements	Reasons
1. $\overline{HJ} \perp \overline{KL}$ and $\overline{HK} \cong \overline{HL}$	1. ?
2. $\angle$s HJK and HJL are rt. $\angle$s	2. ?
3. $\overline{HJ} \cong \overline{HJ}$	3. ?
4. ?	4. HL
5. ?	5. CPCTC

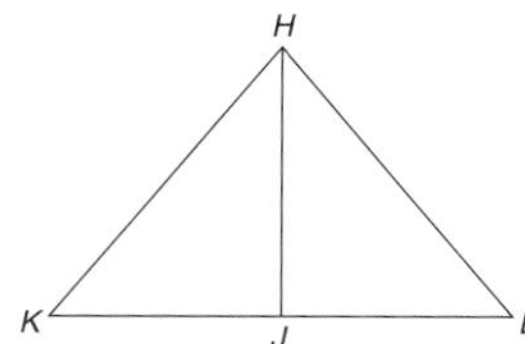

Exercises 11, 12

12. *Given:* $\overrightarrow{HJ}$ bisects $\angle KHL$
$\overline{HJ} \perp \overline{KL}$
Prove: $\angle K \cong \angle L$

PROOF

Statements	Reasons
1. ?	1. Given
2. $\angle JHK \cong \angle JHL$	2. ?
3. $\overline{HJ} \perp \overline{KL}$	3. ?
4. $\angle HJK \cong \angle HJL$	4. ?
5. ?	5. Identity
6. ?	6. ASA
7. $\angle K \cong \angle L$	7. ?

In Exercises 13 to 16, first prove that triangles are congruent, then use CPCTC.

13. *Given:* $\angle P$ and $\angle R$ are right $\angle$s
M is the midpoint of $\overline{PR}$
Prove: $\angle N \cong \angle Q$

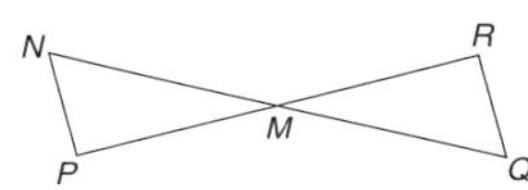

Exercises 13, 14

14. *Given:* M is the midpoint of $\overline{NQ}$
$\overline{NP} \parallel \overline{RQ}$ with transversals $\overline{PR}$ and $\overline{NQ}$
Prove: $\overline{NP} \cong \overline{QR}$

15. *Given:* $\angle 1$ and $\angle 2$ are right $\angle$s
H is the midpoint of $\overline{FK}$
$\overline{FG} \parallel \overline{HJ}$
Prove: $\overline{FG} \cong \overline{HJ}$

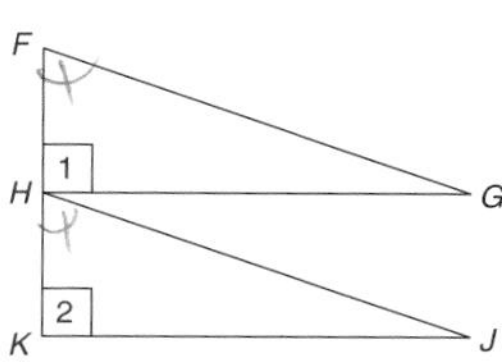

16. *Given:* $\overline{DE} \perp \overline{EF}$ and $\overline{CB} \perp \overline{AB}$
$\overline{AB} \parallel \overline{FE}$
$\overline{AC} \cong \overline{FD}$
Prove: $\overline{EF} \cong \overline{BA}$

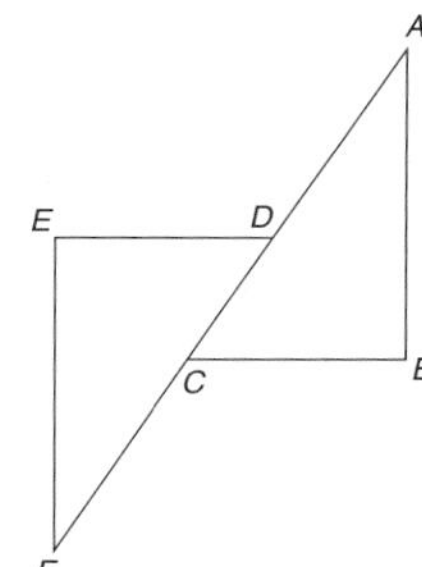

In Exercises 17 to 19, prove the indicated relationship.

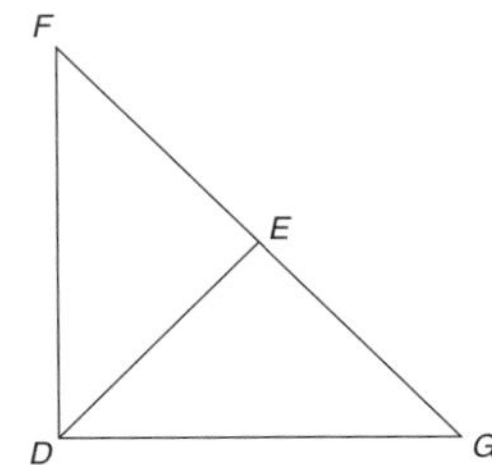

Exercises 17–19

17. *Given:* $\overline{DF} \cong \overline{DG}$ and $\overline{FE} \cong \overline{EG}$
Prove: $\overrightarrow{DE}$ bisects $\angle FDG$

18. *Given:* $\overrightarrow{DE}$ bisects $\angle FDG$
$\angle F \cong \angle G$
Prove: E is the midpoint of $\overline{FG}$

19. *Given:* E is the midpoint of $\overline{FG}$
$\overline{DF} \cong \overline{DG}$
Prove: $\overline{DE} \perp \overline{FG}$

In Exercises 20 to 22, draw the triangles to be shown congruent separately.

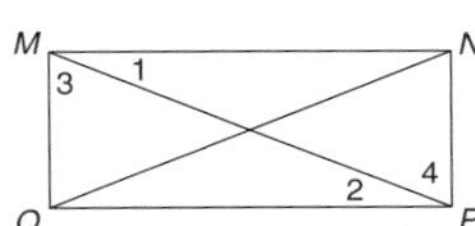

Exercises 20–22

20. *Given:* $\angle MQP$ and $\angle NPQ$ are rt. $\angle$s
$\overline{MQ} \cong \overline{NP}$
Prove: $\overline{MP} \cong \overline{NQ}$
(**HINT:** Show $\triangle MQP \cong \triangle NPQ$.)

21. *Given:* $\angle 1 \cong \angle 2$ and $\overline{MN} \cong \overline{QP}$
Prove: $\overline{MQ} \parallel \overline{NP}$
(**HINT:** Show $\triangle NMP \cong \triangle QPM$.)

22. *Given:* $\overline{MN} \parallel \overline{QP}$ and $\overline{MQ} \parallel \overline{NP}$
Prove: $\overline{MQ} \cong \overline{NP}$
(**HINT:** Show $\triangle MQP \cong \triangle PNM$.)

23. *Given:* $\overrightarrow{RW}$ bisects $\angle SRU$
$\overline{RS} \cong \overline{RU}$
Prove: $\triangle TRU \cong \triangle VRS$
(**HINT:** First show that $\triangle RSW \cong \triangle RUW$.)

Exercise 23

24. *Given:* $\overline{DB} \perp \overline{BC}$ and $\overline{CE} \perp \overline{DE}$
$\overline{AB} \cong \overline{AE}$
Prove: $\triangle BDC \cong \triangle ECD$
(**HINT:** First show that $\triangle ACE \cong \triangle ADB$.)

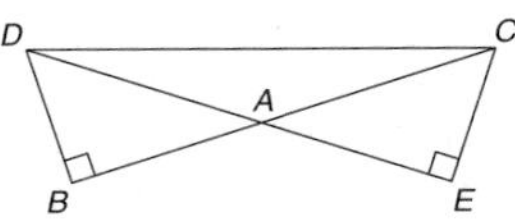

Exercise 24

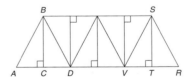

25. In the roof truss shown, $AB = 8$ and $m\angle HAF = 37°$. Find:

a) AH **b)** $m\angle BAD$ **c)** $m\angle ADB$

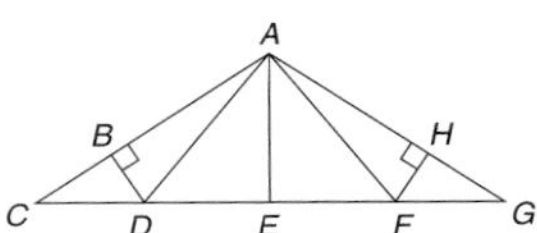

26. In the support system of the bridge shown, $AC = 6$ ft and $m\angle ABC = 28°$. Find:

a) $m\angle RST$ **b)** $m\angle ABD$ **c)** BS

3.3 Isosceles Triangles

In an isosceles triangle, the two sides of equal length are **legs** and the third side is the **base.** The point at which the two legs meet is the **vertex** of the triangle, so the angle formed by the legs (and opposite the base) is the **vertex angle.** The two remaining angles are **base angles.** (See Figure 3.22.) If $\overline{AC} \cong \overline{BC}$ in Figure 3.23, then $\triangle ABC$ would be isosceles

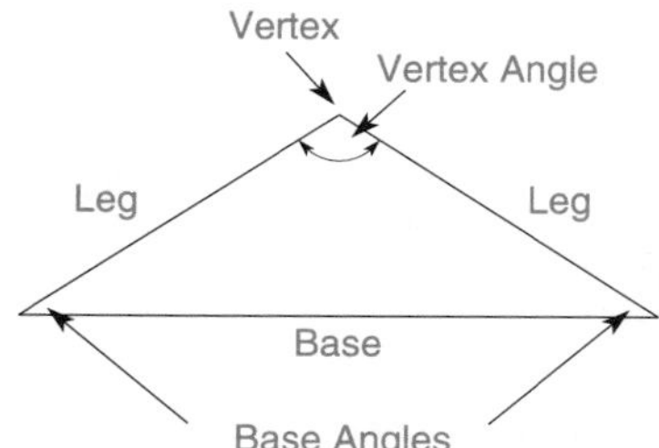

FIGURE 3.22

with legs $\overline{AC}$ and $\overline{BC}$, base $\overline{AB}$, vertex C, vertex angle C, and base angles at A and B. By considering Figure 3.23, we see that the base of an isosceles triangle is not necessarily the "bottom" side.

In any triangle, a number of segments, rays, or lines are related to the triangle. (See Figure 3.23.) Each angle of a triangle has a unique **angle bisector,** and this may be indicated by a ray or segment from the vertex of the bisected angle. Just as an angle bisector begins at the vertex of an angle, so does the **median,** which joins a vertex to the midpoint of the opposite side. Generally, the median from a vertex of a triangle is not the same as the angle bisector from that vertex. An **altitude** is a line segment drawn from a vertex to the opposite side such that it is perpendicular to the opposite side. Finally, the **perpendicular bisector** of a side of a triangle is shown as a line in Figure 3.23. A segment or ray could also perpendicularly bisect a side of the triangle. In Figure 3.23, $\overrightarrow{AD}$ is the angle bisector of $\angle BAC$; $\overline{AE}$ is the altitude from A to $\overline{BC}$; M is the midpoint of $\overline{BC}$; $\overline{AM}$ is the median from A to $\overline{BC}$; and $\overleftrightarrow{FM}$ is the perpendicular bisector of $\overline{BC}$.

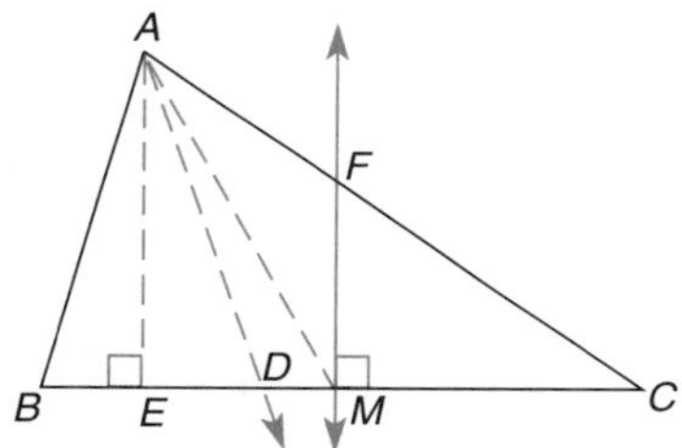

FIGURE 3.23

An altitude can actually lie in the exterior of a triangle. In Figure 3.24, which shows the obtuse triangle $\triangle RST$, the altitude from R must be drawn to an extension of side $\overline{ST}$. Later we will use the length of the altitude $\overline{RH}$ in place of h in the following standard formula for the area of a triangle:

$$A = \frac{1}{2}bh$$

The angle bisector and the median necessarily lie in the interior of the triangle.

Each triangle has three altitudes—one from each vertex. As these are shown for $\triangle ABC$ in Figure 3.25, do the three altitudes seem to meet at a common point?

We now consider a statement that involves the altitudes of congruent triangles.

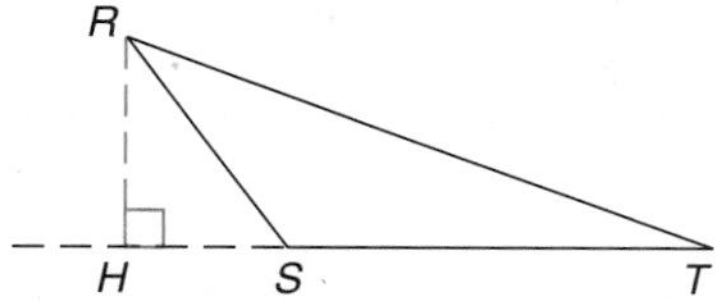

FIGURE 3.24

> **THEOREM 3.3.1:** Corresponding altitudes of congruent triangles are congruent.

Given: $\triangle ABC \cong \triangle RST$
altitudes $\overline{CD}$ to $\overline{AB}$ and $\overline{TV}$ to $\overline{RS}$
(See Figure 3.26.)

Prove: $\overline{CD} \cong \overline{TV}$

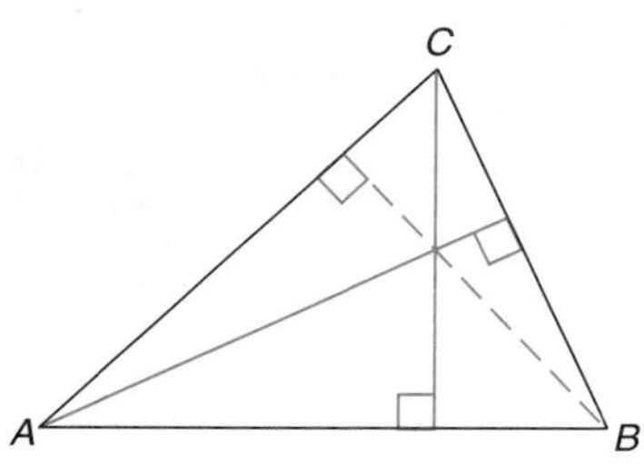

FIGURE 3.25

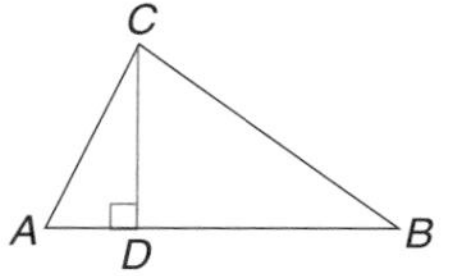

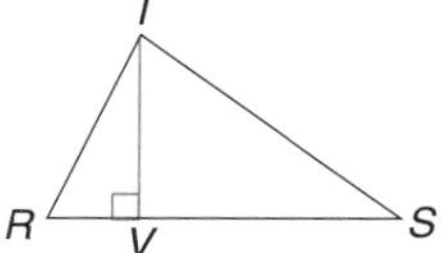

FIGURE 3.26

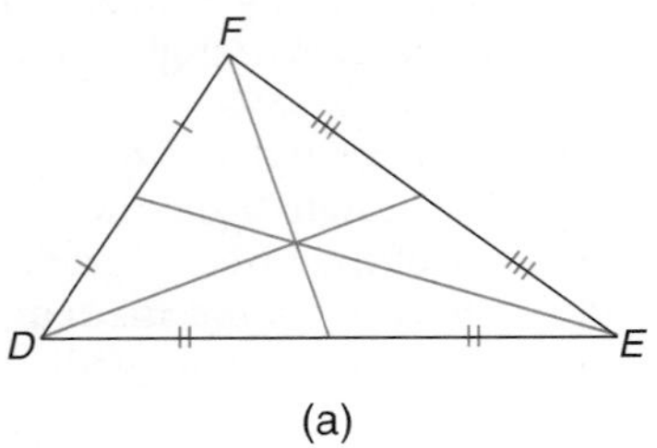

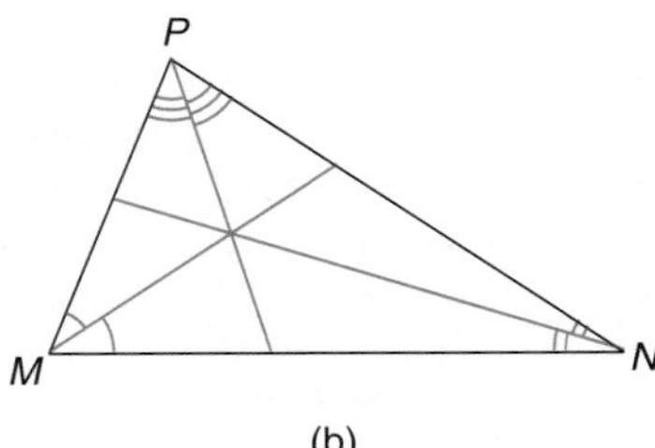

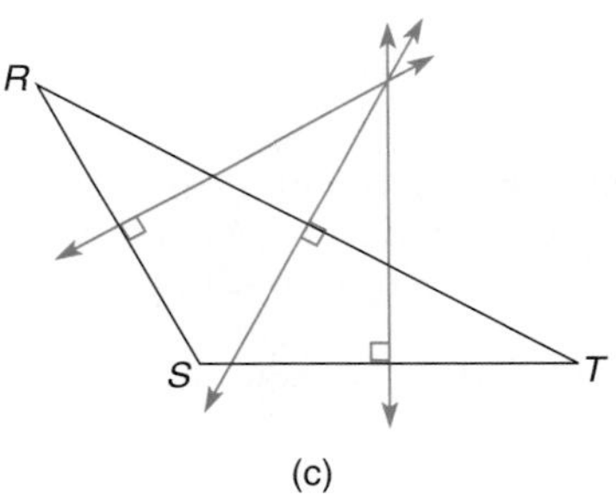

FIGURE 3.27

(Refer to Figure 3.26 on page 119.)

PROOF

Statements	Reasons
1. $\triangle ABC \cong \triangle RST$ altitudes $\overline{CD}$ to $\overline{AB}$ and $\overline{TV}$ to $\overline{RS}$	**1.** Given
2. $\overline{CD} \perp \overline{AB}$ and $\overline{TV} \perp \overline{RS}$	**2.** An altitude of a $\triangle$ is the line segment from one vertex drawn $\perp$ to the opposite side
3. $\angle CDA$ and $\angle TVR$ are right $\angle$s	**3.** If two lines are $\perp$, they form right $\angle$s
4. $\angle CDA \cong \angle TVR$	**4.** All right angles are $\cong$
5. $\overline{AC} \cong \overline{RT}$ and $\angle A \cong \angle R$	**5.** CPCTC (from $\triangle ABC \cong \triangle RST$)
6. $\triangle CDA \cong \triangle TVR$	**6.** AAS
7. $\overline{CD} \cong \overline{TV}$	**7.** CPCTC

Each triangle has three medians—one from each vertex to the midpoint of the opposite side. As the medians are drawn for $\triangle DEF$ in Figure 3.27(a), does it appear that the three medians intersect at a point?

Each triangle has three angle bisectors—one for each of the three angles. As these are shown for $\triangle MNP$ in Figure 3.27(b), does it appear that the three angle bisectors have a point in common?

Each triangle has three perpendicular bisectors for its sides; these are shown for $\triangle RST$ in Figure 3.27(c). Like the altitudes, medians, and angle bisectors, the perpendicular bisectors of the sides also meet at a single point.

The angle bisectors and the medians of a triangle always meet in the interior of the triangle. However, the altitudes and perpendicular bisectors of the sides [see Figure 3.27(c)] can meet in the exterior of the triangle. These points of intersection will be given more attention in Chapter 6.

With the following activity, we open the doors to several further discoveries.

DISCOVER!

Using a sheet of construction paper, cut out an isosceles triangle. Now use your compass to bisect the vertex angle. Fold along the angle bisector to form two smaller triangles. How are the smaller triangles related?

ANSWER

They are congruent.

EXAMPLE 1

Give a formal proof of Theorem 3.3.2.

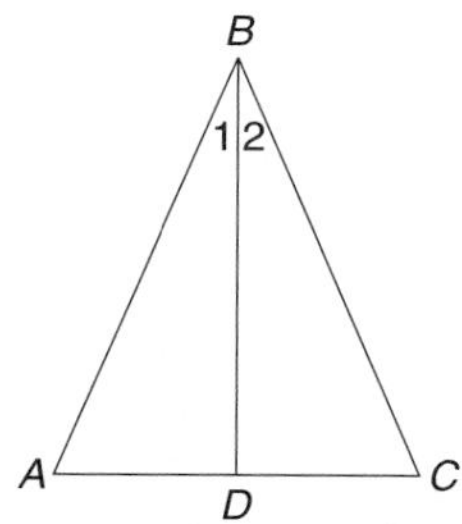

FIGURE 3.28

THEOREM 3.3.2: The bisector of the vertex angle of an isosceles triangle separates the triangle into two congruent triangles.

Given: Isosceles triangle $\triangle ABC$, with $\overline{AB} \cong \overline{BC}$
$\overrightarrow{BD}$ bisects $\angle ABC$
(See Figure 3.28.)

Prove: $\triangle ABD \cong \triangle CBD$

PROOF

Statements	Reasons
1. Isosceles $\triangle ABC$ with $\overline{AB} \cong \overline{BC}$	**1.** Given
2. $\overrightarrow{BD}$ bisects $\angle ABC$	**2.** Given
3. $\angle 1 \cong \angle 2$	**3.** The bisector of an $\angle$ separates it into two $\cong$ $\angle$s
4. $\overline{BD} \cong \overline{BD}$	**4.** Identity
5. $\triangle ABD \cong \triangle CBD$	**5.** SAS

Auxiliary Lines

In geometry, **auxiliary** (helping) lines or segments are often used to solve a problem or to construct a proof. You must account for the unique line, segment, or ray as it is introduced into the existing drawing. That is, each auxiliary figure must be **determined,** but it must not be **underdetermined** or **overdetermined.** A figure is underdetermined when there is more than one possible figure described. On the other extreme, a figure is overdetermined when it is impossible for *all* conditions described to be satisfied.

Consider Figure 3.29 and the following three descriptions, which are coded **D** for determined, **U** for underdetermined, and **O** for overdetermined:

D: Draw a line segment from A perpendicular to $\overline{BC}$ so that the terminal point is on $\overline{BC}$. [*Determined* because the line from A perpendicular to $\overline{BC}$ is unique; see Figure 3.29(a)].

U: Draw a line segment from A to $\overline{BC}$ so that the terminal point is on $\overline{BC}$. [*Underdetermined* because many line segments are possible; see Figure 3.29(b)].

O: Draw a line segment from A perpendicular to $\overline{BC}$ so that it bisects $\overline{BC}$. [*Overdetermined* because the line segment from A drawn perpendicular to $\overline{BC}$ will not contain the midpoint M of $\overline{BC}$; see Figure 3.29(c)].

In Example 2, an auxiliary segment is needed. As you study the proof, note the uniqueness of the segment and its justification in the proof.

(a)

(b)

(c)

FIGURE 3.29

EXAMPLE 2

Give a formal proof of Theorem 3.3.3.

> **THEOREM 3.3.3:** If two sides of a triangle are congruent, then the angles opposite these sides are also congruent.

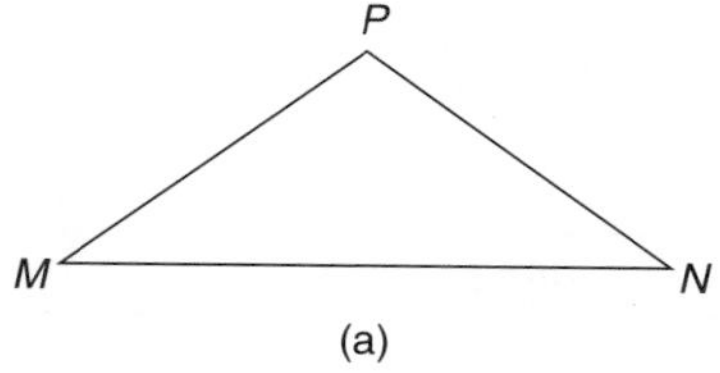

(a)

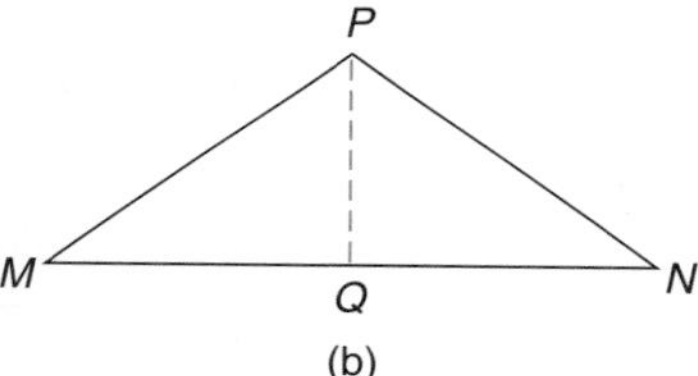

(b)

FIGURE 3.30

Given: Isosceles $\triangle MNP$ with $\overline{MP} \cong \overline{NP}$ (See Figure 3.30.)

Prove: $\angle M \cong \angle N$

NOTE: Figure 3.30(b) shows the auxiliary segment.

PROOF

Statements	Reasons
1. Isosceles $\triangle MNP$ with $\overline{MP} \cong \overline{NP}$	1. Given
2. Draw $\angle$ bisector $\overrightarrow{PQ}$ from P to $\overline{MN}$	2. Every angle has one and only one bisector
3. $\triangle MPQ \cong \triangle NPQ$	3. The bisector of the vertex angle of an isosceles $\triangle$ separates it into two $\cong$ $\triangle$s
4. $\angle M \cong \angle N$	4. CPCTC

Theorem 3.3.3 is sometimes stated, "The base angles of an isosceles triangle are congruent." We apply this theorem in Example 3.

EXAMPLE 3

Find the size of each angle of the isosceles triangle shown in Figure 3.31 if:

a) $m\angle 1 = 36°$

b) The measure of each base $\angle$ is 5° less than twice the measure of the vertex angle

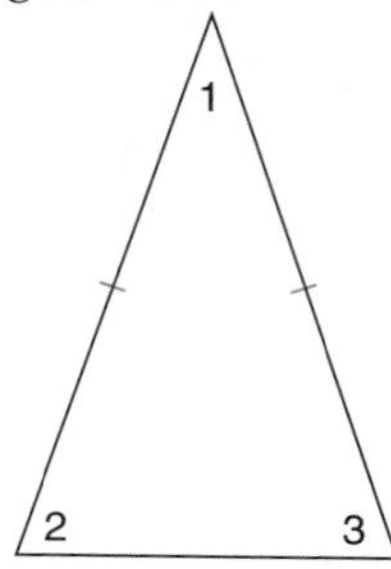

FIGURE 3.31

Solution

a) $m\angle 1 + m\angle 2 + m\angle 3 = 180$. Since $m\angle 1 = 36$ and $\angle 2$ and $\angle 3$ are $\cong$, we have

$$36 + 2(m\angle 2) = 180$$
$$2(m\angle 2) = 144$$
$$m\angle 2 = 72$$

Now $m\angle 1 = 36°$, while $m\angle 2 = m\angle 3 = 72°$.

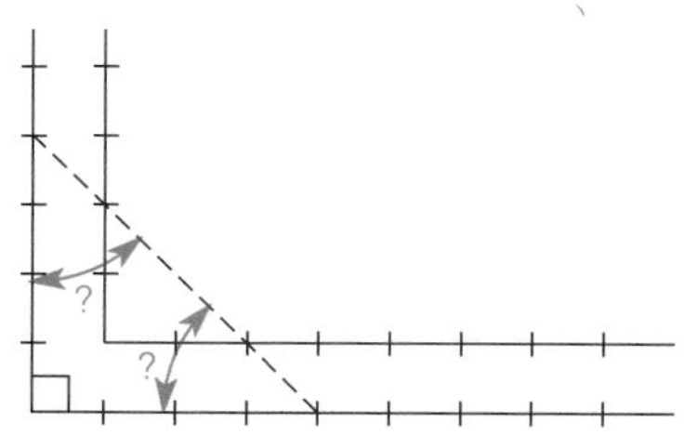

FIGURE 3.32

b) Let the vertex angle measure be given by x. Then the size of each base angle is $2x - 5$. Because the sum of the measures is 180°,

$$x + (2x - 5) + (2x - 5) = 180$$
$$5x - 10 = 180$$
$$5x = 190$$
$$x = 38$$
$$2x - 5 = 2(38) - 5 = 76 - 5 = 71$$

Therefore $m\angle 1 = 38°$ and $m\angle 2 = m\angle 3 = 71°$.

In some instances, a carpenter may want to get a quick, accurate measurement without having to go get his or her tools. Suppose that the carpenter's square shown in Figure 3.32 is handy but that a miter box is not nearby. If two marks are made at lengths of 4 inches from the corner of the square and these are then joined, what size angle is determined? You should see that each angle measures 45°.

Example 4 shows us that the converse of the theorem, "The base angles of an isosceles △ are congruent," is also true. However, see the "Warning" at the left.

■ *Warning*

The converse of an "If, then" statement is not always true. ■

EXAMPLE 4

Give a formal proof of Theorem 3.3.4.

THEOREM 3.3.4: If two angles of a triangle are congruent, then the sides opposite these angles are also congruent.

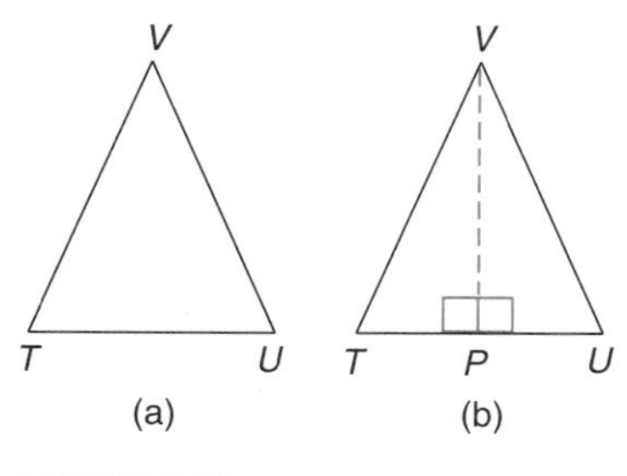

FIGURE 3.33

Given: $\triangle TUV$ with $\angle T \cong \angle U$ in Figure 3.33(a)

Prove: $\overline{VU} \cong \overline{VT}$

PROOF

Statements	Reasons
1. $\triangle TUV$ with $\angle T \cong \angle U$	**1.** Given
2. Draw $\overline{VP}$, the segment from $V \perp \overline{TU}$, as in Figure 3.33(b)	**2.** There is exactly one perpendicular from a point to a line
3. $\angle VPT \cong \angle VPU$	**3.** $\perp$ lines meet to form $\cong$ adjacent $\angle$s
4. $\overline{VP} \cong \overline{VP}$	**4.** Identity
5. $\triangle TPV \cong \triangle UPV$	**5.** AAS
6. $\overline{VU} \cong \overline{VT}$	**6.** CPCTC

When all three sides of a triangle are congruent, the triangle is **equilateral.** If all three angles are congruent, then the triangle is **equiangular.** Theorems 3.3.3 and 3.3.4 can be used to prove the following corollaries.

COROLLARY 3.3.5: An equilateral triangle is also equiangular.

COROLLARY 3.3.6: An equiangular triangle is also equilateral.

DEFINITION: The **perimeter** of a triangle is the sum of the lengths of its sides. Thus, if a, b, and c are the lengths of the three sides, then the perimeter P is given by $P = a + b + c$. (See Figure 3.34.)

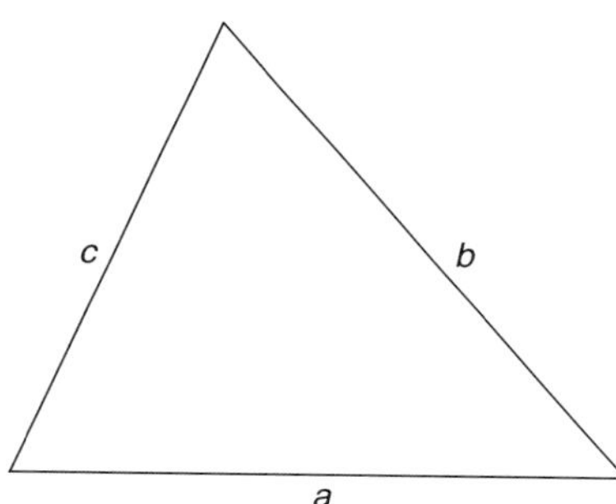

FIGURE 3.34

EXAMPLE 5

Given: $\angle B \cong \angle C$
$AB = 5.3$ and $BC = 3.6$
(See Figure 3.35.)

Find: The perimeter of $\triangle ABC$

Solution If $\angle B \cong \angle C$, then $AC = AB = 5.3$. Therefore

$$P = a + b + c$$
$$P = 3.6 + 5.3 + 5.3$$
$$P = 14.2$$

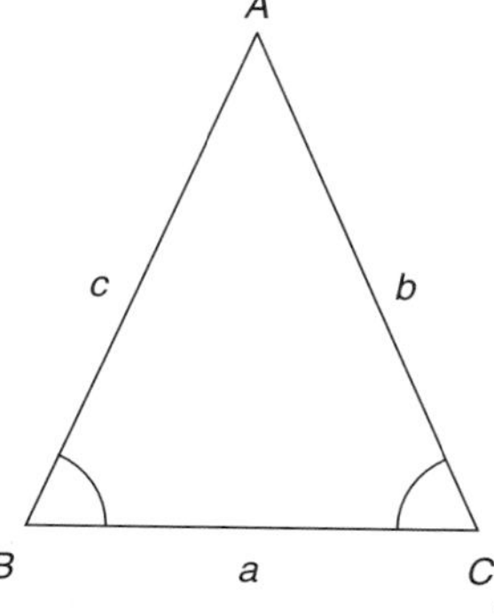

FIGURE 3.35

Many of the properties of triangles that were investigated in earlier sections are summarized in Table 3.1.

TABLE 3.1

Selected Properties of Triangles

	Scalene	Isosceles	Equilateral (equiangular)	Acute	Right	Obtuse
Sides	no two are ≅	exactly two are ≅	all three are ≅	possibly two or three ≅ sides	possibly two ≅ sides	possibly two ≅ sides
Angles	sum of ∠s is 180°	sum of ∠s is 180°; two ∠s ≅	sum of ∠s is 180°; three ≅ 60° ∠s	all ∠s acute; sum of ∠s is 180°; possibly two or three ≅ ∠s	1 right ∠; sum of ∠s is 180°; possibly two ≅ 45° ∠s; acute ∠s are complementary	1 obtuse ∠; sum of ∠s is 180°; possibly two ≅ acute ∠s

3.3 Exercises

In Exercises 1 to 6, describe the segment as determined, underdetermined, or overdetermined. Use the accompanying drawing for reference.

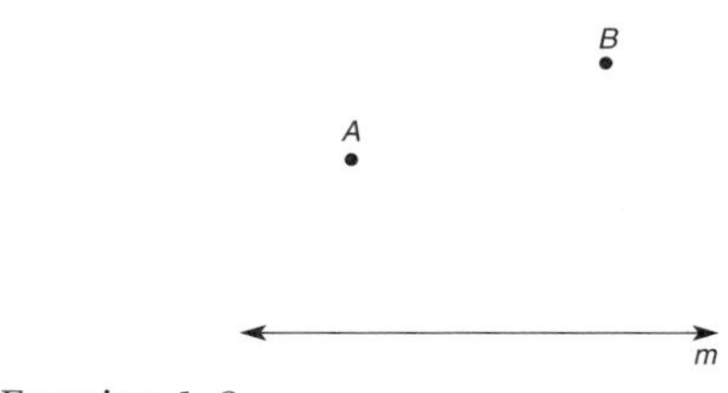

Exercises 1–6

1. Draw a segment through point A.
2. Draw a segment with endpoints A and B.
3. Draw a segment $\overline{AB}$ parallel to line m.
4. Draw segment $\overline{AB}$ perpendicular to m.
5. Draw a segment through A perpendicular to m.
6. Draw $\overline{AB}$ so that line m bisects $\overline{AB}$.
7. A surveyor knows that a lot has the shape of an isosceles triangle. If the vertex angle measures 70° and each equal side is 160 feet long, what measure does each of the base angles have?
8. In concave quadrilateral $ABCD$, the angle at A measures 40°. $\triangle ABD$ is isosceles, and $\overrightarrow{BC}$ bisects $\angle ABD$ while $\overrightarrow{DC}$ bisects $\angle ADB$. What are the measures of $\angle ABC$, $\angle ADC$, and $\angle 1$?

Exercise 8

In Exercises 9 to 14, use arithmetic or algebra as needed to find the measures indicated. Note the use of dashes on equal sides of the given isosceles triangles.

9. Find m$\angle 1$ and m$\angle 2$ if m$\angle 3 = 68°$.

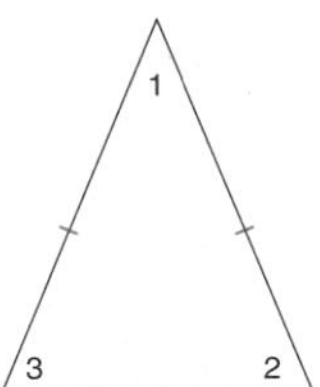

10. If m$\angle 3 = 68°$, find m$\angle 4$, the angle formed by the bisectors of $\angle 3$ and $\angle 2$.

11. Find the measure of $\angle 5$, which is formed by the bisectors of $\angle 1$ and $\angle 3$. Again let m$\angle 3 = 68°$.

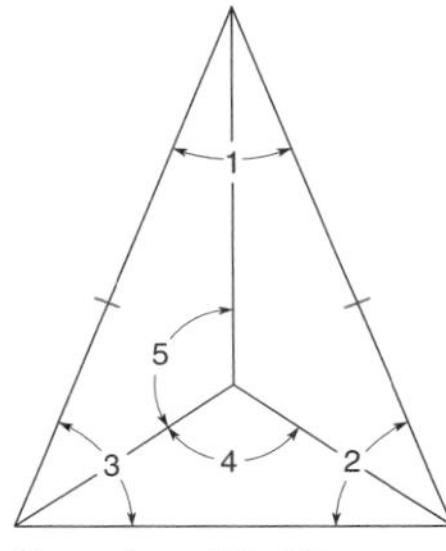

Exercises 10–12

12. Find an expression for the measure of $\angle 5$ if m$\angle 3 = 2x$ and the segments shown bisect the angles of the isosceles triangle.

13. In isosceles $\triangle ABC$ with vertex A (not shown), each base angle is 12° larger than the vertex angle. Find the measure of each angle.

14. In isosceles $\triangle ABC$ (not shown), vertex angle A is 5° more than one-half of base angle B. Find the size of each angle.

In Exercises 15 to 18, suppose that $\overline{BC}$ is the base of isosceles $\triangle ABC$ (not shown).

15. Find the perimeter of $\triangle ABC$ if $AB = 8$ and $BC = 10$.

16. Find AB if the perimeter of $\triangle ABC$ is 36.4 and $BC = 14.6$.

17. Find x if the perimeter of $\triangle ABC$ is 40, $AB = x$, and $BC = x + 4$.

18. Find x if the perimeter of $\triangle ABC$ is 68, $AB = x$, and $BC = 1.4x$.

19. Suppose that $\triangle ABC \cong \triangle DEF$. Also, $\overline{AX}$ bisects $\angle CAB$ and $\overline{DY}$ bisects $\angle FDE$. Are the corresponding angle bisectors of congruent triangles congruent?

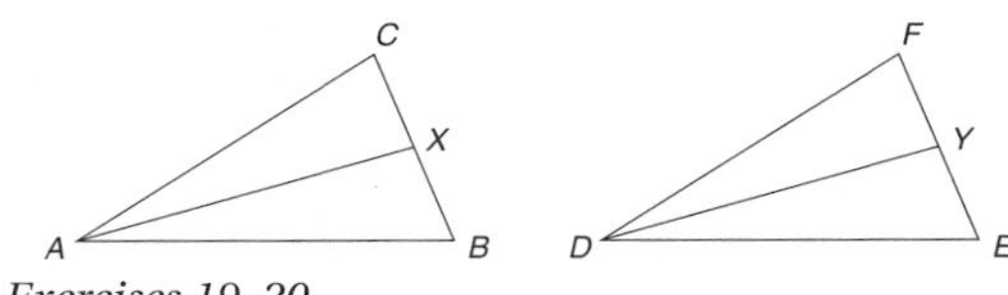

Exercises 19, 20

20. Suppose that $\triangle ABC \cong \triangle DEF$, $\overline{AX}$ is the median from A to $\overline{BC}$, and $\overline{DY}$ is the median from D to $\overline{EF}$. Are the corresponding medians of congruent triangles congruent?

In Exercises 21 and 22, complete each proof.

21. *Given:* $\angle 3 \cong \angle 1$
Prove: $\overline{AB} \cong \overline{AC}$

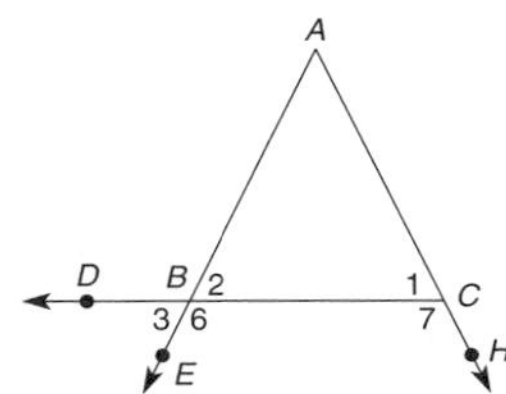

Exercises 21, 22

PROOF

Statements	Reasons
1. $\angle 3 \cong \angle 1$	1. ?
2. ?	2. If two lines intersect, the vertical $\angle$s formed are $\cong$
3. ?	3. Transitive Property of Congruence
4. $\overline{AB} \cong \overline{AC}$	4. ?

22. *Given:* $\overline{AB} \cong \overline{AC}$
Prove: $\angle 6 \cong \angle 7$

PROOF

Statements	Reasons
1. ?	1. Given
2. $\angle 2 \cong \angle 1$	2. ?
3. $\angle 2$ and $\angle 6$ are supplementary; $\angle 1$ and $\angle 7$ are supplementary	3. ?
4. ?	4. If two $\angle$s are supplementary to $\cong$ $\angle$s, they are $\cong$ to each other

In Exercises 23 to 25, complete each proof.

23. *Given:* $\angle 1 \cong \angle 3$
$\overline{RU} \cong \overline{VU}$
Prove: $\triangle STU$ is isosceles
(**HINT:** First show that $\triangle RUS \cong \triangle VUT$.)

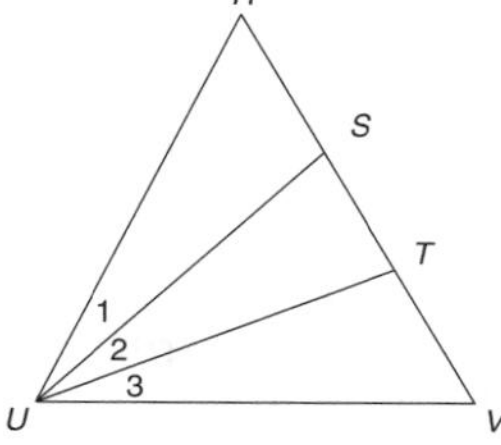

24. *Given:* $\overline{WY} \cong \overline{WZ}$
M is the midpoint of $\overline{YZ}$
$\overline{MX} \perp \overline{WY}$ and
$\overline{MT} \perp \overline{WZ}$
Prove: $\overline{MX} \cong \overline{MT}$

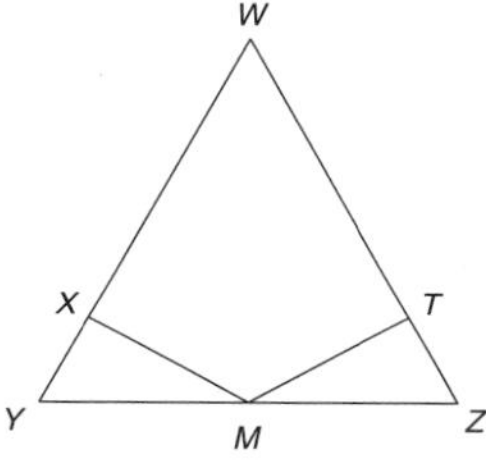

25. *Given:* Isosceles $\triangle MNP$ with vertex P
isosceles $\triangle MNQ$ with vertex Q
Prove: $\triangle MQP \cong \triangle NQP$

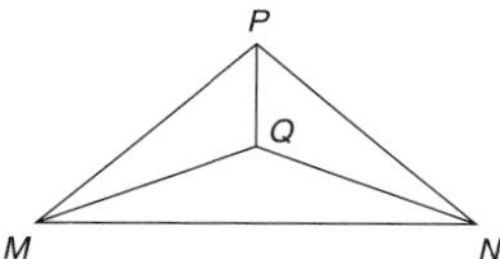

26. In isosceles triangle BAT, $\overline{AB} \cong \overline{AT}$. Also, $\overline{BR} \cong \overline{BT} \cong \overline{AR}$. If $AB = 11$ and $AR = 7$, find the perimeter of $\triangle BAT$.

27. In $\triangle BAT$, $\overline{BR} \cong \overline{BT} \cong \overline{AR}$, while $m\angle RBT = 20°$. Find $m\angle A$.

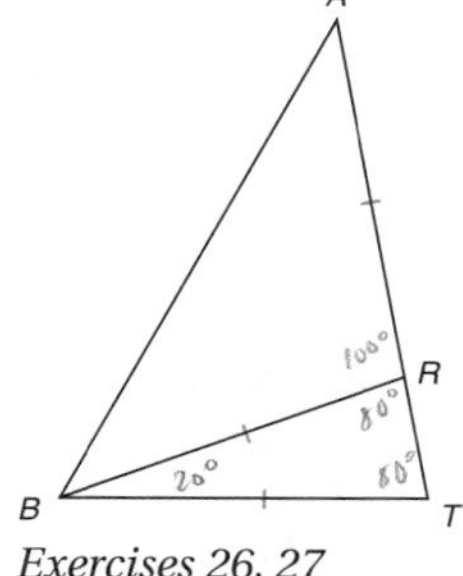

Exercises 26, 27

28. In $\triangle PMN$, $\overline{PM} \cong \overline{PN}$. $\overrightarrow{MB}$ bisects $\angle PMN$, and $\overrightarrow{NA}$ bisects $\angle PNM$. If $m\angle P = 36°$, name all isosceles triangles shown in the drawing.

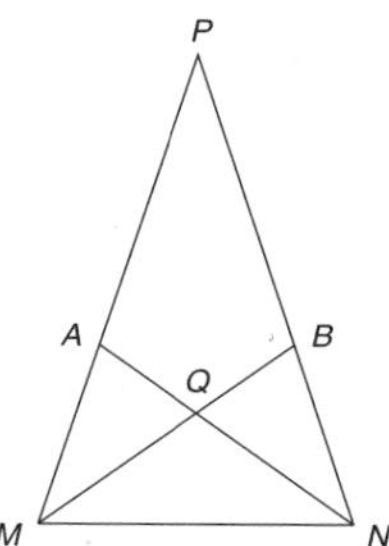

29. $\triangle ABC$ lies in the structural support system of the Ferris wheel. If $m\angle A = 30°$ and $AB = AC = 20$ ft, find the measures of $\angle B$ and $\angle C$.

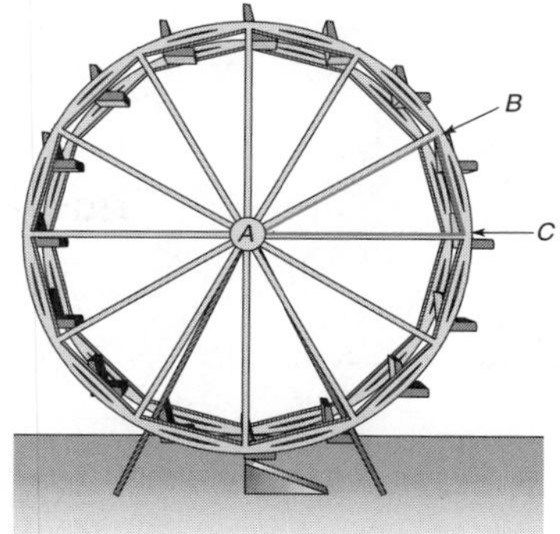

In Exercises 30 to 32, explain why each statement is true.

30. The altitude from the vertex of an isosceles triangle is also the median to the base of the triangle.

31. The bisector of the vertex angle of an isosceles triangle bisects the base.

32. The angle bisectors of the base angles of an isosceles triangle, together with the base, form an isosceles triangle.

3.4 Basic Constructions Justified

In earlier sections, construction methods were introduced that appeared to achieve their goals; however, the methods were presented intuitively. In this section, we justify the construction methods and apply them in further constructions. The justification of the method is a "proof" that demonstrates that the construction accomplished its purpose. See Example 1.

EXAMPLE 1

Justify the method for constructing an angle congruent to a given angle.

Given: $\angle ABC$
$\overline{BD} \cong \overline{BE} \cong \overline{ST} \cong \overline{SR}$ (by construction)
$\overline{DE} \cong \overline{TR}$ (by construction)
(See Figure 3.36.)

Prove: $\angle B \cong \angle S$

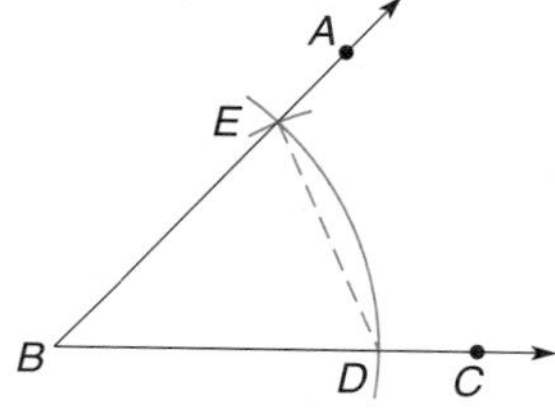

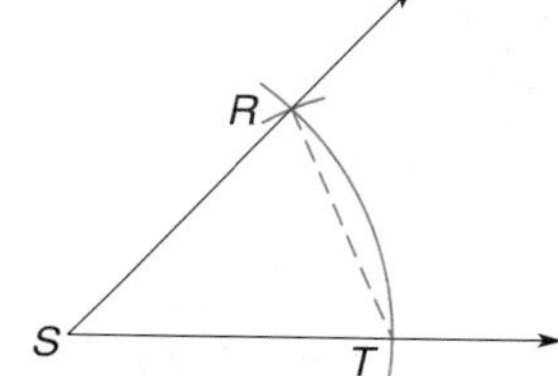

FIGURE 3.36

PROOF

Statements	Reasons
1. $\angle ABC$; $\overline{BD} \cong \overline{BE} \cong \overline{ST} \cong \overline{SR}$	**1.** Given
2. $\overline{DE} \cong \overline{TR}$	**2.** Given
3. $\triangle EBD \cong \triangle RST$	**3.** SSS
4. $\angle B \cong \angle S$	**4.** CPCTC

In Example 2, we will apply the construction method that was justified in Example 1. Our goal is to construct an isosceles triangle that contains an obtuse angle. It is necessary that the congruent sides include the obtuse angle.

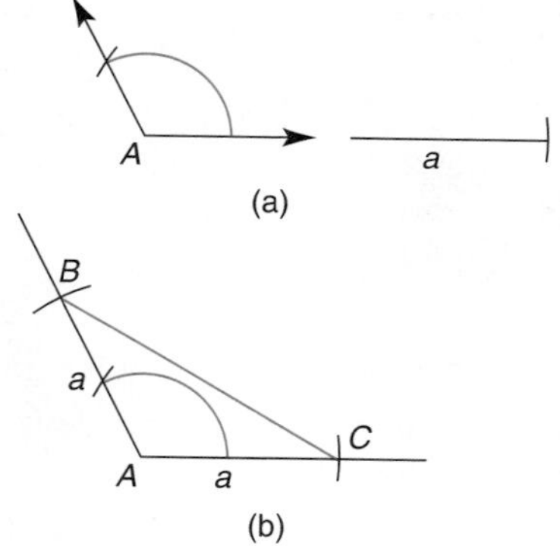

FIGURE 3.37

EXAMPLE 2

Construct an isosceles triangle in which obtuse $\angle A$ is included by two sides of length a [see Figure 3.37(a)].

Solution

Construct an angle congruent to $\angle A$. Then mark off arcs of length a at points B and C as shown in Figure 3.37(b). Join B to C to complete $\triangle ABC$.

In Example 3, we recall the method of construction used to bisect an angle. Although the technique is illustrated, the purpose here is to justify the method.

EXAMPLE 3

Justify the method for constructing the bisector of an angle. Provide the missing reasons in the proof.

Given: $\angle XYZ$
$\overline{YM} \cong \overline{YN}$ (by construction)
$\overline{MW} \cong \overline{NW}$ (by construction)
(See Figure 3.38.)

Prove: $\overrightarrow{YW}$ bisects $\angle XYZ$

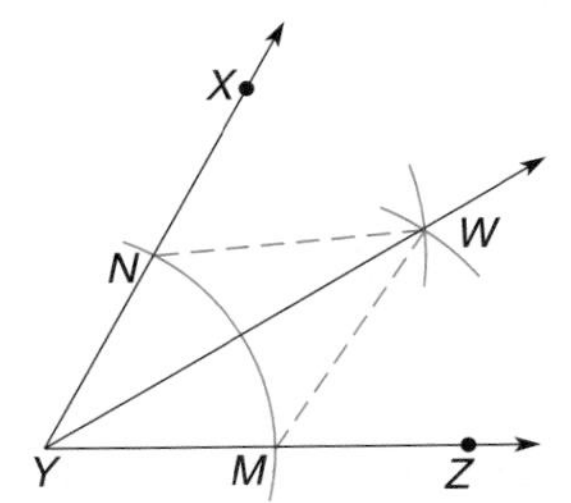

FIGURE 3.38

PROOF

Statements	Reasons
1. $\angle XYZ$; $\overline{YM} \cong \overline{YN}$ and $\overline{MW} \cong \overline{NW}$	1. ?
2. $\overline{YW} \cong \overline{YW}$	2. ?
3. $\triangle YMW \cong \triangle YNW$	3. ?
4. $\angle MYW \cong \angle NYW$	4. ?
5. $\overrightarrow{YW}$ bisects $\angle XYZ$	5. ?

The angle bisector method can be used to construct angles of certain measures. For instance, if a right angle has been constructed, then an angle of measure 45° can be constructed by bisecting the 90° angle. In Example 4, we construct an angle of measure 30°.

EXAMPLE 4

Construct an angle that measures 30°.

Solution

We begin by constructing an equilateral (and therefore equiangular) triangle. To accomplish this, mark off a line segment of length *a* as shown in Figure 3.39(a). From the endpoints of this line segment, mark off arcs using the same radius length *a*. The point of intersection determines the third vertex of this triangle, whose angles measure 60° each [see Figure 3.39(b)]. By constructing the bisector of the angle, we determine an angle that measures 30° [Figure 3.39(c)].

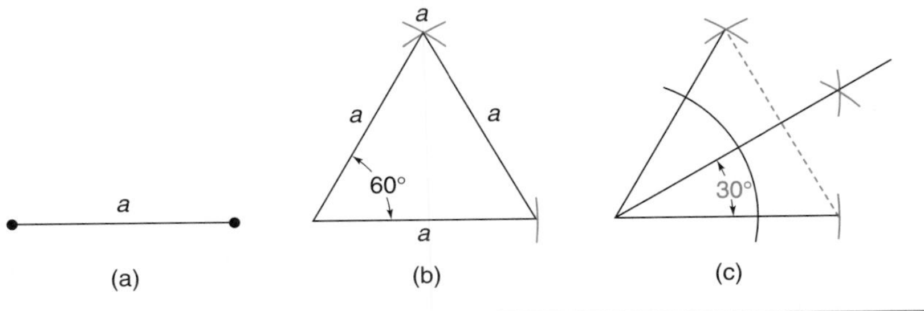

FIGURE 3.39

In Example 5, we justify the method for constructing a line perpendicular to a given line from a point not on that line. In the example, point *P* lies above line ℓ.

EXAMPLE 5

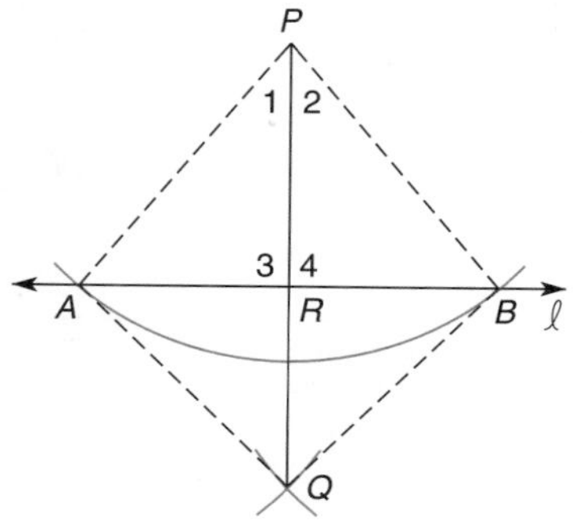

FIGURE 3.40

Given: P not on ℓ
$\overline{PA} \cong \overline{PB}$ (by construction)
$\overline{AQ} \cong \overline{BQ}$ (by construction)
(See Figure 3.40.)

Prove: $\overline{PQ} \perp \overline{AB}$

Provide the missing statements and reasons in the proof.

PROOF

	Statements	Reasons
	1. P not on ℓ $\overline{PA} \cong \overline{PB}$ and $\overline{AQ} \cong \overline{BQ}$	**1.** ?
	2. $\overline{PQ} \cong \overline{PQ}$	**2.** ?
	3. $\triangle PAQ \cong \triangle PBQ$	**3.** ?
	4. $\angle 1 \cong \angle 2$	**4.** ?
	5. $\overline{PR} \cong \overline{PR}$	**5.** ?
(1), (4), (5)	**6.** $\triangle PRA \cong \triangle PRB$	**6.** ?
	7. $\angle 3 \cong \angle 4$	**7.** ?
	8. ?	**8.** If two lines meet to form $\cong$ adjacent $\angle$s, these lines are $\perp$

In Example 6, we recall the method for constructing the line perpendicular to a given line at a point on the line. We illustrate the technique in the example and ask that the student justify the method in Exercise 27. In Example 6, we construct an angle that measures 45°.

EXAMPLE 6

Construct an angle that measures 45°.

Solution

We begin by constructing a line segment perpendicular to line ℓ at point P [Figure 3.41(a)]. Next we bisect one of the right angles that was determined. The bisector forms an angle whose measure is 45° [Figure 3.41(b)].

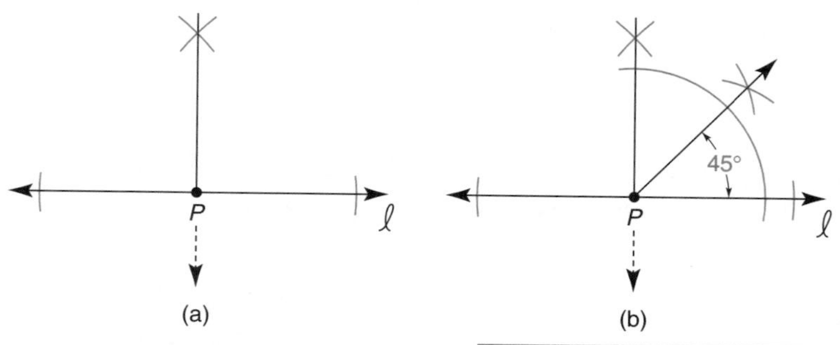

FIGURE 3.41

As we saw in Example 4, constructing an equilateral triangle is fairly simple. It is also possible to construct other regular polygons, such as a square or a regular hexagon. In the following box, we recall some facts that will help us perform such constructions.

> To construct a regular polygon:
>
> **1.** Each interior angle must measure $I = \frac{(n-2)180}{n}$ degrees; alternately, each exterior angle must measure $E = \frac{360}{n}$ degrees.
>
> **2.** All sides must be congruent.

EXAMPLE 7

Construct a regular hexagon having sides of length a.

Solution

We begin by marking off a line segment of length a, as shown in Figure 3.42(a). Each exterior angle of the hexagon ($n = 6$) must measure $E = \frac{360}{6} = 60°$; then each interior angle measures 120°. We construct an equilateral triangle (all sides measure a) so that a 60° exterior angle is formed, as shown in Figure 3.42(b). Again marking off an arc of length a for the second side, we construct another exterior angle of measure 60° as shown in Figure 3.42(c). This procedure is continued until the regular hexagon *ABCDEF* is determined, as shown in Figure 3.42(d).

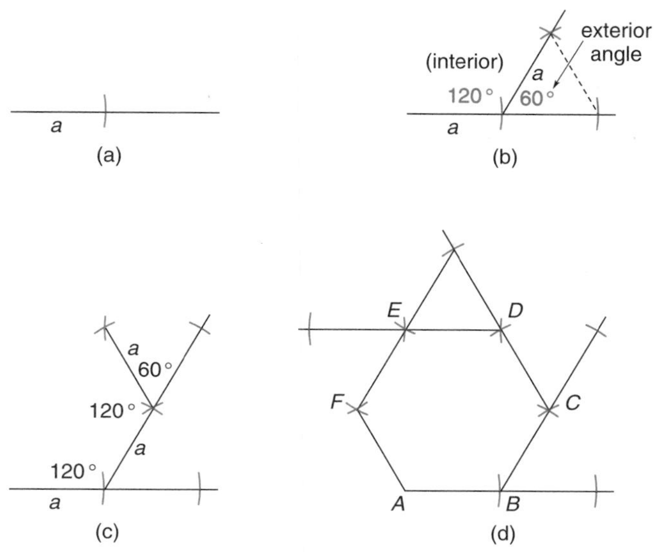

FIGURE 3.42

3.4 Exercises

In Exercises 1 to 6, use line segments of given lengths a, b, and c to perform the constructions.

a

b

c

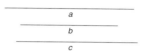

Exercises 1–6

1. Construct a line segment of length $2b$.

2. Construct a line segment of length $b + c$.

3. Construct a line segment of length $\frac{1}{2}c$.

4. Construct a line segment of length $a - b$.

5. Construct a triangle with sides of lengths a, b, and c.

6. Construct an isosceles triangle with a base of length b and legs of length a.

In Exercises 7 to 12, use the angles provided to perform the constructions.

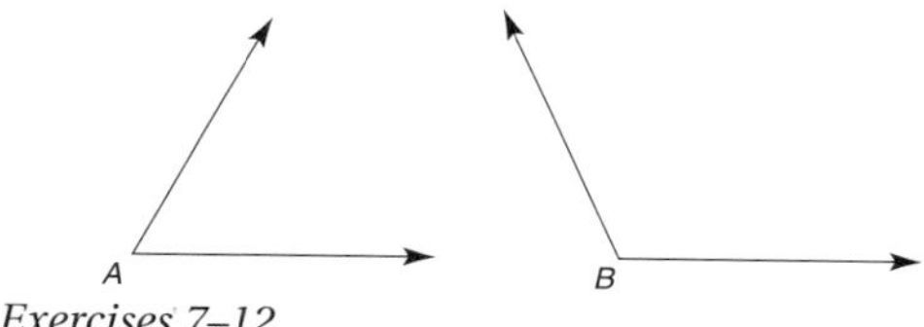

Exercises 7–12

7. Construct an angle that is congruent to acute $\angle A$.

8. Construct an angle that is congruent to obtuse $\angle B$.

9. Construct an angle that has one-half the measure of $\angle A$.

10. Construct an angle that has a measure equal to $m\angle B - m\angle A$.

11. Construct an angle that has twice the measure of $\angle A$.

12. Construct an angle whose measure averages the measures of $\angle A$ and $\angle B$.

In Exercises 13 and 14, use the angles and lengths of sides provided to construct the triangle described.

13. Construct the triangle that has sides of lengths r and t with included angle S.

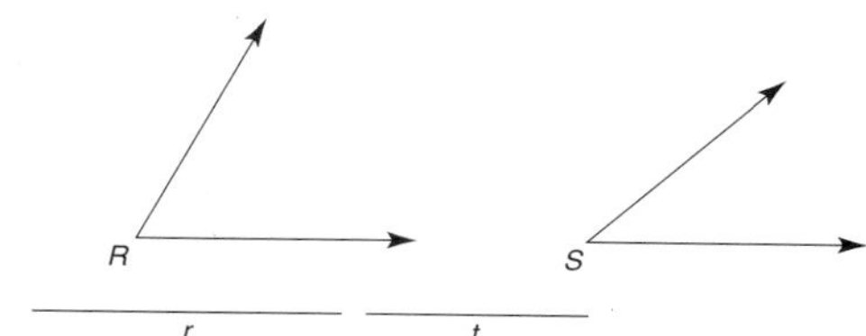

Exercises 13, 14

14. Construct the triangle that has a side of length t included by angles R and S.

In Exercises 15 to 18, construct angles having the given measures.

15. 90° and then 45°

16. 60° and then 30°

17. 30° and then 15°

18. 45° and then 105°
(**HINT:** $105° = 45° + 60°$)

19. Describe how you would construct an angle measuring 22.5°.

20. Describe how you would construct an angle measuring 75°.

21. Construct the complement of the acute angle shown.

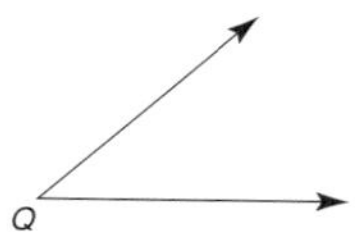

22. Construct the supplement of the obtuse angle shown.

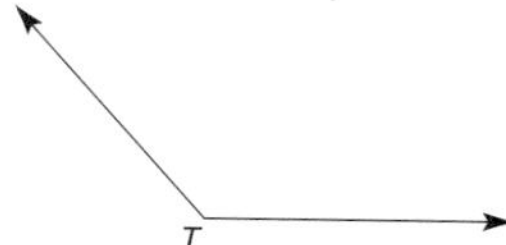

In Exercises 23 and 24, use line segments of lengths a and c as shown.

23. Construct the right triangle with hypotenuse of length c and a leg of length a.

Exercises 23, 24

24. Construct an isosceles triangle with base of length c and altitude of length a.
(**HINT:** The altitude lies on the perpendicular bisector of the base.)

In Exercises 25 and 26, use the given angle and the line segment of length b.

25. Construct the right triangle in which acute angle R has a side (one leg of the triangle) of length b.

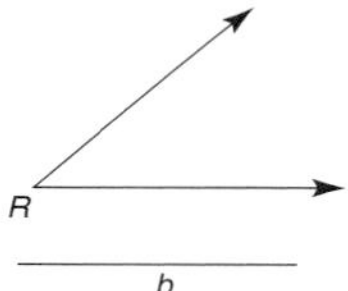

Exercises 25, 26

26. Construct an isosceles triangle with base of length b and congruent base angles having the measure of angle R.

27. Complete the justification of the construction of the line perpendicular to a given line at a point on that line.

Given: Line m, with point P on m
$\overline{PQ} \cong \overline{PR}$ (by construction)
$\overline{QS} \cong \overline{RS}$ (by construction)
Prove: $\overleftrightarrow{SP} \perp m$

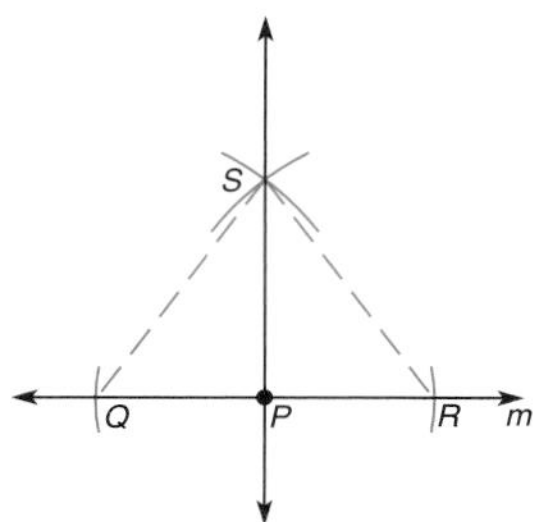

28. Complete the justification of the construction of the perpendicular bisector of a line segment.

Given: $\overline{AB}$ with $\overline{AC} \cong \overline{BC} \cong \overline{AD} \cong \overline{BD}$ (by construction)
Prove: $\overline{AM} \cong \overline{MB}$ and $\overleftrightarrow{CD} \perp \overline{AB}$

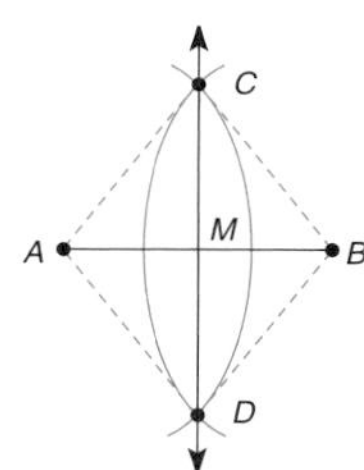

Exercise 28

29. To construct a regular hexagon, what measure would be necessary for each interior angle? Construct an angle of that measure.

30. To construct a regular octagon, what measure would be necessary for each interior angle? Construct an angle of that measure.

31. To construct a regular dodecagon (12 sides), what measure would be necessary for each interior angle? Construct an angle of that measure.

32. Draw an acute triangle and construct the three medians of the triangle. Do the medians appear to meet at a common point?

33. Draw an obtuse triangle and construct the three altitudes of the triangle. Do the altitudes appear to meet at a common point?
(**HINT:** In the construction of two of the altitudes, a side needs to be extended.)

34. Draw a right triangle and construct the angle bisectors of the triangle. Do the angle bisectors appear to meet at a common point?

35. Draw an obtuse triangle and construct the three perpendicular bisectors of its sides. Do the perpendicular bisectors of the three sides appear to meet at a common point?

36. Construct an equilateral triangle and its three altitudes. What does intuition tell you about the three medians, the three angle bisectors, and the three perpendicular bisectors of the sides of that triangle?

37. A carpenter has placed a square over an angle in such a manner that $\overline{AB} \cong \overline{AC}$ and $\overline{BD} \cong \overline{CD}$ (see drawing). What may you conclude about the location of point D?

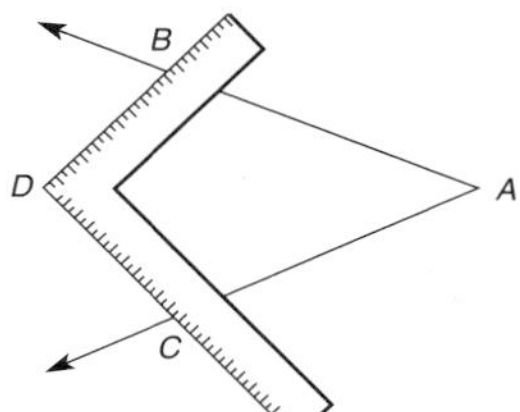

3.5 Inequalities in a Triangle

There are some important inequality relationships among the measured parts of a triangle. To establish some of these, we recall and apply some facts from both algebra and geometry.

DEFINITION: Let a and b be real numbers. $a > b$ (read "a is greater than b") if and only if there is a positive number p for which $a = b + p$.

For instance, $9 > 4$, since there is the positive number 5 for which $9 = 4 + 5$. Because $5 + 2 = 7$, we also know that $7 > 2$ and $7 > 5$. In geometry, let A-B-C on $\overline{AC}$; if $AB + BC = AC$, then $AC > AB$, since BC is a positive number.

Lemmas (Helping Theorems)

We will use the following theorems to help us prove the theorems found later in this section. In their role as "helping" theorems, each of the five boxed statements that follow is called a **lemma**. We will prove the first four lemmas, since their content is geometric.

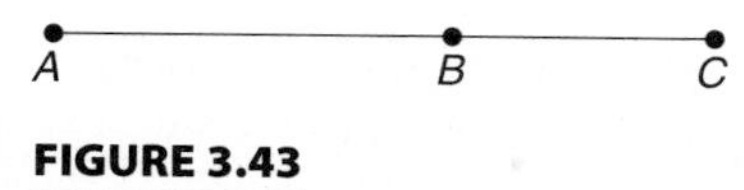

FIGURE 3.43

LEMMA 3.5.1: If B is between A and C on $\overline{AC}$, then $AC > AB$ and $AC > BC$. (The measure of a line segment is greater than the measure of any of its parts. See Figure 3.43.)

Proof

By the Segment-Addition Postulate, $AC = AB + BC$. Since $BC > 0$ (meaning BC is positive), it follows that $AC > AB$. Similarly, $AC > BC$. These relationships follow logically from the definition of $a > b$.

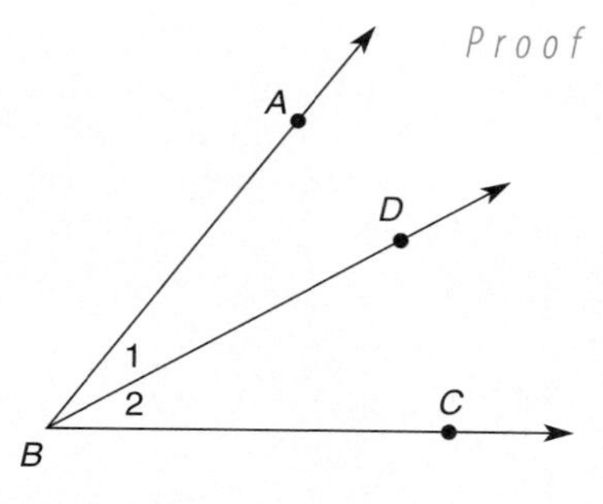

FIGURE 3.44

LEMMA 3.5.2: If $\overrightarrow{BD}$ separates $\angle ABC$ into two parts ($\angle 1$ and $\angle 2$), then $m\angle ABC > m\angle 1$ and $m\angle ABC > m\angle 2$. (The measure of an angle is greater than the measure of any of its parts. See Figure 3.44.)

Proof

By the Angle-Addition Postulate, $m\angle ABC = m\angle 1 + m\angle 2$. Because $m\angle 2 > 0$, it follows that $m\angle ABC > m\angle 1$. Similarly, $m\angle ABC > m\angle 2$.

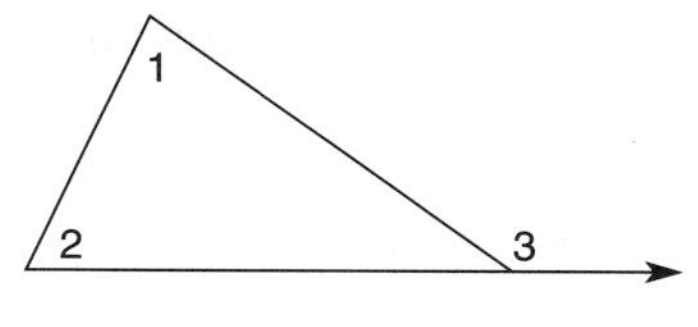

FIGURE 3.45

LEMMA 3.5.3: If $\angle 3$ is an exterior angle of a triangle and $\angle 1$ and $\angle 2$ are the nonadjacent interior angles, then $m\angle 3 > m\angle 1$ and $m\angle 3 > m\angle 2$. (The measure of an exterior angle of a triangle is greater than the measure of either nonadjacent interior angle. See Figure 3.45.)

Proof Because the measure of an exterior angle of a triangle equals the sum of measures of the two nonadjacent interior angles, $m\angle 3 = m\angle 1 + m\angle 2$. It follows that $m\angle 3 > m\angle 1$ and $m\angle 3 > m\angle 2$.

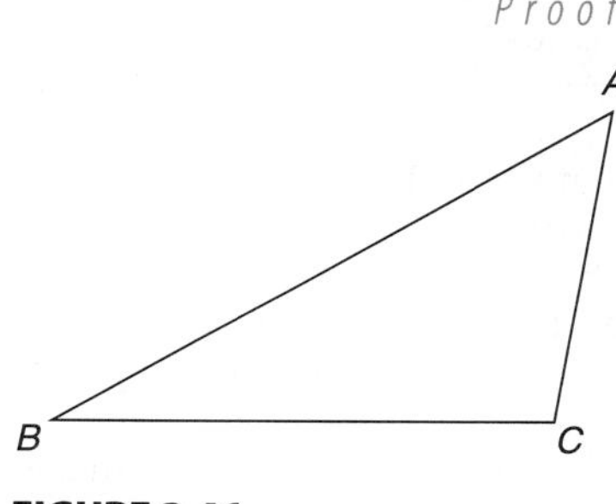

FIGURE 3.46

LEMMA 3.5.4: In $\triangle ABC$, if $\angle C$ is a right angle or an obtuse angle, then $m\angle C > m\angle A$ and $m\angle C > m\angle B$. (If a triangle contains a right or an obtuse angle, then the measure of this angle is greater than the measure of either of the remaining angles. See Figure 3.46.)

Proof In $\triangle ABC$, $m\angle A + m\angle B + m\angle C = 180°$. With $m\angle C \geq 90°$, it follows that $m\angle A + m\angle B \leq 90°$, and each angle ($\angle A$ and $\angle B$) must be acute. Thus $m\angle C > m\angle A$ and $m\angle C > m\angle B$.

The following theorem (also a lemma) is used in Example 1. Its proof (not given) depends upon the definition of "is greater than," which is found on page 134.

LEMMA 3.5.5: **(Addition Property of Inequality)** If $a > b$ and $c > d$, then $a + c > b + d$.

EXAMPLE 1

Give a paragraph proof for the following problem.

Given: $AB > CD$ and $BC > DE$

Prove: $AC > CE$

(See Figure 3.47)

FIGURE 3.47

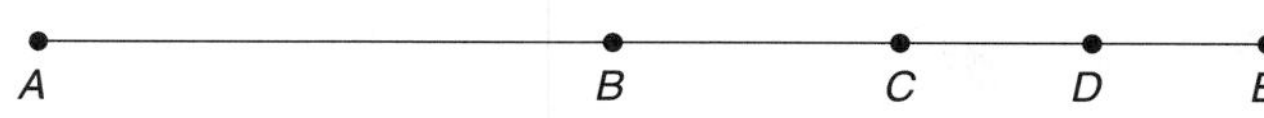

Proof If $AB > CD$ and $BC > DE$, then $AB + BC > CD + DE$ by Lemma 3.5.5. But $AB + BC = AC$ and $CD + DE = CE$ by the Segment-Addition Postulate. Using substitution, it follows that $AC > CE$.

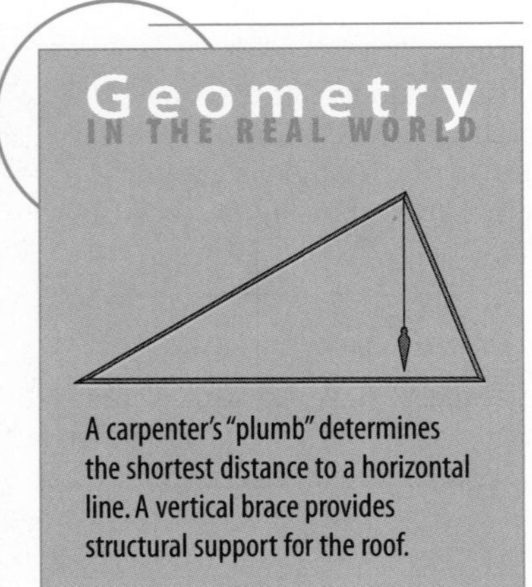

A carpenter's "plumb" determines the shortest distance to a horizontal line. A vertical brace provides structural support for the roof.

The proof in Example 1 could have been written in this more standard format.

PROOF

Statements	Reasons
1. $AB > CD$ and $BC > DE$	**1.** Given
2. $AB + BC > CD + DE$	**2.** Lemma 3.5.5
3. $AB + BC = AC$ and $CD + DE = CE$	**3.** Segment-Addition Postulate
4. $AC > CE$	**4.** Substitution

The paragraph proof and two-column proof of Example 1 are equivalent. In either form, statements must be ordered and justified.

The remaining theorems are the "heart" of this section. Before studying the theorem and its proof, it is a good idea to read each theorem. Many statements of inequality are intuitive; that is, they are easy to believe even though they may not be easily proved.

Study Theorem 3.5.6 and consider Figure 3.48. It appears that $m\angle C > m\angle B$.

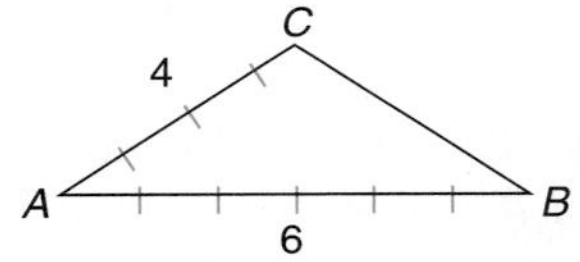

FIGURE 3.48

THEOREM 3.5.6: If one side of a triangle is longer than a second side, then the measure of the angle opposite the longer side is greater than the measure of the angle opposite the shorter side.

EXAMPLE 2

Provide a paragraph proof of Theorem 3.5.6.

Given: $\triangle ABC$, with $AC > BC$ [See Figure 3.49(a).]

Prove: $m\angle B > m\angle A$

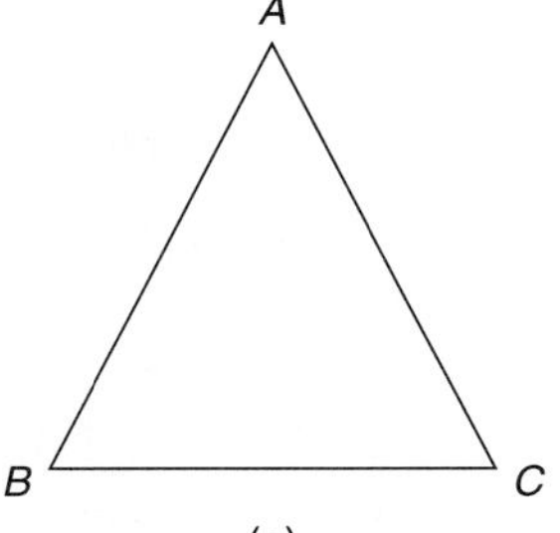

(a)

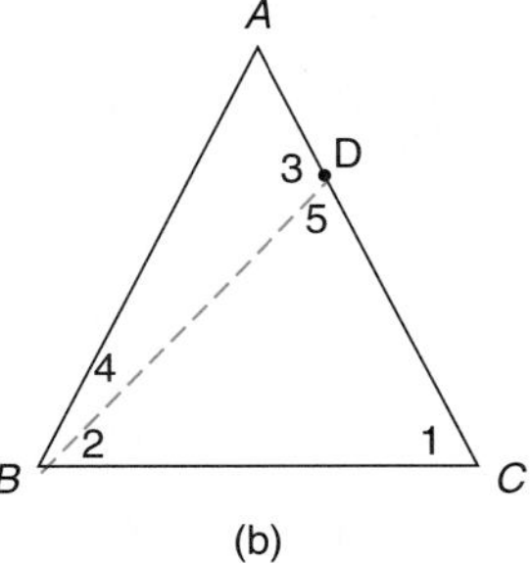

(b)

FIGURE 3.49

Proof: Given $\triangle ABC$ with $AC > BC$, locate point D on $\overline{AC}$ so that $\overline{CD} \cong \overline{BC}$, as in Figure 3.49(b). Now $m\angle 2 = m\angle 5$ in the isosceles triangle, $\triangle BDC$. By Lemma 3.5.2, $m\angle ABC > m\angle 2$; therefore, $m\angle ABC > m\angle 5$ (*) by substitution. But $m\angle 5 > m\angle A$ (*), since $\angle 5$ is an exterior angle of $\triangle ADB$.

Using the two starred statements, we can conclude by the Transitive Property of Inequality that $m\angle ABC > m\angle A$; that is, $m\angle B > m\angle A$ in Figure 3.49(a).

The relationship described in Theorem 3.5.6 extends, of course, to all sides and all angles of a triangle. That is, the largest of the three angles of a triangle is opposite the longest side and the smallest angle is opposite the shortest side.

EXAMPLE 3

Given that the three sides of $\triangle ABC$ are $AB = 4$, $BC = 5$, and $AC = 6$, arrange the angles by size.

Solution

Because $AC > BC > AB$, the largest angle lies opposite $\overline{AC}$, and that is $\angle B$. The angle intermediate in size lies opposite $\overline{BC}$ ($\angle A$), while the smallest angle lies opposite $\overline{AB}$ ($\angle C$). Thus, the order of the angles by size is

$$m\angle B > m\angle A > m\angle C$$

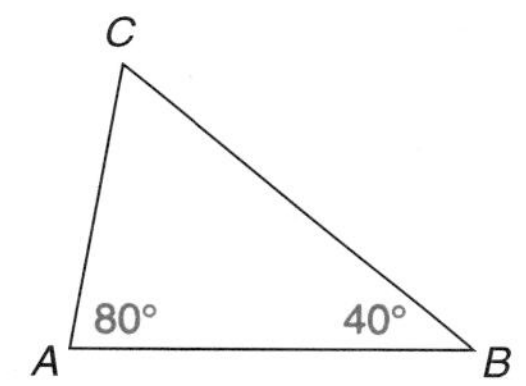

FIGURE 3.50

The converse of Theorem 3.5.6 is also true. It is necessary, however, to use an indirect proof to establish the converse. Recall that this method of proof begins by supposing the exact opposite of what we want to show. Because this assumption leads us to a contradiction, the assumption must be false and the desired claim is therefore true.

Study Theorem 3.5.7 and consider Figure 3.50, in which $m\angle A = 80°$ and $m\angle B = 40°$. It appears that $BC > AC$.

THEOREM 3.5.7: If the measure of one angle of a triangle is greater than the measure of a second angle, then the side opposite the larger angle is longer than the side opposite the smaller angle.

EXAMPLE 4

Prove Theorem 3.5.7 by using an indirect approach in paragraph form.

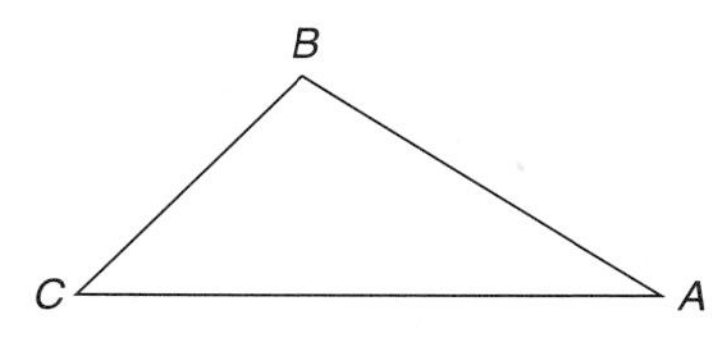

FIGURE 3.51

Given: $\triangle ABC$ with $m\angle B > m\angle A$ (See Figure 3.51.)

Prove: $AC > BC$

Proof: Given $\triangle ABC$ with $m\angle B > m\angle A$, assume that $AC \leq BC$. But if $AC = BC$, then $m\angle B = m\angle A$, which contradicts the hypothesis. Also, if $AC < BC$, then it follows by the previous theorem that $m\angle B < m\angle A$, which also contradicts the hypothesis. Thus the assumed statement must be false, and it follows that $AC > BC$.

EXAMPLE 5

Given $\triangle RST$ in which $m\angle R = 90°$, $m\angle S = 60°$, and $m\angle T = 30°$, write an inequality that compares the lengths of the sides.

Solution

With $m\angle R > m\angle S > m\angle T$, it follows that the sides opposite these $\angle$s are unequal in the same order. That is,

$$ST > RT > SR$$

The following corollary is a consequence of Theorem 3.5.7.

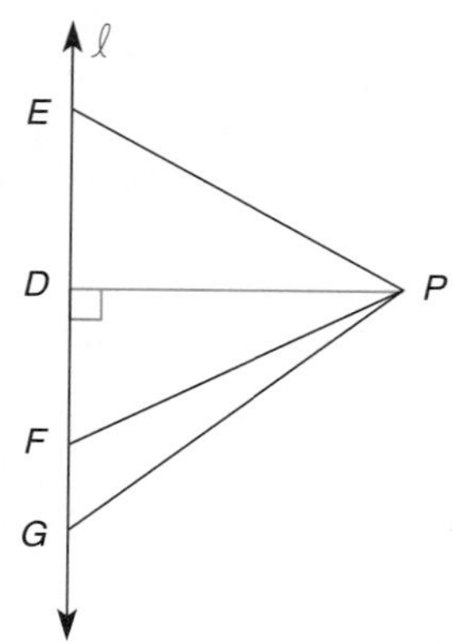

FIGURE 3.52

> COROLLARY 3.5.8: The perpendicular segment from a point to a line is the shortest segment that can be drawn from the point to the line.

In Figure 3.52, $PD < PE$, $PD < PF$, and $PD < PG$. In every case, $\overline{PD}$ is opposite an acute angle of a triangle, while the second segment is always opposite a right angle (necessarily the largest angle of the triangle involved).

Corollary 3.5.8 can easily be extended to three dimensions.

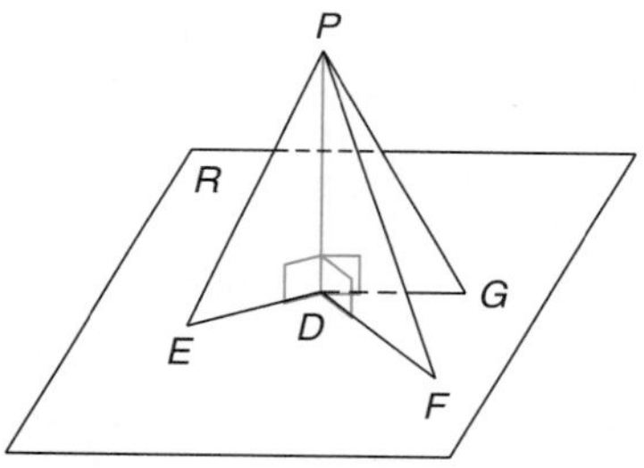

FIGURE 3.53

> COROLLARY 3.5.9: The perpendicular segment from a point to a plane is the shortest segment that can be drawn from the point to the plane.

In Figure 3.53, $\overline{PD}$ has a length less than that of $\overline{PE}$, $\overline{PF}$, or $\overline{PG}$; that is, $\overline{PD}$ is the shortest line segment joining point P to a point in plane R.

Our final theorem shows that no side of a triangle can have a length greater than or equal to the sum of the other two sides. In the proof, the relationship is validated for only one of three possible inequalities. Theorem 3.5.10 is often called the Triangle Inequality. (See Figure 3.54.)

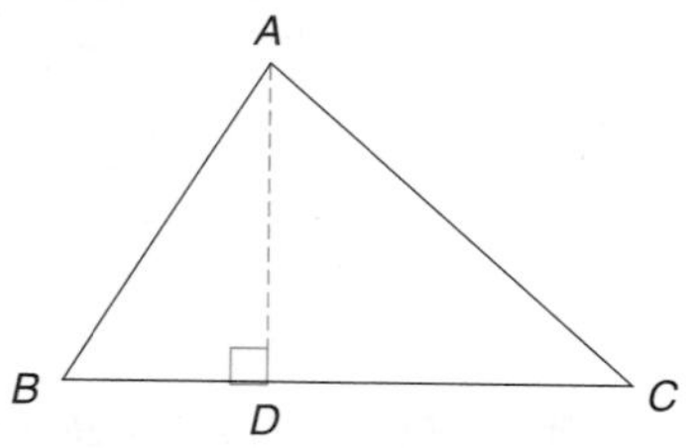

FIGURE 3.54

> THEOREM 3.5.10: **(Triangle Inequality)** The sum of the lengths of any two sides of a triangle is greater than the length of the third side.

Given: $\triangle ABC$

Prove: $BA + CA > BC$

Proof

Draw $\overline{AD} \perp \overline{BC}$. Since the shortest segment from a point to $\overline{AD}$ is the perpendicular segment, $BA > BD$ and $CA > CD$. Using Lemma 3.5.5, we add the inequalities; $BA + CA > BD + CD$. By the Segment-Addition Postulate, the sum $BD + CD$ can be replaced by BC to obtain $BA + CA > BC$.

The following statement is an alternate form of Theorem 3.5.10. If a, b, and c are the lengths of the sides of a triangle and c is the longest side, then $a - b < c < a + b$.

> THEOREM 3.5.10: **(Triangle Inequality)** The length of the longest side of a triangle must be between the sum and difference of the lengths of the other two sides.

EXAMPLE 6

Can a triangle have sides of the following lengths?

a) 3, 4, and 5
b) 3, 4, and 7
c) 3, 4, and 8

Solution

a) Yes, since no side has a length greater than or equal to the sum of the lengths of the other two sides
b) No, because $7 = 3 + 4$
c) No, since $8 > 3 + 4$

From Example 6, you can see that the length of one side cannot be greater than or equal to the sum of the lengths of the other two sides. Considering the alternate form of Theorem 3.5.10, we see that $4 - 3 < 5 < 4 + 3$ in part (a). When 5 is replaced by 7 [as in part (b)] or 8 [as in part (c)], this inequality becomes a false statement.

Our final example illustrates a practical application of inequality relationships in triangles.

EXAMPLE 7

On a map, firefighters are located at points *A* and *B*. A fire has broken out at point *C*. Which group of firefighters is nearer the location of the fire? (See Figure 3.55.)

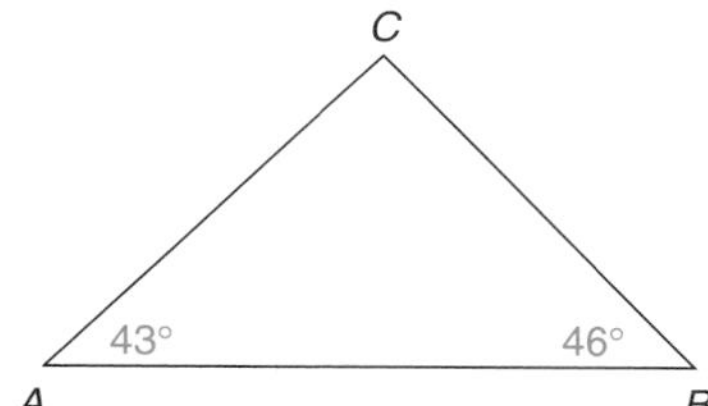

FIGURE 3.55

Solution

With $m\angle A = 43°$ and $m\angle B = 46°$, it follows that $AC > BC$. Since the distance from *B* to *C* is less than the distance from *A* to *C*, the firefighters at site *B* should be sent to the fire located at *C*.

3.5 Exercises

In Exercises 1 to 10, classify each statement as true or false.

1. $\overline{AB}$ is the longest side of $\triangle ABC$.

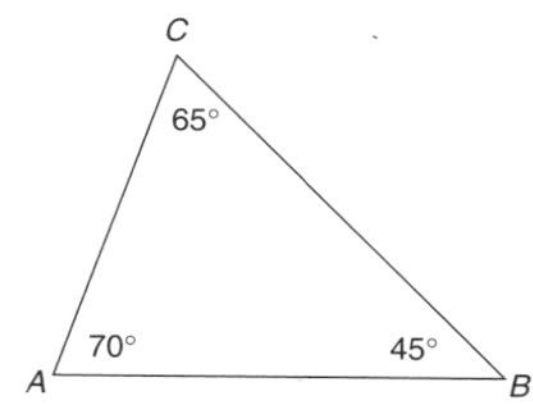

Exercises 1, 2

2. $AB < BC$
3. $DB > AB$

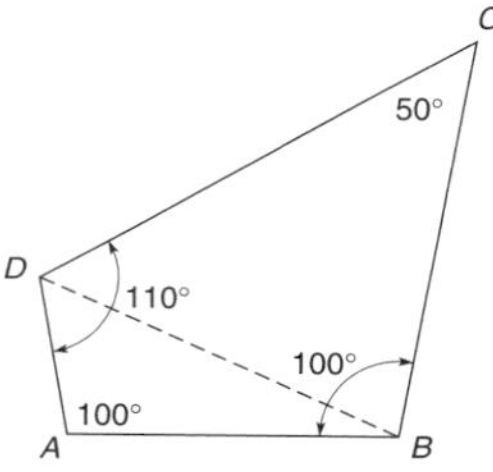

Exercises 3, 4

4. Because $m\angle A = m\angle B$, it follows that $DA = DC$.
5. $m\angle A + m\angle B = m\angle C$

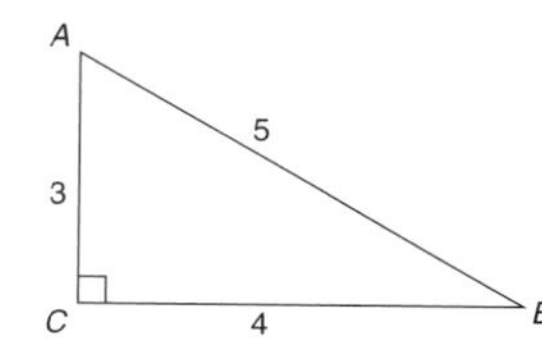

Exercises 5, 6

6. $m\angle A > m\angle B$
7. $DF > DE + EF$

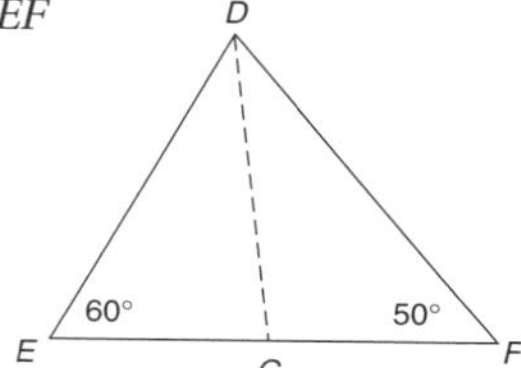

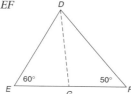

Exercises 7, 8

8. If $\overrightarrow{DG}$ is the bisector of $\angle EDF$, then $DG > DE$.
9. $DA > AC$

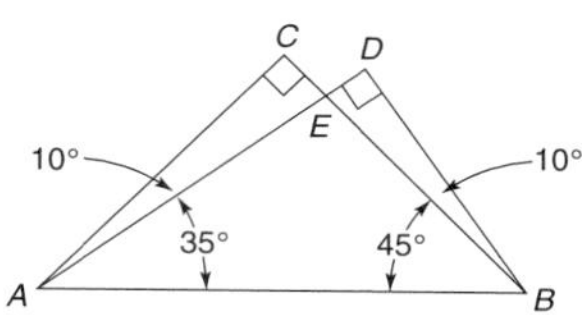

Exercises 9, 10

10. $CE = ED$
11. If possible, draw a triangle whose angles measure:
    - **a)** 100°, 100°, and 60°
    - **b)** 45°, 45°, and 90°
12. If possible, draw a triangle whose sides measure:
    - **a)** 7, 7, and 14
    - **b)** 6, 7, and 14
    - **c)** 6, 7, and 8
13. NASA in Huntsville, Alabama (at point H) has called a manufacturer for parts needed as soon as possible. NASA will, in fact, send a courier for the necessary equipment. The manufacturer has two distribution centers located in nearby Tennessee—one in Nashville (at point N) and the other in Jackson (at point J). Using the angle measurements indicated on the map below, to which town should the courier be dispatched to obtain the needed parts?

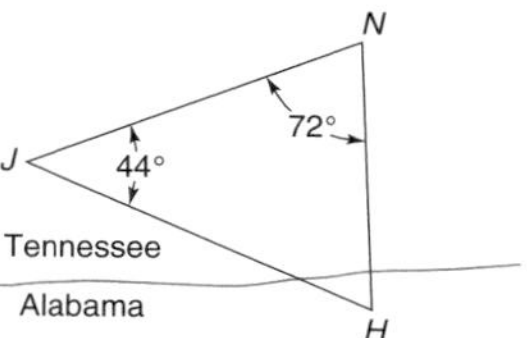

14. A tornado has just struck a small Kansas community at point T. There are Red Cross units stationed in both Salina (at point S) and Wichita (at point W). Using the angle measurements indicated on the map on page 141, which Red Cross unit would reach the victims first? (Assume that units have the same mode of travel available.)

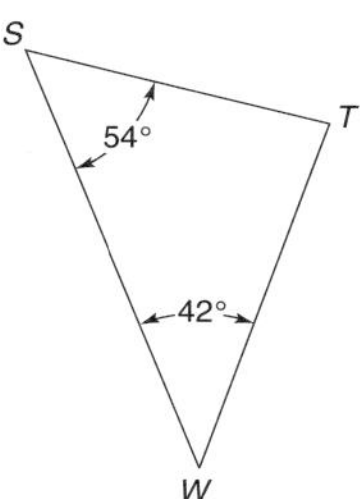

Exercise 14

In Exercises 15 and 16, complete each proof.

15. *Given:* m∠*ABC* > m∠*DBE*
m∠*CBD* > m∠*EBF*
Prove: m∠*ABD* > m∠*DBF*

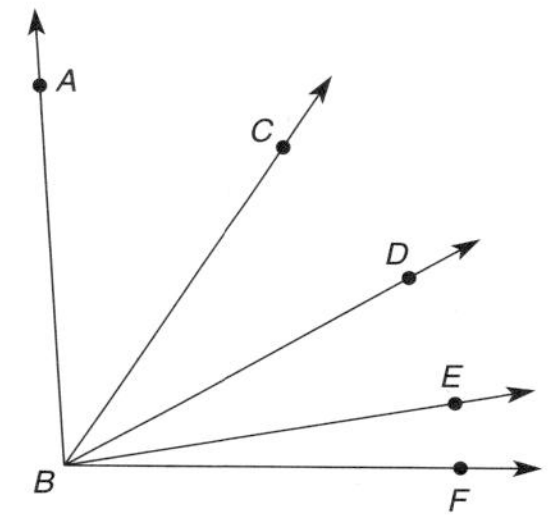

PROOF

Statements	Reasons
1. ?	**1.** Given
2. m∠*ABC* + m∠*CBD* > m∠*DBE* + m∠*EBF*	**2.** Addition Property of Inequality
3. m∠*ABD* = m∠*ABC* + m∠*CBD* and m∠*DBF* = m∠*DBE* + m∠*EBF*	**3.** ?
4. ?	**4.** Substitution

16. *Given:* Equilateral △*ABC* and *D-B-C*
Prove: *DA* > *AC*

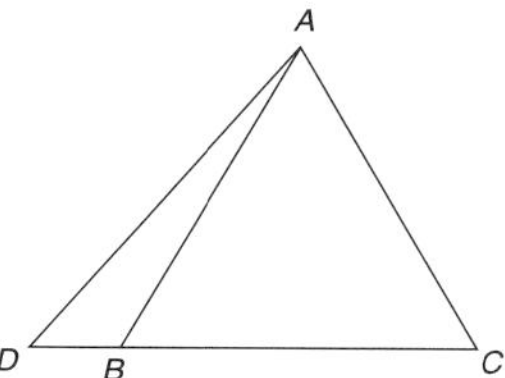

PROOF

Statements	Reasons
1. ?	**1.** Given
2. △*ABC* is equiangular, so m∠*ABC* = m∠*C*	**2.** ?
3. m∠*ABC* > m∠*D* (∠*D* of △*ABD*)	**3.** The measure of an ext. ∠ of a △ is greater than the measure of either nonadjacent int. ∠
4. ?	**4.** Substitution
5. ?	**5.** ?

In Exercises 17 and 18, construct proofs.

17. *Given:* Quadrilateral *RSTU* with diagonal $\overline{US}$
∠*R* and ∠*TUS* are right ∠s
Prove: *TS* > *UR*

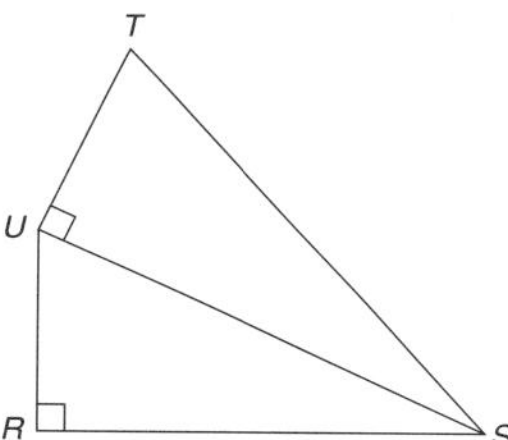

18. *Given:* Quadrilateral *ABCD* with $\overline{AB} \cong \overline{DE}$
Prove: *DC* > *AB*

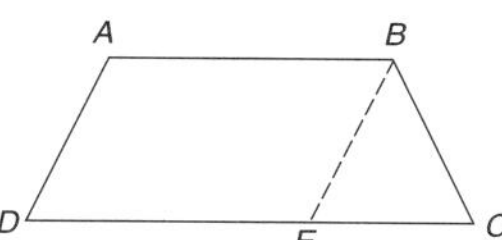

19. For △*ABC* and △*DEF* (not shown), suppose that $\overline{AC} \cong \overline{DF}$ and $\overline{AB} \cong \overline{DE}$, but that m∠*A* < m∠*D*. Draw a conclusion regarding $\overline{BC}$ and $\overline{EF}$.

20. In △*MNP* (not shown), $\overrightarrow{MQ}$ bisects ∠*NMP* and *MN* < *MP*. Draw a conclusion about the relative lengths of $\overline{NQ}$ and $\overline{QP}$.

In Exercises 21 to 24, apply a form of Theorem 3.5.10.

21. The sides of a triangle have lengths of 4, 6, and *x*. Write an inequality that states the possible values of *x*.

22. The sides of a triangle have lengths of 7, 13, and *x*. As in Exercise 21, write an inequality that describes possible values of *x*.

23. If the lengths of two sides of a triangle are represented by $2x + 5$ and $3x + 7$ (in which x is positive), describe in terms of x the possible lengths of the third side whose length is represented by y.

24. Prove by the indirect method: "The diagonal of a square is not equal in length to the length of any of the sides of the square."

25. Prove by the indirect method:
Given: $\triangle MPN$ is not isosceles
Prove: $PM \neq PN$

In Exercises 26 and 27, prove each theorem.

26. The length of the median from the vertex of an isosceles triangle is less than the length of either of the legs.

27. The length of an altitude of a triangle that does not contain a right angle is less than the length of either side containing the same vertex as the altitude.

A Look Beyond: Historical Sketch of Archimedes

While Euclid (see *A Look Beyond,* Chapter 1) was a great teacher and wrote so that the majority might understand the principles of geometry, Archimedes wrote only for the very well-educated mathematicians and scientists of his day. Archimedes (287 B.C.–212 B.C.) wrote on such topics as the measure of the circle, the quadrature of the parabola, and spirals. In his works, Archimedes found a very good approximation of π. His other geometric works included investigations of conic sections and spirals, and he also wrote about physics. He was a great inventor and is probably remembered more for his inventions than for his writing.

Several historical events concerning the life of Archimedes have been substantiated, and one account involves his detection of a dishonest goldsmith. In that story, Archimedes was called upon to determine whether the crown that had been ordered by the king was constructed entirely of gold. By applying the principle of hydrostatics (which he had discovered), Archimedes established that the goldsmith had not constructed the crown entirely of gold. (The principle of hydrostatics states that an object placed in a fluid displaces an amount of fluid equal in weight to the amount of weight the object loses while submerged.)

One of his inventions is known as Archimedes' screw. This device allows water to flow from one level to a higher level so that, for example, holds of ships can be emptied of water. Archimedes' screw was used in Egypt to drain fields when the Nile River overflowed its banks.

When Syracuse (where Archimedes lived) came under siege by the Romans, Archimedes designed a long-range catapult that was so effective that Syracuse was able to fight off the powerful Roman army for three years before being overcome.

One report concerning the inventiveness of Archimedes has been treated as false, because his result has not been duplicated. It was said that he designed a wall of mirrors that could focus and reflect the sun's heat with such intensity as to set fire to Roman ships at sea. Because recent experiments with concave mirrors have failed to produce such intense heat, the account is difficult to believe.

Archimedes eventually died at the hands of a Roman soldier, even though the Roman army had been given orders not to harm him. After his death, the Romans honored his brilliance with a tremendous monument displaying the figure of a sphere inscribed in a right circular cylinder.

Summary

A Look Back at Chapter 3

In this chapter, we considered several methods for proving triangles congruent. Inequality relationships for the sides and angles of a triangle were also investigated.

A Look Ahead to Chapter 4

In the next chapter, we use properties of triangles to develop the properties of quadrilaterals. We consider several special types of quadrilaterals, including the parallelogram, kite, rhombus, and trapezoid.

Important Terms and Concepts of Chapter 3

3.1 Congruent Triangles
SSS, SAS, ASA, AAS
Included Angle, Included Side
Reflexive Property of Congruence (Identity)
Symmetric, Transitive Properties of Congruence

3.2 CPCTC
Hypotenuse and Legs of a Right Triangle
HL

3.3 Isosceles Triangles
Vertex, Legs, and Base of an Isosceles Triangle
Base Angles, Vertex Angle
Angle Bisector
Median
Altitude
Perpendicular Bisector
Auxiliary Line
Determined, Underdetermined, Overdetermined Figures
Equilateral, Equiangular Triangles
Perimeter

3.4 Justifying Constructions

3.5 Lemma
Triangle Inequality

A Look Beyond: *Historical Sketch of Archimedes*

Review Exercises

1. *Given:* $\angle AEB \cong \angle DEC$
$\overline{AE} \cong \overline{ED}$
Prove: $\triangle AEB \cong \triangle DEC$

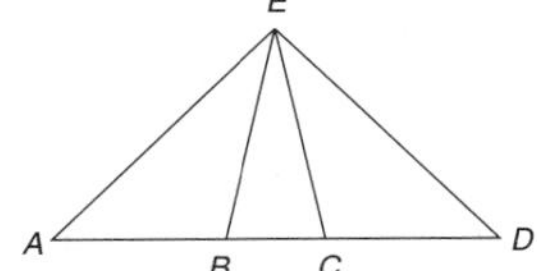

2. *Given:* $\overline{AB} \cong \overline{EF}$
$\overline{AC} \cong \overline{DF}$
$\angle 1 \cong \angle 2$
Prove: $\angle B \cong \angle E$

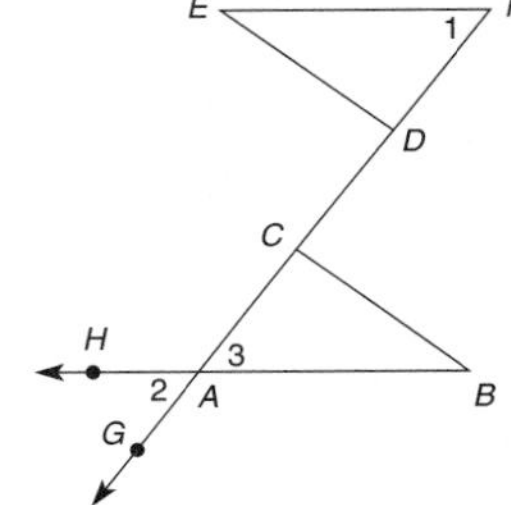

3. *Given:* $\overline{AD}$ bisects $\overline{BC}$
$\overline{AB} \perp \overline{BC}$
$\overline{DC} \perp \overline{BC}$
Prove: $\overline{AE} \cong \overline{ED}$

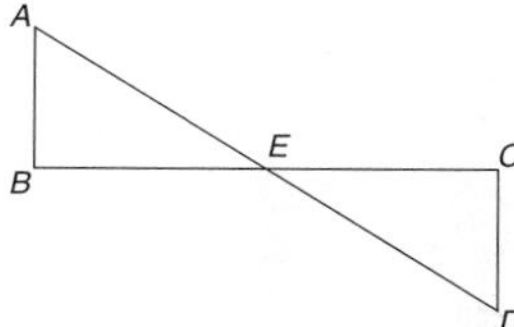

4. *Given:* $\overline{OA} \cong \overline{OB}$
$\overline{OC}$ is the median to $\overline{AB}$
Prove: $\overline{OC} \perp \overline{AB}$

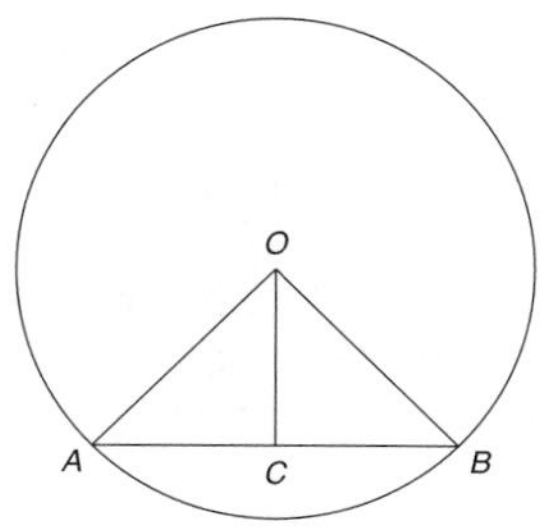

5. *Given:* $\overline{AB} \cong \overline{DE}$
$\overline{AB} \parallel \overline{DE}$
$\overline{AC} \cong \overline{DF}$
Prove: $\overline{BC} \parallel \overline{FE}$

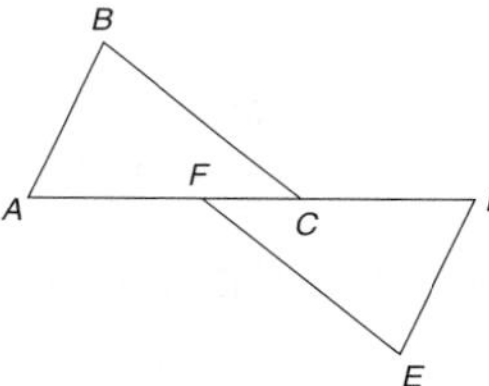

6. *Given:* B is the midpoint of $\overline{AC}$
$\overline{BD} \perp \overline{AC}$
Prove: $\triangle ADC$ is isosceles

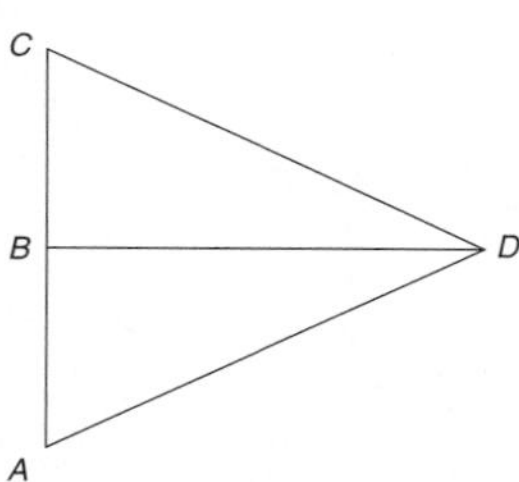

7. *Given:* $\overline{JM} \perp \overline{GM}$ and $\overline{GK} \perp \overline{KJ}$
$\overline{GH} \cong \overline{HJ}$
Prove: $\overline{GM} \cong \overline{JK}$

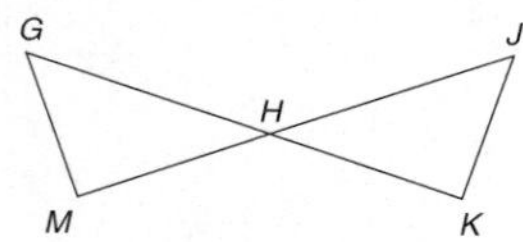

8. *Given:* $\overline{TN} \cong \overline{TR}$
$\overline{TO} \perp \overline{NP}$
$\overline{TS} \perp \overline{PR}$
$\overline{TO} \cong \overline{TS}$
Prove: $\angle N \cong \angle R$

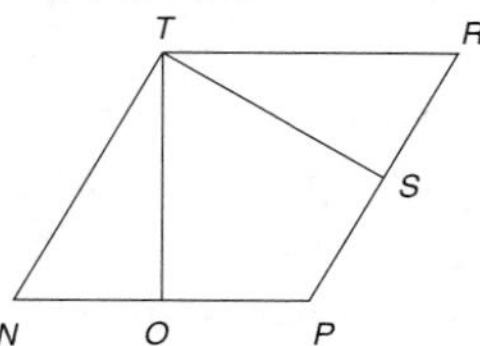

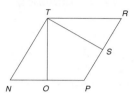

9. *Given:* $\overline{YZ}$ is the base of an isosceles triangle
$\overrightarrow{XA} \parallel \overline{YZ}$
Prove: $\angle 1 \cong \angle 2$

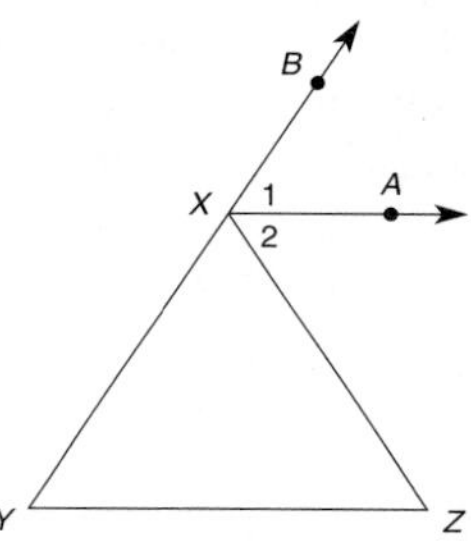

10. *Given:* $\overline{AB} \parallel \overline{DC}$
$\overline{AB} \cong \overline{DC}$
C is the midpoint of $\overline{BE}$
Prove: $\overline{AC} \parallel \overline{DE}$

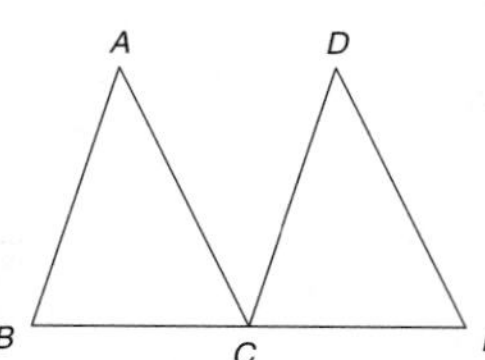

11. *Given:* $\angle BAD \cong \angle CDA$
$\overline{AB} \cong \overline{CD}$
Prove: $\overline{AE} \cong \overline{ED}$
(**HINT:** Prove $\triangle BAD \cong \triangle CDA$ first.)

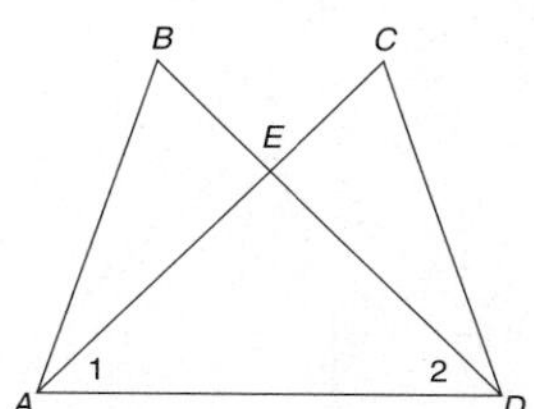

12. *Given:* $\overline{BE}$ is the altitude to $\overline{AC}$
$\overline{AD}$ is the altitude to $\overline{CE}$
$\overline{BC} \cong \overline{CD}$
Prove: $\overline{BE} \cong \overline{AD}$
(**HINT:** Prove $\triangle CBE \cong \triangle CDA$.)

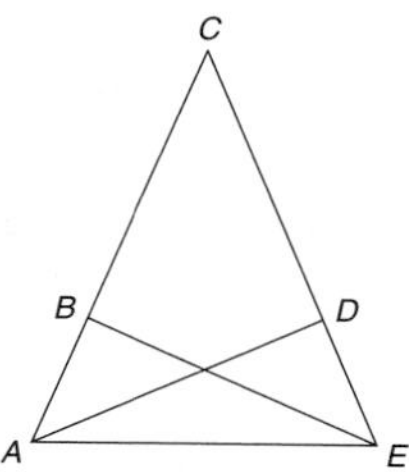

13. *Given:* $\overline{AB} \cong \overline{CD}$
$\angle BAD \cong \angle CDA$
Prove: $\triangle AED$ is isosceles
(**HINT:** Prove $\angle CAD \cong \angle BDA$ by CPCTC.)

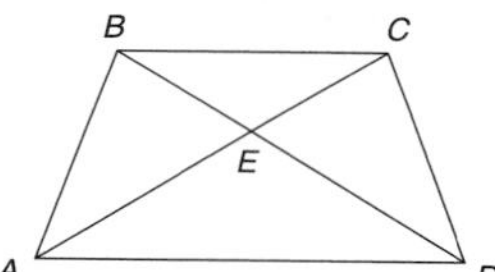

14. *Given:* $\overrightarrow{AC}$ bisects $\angle BAD$
Prove: $AD > CD$

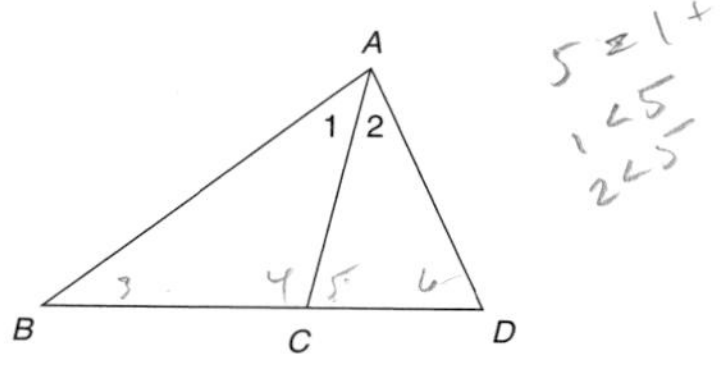

15. In $\triangle PQR$ (not shown), $m\angle P = 67°$ and $m\angle Q = 23°$.

a) Name the shortest side.
b) Name the longest side.

16. In $\triangle ABC$ (not shown), $m\angle A = 40°$ and $m\angle B = 65°$. List the sides in order of their lengths, starting with the smallest side.

17. In $\triangle PQR$ (not shown), $PQ = 1.5$, $PR = 2$, and $QR = 2.5$. List the angles in order of size, starting with the smallest angle.

18. Name the longest line segment in the figure.

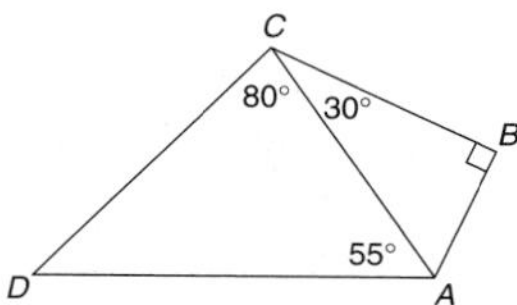

19. Which of the following can be the lengths of the sides of a triangle?

a) 3, 6, 9 **b)** 4, 5, 8 **c)** 2, 3, 8

20. Two sides of a triangle have lengths 15 and 20. The length of the third side can be any number between ___ and ___.

21. *Given:* $\overline{DB} \perp \overline{AC}$
$\overline{AD} \cong \overline{DC}$
$m\angle C = 70°$
Find: $m\angle ADB$

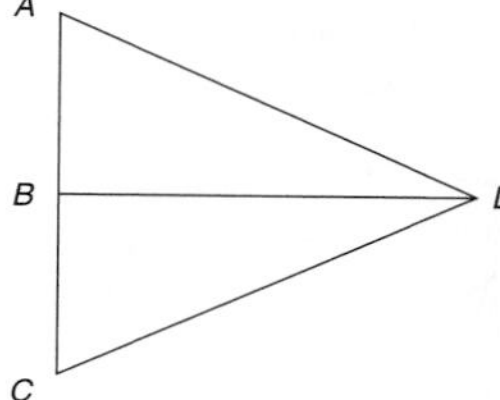

22. *Given:* $\overline{AB} \cong \overline{BC}$
$\angle DAC \cong \angle BCD$
$m\angle B = 50°$
Find: $m\angle ADC$

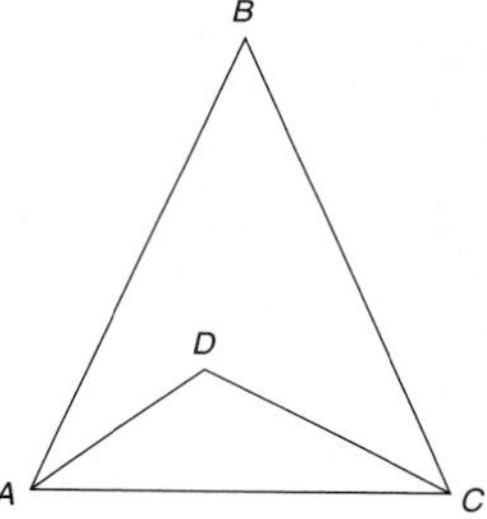

23. *Given:* $\triangle ABC$ is isosceles with base $\overline{AB}$
$m\angle 2 = 3x + 10$
$m\angle 4 = \frac{5}{2}x + 18$
Find: $m\angle C$

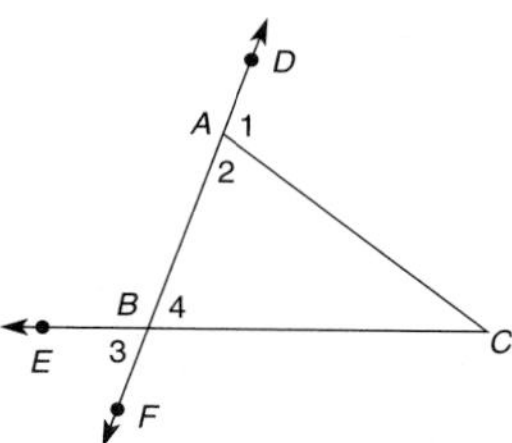

Exercises 23, 24

24. *Given:* $\triangle ABC$ with perimeter 40
$AB = 10$
$BC = x + 6$
$AC = 2x - 3$
Find: Whether $\triangle ABC$ is scalene, isosceles, or equilateral

25. *Given:* $\triangle ABC$ is isosceles with base $\overline{AB}$
$AB = y + 7$
$BC = 3y + 5$
$AC = 9 - y$
Find: Whether $\triangle ABC$ is also equilateral

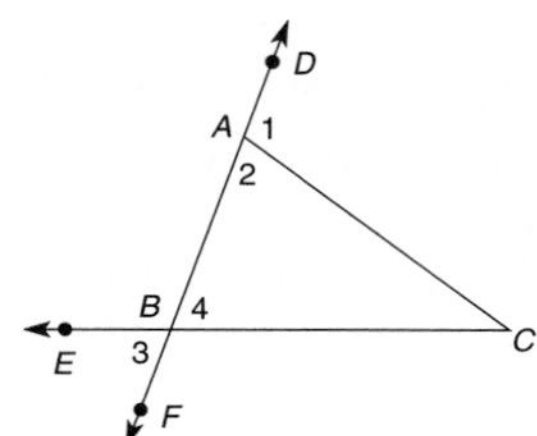

Exercises 25, 26

26. *Given:* $\overline{AC}$ and $\overline{BC}$ are the legs of isosceles $\triangle ABC$
$m\angle 1 = 5x$
$m\angle 3 = 2x + 12$
Find: $m\angle 2$

27. Construct an angle that measures 75°.

28. Construct a right triangle that has acute angle A and hypotenuse of length c.

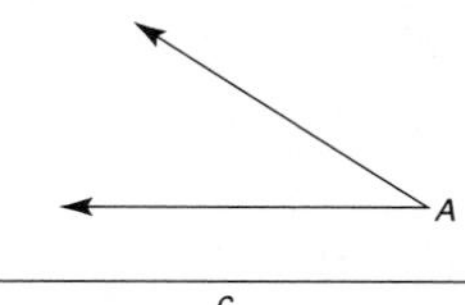

29. Construct a second isosceles triangle in which the base angles are half as large as the base angles of the given isosceles triangle.

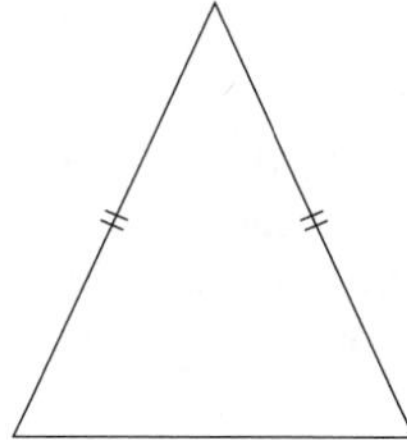

Chapter 4
Quadrilaterals

In the design of most homes and buildings, the geometric shape that dominates is the quadrilateral, a figure with four sides. In this chapter, we consider special types of quadrilaterals and their characteristics. You may already be familiar with some of these quadrilaterals. Do you see any rectangles or squares in the picture below? Two sides of the attached garage take the shape of a trapezoid.

CHAPTER OUTLINE

4.1 Properties of a Parallelogram

A **quadrilateral** is a polygon that has four sides. Unless otherwise stated, the term "quadrilateral" refers to a figure such as *ABCD* in Figure 4.1(a), in which the line segment sides lie within a single plane. When the sides of the quadrilateral are not coplanar, as with *MNPQ* in Figure 4.1(b), the quadrilateral is said to be **skew.** Thus, *MNPQ* is a skew quadrilateral. In this textbook, we deal almost exclusively with quadrilaterals whose sides are coplanar.

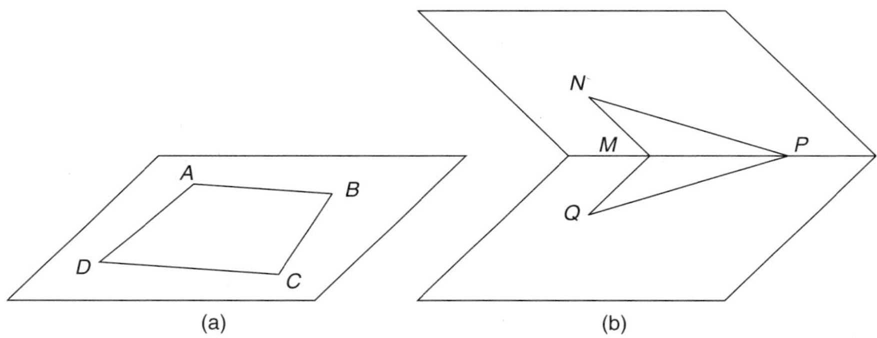

FIGURE 4.1

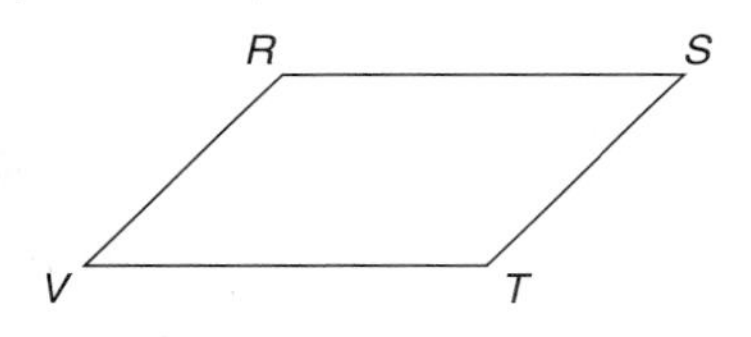

FIGURE 4.2

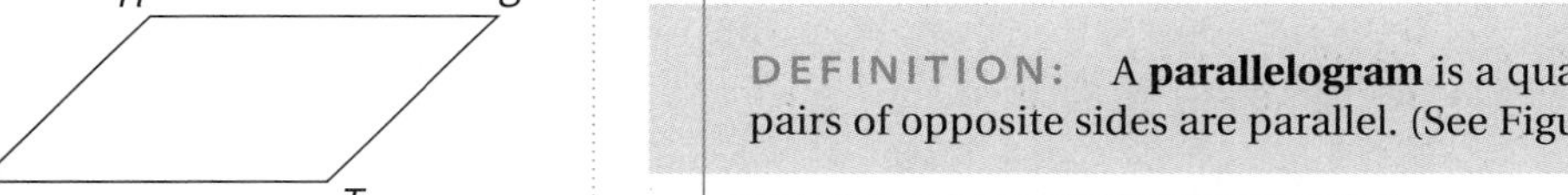

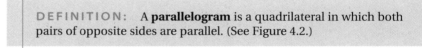

DEFINITION: A **parallelogram** is a quadrilateral in which both pairs of opposite sides are parallel. (See Figure 4.2.)

Because the symbol for parallelogram is ▱, the quadrilateral in Figure 4.2 is ▱*RSTV.*

The following activity guides us toward many of the theorems of this section.

DISCOVER!

From a standard sheet of construction paper, cut out a parallelogram as shown. Then cut along one diagonal. How are the two triangles that are formed related?

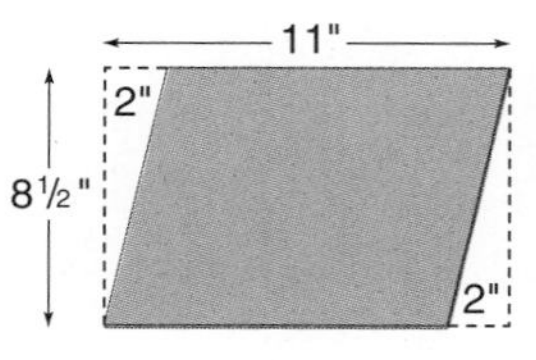

ANSWER

They are congruent.

EXAMPLE 1

Give a formal proof of Theorem 4.1.1.

> **THEOREM 4.1.1:** A diagonal of a parallelogram separates it into two congruent triangles.

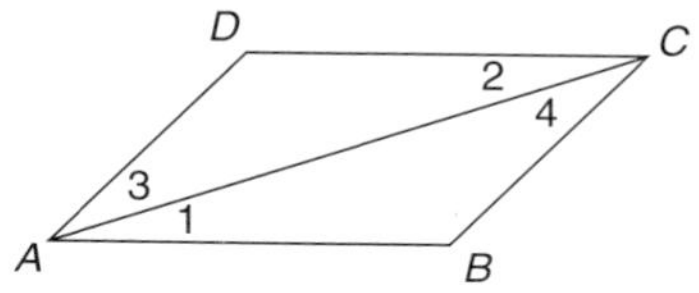

FIGURE 4.3

Given: $\square ABCD$ with diagonal $\overline{AC}$ (See Figure 4.3.)
Prove: $\triangle ACD \cong \triangle CAB$

PROOF

Statements	Reasons
1. $\square ABCD$	**1.** Given
2. $\overline{AB} \parallel \overline{CD}$	**2.** The opposite sides of a $\square$ are $\parallel$ (definition)
3. $\angle 1 \cong \angle 2$	**3.** If two $\parallel$ lines are cut by a transversal, the alternate interior $\angle$s are congruent
4. $\overline{AD} \parallel \overline{BC}$	**4.** Same as reason 2
5. $\angle 3 \cong \angle 4$	**5.** Same as reason 3
6. $\overline{AC} \cong \overline{AC}$	**6.** Identity
7. $\triangle ACD \cong \triangle CAB$	**7.** ASA

Three corollaries of Theorem 4.1.1 follow. The student should make a drawing to illustrate each corollary.

> **COROLLARY 4.1.2:** Opposite angles of a parallelogram are congruent.

> **COROLLARY 4.1.3:** Opposite sides of a parallelogram are congruent.

> **COROLLARY 4.1.4:** Diagonals of a parallelogram bisect each other.

Recall Theorem 2.1.4: "If two parallel lines are cut by a transversal, then the interior angles on the same side of the transversal are supplementary." A corollary of that theorem is stated next.

COROLLARY 4.1.5: Consecutive angles of a parallelogram are supplementary.

Example 2 illustrates the fact that two parallel lines are everywhere equidistant. In general, the phrase "distance between two parallel lines" refers to the length of the perpendicular segment between the two parallel lines.

EXAMPLE 2

FIGURE 4.4

Given: $\overleftrightarrow{AB} \parallel \overleftrightarrow{CD}$
$\overline{AC} \perp \overleftrightarrow{CD}$ and $\overline{BD} \perp \overleftrightarrow{CD}$
(See Figure 4.4.)

Prove: $\overline{AC} \cong \overline{BD}$

PROOF

Statements	Reasons
1. $\overleftrightarrow{AB} \parallel \overleftrightarrow{CD}$	1. Given
2. $\overline{AC} \perp \overleftrightarrow{CD}$ and $\overline{BD} \perp \overleftrightarrow{CD}$	2. Given
3. $\overline{AC} \parallel \overline{BD}$	3. If two lines are $\perp$ to the same line, they are parallel
4. $ABDC$ is a ▱	4. If both pairs of opposite sides of a quadrilateral are $\parallel$, the quadrilateral is a ▱
5. $\overline{AC} \cong \overline{BD}$	5. Opposite sides of a ▱ are congruent

In Example 2, we used the definition of a parallelogram to prove that a particular quadrilateral was a parallelogram, but there are other ways of establishing that a given quadrilateral is a parallelogram. We will investigate those methods in Section 4.2.

Next we consider an inequality relationship for the parallelogram. In order to develop this relationship, we need to investigate an inequality involving two triangles.

In $\triangle ABC$ and $\triangle DEF$ of Figure 4.5, $\overline{AB} \cong \overline{DE}$ and $\overline{BC} \cong \overline{EF}$. If $m\angle B > m\angle E$, then $AC > DF$.

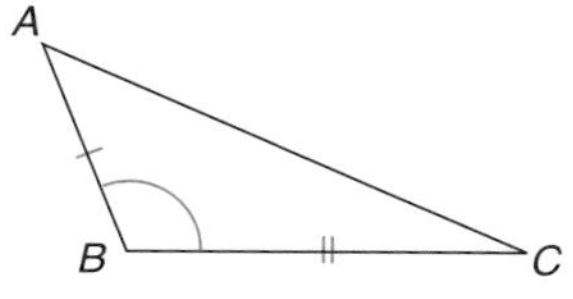

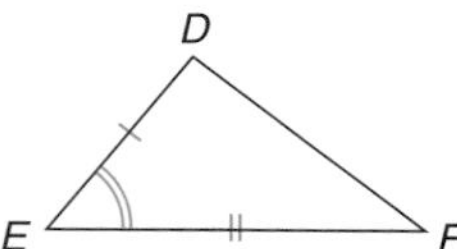

FIGURE 4.5

We will use, but not prove, the following relationship:

Given: $\overline{AB} \cong \overline{DE}$ and
$\overline{BC} \cong \overline{EF}$; $m\angle B > m\angle E$ (See Figure 4.5.)

Prove: $AC > DF$

The corresponding theorem follows:

DISCOVER!

Draw $\square ABCD$ so that $m\angle A > m\angle B$. Which diagonal has the greater length?

ANSWER
$\overline{BD}$

THEOREM 4.1.6: If two sides of one triangle are congruent to two sides of a second triangle and the included angle of the first triangle is greater than the included angle of the second, then the length of the side opposite the included angle of the first triangle is greater than the length of the side opposite the included angle of the second.

Now we can compare the lengths of the diagonals of a parallelogram. For a parallelogram having no right angles, two consecutive angles are unequal but supplementary; thus, one angle of the parallelogram will be acute while the next angle will be obtuse. In Figure 4.6(a), $\square ABCD$ has acute angle A and obtuse angle D. In Figure 4.6(b), diagonal $\overline{AC}$ lies opposite the obtuse angle ADC in $\triangle ACD$ and diagonal $\overline{BD}$ lies opposite the acute angle DAB in $\triangle ABD$. In Figures 4.6(c) and (d), we have taken $\triangle ACD$ and $\triangle ABD$ from $\square ABCD$ of Figure 4.6(b). In the two triangles, the pairs of sides that include $\angle ADC$ and $\angle DAB$ are congruent by Identity ($\overline{AD} \cong \overline{AD}$) and the fact that opposite sides of a parallelogram are congruent ($\overline{DC} \cong \overline{AB}$). Using Theorem 4.1.6, we see that $AC > BD$ because $\overline{AC}$ lies opposite an obtuse angle of $\triangle ADC$ while $\overline{BD}$ lies opposite an acute angle of $\triangle DAB$.

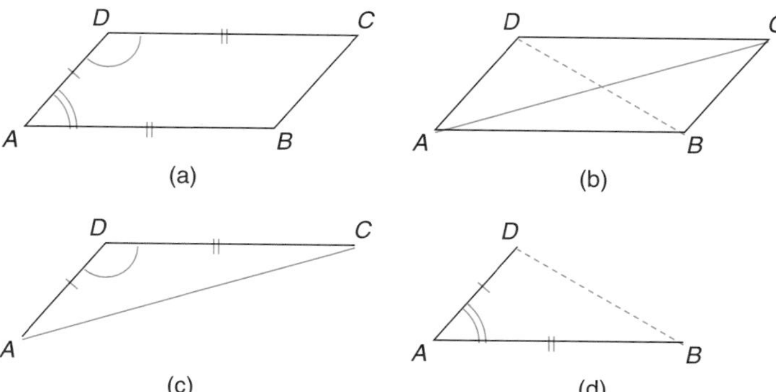

FIGURE 4.6

Based upon the preceding discussion, we have the following theorem.

THEOREM 4.1.7: In a parallelogram with unequal pairs of consecutive angles, the longer diagonal lies opposite the obtuse angle.

EXAMPLE 3

In parallelogram $RSTV$ (not shown), $m\angle R = 67°$.

a) Find the measure of $\angle S$.
b) Determine which diagonal ($\overline{RT}$ or $\overline{SV}$) has the greater length.

Solution

a) $m\angle S = 180° - 67° = 113°$
b) Since $\angle S$ is obtuse, the diagonal opposite this angle is longer; that is, $\overline{RT}$ is the longer diagonal.

We use an indirect approach to solve Example 4.

EXAMPLE 4

In parallelogram $ABCD$ (not shown), $\overline{AC}$ and $\overline{BD}$ are diagonals and $AC > BD$. Determine which angles of the parallelogram are obtuse and which angles are acute.

Solution

Since the longer diagonal $\overline{AC}$ lies opposite angles B and D, these angles are obtuse. The remaining angles A and C are necessarily acute.

Our next example uses algebra to relate angle sizes and diagonal lengths.

EXAMPLE 5

In $\square MNPQ$ in Figure 4.7, $m\angle M = 2(x + 10)$ and $m\angle Q = 3x - 10$. Determine which diagonal would be longer, $\overline{QN}$ or $\overline{MP}$.

Solution

Consecutive angles M and Q are supplementary, so $m\angle M + m\angle Q = 180$.

FIGURE 4.7

$$2(x + 10) + (3x - 10) = 180$$
$$2x + 20 + 3x - 10 = 180$$
$$5x + 10 = 180 \rightarrow 5x = 170 \rightarrow x = 34$$

Then $m\angle M = 2(34 + 10) = 88°$ while $m\angle Q = 3(34) - 10 = 92°$. Because $m\angle Q > m\angle M$, diagonal $\overline{MP}$ (opposite $\angle Q$) would be longer than $\overline{QN}$.

Speed and Direction of Aircraft

For the application in Example 6, we indicate the velocity of an airplane or the wind by drawing a directed arrow. In each case, a scale is used on a grid in which a North-South line meets an East-West line at right angles. Consider the sketches in Figure 4.8 and their explanations.

In some scientific applications, such as Example 6, a parallelogram can be used to determine the solution to the problem. For instance, the Parallelogram Law enables us to determine the resulting speed and direction of an airplane when the velocity of the airplane and of the wind are considered together. In Figure 4.9, the arrows representing the two velocities are placed head-to-tail from the point of origin. Because the order of the two velocities is not important, the drawing leads to a parallelogram. In the parallelogram, it is the length and direction of the diagonal that solves the problem. In Example 6, it is very important to be accurate when scaling the drawing. Otherwise, the ruler and protractor will give poor results.

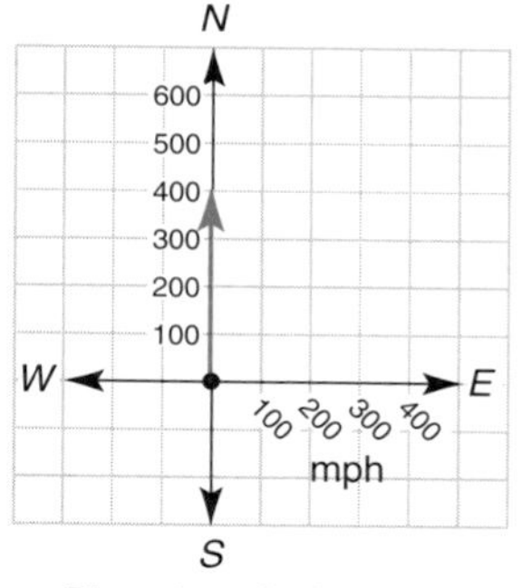

Plane travels due north at 400 mph

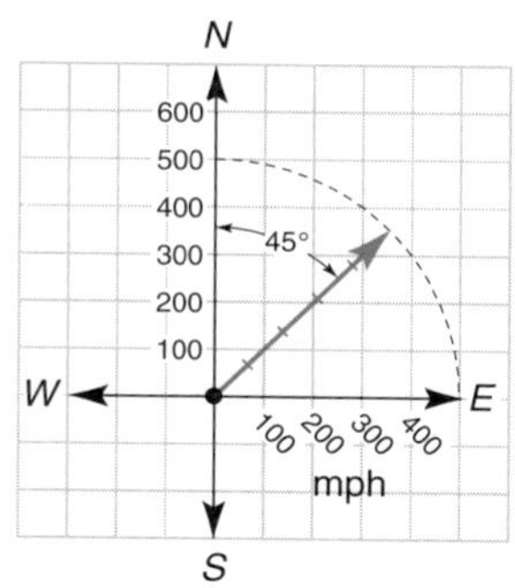

Plane travels at 500 mph in the direction N 45° E

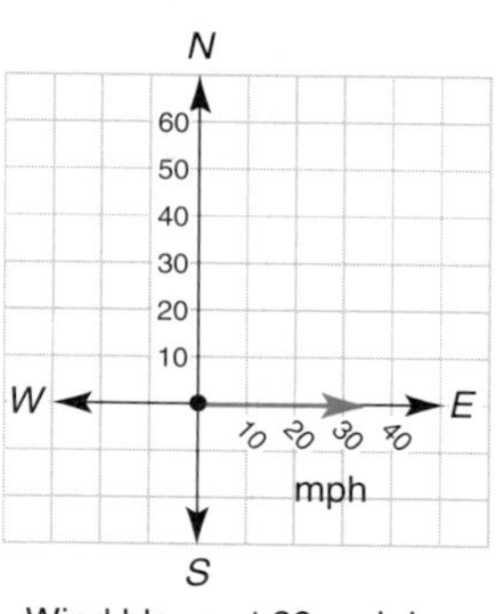

Wind blows at 30 mph in the direction west to east

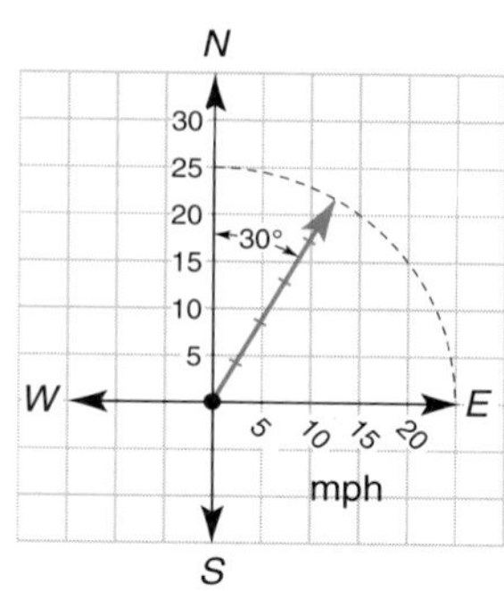

Wind blows at 25 mph in the direction N 30° E

FIGURE 4.8

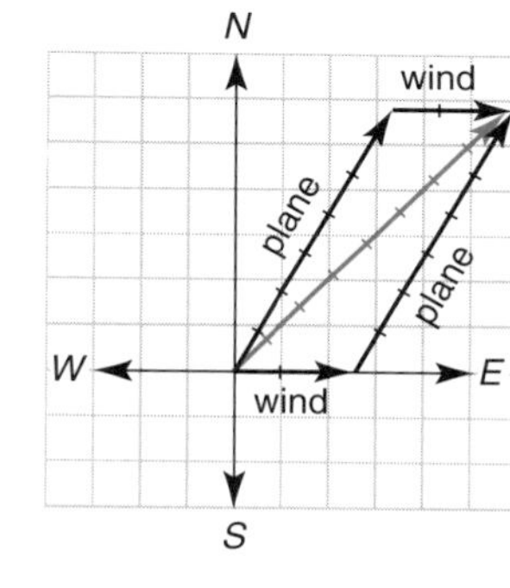

FIGURE 4.9

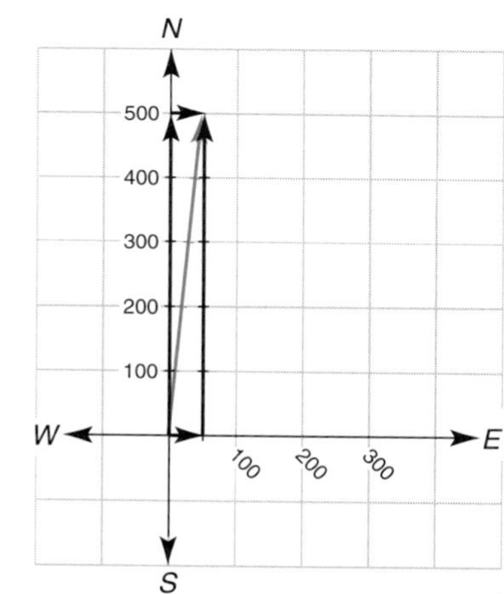

FIGURE 4.10

NOTE: In Example 6, kph means *kilometers per hour.*

EXAMPLE 6

An airplane travels due north at 500 kph. If the wind blows at 50 kph from west to east, what is the resulting speed and direction of the plane?

Solution

Using a ruler to measure the diagonal of the parallelogram, the length corresponds to a speed of approximately 505 kph. Using a protractor, the direction is approximately N 6° E. (See Figure 4.10.)

NOTE: By methods introduced in later sections, the actual speed is approximately 502.5 kph and the direction is N 5.7° E.

4.1 Exercises

1. *ABCD* is a parallelogram.
   a) Using a ruler, compare the lengths of sides $\overline{AB}$ and $\overline{DC}$.
   b) Using a ruler, compare the lengths of sides $\overline{AD}$ and $\overline{BC}$.

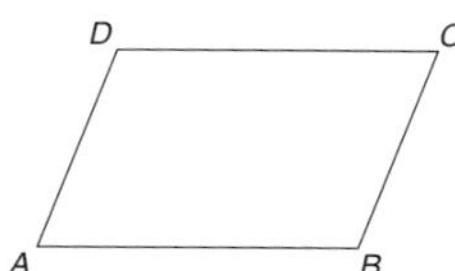

Exercises 1, 2

2. *ABCD* is a parallelogram.
   a) Using a protractor, compare the measures of $\angle A$ and $\angle C$.
   b) Using a protractor, compare the measures of $\angle B$ and $\angle D$.
3. *MNPQ* is a parallelogram. Suppose that $MQ = 5$, $MN = 8$, and $m\angle M = 110°$. Find:
   a) QP b) NP
   c) $m\angle Q$ d) $m\angle P$

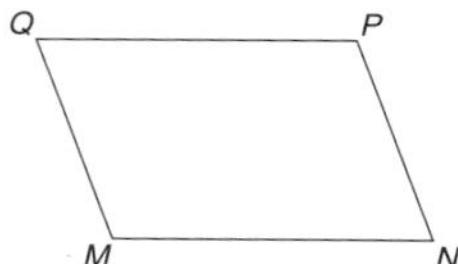

Exercises 3, 4

4. *MNPQ* is a parallelogram. Suppose that $MQ = 12.7$, $MN = 17.9$, and $m\angle M = 122°$. Find:
   a) QP b) NP
   c) $m\angle Q$ d) $m\angle P$
5. Given that $AB = 3x + 2$, $BC = 4x + 1$, and $CD = 5x - 2$, find the length of each side of $\square ABCD$.
6. Given that $m\angle A = 2x + 3$ and $m\angle C = 3x - 27$, find the measure of each angle of $\square ABCD$.
7. Given that $m\angle A = 2x + 3$ and $m\angle B = 3x - 23$, find the measure of each angle of $\square ABCD$.

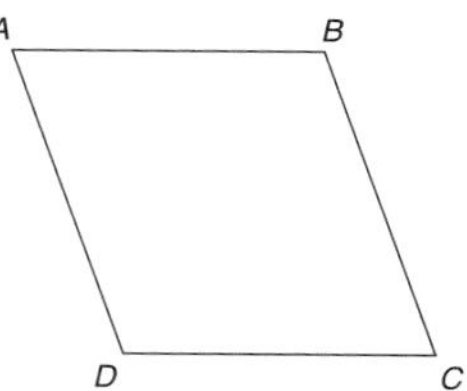

Exercises 5–8

8. Given that $m\angle A = 2x + y$, $m\angle B = 2x + 3y - 20$, and $m\angle C = 3x - y + 16$, find the measure of each angle of $\square ABCD$.
9. In quadrilateral *RSTV*, the midpoints of consecutive sides are joined in order. Try drawing other quadrilaterals and joining their midpoints. What can you conclude about the resulting quadrilateral in each case?

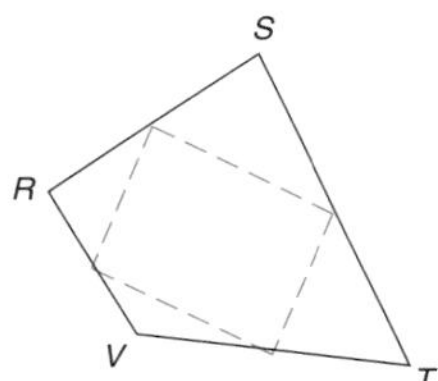

10. In quadrilateral *ABCD*, the midpoints of opposite sides are joined to form two intersecting segments. Try drawing other quadrilaterals and join their opposite midpoints. What can you conclude about these segments in each case?

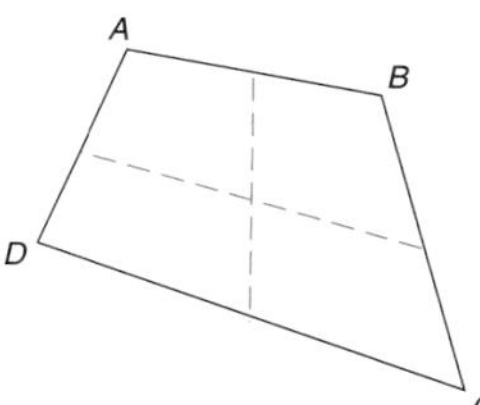

11. Quadrilateral *ABCD* has $\overline{AB} \cong \overline{DC}$ and $\overline{AD} \cong \overline{BC}$. Using intuition, what type of quadrilateral is *ABCD*?

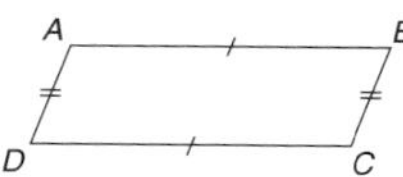

12. Quadrilateral *RSTV* has $\overline{RS} \cong \overline{TV}$ and $\overline{RS} \parallel \overline{TV}$. Using intuition, what type of quadrilateral is *RSTV*?

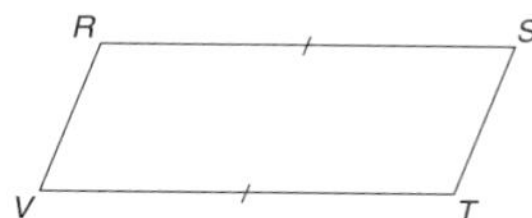

In Exercises 13 to 16, use the definition of parallelogram to complete each proof.

13. *Given:* $\overline{RS} \parallel \overline{VT}$, $\overline{RV} \perp \overline{VT}$, and $\overline{ST} \perp \overline{VT}$
Prove: *RSTV* is a parallelogram

PROOF

Statements	Reasons
1. $\overline{RS} \parallel \overline{VT}$	**1.** ?
2. ?	**2.** Given
3. ?	**3.** If two lines are ⊥ to the same line, they are ∥ to each other
4. ?	**4.** If both pairs of opposite sides of a quadrilateral are ∥, the quad. is a ▱

14. *Given:* $\overline{WX} \parallel \overline{ZY}$ and ∠s *Z* and *Y* are supplementary
Prove: *WXYZ* is a parallelogram

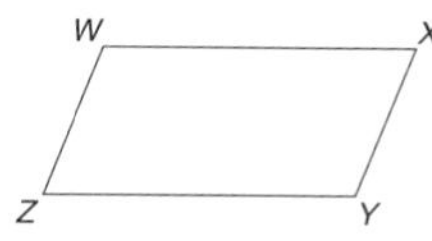

PROOF

Statements	Reasons
1. $\overline{WX} \parallel \overline{ZY}$	**1.** ?
2. ?	**2.** Given
3. ?	**3.** If two lines are cut by a transversal so that int. ∠s on the same side of the trans. are supplementary, these lines are ∥
4. ?	**4.** If both pairs of opposite sides of a quadrilateral are ∥, the quad. is a ▱.

15. *Given:* Parallelogram *RSTV*; also $\overline{XY} \parallel \overline{VT}$
Prove: $\angle 1 \cong \angle S$
Plan: First show that *RSYX* is a parallelogram.

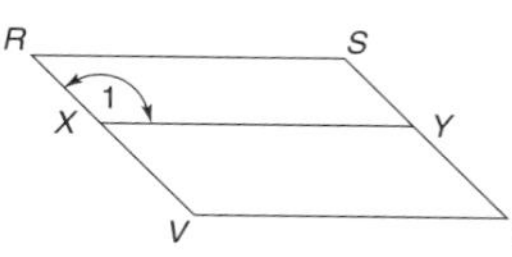

16. *Given:* Parallelogram *ABCD* with $\overline{DE} \perp \overline{AB}$ and $\overline{FB} \perp \overline{AB}$
Prove: $\overline{DE} \cong \overline{FB}$
Plan: First show that *DEBF* is a parallelogram.

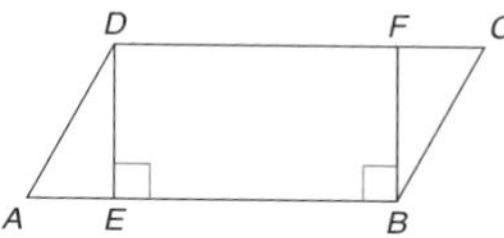

In Exercises 17 to 20, write a formal proof of each theorem (corollary).

17. The opposite angles of a parallelogram are congruent.

18. The opposite sides of a parallelogram are congruent.

19. The diagonals of a parallelogram bisect each other.

20. The consecutive angles of a parallelogram are supplementary.

21. The bisectors of two consecutive angles of ▱*HJKL* are shown. What can you conclude about ∠*P*?

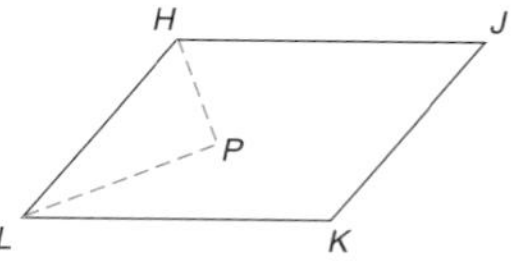

22. When the bisectors of two consecutive angles of a parallelogram meet at a point on the remaining side, what may you conclude about △*DEC*? About △*ADE*? About △*BCE*?

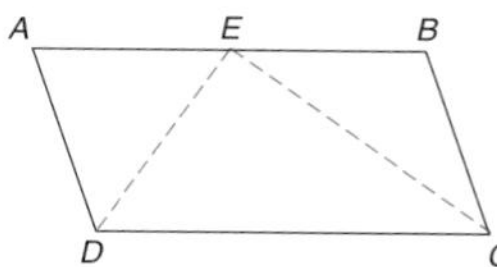

23. Draw parallelogram *RSTV* with $m\angle R = 70°$ and $m\angle S = 110°$. Which diagonal of ▱*RSTV* has the greater length?

24. Draw parallelogram *RSTV* so that the diagonals have the lengths $RT = 5$ and $SV = 4$. Which two angles of ▱*RSTV* have the greater measure?

25. The following problem is based upon the Parallelogram Law. In the scaled drawing, each unit corresponds to 50 miles per hour. A small airplane travels due east at 250 mph. The wind is blowing at 50 mph in the direction due north. Using the scale provided, determine the approximate length of the indicated diagonal and use it to determine the speed of the airplane in miles per hour.

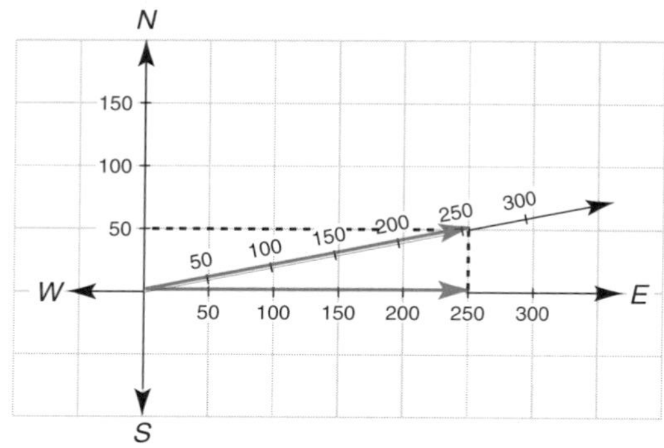

Exercises 25, 26

26. In the drawing, the bearing (direction) in which the airplane travels is described as North x degrees East, where x is the measure of the angle from the north axis toward the east axis. Using a protractor, find the approximate bearing of the airplane.

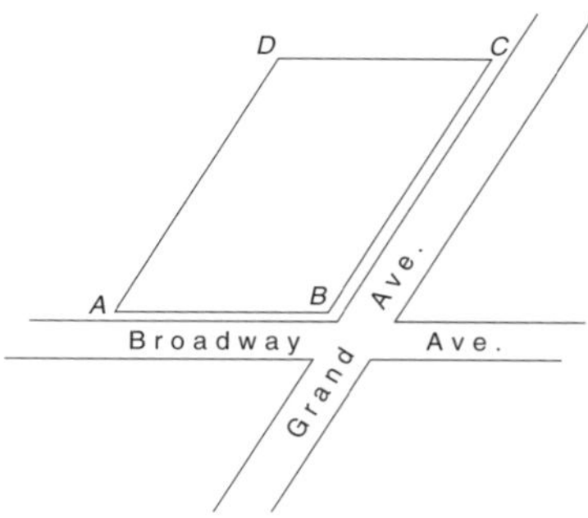

Exercises 27, 28

27. Two streets meet to form an obtuse angle at point B. On that corner, the newly poured foundation for a building takes the shape of a parallelogram. Which diagonal, $\overline{AC}$ or $\overline{BD}$, is longer?

28. To test the accuracy of the foundation's measurements, lines (strings) are joined from opposite corners of the building's foundation. How should the strings that are represented by $\overline{AC}$ and $\overline{BD}$ be related?

4.2 The Parallelogram and Kite

The quadrilaterals discussed in this section have two pairs of congruent sides.

The Parallelogram

Because the hypothesis of each theorem in Section 4.1 included a given parallelogram, our goal was to develop the properties of parallelograms. In this section, Theorems 4.2.1–4.2.3 take the form, "If . . . , then this quadrilateral is a parallelogram." In this section, we find that quadrilaterals having certain characteristics must be parallelograms. For instance, one set of characteristics that determines a parallelogram is "a quadrilateral has one pair of *opposite* sides that are both congruent and parallel."

EXAMPLE 1

Give a formal proof of Theorem 4.2.1.

> **THEOREM 4.2.1:** If two sides of a quadrilateral are both congruent and parallel, then the quadrilateral is a parallelogram.

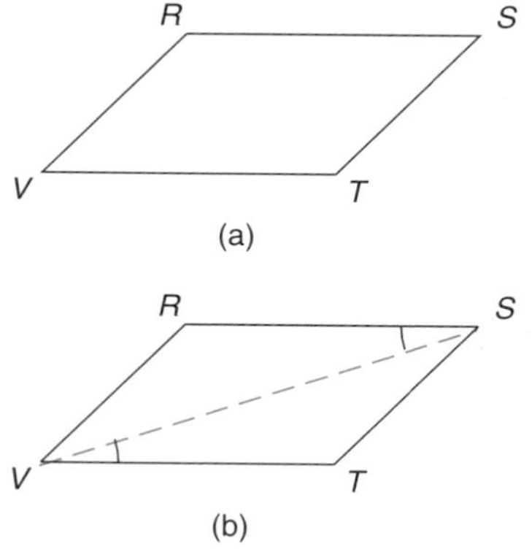

FIGURE 4.11

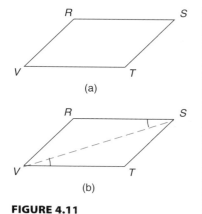

Given: In Figure 4.11(a), $\overline{RS} \parallel \overline{VT}$ and $\overline{RS} \cong \overline{VT}$
Prove: $RSTV$ is a ▱

PROOF

Statements	Reasons
1. $\overline{RS} \parallel \overline{VT}$ and $\overline{RS} \cong \overline{VT}$	1. Given
2. Draw diagonal $\overline{VS}$, as in Figure 4.11(b)	2. Exactly one line passes through two points
3. $\overline{VS} \cong \overline{VS}$	3. Identity
4. $\angle RSV \cong \angle SVT$	4. If two ∥ lines are cut by a transversal, alternate interior ∠s are ≅
5. $\triangle RSV \cong \triangle TVS$	5. SAS
6. $\therefore \angle RVS \cong \angle VST$	6. CPCTC
7. $\overline{RV} \parallel \overline{ST}$	7. If two lines are cut by a transversal so that alternate interior ∠s are ≅, these lines are ∥
8. $RSTV$ is a ▱	8. If both pairs of opposite sides of a quadrilateral are ∥, the quadrilateral is a parallelogram

Consider the following activity. Through it, we discover another type of quadrilateral that must be a parallelogram.

DISCOVER!

Take two straws and cut each straw into two pieces so that the lengths of the pieces of one straw match those of the second. Now form a quadrilateral by placing the pieces end-to-end, but so that congruent sides lie in opposite positions. What type of quadrilateral is always formed?

ANSWER
A parallelogram

The preceding activity led to a parallelogram and also leads to the following theorem. Proof of the theorem is left to the student.

THEOREM 4.2.2: If both pairs of opposite sides of a quadrilateral are congruent, then it is a parallelogram.

Another quality of quadrilaterals that determines a parallelogram is stated in Theorem 4.2.3. Its proof is also left to the student.

THEOREM 4.2.3: If the diagonals of a quadrilateral bisect each other, then the quadrilateral is a parallelogram.

When a figure is drawn to represent the hypothesis of a theorem, we should not include more conditions than the hypothesis states. For instance, if we drew two diagonals that not only bisected each other but were also of equal lengths, the quadrilateral would be the special type of parallelogram known as a **rectangle.** We will deal with rectangles in the next section.

The Kite

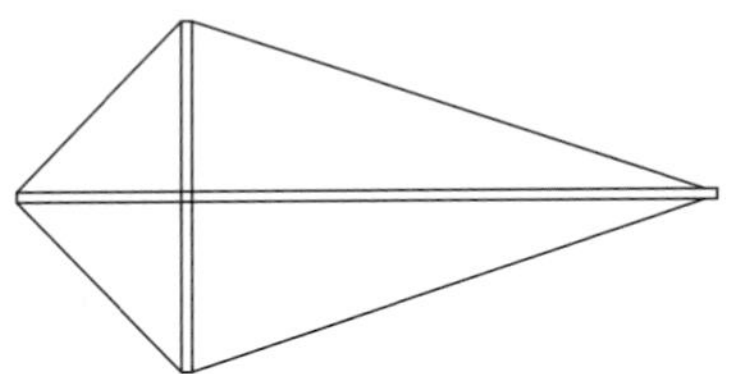

FIGURE 4.12

The next quadrilateral we consider is known as a *kite.* This quadrilateral gets its name from the child's toy pictured in Figure 4.12. In the construction of the kite, there are two pairs of congruent *adjacent* sides. This leads to the formal definition of a kite.

DEFINITION: A **kite** is a quadrilateral with two distinct pairs of congruent adjacent sides.

THEOREM 4.2.4: In a kite, one pair of opposite angles is congruent.

In Example 2, we verify Theorem 4.2.4 by proving that $\angle B \cong \angle D$. With congruent sides as marked, $\angle A \not\cong \angle C$.

EXAMPLE 2

Complete the proof of Theorem 4.2.4.

Given: Kite *ABCD* with congruent sides as marked. [See Figure 4.13(a).]

Prove: $\angle B \cong \angle D$

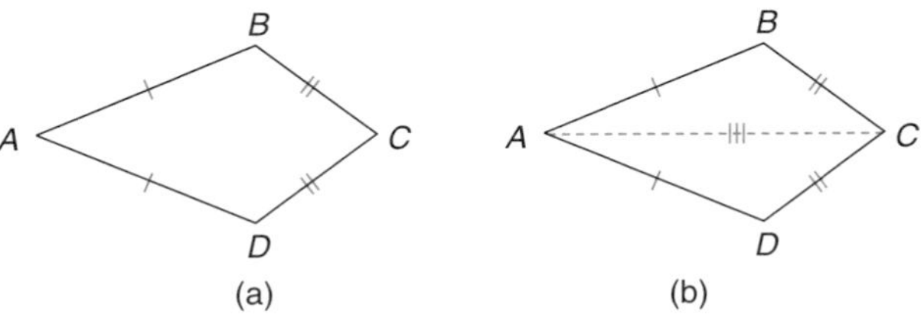

FIGURE 4.13

PROOF

Statements	Reasons
1. Kite $ABCD$	**1.** ?
2. $\overline{BC} \cong \overline{CD}$ and $\overline{AB} \cong \overline{AD}$	**2.** A kite has two pairs of ≅ adjacent sides
3. Draw $\overline{AC}$ [Figure 4.13(b)]	**3.** Through two points, there is exactly one line
4. $\overline{AC} \cong \overline{AC}$	**4.** ?
5. $\triangle ACD \cong \triangle ACB$	**5.** ?
6. ?	**6.** CPCTC

Two additional theorems involving the kite are found in Exercises 25 and 26 of this section.

When observing an old barn or shed, we often see that it has begun to lean. While a triangle is rigid in shape [Figure 4.14(a)] and bends only when broken, a quadrilateral [Figure 4.14(b)] does *not* provide the same level of strength and stability. In observing the construction of a house, bridge, or building, note the use of wooden or metal triangles as braces. The brace in the swing set in Figure 4.14(c) suggests the following theorem.

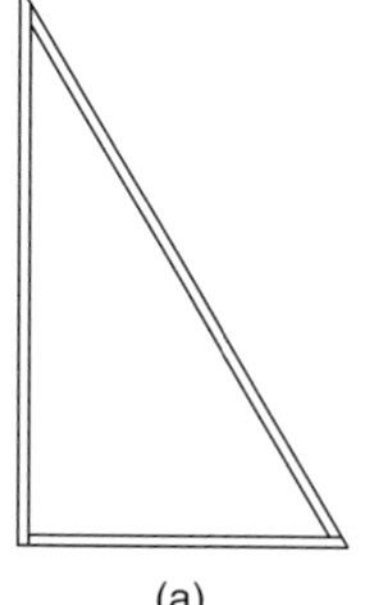

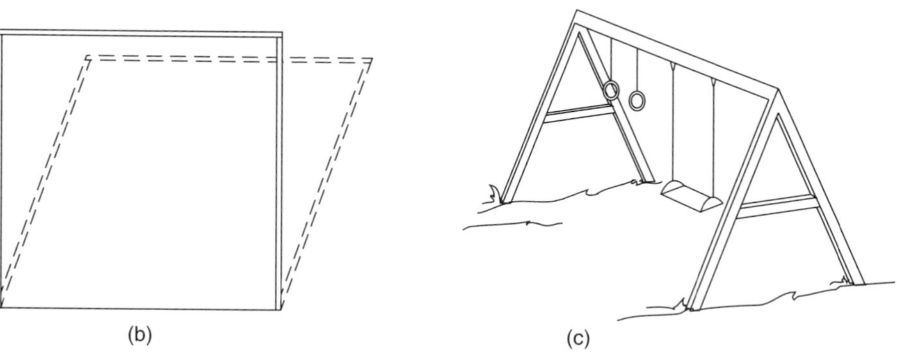

FIGURE 4.14

> **THEOREM 4.2.5:** The segment that joins the midpoints of two sides of a triangle is parallel to the third side and has a length equal to one-half the length of the third side.

Refer to Figure 4.15(a); Theorem 4.2.5 claims that $\overline{MN} \parallel \overline{BC}$ and $MN = \frac{1}{2}(BC)$. We will prove the first part of this theorem but leave the second part as an exercise.

> The segment that joins the midpoints of two sides of a triangle is parallel to the third side of the triangle.

Given: In Figure 4.15(a), $\triangle ABC$ with M and N the midpoints of $\overline{AB}$ and $\overline{AC}$
Prove: $\overline{MN} \parallel \overline{BC}$

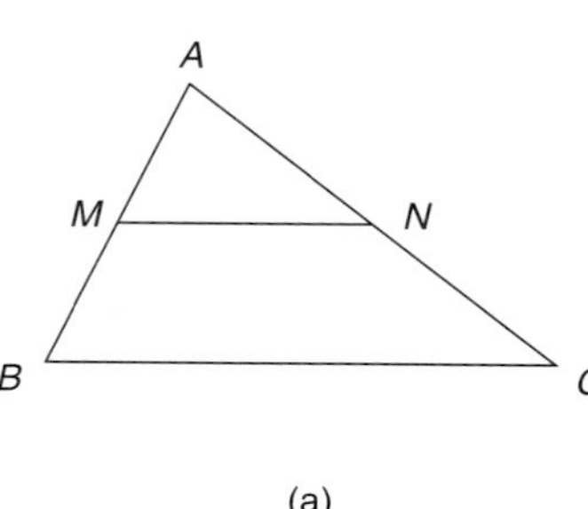

(a)

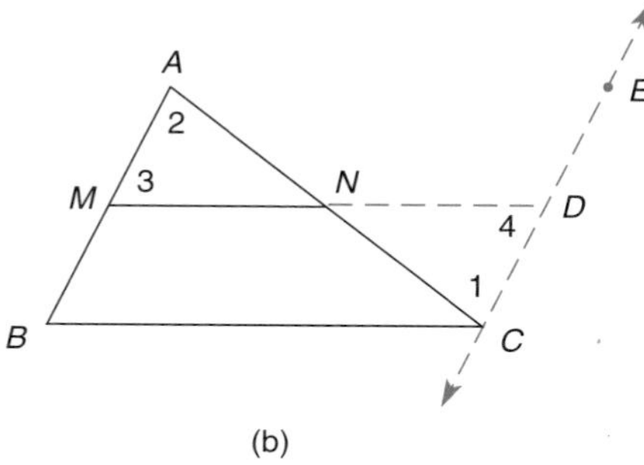

(b)

FIGURE 4.15

PROOF

Statements	Reasons
1. $\triangle ABC$, with midpoints M and N of $\overline{AB}$ and $\overline{AC}$, respectively	1. Given
2. Through C, construct $\overleftrightarrow{CE} \parallel \overline{AB}$, as in Figure 4.15(b)	2. Parallel Postulate
3. Extend $\overline{MN}$ to meet $\overleftrightarrow{CE}$ at D, as in Figure 4.15(b)	3. Exactly one line passes through two points
4. $\overline{AM} \cong \overline{MB}$ and $\overline{AN} \cong \overline{NC}$	4. The midpoint of a segment divides it into $\cong$ segments
5. $\angle 1 \cong \angle 2$ and $\angle 4 \cong \angle 3$	5. If two $\parallel$ lines are cut by a transversal, alternate interior $\angle$s are $\cong$
6. $\triangle ANM \cong \triangle CND$	6. AAS
7. $\overline{AM} \cong \overline{DC}$	7. CPCTC
8. $\overline{MB} \cong \overline{DC}$	8. Transitive (both are $\cong$ to $\overline{AM}$)
9. Quadrilateral $BMDC$ is a ▱	9. If two sides of a quadrilateral are both $\cong$ and $\parallel$, the quadrilateral is a parallelogram
10. $\overline{MN} \parallel \overline{BC}$	10. Opposite sides of a ▱ are $\parallel$

Theorem 4.2.5 also asserts that the segment formed by joining the midpoints of two sides of a triangle has a length equal to one-half the length of the third side. This part of the theorem is used in Examples 3 and 4.

DISCOVER!

Sketch regular hexagon *ABCDEF*. Draw diagonals $\overline{AE}$ and $\overline{CE}$. What type of quadrilateral is *ABCE*?

ANSWER
Kite

EXAMPLE 3

In $\triangle RST$ in Figure 4.16, M and N are the midpoints of $\overline{RS}$ and $\overline{RT}$, respectively.

a) If $ST = 12.7$, find MN.
b) If $MN = 15.8$, find ST.

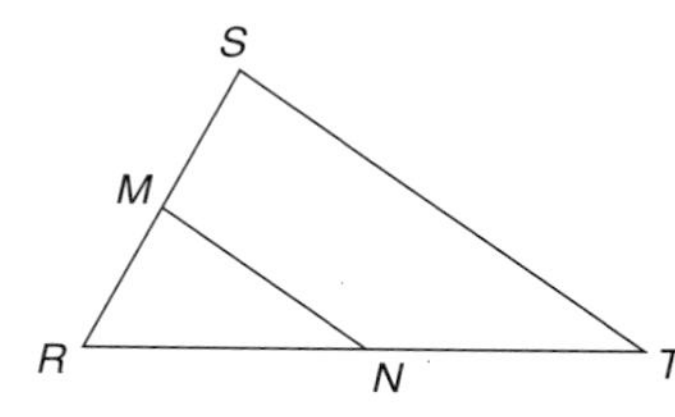

FIGURE 4.16

Solution

a) $MN = \frac{1}{2}(ST)$, so $MN = \frac{1}{2}(12.7) = 6.35$.
b) $MN = \frac{1}{2}(ST)$, so $15.8 = \frac{1}{2}(ST)$.
Multiplying by 2, $ST = 31.6$.

EXAMPLE 4

Given: $\triangle ABC$ in Figure 4.17, with D the midpoint of $\overline{AC}$ and E the midpoint of $\overline{BC}$; $DE = 2x + 1$; $AB = 5x - 1$

Find: x, DE, and AB

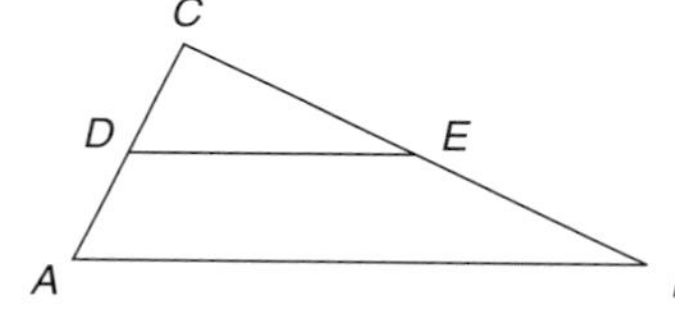

FIGURE 4.17

Solution

By Theorem 4.2.6,

$$DE = \frac{1}{2}(AB)$$

so

$$2x + 1 = \frac{1}{2}(5x - 1)$$

Multiplying by 2, we have

$$4x + 2 = 5x - 1$$
$$3 = x$$

Therefore, $DE = 2 \cdot 3 + 1 = 7$. Similarly, $AB = 5 \cdot 3 - 1 = 14$.

In the final example of this section, we consider the design of a product. Also see related Exercises 15 and 16 of this section.

EXAMPLE 5

In an efficiency apartment, there is a bed that folds down from the wall. In the vertical position, the design shows drop-down legs of equal length; that is, $AB = CD$ [see Figure 4.18(a)]. When the bed is lowered to a horizontal position, determine the type of quadrilateral $ABDC$ that is shown in Figure 4.18(b).

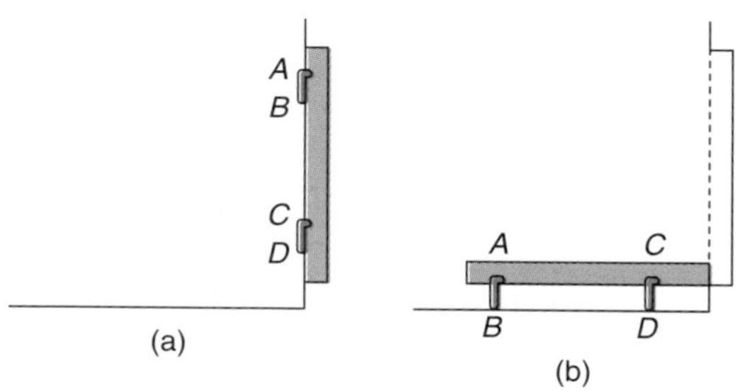

FIGURE 4.18

Solution

See Figure 4.18(a): Because $AB = CD$, it follows that $AB + BC = BC + CD$; here, BC was added to each side of the equation. But $AB + BC = AC$ and $BC + CD = BD$. Thus $AC = BD$ by substitution.

In Figure 4.18(b), we see that $AB = CD$ and $AC = BD$. Because both pairs of opposite sides of the quadrilateral are congruent, $ABDC$ is a parallelogram.

NOTE: In Section 4.3, we will also show that $ABDC$ of Figure 4.18(b) is a *rectangle* (a special type of parallelogram).

4.2 Exercises

1. a) As shown, must quadrilateral $ABCD$ be a parallelogram?
   b) Given the lengths of the sides as shown, is the measure of $\angle A$ unique?

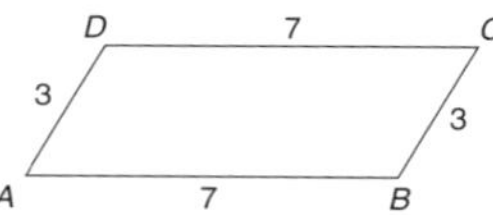

2. a) As shown, must $RSTV$ be a parallelogram?
   b) With measures as indicated, is it necessary that $RS = 8$?

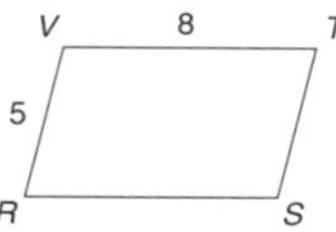

3. In the drawing, suppose that $\overline{WY}$ and $\overline{XZ}$ bisect each other. What type of quadrilateral is $WXYZ$?

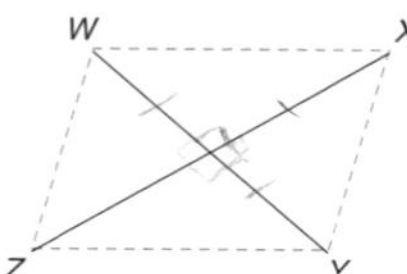

Exercises 3, 4

4. In the drawing, suppose that $\overline{ZX}$ is the perpendicular bisector of $\overline{WY}$. What type of quadrilateral is $WXYZ$?

5. A carpenter lays out boards of lengths 8 ft, 8 ft, 4 ft, and 4 ft by placing them end-to-end.
   a) If these are joined at the ends to form a quadrilateral that has the 8-ft pieces connected in order, what type of quadrilateral is formed?
   b) If these are joined at the ends to form a quadrilateral that has the 4- and 8-ft pieces alternating, what type of quadrilateral is formed?

6. A carpenter joins four boards of lengths 6 ft, 6 ft, 4 ft, and 4 ft, in that order, to form quadrilateral $ABCD$ as shown.
   a) What type of quadrilateral is formed?
   b) How are angles B and D related?

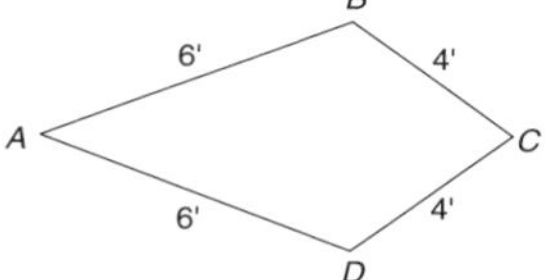

7. In parallelogram $ABCD$ (not shown), $AB = 8$, $m\angle B = 110°$, and $BC = 5$. Which diagonal has the greater length?

8. In kite $WXYZ$, the measures of selected angles are shown. Which diagonal of the kite has the greater length?

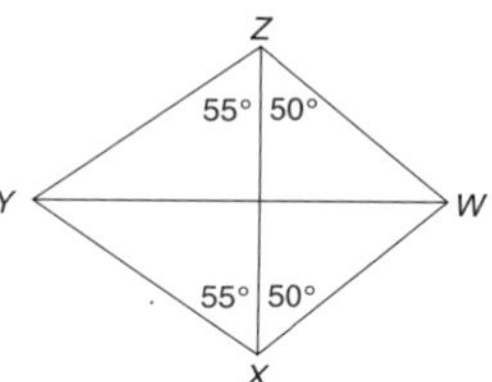

9. In $\triangle ABC$, M and N are midpoints of $\overline{AC}$ and $\overline{BC}$, respectively. If $AB = 12.36$, how long is $\overline{MN}$?

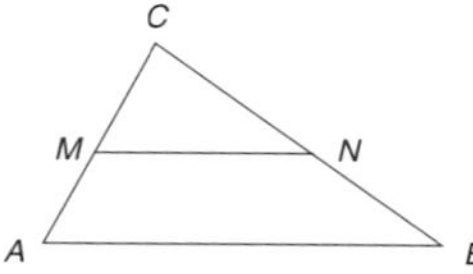

Exercises 9, 10

10. In $\triangle ABC$, M and N are midpoints of $\overline{AC}$ and $\overline{BC}$, respectively. If $MN = 7.65$, how long is $\overline{AB}$?

In Exercises 11–14, assume that X, Y, and Z are midpoints of the sides of △RST.

11. If $RS = 12$, $ST = 14$, and $RT = 16$, find:

a) XY **b)** XZ **c)** YZ

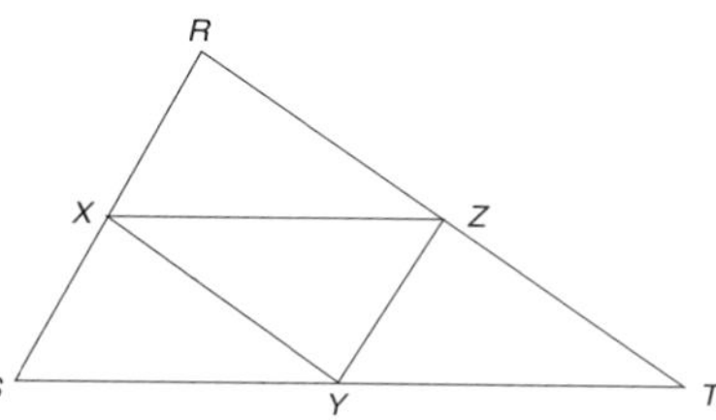

Exercises 11–14

12. If $XY = 6$, $YZ = 8$, and $XZ = 10$, find:

a) RS **b)** ST **c)** RT

13. If the perimeter (sum of the lengths of all three sides) of $\triangle RST$ is 20, what is the perimeter of $\triangle XYZ$?

14. If the perimeter (sum of the lengths of all three sides) of $\triangle XYZ$ is 12.7, what is the perimeter of $\triangle RST$?

15. For compactness, the drop-down wheels of a stretcher are folded under it as shown. To obtain a surface parallel to the ground when the wheels are dropped, what relationship must exist between $\overline{AB}$ and $\overline{CD}$?

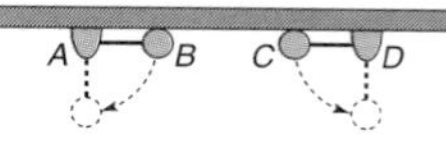

16. For compactness, the drop-down legs of an ironing board fold up under the board. A sliding mechanism at point A and the legs being connected at common midpoint M cause the board's upper surface to be parallel to the floor. How are $\overline{AB}$ and $\overline{CD}$ related?

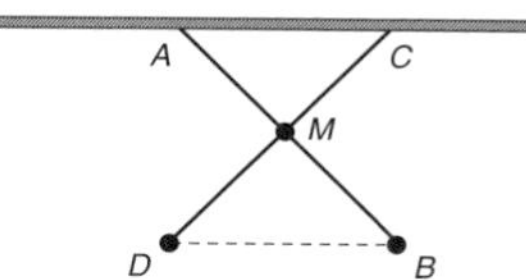

In Exercises 17 to 22, complete each proof.

17. *Given:* $\angle 1 \cong \angle 2$ and $\angle 3 \cong \angle 4$
Prove: $MNPQ$ is a kite

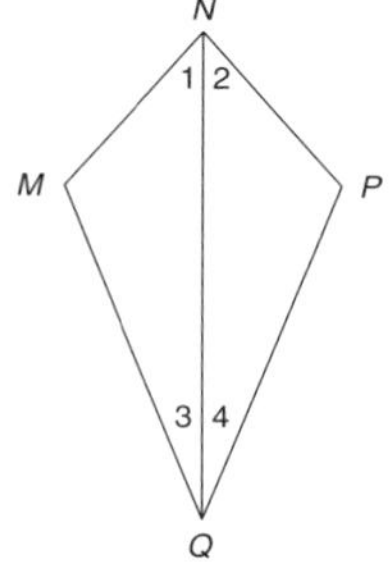

PROOF

Statements	Reasons
1. $\angle 1 \cong \angle 2$ and $\angle 3 \cong \angle 4$	**1.** ?
2. $\overline{NQ} \cong \overline{NQ}$	**2.** ?
3. ?	**3.** ASA
4. $\overline{MN} \cong \overline{PN}$ and $\overline{MQ} \cong \overline{PQ}$	**4.** ?
5. ?	**5.** If a quadrilateral has two pairs of ≅ adjacent sides, it is a kite

18. *Given:* Quadrilateral $ABCD$, with midpoints E, F, G, and H of the sides
Prove: $\overline{EF} \parallel \overline{HG}$

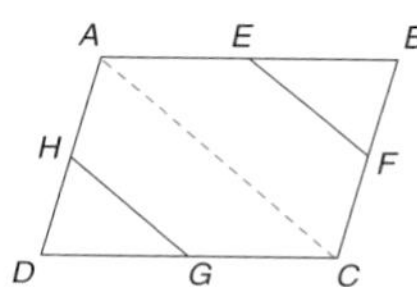

PROOF

Statements	Reasons
1. ?	**1.** Given
2. Draw $\overline{AC}$	**2.** Through two points, there is one line
3. In $\triangle ABC$, $\overline{EF} \parallel \overline{AC}$ and in $\triangle ADC$, $\overline{HG} \parallel \overline{AC}$	**3.** ?
4. ?	**4.** If two lines are ∥ to the same line, these lines are ∥ to each other

19. *Given:* M-Q-T and P-Q-R so that $MNPQ$ and $QRST$ are ▱s
Prove: $\angle N \cong \angle S$

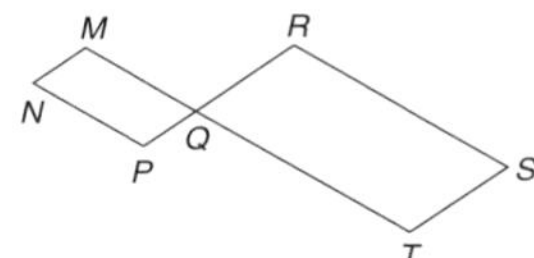

20. *Given:* ▱$WXYZ$ with diagonals $\overline{WY}$ and $\overline{XZ}$
Prove: $\triangle WMX \cong \triangle YMZ$

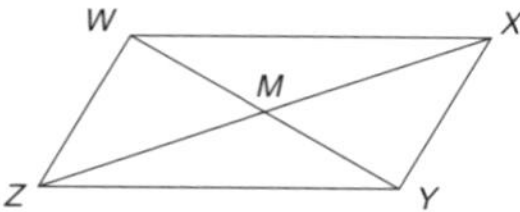

21. *Given:* Kite $HJKL$ with diagonal $\overline{HK}$
Prove: $\overrightarrow{HK}$ bisects $\angle LHJ$

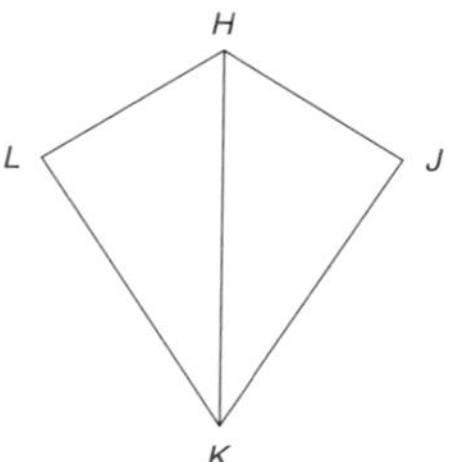

22. *Given:* ▱$MNPQ$, with T the midpoint of $\overline{MN}$ and S the midpoint of $\overline{QP}$
Prove: $\triangle QMS \cong \triangle NPT$, and $MSPT$ is a ▱

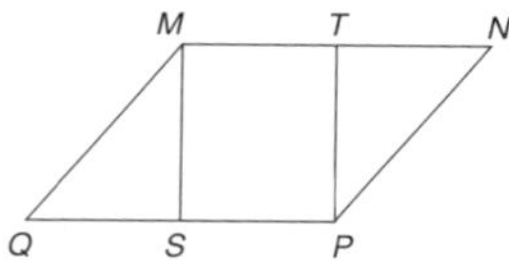

In Exercises 23 to 26, write a formal proof of each theorem or corollary.

23. If both pairs of opposite sides of a quadrilateral are congruent, then the quadrilateral is a parallelogram.

24. If the diagonals of a quadrilateral bisect each other, then the quadrilateral is a parallelogram.

25. In a kite, one diagonal is the perpendicular bisector of the other diagonal.

26. One diagonal of a kite bisects two of the angles of the kite.

In Exercises 27 to 29, $\triangle RST$ has M and N for midpoints of sides $\overline{RS}$ and $\overline{RT}$, respectively.

27. *Given:* $MN = 2y - 3$
$ST = 3y$
Find: y, MN, and ST

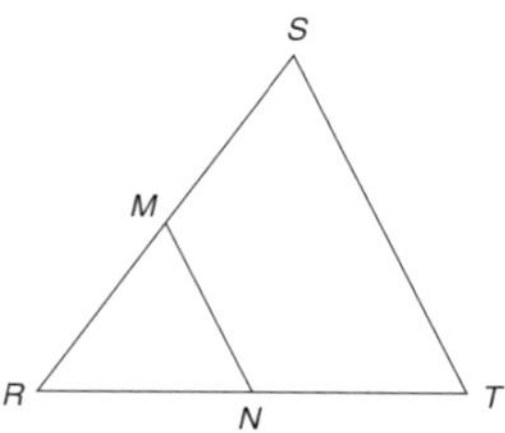

Exercises 27–29

28. *Given:* $MN = x^2 + 5$
$ST = x(2x + 5)$
Find: x, MN, and ST

29. *Given:* $RM = RN = 2x + 1$;
$ST = 5x - 3$; $m\angle R = 60°$
Find: x, RM, and ST

30. $RSTV$ is a kite, with $\overline{RS} \perp \overline{ST}$ and $\overline{RV} \perp \overline{VT}$. If $m\angle STV = 40°$, how large is the angle formed by the bisectors of $\angle RST$ and $\angle STV$? The bisectors of $\angle SRV$ and $\angle RST$?

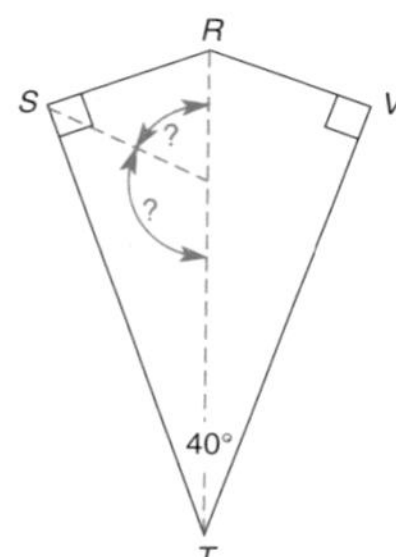

31. ➢ Prove that the segment that joins the midpoints of two sides of a triangle has a length equal to one-half the length of the third side.
(**HINT:** In the drawing, $\overline{MN}$ is extended to D, a point on $\overline{CD}$. Also, $\overline{CD}$ is parallel to $\overline{AB}$.)

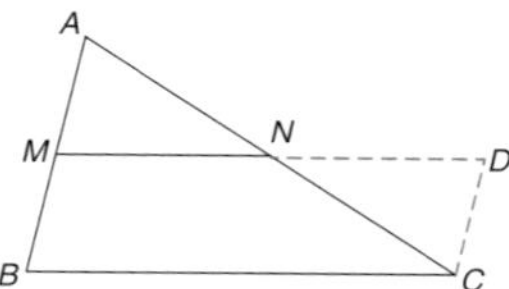

32. ➢ Prove that, when the midpoints of consecutive sides of a quadrilateral are joined in order, the resulting quadrilateral is a parallelogram.

4.3 The Rectangle, Square, and Rhombus

The Rectangle

In this section, we investigate special parallelograms. The first of these is the rectangle, which is defined as follows.

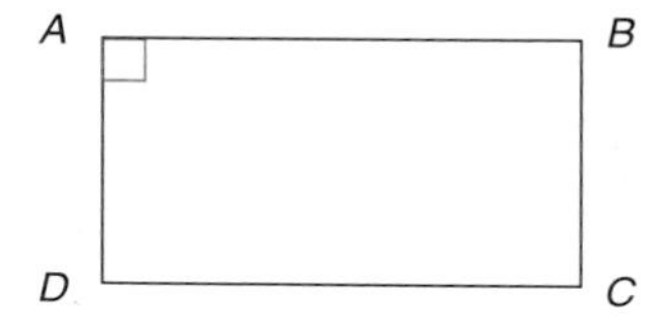

FIGURE 4.19

> DEFINITION: A **rectangle** is a parallelogram that has a right angle. (See Figure 4.19.)

Rectangle *ABCD* is shown in Figure 4.19. An abbreviation for the word rectangle is "rect.". Because a rectangle is a parallelogram by definition, the following statement is easily proven by applying Corollaries 4.1.3 and 4.1.5. Its proof is left to the student.

> COROLLARY 4.3.1: All angles of a rectangle are right angles.

The following theorem is true for rectangles, but not for parallelograms in general.

> THEOREM 4.3.2: The diagonals of a rectangle are congruent.

NOTE: To follow the flow of the following proof, it may be best to draw triangles *NMQ* and *PQM* separately.

EXAMPLE 1

Complete a proof of Theorem 4.3.2.

Given: Rectangle *MNPQ* with diagonals $\overline{MP}$ and $\overline{NQ}$ (See Figure 4.20.)

Prove: $\overline{MP} \cong \overline{NQ}$

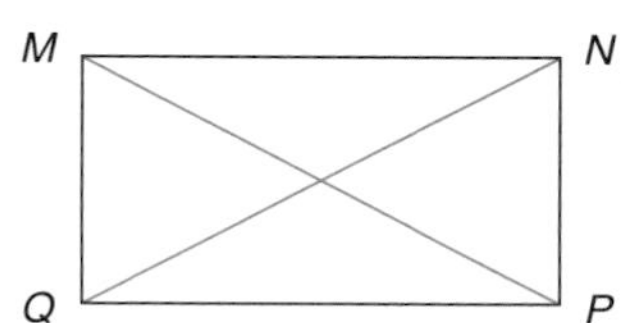

FIGURE 4.20

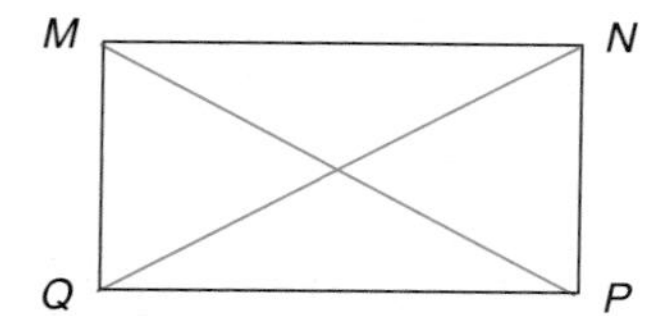

FIGURE 4.20

PROOF

Statements	Reasons
1. Rectangle $MNPQ$ with diagonals $\overline{MP}$ and $\overline{NQ}$	1. Given
2. $MNPQ$ is a ▱	2. By definition, a rectangle is a ▱ with a right angle
3. $\overline{MN} \cong \overline{QP}$	3. Opposite sides of a ▱ are ≅
4. $\overline{MQ} \cong \overline{MQ}$	4. Identity
5. $\angle NMQ$ and $\angle PQM$ are right ∠s	5. By Corollary 4.3.1, the four ∠s of a rectangle are right ∠s
6. $\angle NMQ \cong \angle PQM$	6. All right ∠s are ≅
7. $\triangle NMQ \cong \triangle PQM$	7. SAS
8. $\overline{MP} \cong \overline{NQ}$	8. CPCTC

The Square

All rectangles are parallelograms; some parallelograms are rectangles; and some rectangles are *squares.*

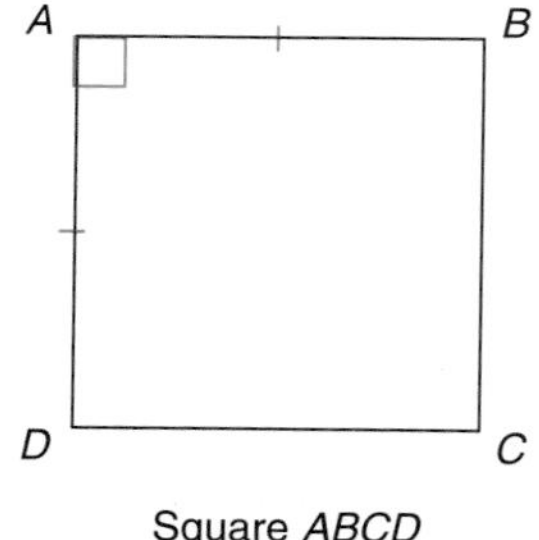

Square *ABCD*

FIGURE 4.21

DEFINITION: A **square** is a rectangle that has two congruent adjacent sides. (See Figure 4.21.)

COROLLARY 4.3.3: All sides of a square are congruent.

Because a square is a type of rectangle, it has four right angles and its diagonals are congruent. Because a square is also a parallelogram, its opposite sides are parallel. For any square, we can show that the diagonals are perpendicular.

The Rhombus

The next type of quadrilateral we consider is the rhombus.

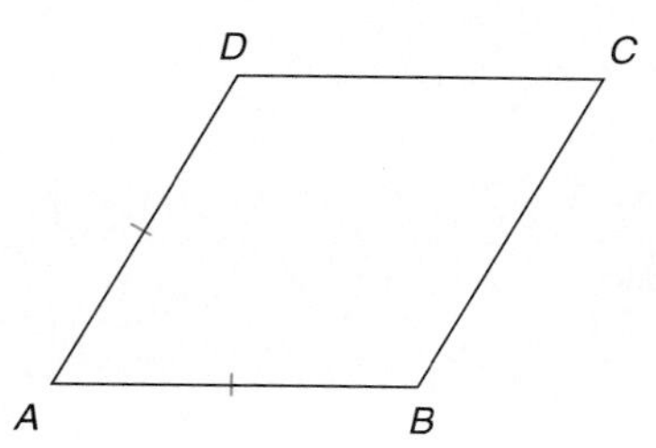

FIGURE 4.22

DEFINITION: A **rhombus** is a parallelogram with two congruent adjacent sides.

In Figure 4.22, the adjacent sides $\overline{AB}$ and $\overline{AD}$ of rhombus *ABCD* are marked congruent. Because a rhombus is a type of parallelogram, it is also necessary that $\overline{AB} \cong \overline{DC}$ and $\overline{AD} \cong \overline{BC}$.

COROLLARY 4.3.4: All sides of a rhombus are congruent.

We will use Corollary 4.3.4 in the proof of the following theorem.

THEOREM 4.3.5: The diagonals of a rhombus are perpendicular.

EXAMPLE 2

Give a formal proof of Theorem 4.3.5.

Given: Rhombus $ABCD$, with diagonals $\overline{AC}$ and $\overline{DB}$. (See Figure 4.23.)

Prove: $\overline{AC} \perp \overline{DB}$

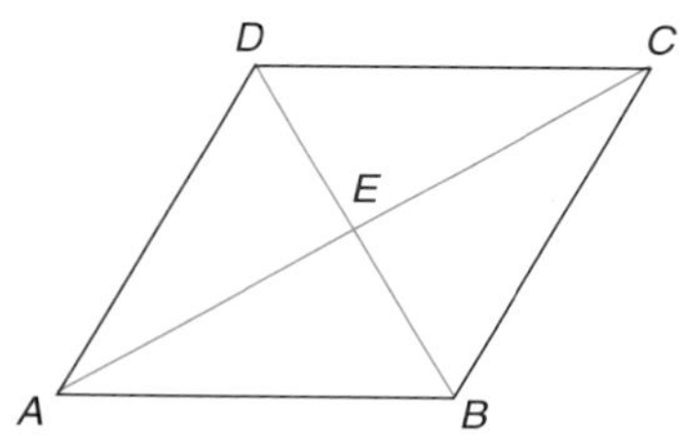

FIGURE 4.23

PROOF

Statements	Reasons
1. Rhombus $ABCD$ with diagonals $\overline{AC}$ and $\overline{DB}$	1. Given
2. $ABCD$ is a ▱	2. A rhombus is a ▱ with two $\cong$ adjacent sides
3. $\overline{DB}$ bisects $\overline{AC}$	3. Diagonals of a ▱ bisect each other
4. $\overline{AE} \cong \overline{EC}$	4. If a segment is bisected, it is divided into two $\cong$ segments
5. $\overline{AD} \cong \overline{DC}$	5. All sides of a rhombus are $\cong$
6. $\overline{DE} \cong \overline{DE}$	6. Identity
7. $\triangle ADE \cong \triangle CDE$	7. SSS
8. $\angle DEA \cong \angle DEC$	8. CPCTC
9. $\overline{AC} \perp \overline{DB}$	9. If two lines meet to form $\cong$ adjacent $\angle$s, the lines are $\perp$

DISCOVER!

Sketch regular hexagon *RSTVWX*. Draw diagonals $\overline{RT}$ and $\overline{XV}$. What type of quadrilateral is *RTVX*?

ANSWER

Rectangle

An alternate definition of "square" is, "A square is a rhombus whose adjacent sides form a right angle." Therefore, a further property of a square is that its diagonals are perpendicular.

The following theorem, which deals with right triangles, is also useful in applications involving quadrilaterals that have right angles. In antiquity, the theorem claimed that "the square upon the hypotenuse equals the sum of the squares upon the legs of the right triangle." See Figure 4.24(a) on the next page. This interpretation involves the area concept, which we study in a later chapter. Our interpretation of the Pythagorean Theorem uses number (length) relationships.

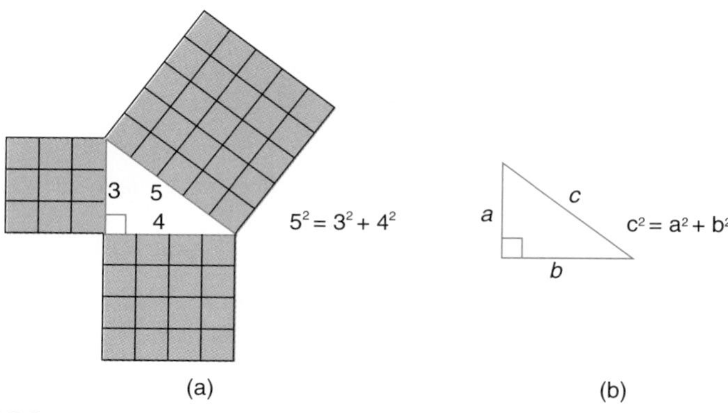

FIGURE 4.24

The Pythagorean Theorem

The Pythagorean Theorem will be proved in Section 5.3. Stated without proof at this time, we apply the "rule of Pythagoras" in Examples 3, 4, and 5.

> **The Pythagorean Theorem** In a right triangle with hypotenuse of length c and legs of lengths a and b, it follows that $c^2 = a^2 + b^2$.

The Pythagorean Theorem can be applied to any right triangle or figure containing a right triangle. In Example 3, we seek the length of the hypotenuse in a right triangle whose lengths of legs are known. When using the Pythagorean Theorem, c represents the length of the hypotenuse; however, either leg can be chosen for length a (or b).

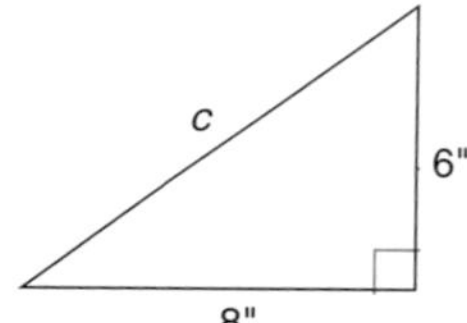

FIGURE 4.25

EXAMPLE 3

What is the length of the hypotenuse of a right triangle whose legs measure 6 in. and 8 in.? (See Figure 4.25.)

Solution

$$c^2 = a^2 + b^2$$
$$c^2 = 6^2 + 8^2$$
$$c^2 = 36 + 64 \rightarrow c^2 = 100 \rightarrow c = 10 \text{ in.}$$

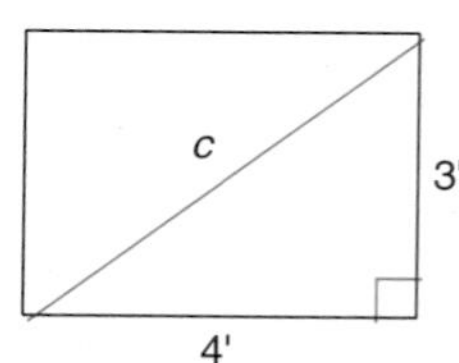

FIGURE 4.26

In the following example, the diagonal of a rectangle separates it into two right triangles. As shown in Figure 4.26, the diagonal of the rectangle is the hypotenuse of each triangle.

EXAMPLE 4

What is the length of the diagonal in a rectangle whose sides measure 3 ft and 4 ft?

Solution For the triangle in Figure 4.26, $c^2 = a^2 + b^2$ becomes $c^2 = 3^2 + 4^2$ or $c^2 = 9 + 16$. Then $c^2 = 25$, so $c = 5$. The length of the diagonal is 5 ft.

In Example 5, we use the fact that a rhombus is a parallelogram to justify that its diagonals bisect each other. Using Theorem 4.3.5, the diagonals of the rhombus are also perpendicular.

EXAMPLE 5

What is the length of each side of a rhombus whose diagonals measure 10 cm and 24 cm? (See Figure 4.27.)

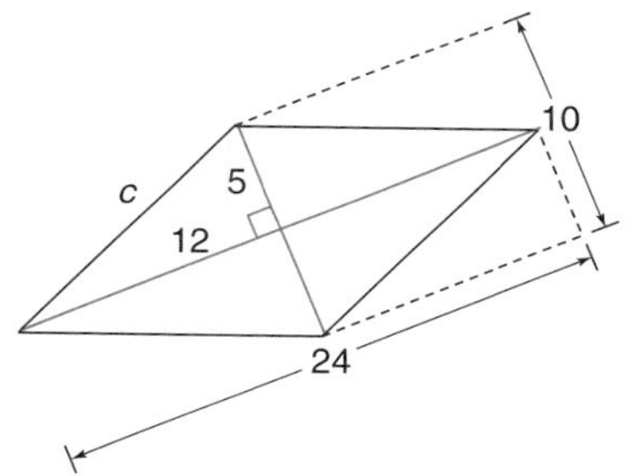

FIGURE 4.27

Solution The diagonals of a rhombus are perpendicular bisectors of each other. Thus, the diagonals separate the rhombus shown into four congruent right triangles with sides of lengths 5 cm and 12 cm. For each triangle, $c^2 = a^2 + b^2$ becomes $c^2 = 5^2 + 12^2$ or $c^2 = 25 + 144$. Then $c^2 = 169$, so $c = 13$. The length of each side is 13 cm.

EXAMPLE 6

On a softball diamond (actually a square), the distance along base paths is 60 ft. Using the triangle in Figure 4.28, find the distance from home plate to second base.

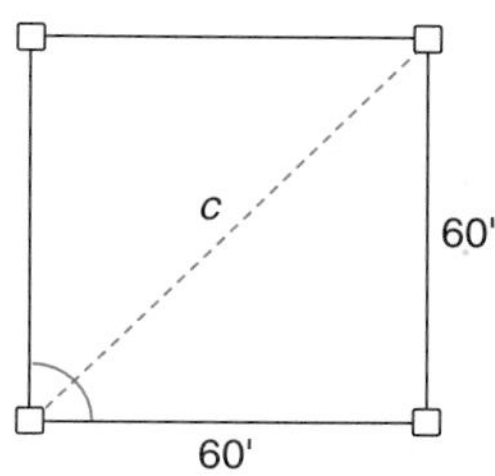

FIGURE 4.28

Solution Using $c^2 = a^2 + b^2$, we have

$$c^2 = 60^2 + 60^2$$
$$c^2 = 7200$$

Then $c = \sqrt{7200}$ or $c \approx 84.85$ ft.

4.3 Exercises

1. If diagonal $\overline{DB}$ is congruent to each side of rhombus $ABCD$, what is the measure of $\angle A$? Of $\angle ABC$?

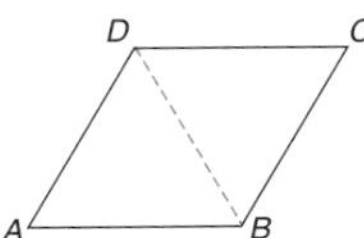

2. If the diagonals of a parallelogram are perpendicular, what may you conclude about the parallelogram? (**HINT:** Make a number of drawings in which you use only the information suggested.)
3. If the diagonals of a parallelogram are congruent, what may you conclude about the parallelogram?
4. If the diagonals of a parallelogram are perpendicular and congruent, what may you conclude about the parallelogram?
5. If the diagonals of a quadrilateral are perpendicular bisectors of each other (but not congruent), what may you conclude about the quadrilateral?
6. If the diagonals of a rhombus are congruent, what may you conclude about the rhombus?
7. A line segment joins the midpoints of two opposite sides of a rectangle as shown. What may you conclude regarding $\overline{MN}$ and MN?

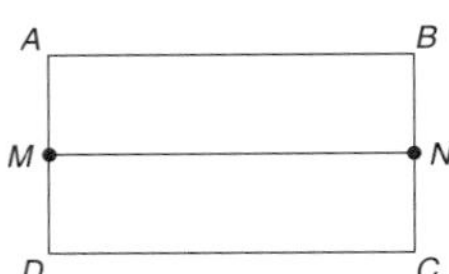

In Exercises 8 to 10, use the properties of rectangles to solve each problem. Rectangle ABCD is shown in the figure.

8. *Given:* $AB = 5$ and $BC = 12$
   Find: CD, AD, and AC (not shown)
9. *Given:* $AB = 2x + 7$, $BC = 3x + 4$, and $CD = 3x + 2$
   Find: x and DA
10. *Given:* $AB = x + y$, $BC = x + 2y$, $CD = 2x - y - 1$, and $DA = 3x - 3y + 1$
    Find: x and y
11. *Given:* Rectangle $ABCD$ (not shown) with $AB = 8$ and $BC = 6$; M and N are the midpoints of sides $\overline{AB}$ and $\overline{BC}$, respectively
    Find: MN
12. *Given:* Rhombus $RSTV$ (not shown) with diagonals $\overline{RT}$ and $\overline{SV}$ so that $RT = 8$ and $SV = 6$
    Find: RS, the length of a side

In Exercises 13 and 14, supply the missing statements and reasons.

13. *Given:* Quadrilateral $PQST$ with midpoints A, B, C, and D of the sides
    Prove: $ABCD$ is a ▱

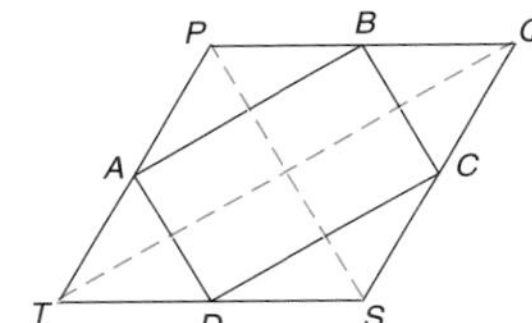

PROOF

Statements	Reasons
1. Quadrilateral $PQST$ with midpoints A, B, C, and D of the sides	1. ?
2. Draw $\overline{TQ}$	2. Through two points, there is one line
3. $\overline{AB} \parallel \overline{TQ}$ in $\triangle TPQ$	3. The line joining the midpoints of two sides of a triangle is ∥ to the third side
4. $\overline{DC} \parallel \overline{TQ}$ in $\triangle TSQ$	4. ?
5. $\overline{AB} \parallel \overline{DC}$	5. ?
6. Draw $\overline{PS}$	6. ?
7. $\overline{AD} \parallel \overline{PS}$ in $\triangle TSP$	7. ?
8. $\overline{BC} \parallel \overline{PS}$ in $\triangle PSQ$	8. ?
9. $\overline{AD} \parallel \overline{BC}$	9. ?
10. ?	10. If both pairs of opposite sides of a quadrilateral are ∥, the quad. is a ▱

14. *Given:* Rectangle $WXYZ$ with diagonals $\overline{WY}$ and $\overline{XZ}$
Prove: $\angle 1 \cong \angle 2$

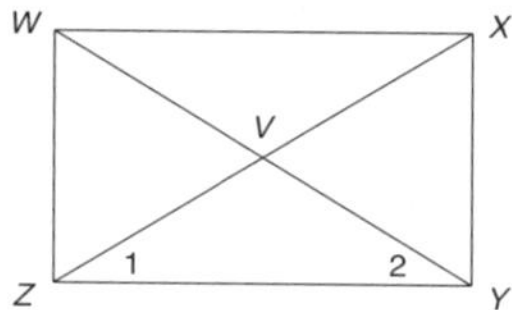

PROOF

Statements	Reasons
1. ?	**1.** Given
2. ?	**2.** The diagonals of a rectangle are ≅
3. $\overline{WZ} \cong \overline{XY}$	**3.** The opposite sides of a rectangle (▱) are ≅
4. $\overline{ZY} \cong \overline{ZY}$	**4.** ?
5. $\triangle XZY \cong \triangle WYZ$	**5.** ?
6. ?	**6.** ?

In Exercises 15 to 17, explain why each statement is true.

15. All angles of a rectangle are right angles.

16. All sides of a rhombus are congruent.

17. All sides of a square are congruent.

In Exercises 18 to 23, write a formal proof of each theorem.

18. The diagonals of a square are perpendicular.

19. A diagonal of a rhombus bisects two angles of the rhombus.

20. If the diagonals of a parallelogram are congruent, the parallelogram is a rectangle.

21. If the diagonals of a parallelogram are perpendicular, the parallelogram is a rhombus.

22. If the diagonals of a parallelogram are congruent and perpendicular, the parallelogram is a square.

23. If the midpoints of the sides of a rectangle are joined in order, the quadrilateral formed is a rhombus.

In Exercises 24 and 25, you will need to use the square root ($\sqrt{\ }$) function of your calculator.

24. A wall that is 12 ft long by 8 ft high has a triangular brace along the diagonal. Use a calculator to approximate the length of the brace to the nearest tenth of a foot.

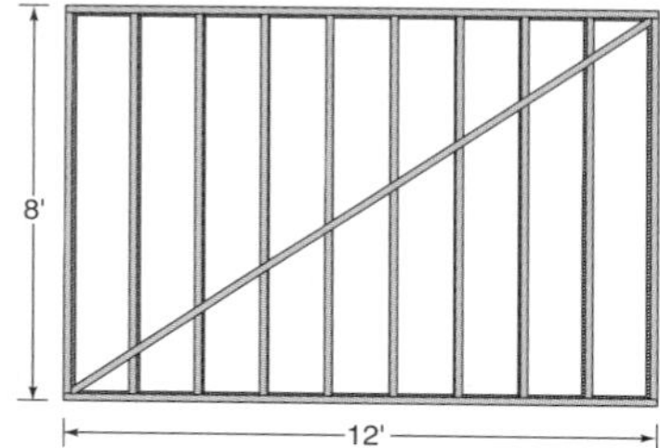

25. A walk-up ramp moves horizontally 20 ft while rising 4 ft. Use a calculator to approximate its length to the nearest tenth of a foot.

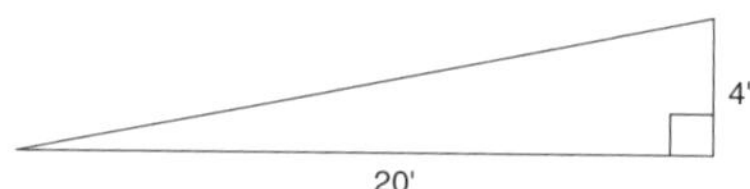

26. **a)** Argue that the midpoint of the hypotenuse of a right triangle is equidistant from the three vertices of the triangle. Use the fact that the congruent diagonals of a rectangle bisect each other. Be sure to provide a drawing.

b) Use the relationship from part (a) to find CM, the length of the median to the hypotenuse of right $\triangle ABC$, in which $m\angle C = 90°$, $AC = 6$, and $BC = 8$.

4.4 The Trapezoid

DEFINITION: A **trapezoid** is a quadrilateral with exactly two parallel sides.

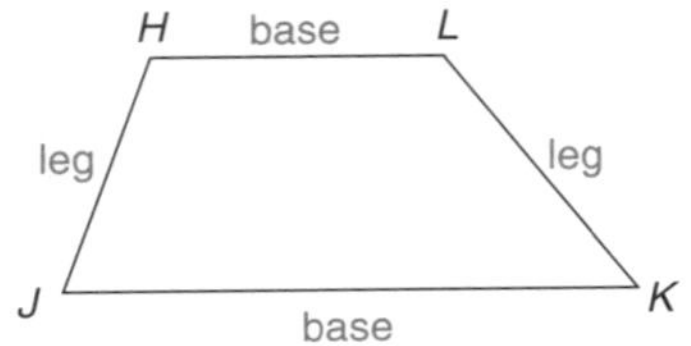

FIGURE 4.29

Figure 4.29 shows trapezoid $HJKL$, in which $\overline{HL} \parallel \overline{JK}$. The parallel sides $\overline{HL}$ and $\overline{JK}$ are **bases,** and the nonparallel sides $\overline{HJ}$ and $\overline{LK}$ are **legs.** Because $\angle J$ and $\angle K$ each have $\overline{JK}$ for a side, they are **base angles** of the trapezoid; $\angle H$ and $\angle L$ are also base angles since $\overline{HL}$ is a base.

When the midpoints of the two legs of a trapezoid are joined, the resulting segment is known as the **median** of the trapezoid. Given that M and N are the midpoints of the legs $\overline{HJ}$ and $\overline{LK}$ in trapezoid $HJKL$, $\overline{MN}$ is the median of the trapezoid. [See Figure 4.30(a).]

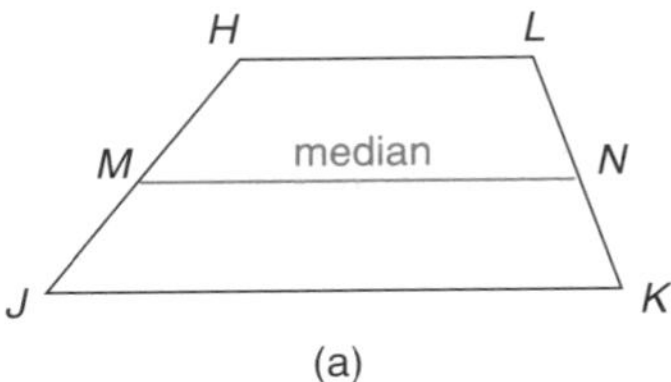

(a)

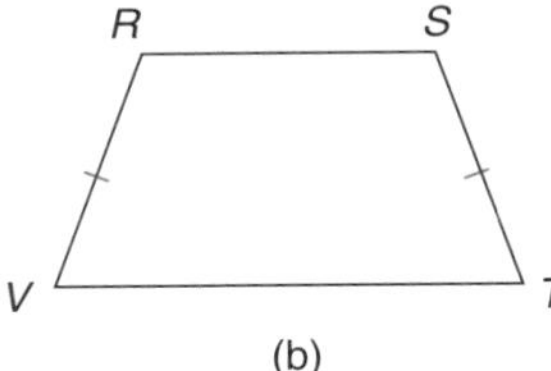

(b)

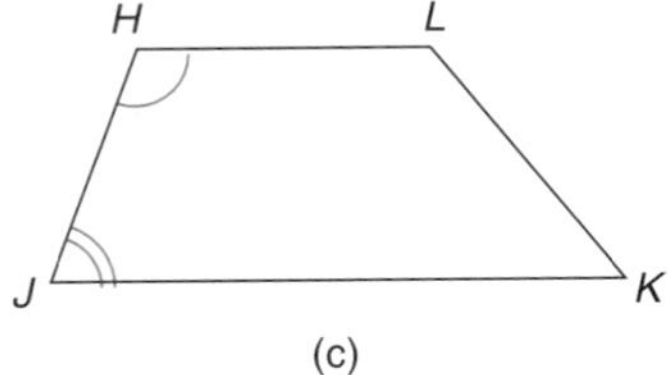

(c)

FIGURE 4.30

If the two legs of a trapezoid are congruent, the trapezoid is known as an **isosceles trapezoid.** In Figure 4.30(b), $RSTV$ is an isosceles trapezoid because $\overline{RV} \cong \overline{ST}$ and $\overline{RS} \parallel \overline{VT}$.

Every trapezoid contains two pairs of consecutive interior angles that are supplementary. Each of these pairs of angles is formed when parallel lines are cut by a transversal. In Figure 4.30(c), angles H and J are supplementary, as are angles L and K.

EXAMPLE 1

In Figure 4.29, suppose that $m\angle H = 107°$ and $m\angle K = 58°$. Find $m\angle J$ and $m\angle L$.

Solution

Because $\overline{HL} \parallel \overline{JK}$, $\angle$s H and J are supplementary angles, as are $\angle$s L and K. Then $m\angle H + m\angle J = 180$ and $m\angle L + m\angle K = 180$. Substitution leads to $107 + m\angle J = 180$ and $m\angle L + 58 = 180$, so $m\angle J = 73°$ and $m\angle L = 122°$.

DISCOVER!

Using construction paper, cut out two trapezoids that are copies of each other. To accomplish this, hold two pieces of paper together and cut once. Take the second trapezoid and turn it so that a pair of congruent legs coincide. What type of quadrilateral has been formed?

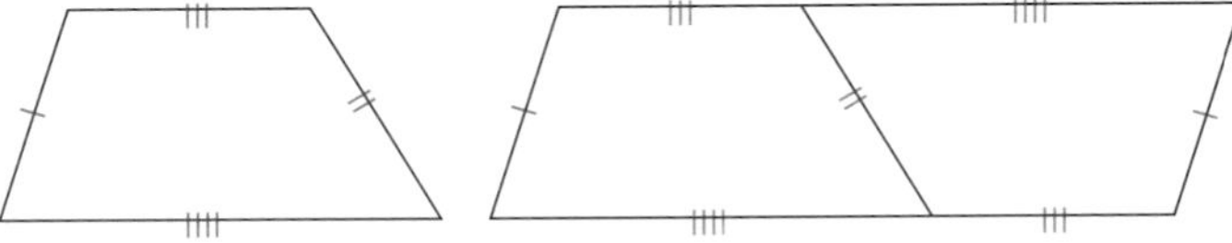

ANSWER
Parallelogram

The preceding activity may provide insight for a number of theorems involving the trapezoid.

THEOREM 4.4.1: The base angles of an isosceles trapezoid are congruent.

EXAMPLE 2

Give a formal proof of Theorem 4.4.1.

Given: In Figure 4.31(a), trapezoid $RSTV$ with $\overline{RV} \cong \overline{ST}$ and $\overline{RS} \parallel \overline{VT}$

Prove: $\angle V \cong \angle T$ and $\angle R \cong \angle S$

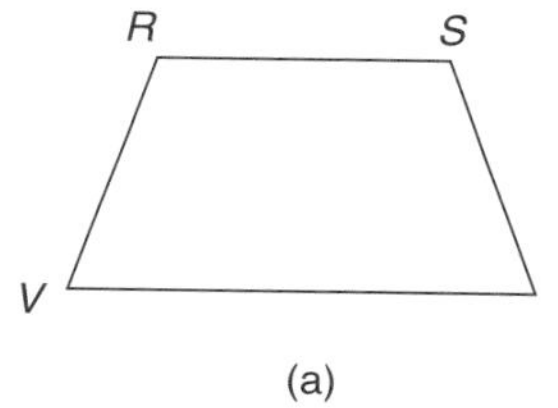

(a)

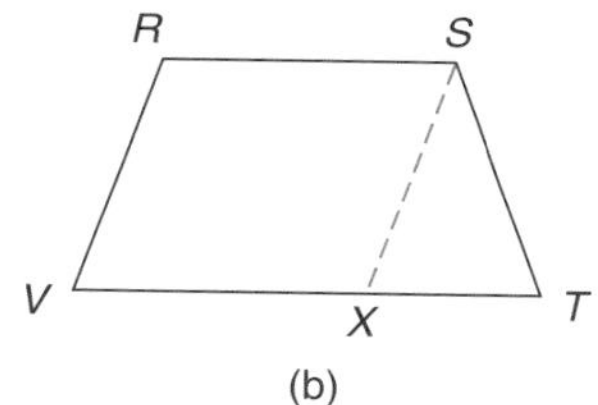

(b)

FIGURE 4.31

PROOF

Statements	Reasons
1. Trapezoid $RSTV$ with $\overline{RV} \cong \overline{ST}$ and $\overline{RS} \parallel \overline{VT}$	1. Given
2. Draw $\overline{SX} \parallel \overline{RV}$, as in Figure 4.31(b)	2. Parallel Postulate
3. $RSXV$ is a ▱	3. If both pairs of opposite sides of a quadrilateral are ∥, it is a ▱
4. $\overline{RV} \cong \overline{SX}$	4. Opposite sides of a ▱ are ≅
5. $\overline{ST} \cong \overline{SX}$	5. Transitive
6. $\angle SXT \cong \angle T$	6. If two sides of a △ are ≅, the angles opposite those sides are also ≅
7. $\angle V \cong \angle SXT$	7. If two ∥ lines are cut by a transversal, corresponding ∠s are ≅
8. $\angle V \cong \angle T$	8. Transitive
9. $\angle V$ is supplementary to $\angle VRS$ and $\angle T$ is supplementary to $\angle TSR$	9. If two ∥ lines are cut by a transversal, interior ∠s on the same side of the transversal are supplementary
10. $\angle VRS \cong \angle TSR$	10. Supplements of ≅ ∠s are ≅

Geometry IN THE REAL WORLD

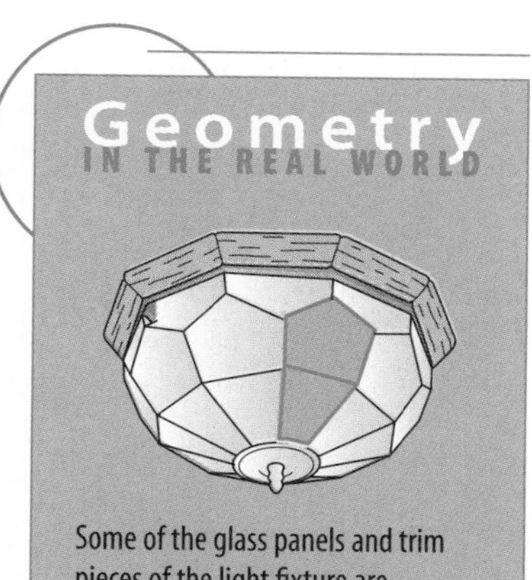

Some of the glass panels and trim pieces of the light fixture are isosceles trapezoids. Other glass panels are pentagons.

The following statement is a corollary of Theorem 4.4.1. Its proof is left to the student.

> COROLLARY 4.4.2: The diagonals of an isosceles trapezoid are congruent.

EXAMPLE 3

Given isosceles trapezoid $ABCD$ with $\overline{AB} \parallel \overline{DC}$ (see Figure 4.32):

a) Find the measures of the angles of $ABCD$ if $m\angle A = 12x + 30$ and $m\angle B = 10x + 46$.
b) Using the result from part(a), find the length of each diagonal (not shown) if $AC = 2x - 5$.

Solution

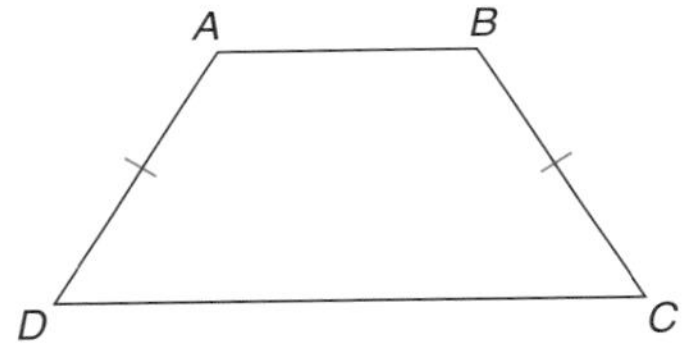

a) Since $m\angle A = m\angle B$, $12x + 30 = 10x + 46$. So $2x = 16$ and $x = 8$. Then $m\angle A = 12(8) + 30$ or $126°$ and $m\angle B = 10(8) + 46$ or $126°$. Subtracting $(180 - 126 = 54)$, we determine the supplements of $\angle$s A and B. That is $m\angle C = m\angle D = 54°$.
b) Since $x = 8$, $AC = 2(8) - 5$ or 11. By Corollary 4.4.2, $\overline{AC} \cong \overline{BD}$ so $BD = 11$ also.

A B D C

FIGURE 4.32

The proof of the following theorem is left as an exercise. We apply Theorem 4.4.3 in Examples 4 and 5.

> THEOREM 4.4.3: The length of the median of a trapezoid equals one-half the sum of the lengths of the two bases.

EXAMPLE 4

In trapezoid $RSTV$ in Figure 4.33, $\overline{RS} \parallel \overline{VT}$ and M and N are the midpoints of $\overline{RV}$ and $\overline{TS}$, respectively. Find the length of median $\overline{MN}$ if $RS = 12$ and $VT = 18$.

Solution

Using Theorem 4.4.3, $MN = \frac{1}{2}(RS + VT)$, so $MN = \frac{1}{2}(12 + 18)$ or $MN = \frac{1}{2}(30)$. Thus, $MN = 15$.

EXAMPLE 5

In trapezoid $RSTV$, $\overline{RS} \parallel \overline{VT}$ and M and N are the midpoints of $\overline{RV}$ and $\overline{TS}$, respectively (see Figure 4.33). Find MN, RS, and VT if $RS = 2x$, $MN = 3x - 5$, and $VT = 2x + 10$.

Solution

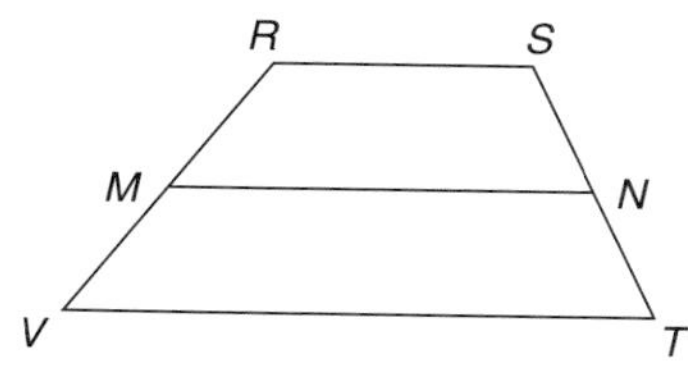

FIGURE 4.33

Using Theorem 4.4.3, $MN = \frac{1}{2}(RS + VT)$, so

$$3x - 5 = \frac{1}{2}[2x + (2x + 10)] \quad \text{or} \quad 3x - 5 = \frac{1}{2}(4x + 10)$$

Then $3x - 5 = 2x + 5$ and $x = 10$. Now $RS = 2x = 2(10)$, so $RS = 20$. Also, $MN = 3x - 5 = 3(10) - 5$; therefore, $MN = 25$. Finally, $VT = 2x + 10$; therefore, $VT = 2(10) + 10 = 30$.

NOTE: As a check, $MN = \frac{1}{2}(RS + VT)$ leads to the true statement $25 = \frac{1}{2}(20 + 30)$.

> **THEOREM 4.4.4:** The median of a trapezoid is parallel to each base.

The proof of Theorem 4.4.4 is left as an exercise.

Theorems 4.4.5–4.4.7 enable us to show that a quadrilateral with certain characteristics is a trapezoid or perhaps even an isosceles trapezoid. We state these theorems as follows:

> **THEOREM 4.4.5:** If exactly two consecutive angles of a quadrilateral are supplementary, the quadrilateral is a trapezoid.

> **THEOREM 4.4.6:** If two base angles of a trapezoid are congruent, the trapezoid is an isosceles trapezoid.

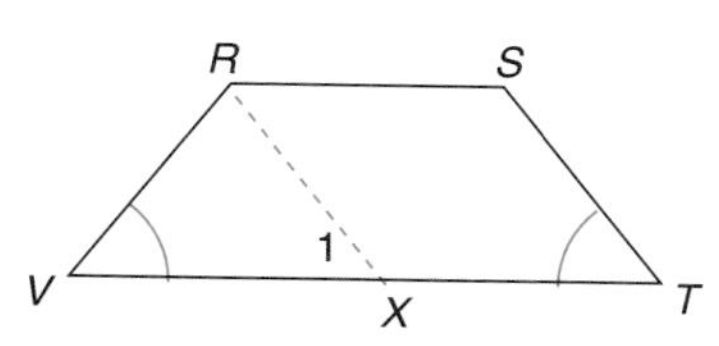

FIGURE 4.34

Theorem 4.4.5 is easily proven by applying Theorem 2.3.4. However, we consider a plan for proving Theorem 4.4.6. See Figure 4.34.

Given: Trapezoid $RSTV$ with $\overline{RS} \parallel \overline{VT}$ and $\angle V \cong \angle T$

Prove: $RSTV$ is an isosceles trapezoid

Plan: Draw auxiliary line $\overline{RX}$ parallel to $\overline{ST}$. Now show that $\angle V \cong \angle 1$, so $\overline{RV} \cong \overline{RX}$ in $\triangle RXV$. But $\overline{RX} \cong \overline{ST}$ in parallelogram $RXTS$, so $\overline{RV} \cong \overline{ST}$ and $RSTV$ is isosceles.

> **THEOREM 4.4.7:** If the diagonals of a trapezoid are congruent, the trapezoid is an isosceles trapezoid.

For several reasons, our final theorem is a challenge to prove. Looking ahead at Figure 4.35, one sees trapezoids such as *ABED* and *BCFE*. However, the proof (whose "plan" we provide) uses auxiliary lines, parallelograms, and congruent triangles.

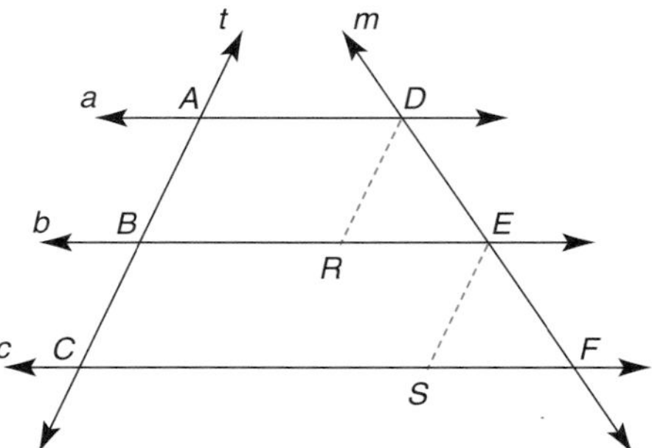

FIGURE 4.35

> **THEOREM 4.4.8:** If three (or more) parallel lines intercept congruent segments on one transversal, then they intercept congruent segments on any transversal.

Given: Parallel lines *a, b,* and *c* cut by transversal *t* so that $\overline{AB} \cong \overline{BC}$; also transversal *m* in Figure 4.35

Prove: $\overline{DE} \cong \overline{EF}$

Plan: Through *D* and *E*, draw $\overline{DR} \parallel \overline{AB}$ and $\overline{ES} \parallel \overline{AB}$. Now show that $\overline{DR} \cong \overline{ES}$ (since $\cong$ to $\overline{AB}$ and $\overline{BC}$ as opposite sides of ▱). By *AAS*, show $\triangle DER \cong \triangle EFS$; then $\overline{DE} \cong \overline{EF}$ by CPCTC.

EXAMPLE 6

In Figure 4.35, $a \parallel b \parallel c$. If $AB = BC = 7.2$ and $DE = 8.4$, find *EF*.

Solution Using Theorem 4.4.8, $EF = 8.4$.

4.4 Exercises

1. Find the measures of the remaining angles of trapezoid *ABCD* (not shown) if $\overline{AB} \parallel \overline{DC}$ and $m\angle A = 58°$ while $m\angle C = 125°$.
2. Find the measures of the remaining angles of trapezoid *ABCD* (not shown) if $\overline{AB} \parallel \overline{DC}$ and $m\angle B = 63°$ while $m\angle D = 118°$.
3. If the diagonals of a trapezoid are congruent, what may you conclude about the trapezoid?
4. If two of the base angles of a trapezoid are congruent, what type of trapezoid is it?
5. What type of quadrilateral is formed when the midpoints of the sides of an isosceles trapezoid are joined?

6. Without writing a formal proof, explain why $MN = \frac{1}{2}(AB + DC)$.

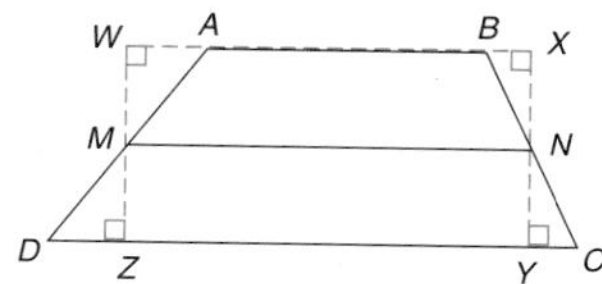

In Exercises 7 to 12, the drawing shows trapezoid ABCD with $\overline{AB} \parallel \overline{DC}$; also, M and N are midpoints of $\overline{AD}$ and $\overline{BC}$, respectively.

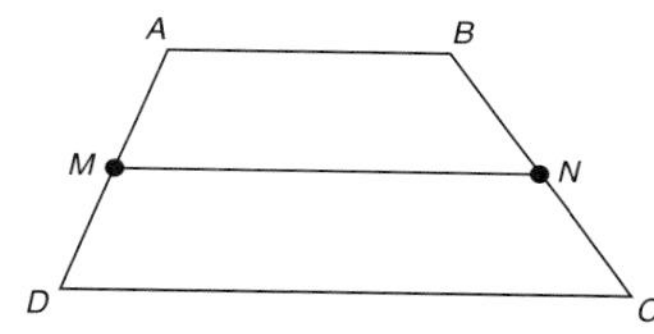

Exercises 7–12

7. *Given:* $AB = 7.3$ and $DC = 12.1$
Find: MN

8. *Given:* $MN = 6.3$ and $DC = 7.5$
Find: AB

9. *Given:* $AB = 8.2$ and $MN = 9.5$
Find: DC

10. *Given:* $AB = 7x + 5$,
$DC = 4x - 2$,
$MN = 5x + 3$
Find: x

11. *Given:* $AB = 6x + 5$,
$DC = 8x - 1$
Find: MN, in terms of x

12. *Given:* $AB = x + 3y + 4$ and $DC = 3x + 5y - 2$
Find: MN, in terms of x and y

13. *Given:* $ABCD$ is an isosceles trapezoid
Prove: $\triangle ABE$ is isosceles

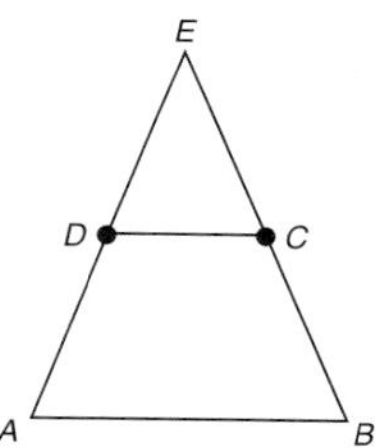

Exercises 13, 14

14. *Given:* Isosceles $\triangle ABE$ with $\overline{AE} \cong \overline{BE}$; also, D and C are midpoints of $\overline{AE}$ and $\overline{BE}$, respectively
Prove: $ABCD$ is an isosceles trapezoid

15. In isosceles trapezoid $MNPQ$ with $\overline{MN} \parallel \overline{QP}$, diagonal $\overline{MP} \perp \overline{MQ}$. If $PQ = 13$ and $NP = 5$, how long is diagonal $\overline{MP}$?

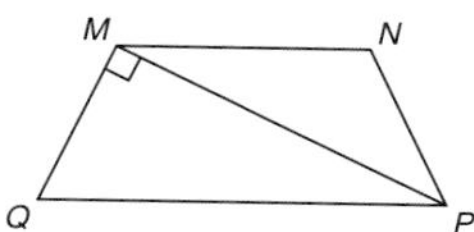

16. In trapezoid $RSTV$, $\overline{RV} \parallel \overline{ST}$, $m\angle SRV = 90°$, and M and N are midpoints of the nonparallel sides. If $ST = 13$, $RV = 17$, and $RS = 16$, how long is $\overline{RN}$?

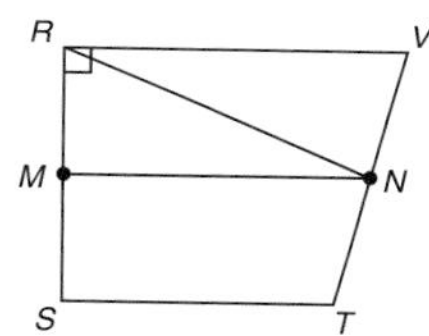

17. Each vertical section of a suspension bridge is in the shape of a trapezoid. For additional support, a vertical cable is placed midway as shown. If the two vertical columns shown have heights of 20 ft and 24 ft and the section is 10 ft wide, what will the height of the cable be?

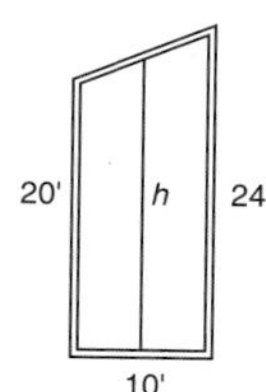

18. The state of Nevada approximates the shape of a trapezoid with these dimensions for boundaries: 340 miles on the north, 515 miles on the east, 435 miles on the south, and 225 miles on the west. If A and B are points located midway across the north and south boundaries, what is the distance from A to B?

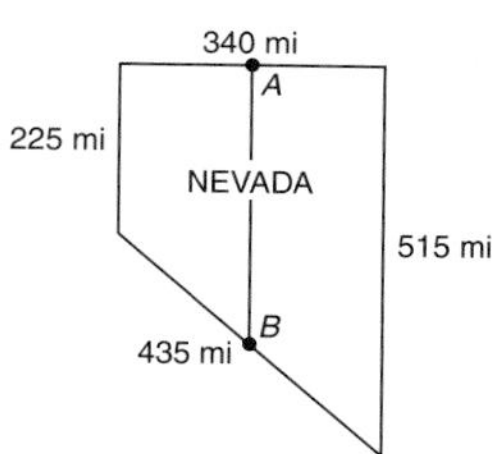

19. In the figure, $a \parallel b \parallel c$ and B is the midpoint of $\overline{AC}$. If $AB = 2x + 3$, $BC = x + 7$, and $DE = 3x + 2$, find the length of $\overline{EF}$.

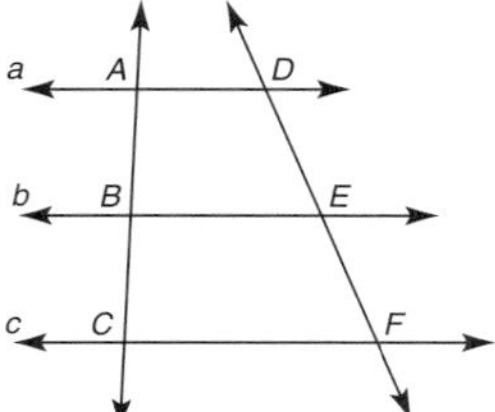

Exercises 19, 20

20. In the figure, $a \parallel b \parallel c$ and B is the midpoint of $\overline{AC}$. If $AB = 2x + 3y$, $BC = x + y + 7$, $DE = 2x + 3y + 3$, and $EF = 5x - y + 2$, find x and y.

In Exercises 21 to 27, complete a formal proof.

21. The diagonals of an isosceles trapezoid are congruent.

22. The median of a trapezoid is parallel to each base.

23. If two consecutive angles of a quadrilateral are supplementary, the quadrilateral is a trapezoid.

24. If two base angles of a trapezoid are congruent, the trapezoid is an isosceles trapezoid.

25. If three parallel lines intercept congruent segments on one transversal, then they intercept congruent segments on any transversal.

26. If the midpoints of the sides of an isosceles trapezoid are joined in order, then the quadrilateral formed is a rhombus.

27. *Given:* $\overline{EF}$ is the median of trapezoid $ABCD$
Prove: $EF = \frac{1}{2}(AB + DC)$
(**HINT:** Using Theorem 4.4.8, show that M is the midpoint of $\overline{AC}$. For $\triangle ADC$ and $\triangle CBA$, apply Theorem 4.2.5.)

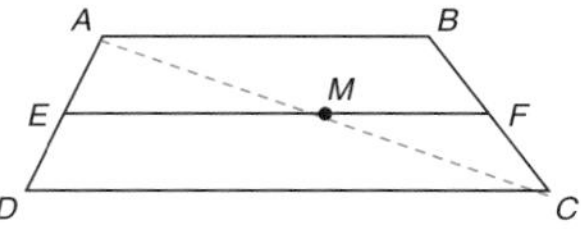

28. *Given:* $\overline{AB} \parallel \overline{DC}$
$m\angle A = m\angle B = 56°$
$\overline{CE} \parallel \overline{DA}$ and $\overrightarrow{CF}$ bisects $\angle DCB$
Find: $m\angle FCE$

A Look Beyond: Historical Sketch of Thales

One of the most significant contributors to the development of geometry was the Greek mathematician Thales of Miletus (625 B.C.–547 B.C.). Thales is credited with being the "Father of Geometry" because he was the first person to organize geometric thought and utilize the deductive method as a means of verifying propositions (theorems). It is not surprising that Thales made original discoveries in geometry. Just as significant as his discoveries was Thales' persistence in verifying the claims of his predecessors. In this textbook, you will find that propositions such as these are only a portion of those that can be attributed to Thales:

Chapter 1: If two straight lines intersect, the opposite (vertical) angles formed are equal.
Chapter 3: The base angles of an isosceles triangle are equal.
Chapter 5: The sides of similar triangles are proportional.
Chapter 6: An angle inscribed in a semicircle is a right angle.

Thales' knowledge of geometry was matched by the wisdom that he displayed in everyday affairs. For example, he is known to have measured the height of the Great Pyramid of Egypt by comparing the lengths of the shadows cast by the pyramid and by his own staff. Thales also used his insights into geometry to measure the distances from the land to ships at sea.

Perhaps the most interesting story concerning Thales was one related by Aesop (famous for fables). It seems that Thales was on his way to market with his beasts of burden carrying saddlebags filled with salt. Quite by accident, one of the mules discovered that rolling in the stream greatly reduced this load; of course, the lesser weight was due to the dissolving of salt in the saddlebags. On subsequent trips, the same mule continued to lighten his load by rolling in the water. Thales soon realized the need to do something (anything!) to modify the mule's behavior. When preparing for the next trip, Thales filled the offensive mule's saddlebags with sponges. When the mule took his usual dive, he found that his load was heavier than ever. Soon the mule realized the need to keep the saddlebags out of the water. In this way, it is said that Thales discouraged the mule from allowing the precious salt to dissolve during later trips to market.

Summary

A Look Back at Chapter 4

The goal of this chapter has been to develop the properties of quadrilaterals, including special types of quadrilaterals such as the parallelogram, rectangle, and trapezoid. Table 4.1 (below), which summarizes the properties of quadrilaterals, is continued on page 180.

A Look Ahead to Chapter 5

In the next chapter, similarity will be defined for all polygons, with an emphasis on triangles. The Pythagorean Theorem, which we mentioned in Chapter 4, will be proved in Chapter 5. Special right triangles will be discussed.

Important Terms and Concepts of Chapter 4

4.1 Quadrilateral
Skew
Parallelogram
Diagonals of a Parallelogram

4.2 Quadrilaterals that are Parallelograms
Rectangle
Kite

4.3 Rectangle
Square
Rhombus
Pythagorean Theorem

4.4 Trapezoid (Bases, Legs, Base Angles, Median)
Isosceles Trapezoid

A Look Beyond: Historical Sketch of Thales

TABLE 4.1
Properties of Quadrilaterals

	parallelogram	rectangle	rhombus	square	kite	trapezoid	isosceles trapezoid
congruent sides	both pairs of opposite sides	both pairs of opposite sides	all four sides	all four sides	both pairs of adjacent sides	possible; also see isosceles trapezoid	pair of legs

(continued)

TABLE 4.1 (continued)

Properties of Quadrilaterals

	parallelogram	*rectangle*	*rhombus*	*square*	*kite*	*trapezoid*	*isosceles trapezoid*
parallel sides	both pairs of opposite sides 	both pairs of opposite sides 	both pairs of opposite sides 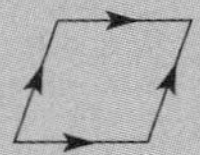	both pairs of opposite sides 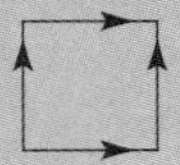	generally none	pair of bases 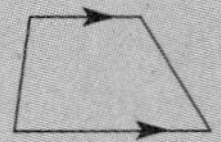	pair of bases 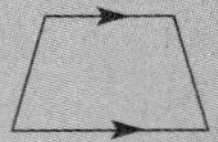
perpendicular sides	if parallelo-gram is a rectangle or square	consecutive pairs 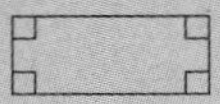	if rhombus is a square	consecutive pairs 	possible 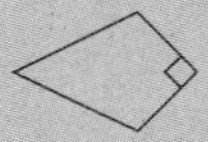	possible 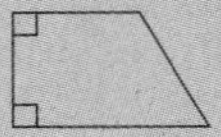	generally none
congruent angles	both pairs of opposite angles 	all four angles 	both pairs of opposite angles 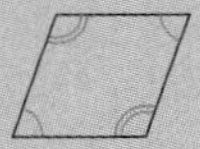	all four angles 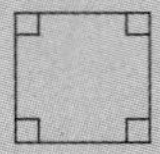	one pair of opposite angles 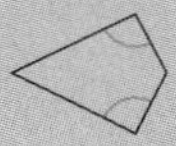	possible; also see isosceles trapezoid 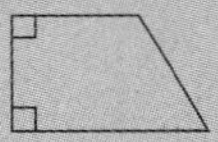	each pair of base angles 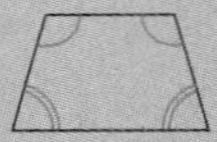
supplementary angles	all pairs of consecutive angles 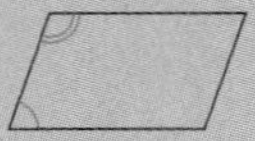	any two angles 	all pairs of consecutive angles 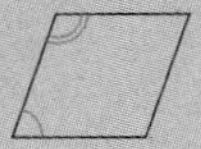	any two angles 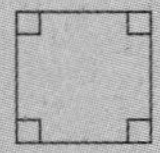	possibly two pairs 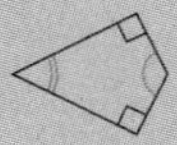	each pair of leg angles 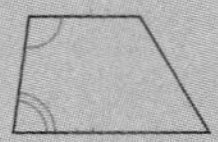	each pair of leg angles 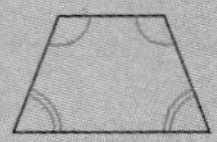
diagonal relationships	bisect each other 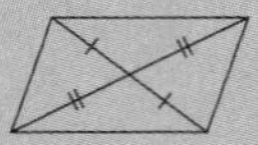	congruent; bisect each other 	perpendicu-lar; bisect each other and interior angles 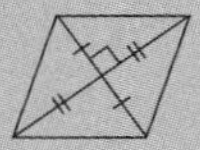	congruent; perpendicu-lar; bisect each other and interior angles 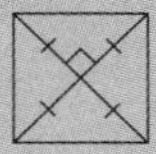	perpendicu-lar; one bisects other and two interior angles 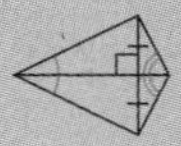	intersect 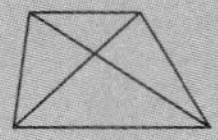	congruent 

Review Exercises

State whether the statements in Review Exercises 1 to 12 are always true (A), sometimes true (S), or never true (N).

1. A square is a rectangle.
2. If two of the angles of a trapezoid are congruent, then the trapezoid is isosceles.
3. The diagonals of a trapezoid bisect each other.
4. The diagonals of a parallelogram are perpendicular.
5. A rectangle is a square.
6. The diagonals of a square are perpendicular.
7. Two consecutive angles of a parallelogram are supplementary.
8. Opposite angles of a rhombus are congruent.
9. The diagonals of a rectangle are congruent.
10. The four sides of a kite are congruent.
11. The diagonals of a parallelogram are congruent.
12. The diagonals of a kite are perpendicular bisectors of each other.
13. *Given:* $\square ABCD$
    $CD = 2x + 3$
    $BC = 5x - 4$
    Perimeter of $\square ABCD = 96$ cm
    Find: The lengths of the sides of $\square ABCD$

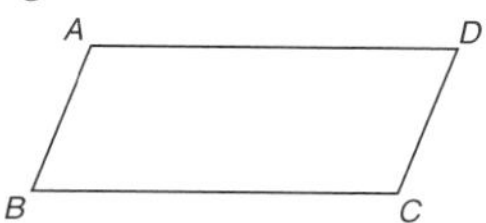

Exercises 13, 14

14. *Given:* $\square ABCD$
    $m\angle A = 2x + 6$
    $m\angle B = x + 24$
    Find: $m\angle C$
15. The diagonals of $\square ABCD$ (not shown) are perpendicular. If one diagonal has a length of 10 and the other diagonal has a length of 24, find the perimeter of the parallelogram.
16. *Given:* $\square MNOP$
    $m\angle M = 4x$
    $m\angle O = 2x + 50$
    Find: $m\angle M$ and $m\angle P$

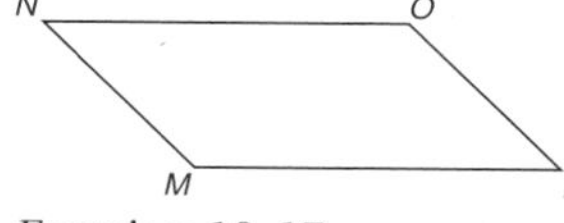

Exercises 16, 17

17. Using the information from Exercise 16, which diagonal ($\overline{MO}$ or $\overline{PN}$) would be longer?
18. In quadrilateral $ABCD$, M is the midpoint of $\overline{BD}$ and $\overline{AC} \perp \overline{DB}$ at M. What special type of quadrilateral is $ABCD$?
19. In isosceles trapezoid $DEFG$, $\overline{DE} \parallel \overline{GF}$ and $m\angle D = 108°$. Find the measures of the other angles in the trapezoid.
20. One base of a trapezoid has a length of 12.3 cm while the other base is 17.5 cm. Find the length of the median of the trapezoid.
21. In trapezoid $MNOP$, $\overline{MN} \parallel \overline{PO}$ and R and S are the midpoints of $\overline{MP}$ and $\overline{NO}$, respectively. Find the lengths of the bases if $RS = 15$, $MN = 3x + 2$, and $PO = 2x - 7$.

In Review Exercises 22 to 24, M and N are the midpoints of $\overline{FJ}$ and $\overline{FH}$, respectively.

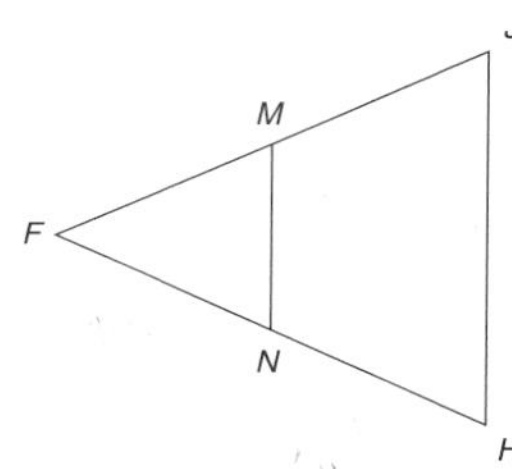

Exercises 22–24

22. *Given:* Isosceles $\triangle FJH$ with
    $\overline{FJ} \cong \overline{FH}$
    $FM = 2y + 3$
    $NH = 5y - 9$
    $JH = 2y$
    Find: The perimeter of $\triangle FMN$
23. *Given:* $JH = 12$
    $m\angle J = 80$
    $m\angle F = 60$
    Find: MN, $m\angle FMN$, $m\angle FNM$
24. *Given:* $MN = x^2 + 6$
    $JH = 2x(x + 2)$
    Find: x, MN, JH
25. *Given:* $ABCD$ is a $\square$
    $\overline{AF} \cong \overline{CE}$
    Prove: $\overline{DF} \parallel \overline{EB}$

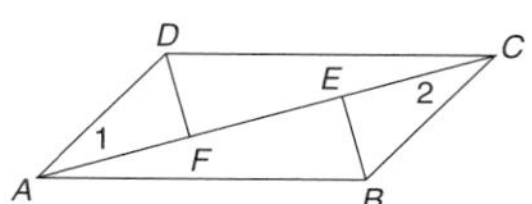

Exercise 25

26. *Given:* $ABEF$ is a rectangle
$BCDE$ is a rectangle
$\overline{FE} \cong \overline{ED}$
Prove: $\overline{AE} \cong \overline{BD}$ and $\overline{AE} \parallel \overline{BD}$

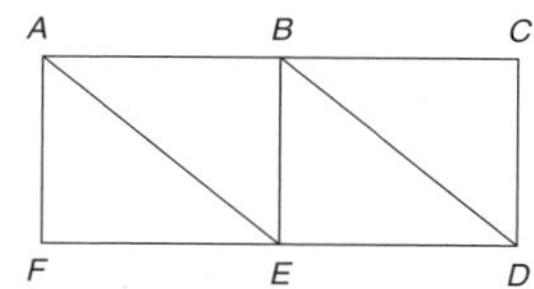

27. *Given:* $\overline{DE}$ is a median in $\triangle ADC$
$\overline{BE} \cong \overline{FD}$
$\overline{EF} \cong \overline{FD}$
Prove: $ABCF$ is a ▱

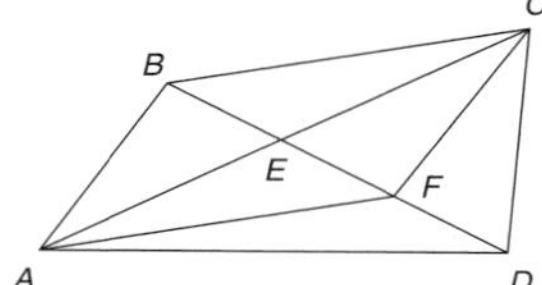

28. *Given:* $\triangle FAB \cong \triangle HCD$
$\triangle EAD \cong \triangle GCB$
Prove: $ABCD$ is a ▱

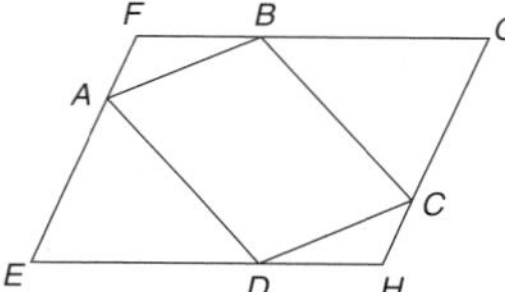

29. *Given:* $ABCD$ is a parallelogram
$\overline{DC} \cong \overline{BN}$
$\angle 3 \cong \angle 4$
Prove: $ABCD$ is a rhombus

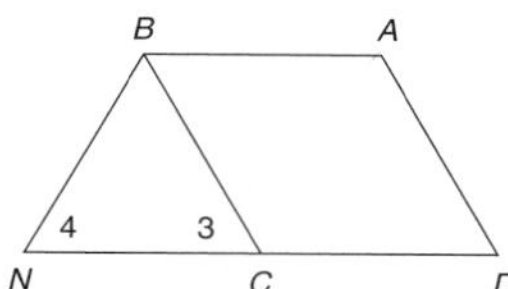

30. *Given:* $\triangle TWX$ is isosceles, with base $\overline{WX}$
$\overline{RY} \parallel \overline{WX}$
Prove: $RWXY$ is an isosceles trapezoid

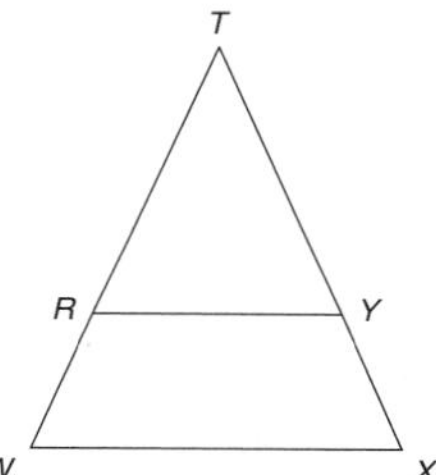

31. Construct a rhombus, given these lengths for the diagonals.

Chapter 5 Similar Triangles

The picture on the right is an enlargement of the one on the left. The proportionality of the person's features tells us of the relationship between the two portraits. Likewise the rectangular frame on the right is an enlargement of the other frame. In geometry, we say that the two rectangles (shapes of the frames) are similar. In this chapter, we study similar figures while emphasizing the similarity of triangles.

The applications found in this chapter and later chapters sometimes lead to quadratic equations. A review of methods for solving quadratic equations can be found in Appendix A of this textbook.

CHAPTER OUTLINE

5.1 Ratios, Rates, and Proportions

The concepts and techniques discussed in Section 5.1 are often necessary for the geometry applications found later in Chapter 5 and beyond.

A **ratio** is the quotient $\frac{a}{b}$ (where $b \neq 0$) that provides a comparison between numbers a and b. Since every fraction indicates a division, every fraction represents a ratio. Read "a to b," the ratio is sometimes written in the form $a:b$.

It is generally preferable to provide the ratio in simplest form, so the ratio 6 to 8 would be reduced (in fraction form) from $\frac{6}{8}$ to $\frac{3}{4}$. If units of measure are involved, these units must be **commensurable** (convertible to the same unit of measure). When simplifying the ratio of two quantities that are expressed in the same unit, you eliminate the common unit in the process. If two quantities cannot be compared because no common unit of measure is possible, the quantities are **incommensurable.**

EXAMPLE 1

Find the best form of each ratio:

a) 12 to 20
b) 12 in. to 24 in.
c) 12 in. to 3 ft (NOTE: 1 ft = 12 in.)
d) 5 lb to 20 oz (NOTE: 1 lb = 16 oz)
e) 5 lb to 2 ft
f) 4 m to 30 cm (NOTE: 1 m = 100 cm)

Solution

a) $\frac{12}{20} = \frac{3}{5}$

b) $\frac{12 \text{ in.}}{24 \text{ in.}} = \frac{12}{24} = \frac{1}{2}$

c) $\frac{12 \text{ in.}}{3 \text{ ft}} = \frac{12 \text{ in.}}{3(12 \text{ in.})} = \frac{12 \text{ in.}}{36 \text{ in.}} = \frac{1}{3}$

d) $\frac{5 \text{ lb}}{20 \text{ oz}} = \frac{5(16 \text{ oz})}{20 \text{ oz}} = \frac{80 \text{ oz}}{20 \text{ oz}} = \frac{4}{1}$

e) $\frac{5 \text{ lb}}{2 \text{ ft}}$ is incommensurable!

f) $\frac{4 \text{ m}}{30 \text{ cm}} = \frac{4(100 \text{ cm})}{30 \text{ cm}} = \frac{400 \text{ cm}}{30 \text{ cm}} = \frac{40}{3}$

Geometry IN THE REAL WORLD

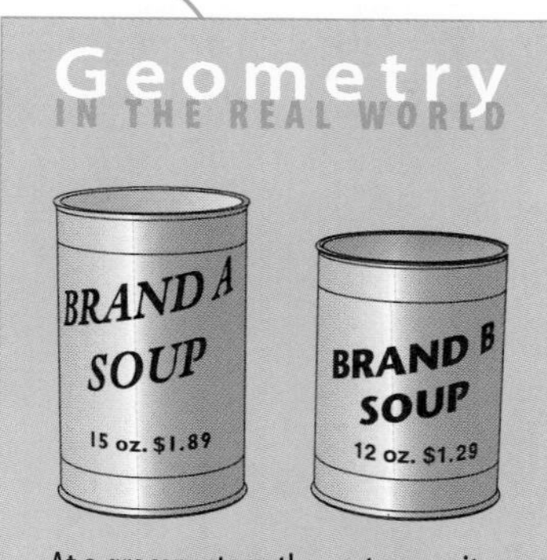

At a grocery store, the cost per unit is a rate that allows the consumer to know which brand name is more expensive.

A **rate** is a quotient that compares two quantities that are incommensurable. If an automobile can travel 300 miles of an interstate on a full tank of 10 gallons of gasoline, then its consumption *rate* is $\frac{300 \text{ miles}}{10 \text{ gallons}}$. In simplified form, the consumption rate is $\frac{30 \text{ mi}}{\text{gal}}$, or "30 miles per gallon or 30 mpg."

A **proportion** is a statement that two ratios or two rates are equal. Thus $\frac{a}{b} = \frac{c}{d}$ is a proportion and may be read as "a is to b as c is to d." In the order read, a is the *first term* of the proportion, b is the *second term*, c is the *third term*, and d is the *fourth term*. The first and last terms (a and d) of the proportion are the **extremes** while the second and third terms (b and c) are the **means.**

The following property is convenient for solving many proportions.

PROPERTY 1: (Means-Extremes Property)
In a proportion, the product of the means equals the product of the extremes; that is, if $\frac{a}{b} = \frac{c}{d}$ (where $b \neq 0$ and $d \neq 0$), then $a \cdot d = b \cdot c$.

In the false proportion $\frac{9}{12} = \frac{2}{3}$, it is obvious that $9 \cdot 3 \neq 12 \cdot 2$; on the other hand, the truth of the statement $\frac{9}{12} = \frac{3}{4}$ is evident from the fact that $9 \cdot 4 = 12 \cdot 3$. Henceforth, any proportion given in this text is intended to be a true proportion.

EXAMPLE 2

Use the Means-Extremes Property to solve each proportion for x.

a) $\frac{x}{8} = \frac{5}{12}$

b) $\frac{x+1}{9} = \frac{x-3}{3}$

c) $\frac{3}{x} = \frac{x}{2}$

d) $\frac{x+3}{3} = \frac{9}{x-3}$

e) $\frac{x+2}{5} = \frac{4}{x-1}$

Solution

a) $x \cdot 12 = 8 \cdot 5$ (Means-Extremes Property)

$12x = 40$

$x = \frac{40}{12} = \frac{10}{3}$

b) $3(x + 1) = 9(x - 3)$ (Means-Extremes Property)

$3x + 3 = 9x - 27$

$30 = 6x$

$x = 5$

c) $3 \cdot 2 = x \cdot x$ (Means-Extremes Property)

$x^2 = 6$

$x = \pm\sqrt{6} \approx \pm 2.45$

d) $(x + 3)(x - 3) = 3 \cdot 9$ (Means-Extremes Property)

$x^2 - 9 = 27$

$x^2 - 36 = 0$

$(x + 6)(x - 6) = 0$ (using factoring)

$x + 6 = 0$ or $x - 6 = 0$

$x = -6$ or $x = 6$

e) $(x + 2)(x - 1) = 5 \cdot 4$ (Means-Extremes Property)

$x^2 + x - 2 = 20$

$x^2 + x - 22 = 0$

$x = \frac{-b \pm \sqrt{b^2 - 4ac}}{2a}$ (using Quadratic Formula; see Appendix A)

$= \frac{-1 \pm \sqrt{(1)^2 - 4(1)(-22)}}{2(1)}$

$= \frac{-1 \pm \sqrt{1 + 88}}{2}$

$= \frac{-1 \pm \sqrt{89}}{2}$

≈ 4.22 or -5.22

In application problems involving proportions, it is essential to order the related quantities in each ratio or rate. The following example illustrates the care that must be taken in forming the proportion for an application.

EXAMPLE 3

If an automobile can travel 90 mi on 4 gal of gasoline, how far can it travel on 6 gal of gasoline?

Solution By form,

$$\frac{\text{number miles first trip}}{\text{number gallons first trip}} = \frac{\text{number miles second trip}}{\text{number gallons second trip}}$$

Where x represents the number of miles traveled on the second trip, we have

$$\frac{90}{4} = \frac{x}{6}$$
$$4x = 540$$
$$x = 135$$

Thus the car can travel 135 mi on 6 gal of gasoline.

In $\frac{a}{b} = \frac{b}{c}$, where the second and third terms of the proportion are identical, the value of b is known as the **geometric mean** of a and c. For example, 6 and -6 are the geometric means of 4 and 9 because $\frac{4}{6} = \frac{6}{9}$ and $\frac{4}{-6} = \frac{-6}{9}$. Because applications in geometry generally require positive solutions, we usually seek only the positive geometric mean of a and c.

EXAMPLE 4

In Figure 5.1, AD is the geometric mean of BD and DC. If $BC = 10$ and $BD = 4$, determine AD.

Solution $\frac{BD}{AD} = \frac{AD}{DC}$. Because $DC = BC - BD$, we know that $DC = 10 - 4 = 6$. Therefore

$$\frac{4}{x} = \frac{x}{6}$$

in which x is the length of $\overline{AD}$. Applying the Means-Extremes Property, we get

$$x^2 = 24$$
$$x = \pm\sqrt{24} = \pm\sqrt{4 \cdot 6} = \pm\sqrt{4} \cdot \sqrt{6} = \pm 2\sqrt{6}$$

To have a permissible length for $\overline{AD}$, the geometric mean is the positive solution. Thus $AD = 2\sqrt{6}$ or $AD \approx 4.90$.

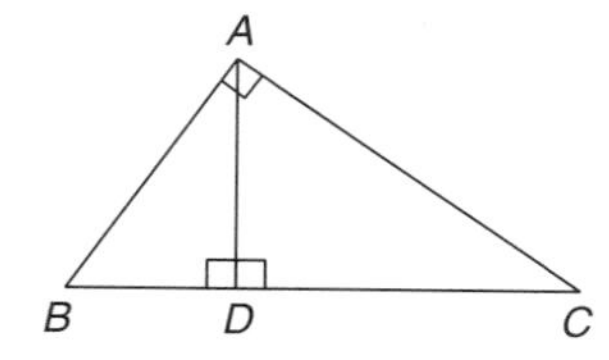

FIGURE 5.1

An **extended ratio** compares more than two quantities and must be expressed in a form such as a:b:c or d:e:f:g. If you know that the angles of a triangle are 90°, 60°, and 30°, then the ratio that compares these measures is 90:60:30 or 3:2:1 (since 90, 60, and 30 have the greatest common factor of 30). Unknown quantities in the ratio a:b:c:d: . . . are generally represented by variable expressions such as ax, bx, cx, dx, . . .

EXAMPLE 5

Suppose that the perimeter of a quadrilateral is 70 and that the sides are in the ratio of 2:3:4:5. Find the measure of each side.

Solution

Let the lengths of the sides be represented by $2x$, $3x$, $4x$, and $5x$. Then

$$\begin{aligned} 2x + 3x + 4x + 5x &= 70 \\ 14x &= 70 \\ x &= 5 \end{aligned}$$

The lengths of sides are 10, 15, 20, and 25, because

$$2x = 10,\ 3x = 15,\ 4x = 20, \text{ and } 5x = 25$$

It is possible to solve certain problems in more ways than one, as is illustrated in the next example. However, the solution is unique and will not be affected by the method chosen.

EXAMPLE 6

Two complementary angles are in the ratio 2 to 3. Find the measure of each angle.

Solution

Let the first of the complementary angles have measure x; then the second has the measure $90 - x$. Thus we have

$$\frac{x}{90 - x} = \frac{2}{3}$$

Using the Means-Extremes Property, we have

$$\begin{aligned} 3x &= 2(90 - x) \\ 3x &= 180 - 2x \\ 5x &= 180 \\ x &= 36 \\ 90 - x &= 54 \end{aligned}$$

The angles have measures of 36° and 54°.

Alternate Solution

Let the measures of the two angles be $2x$ and $3x$, since they are in the ratio 2:3. Because the angles are complementary,

$$\begin{aligned} 2x + 3x &= 90 \\ 5x &= 90 \\ x &= 18 \end{aligned}$$

Now $2x = 36$ and $3x = 54$, so the measures of the two angles are 36° and 54°.

Some additional properties of proportions follow. Because they are not cited as often as the Means-Extremes Property, they are not given titles.

PROPERTY 2: In a proportion, the means or the extremes (or both the means and the extremes) may be interchanged; that is, if $\frac{a}{b} = \frac{c}{d}$ (where a, b, c, and d are nonzero), then $\frac{a}{c} = \frac{b}{d}$, $\frac{d}{b} = \frac{c}{a}$, and $\frac{d}{c} = \frac{b}{a}$.

NOTE: The last proportion is the inverted form of the given proportion. This property can easily be remembered by considering that $a \cdot d = b \cdot c$ (product of means = product of extremes) in all proportions.

If given the proportion $\frac{2}{3} = \frac{8}{12}$, Property 2 enables us to draw conclusions such as:

1. $\frac{2}{8} = \frac{3}{12}$ (means switched);
2. $\frac{12}{3} = \frac{8}{2}$ (extremes switched); and
3. $\frac{3}{2} = \frac{12}{8}$ (both sides inverted).

PROPERTY 3: If $\frac{a}{b} = \frac{c}{d}$ (where $b \neq 0$ and $d \neq 0$), then $\frac{a+b}{b} = \frac{c+d}{d}$ and $\frac{a-b}{b} = \frac{c-d}{d}$.

NOTE: When each denominator is added to (subtracted from) each numerator on each side of a proportion, another true proportion is obtained.

If given the proportion $\frac{2}{3} = \frac{8}{12}$, Property 3 enables us to draw such conclusions as:

1. $\frac{2+3}{3} = \frac{8+12}{12}$ (each side simplifies to $\frac{5}{3}$) and
2. $\frac{2-3}{3} = \frac{8-12}{12}$ (each side equals $-\frac{1}{3}$).

Just as there are extended ratios, there are also **extended proportions,** such as

$$\frac{a}{b} = \frac{c}{d} = \frac{e}{f} = \dots$$

Suggested by different numbers of servings of a particular recipe, the statement below is an extended proportion comparing numbers of eggs to numbers of cups of milk:

$$\frac{2 \text{ eggs}}{3 \text{ cups}} = \frac{4 \text{ eggs}}{6 \text{ cups}} = \frac{6 \text{ eggs}}{9 \text{ cups}}$$

EXAMPLE 7

In the triangles shown in Figure 5.2, $\frac{AB}{DE} = \frac{AC}{DF} = \frac{BC}{EF}$. Find the lengths of $\overline{DF}$ and $\overline{EF}$.

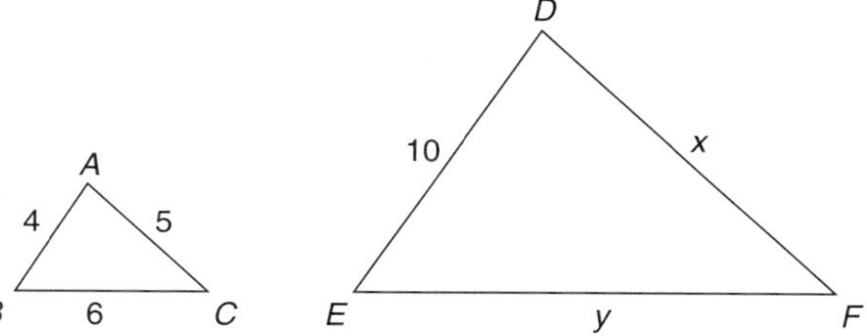

FIGURE 5.2

Solution Substituting, we have

$$\frac{4}{10} = \frac{5}{x} = \frac{6}{y}$$

From the equation

$$\frac{4}{10} = \frac{5}{x}$$

it follows that $4x = 50$ and that $x = DF = 12.5$. Using the equation

$$\frac{4}{10} = \frac{6}{y}$$

we find that $4y = 60$, so $y = EF = 15$.

5.1 Exercises

In Exercises 1 to 4, give the ratios in simplified form.

1. **a)** 12 to 15 **b)** 12 in. to 15 in. **c)** 1 ft to 18 in. **d)** 1 ft to 18 oz
2. **a)** 20 to 36 **b)** 24 oz to 52 oz **c)** 20 oz to 2 lb **d)** 2 lb to 20 oz
3. **a)** 15:24 **b)** 2 ft:2 yd **c)** 2 m:150 cm **d)** 2 m:1 lb
4. **a)** 24:32 **b)** 12 in.:2 yd **c)** 150 cm:2 m **d)** 1 gal:24 mi

In Exercises 5 to 14, find the value of x in each proportion.

5. **a)** $\frac{x}{4} = \frac{9}{12}$ **b)** $\frac{7}{x} = \frac{21}{24}$
6. **a)** $\frac{x-1}{10} = \frac{3}{5}$ **b)** $\frac{x+1}{6} = \frac{10}{12}$

7. a) $\frac{x-3}{8} = \frac{x+3}{24}$ b) $\frac{x+1}{6} = \frac{4x-1}{18}$

8. a) $\frac{9}{x} = \frac{x}{16}$ b) $\frac{32}{x} = \frac{x}{2}$

9. a) $\frac{x}{4} = \frac{7}{x}$ b) $\frac{x}{6} = \frac{3}{x}$

10. a) $\frac{x+1}{3} = \frac{10}{x+2}$ b) $\frac{x-2}{5} = \frac{12}{x+2}$

11. a) $\frac{x+1}{x} = \frac{10}{2x}$ b) $\frac{2x+1}{x+1} = \frac{14}{3x-1}$

12. a) $\frac{x+1}{2} = \frac{7}{x-1}$ b) $\frac{x+1}{3} = \frac{5}{x-2}$

13. a) $\frac{x+1}{x} = \frac{2x}{3}$ b) $\frac{x+1}{x-1} = \frac{2x}{5}$

14. a) $\frac{x+1}{x} = \frac{x}{x-1}$ b) $\frac{x+2}{x} = \frac{2x}{x-2}$

15. Sarah ran the 300-m hurdles in 47.7 seconds. In meters per second, find the rate at which Sarah ran. Give the answer to the nearest tenth of a meter per second.

16. Fran has been hired to sew the dance troupe's dresses for the school musical. If $13\frac{1}{3}$ yd of material are needed for the four dresses, find the rate that describes the amount of material needed for each dress.

In Exercises 17 to 22, use proportions to solve each problem.

17. A recipe calls for 4 eggs and 3 cups of milk. To prepare for a larger number of guests, a cook uses 14 eggs. How many cups of milk are needed?

18. If a school secretary copies 168 worksheets for a class of 28 students, how many must be prepared for a class of 32 students?

19. An electrician installs 20 electrical outlets in a new six-room house. Assuming proportionality, how many outlets should be installed in a new construction having seven rooms? (Round up to an integer.)

20. The secretarial pool (15 secretaries in all) on one floor of a corporate complex has access to four copy machines. If there are 23 secretaries on a different floor, approximately what number of copy machines should be available? (Assume a proportionality.)

21. Assume that AD is the geometric mean of BD and DC in $\triangle ABC$ shown at the top of the next column.

   a) Find AD if $BD = 6$ and $DC = 8$.
   b) Find BD if $AD = 6$ and $DC = 8$.

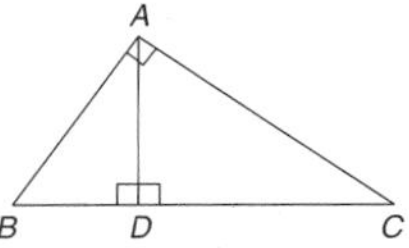

Exercises 21, 22

22. In the drawing, assume that AB is the geometric mean of BD and BC.

   a) Find AB if $BD = 6$ and $DC = 10$.
   b) Find DC if $AB = 10$ and $BC = 15$.

23. The salaries of a secretary, a salesperson, and a vice president for a retail sales company are in the ratio 2:3:5. If their combined annual salaries amount to \$92,500, what is the annual salary of each?

24. If the angles of a quadrilateral are in the ratio of 2:3:4:6, find the measure of each angle.

25. The measures of two complementary angles are in the ratio 4:5. Find the measure of each angle, using the two methods shown in Example 6.

26. The measures of two supplementary angles are in the ratio of 2:7. Find the measure of each angle, using the two methods of Example 6.

27. If 1 inch equals 2.54 centimeters, use a proportion to convert 12 inches to centimeters.
(**HINT:** $\frac{2.54 \text{ cm}}{1 \text{ in.}} = \frac{x \text{ cm}}{12 \text{ in.}}$)

28. If 1 kilogram equals 2.2 pounds, use a proportion to convert 12 pounds to kilograms.

29. For the quadrilaterals shown, $\frac{MN}{WX} = \frac{NP}{XY} = \frac{PQ}{YZ} = \frac{MQ}{WZ}$. If $MN = 7$, $WX = 3$, and $PQ = 6$, find YZ.

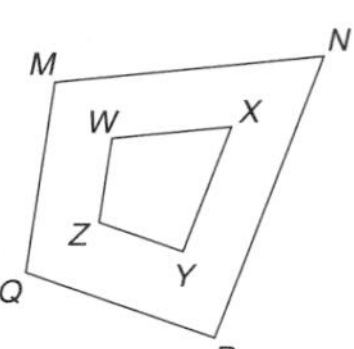

Exercises 29, 30

30. For this exercise, use the drawing and extended ratio of Exercise 29. If $NP = 2 \cdot XY$ and $WZ = 3\frac{1}{2}$, find MQ.

31. Two numbers a and b are in the ratio 3:4. If the first number is decreased by 2 and the second is decreased by 1, they are in the ratio 2:3. Find a and b.

32. If the ratio of the measure of the complement of an angle to the measure of its supplement is 1:4, what is the measure of the angle?

33. On a blueprint, a 1-in. scale corresponds to 3 ft. To show a room with actual dimensions 12 ft wide by 14 ft long, what dimensions should be shown on the blueprint?

5.2 Similar Triangles and Polygons

When two geometric figures have exactly the same shape, they are **similar;** the symbol for "is similar to" is $\sim$. When two figures have the same shape ($\sim$) and all corresponding parts have equal ($=$) measures, the two figures are **congruent** ($\cong$). Notice that the symbol for congruence combines the symbols for similarity and equality.

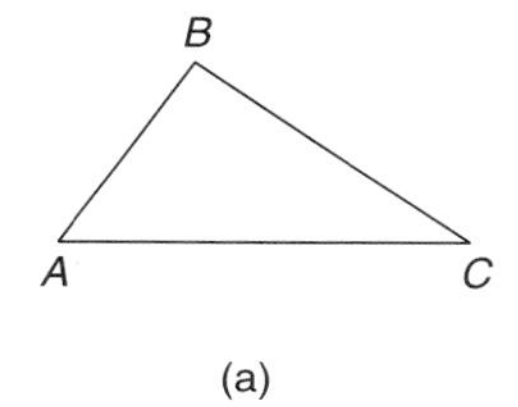

(a)

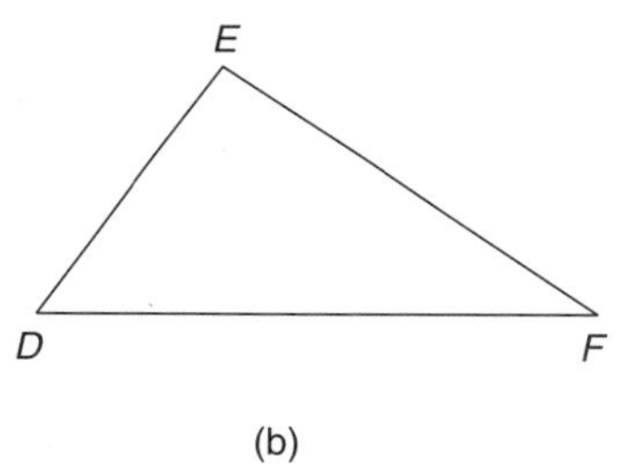

(b)

FIGURE 5.3

While two-dimensional figures can be similar, like $\triangle ABC$ and $\triangle DEF$ in Figure 5.3, it is also possible for three-dimensional figures to be similar; similar orange juice containers are shown in Figures 5.4(a) and (b). Informally, two figures are termed "similar" if one is an enlargement of the other. Thus, a tuna fish can and an orange juice can are *not* similar even if both are right-circular cylinders [see Figures 5.4(b) and (c)].

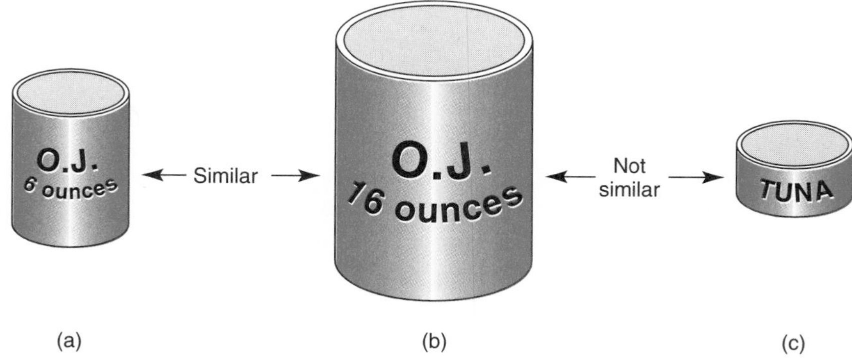

FIGURE 5.4

Generally our discussion of similarity will be limited to plane figures.

For two polygons to be similar, one requirement is that each angle of one polygon must be congruent to the corresponding angle of the other. While this congruence of angles is necessary for similarity of polygons, it alone is not sufficient to establish similarity. The vertices of the congruent angles are **corresponding vertices** of the similar polygons. If $\angle A$ in one polygon is congruent to $\angle M$ in the second polygon, then vertex A corresponds to vertex M, and this is symbolized $A \leftrightarrow M$; we can indicate that $\angle A$ corresponds to $\angle M$ by writing $\angle A \leftrightarrow \angle M$. A pair of angles like $\angle A$ and $\angle M$ are **corresponding angles,** and the sides determined by consecutive and corresponding vertices are **corresponding sides** of the similar polygons.

EXAMPLE 1

Given that quadrilaterals *ABCD* and *HJKL* are similar, with congruent angles indicated in Figure 5.5, name the vertices, angles, and sides that correspond to each other.

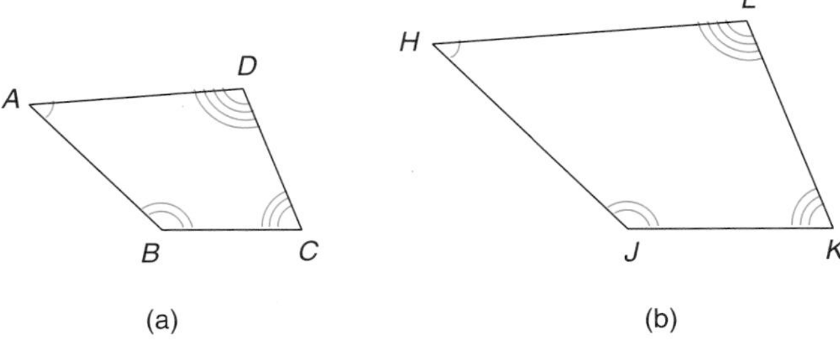

FIGURE 5.5

Solution Since $\angle A \cong \angle H$, it follows that

$$A \leftrightarrow H \quad \text{and} \quad \angle A \leftrightarrow \angle H$$

Similarly,

$$B \leftrightarrow J \quad \text{and} \quad \angle B \leftrightarrow \angle J$$
$$C \leftrightarrow K \quad \text{and} \quad \angle C \leftrightarrow \angle K$$
$$D \leftrightarrow L \quad \text{and} \quad \angle D \leftrightarrow \angle L$$

By associating pairs of consecutive and corresponding vertices, the corresponding sides are included between the corresponding angles.

$$\overline{AB} \leftrightarrow \overline{HJ}, \quad \overline{BC} \leftrightarrow \overline{JK}, \quad \overline{CD} \leftrightarrow \overline{KL}, \quad \overline{AD} \leftrightarrow \overline{HL}$$

DEFINITION: Two polygons are **similar** if and only if two conditions are satisfied:

1. All pairs of corresponding angles are congruent.
2. All pairs of corresponding sides are proportional.

Geometry IN NATURE

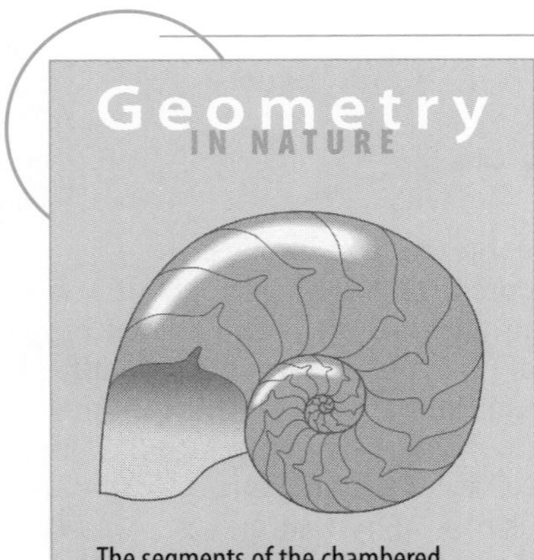

The segments of the chambered nautilus are similar (not congruent) in shape.

The second condition for similarity requires that the following extended proportion exist for the sides of the similar quadrilaterals of Example 1.

$$\frac{AB}{HJ} = \frac{BC}{JK} = \frac{CD}{KL} = \frac{AD}{HL}$$

Notice that *both* conditions for similarity are necessary! While condition 1 is satisfied for square *EFGH* and rectangle *RSTU* [see Figures 5.6(a) and (b)], the figures are not similar—that is, one is not an enlargement of the other—because the extended proportion is not true. On the other hand, condition 2 is satisfied for square *EFGH* and rhombus *WXYZ* [see Figures 5.6(a) and (c)], but the figures are not similar because the angles are not congruent.

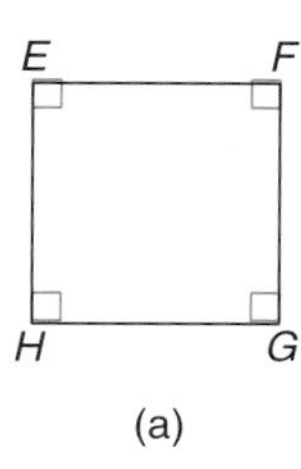

(a)

(b)

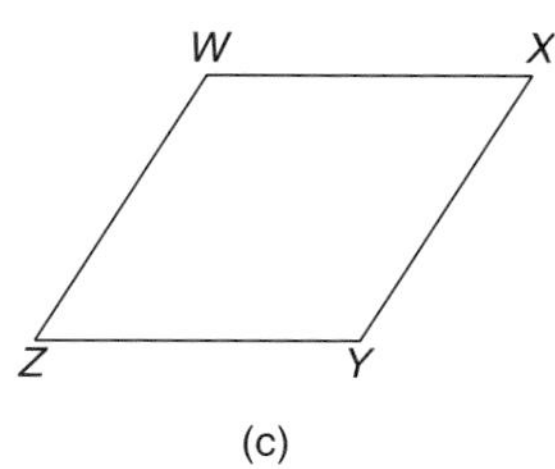

(c)

FIGURE 5.6

The practice of naming corresponding vertices in consecutive order for the two polygons is most convenient! For instance, if pentagon $ABCDE$ is similar to pentagon $MNPQR$, then we know that $A \leftrightarrow M$, $B \leftrightarrow N$, $C \leftrightarrow P$, $D \leftrightarrow Q$, $E \leftrightarrow R$, $\angle A \cong \angle M$, $\angle B \cong \angle N$, $\angle C \cong \angle P$, $\angle D \cong \angle Q$, and $\angle E \cong \angle R$. Because the vertices correspond, we also know that

$$\frac{AB}{MN} = \frac{BC}{NP} = \frac{CD}{PQ} = \frac{DE}{QR} = \frac{EA}{RM}$$

EXAMPLE 2

If $\triangle ABC \sim \triangle DEF$ in Figure 5.7, use the indicated measures to find the measures of the remaining parts of each of the triangles.

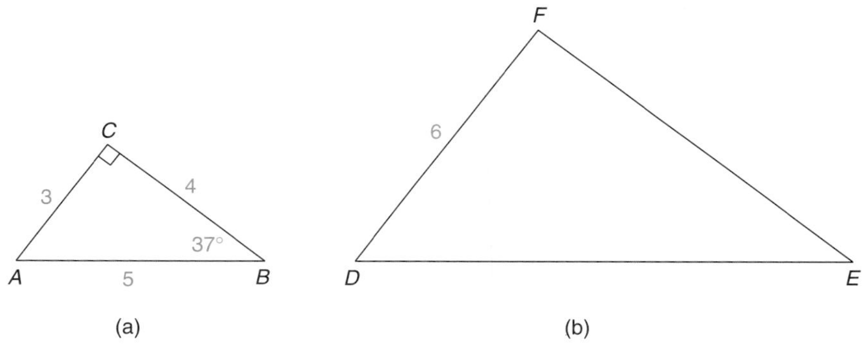

FIGURE 5.7

Solution Because the sum of the measures of the angles of a triangle is 180°,

$$m\angle A = 180 - (90 + 37) = 53°$$

Due to the similarity and the corresponding vertices,

$$m\angle D = 53°, \qquad m\angle E = 37°, \qquad m\angle F = 90°$$

The proportion that relates the lengths of the sides is

$$\frac{AC}{DF} = \frac{CB}{FE} = \frac{AB}{DE} \quad \text{so} \quad \frac{3}{6} = \frac{4}{FE} = \frac{5}{DE}$$

From $\frac{3}{6} = \frac{4}{FE}$, we see that

$$3 \cdot FE = 6 \cdot 4 = 24$$
$$FE = 8$$

From $\frac{3}{6} = \frac{5}{DE}$, we see that

$$3 \cdot DE = 6 \cdot 5 = 30$$
$$DE = 10$$

In a proportion, the ratios can *all* be inverted; thus Example 2 could have been solved by using the ratio

$$\frac{DF}{AC} = \frac{FE}{CB} = \frac{DE}{AB}$$

In an extended proportion, the equal ratios must all be equal to the same constant value. By designating this number (often called the "constant of proportionality") by k, we see that

$$\frac{DF}{AC} = k \qquad \frac{FE}{CB} = k \qquad \frac{DE}{AB} = k$$

It follows that $DF = k \cdot AC$, $FE = k \cdot CB$, and $DE = k \cdot AB$. In Example 2, this constant of proportionality had the value $k = 2$, which means that each side of the larger triangle was double the length of each side of the smaller triangle.

Our definition of similar polygons (and therefore of similar triangles) is almost impossible to use as a method of proof. Fortunately, some easier methods are available for proving triangles similar. If two triangles are carefully sketched or constructed so that their angles are congruent, they will appear to be similar, as shown in Figure 5.8.

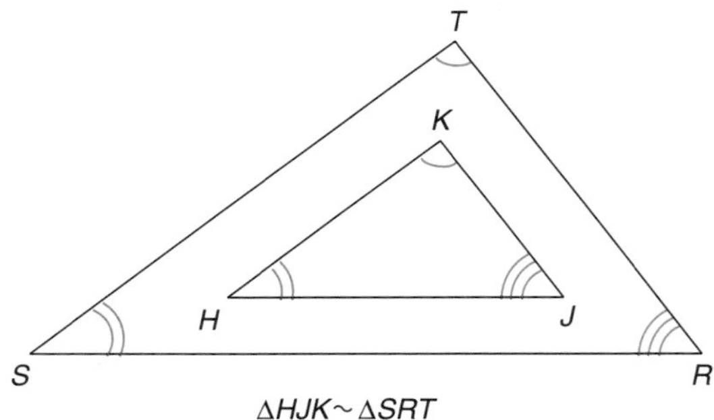

FIGURE 5.8

POSTULATE 15:
If the three angles of one triangle are congruent to the three angles of a second triangle, then the triangles are similar (AAA).

Corollary 5.2.1 follows from knowing that if two angles of one triangle are congruent to two angles of another triangle, then the third angles *must* also be congruent.

COROLLARY 5.2.1: If two angles of one triangle are congruent to two angles of another triangle, then the triangles are similar (AA).

EXAMPLE 3

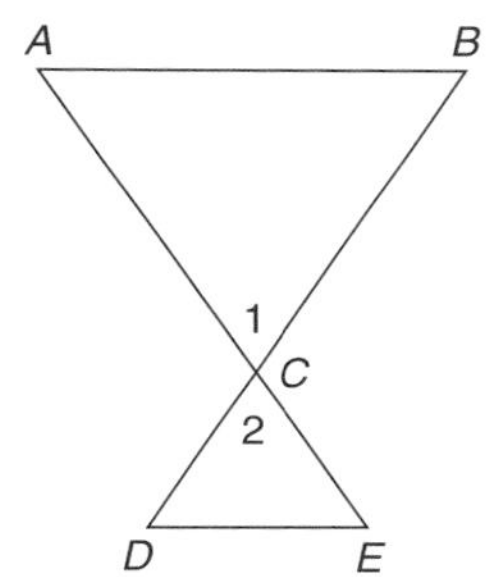

FIGURE 5.9

Provide a two-column proof of the following problem.

Given: $\overline{AB} \parallel \overline{DE}$ in Figure 5.9.

Prove: $\triangle ABC \sim \triangle EDC$

PROOF

Statements	Reasons
1. $\overline{AB} \parallel \overline{DE}$	**1.** Given
2. $\angle A \cong \angle E$	**2.** If two $\parallel$ lines are cut by a transversal, the alternate interior angles are $\cong$
3. $\angle 1 \cong \angle 2$	**3.** Vertical angles are $\cong$
4. $\triangle ABC \sim \triangle EDC$	**4.** AA

In many instances, we wish to prove some relationship beyond the similarity of triangles. This is possible through the definition of "similarity." The following part of the definition of similarity is often cited as a reason in a proof. This fact, abbreviated CSSTP, is used in Example 4. Although the statement involves triangles, we realize that the corresponding sides of any two similar polygons are proportional.

CSSTP: Corresponding sides of similar triangles are proportional.

EXAMPLE 4

Complete the following two-column proof.

Given: $\angle ADE \cong \angle B$ in Figure 5.10 on the next page

Prove: $\frac{DE}{BC} = \frac{AE}{AC}$

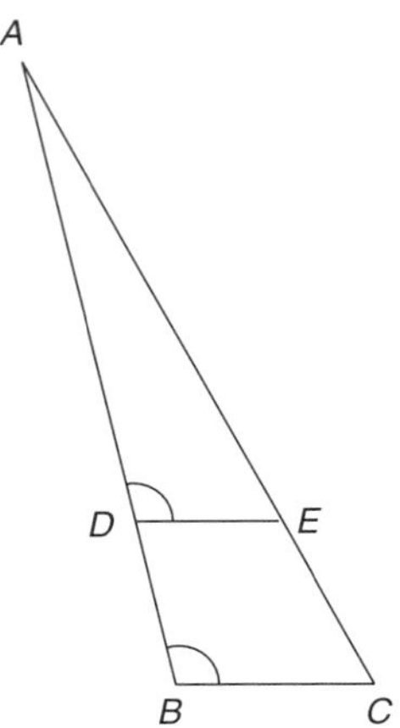

FIGURE 5.10

PROOF

Statements	Reasons
1. $\angle ADE \cong \angle B$	1. Given
2. $\angle A \cong \angle A$	2. Identity
3. $\triangle ADE \sim \triangle ABC$	3. AA
4. $\frac{DE}{BC} = \frac{AE}{AC}$	4. CSSTP

NOTE: In this proof, DE appears above BC because the sides with these names lie opposite $\angle A$ in the two similar triangles. AE and AC are the lengths of sides opposite the congruent and corresponding angles $\angle ADE$ and $\angle B$. That is, corresponding sides of similar triangles always lie opposite corresponding angles.

> **THEOREM 5.2.2:** The lengths of the corresponding altitudes of similar triangles have the same ratio as the lengths of any pair of corresponding sides.

The proof of this theorem is left to the student; see Exercise 21. Note that this proof also requires the use of CSSTP.

EXAMPLE 5

$\angle ADE \cong \angle B$ in Figure 5.11. If $DE = 3$, $AC = 16$, and $EC = BC$, find the length BC.

Solution

From the similar triangles, we proved (in Example 4) that $\frac{DE}{BC} = \frac{AE}{AC}$. With $AC = AE + EC$ and letting the lengths of the congruent segments ($\overline{EC}$ and $\overline{BC}$) be denoted by x, we have

$$16 = AE + x \qquad \text{so} \qquad AE = 16 - x$$

Substituting into the proportion, we have

$$\frac{3}{x} = \frac{16 - x}{16}$$

It follows that

$$\begin{aligned} x(16 - x) &= 3 \cdot 16 \\ 16x - x^2 &= 48 \\ x^2 - 16x + 48 &= 0 \\ (x - 4)(x - 12) &= 0 \end{aligned}$$

Now x (or BC) equals 4 or 12. Each length is acceptable, but the drawings differ as illustrated in Figure 5.12.

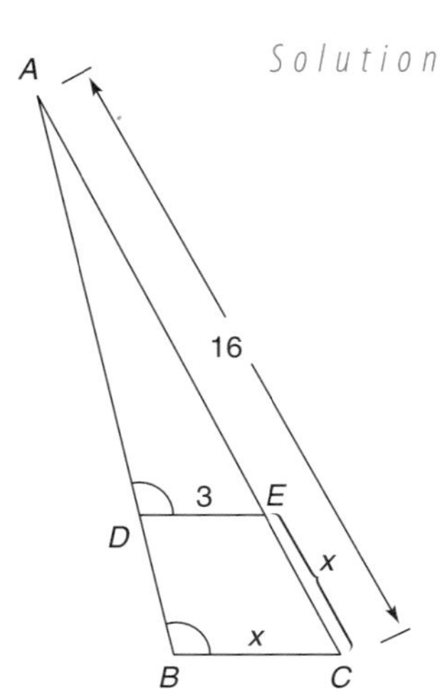

FIGURE 5.11

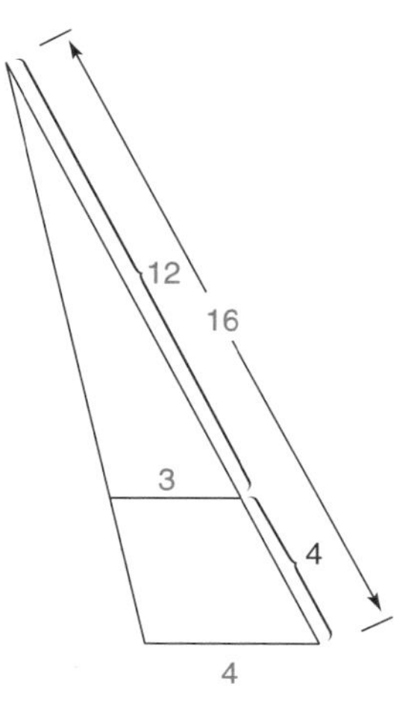

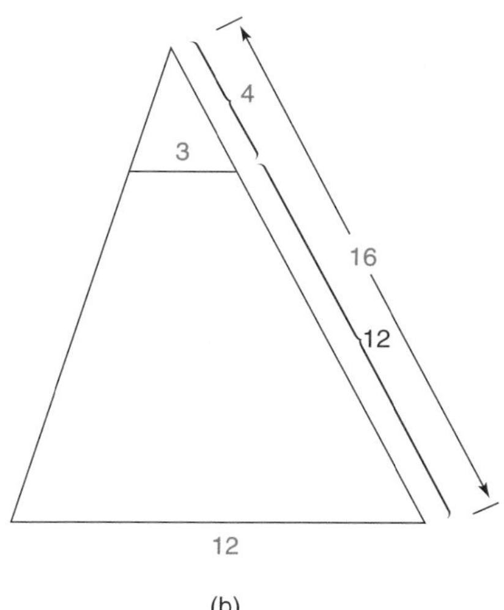

FIGURE 5.12

In Example 6, you are asked to prove that the product of lengths of two segments equals the product of lengths of two other segments. Here is a plan for establishing such a relationship:

1. Use AA to show that two triangles are similar, like $\triangle ABC \sim \triangle DEF$.
2. Use CSSTP to form a proportion by choosing the ratios from

$$\frac{AB}{DE} = \frac{BC}{EF} = \frac{AC}{DF}$$

3. Use the Means-Extremes Property to obtain equal products, such as

$$AB \cdot EF = DE \cdot BC$$

when $\frac{AB}{DE} = \frac{BC}{EF}$ is the proportion chosen in step 2.

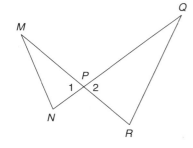

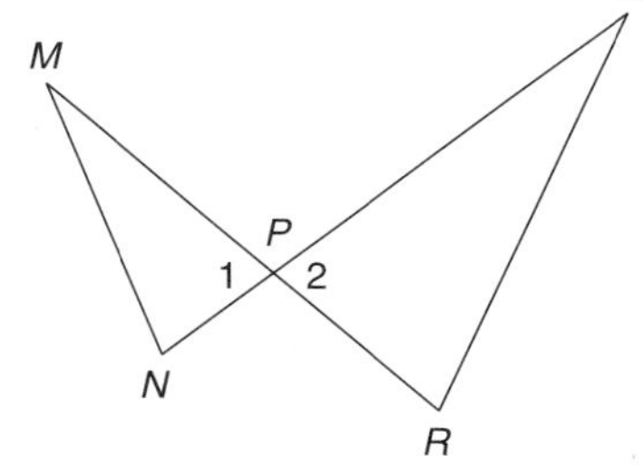

FIGURE 5.13

EXAMPLE 6

Use a paragraph proof for this problem.

Given: $\angle M \cong \angle Q$ in Figure 5.13.

Prove: $NP \cdot QR = RP \cdot MN$

Proof: By hypothesis, $\angle M \cong \angle Q$. Also, $\angle 1 \cong \angle 2$ by the fact that vertical angles are congruent. Now $\triangle MPN \sim \triangle QPR$ by AA. Using CSSTP, $\frac{NP}{RP} = \frac{MN}{QR}$. Then $NP \cdot QR = RP \cdot MN$ by the Means-Extremes Property.

NOTE: In the proof, the sides selected for the proportion were carefully chosen. The statement to be proved suggested that we include *NP*, *QR*, *RP*, and *MN* in the proportion.

The following example uses a method called *shadow reckoning*. Not a new technique, this method of calculating a length dates back more than 2500 years. It was used by Thales to estimate the height of the pyramids in Egypt.

EXAMPLE 7

Darnell is curious about the height of a flagpole that stands in front of his school. Darnell, who is 6 ft tall, casts a shadow that he paces off at 9 ft. He walks the length of the shadow of the flagpole, a distance of 30 ft. How tall is the flagpole?

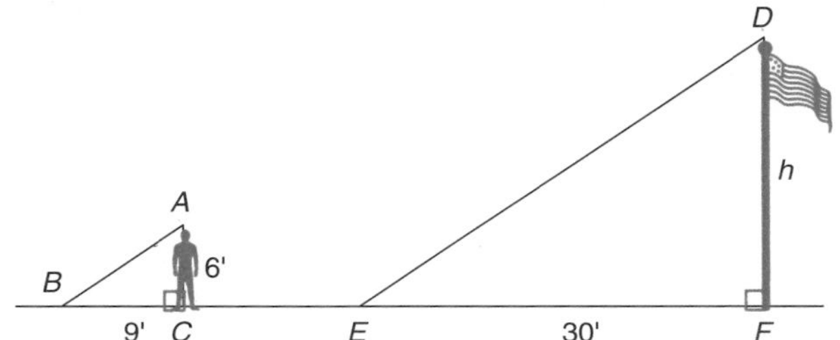

FIGURE 5.14

Solution

In Figure 5.14, $\triangle ABC \sim \triangle DEF$. From similar triangles, we know that $\frac{AC}{DF} = \frac{BC}{EF}$ or $\frac{AC}{BC} = \frac{DF}{EF}$ by interchanging the means.

Where h is the height of the flagpole, substitution into the second proportion leads to

$$\frac{6}{9} = \frac{h}{30} \rightarrow 9h = 180 \rightarrow h = 20$$

The height of the flagpole is 20 ft.

5.2 Exercises

In Exercises 1 and 2, refer to the drawing.

1. a) Given that $A \leftrightarrow X$, $B \leftrightarrow T$, and $C \leftrightarrow N$, write a statement claiming that the triangles shown are similar.
   b) If $A \leftrightarrow N$, $C \leftrightarrow X$, and $B \leftrightarrow T$, write a statement claiming that the triangles shown are similar.

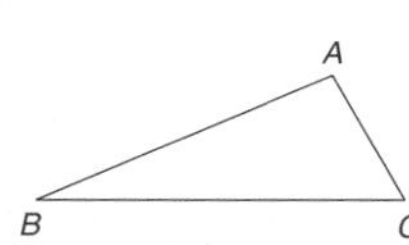

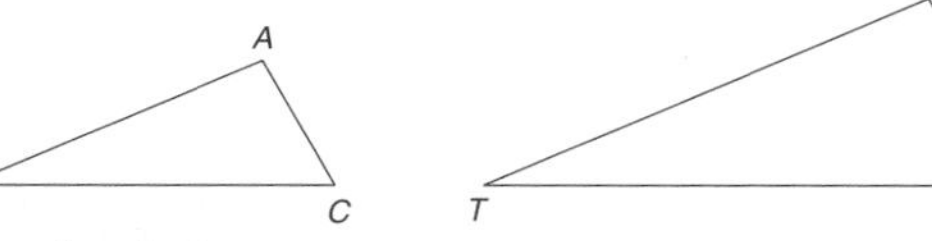

Exercises 1, 2

2. a) If $\triangle ABC \sim \triangle XTN$, which angle of $\triangle ABC$ corresponds to $\angle N$ of $\triangle XTN$?
   b) If $\triangle ABC \sim \triangle XTN$, which side of $\triangle XTN$ corresponds to side $\overline{AC}$ of $\triangle ABC$?

3. A **sphere** is the three-dimensional surface that contains all points in space lying at a fixed distance from a point known as the center of the sphere. Consider the two spheres shown. Are these two spheres similar? Are any two spheres similar? Explain.

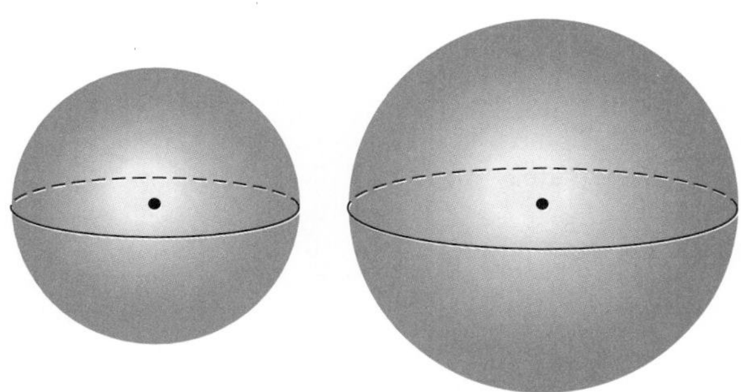

4. Given that rectangle $ABCE$ is similar to rectangle $MNPR$ and that $\triangle CDE \sim \triangle PQR$, what can you conclude regarding pentagon $ABCDE$ and pentagon $MNPQR$?

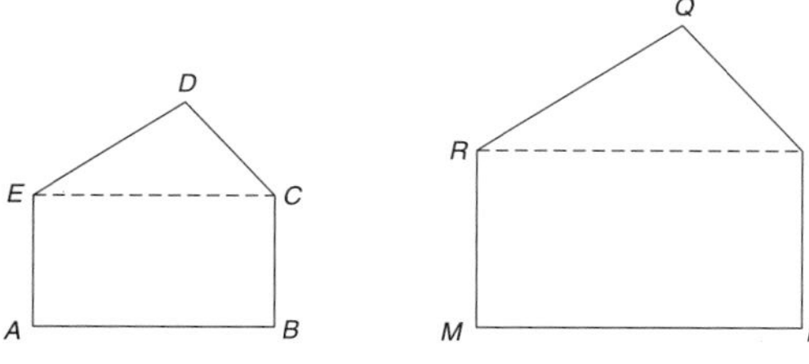

5. In $\triangle RST$ and $\triangle UVW$, $WU = \frac{3}{2} \cdot TR$, $WV = \frac{3}{2} \cdot TS$, and $UV = \frac{3}{2} \cdot RS$. Use intuition to draw a conclusion about the two triangles. If your conclusion includes a similarity relationship, write an extended proportion relating the lengths of the sides of the two triangles.

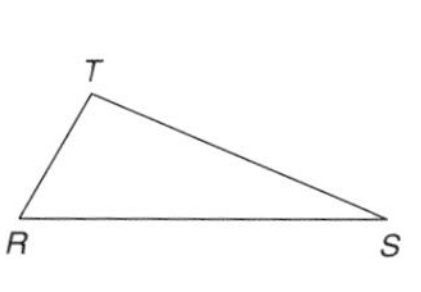

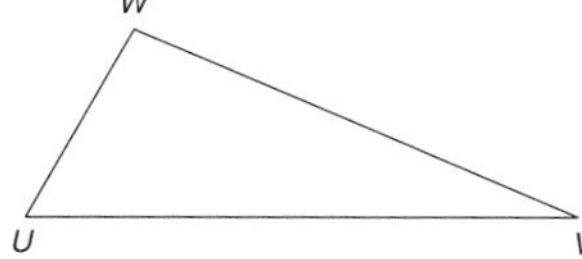

6. In $\triangle DGH$ and $\triangle DEF$, $DE = 3 \cdot DG$ and $DF = 3 \cdot DH$. Use intuition to form a conclusion about the triangle relationship. If the conclusion involves a similarity relationship, state an extended proportion relating the lengths of the sides of the two triangles.

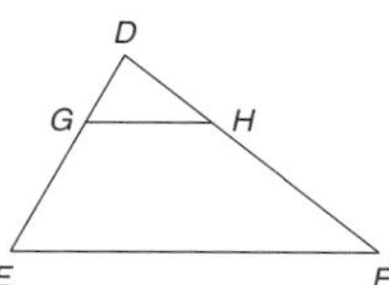

7. *Given:* $\triangle MNP \sim \triangle QRS$, $m\angle M = 56°$, $m\angle R = 82°$, $MN = 9$, $QR = 6$, $RS = 7$, $MP = 12$
   Find: **a)** $m\angle N$ **c)** NP
   b) $m\angle P$ **d)** QS

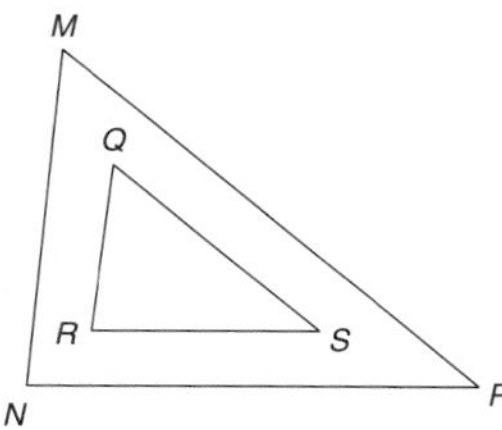

8. *Given:* $\triangle ABC \sim \triangle PRC$, $m\angle A = 67°$, $PC = 5$, $CR = 12$, $PR = 13$, $AB = 26$
   Find: **a)** $m\angle B$ **c)** AC
   b) $m\angle RPC$ **d)** CB

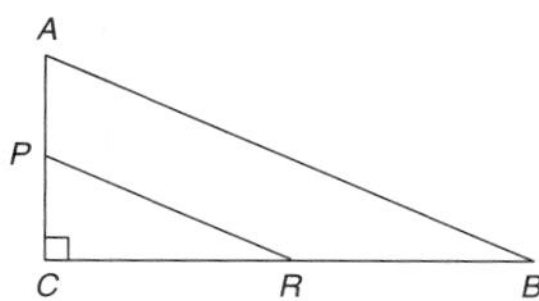

9. a) Does the similarity relationship have a **reflexive** property for triangles (and polygons in general)?
   b) Is there a **symmetric** property for the similarity of triangles (and polygons)?
   c) Is there a **transitive** property for the similarity of triangles (and polygons)?

10. Using the names of properties from Exercise 9, identify the property illustrated by each statement:
    a) If $\triangle 1 \sim \triangle 2$, then $\triangle 2 \sim \triangle 1$
    b) If $\triangle 1 \sim \triangle 2$, $\triangle 2 \sim \triangle 3$, and $\triangle 3 \sim \triangle 4$, then $\triangle 1 \sim \triangle 4$
    c) $\triangle 1 \sim \triangle 1$

In Exercises 11 to 14, complete each proof.

11. *Given:* $\overline{MN} \perp \overline{NP}$, $\overline{QR} \perp \overline{RP}$
Prove: $\triangle MNP \sim \triangle QRP$

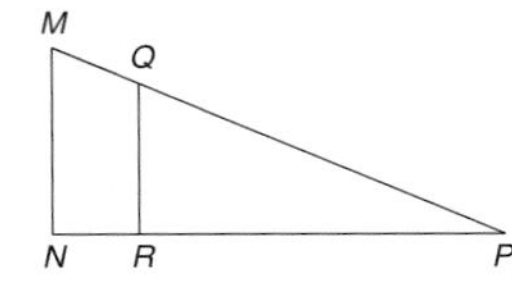

Exercises 11, 12

PROOF

Statements	Reasons
1. ?	**1.** Given
2. $\angle$s N and QRP are right $\angle$s	**2.** ?
3. ?	**3.** All right $\angle$s are $\cong$
4. $\angle P \cong \angle P$	**4.** ?
5. ?	**5.** ?

12. *Given:* $\overline{MN} \parallel \overline{QR}$
Prove: $\triangle MNP \sim \triangle QRP$

PROOF

Statements	Reasons
1. ?	**1.** Given
2. $\angle M \cong \angle RQP$	**2.** ?
3. ?	**3.** If two $\parallel$ lines are cut by a transversal, the corresponding $\angle$s are $\cong$
4. ?	**4.** ?

13. *Given:* $\angle H \cong \angle F$
Prove: $\triangle HJK \sim \triangle FGK$

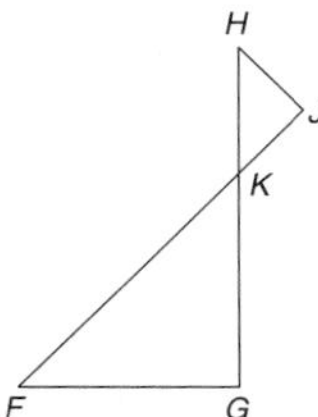

Exercises 13, 14

PROOF

Statements	Reasons
1. ?	**1.** Given
2. $\angle HKJ \cong \angle FKG$	**2.** ?
3. ?	**3.** ?

14. *Given:* $\overline{HJ} \perp \overline{JF}$, $\overline{HG} \perp \overline{FG}$
Prove: $\triangle HJK \sim \triangle FGK$

PROOF

Statements	Reasons
1. ?	**1.** Given
2. $\angle$s G and J are right $\angle$s	**2.** ?
3. $\angle G \cong \angle J$	**3.** ?
4. $\angle HKJ \cong \angle GKF$	**4.** ?
5. ?	**5.** ?

In Exercises 15 and 16, provide two-column proofs.

15. *Given:* $\overline{AB} \parallel \overline{DF}$, $\overline{BD} \parallel \overline{FG}$
Prove: $\triangle ABC \sim \triangle EFG$

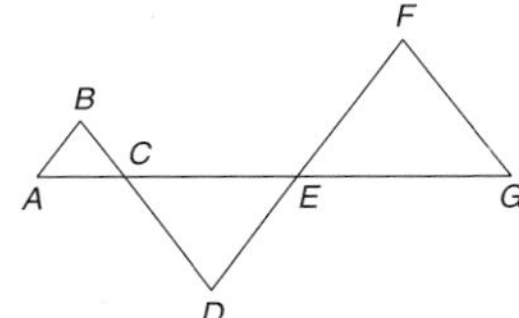

Exercises 15, 16

16. *Given:* $\overline{AB} \cong \overline{BC}$, $\overline{CD} \cong \overline{DE}$, and $\overline{EF} \cong \overline{FG}$
Prove: $\triangle ABC \sim \triangle EFG$

In Exercises 17 to 20, provide paragraph proofs.

17. *Given:* $\angle RVU \cong \angle S$, $\angle RUV \cong \angle Q$
Prove: $\triangle PQR \sim \triangle STR$

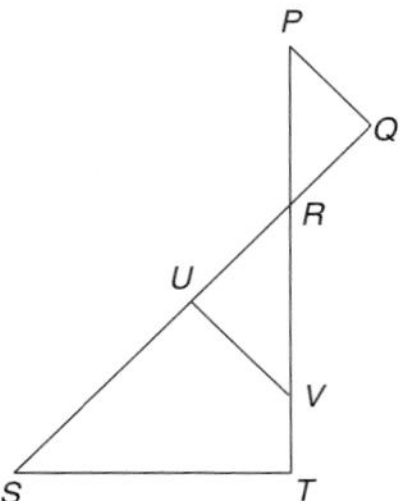

18. *Given:* $\overline{RS} \perp \overline{AB}$, $\overline{CB} \perp \overline{AC}$
Prove: $\triangle BSR \sim \triangle BCA$

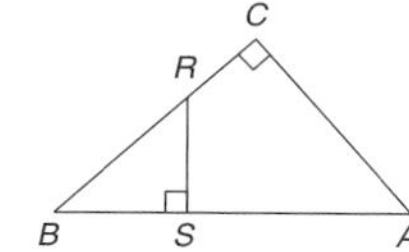

19. *Given:* $\overline{RS} \parallel \overline{UV}$
Prove: $\frac{RT}{VT} = \frac{RS}{VU}$

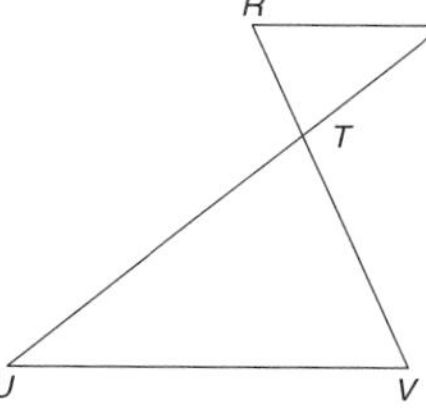

20. *Given:* $\overline{AB} \parallel \overline{DC}$, $\overline{AC} \parallel \overline{DE}$
Prove: $\frac{AB}{DC} = \frac{BC}{CE}$

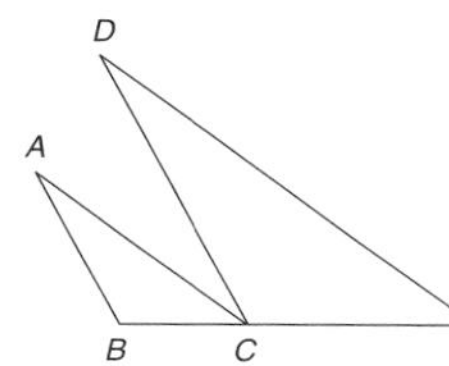

21. Use a two-column proof to prove the theorem, "The lengths of the corresponding altitudes of similar triangles have the same ratio as the lengths of any pair of corresponding sides."
Given: $\triangle DEF \sim \triangle MNP$; $\overline{DG}$ and $\overline{MQ}$ are altitudes
Prove: $\frac{DG}{MQ} = \frac{DE}{MN}$

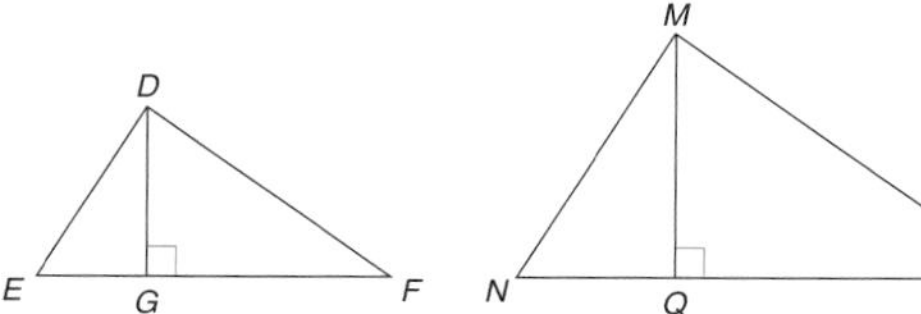

22. Provide a paragraph proof for the following problem.
Given: $\overline{RS} \parallel \overline{YZ}$, $\overline{RU} \parallel \overline{XZ}$
Prove: $RS \cdot ZX = ZY \cdot RT$

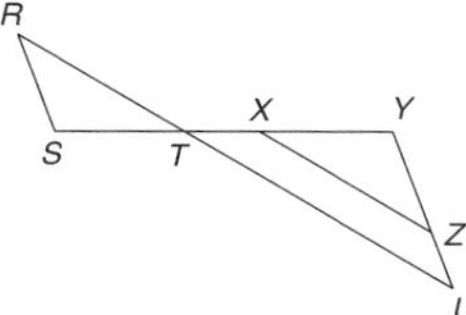

In Exercises 23 to 26, $\triangle ABC \sim \triangle DBE$.

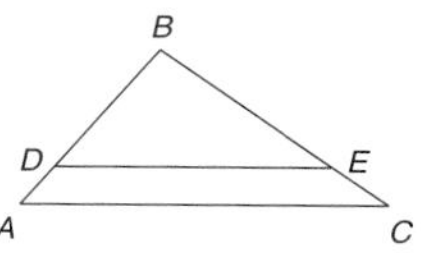

Exercises 23–26

23. *Given:* $AC = 8$, $DE = 6$, $CB = 6$
Find: EB
(**HINT:** Let $EB = x$, and solve an equation.)

24. *Given:* $AC = 10$, $CB = 12$
E the midpoint of $\overline{CB}$
Find: DE

25. *Given:* $AC = 10$, $DE = 8$, $AD = 4$
Find: DB

26. *Given:* $CB = 12$, $CE = 4$, $AD = 5$
Find: DB

In Exercises 27 to 30, $\triangle ADE \sim \triangle ABC$.

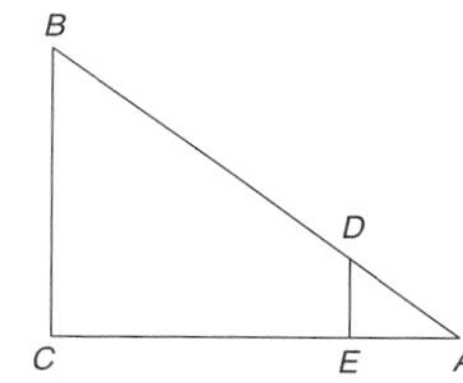

Exercises 27–30

27. *Given:* $DE = 4$, $AE = 6$, $EC = BC$
Find: BC

28. *Given:* $DE = 5$, $AD = 8$, $DB = BC$
Find: AB
(**HINT:** Find DB first.)

29. *Given:* $DE = 4$, $AC = 20$, $EC = BC$
Find: BC

30. *Given:* $AD = 4$, $AC = 18$, $DB = AE$
Find: AE

31. A person who is walking away from a 10-ft lamppost casts a shadow 6 ft long. If the person is at a distance of 10 ft from the lamppost at that moment, what is the person's height?

32. With 100 ft of string out, a kite is 64 ft above ground level. When the girl flying the kite pulls in 40 ft of string, the angle formed by the string and the ground does not change. What is the height of the kite above the ground after the 40 ft of string have been taken in?

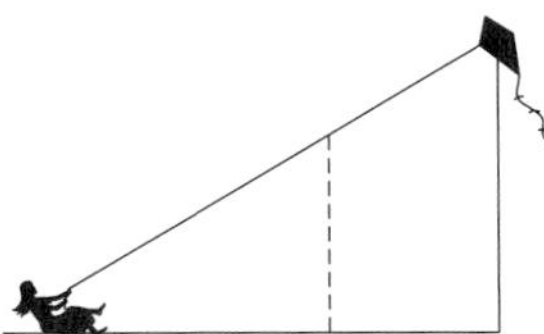

33. While admiring a rather tall tree, Fred notes that the shadow of his 6-ft frame has a length of 3 paces. On the level ground, he walks off the complete shadow of the tree in 37 paces. How tall is the tree?

34. As a garage door closes, light is cast 6 ft beyond the base of the door (as shown in the accompanying drawing) by a light fixture that is set in the garage ceiling 10 ft back from the door. If the ceiling of the garage is 10 ft above the floor, how far is the garage door above the floor at the time that light is cast 6 ft beyond the door?

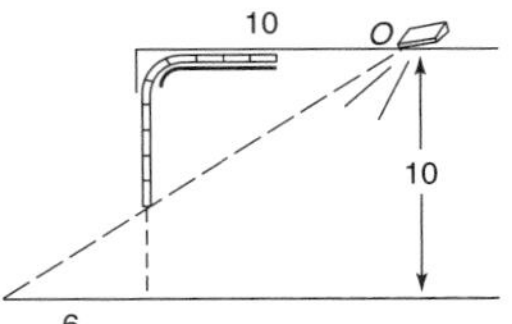

35. Prove that the altitude drawn to the hypotenuse of a right triangle separates the right triangle into two right triangles that are similar to each other and to the original right triangle.

36. Prove that the line segment joining the midpoints of two sides of a triangle determine a triangle that is similar to the original triangle.

5.3 The Pythagorean Theorem

The following theorem which was proved in Exercise 35 of Section 5.2, will enable us to prove the well-known Pythagorean Theorem.

> **THEOREM 5.3.1:** The altitude drawn to the hypotenuse of a right triangle separates the right triangle into two right triangles that are similar to each other and to the original right triangle.

Theorem 5.3.1 is illustrated by Figure 5.15, in which the right triangle $\triangle ABC$ has its right angle at vertex C so that $\overline{CD}$ is the altitude to hypotenuse $\overline{AB}$. The smaller triangles are shown in Figures 5.15(b) and (c), and the original triangle is shown in Figure 5.15(d). Notice the matched arcs indicating congruent angles.

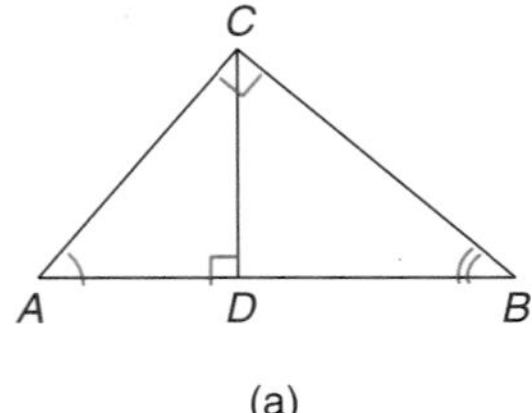

(a)

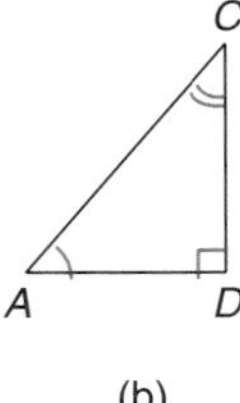

(b)

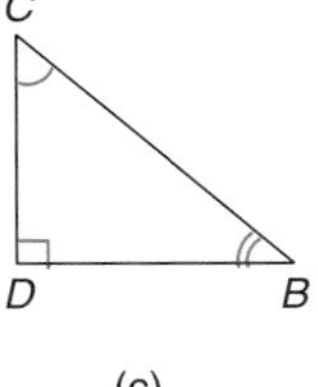

(c)

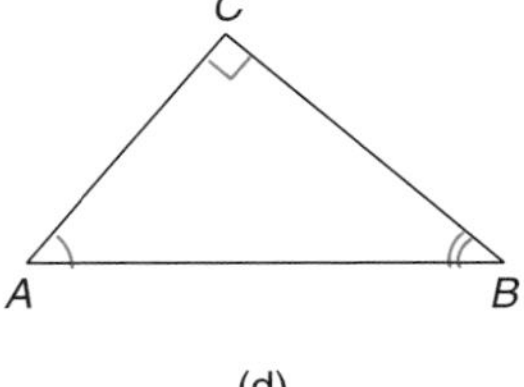

(d)

FIGURE 5.15

Two less obvious claims are now stated as theorems, and proofs of these are provided. In Figure 5.15(a), $\overline{AD}$ and $\overline{DB}$ are known as segments (parts) of the hypotenuse $\overline{AB}$. Furthermore, $\overline{AD}$ is the segment of the hypotenuse *adjacent* to leg $\overline{AC}$ while $\overline{BD}$ is the segment of the hypotenuse *adjacent* to leg $\overline{BC}$.

> **THEOREM 5.3.2:** The length of the altitude to the hypotenuse of a right triangle is the geometric mean of the lengths of the segments of the hypotenuse.

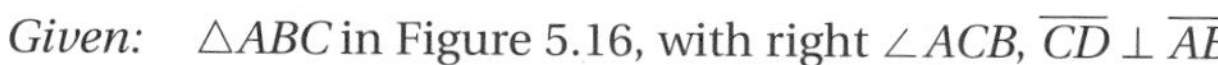

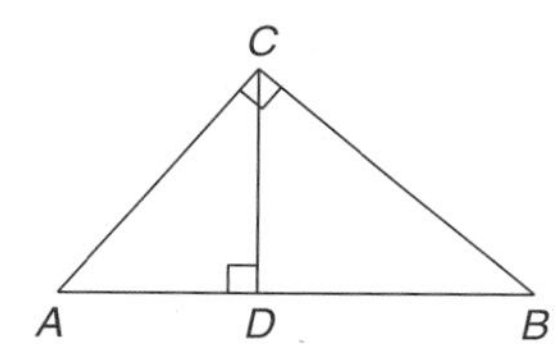

FIGURE 5.16

Given: $\triangle ABC$ in Figure 5.16, with right $\angle ACB$, $\overline{CD} \perp \overline{AB}$

Prove: $\frac{AD}{CD} = \frac{CD}{DB}$

Proof: With right $\triangle ABC$ and $\overline{CD} \perp \overline{AB}$, we can use the fact that the altitude drawn to the hypotenuse of a right triangle separates it into two right triangles that are similar to each other. Because $\triangle ADC \sim \triangle CDB$, it follows by CSSTP that

$$\frac{AD}{CD} = \frac{CD}{DB} \text{ [See Figures 5.15(b) and (c).]}$$

In general, the phrase "segments of the hypotenuse" refers to the parts of the hypotenuse determined by the altitude drawn from the vertex of a right triangle to the hypotenuse. See Theorem 5.3.3 and its proof, in which $\overline{AD}$ and $\overline{DB}$ are the segments of hypotenuse $\overline{AB}$ of right triangle ABC determined by altitude $\overline{CD}$.

> **THEOREM 5.3.3:** The length of each leg of a right triangle is the geometric mean of the length of the hypotenuse and the length of the segment of the hypotenuse adjacent to that leg.

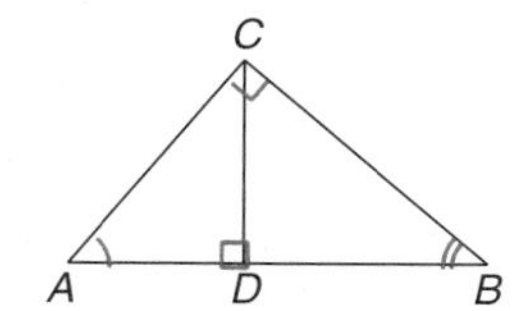

FIGURE 5.17

Given: $\triangle ABC$ with right $\angle ACB$; $\overline{CD} \perp \overline{AB}$. (See Figure 5.17.)

Prove: $\frac{AB}{AC} = \frac{AC}{AD}$

Proof: With right $\triangle ABC$ and $\overline{CD} \perp \overline{AB}$, we know that $\triangle ADC \sim \triangle ACB$ because the altitude drawn to the hypotenuse of a right triangle forms two right triangles that are similar to the original right triangle. (See Figure 5.18.) In turn, $\frac{AB}{AC} = \frac{AC}{AD}$. (Similarly, it can be shown that $\frac{AB}{CB} = \frac{CB}{DB}$.)

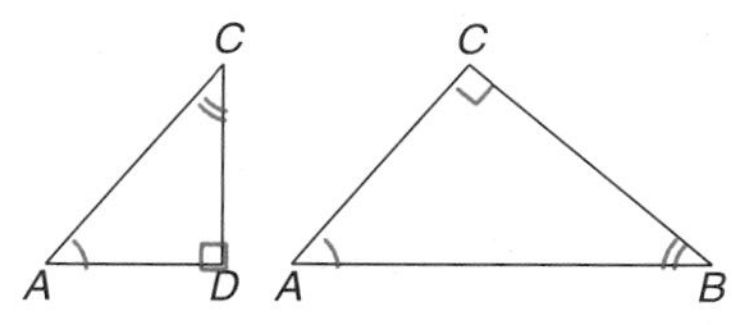

FIGURE 5.18

NOTE: Although $\overline{AD}$ and $\overline{DB}$ are both segments of the hypotenuse, $\overline{AD}$ is the segment adjacent to (next to) $\overline{AC}$.

The preceding theorem opens the doors to a proof of the famous Pythagorean Theorem, one of the most often applied relationships in geometry. While the theorem's title gives credit to the Greek geometer Pythagoras, many other proofs are known, and the ancient Chinese were aware of the relationship before the time of Pythagoras.

THEOREM 5.3.4: **(Pythagorean Theorem)** The square of the length of the hypotenuse of a right triangle is equal to the sum of the squares of the lengths of the legs.

(a)

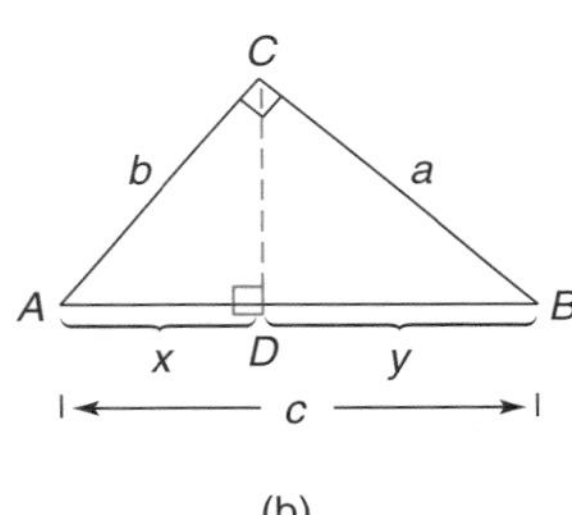

(b)

FIGURE 5.19

Thus, where c is the length of the hypotenuse and a and b are the lengths of the legs, $c^2 = a^2 + b^2$.

Given: In Figure 5.19(a), $\triangle ABC$ with right $\angle C$

Prove: $c^2 = a^2 + b^2$

Proof: Draw $\overline{CD} \perp \overline{AB}$, as shown in Figure 5.19(b).
Denote $AD = x$ and $DB = y$. By Theorem 5.3.3,

$$\frac{c}{b} = \frac{b}{x} \quad \text{and} \quad \frac{c}{a} = \frac{a}{y}$$

Therefore $\quad b^2 = cx \quad$ and $\quad a^2 = cy$

Using the Addition Property of Equality, we have

$$a^2 + b^2 = cy + cx = c(y + x)$$

But $y + x = x + y = AD + DB = AB = c$. Thus

$$a^2 + b^2 = c(c) = c^2$$

EXAMPLE 1

Given $\triangle RST$ with right $\angle S$ in Figure 5.20, find:

a) RT if $RS = 3$ and $ST = 4$
b) RT if $RS = 4$ and $ST = 6$
c) RS if $RT = 13$ and $ST = 12$
d) ST if $RS = 6$ and $RT = 9$

Solution

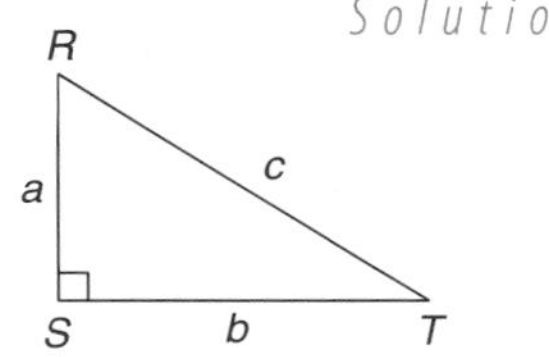

FIGURE 5.20

With right $\angle S$, the hypotenuse is $\overline{RT}$. Then $RT = c$, $RS = a$, and $ST = b$.

a) $3^2 + 4^2 = c^2 \rightarrow 9 + 16 = c^2$
$c^2 = 25$
$c = 5$; $RT = 5$

b) $4^2 + 6^2 = c^2 \rightarrow 16 + 36 = c^2$
$c^2 = 52$

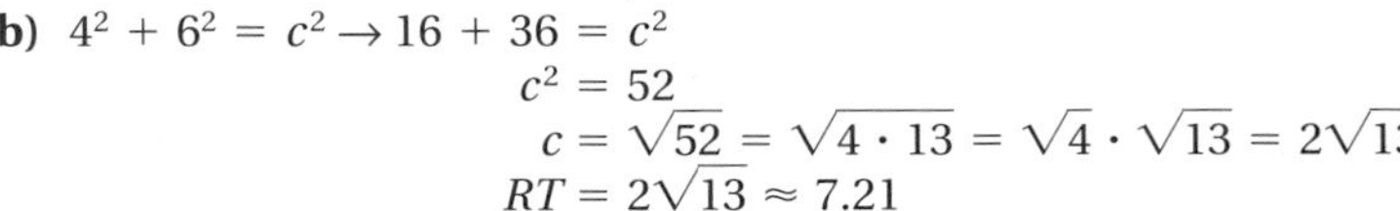

$c = \sqrt{52} = \sqrt{4 \cdot 13} = \sqrt{4} \cdot \sqrt{13} = 2\sqrt{13}$
$RT = 2\sqrt{13} \approx 7.21$

DISCOVER!

A video titled "The Rule of Pythagoras" is available through Project Mathematics at Cal Tech University in Pasadena, CA. It is well worth watching!

c) $a^2 + 12^2 = 13^2 \rightarrow a^2 + 144 = 169$
$a^2 = 25$
$a = 5$; $RS = 5$

d) $6^2 + b^2 = 9^2 \rightarrow 36 + b^2 = 81$
$b^2 = 45$
$b = \sqrt{45} = \sqrt{9 \cdot 5} = \sqrt{9} \cdot \sqrt{5} = 3\sqrt{5}$
$ST = 3\sqrt{5} \approx 6.71$

The converse of the Pythagorean Theorem is also true.

THEOREM 5.3.5: **(Converse of Pythagorean Theorem)** If a, b, and c are the lengths of the three sides of a triangle, with c the length of the longest side, and if $c^2 = a^2 + b^2$, then the triangle is a right triangle with the right angle opposite the side of length c.

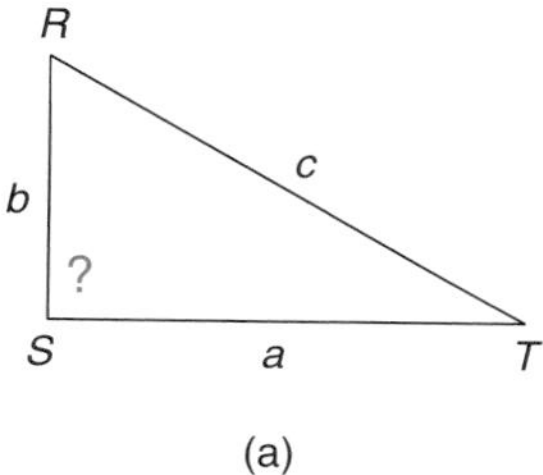

(a)

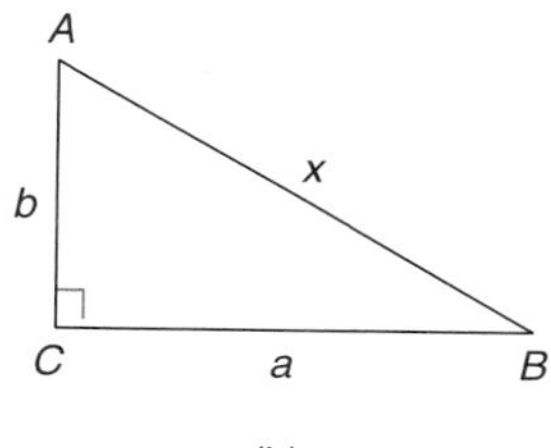

(b)

FIGURE 5.21

Given: $\triangle RST$ [Figure 5.21(a)] with sides a, b, and c so that $c^2 = a^2 + b^2$

Prove: $\triangle RST$ is a right triangle

Proof: We are given $\triangle RST$ for which $c^2 = a^2 + b^2$. Construct the right $\triangle ABC$, which has legs of lengths a and b and a hypotenuse of length x. [See Figure 5.21(b).] By the Pythagorean Theorem, $x^2 = a^2 + b^2$. By substitution, $x^2 = c^2$ and $x = c$. Thus $\triangle RTS \cong \triangle ABC$ by SSS. Then $\angle S$, opposite the side of length c, must be $\cong$ to $\angle C$, the right $\angle$ of $\triangle ABC$. Then $\angle S$ is a right $\angle$, and $\triangle RST$ is a right triangle.

EXAMPLE 2

Which of the following can be the lengths of sides of a right triangle?

a) $a = 5$, $b = 12$, $c = 13$
b) $a = 15$, $b = 8$, $c = 17$
c) $a = 7$, $b = 9$, $c = 10$
d) $a = \sqrt{2}$, $b = \sqrt{3}$, $c = \sqrt{5}$

Solution

a) Because $5^2 + 12^2 = 13^2$ (that is, $25 + 144 = 169$), this triangle is a right triangle.
b) Because $15^2 + 8^2 = 17^2$ (that is, $225 + 64 = 289$), this triangle is a right triangle.
c) $7^2 + 9^2 = 49 + 81 = 130$, which is not 10^2 (that is, 100), so this triangle is not a right triangle.
d) Because $(\sqrt{2})^2 + (\sqrt{3})^2 = (\sqrt{5})^2$ (that is, $2 + 3 = 5$), this triangle is a right triangle.

EXAMPLE 3

A ladder 12 ft long is leaning against a wall so that its base is 4 ft from the wall at ground level (see Figure 5.22). How far up the wall does the ladder reach?

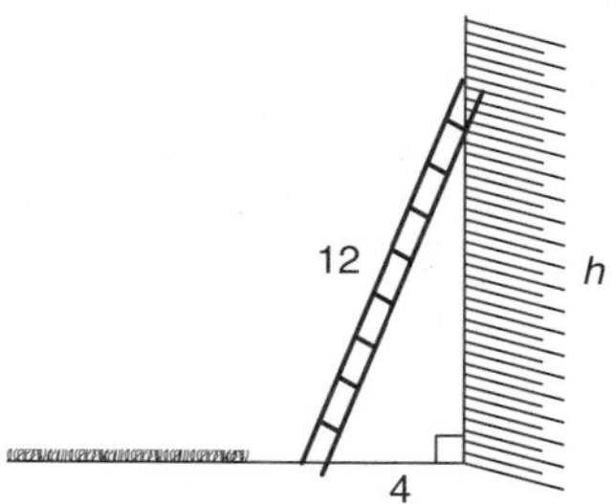

FIGURE 5.22

Solution The desired height is represented by h, so we have

$$\begin{aligned} 4^2 + h^2 &= 12^2 \\ 16 + h^2 &= 144 \\ h^2 &= 128 \\ h &= \sqrt{128} = \sqrt{64 \cdot 2} = \sqrt{64} \cdot \sqrt{2} = 8\sqrt{2} \end{aligned}$$

The height is represented exactly by $h = 8\sqrt{2}$, which is approximately 11.31 ft.

EXAMPLE 4

One diagonal of a rhombus has the same length, 10 cm, as each side (see Figure 5.23). How long is the other diagonal?

Solution Because the diagonals are perpendicular bisectors of each other, four right $\triangle$s are formed. For each right $\triangle$, a side of the rhombus is the hypotenuse. Half of each diagonal is a leg of each right triangle. Therefore

$$\begin{aligned} 5^2 + b^2 &= 10^2 \\ 25 + b^2 &= 100 \\ b^2 &= 75 \\ b &= \sqrt{75} = \sqrt{25 \cdot 3} = \sqrt{25} \cdot \sqrt{3} = 5\sqrt{3} \end{aligned}$$

Thus the length of the whole diagonal is $10\sqrt{3}$ cm $\approx$ 17.32 cm.

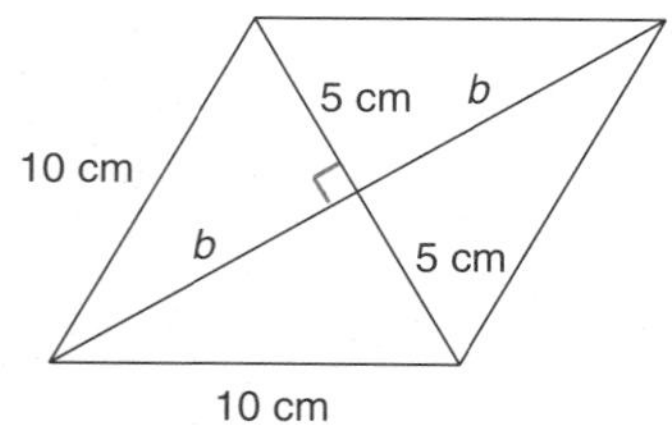

FIGURE 5.23

While Example 5 also uses the Pythagorean Theorem, it is considerably more complicated than Example 4. Indeed, it is one of those situations that may require some insight to solve. Note that the triangle described in Example 5 is *not* a right triangle because $4^2 + 5^2 \neq 6^2$.

EXAMPLE 5

A triangle has sides of lengths 4, 5, and 6, as shown in Figure 5.24. Find the length of the altitude to the side of length 6.

Solution

The altitude to the side of length 6 separates that side into two parts whose lengths are given by x and $6 - x$. Using the two right triangles formed, we have by the Pythagorean Theorem:

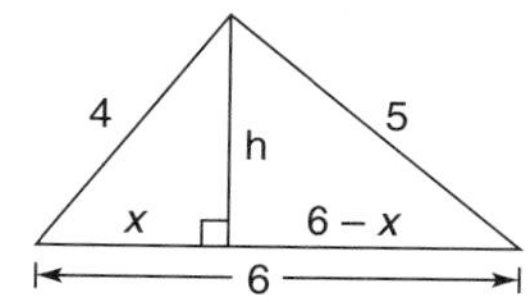

FIGURE 5.24

$$x^2 + h^2 = 4^2 \qquad \text{and} \qquad (6 - x)^2 + h^2 = 5^2$$

Subtracting the first equation from the second, we can calculate x.

$$\begin{aligned} 36 - 12x + x^2 + h^2 &= 25 \\ x^2 + h^2 &= 16 \\ \hline 36 - 12x &= 9 \qquad \text{(subtraction)} \\ -12x &= -27 \\ x &= \frac{27}{12} = \frac{9}{4} \end{aligned}$$

Now we use $x = \frac{9}{4}$ to find h:

$$\begin{aligned} x^2 + h^2 &= 4^2 \\ \left(\frac{9}{4}\right)^2 + h^2 &= 4^2 \\ \frac{81}{16} + h^2 &= 16 \\ \frac{81}{16} + h^2 &= \frac{256}{16} \\ h^2 &= \frac{175}{16} \\ h &= \frac{\sqrt{175}}{4} = \frac{\sqrt{25 \cdot 7}}{4} = \frac{\sqrt{25} \cdot \sqrt{7}}{4} = \frac{5\sqrt{7}}{4} \approx 3.31 \end{aligned}$$

It is now possible to prove the HL method for the congruence of triangles.

THEOREM 5.3.6: If the hypotenuse and a leg of one right triangle are congruent to the hypotenuse and a leg of a second right triangle, then the triangles are congruent (HL).

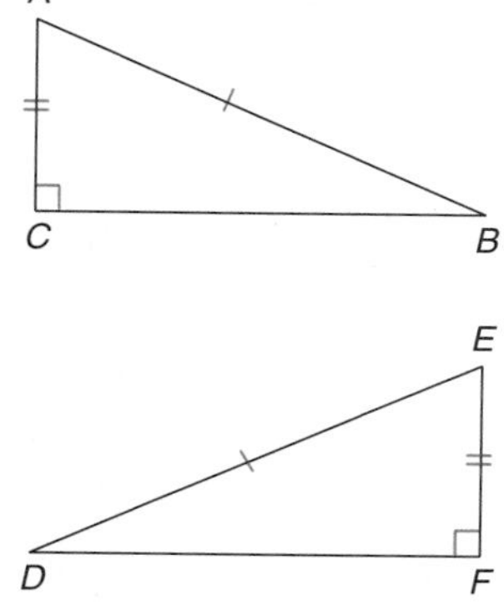

FIGURE 5.25

Given: Right $\triangle ABC$ with right $\angle C$ and right $\triangle DEF$ with right $\angle F$ (see Figure 5.25); $\overline{AB} \cong \overline{DE}$ and $\overline{AC} \cong \overline{EF}$

Prove: $\triangle ABC \cong \triangle EDF$

Proof: With right $\angle C$, the hypotenuse of $\triangle ABC$ is $\overline{AB}$; similarly, $\overline{DE}$ is the hypotenuse of right $\triangle EDF$. Because $\overline{AB} \cong \overline{DE}$, we denote the common length by c; that is, $AB = DE = c$. Because $\overline{AC} \cong \overline{EF}$, we also have $AC = EF = a$. Then

$$a^2 + (BC)^2 = c^2 \quad \text{and} \quad a^2 + (DF)^2 = c^2$$

which leads to

$$BC = \sqrt{c^2 - a^2} \quad \text{and} \quad DF = \sqrt{c^2 - a^2}$$

Then $BC = DF$ so that $\overline{BC} \cong \overline{DF}$. Hence, $\triangle ABC \cong \triangle EDF$ by SSS.

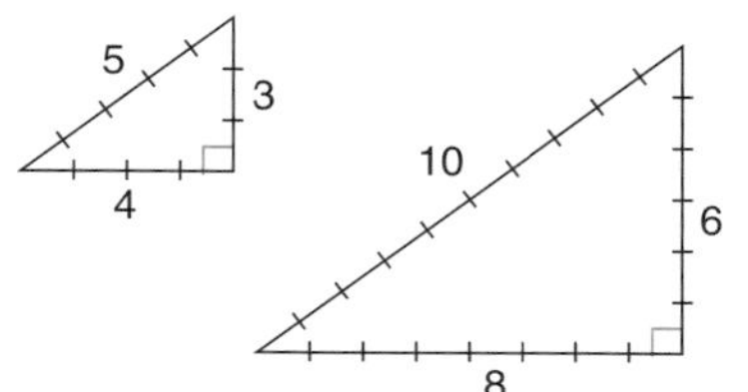

FIGURE 5.26

Our work with the Pythagorean Theorem would be incomplete if we did not address two concerns. The first, Pythagorean Triples, involves natural (or counting) numbers as possible choices of a, b, and c. The second leads to a classification of triangles according to the lengths of their sides; see Theorem 5.3.7.

A **Pythagorean triple** is a set of three natural numbers (a, b, c) for which $a^2 + b^2 = c^2$. Three sets of Pythagorean triples encountered in this section are (3, 4, 5), (5, 12, 13), and (8, 15, 17). These numbers will always fit the sides of a right triangle.

Natural-number multiples of any of these triples will also constitute Pythagorean triples. For example, doubling (3, 4, 5) yields (6, 8, 10), which is also a Pythagorean triple (see Figure 5.26).

The Pythagorean triple (3, 4, 5) also leads to (9, 12, 15), (12, 16, 20), and (15, 20, 25). The Pythagorean triple (5, 12, 13) leads to triples such as (10, 24, 26) and (15, 36, 39). Basic Pythagorean triples that are used less frequently include (7, 24, 25), (9, 40, 41), and (20, 21, 29).

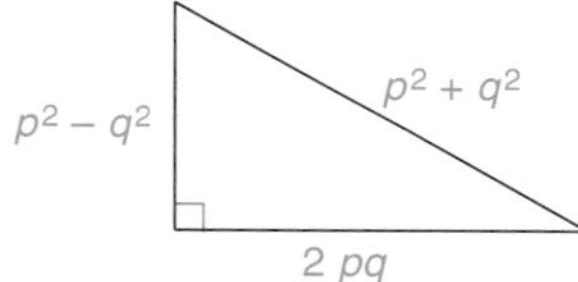

FIGURE 5.27

Pythagorean triples can be generated by using any of several formulas. One formula uses $2pq$ for one leg, $p^2 - q^2$ for the other leg, and $p^2 + q^2$ for the hypotenuse, where p and q are natural numbers and $p > q$. (See Figure 5.27.)

The following table lists some Pythagorean triples corresponding to choices for p and q. The boldface triples are basic triples, also known as primitive triples. While no exercises for this section include the phrase "Pythagorean triple," using these triples and their multiples will save you considerable time and effort. In the final column, the resulting triple is provided in order from a (small) to c (large).

p	q	a (or b) $p^2 - q^2$	b (or a) $2pq$	c $p^2 + q^2$	(a, b, c)
2	1	3	4	5	**(3, 4, 5)**
3	1	8	6	10	(6, 8, 10)
3	2	5	12	13	**(5, 12, 13)**
4	1	15	8	17	**(8, 15, 17)**
4	3	7	24	25	**(7, 24, 25)**
5	1	24	10	26	(10, 24, 26)
5	2	21	20	29	**(20, 21, 29)**
5	3	16	30	34	(16, 30, 34)
5	4	9	40	41	**(9, 40, 41)**

The Converse of the Pythagorean Theorem allows us to recognize a right triangle by knowing the lengths of sides. A variation on the Converse allows us to recognize whether a triangle is acute or obtuse. This theorem is stated without proof.

THEOREM 5.3.7: Let a, b, and c represent the lengths of the three sides of a triangle, with c the length of the longest side.

1. If $c^2 > a^2 + b^2$, then the triangle is obtuse and the obtuse angle lies opposite the side of length c.
2. If $c^2 < a^2 + b^2$, then the triangle is acute.

EXAMPLE 6

Determine the type of triangle represented if the lengths of its sides are as follows:

a) 4, 5, 7 **b)** 6, 7, 8 **c)** 9, 12, 15 **d)** 3, 4, 9

Solution

a) Because c is the longest side, $c = 7$, and we have $7^2 > 4^2 + 5^2$ or $49 > 16 + 25$; the triangle is obtuse.

b) Choosing $c = 8$, we have $8^2 < 6^2 + 7^2$ or $64 < 36 + 49$; the triangle is acute.

c) Choosing $c = 15$, we have $15^2 = 9^2 + 12^2$ or $225 = 81 + 144$; the triangle is a right triangle.

d) Because $9 > 3 + 4$, no triangle is possible. (**RECALL:** The sum of the lengths of two sides of a triangle must be greater than the length of the third side.)

5.3 Exercises

1. By naming the vertices in order, state three different triangles that are similar to each other.

2. Write a proportion in which

a) SV is used as a geometric mean.
b) RS is a geometric mean.
c) TS is a geometric mean.

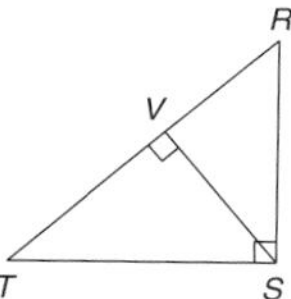

Exercises 1–4

3. Use Theorem 5.3.2 to find RV if $SV = 6$ and $VT = 8$.

4. Use Theorem 5.3.3 to find RT if $RS = 6$ and $VR = 4$.

5. Find the length of $\overline{DF}$ if:

a) $DE = 8$ and $EF = 6$
b) $DE = 5$ and $EF = 3$

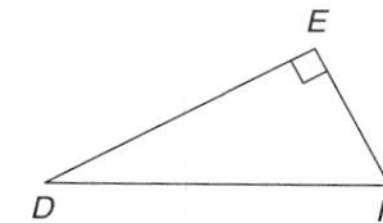

Exercises 5–8

6. Find the length of $\overline{DE}$ if:

a) $DF = 13$ and $EF = 5$
b) $DF = 12$ and $EF = 6\sqrt{3}$

7. Find EF if:

a) $DF = 17$ and $DE = 15$
b) $DF = 12$ and $DE = 8\sqrt{2}$

8. Find DF if:

a) $DE = 12$ and $EF = 5$
b) $DE = 12$ and $EF = 6$

9. Determine the type of triangle represented if the lengths of its sides are:

a) $a = 4$, $b = 3$, and $c = 5$
b) $a = 4$, $b = 5$, and $c = 6$
c) $a = 2$, $b = \sqrt{3}$, and $c = \sqrt{7}$
d) $a = 3$, $b = 8$, and $c = 15$

10. Determine the type of triangle represented if the lengths of its sides are:

a) $a = 1.5$, $b = 2$, and $c = 2.5$
b) $a = 20$, $b = 21$, and $c = 29$
c) $a = 10$, $b = 12$, and $c = 16$
d) $a = 5$, $b = 7$, and $c = 9$

11. A guy wire 25 ft long supports an antenna at a point that is 20 ft above the base of the antenna. How far from the base of the antenna is the guy wire secured?

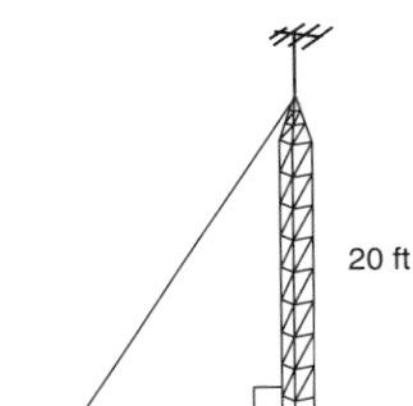

12. A strong wind holds a kite 30 ft above the earth in a position 40 ft across the ground. How much string does the girl have out (to the kite)?

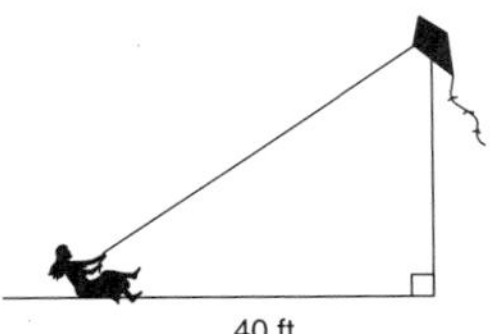

13. A boat is 6 m below the level of a pier and 12 m from the pier as measured across the water. How much rope is needed to reach the boat?

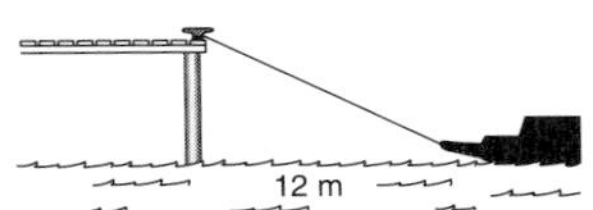

14. A hot air balloon is held in place by the ground crew at a point that is 21 ft from a point directly beneath the balloon. If the rope is of length 29 ft, how far above ground level is the balloon?

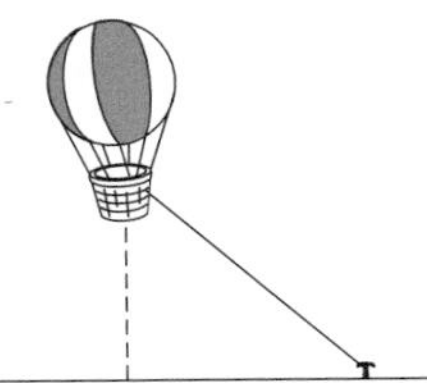

15. A rectangle has a width of 16 cm and a diagonal of length 20 cm. How long is the rectangle?

16. A right triangle has legs of lengths x and $2x + 2$ and a hypotenuse of length $2x + 3$. What are the lengths of its sides?

17. A rectangle has base $x + 3$, altitude $x + 1$, and diagonals $2x$ each. What are the lengths of its base, altitude, and diagonals?

18. The diagonals of a rhombus measure 6 m and 8 m. How long are each of the congruent sides?

19. Each side of a rhombus measures 12 in. If one diagonal is 18 in. long, how long is the other diagonal?

20. An isosceles right triangle has a hypotenuse of length 10 cm. How long is each leg?

21. Each leg of an isosceles right triangle has a length of $6\sqrt{2}$ in. What is the length of the hypotenuse?

22. In right $\triangle ABC$ with right $\angle C$, $AB = 10$ and $BC = 8$. Find the length of $\overline{MB}$ if M is the midpoint of $\overline{AC}$.

23. In right $\triangle ABC$ with right $\angle C$, $AB = 17$ and $BC = 15$. Find the length of $\overline{MN}$ if M and N are the midpoints of $\overline{AB}$ and $\overline{BC}$, respectively.

24. Find the length of the altitude to the 10-in. side of a triangle whose sides are 6, 8, and 10 in. in length.

25. Find the length of the altitude to the 26-in. side of a triangle whose sides are 10, 24, and 26 in. in length.

26. In quadrilateral $ABCD$, $\overline{BC} \perp \overline{AB}$ and $\overline{DC} \perp$ diagonal $\overline{AC}$. If $AB = 4$, $BC = 3$, and $DC = 12$, determine DA.

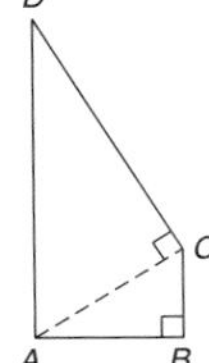

27. In quadrilateral $RSTU$, $\overline{RS} \perp \overline{ST}$ and $\overline{UT} \perp$ diagonal $\overline{RT}$. If $RS = 6$, $ST = 8$, and $RU = 15$, find UT.

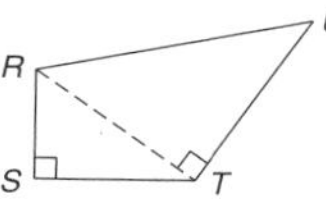

28. *Given:* $\triangle ABC$ is not a right $\triangle$
Prove: $a^2 + b^2 \neq c^2$

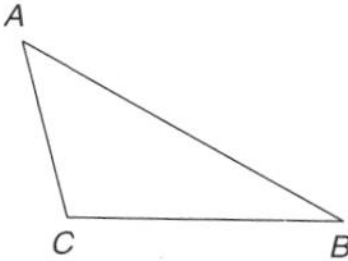

29. ➢ If $a = p^2 - q^2$, $b = 2pq$, and $c = p^2 + q^2$, show that $c^2 = a^2 + b^2$.

30. Given that the line segment shown has length 1, construct a line segment whose length is $\sqrt{2}$.

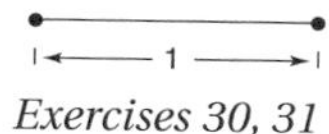

Exercises 30, 31

31. Using the same line segment as in Exercise 30, construct a segment of length 2 and then a second segment of length $\sqrt{5}$.

32. When the rectangle in the accompanying drawing (whose dimensions are 16 by 9) is cut into pieces and rearranged, a square can be formed. What is the perimeter of this square?

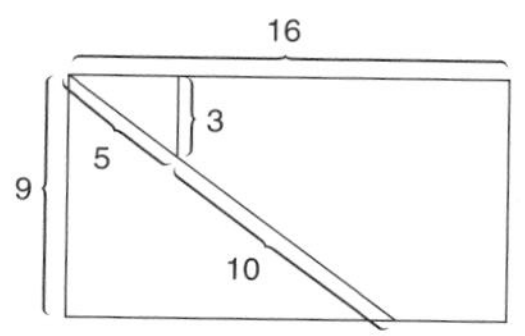

33. A, C, and F are three of the vertices of the cube shown in the accompanying figure. Given that each face of the cube is a square, what is the measure of angle ACF?

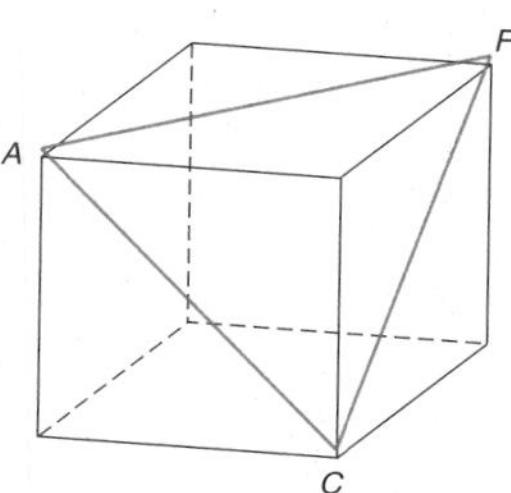

34. ➢ Find the length of the altitude to the 8-in. side of a triangle whose sides are 4, 6, and 8 in. long. (**HINT:** Use the Pythagorean Theorem twice.)

35. In the figure, square $RSTV$ has its vertices on the sides of square $WXYZ$ as shown. If $ZT = 5$ and $TY = 12$, find TS. Also find RT.

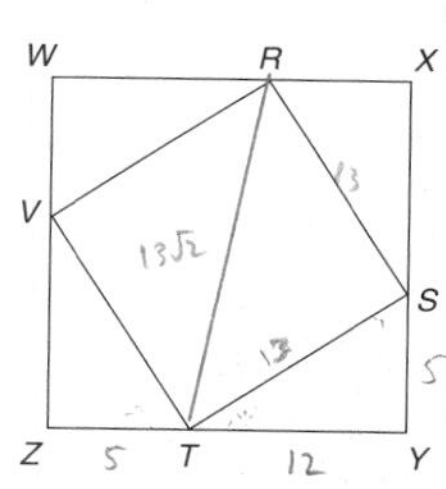

36. Prove that if (a, b, c) is a Pythagorean triple and n is a natural number, then (na, nb, nc) is also a Pythagorean triple.

5.4 Special Right Triangles

Many of the calculations that we do in this section involve square root radicals. To better understand some of these calculations, it may be necessary to review the Properties of Square Roots in Appendix A.

The 45°-45°-90° Right Triangle

Certain right triangles occur so often that they deserve more attention than others. The special right triangles we will consider have angle measures of 45°, 45°, and 90° or of 30°, 60°, and 90°.

In the 45-45-90 triangle, the legs are opposite the congruent angles and are also congruent. Rather than using a and b to represent the lengths of the legs, we use a for both lengths as shown in Figure 5.28. It then follows by the Pythagorean Theorem that

$$\begin{aligned} c^2 &= a^2 + a^2 \\ c^2 &= 2a^2 \\ c &= \sqrt{2a^2} \\ c &= \sqrt{2} \cdot \sqrt{a^2} \\ c &= a\sqrt{2} \end{aligned}$$

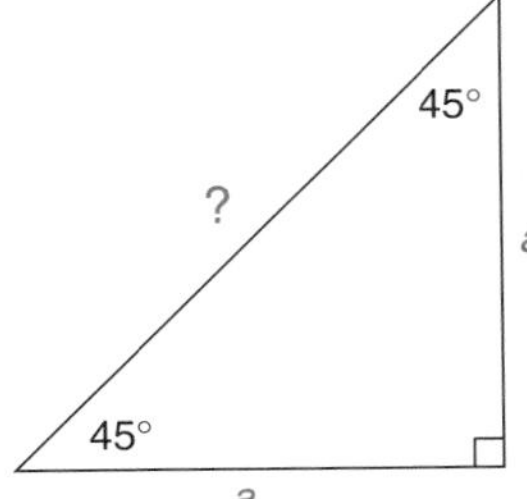

FIGURE 5.28

> **THEOREM 5.4.1:** **(45-45-90 Theorem)** In a triangle whose angles measure 45°, 45°, and 90°, the hypotenuse has a length equal to the product of $\sqrt{2}$ and the length of either leg.

FIGURE 5.29

It is better to memorize the sketch in Figure 5.29 than to repeat the steps preceding the 45-45-90 Theorem.

EXAMPLE 1

Find the lengths of the missing sides in each triangle in Figure 5.30.

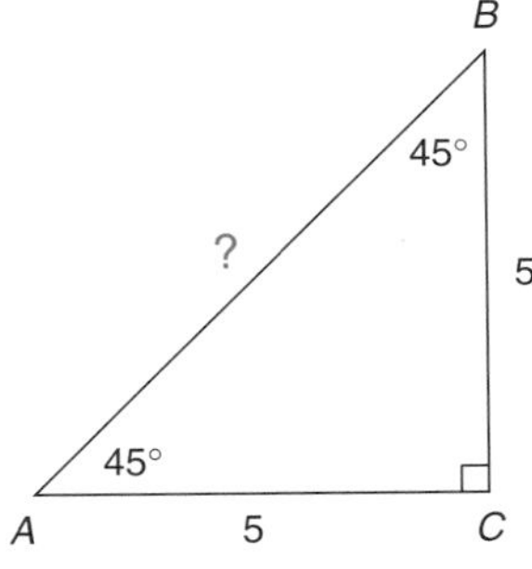

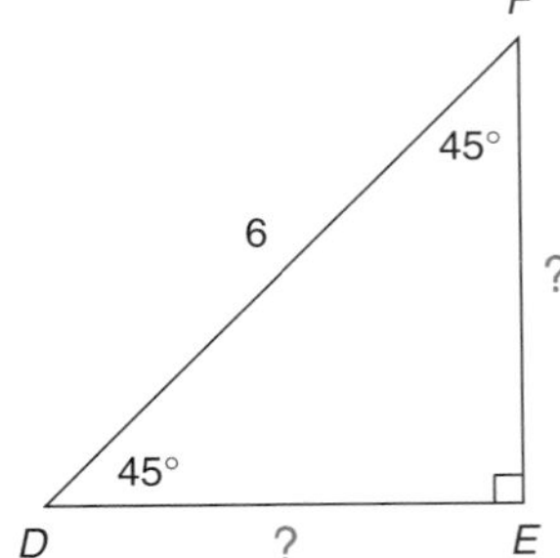

FIGURE 5.30

Solution

a) The length of hypotenuse $\overline{AB}$ is $5\sqrt{2}$, the product of $\sqrt{2}$ and the length of either of the equal legs.

b) Let a denote the length of $\overline{DE}$ and of $\overline{EF}$. The length of hypotenuse $\overline{DF}$ is $a\sqrt{2}$. Then $a\sqrt{2} = 6$, so $a = \frac{6}{\sqrt{2}}$. Simplifying,

$$\begin{aligned} a &= \frac{6}{\sqrt{2}} \cdot \frac{\sqrt{2}}{\sqrt{2}} \\ &= \frac{6\sqrt{2}}{2} \\ &= 3\sqrt{2} \end{aligned}$$

Therefore $DE = EF = 3\sqrt{2} \approx 4.24$.

NOTE: If we use the Pythagorean Theorem to solve Example 1, the solution in (a) can be found by solving the equation $5^2 + 5^2 = c^2$ and the solution in (b) can be found by solving $a^2 + a^2 = 6^2$.

EXAMPLE 2

Each side of a square has a length of $\sqrt{5}$. Find the length of a diagonal.

Solution

The square shown in Figure 5.31(a) is separated into two 45°-45°-90° triangles. With each of the congruent legs represented by a in Figure 5.31(b), we see that $a = \sqrt{5}$ and the diagonal (hypotenuse length) is $a \cdot \sqrt{2} = \sqrt{5} \cdot \sqrt{2}$, so $a = \sqrt{10} \approx 3.16$.

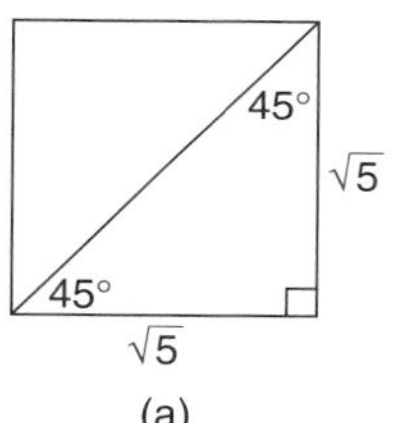

(a)

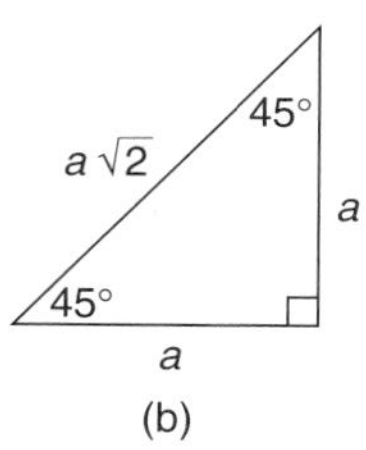

(b)

FIGURE 5.31

The 30°-60°-90° Right Triangle

The second special triangle is the 30-60-90 triangle, shown in Figure 5.32(a). Its sides are related to each other in a way that can be seen by reflecting (as a mirror might) the triangle across the side included by the 30° and 90° angles. As shown in Figure 5.32(b), the reflection produces an equiangular and therefore equilateral triangle in which the length of the side that was the hypotenuse of the 30-60-90 triangle is twice the length of the leg opposite the 30° angle. Extending the shorter leg while allowing for the 30° angle ensures that the two smaller triangles are congruent.

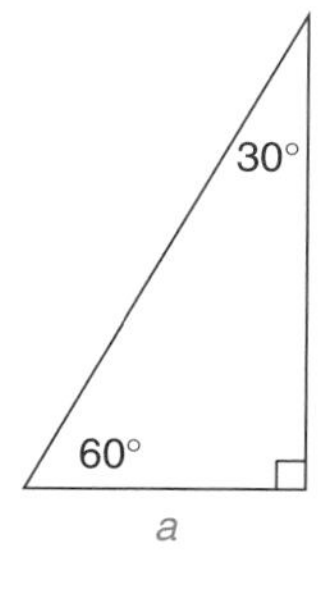

(a)

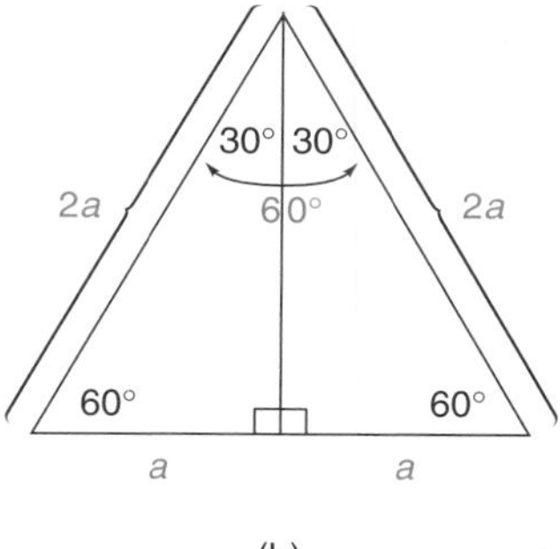

(b)

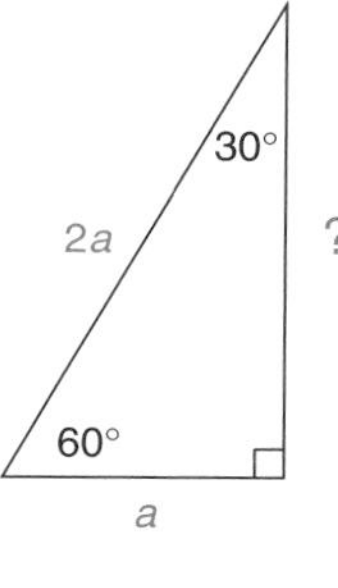

(c)

FIGURE 5.32

Take another look at the 30-60-90 triangle in Figure 5.32(c). Since we know that the length of the hypotenuse is twice the length of the shorter leg, we can again use the Pythagorean Theorem to represent the length of the longer leg. Where b is the length of the longer leg, we have

$$\begin{aligned} c^2 &= a^2 + b^2 \\ (2a)^2 &= a^2 + b^2 \\ 4a^2 &= a^2 + b^2 \\ 3a^2 &= b^2, \text{ so } b^2 = 3a^2 \end{aligned}$$

Thus

$$\begin{aligned} b &= \sqrt{3a^2} \\ b &= \sqrt{3} \cdot \sqrt{a^2} \\ b &= a\sqrt{3} \end{aligned}$$

THEOREM 5.4.2: (30-60-90 Theorem) In a triangle whose angles measure 30°, 60°, and 90°, the hypotenuse has a length equal to twice the length of the shorter leg, while the length of the longer leg is the product of $\sqrt{3}$ and the length of the shorter leg.

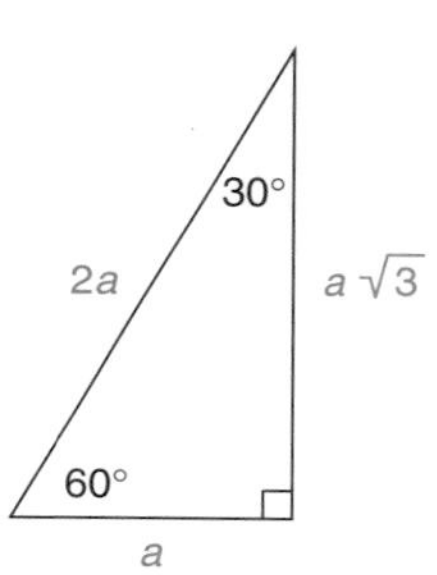

FIGURE 5.33

It would be best to memorize the sketch in Figure 5.33. So that you more easily recall which expression is used for each side, remember that the lengths of the sides follow the same order as the angles opposite them. Thus:

opposite the 30° ∠ (smallest angle) is a (shortest side)
opposite the 60° ∠ (middle angle) is $a\sqrt{3}$ (middle side)
opposite the 90° ∠ (largest angle) is $2a$ (longest side)

EXAMPLE 3

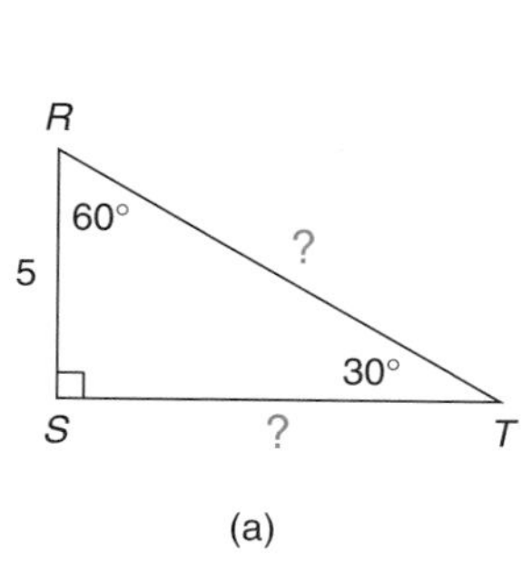

(a)

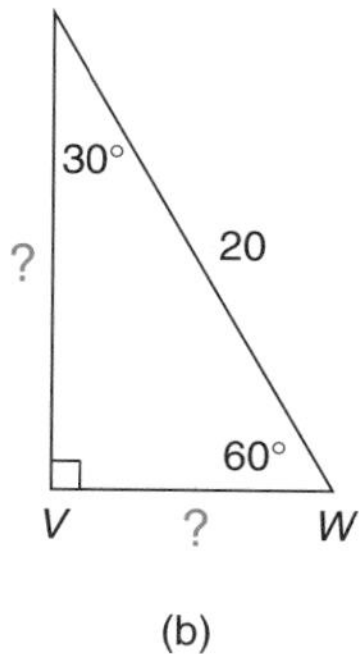

(b)

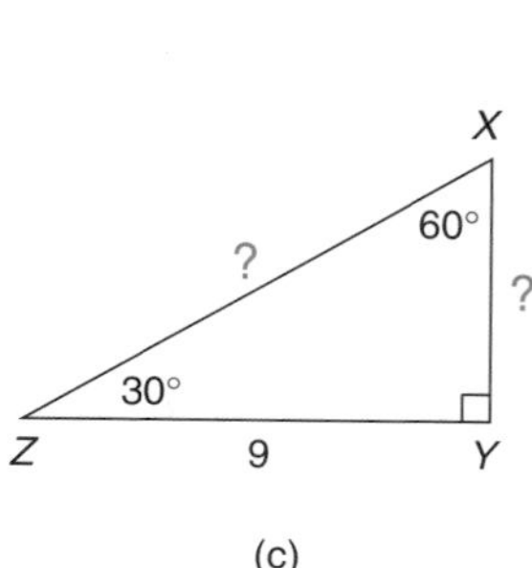

(c)

FIGURE 5.34

Find the lengths of the missing sides of each triangle in Figure 5.34.

Solution

a) $RT = 2 \cdot RS = 2 \cdot 5 = 10$
$ST = RS\sqrt{3} = 5\sqrt{3} \approx 8.66$

b) $UW = 2 \cdot VW \rightarrow 20 = 2 \cdot VW \rightarrow VW = 10$
$UV = VW\sqrt{3} = 10\sqrt{3} \approx 17.32$

c) $ZY = XY\sqrt{3} \rightarrow 9 = XY \cdot \sqrt{3} \rightarrow XY = \frac{9}{\sqrt{3}} = \frac{9}{\sqrt{3}} \cdot \frac{\sqrt{3}}{\sqrt{3}}$
$= \frac{9\sqrt{3}}{3} = 3\sqrt{3}$

$XZ = 2 \cdot XY = 2 \cdot 3\sqrt{3} = 6\sqrt{3} \approx 10.39$

EXAMPLE 4

Each side of an equilateral triangle measures 6 in. Find the length of an altitude of the triangle.

Solution

The equilateral triangle shown in Figure 5.35(a) is separated into two 30°-60°-90° triangles by the altitude. In the 30°-60°-90° triangle in Figure 5.35(b), the side of the equilateral triangle becomes the hypotenuse, so $2a = 6$ and $a = 3$. The altitude lies opposite the 60° angle of the 30°-60°-90° triangle, so its length is $a\sqrt{3}$ or $3\sqrt{3}$ in. $\approx$ 5.20 in.

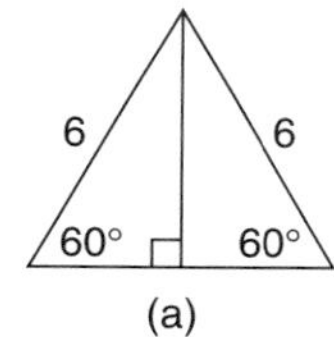

(a)

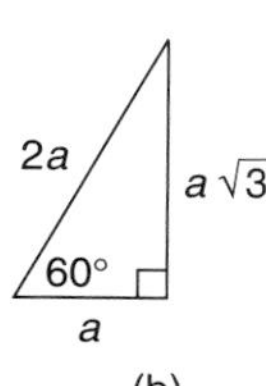

(b)

FIGURE 5.35

The converse of Theorem 5.4.1 is true and is described in the following theorem.

> **THEOREM 5.4.3:** If the length of the hypotenuse of a right triangle equals the product of $\sqrt{2}$ and the length of either leg, then the angles of the triangle measure 45°, 45°, and 90°.

Proof

In Figure 5.36, we represent the length of the hypotenuse by $h = a\sqrt{2}$, where a is the length of either leg. In a right triangle, the angles that lie opposite the congruent legs are also congruent. In a right triangle, the acute angles are complementary, so each of the congruent acute angles measures 45°.

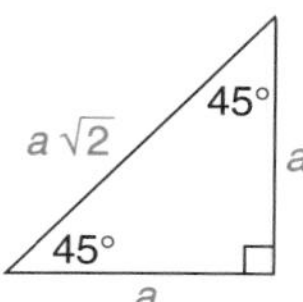

FIGURE 5.36

EXAMPLE 5

In right $\triangle RST$, $RS = ST$. (See Figure 5.37.) What are the measures of the angles of the triangle? If $RT = 12\sqrt{2}$, what is the length of $\overline{RS}$ (or $\overline{ST}$)?

Solution

The longest side is the hypotenuse $\overline{RT}$, so the right angle is $\angle S$ and $m\angle S = 90°$. Since $\overline{RS} \cong \overline{ST}$, the congruent acute angles are $\angle$s R and T and $m\angle R = m\angle T = 45°$. Because $RT = 12\sqrt{2}$, $RS = ST = 12$.

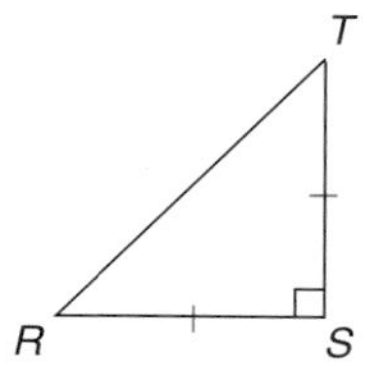

FIGURE 5.37

The converse of Theorem 5.4.2 is also true and can be proven by the indirect method. Rather than construct the proof, we state and apply this theorem. See Figure 5.38.

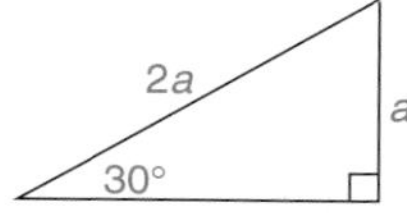

FIGURE 5.38

> **THEOREM 5.4.4:** If the length of the hypotenuse of a right triangle is twice the length of one leg of the triangle, then the angle of the triangle opposite that leg measures 30°.

An equivalent form of this theorem is stated as follows:

> If one leg of a right triangle has a length equal to one-half the length of the hypotenuse, then the angle of the triangle opposite that leg measures 30° (see Figure 5.39).

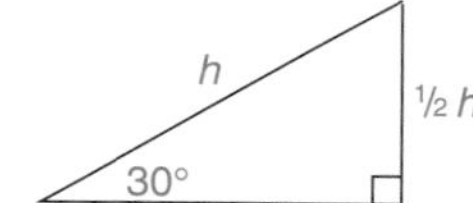

FIGURE 5.39

EXAMPLE 6

In right $\triangle ABC$ with right $\angle C$, $AB = 24.6$ and $BC = 12.3$ (see Figure 5.40). What are the measures of the angles of the triangle? Also, what is the length of $\overline{AC}$?

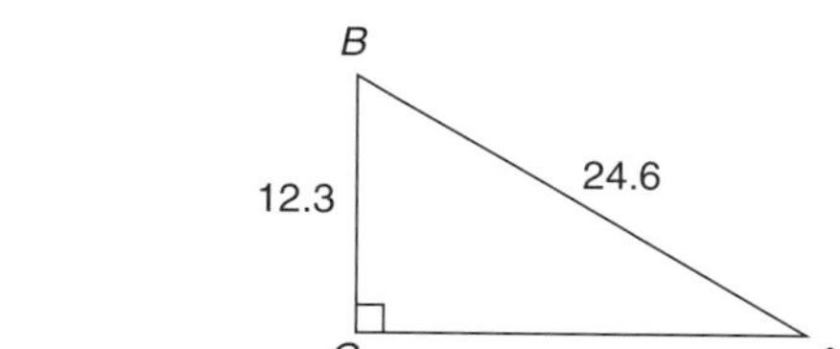

FIGURE 5.40

Solution

Since $\angle C$ is a right angle, $m\angle C = 90°$ and $\overline{AB}$ is the hypotenuse. Since $BC = \frac{1}{2}(AB)$, the angle opposite $\overline{BC}$ measures 30°. Thus, $m\angle A = 30°$ and $m\angle B = 60°$.

Because $\overline{AC}$ lies opposite the 60° angle, $AC = (12.3)\sqrt{3} \approx 21.3$.

5.4 Exercises

In Exercises 1 to 14, find the missing lengths. Give your answers in both simplest radical form and as approximations to two decimal places.

1. *Given:* Right $\triangle XYZ$ with $m\angle X = 45°$ and $XZ = 8$
Find: YZ and XY

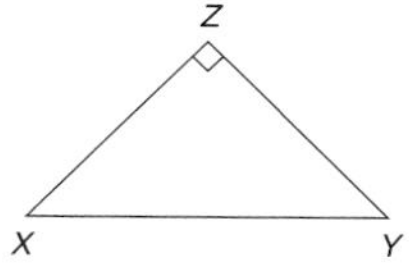

Exercises 1, 2

2. *Given:* Right $\triangle XYZ$ with $\overline{XZ} \cong \overline{YZ}$ and $XY = 10$
Find: XZ and YZ

3. *Given:* Right $\triangle DEF$ with $m\angle E = 60°$ and $DE = 5$
Find: DF and FE

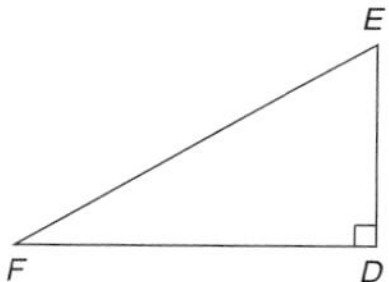

Exercises 3, 4

4. *Given:* Right $\triangle DEF$ with $m\angle E = 2 \cdot m\angle F$ and $EF = 12\sqrt{3}$
Find: DE and DF

5. *Given:* Rectangle $HJKL$ with diagonals $\overline{HK}$ and $\overline{JL}$
$m\angle HKL = 30°$
Find: HL, HK, and MK

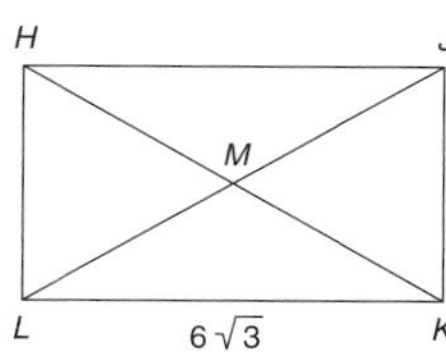

6. *Given:* Right $\triangle RST$ with $RT = 6\sqrt{2}$ and $m\angle STV = 150°$
Find: RS and ST

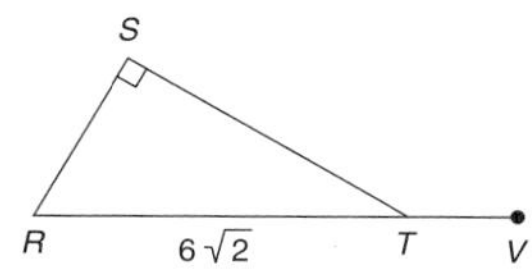

7. *Given:* $\triangle ABC$ with $m\angle A = m\angle B = 45°$ and $BC = 6$
Find: AC and AB

8. *Given:* Right $\triangle MNP$ with $MP = PN$ and $MN = 10\sqrt{2}$
Find: PM and PN

9. *Given:* $\triangle RST$ with $m\angle T = 30°$, $m\angle S = 60°$, and $ST = 12$
Find: RS and RT

10. *Given:* $\triangle XYZ$ with $\overline{XY} \cong \overline{XZ} \cong \overline{YZ}$
$\overline{ZW} \perp \overline{XY}$
$YZ = 6$
Find: ZW

11. *Given:* Square $ABCD$ with diagonals $\overline{DB}$ and $\overline{AC}$ intersecting at E
$DC = 5\sqrt{3}$
Find: DB

12. *Given:* $\triangle NQM$ with $\angle$s as shown in the drawing
$\overline{MP} \perp \overline{NQ}$
Find: NM, MP, MQ, PQ, and NQ

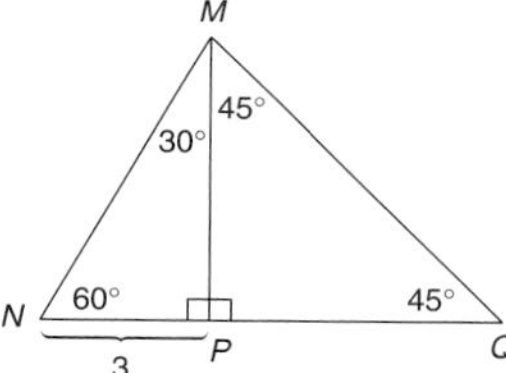

13. *Given:* $\triangle XYZ$ with $\angle$s as shown in the drawing
Find: XY
(**HINT:** Compare this drawing to the one for Exercise 12.)

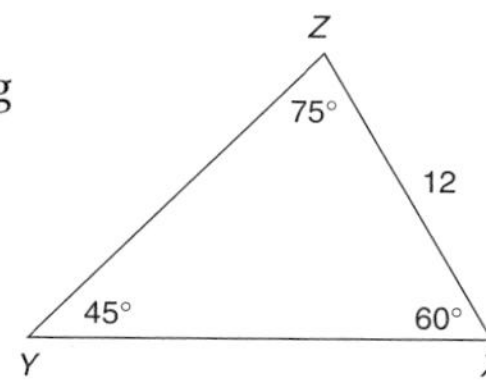

14. *Given:* Rhombus $ABCD$ in which diagonals $\overline{AC}$ and $\overline{DB}$ intersect at E
$DB = AB = 8$
Find: AC

15. A carpenter is working with a board that is $3\frac{3}{4}$ in. wide. After marking off a point down the side of length $3\frac{3}{4}$ in., the carpenter makes a cut along $\overline{BC}$ with a saw. What is the measure of the angle ($\angle ACB$) that is formed?

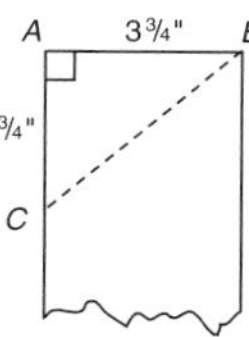

16. To unload groceries from a delivery truck at the Piggly Wiggly Market, an 8-ft ramp that rises 4 ft to the door of the trailer is used. What is the measure of the indicated angle ($\angle D$)?

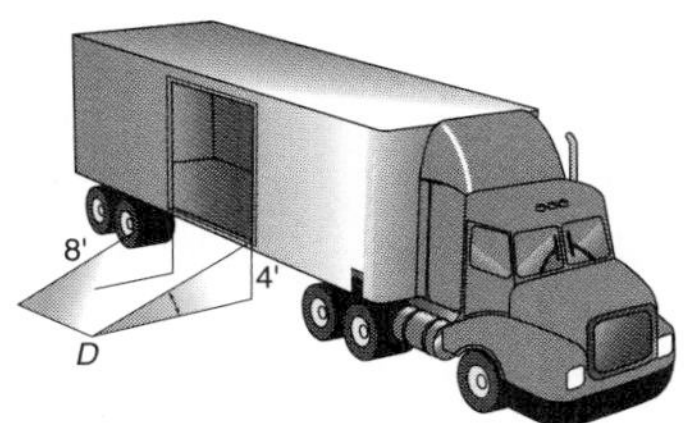

17. A jogger runs along two sides of an open rectangular lot. If the first side of the lot is 200 ft long and the diagonal distance across the lot is 400 ft, what is the measure of the angle formed by the 200-ft and 400-ft dimensions? To the nearest foot, how much farther does the jogger run by traveling the two sides of the block rather than the diagonal distance across the lot?

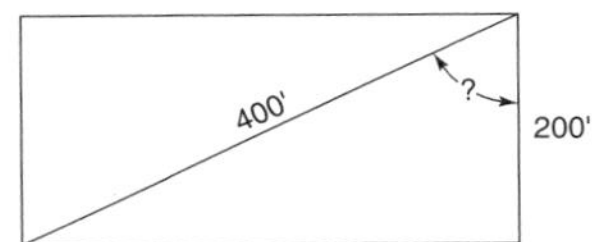

18. Thelma's boat leaves the dock at the same time that Gina's boat leaves the dock. Thelma's boat travels due east at 12 mph. Gina's boat travels at 24 mph in the direction N 30° E. To the nearest tenth of a mile, how far apart will the boats be in half an hour?

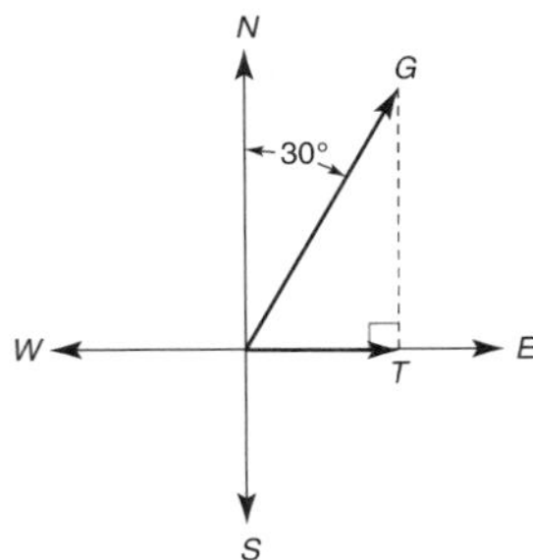

In Exercises 19–25, give both exact solutions and approximate solutions to two decimal places.

19. *Given:* In $\triangle ABC$, $\overrightarrow{AD}$ bisects $\angle BAC$
$m\angle B = 30°$ and $AB = 12$
Find: DC and DB

20. *Given:* In $\triangle ABC$, $\overrightarrow{AD}$ bisects $\angle BAC$
$AB = 20$ and $AC = 10$
Find: DC and DB

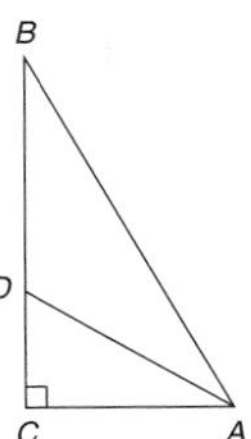

Exercises 19, 20

21. *Given:* $\triangle MNQ$ is equiangular and $NR = 6$
$\overrightarrow{NR}$ bisects $\angle MNQ$
$\overrightarrow{QR}$ bisects $\angle MQN$
Find: NQ

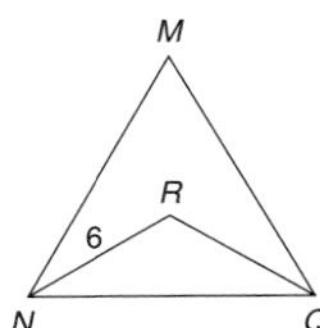

22. *Given:* $\triangle STV$ is an isosceles right triangle
M and N are midpoints of $\overline{ST}$ and $\overline{SV}$
Find: MN

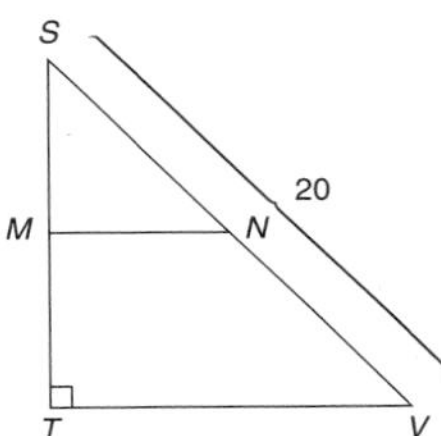

23. *Given:* Right $\triangle ABC$ with $m\angle C = 90°$ and
$m\angle BAC = 60°$; point D on $\overline{BC}$; $\overrightarrow{AD}$ bisects $\angle BAC$
and $AB = 12$
Find: BD

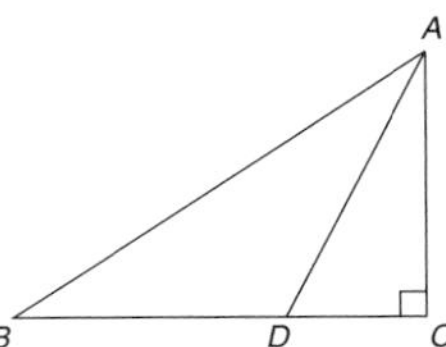

Exercises 23, 24

24. *Given:* Right $\triangle ABC$ with $m\angle C = 90°$ and
$m\angle BAC = 60°$; point D on $\overline{BC}$; $\overrightarrow{AD}$ bisects $\angle BAC$
and $AC = 2\sqrt{3}$
Find: BD

25. *Given:* $\triangle ABC$ with $m\angle A = 45°$, $m\angle B = 30°$, and $BC = 12$
Find: AB
(**HINT:** Use altitude $\overline{CD}$ from C to $\overline{AB}$ as an auxiliary line.)

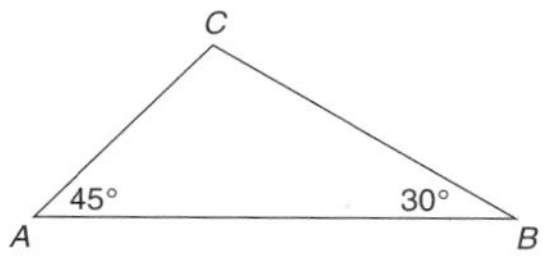

26. ➢ *Given:* Isosceles trapezoid $MNPQ$ with $QP = 12$ and $m\angle M = 120°$; the bisectors of $\angle$s MQP and NPQ meet at point T on $\overline{MN}$
Find: The perimeter of $MNPQ$

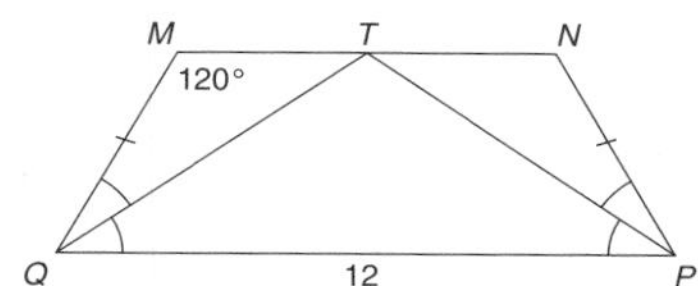

5.5 Segments Divided Proportionally

In this section, we begin with an informal description of the phrase "divided proportionally." Suppose that three children have been provided with a joint savings account by their parents. Equal monthly deposits have been made to the account for each child since birth. If the ages of the children are 2, 4, and 6 (assume exactness of ages for simplicity) and the total in the account is \$7200, then the amount that each child should receive can be found by solving the equation

$$2x + 4x + 6x = 7200$$

Solving leads to the solution \$1200 for the 2-year-old, \$2400 for the 4-year-old, and \$3600 for the 6-year-old. We may say that the amount has been divided proportionally. Expressed as a proportion, this is

$$\frac{1200}{2} = \frac{2400}{4} = \frac{3600}{6}$$

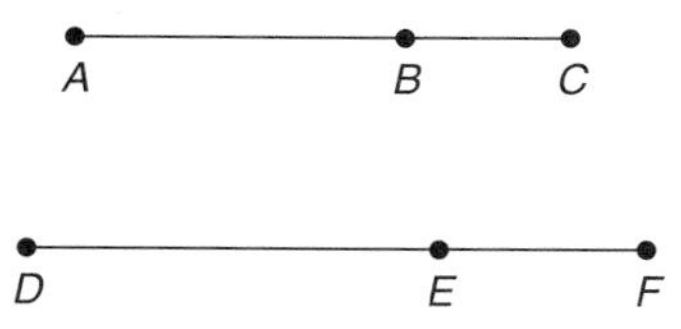

FIGURE 5.41

In Figure 5.41, $\overline{AC}$ and $\overline{DF}$ are divided proportionally at points B and E if $\frac{AB}{DE} = \frac{BC}{EF}$ $\left(\text{or } \frac{AB}{BC} = \frac{DE}{EF}\right)$.

Of course, a pair of segments may be divided proportionally by several points, as shown in Figure 5.42. In this case, $\overline{RW}$ and $\overline{HM}$ are divided proportionally when

$$\frac{RS}{HJ} = \frac{ST}{JK} = \frac{TV}{KL} = \frac{VW}{LM}$$

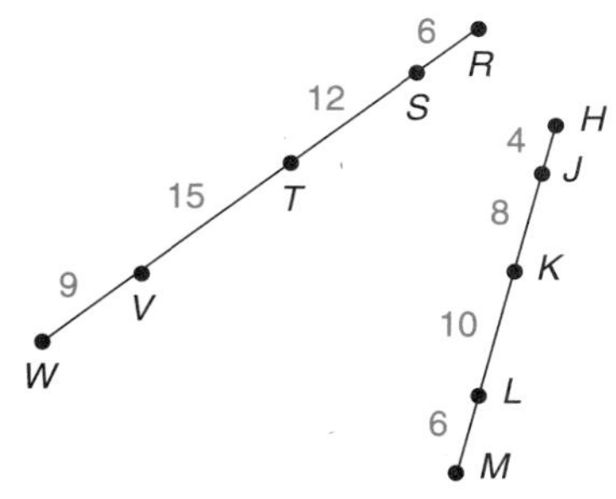

FIGURE 5.42

EXAMPLE 1

In Figure 5.43 on the next page, points D and E divide $\overline{AB}$ and $\overline{AC}$ proportionally. If $AD = 4$, $DB = 7$, and $EC = 6$, find AE.

Solution

$\frac{AD}{AE} = \frac{DB}{EC}$, so $\frac{4}{x} = \frac{7}{6}$, where $x = AE$. Then $7x = 24$, so $x = AE = \frac{24}{7} = 3\frac{3}{7}$.

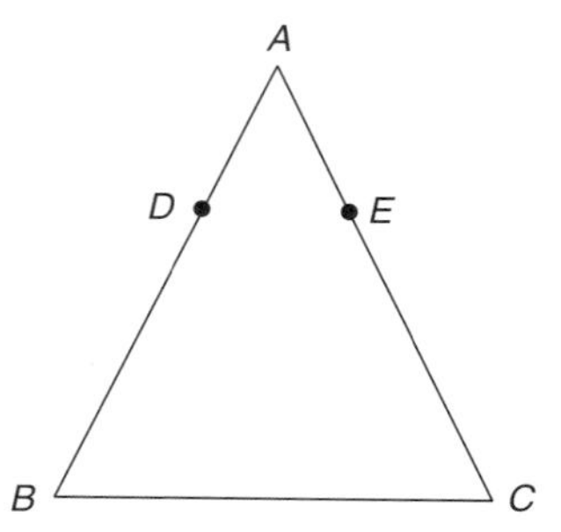

FIGURE 5.43

A property that will be proved in Exercise 25 of this section is

$$\text{If } \frac{a}{b} = \frac{c}{d}, \text{ then } \frac{a+c}{b+d} = \frac{a}{b} = \frac{c}{d}$$

In words, we may restate this property as follows:

The fraction whose numerator and denominator are determined, respectively, by adding numerators and denominators of equal fractions is equal to each of those equal fractions.

Here is a numerical example of this claim:

$$\text{If } \frac{2}{3} = \frac{4}{6}, \text{ then } \frac{2+4}{3+6} = \frac{2}{3} = \frac{4}{6}$$

In Example 2, the preceding property is necessary as a reason.

EXAMPLE 2

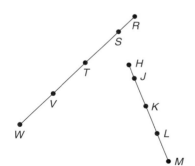

Given: $\overline{RW}$ and $\overline{HM}$ are divided proportionally at the points shown in Figure 5.44.

Prove: $\frac{RT}{HK} = \frac{TW}{KM}$

Proof: $\overline{RW}$ and $\overline{HM}$ are divided proportionally so that

$$\frac{RS}{HJ} = \frac{ST}{JK} = \frac{TV}{KL} = \frac{VW}{LM}$$

Using the property that if $\frac{a}{b} = \frac{c}{d}$, then $\frac{a+c}{b+d} = \frac{a}{b} = \frac{c}{d}$, we have

$$\frac{RS}{HJ} = \frac{RS+ST}{HJ+JK} = \frac{TV+VW}{KL+LM} = \frac{TV}{KL}$$

Because $RS + ST = RT$, $HJ + JK = HK$, $TV = VW = TW$, and $KL + LM = KM$,

$$\frac{RT}{HK} = \frac{TW}{KM}$$

FIGURE 5.44

Two properties that were introduced earlier (Property 3 of Section 5.1) are now recalled.

$$\text{If } \frac{a}{b} = \frac{c}{d}, \text{ then } \frac{a \pm b}{b} = \frac{c \pm d}{d}$$

The subtraction part of the property is needed for the proof of Theorem 5.5.1.

THEOREM 5.5.1: If a line is parallel to one side of a triangle and intersects the other two sides, then it divides these sides proportionally.

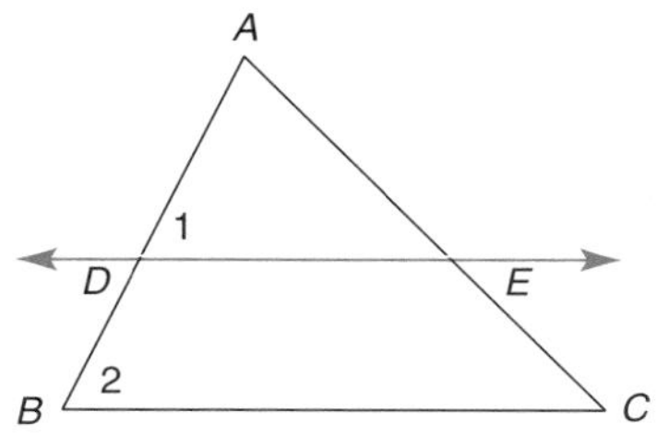

FIGURE 5.45

Given: In Figure 5.45, $\triangle ABC$ with $\overleftrightarrow{DE} \parallel \overline{BC}$ and with $\overleftrightarrow{DE}$ intersecting $\overline{AB}$ at D and $\overline{AC}$ at E

Prove: $\frac{AD}{DB} = \frac{AE}{EC}$

Proof: Since $\overleftrightarrow{DE} \parallel \overline{BC}$, $\angle 1 \cong \angle 2$. With $\angle A$ as a common angle for $\triangle ADE$ and $\triangle ABC$, it follows by AA that these triangles are similar. Now

$$\frac{AB}{AD} = \frac{AC}{AE} \qquad \text{by CSSTP}$$

By Property 3 of Section 5.1,

$$\frac{AB - AD}{AD} = \frac{AC - AE}{AE}$$

Because $AB - AD = DB$ and $AC - AE = EC$, the proportion becomes

$$\frac{DB}{AD} = \frac{EC}{AE}$$

Inverting both fractions gives the desired conclusion:

$$\frac{AD}{DB} = \frac{AE}{EC}$$

COROLLARY 5.5.2: When three (or more) parallel lines are cut by a pair of transversals, the transversals are divided proportionally by the parallel lines.

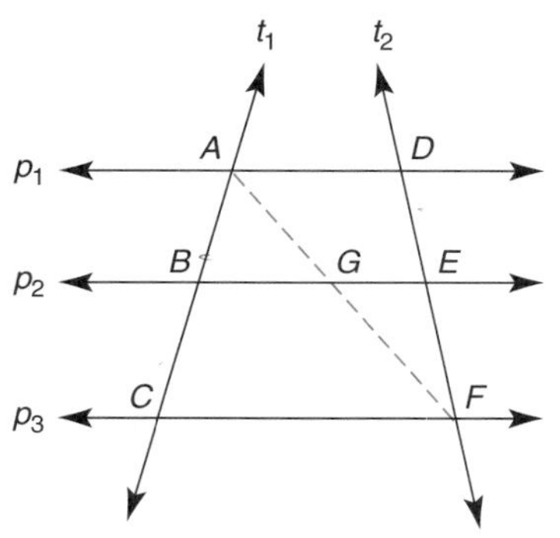

FIGURE 5.46

Corollary 5.5.2 is now illustrated, and its proof is briefly described. This corollary claims, for example, that $\frac{AB}{BC} = \frac{DE}{EF}$ as shown in Figure 5.46, where $p_1 \parallel p_2 \parallel p_3$. The proof requires the use of an auxiliary segment $\overline{AF}$. Then

$$\frac{AB}{BC} = \frac{AG}{GF} \quad \text{and} \quad \frac{AG}{GF} = \frac{DE}{EF} \qquad \text{by Theorem 5.5.1}$$

Consequently,

$$\frac{AB}{BC} = \frac{DE}{EF} \qquad \text{Transitive Property of Equality}$$

NOTE: By interchanging the means, we can write this proportion in the form $\frac{AB}{DE} = \frac{BC}{EF}$.

EXAMPLE 3

Given parallel lines p_1, p_2, p_3, and p_4 cut by t_1 and t_2 so that $AB = 4$, $EF = 3$, $BC = 2$, and $GH = 5$, find FG and CD. (See Figure 5.47.)

Solution

Because the transversals are divided proportionally,

$$\frac{AB}{EF} = \frac{BC}{FG} = \frac{CD}{GH}$$

so $$\frac{4}{3} = \frac{2}{FG} = \frac{CD}{5}$$

Then $4 \cdot FG = 6$ and $3 \cdot CD = 20$

$$FG = \frac{3}{2} = 1\tfrac{1}{2} \quad \text{and} \quad CD = \frac{20}{3} = 6\tfrac{2}{3}$$

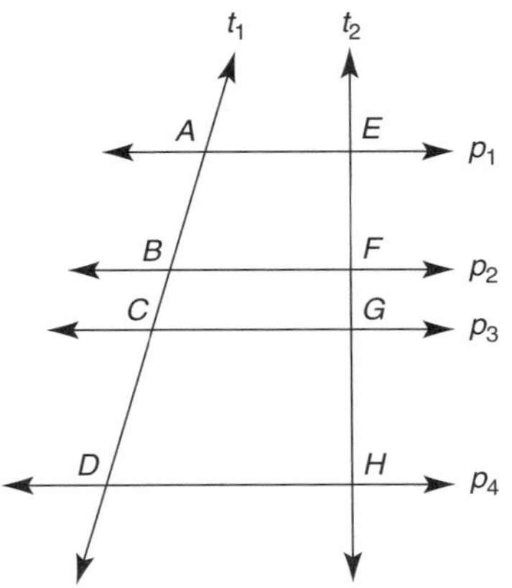

FIGURE 5.47

The following activity leads us to the relationship described in Theorem 5.5.3.

DISCOVER!

On a piece of paper, draw or construct $\triangle ABC$ whose sides measure $AB = 4$, $BC = 6$, and $AC = 5$. Then construct the angle bisector $\overrightarrow{BD}$ of $\angle B$. How does $\frac{AB}{AD}$ compare to $\frac{BC}{DC}$?

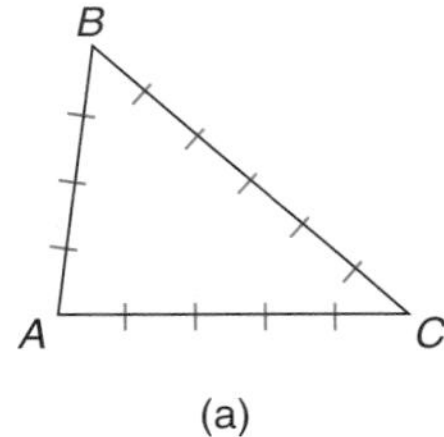

(a)

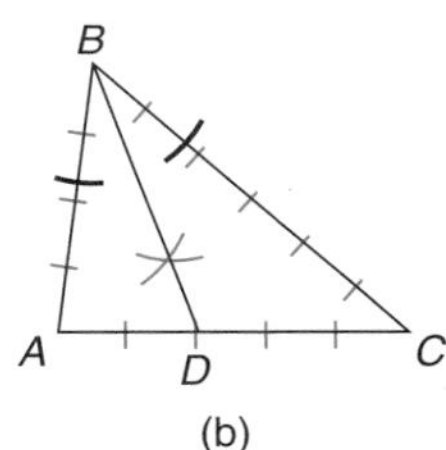

(b)

ANSWER

Though not by chance, it may come as a surprise that $\frac{AB}{AD} = \frac{BC}{DC}$ (that is, $\frac{4}{2} = \frac{6}{3}$) and $\frac{AB}{BC} = \frac{AD}{DC}$ $\left(\frac{4}{6} = \frac{2}{3}\right)$. It seems that the bisector of an angle included by two sides of a triangle separates the third side into segments whose lengths are proportional to the lengths of the two sides forming the angle.

The proof of Theorem 5.5.3 requires the use of Theorem 5.5.1.

THEOREM 5.5.3: If a ray bisects one angle of a triangle, then it divides the opposite side into segments whose lengths are proportional to the lengths of the two sides which form that angle.

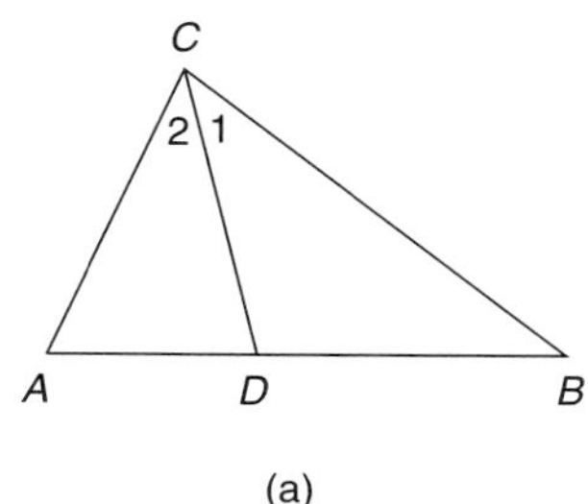

(a)

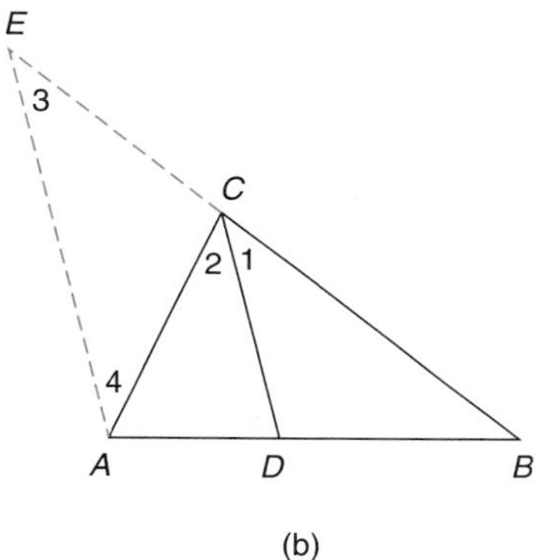

(b)

FIGURE 5.48

Given: $\triangle ABC$ in Figure 5.48(a), in which $\overrightarrow{CD}$ bisects $\angle ACB$

Prove: $\frac{AD}{AC} = \frac{DB}{CB}$

Proof: We begin by extending $\overline{BC}$ beyond C (there is only one line through B and C) to meet the line through A drawn parallel to $\overline{DC}$. [See Figure 5.48(b).] Let E be the point of intersection (these lines must intersect; otherwise $\overline{AE}$ would have two parallels, $\overline{BC}$ and $\overline{CD}$, through point C).

Since $\overline{CD} \parallel \overline{EA}$ we have

$$\frac{EC}{AD} = \frac{CB}{DB}$$

by Theorem 5.5.1. Now $\angle 1 \cong \angle 2$ (due to the angle bisector), while $\angle 1 \cong \angle 3$ (since they are corresponding angles for parallel lines) and $\angle 2 \cong \angle 4$ (alternate interior angles for parallel lines). By the Transitive Property, $\angle 3 \cong \angle 4$ so $\triangle ACE$ is isosceles with $\overline{EC} \cong \overline{AC}$. Using substitution, the proportion becomes

$$\frac{AC}{AD} = \frac{CB}{DB} \quad \text{or} \quad \frac{AD}{AC} = \frac{DB}{CB} \quad \text{by inversion}$$

The "Prove" of the preceding theorem indicates that one form of the proportionality described is given informally by

$$\frac{\text{segment at left}}{\text{side at left}} = \frac{\text{segment at right}}{\text{side at right}}$$

Equivalently, the proportion could state

$$\frac{\text{segment at left}}{\text{segment at right}} = \frac{\text{side at left}}{\text{side at right}}$$

Other forms of the proportion are also possible!

EXAMPLE 4

For $\triangle XYZ$ in Figure 5.49, $XY = 3$ and $YZ = 5$. If $\overrightarrow{YW}$ bisects $\angle XYZ$ and $XW = 2$, find XZ.

Solution

Let $WZ = x$. We know that $\frac{YX}{XW} = \frac{YZ}{WZ}$, so $\frac{3}{2} = \frac{5}{x}$.

Therefore

$$3x = 10$$
$$x = \frac{10}{3} = 3\tfrac{1}{3}$$

Then $WZ = 3\frac{1}{3}$.

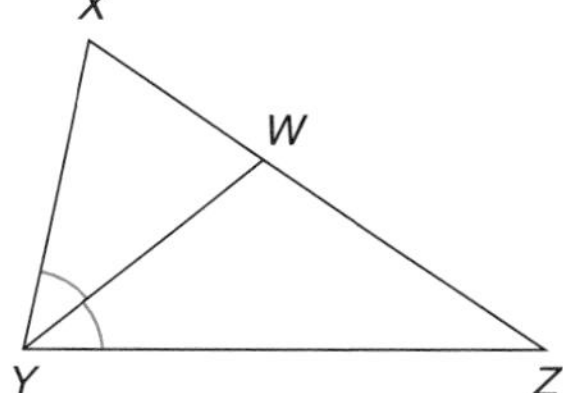

FIGURE 5.49

Because $XZ = XW + WZ$, we have $XZ = 2 + 3\frac{1}{3} = 5\frac{1}{3}$.

EXAMPLE 5

In Figure 5.49 on page 223, $\triangle XYZ$ has sides of lengths $XY = 3$, $YZ = 4$, and $XZ = 5$. If $\overrightarrow{YW}$ bisects $\angle XYZ$, find XW and WZ.

Solution

Let $XW = y$; then $WZ = 5 - y$, and $\frac{XY}{YZ} = \frac{XW}{WZ}$ becomes $\frac{3}{4} = \frac{y}{5-y}$. From this proportion we can find y.

$$3(5 - y) = 4y$$
$$15 - 3y = 4y$$
$$15 = 7y$$
$$y = \frac{15}{7}$$

Then $XW = \frac{15}{7} = 2\frac{1}{7}$ while $WZ = 5 - 2\frac{1}{7} = 2\frac{6}{7}$.

5.5 Exercises

1. In preparing a recipe, 5 oz of ingredient A, 4 oz of ingredient B, and 6 oz of ingredient C are used. If 90 oz of this dish are needed, how many ounces of each ingredient should be used?

2. In a chemical mixture, 2 g of chemical A are used for each gram of chemical B, and 3 g of chemical C are needed for each gram of B. If 72 g of the mixture are prepared, what amount (in grams) of each chemical is needed?

3. Given that $\frac{AB}{EF} = \frac{BC}{FG} = \frac{CD}{GH}$, do the following proportions hold?

   a) $\frac{AC}{EG} = \frac{CD}{GH}$

   b) $\frac{AB}{EF} = \frac{BD}{FH}$

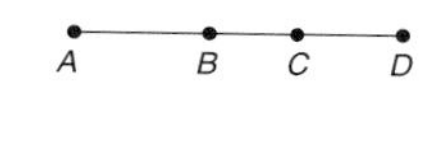

4. Given that $\overleftrightarrow{XY} \parallel \overline{TS}$, do the following proportions hold?

   a) $\frac{TX}{XR} = \frac{RY}{YS}$

   b) $\frac{TR}{XR} = \frac{SR}{YR}$

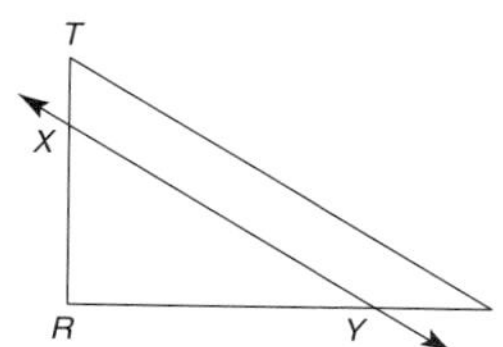

5. *Given:* $\ell_1 \parallel \ell_2 \parallel \ell_3 \parallel \ell_4$, $AB = 5$, $BC = 4$, $CD = 3$, $EH = 10$
   Find: EF, FG, and GH

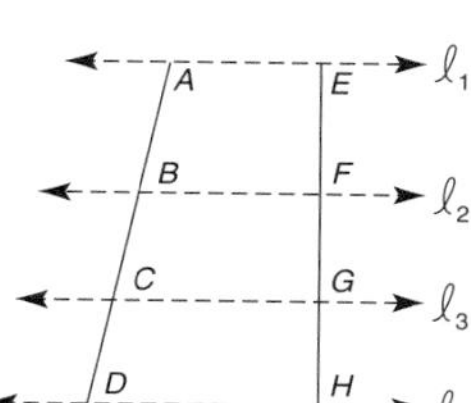

Exercises 5, 6

6. *Given:* $\ell_1 \parallel \ell_2 \parallel \ell_3 \parallel \ell_4$, $AB = 7$, $BC = 5$, $CD = 4$, $EF = 6$
   Find: FG, GH, EH

7. *Given:* $\ell_1 \parallel \ell_2 \parallel \ell_3$, $AB = 4$, $BC = 5$, $DE = x$, $EF = 12 - x$
   Find: x, DE, EF

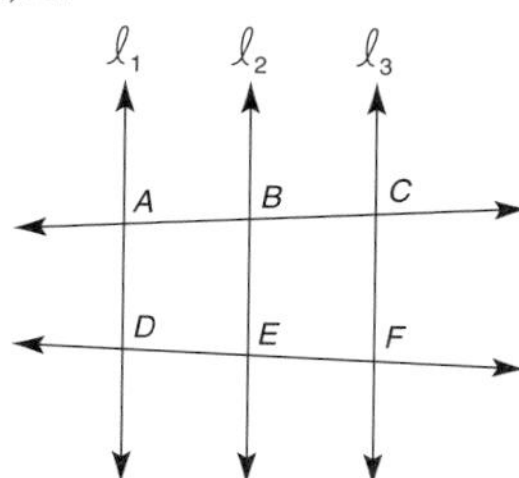

Exercises 7, 8

8. *Given:* $\ell_1 \parallel \ell_2 \parallel \ell_3$, $AB = 5$, $BC = x$, $DE = x - 2$, $EF = 7$
   Find: x, BC, DE

9. *Given:* $\overleftrightarrow{DE} \parallel \overline{BC}$, $AD = 5$, $DB = 12$, $AE = 7$
Find: EC

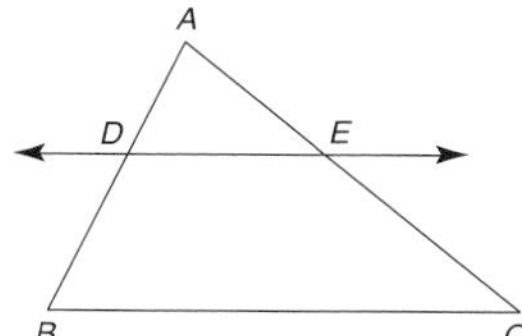

Exercises 9–12

10. *Given:* $\overleftrightarrow{DE} \parallel \overline{BC}$, $AD = 6$, $DB = 10$, $AC = 20$
Find: EC

11. *Given:* $\overleftrightarrow{DE} \parallel \overline{BC}$, $AD = a - 1$, $DB = 2a + 2$, $AE = a$, $EC = 4a - 5$
Find: a and AD

12. *Given:* $\overleftrightarrow{DE} \parallel \overline{BC}$, $AD = 5$, $DB = a + 3$, $AE = a + 1$, $EC = 3(a - 1)$
Find: a and EC

13. *Given:* $\overrightarrow{RW}$ bisects $\angle SRT$
Do the following equalities hold?

a) $SW = WT$
b) $\dfrac{RS}{RT} = \dfrac{SW}{WT}$

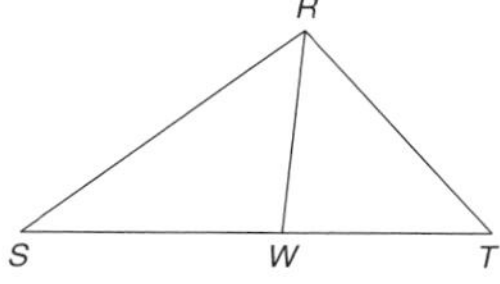

Exercises 13, 14

14. *Given:* $\overrightarrow{RW}$ bisects $\angle SRT$
Do the following equalities hold?

a) $\dfrac{RS}{SW} = \dfrac{RT}{WT}$
b) $m\angle S = m\angle T$

15. *Given:* $\overrightarrow{UT}$ bisects $\angle WUV$, $WU = 8$, $UV = 12$, $WT = 6$
Find: TV

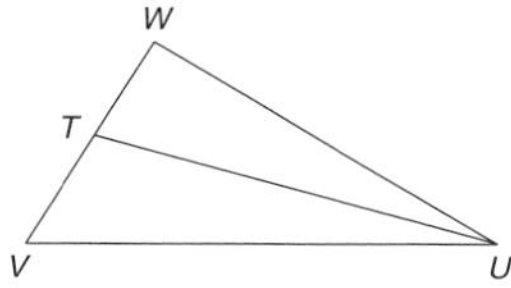

Exercises 15, 16

16. *Given:* $\overrightarrow{UT}$ bisects $\angle WUV$, $WU = 9$, $UV = 12$, $WV = 9$
Find: WT

17. *Given:* $\overrightarrow{NQ}$ bisects $\angle MNP$, $NP = MQ$, $QP = 8$, $MN = 12$
Find: NP

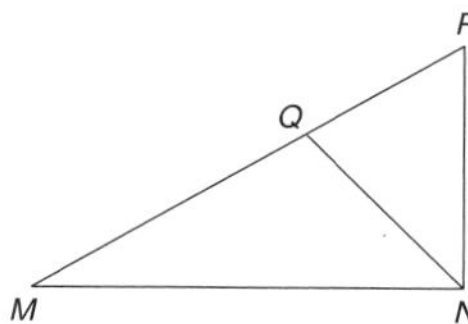

18. *Given:* In $\triangle ABC$, $\overrightarrow{AD}$ bisects $\angle BAC$
$AB = 20$ and $AC = 16$
Find: DC and DB

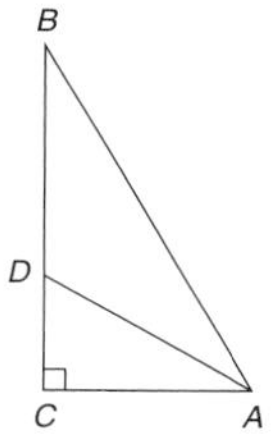

19. In $\triangle ABC$, $\angle ACB$ is trisected by $\overrightarrow{CD}$ and $\overrightarrow{CE}$ so that $\angle 1 \cong \angle 2 \cong \angle 3$. Write two different proportions that follow from this information.

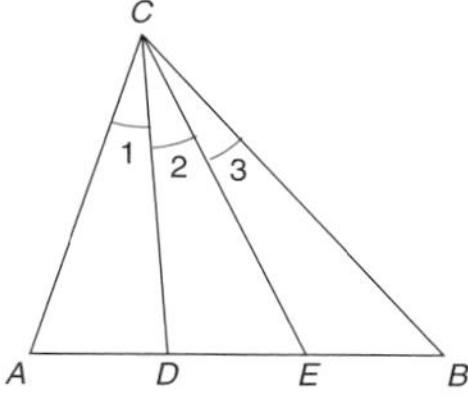

20. In $\triangle ABC$, $m\angle CAB = 80°$, $m\angle ACB = 60°$, and $m\angle ABC = 40°$. With the angle bisectors as shown, which line segment is longer?

a) $\overline{AE}$ or $\overline{EC}$? **b)** $\overline{CD}$ or $\overline{DB}$? **c)** $\overline{AF}$ or $\overline{FB}$?

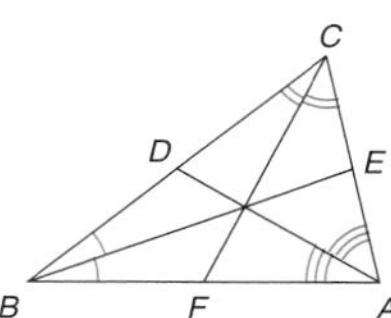

21. In right $\triangle RST$ (not shown) with right $\angle S$, $\overrightarrow{RV}$ bisects $\angle SRT$ so that V lies on side $\overline{ST}$. If $RS = 6$ and $RT = 12$, find SV and VT.

22. *Given:* AC is the geometric mean between AD and AB.
$AD = 4$, $DB = 6$
Find: AC

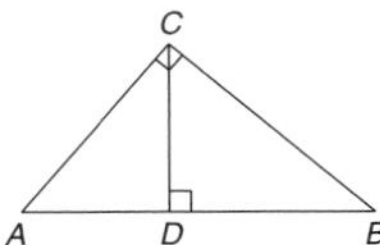

23. *Given:* $\overrightarrow{RV}$ bisects $\angle SRT$, $RS = x - 6$, $SV = 3$, $RT = 2 - x$, $VT = x + 2$
Find: x
(**HINT:** You will need to apply the Quadratic Formula.)

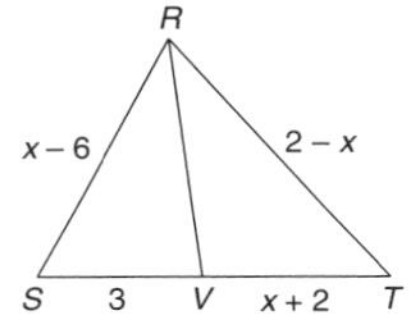

24. *Given:* $\overrightarrow{MR}$ bisects $\angle NMP$, $MN = 2x$, $NR = x$, $RP = x + 1$, and $MP = 3x - 1$.
Find: x

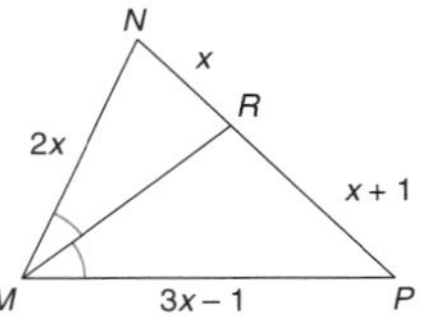

25. Complete the proof of this property:

$$\text{If } \frac{a}{b} = \frac{c}{d}, \text{ then } \frac{a+c}{b+d} = \frac{a}{b} \text{ and } \frac{a+c}{b+d} = \frac{c}{d}$$

PROOF

Statements	Reasons
1. $\frac{a}{b} = \frac{c}{d}$	**1.** ?
2. $b \cdot c = a \cdot d$	**2.** ?
3. $ab + bc = ab + ad$	**3.** ?
4. $b(a + c) = a(b + d)$	**4.** ?
5. $\frac{a+c}{b+d} = \frac{a}{b}$	**5.** Means-Extremes Property (symmetric form)
6. $\frac{a+c}{b+d} = \frac{c}{d}$	**6.** ?

26. *Given:* $\triangle RST$, with $\overleftrightarrow{XY} \parallel \overline{RT}$, $\overleftrightarrow{YZ} \parallel \overline{RS}$
Prove: $\frac{RX}{XS} = \frac{ZT}{RZ}$

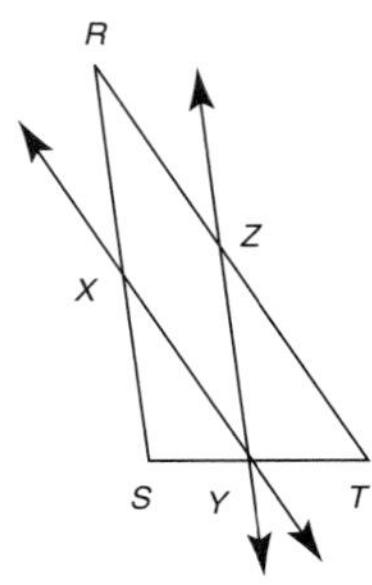

27. Use Theorem 5.5.1 and the drawing to complete the proof of this theorem: "If a line is parallel to one side of a triangle and passes through the midpoint of a second side, then it will pass through the midpoint of the third side."
Given: $\triangle RST$ with M the midpoint of $\overline{RS}$; $\overleftrightarrow{MN} \parallel \overline{ST}$
Prove: N is the midpoint of $\overline{RT}$

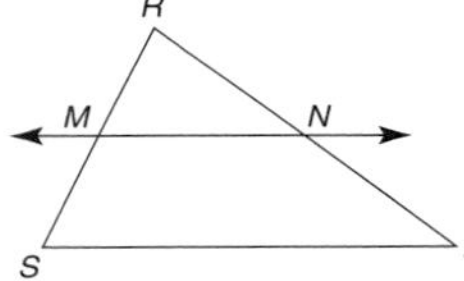

28. Use Exercises 27 and the drawing to complete the proof of this theorem: "The length of the median of a trapezoid is one-half the sum of lengths of the two bases."
Given: Trapezoid $ABCD$ with median $\overline{MN}$
Prove: $MN = \frac{1}{2}(AB + CD)$

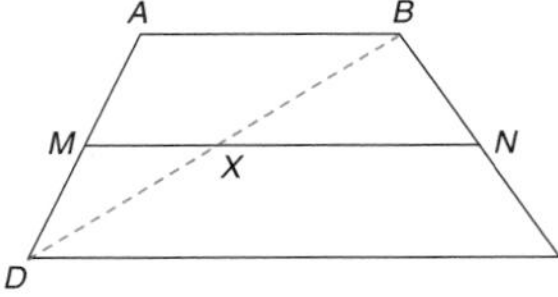

29. Use Theorem 5.5.3 to complete the proof of this theorem: "If the bisector of an angle of a triangle also bisects the opposite side, then the triangle is an isosceles triangle."
Given: $\triangle XYZ$; $\overrightarrow{YW}$ bisects $\angle XYZ$; $\overline{WX} \cong \overline{WZ}$
Prove: $\triangle XYZ$ is isosceles
(**HINT:** Use a proportion to show that $YX = YZ$.)

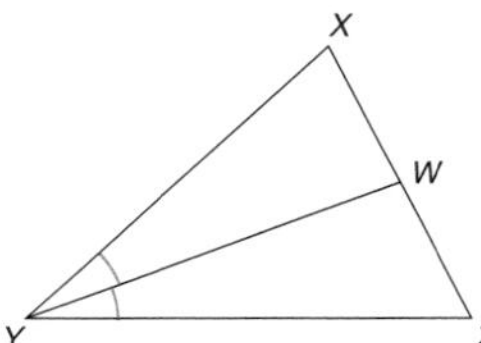

30. In right $\triangle ABC$ (not shown) with right $\angle C$, $\overrightarrow{AD}$ bisects $\angle BAC$ so that D lies on side $\overline{CB}$. If $AC = 6$ and $DC = 3$, find BD and AB.
(**HINT:** Let $BD = x$ and $AB = 2x$. Then use the Pythagorean Theorem.)

31. *Given:* $\triangle ABC$ (not shown) is isosceles with $m\angle ABC = m\angle C = 72°$; $\overrightarrow{BD}$ bisects $\angle ABC$ and $AB = 1$
Find: BC

32. *Given:* $\triangle RST$ with right $\angle RST$; $m\angle R = 30°$ and $ST = 6$; $\angle RST$ is trisected by $\overrightarrow{SM}$ and $\overrightarrow{SN}$
Find: TN, NM, and MR

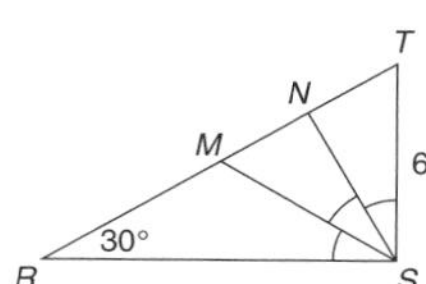

A Look Beyond: An Unusual Application of Similar Triangles

The following problem is one that can be solved in many ways. If methods of calculus are applied, the solution is found through many complicated and tedious calculations. The simplest solution, which follows, utilizes geometry and similar triangles.

Problem: A hiker is at a location 450 ft downstream from his campsite. He is 200 ft away from the straight stream, and his tent is 100 ft away, as shown in Figure 5.50(a). Across the flat field, he sees that a spark from his campfire has ignited the tent. Taking the empty bucket he is carrying, he runs to the river to get water and then on to the tent. To what point on the river should he run to minimize the distance he travels?

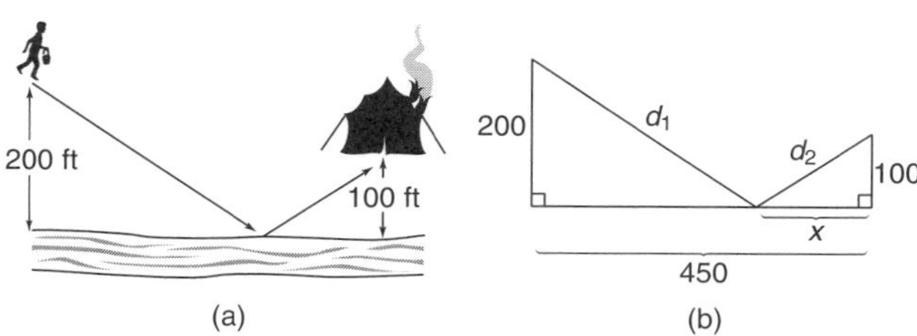

FIGURE 5.50

We wish to determine x in Figure 5.50(b) so that the total distance $D = d_1 + d_2$ is as small as possible. Consider three possible choices of this point on the river. These are suggested by dashed, dotted, and solid lines in Figure 5.51(a). Also consider the reflections of the triangles across the river. [See Figure 5.51(b).]

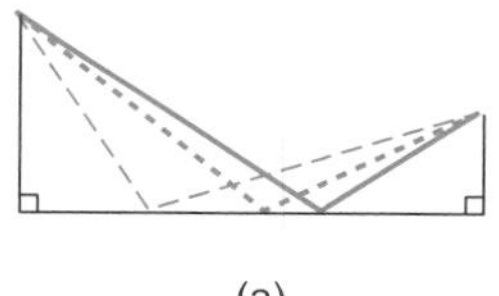

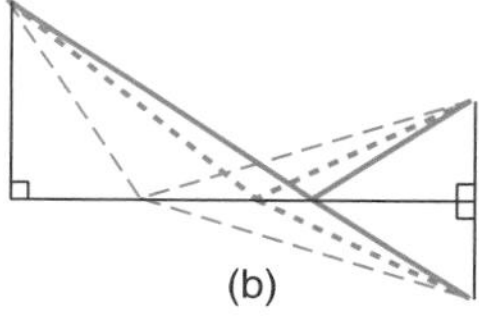

FIGURE 5.51

The minimum distance D occurs where the segments of lengths d_1 and d_2 form a straight line. That is, the configuration with the solid line segments minimizes the distance. In that case, the triangle at left and the reflected triangle at right are similar. (See Figure 5.52.)

Thus $$\frac{200}{100} = \frac{450 - x}{x}$$

$$200x = 100(450 - x)$$
$$200x = 45{,}000 - 100x$$
$$300x = 45{,}000$$
$$x = 150$$

Thus the desired point on the river is 300 ft (determined by $450 - x$) upstream from the hiker's location.

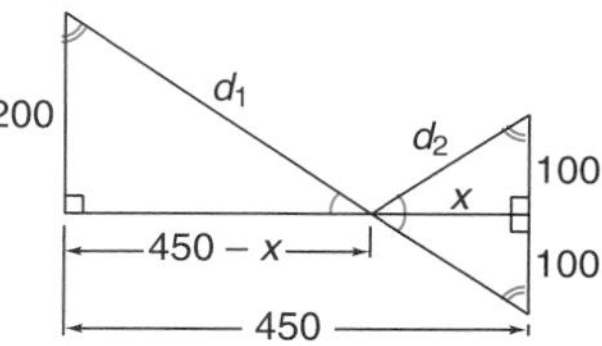

FIGURE 5.52

Summary

A Look Back at Chapter 5

The goal of this chapter has been to define similarity for all polygons. We postulated a method for proving triangles similar and showed that proportions are a consequence of similar triangles, a line parallel to one side of a triangle, and a ray bisecting one angle of a triangle. The Pythagorean Theorem and its converse were proved. The 30-60-90 triangle, the 45-45-90 triangle, and other special right triangles with sides forming Pythagorean triples were discussed.

A Look Ahead to Chapter 6

In the next chapter, we will begin our work with the circle. Segments and lines of the circle will be defined, as will special angles in a circle. Several theorems dealing with the measurements of these angles and line segments will be proved. Our work with constructions will enable us to deal with the locus of points and the concurrence of lines.

Important Terms and Concepts of Chapter 5

5.1 Ratio, Rate, Proportion
Commensurable, Incommensurable
Extremes, Means of a Proportion
Means-Extremes Property
Geometric Mean
Extended Ratio, Extended Proportion

5.2 Similar Polygons
Congruent Polygons
Corresponding Vertices, Angles, and Sides
AAA, AA, and CSSTP

5.3 Pythagorean Theorem and Converse
Pythagorean triple

5.4 45-45-90 Triangle
30-60-90 Triangle

5.5 Segments Divided Proportionally

A Look Beyond: An Unusual Application of Similar Triangles

Review Exercises

Answer true or false for Review Exercises 1 to 7.

1. The ratio of 12 hr to 1 day is 2 to 1.
2. If the numerator and the denominator of a ratio are multiplied by 4, the new ratio equals the given ratio.
3. The value of a ratio must be less than one.
4. The three numbers 6, 14, and 22 are in a ratio of 3:7:11.
5. To correctly express a ratio, the terms must have the same unit of measure.
6. The ratio 3:4 is the same as the ratio 4:3.
7. If the second and third terms of a proportion are equal, then either is the geometric mean of the first and fourth terms.
8. Find the value(s) of x in each proportion:

   a) $\frac{x}{6} = \frac{3}{x}$

   b) $\frac{x-5}{3} = \frac{2x-3}{7}$

   c) $\frac{6}{x+4} = \frac{2}{x+2}$

   d) $\frac{x+3}{5} = \frac{x+5}{7}$

   e) $\frac{x-2}{x-5} = \frac{2x+1}{x-1}$

   f) $\frac{x(x+5)}{4x+4} = \frac{9}{5}$

   g) $\frac{x-1}{x+2} = \frac{10}{3x-2}$

   h) $\frac{x+7}{2} = \frac{x+2}{x-2}$

Use proportions to solve Review Exercises 9 to 11.

9. Four containers of fruit juice cost \$2.52. How much do six containers cost?

10. Two packages of M&Ms cost 69¢. How many packages can you buy for \$2.25?

11. A rug measuring 20 square meters costs \$132. How much would a 12 square meter rug of the same material cost?

12. The ratio of the measures of the sides of a quadrilateral is 2:3:5:7. If the perimeter is 68, find the length of each side.

13. The length and width of a rectangle are 18 and 12, respectively. A similar rectangle has length 27. What is its width?

14. The sides of a triangle are 6, 8, and 9. The shortest side of a similar triangle is 15. How long are its other sides?

15. The ratio of the measure of the supplement of an angle to that of the complement of the angle is 5:2. Find the measure of the supplement.

16. *Given:* $ABCD$ is a parallelogram
$\overline{DB}$ intersects $\overline{AE}$ at point F
Prove: $\frac{AF}{EF} = \frac{AB}{DE}$

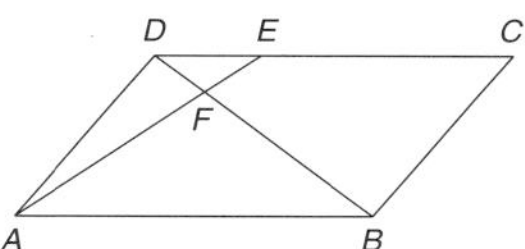

17. *Given:* $\angle 1 \cong \angle 2$
Prove: $\frac{AB}{AC} = \frac{BE}{CD}$

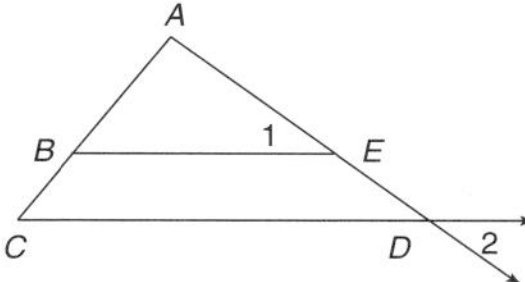

18. *Given:* $\triangle ABC \sim \triangle DEF$ (not shown)
$m\angle A = 50°$, $m\angle E = 33°$
$m\angle D = 2x + 40$
Find: x, $m\angle F$

19. *Given:* In $\triangle ABC$ and $\triangle DEF$ (not shown)
$\angle B \cong \angle F$ and $\angle C \cong \angle E$
$AC = 9$, $DE = 3$, $DF = 2$, $FE = 4$
Find: AB, BC

For Review Exercises 20 to 22, $\overline{DE} \parallel \overline{AC}$.

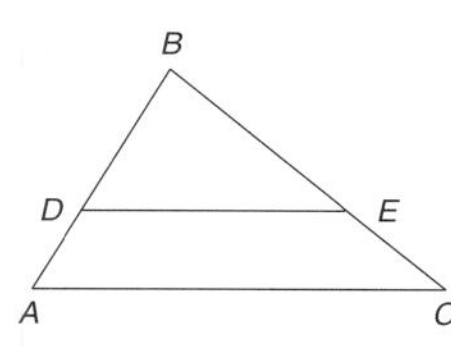

Exercises 20–22

20. $BD = 6$, $BE = 8$, $EC = 4$, $AD = ?$

21. $AD = 4$, $BD = 8$, $DE = 3$, $AC = ?$

22. $AD = 2$, $AB = 10$, $BE = 5$, $BC = ?$

For Review Exercises 23 to 25, $\overrightarrow{GJ}$ bisects $\angle FGH$.

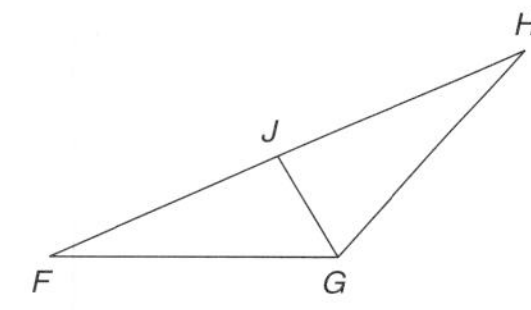

Exercises 23–25

23. *Given:* $FG = 10$, $GH = 8$, $FJ = 7$
Find: JH

24. *Given:* $GF:GH = 1:2$, $FJ = 5$
Find: JH

25. *Given:* $FG = 8$, $HG = 12$, $FH = 15$
Find: FJ

26. *Given:* $\overleftrightarrow{EF} \parallel \overleftrightarrow{GO} \parallel \overleftrightarrow{HM} \parallel \overleftrightarrow{JK}$, with transversals $\overline{FJ}$ and $\overline{EK}$
$FG = 2$, $GH = 8$, $HJ = 5$, $EM = 6$
Find: EO, EK

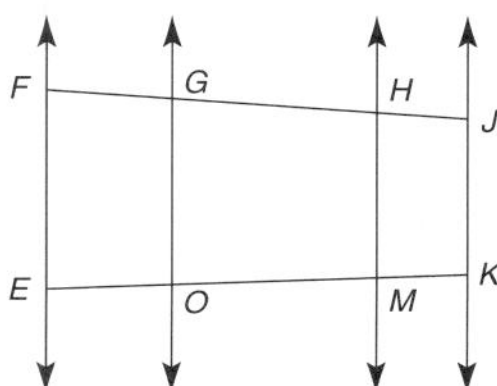

27. Prove that if a line bisects one side of a triangle and is parallel to a second side, then it bisects the third side.

28. Prove that the diagonals of a trapezoid divide themselves proportionally.

29. *Given:* $\triangle ABC$ with right $\angle BAC$
$\overline{AD} \perp \overline{BC}$

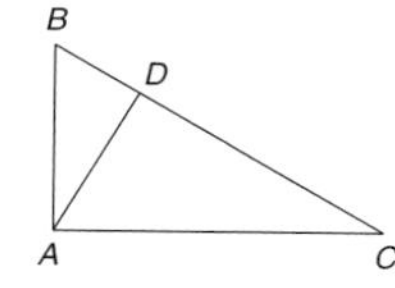

a) $BD = 3$, $AD = 5$, $DC = ?$
b) $AC = 10$, $DC = 4$, $BD = ?$
c) $BD = 2$, $BC = 6$, $BA = ?$
d) $BD = 3$, $AC = 3\sqrt{2}$, $DC = ?$

30. *Given:* $\triangle ABC$ with right $\angle ABC$
$\overline{BD} \perp \overline{AC}$

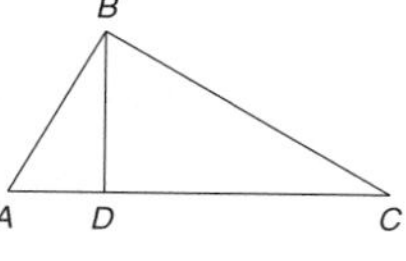

a) $BD = 12$, $AD = 9$, $DC = ?$
b) $DC = 5$, $BC = 15$, $AD = ?$
c) $AD = 2$, $DC = 8$, $AB = ?$
d) $AB = 2\sqrt{6}$, $DC = 2$, $AD = ?$

31. In the drawings shown, find x.

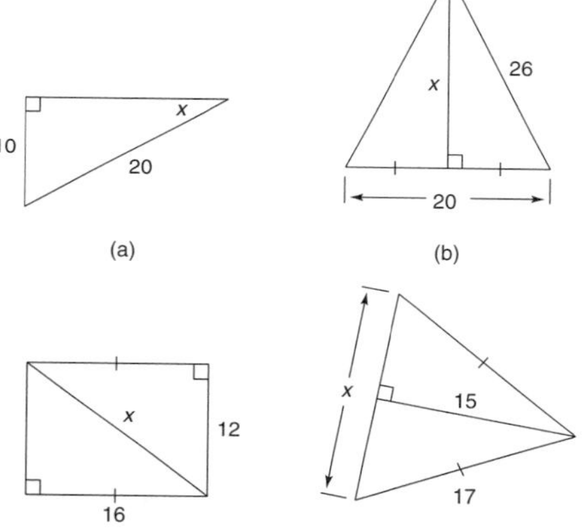

32. *Given:* $ABCD$ is a rectangle
E is the midpoint of $\overline{BC}$
$AB = 16$, $CF = 9$, $AD = 24$
Find: AE, EF, AF

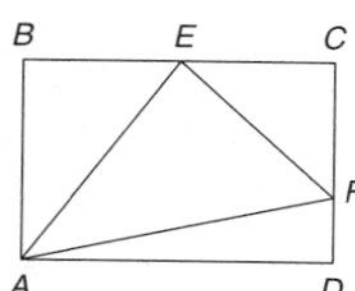

33. Find the length of a diagonal of a square whose side is 4 in. long.

34. Find the length of a side of a square whose diagonal is 6 cm long.

35. Find the length of a side of a rhombus whose diagonals are 48 cm and 14 cm long.

36. Find the length of an altitude of an equilateral triangle whose side is 10 in. long.

37. Find the length of a side of an equilateral triangle if an altitude is 6 in. long.

38. The lengths of three sides of a triangle are 13 cm, 14 cm, and 15 cm. Find the length of the altitude to the 14-cm side.

39. In the drawings, find x and y.

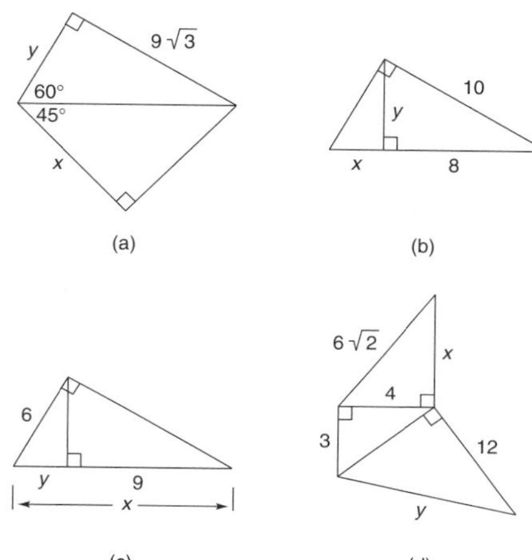

40. An observation aircraft flying at a height of 12 km has detected a Brazilian ship at a distance of 20 km from the aircraft and in line with an American ship that is 13 km from the aircraft. How far apart are the U.S. and Brazilian ships?

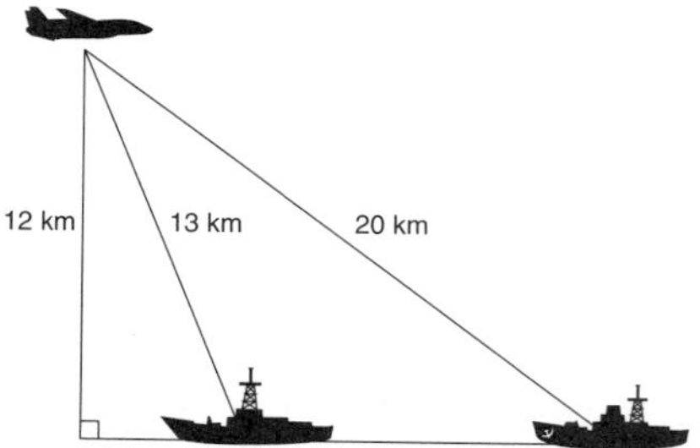

41. Tell whether each set of numbers represents the lengths of the sides of an acute triangle, of an obtuse triangle, of a right triangle, or of no triangle:

a) 12, 13, 14
b) 11, 5, 18
c) 9, 15, 18
d) 6, 8, 10
e) 8, 7, 16
f) 8, 7, 6
g) 9, 13, 8
h) 4, 2, 3

Chapter 6
Circles

When we consider something as simple as a pancake, as functional as a gear or pulley, or as attractive as a wire-spoke wheel, we generally think of a circle. In this chapter, we deal with circles and develop their properties, which are logical consequences of the properties that have been developed in previous chapters. The gears and pulleys shown illustrate the use of the circle in mechanical engineering.

CHAPTER OUTLINE

6.1 Circles and Related Segments and Angles

In this chapter, we will introduce terminology related to the circle, some methods of measurement, and many properties of the circle.

Warning

If the phrase "in a plane" is omitted from the definition of a circle, the result is the definition of a sphere.

DEFINITION: A **circle** is the set of all points in a plane that are at a fixed distance from a given point known as the *center* of the circle.

A circle is named after its center point. The symbol for circle is $\odot$, so the circle in Figure 6.1 is $\odot P$. Points A, B, C, and D are points *of* (or *on*) the circle. Points P (the center) and R are in the *interior* of circle P; points G and H are in the *exterior* of the circle.

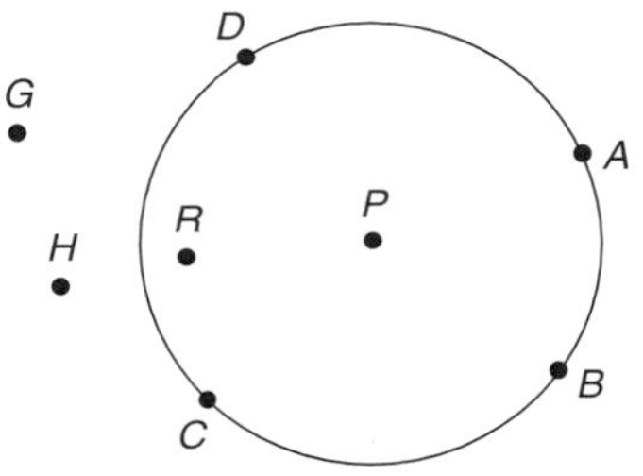

FIGURE 6.1

In $\odot Q$ of Figure 6.2, $\overline{SQ}$ is a radius of the circle. A **radius** is a segment that joins the center of the circle to a point on the circle. $\overline{SQ}$, $\overline{TQ}$, $\overline{VQ}$, and $\overline{WQ}$ are **radii** (plural of "radius") of $\odot Q$. By definition, $SQ = TQ = VQ = WQ$.

The following statement is a consequence of the definition of a circle.

All radii of a circle are congruent.

A segment that joins two points of a circle (like $\overline{SW}$ in Figure 6.2) is a **chord** of the circle. A **diameter** of a circle is a chord that contains the center of the circle; in Figure 6.2, $\overline{TW}$ is a diameter of $\odot Q$.

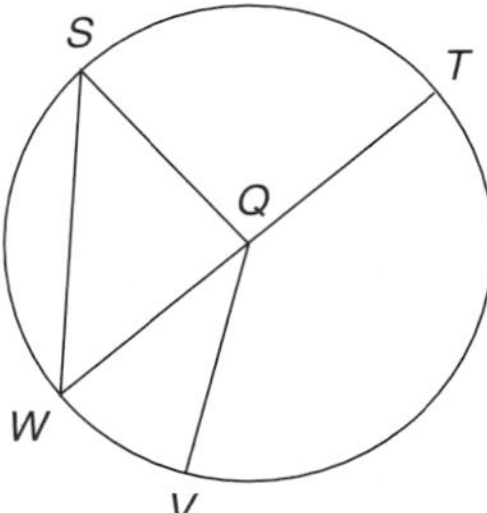

FIGURE 6.2

DEFINITION: **Congruent circles** are two or more circles that have congruent radii.

Circles P and Q in Figure 6.3 are congruent because their radii have equal lengths. We can slide $\odot P$ to the right to coincide with $\odot Q$.

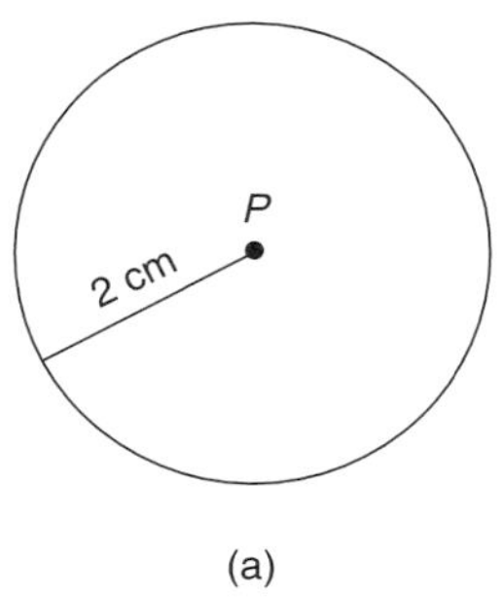

(a)

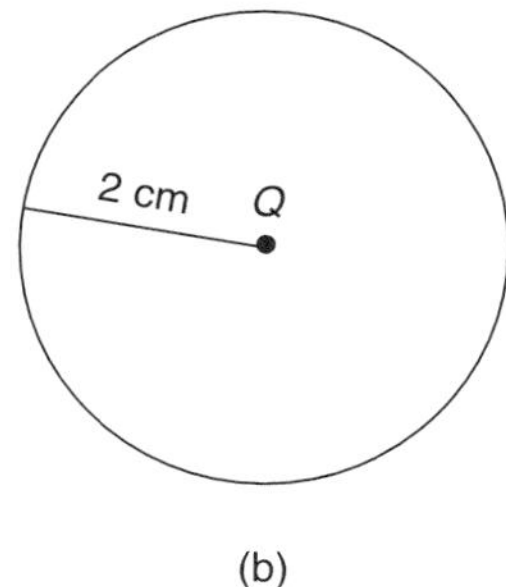

(b)

FIGURE 6.3

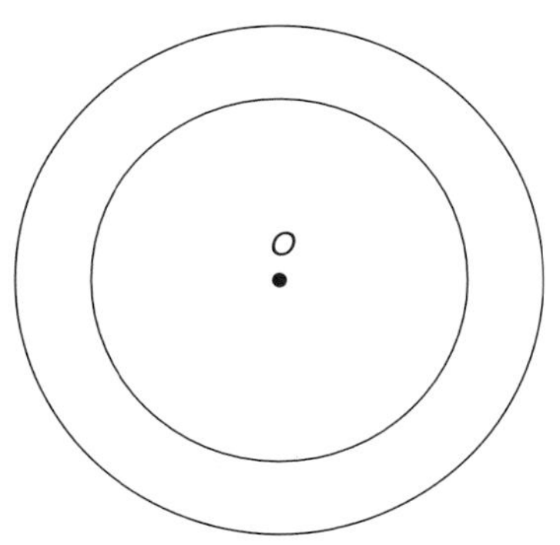

FIGURE 6.4

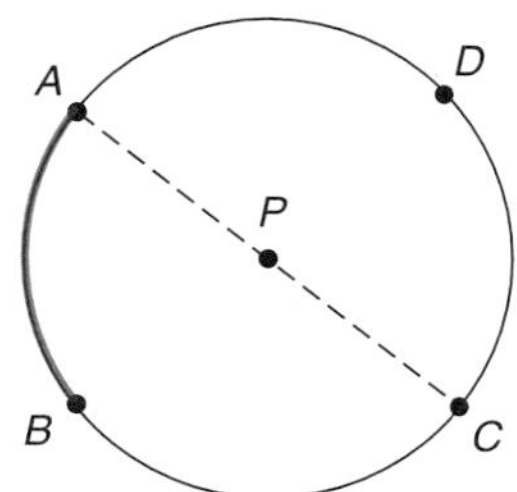

FIGURE 6.5

> DEFINITION: **Concentric circles** are coplanar circles that have a common center.

The concentric circles in Figure 6.4 have the common center O.

In $\odot P$ in Figure 6.5, the part of the circle shown from point A to point B is **arc** AB, as symbolized by $\overset{\frown}{AB}$. If $\overline{AC}$ is a diameter, then $\overset{\frown}{ABC}$ (three letters are used for clarity) is a **semicircle.** In Figure 6.5, a **minor arc** like $\overset{\frown}{AB}$ is part of a semicircle; a **major arc** such as $\overset{\frown}{ABCD}$ (also denoted by $\overset{\frown}{ABD}$ or $\overset{\frown}{ACD}$) is more than a semicircle but less than the entire circle.

> DEFINITION: A **central angle** of a circle is an angle whose vertex is the center of the circle and whose sides are radii of the circle.

In Figure 6.6, $\angle NOP$ is a central angle of $\odot O$. The **intercepted arc** of $\angle NOP$ is $\overset{\frown}{NP}$. The intercepted arc of an angle is determined by the two points of intersection of the angle with the circle and all points of the arc in the interior of the angle.

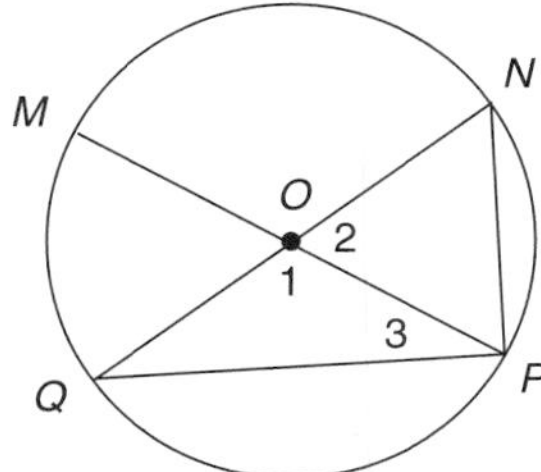

FIGURE 6.6

In Example 1, we "check" the terminology just introduced.

EXAMPLE 1

In Figure 6.6, $\overline{MP}$ and $\overline{NQ}$ intersect at O, the center of the circle. Name:

a) All four radii (shown)
b) Both diameters (shown)
c) All four chords (shown)
d) One central angle
e) One minor arc
f) One semicircle
g) One major arc
h) Intercepted arc of $\angle MON$
i) Central angle that intercepts $\overset{\frown}{NP}$

Solution

a) $\overline{OM}$, $\overline{OQ}$, $\overline{OP}$, and $\overline{ON}$
b) $\overline{MP}$ and $\overline{QN}$
c) $\overline{MP}$, $\overline{QN}$, $\overline{QP}$, and $\overline{NP}$
d) $\angle QOP$ (other answers are possible)
e) $\overset{\frown}{NP}$ (other answers are possible)
f) $\overset{\frown}{MQP}$ (other answers are possible)
g) $\overset{\frown}{MQN}$ (can be named $\overset{\frown}{MQPN}$; other answers are possible)
h) $\overset{\frown}{MN}$ (lies in the interior of $\angle MON$)
i) $\angle NOP$ (also called $\angle 2$)

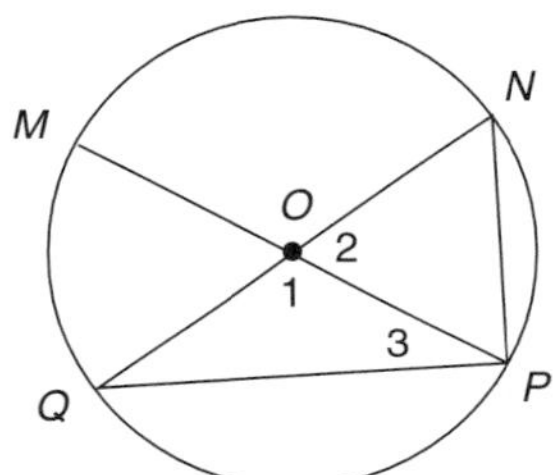

FIGURE 6.6

The following statement is a consequence of the Segment-Addition Postulate.

> In a circle, the length of a diameter is twice that of a radius.

EXAMPLE 2

$\overline{QN}$ is a diameter of $\odot O$ in Figure 6.6 and $PN = ON = 12$. Find the length of chord $\overline{QP}$.

Solution

Because $PN = ON$ and $ON = OP$, $\triangle NOP$ is equilateral. Then $m\angle 2 = m\angle N = m\angle NPO = 60°$. Also, $OP = OQ$, so $\triangle POQ$ is isosceles with $m\angle 1 = 120°$, since this angle is supplementary to $\angle 2$. Now $m\angle Q = m\angle 3 = 30°$ since the sum of the measures of the angles of $\triangle POQ$ is 180°. If $m\angle N = 60°$ and $m\angle Q = 30°$, then $\triangle NPQ$ is a right $\triangle$ whose angle measures are 30°, 60°, and 90°. It follows that $QP = PN \cdot \sqrt{3} = 12\sqrt{3}$.

> **THEOREM 6.1.1:** A radius that is perpendicular to a chord bisects the chord.

FIGURE 6.7

Given: $\overline{OD} \perp \overline{AB}$ in $\odot O$ (See Figure 6.7.)
Prove: $\overline{OD}$ bisects $\overline{AB}$

Proof: $\overline{OD} \perp \overline{AB}$ in $\odot O$. Draw radii $\overline{OA}$ and $\overline{OB}$. Now $\overline{OA} \cong \overline{OB}$ since all radii of a circle are $\cong$. Because $\angle 1$ and $\angle 2$ are right $\angle$s and $\overline{OC} \cong \overline{OC}$, we see that $\triangle OCA \cong \triangle OCB$ by *HL*. Then $\overline{AC} \cong \overline{CB}$ by CPCTC, so $\overline{OD}$ bisects $\overline{AB}$.

Angle and Arc Relationships in the Circle

In Figure 6.8, the sum of the measures of the angles about point O (angles determined by perpendicular diameters $\overline{AC}$ and $\overline{BD}$) is 360°. Similarly, the circle can be separated into 360 equal arcs, *each of which measures 1 degree of arc measure;* that is, each arc would be intercepted by a central angle measuring 1°. Our description of arc measure leads to the following postulate.

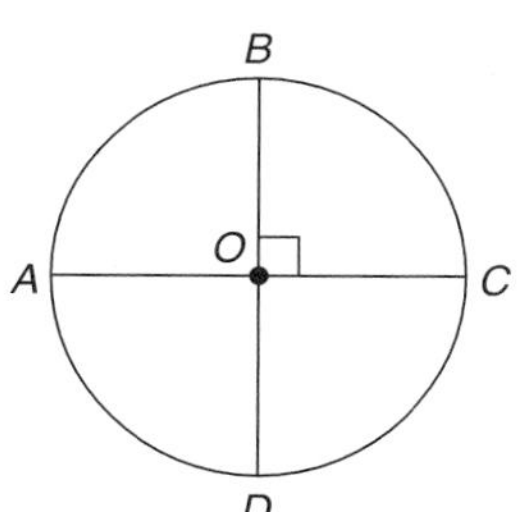

FIGURE 6.8

POSTULATE 16: (Central Angle Postulate)
In a circle, the degree measure of a central angle is equal to the degree measure of its intercepted arc.

In Figure 6.8, $m\overset{\frown}{AB} = 90°$, $m\overset{\frown}{BCD} = 180°$, and $m\overset{\frown}{AD} = 90°$. It follows that $m\overset{\frown}{AB} + m\overset{\frown}{BCD} + m\overset{\frown}{AD} = 360°$. Consequently, we have the following generalization.

The sum of the measures of the consecutive arcs that form a circle is exactly 360°.

In $\odot Y$ [Figure 6.9(a)], if $m\angle XYZ = 76°$, then $m\overset{\frown}{XZ} = 76°$ by the Central Angle Postulate. If two arcs have equal degree measures [Figures 6.9(b) and (c)] but are parts of two circles with unequal radii, then these arcs will not coincide. This observation leads to the following definition.

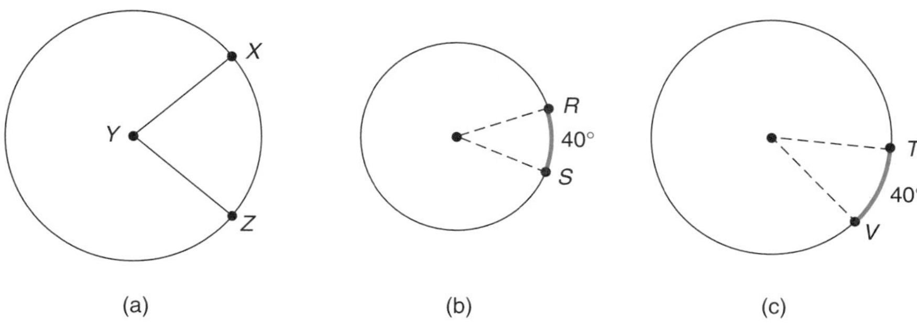

FIGURE 6.9

DEFINITION: In a circle or congruent circles, **congruent arcs** are arcs with equal measures.

To fully appreciate this definition, consider the concentric circles (having the same center) in Figure 6.10. Here the degree measure of $\angle AOB$ of the smaller circle is the same as the degree measure of $\angle COD$ of the larger circle. Even though $m\overset{\frown}{AB} = m\overset{\frown}{CD}$, $\overset{\frown}{AB} \not\cong \overset{\frown}{CD}$ since the arcs would not coincide.

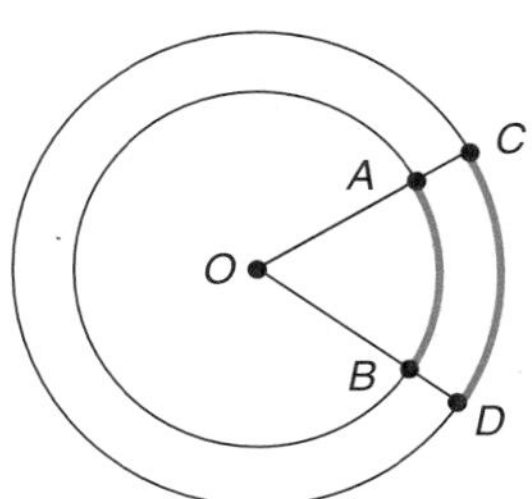

FIGURE 6.10

EXAMPLE 3

In $\odot O$ in Figure 6.11, $\overrightarrow{OE}$ bisects $\angle AOD$. Using the measures indicated, find:

a) $m\overset{\frown}{AB}$ **b)** $m\overset{\frown}{BC}$ **c)** $m\overset{\frown}{BD}$ **d)** $m\angle AOD$
e) $m\overset{\frown}{AE}$ **f)** $m\overset{\frown}{ACE}$ **g)** whether $\overset{\frown}{AE} \cong \overset{\frown}{ED}$

Solution

a) 105° **b)** 70° **c)** 105° **d)** 150°, from 360 − (105 + 70 + 35) **e)** 75° since the corresponding central angle ($\angle AOE$) is the result of bisecting $\angle AOD$, which was found to be 150° **f)** 285° (from 360 − 75, the measure of $\overset{\frown}{AE}$) **g)** The arcs are congruent since both measure 75° and both are found in the same circle.

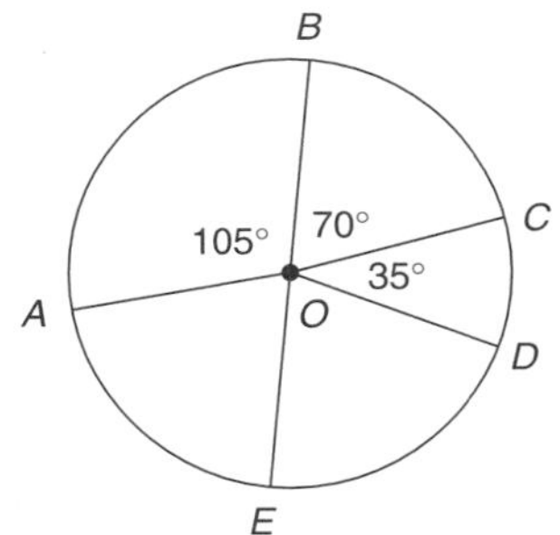

FIGURE 6.11

In Example 3(c), notice that $m\overset{\frown}{BD} = m\overset{\frown}{BC} + m\overset{\frown}{CD}$. Since D lies between B and A so that the union of $\overset{\frown}{BD}$ and $\overset{\frown}{DA}$ is a major arc, $m\overset{\frown}{BDA} = m\overset{\frown}{BD} + m\overset{\frown}{DA}$. With this understanding, we have the following postulate.

> **POSTULATE 17: (Arc-Addition Postulate)**
> If B lies between A and C on a circle, then $m\overset{\frown}{AB} + m\overset{\frown}{BC} = m\overset{\frown}{ABC}$.

The drawing in Figure 6.12(a) further supports the claim in Postulate 17.

Given points A, B, and C on $\odot O$ as shown in Figure 6.12(a), suppose that radii $\overline{OA}$, $\overline{OB}$, and $\overline{OC}$ are drawn. Because

$$m\angle AOB + m\angle BOC = m\angle AOC$$

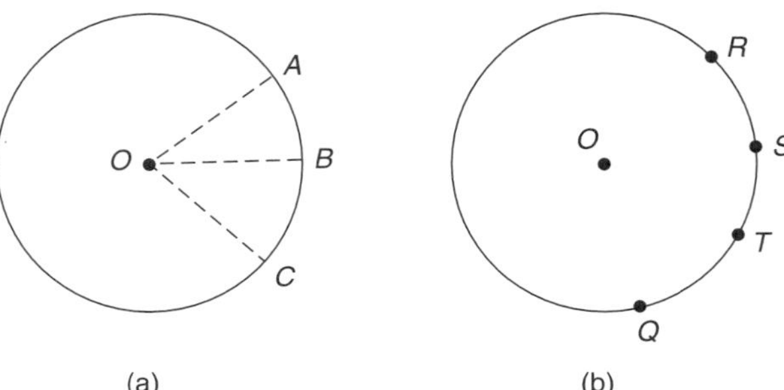

FIGURE 6.12

by the Angle-Addition Postulate, it follows that

$$m\widehat{AB} + m\widehat{BC} = m\widehat{ABC}$$

The reason for writing $\widehat{ABC}$, rather than $\widehat{AC}$, in stating the Arc-Addition Postulate is to avoid confusing minor arc $\widehat{AC}$ with the major arc with endpoints at A and C. It is easy to show that $m\widehat{ABC} - m\widehat{BC} = m\widehat{AB}$.

The Arc-Addition Postulate can easily be extended to include more than two arcs. In Figure 6.12(b), $m\widehat{RS} + m\widehat{ST} + m\widehat{TQ} = m\widehat{RSTQ}$.

If $m\widehat{RS} = m\widehat{ST}$ in Figure 6.12(b), then point S is the **midpoint** of $\widehat{RT}$ and $\widehat{RT}$ is **bisected** at point S.

In Example 4, we use the fact that the entire circle measures 360°.

EXAMPLE 4

Determine the measure of the angle formed by the hands of a clock at 3:12 P.M. (See Figure 6.13.)

Solution

The minute hand moves through 12 minutes, which is $\frac{12}{60}$ or $\frac{1}{5}$ of an hour. Thus, the minute hand points in a direction whose angle measure from the vertical is $\frac{1}{5}(360°)$ or 72°. At exactly 3 P.M., the hour hand would form an angle of 90° with the vertical. However, gears inside the clock turn the hour hand through $\frac{1}{5}$ of the 30° arc from the 3 toward the 4; that is, the hour hand moves another $\frac{1}{5}(30°)$ or 6° to form an angle of 96° with the vertical. The angle between the hands must measure 96° − 72° or 24°.

FIGURE 6.13

As we have seen, the measure of an arc can be used to measure the corresponding central angle. The measure of an arc can also be used to measure other types of angles related to the circle, including the inscribed angle.

> **DEFINITION:** An **inscribed angle** of a circle is an angle whose vertex is a point on the circle and whose sides are chords of the circle.

The word "inscribed" is often linked to the word "inside."

DISCOVER!

In Figure 6.14, $\angle B$ is the inscribed angle whose sides are chords $\overline{BA}$ and $\overline{BC}$.

a) Use a protractor to find the measure of central $\angle AOC$. **b)** Also find the measure of $\widehat{AC}$. **c)** Finally, measure inscribed $\angle B$. **d)** How is the measure of inscribed $\angle B$ related to the measure of its intercepted arc $\widehat{AC}$?

ANSWERS

a) 56° b) 56° c) 28° d) $m\angle B = \frac{1}{2}m\widehat{AC}$

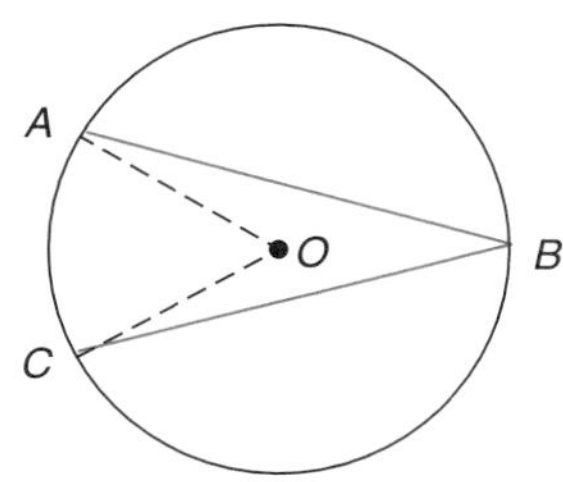

FIGURE 6.14

The above relationship between the measure of an inscribed angle and its intercepted arc is true in general.

THEOREM 6.1.2: The measure of an inscribed angle of a circle is one-half the measure of its intercepted arc.

The proof of Theorem 6.1.2 must be divided into three cases:

CASE 1. One side of the inscribed angle is a diameter.
CASE 2. The diameter to the vertex of the inscribed angle lies in the interior of the angle.
CASE 3. The diameter to the vertex of the inscribed angle lies in the exterior of the angle.

The proof of case 1 follows, but proofs of the other cases are left as exercises for the student. Drawings for cases 2 and 3 are found in Figure 6.15.

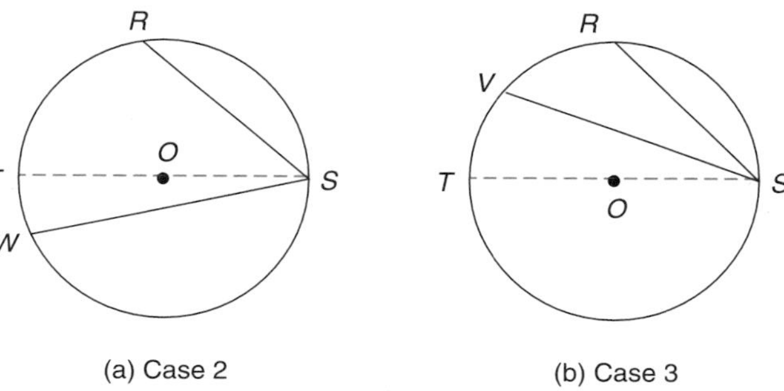

FIGURE 6.15

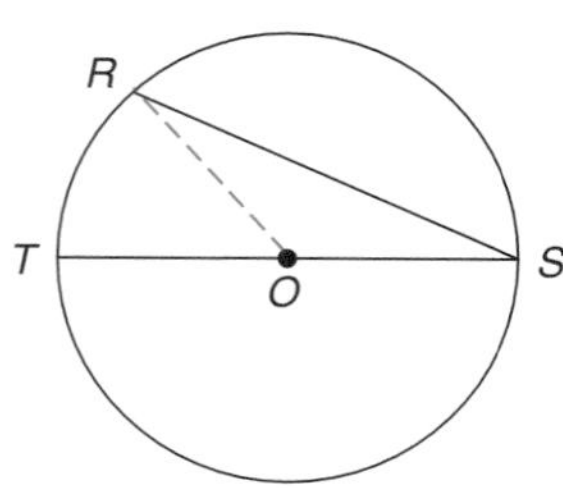

FIGURE 6.16

Given: $\odot O$ with inscribed $\angle RST$ and diameter $\overline{ST}$ (Figure 6.16)

Prove: $m\angle S = \frac{1}{2} m\overset{\frown}{RT}$

Proof of case (1): We begin by constructing radius $\overline{RO}$. Then $m\angle ROT = m\overset{\frown}{RT}$ since the central angle has a measure equal to its intercepted arc. With $\overline{OR} \cong \overline{OS}$, $\triangle ROS$ is isosceles and $m\angle R = m\angle S$. Now the exterior angle of the triangle is $\angle ROT$, so

$$m\angle ROT = m\angle R + m\angle S$$

Since $m\angle R = m\angle S$, $m\angle ROT = 2(m\angle S)$.
Then $m\angle S = \frac{1}{2} m\angle ROT$.
With $m\angle ROT = m\overset{\frown}{RT}$, we have $m\angle S = \frac{1}{2} m\overset{\frown}{RT}$ by substitution.

While proofs in this chapter generally take the less formal paragraph form, it remains necessary to justify each statement of the proof.

THEOREM 6.1.3: In a circle (or in congruent circles), congruent minor arcs have congruent central angles. (See Figure 6.17.)

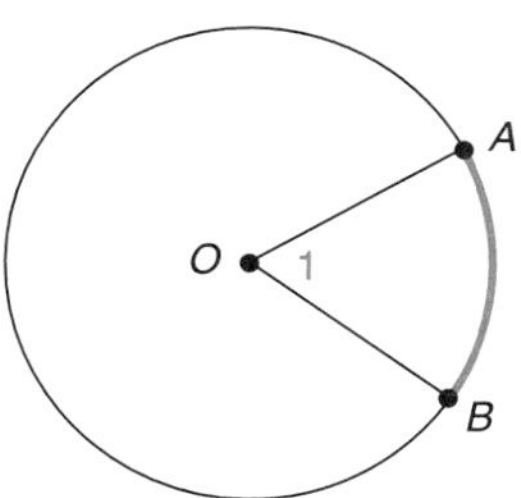

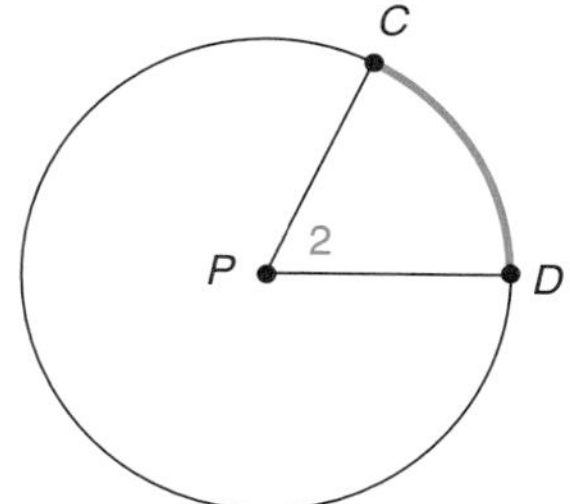

If $\overset{\frown}{AB} \cong \overset{\frown}{CD}$ in congruent circles O and P, then $\angle 1 \cong \angle 2$ by Theorem 6.1.3.

FIGURE 6.17

We suggest that the student make a drawing to illustrate each of the next three theorems. Some of the proofs depend upon auxiliary radii. See Exercise 15.

THEOREM 6.1.4: In a circle (or in congruent circles), congruent central angles have congruent arcs.

THEOREM 6.1.5: In a circle (or in congruent circles), congruent chords have congruent minor (major) arcs.

THEOREM 6.1.6: In a circle (or in congruent circles), congruent arcs have congruent chords.

Based upon an earlier definition, we define the distance from the center of a circle to a chord to be the length of the perpendicular segment joining the center to that chord.

Congruent triangles are used to prove the next two theorems.

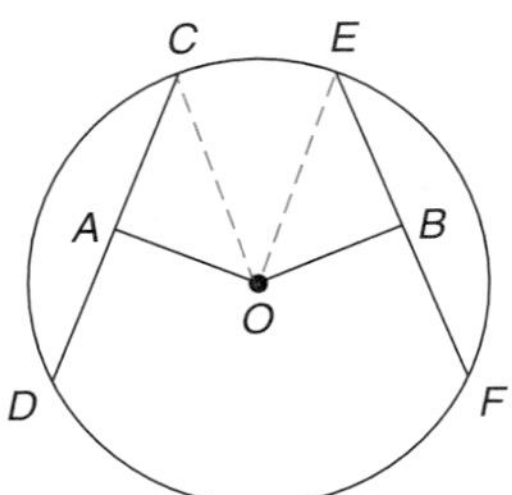

FIGURE 6.18

THEOREM 6.1.7: Chords that are at the same distance from the center of a circle are congruent.

Given: $\overline{OA} \perp \overline{CD}$ and $\overline{OB} \perp \overline{EF}$ in $\odot O$ (See Figure 6.18.)
$\overline{OA} \cong \overline{OB}$

Prove: $\overline{CD} \cong \overline{EF}$

Proof: Draw radii $\overline{OC}$ and $\overline{OE}$. With $\overline{OA} \perp \overline{CD}$ and $\overline{OB} \perp \overline{EF}$, $\angle OAC$ and $\angle OBE$ are right $\angle$s. $\overline{OA} \cong \overline{OB}$ is given and $\overline{OC} \cong \overline{OE}$ since all radii of a circle are congruent. Since $\triangle OAC$ and $\triangle OBE$ are right triangles, $\triangle OAC \cong \triangle OBE$ by HL.

By CPCTC, $\overline{CA} \cong \overline{BE}$ so that $CA = BE$. Then $2(CA) = 2(BE)$. But $2(CA) = CD$ since A is the midpoint of chord $\overline{CD}$. ($\overline{OA}$ bisects chord $\overline{CD}$ because $\overline{OA}$ is part of a radius. See Theorem 6.1.1). Likewise, $2(BE) = EF$, and it follows that

$$CD = EF \text{ and } \overline{CD} \cong \overline{EF}$$

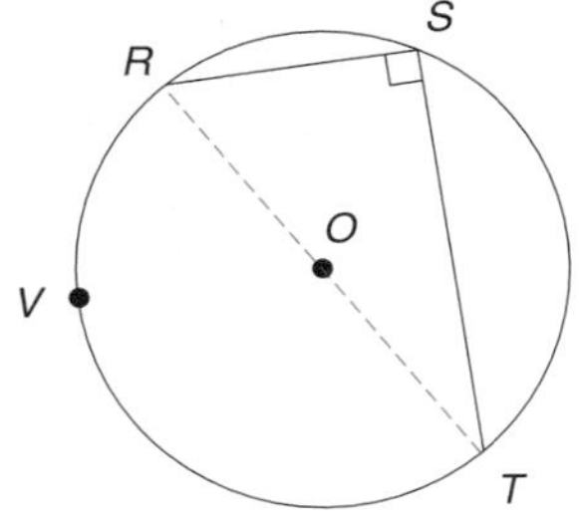

FIGURE 6.19

THEOREM 6.1.8: Congruent chords are located at the same distance from the center of a circle.

The student should make a drawing to illustrate Theorem 6.1.8.

Proofs of the remaining theorems are left as exercises.

THEOREM 6.1.9: An angle inscribed in a semicircle is a right angle.

In Figure 6.19, $\angle S$ is inscribed in the semicircle $\widehat{RST}$. Notice that $\angle S$ also intercepts semicircle $\widehat{RVT}$. Therefore, $\angle S$ is a right angle.

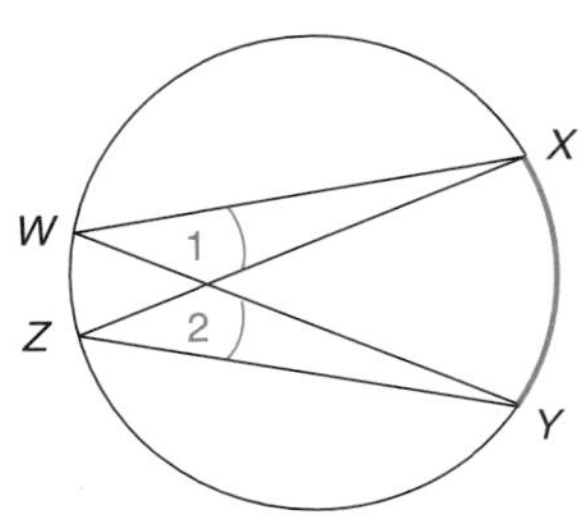

FIGURE 6.20

THEOREM 6.1.10: If two inscribed angles intercept the same arc, then these angles are congruent.

In Figure 6.20, notice that $\angle 1$ and $\angle 2$ both intercept $\widehat{XY}$. Then $\angle 1 \cong \angle 2$.

6.1 Exercises

1. *Given:* $\overline{AO} \perp \overline{OB}$ and $\overline{OC}$ bisects $\widehat{ACB}$ in $\odot O$
   Find: **a)** $m\widehat{AB}$
   b) $m\widehat{ACB}$
   c) $m\widehat{BC}$
   d) $m\angle AOC$

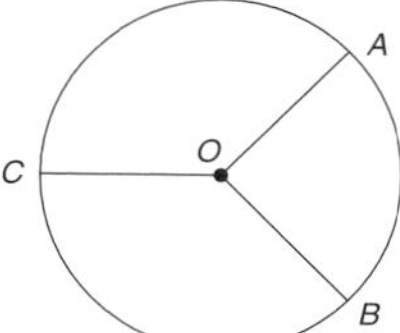

2. *Given:* $ST = \frac{1}{2}(SR)$ in $\odot Q$
   $\overline{SR}$ is a diameter
   Find: **a)** $m\widehat{ST}$
   b) $m\widehat{TR}$
   c) $m\widehat{STR}$
   d) $m\angle S$
   (**HINT:** Draw $\overline{QT}$.)

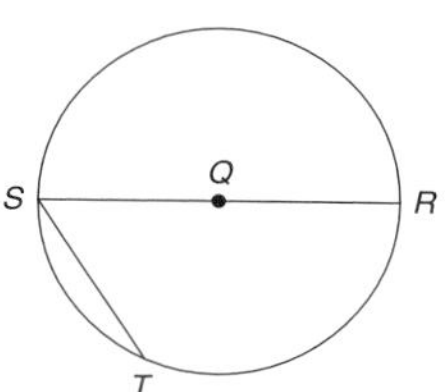

3. *Given:* $\odot Q$ in which $m\widehat{AB}:m\widehat{BC}:m\widehat{CA} = 2:3:4$
   Find: **a)** $m\widehat{AB}$
   b) $m\widehat{BC}$
   c) $m\widehat{CA}$
   d) $m\angle 1$ ($\angle AQB$)
   e) $m\angle 2$ ($\angle CQB$)
   f) $m\angle 3$ ($\angle CQA$)
   g) $m\angle 4$ ($\angle CAQ$)
   h) $m\angle 5$ ($\angle QAB$)
   i) $m\angle 6$ ($\angle QBC$)

4. *Given:* $m\angle DOE = 76°$ and $m\angle EOG = 82°$ in $\odot O$
   $\overline{EF}$ is a diameter
   Find: **a)** $m\widehat{DE}$
   b) $m\widehat{DF}$
   c) $m\angle F$
   d) $m\angle DGE$
   e) $m\angle EHG$
   f) whether $m\angle EHG = \frac{1}{2}(m\widehat{EG} + m\widehat{DF})$

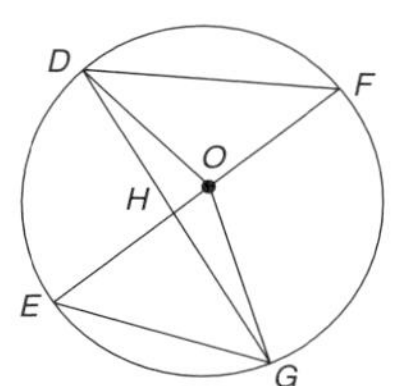

5. *Given:* $\odot O$ with $\overline{AB} \cong \overline{AC}$ and $m\angle BOC = 72°$
   Find: **a)** $m\widehat{BC}$
   b) $m\widehat{AB}$
   c) $m\angle A$
   d) $m\angle ABC$
   e) $m\angle ABO$

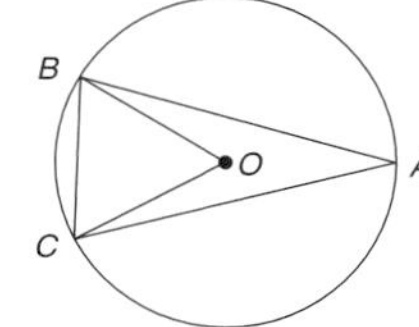

6. In $\odot O$ (not shown), $\overline{OA}$ is a radius, $\overline{AB}$ is a diameter, and $\overline{AC}$ is a chord.
   a) How does OA compare to AB?
   b) How does AC compare to AB?
   c) How does AC compare to OA?

7. *Given:* $\overline{OC} \perp \overline{AB}$ and $OC = 6$ in $\odot O$
   Find: **a)** AB
   b) BC

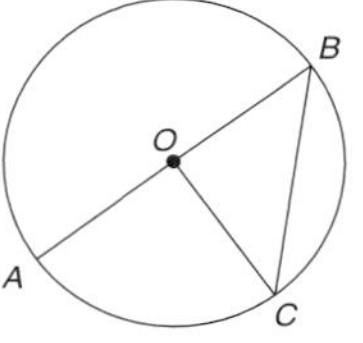

Exercise 7

8. *Given:* Concentric circles with center Q
   $SR = 3$ and $RQ = 4$
   $\overline{QS} \perp \overline{TV}$ at R
   Find: **a)** RV
   b) TV

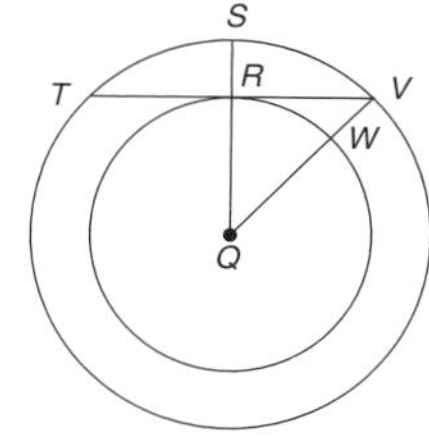

Exercises 8, 9

9. *Given:* Concentric circles with center Q
   $TV = 8$ and $VW = 2$
   $\overline{RQ} \perp \overline{TV}$
   Find: RQ (**HINT:** Let $RQ = x$.)

10. $\overline{AB}$ is the **common chord** of $\odot O$ and $\odot Q$. If $AB = 12$ and each circle has a radius of length 10, how long is $\overline{OQ}$?

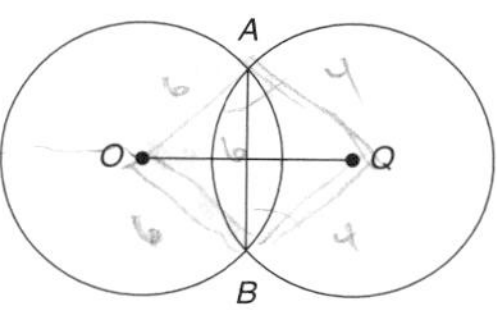

Exercises 10, 11

11. Circles O and Q have the common chord $\overline{AB}$. If $AB = 6$, $\odot O$ has a radius of length 4, and $\odot Q$ has a radius of length 6, how long is $\overline{OQ}$?

12. Suppose that a circle is divided into three congruent arcs by points A, B, and C. What is the measure of each arc? What type of figure results when A, B, and C are joined by segments?

13. Suppose that a circle is divided by points A, B, C, and D into four congruent arcs. What is the measure of each arc? If these points are joined in order, what type of quadrilateral results?

14. Following the pattern of Exercises 12 and 13, what type of figure results from dividing the circle equally by five points and joining those points in order? What type of polygon is formed by joining consecutively the n points that separate the circle into n congruent arcs?

15. Consider a circle or congruent circles, and explain why each statement is true:

    a) Congruent arcs have congruent central angles.
    b) Congruent central angles have congruent arcs.
    c) Congruent chords have congruent arcs.
    d) Congruent arcs have congruent chords.
    e) Congruent central angles have congruent chords.
    f) Congruent chords have congruent central angles.

16. State the measure of the angle formed by the minute hand and the hour hand of a clock when the time is

    a) 1:30 P.M. b) 2:20 A.M.

17. State the measure of the angle formed by the hands of the clock at

    a) 6:30 P.M. b) 5:40 A.M.

18. Five points are equally spaced on a circle. A five-pointed star (pentagram) is formed by joining nonconsecutive points two at a time. What is the degree measure of an arc determined by two consecutive points?

19. A ceiling fan has five equally spaced blades. What is the measure of the angle formed by two consecutive blades?

20. Repeat Exercise 19, but with the fan having 6 blades.

21. An amusement park ride (the "Octopus") has eight support arms that are equally spaced about a circle. What is the measure of the central angle formed by two consecutive arms?

In Exercises 22 and 23, complete each proof.

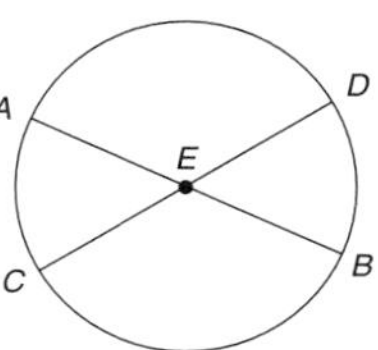

22. *Given:* Diameters $\overline{AB}$ and $\overline{CD}$ intersecting at E in $\odot E$
    Prove: $\overset{\frown}{AC} \cong \overset{\frown}{DB}$

PROOF

Statements	Reasons
1. ?	1. Given
2. $\angle AEC \cong \angle DEB$	2. ?
3. $m\angle AEC = m\angle DEB$	3. ?
4. $m\angle AEC = m\overset{\frown}{AC}$ and $m\angle DEB = m\overset{\frown}{DB}$	4. ?
5. $m\overset{\frown}{AC} = m\overset{\frown}{DB}$	5. ?
6. ?	6. If two arcs of a circle have the same measure, they are $\cong$

23. *Given:* $\overline{MN} \parallel \overline{OP}$ in $\odot O$
    Prove: $m\overset{\frown}{MQ} = 2(m\overset{\frown}{NP})$

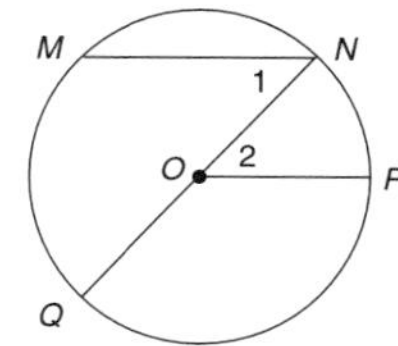

PROOF

Statements	Reasons
1. ?	1. Given
2. $\angle 1 \cong \angle 2$	2. ?
3. $m\angle 1 = m\angle 2$	3. ?
4. $m\angle 1 = \frac{1}{2}(m\overset{\frown}{MQ})$	4. ?
5. $m\angle 2 = m\overset{\frown}{NP}$	5. ?
6. $\frac{1}{2}(m\overset{\frown}{MQ}) = m\overset{\frown}{NP}$	6. ?
7. $m\overset{\frown}{MQ} = 2(m\overset{\frown}{NP})$	7. Multiplication Prop. of Equality

In Exercises 24 to 29, write a paragraph proof.

24. *Given:* $\overline{RS}$ and $\overline{TV}$ are diameters of $\odot W$
Prove: $\triangle RST \cong \triangle VTS$

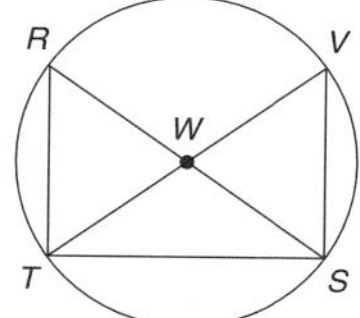

25. *Given:* Chords $\overline{AB}$, $\overline{BC}$, $\overline{CD}$, and $\overline{AD}$ in $\odot O$
Prove: $\triangle ABE \sim \triangle CDE$

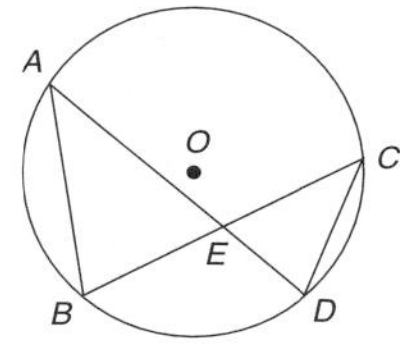

26. Congruent chords are at the same distance from the center of a circle.

27. A radius perpendicular to a chord bisects the arc of that chord.

28. An angle inscribed in a semicircle is a right angle.

29. If two inscribed angles intercept the same arc, then these angles are congruent.

30. If $\overleftrightarrow{MN} \parallel \overleftrightarrow{PQ}$ in $\odot O$, explain why $MNPQ$ is an isosceles trapezoid. (**HINT:** Draw a diagonal.)

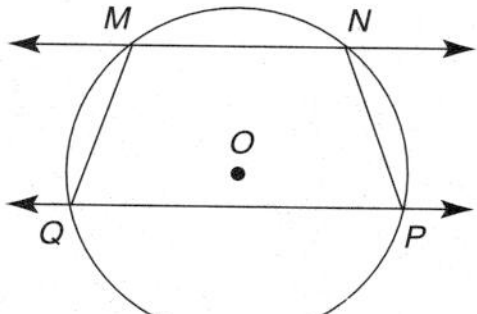

31. If $\overset{\frown}{ST} \cong \overset{\frown}{TV}$, explain why $\triangle STV$ is an isosceles triangle.

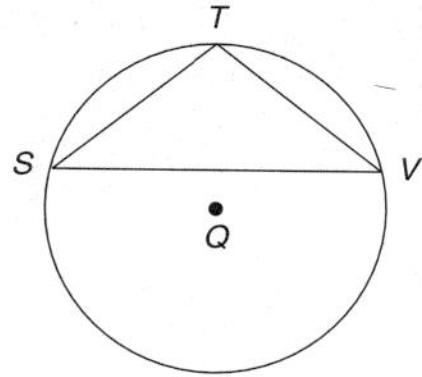

32. ➢ Use a paragraph proof to complete this exercise.
Given: $\odot O$ with chords $\overline{AB}$ and $\overline{BC}$
radii $\overline{AO}$ and $\overline{OC}$
Prove: $m\angle ABC < m\angle AOC$

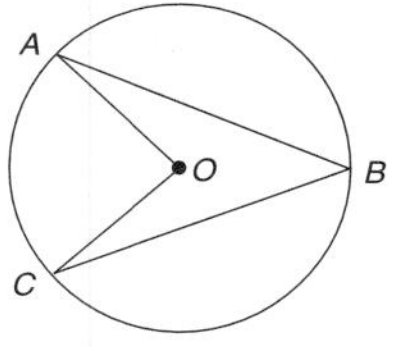

33. Prove case 2 of Theorem 6.1.2.

34. Prove case 3 of Theorem 6.1.2.

6.2 More Angle Measures in the Circle

We begin this section by considering lines, rays, and segments that are related to the circle.

DEFINITION: A **tangent** is a line that intersects a circle at exactly one point; the point of intersection is the **point of contact** or **point of tangency.**

The term "tangent" also applies to a segment or ray that is part of a tangent line to a circle. In each case, the tangent touches the circle at one point.

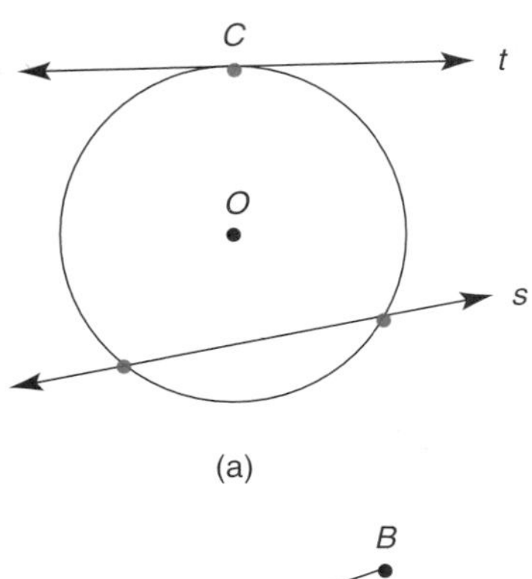

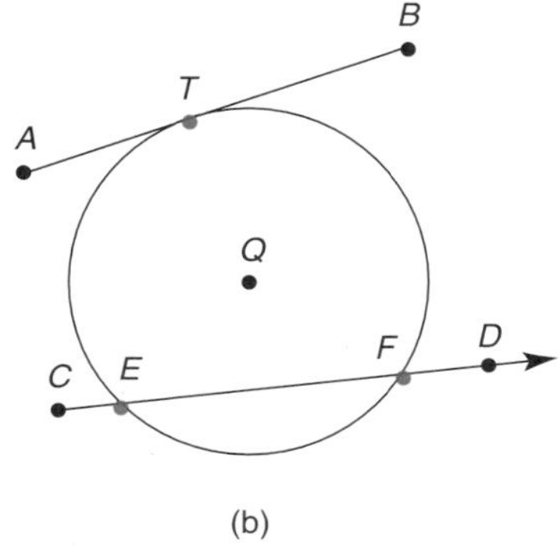

FIGURE 6.21

> DEFINITION: A **secant** is a line (or segment or ray) that intersects a circle at exactly two points.

In Figure 6.21(a), line s is a secant to $\odot O$; also, line t is a tangent to $\odot O$ and point C is its point of contact. In Figure 6.21(b), $\overline{AB}$ is a tangent to $\odot Q$ and point T is its point of tangency; $\overrightarrow{CD}$ is a secant with intersections at points E and F.

> DEFINITION: A polygon is **inscribed in a circle** if its vertices are points on the circle and its sides are chords of the circle. Equivalently, the circle is said to be **circumscribed about the polygon.**

In Figure 6.22, $\triangle ABC$ is inscribed in $\odot O$ and quadrilateral $RSTV$ is inscribed in $\odot Q$. Conversely, $\odot O$ is circumscribed about $\triangle ABC$ and $\odot Q$ is circumscribed about quadrilateral $RSTV$. Notice that $\overline{AB}$, $\overline{BC}$, and $\overline{AC}$ are chords of $\odot O$ and $\overline{RS}$, $\overline{ST}$, $\overline{TV}$, and $\overline{RV}$ are chords of $\odot Q$.

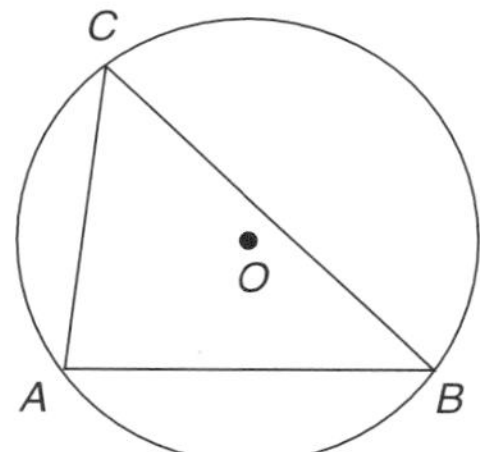

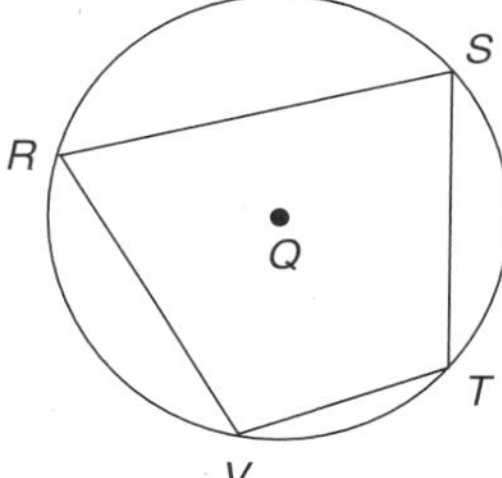

FIGURE 6.22

DISCOVER!

Draw any circle and call it $\odot O$. Now choose four points on $\odot O$ (call these A, B, C, and D). Join these points to form quadrilateral $ABCD$ inscribed in $\odot O$. Measure each of the inscribed angles ($\angle A$, $\angle B$, $\angle C$, and $\angle D$).

a) Find the sum $m\angle A + m\angle C$.
b) How are $\angle$s A and C related?
c) Find the sum $m\angle B + m\angle D$.
d) How are $\angle$s B and D related?

ANSWERS

a) 180° b) supplementary c) 180° d) supplementary

The preceding *Discover!* prepares the way for the next theorem.

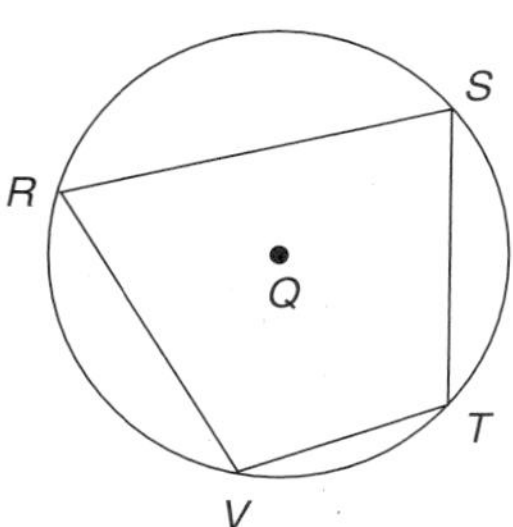

FIGURE 6.23

THEOREM 6.2.1: If a quadrilateral is inscribed in a circle, the opposite angles are supplementary.

The proof of Theorem 6.2.1 follows.

Given: $RSTV$ is inscribed in $\odot Q$ (See Figure 6.23.)

Prove: $\angle R$ and $\angle T$ are supplementary

Proof: From Section 6.1, an inscribed angle is equal in measure to one-half the measure of its intercepted arc. Since $m\angle R = \frac{1}{2}m\widehat{STV}$ and $m\angle T = \frac{1}{2}m\widehat{SRV}$, it follows that

$$\begin{aligned} m\angle R + m\angle T &= \frac{1}{2}m\widehat{STV} + \frac{1}{2}m\widehat{SRV} \\ &= \frac{1}{2}(m\widehat{STV} + m\widehat{SRV}) \end{aligned}$$

Since $\widehat{STV}$ and $\widehat{SRV}$ form the entire circle, $m\widehat{STV} + m\widehat{SRV} = 360°$. By substitution,

$$m\angle R + m\angle T = \frac{1}{2}(360°) = 180°$$

By definition, $\angle R$ and $\angle T$ are supplementary.

DEFINITION: A polygon is **circumscribed about a circle** if all sides of the polygon are segments tangent to the circle; also, the circle is said to be **inscribed in the polygon.**

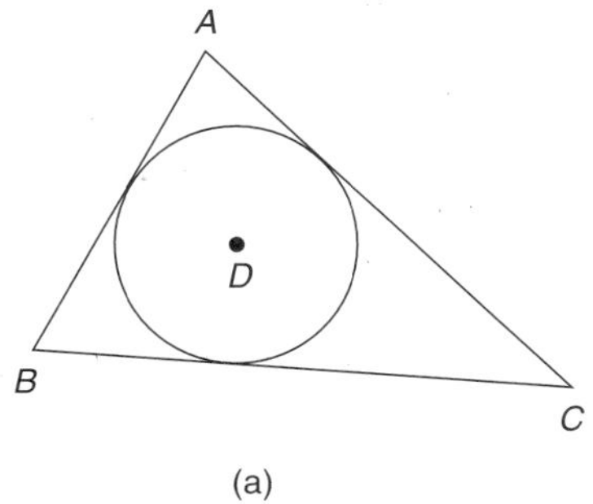

(a)

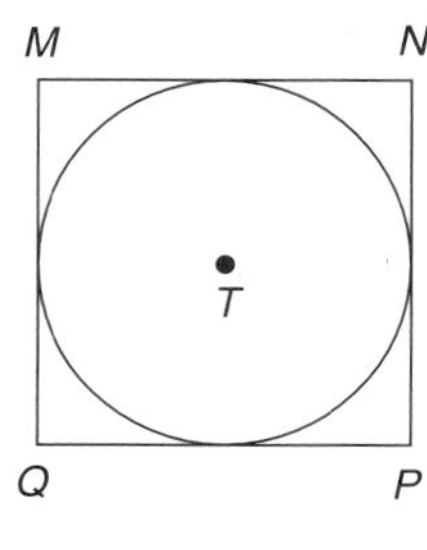

(b)

FIGURE 6.24

In Figure 6.24(a), $\triangle ABC$ is circumscribed about $\odot D$. In Figure 6.24(b), square $MNPQ$ is circumscribed about $\odot T$. Furthermore, $\odot D$ is inscribed in $\triangle ABC$ while $\odot T$ is inscribed in square $MNPQ$. Notice that $\overline{AB}$, $\overline{AC}$, and $\overline{BC}$ are tangents to $\odot D$ while $\overline{MN}$, $\overline{NP}$, $\overline{PQ}$, and $\overline{MQ}$ are tangents to $\odot T$.

We know that a central angle has a measure equal to its intercepted arc and that an inscribed angle has a measure one-half its intercepted arc. Now we consider another type of angle in the circle.

THEOREM 6.2.2: The measure of an angle formed by two chords that intersect within a circle is one-half the sum of the measures of the arcs intercepted by the angle and its vertical angle.

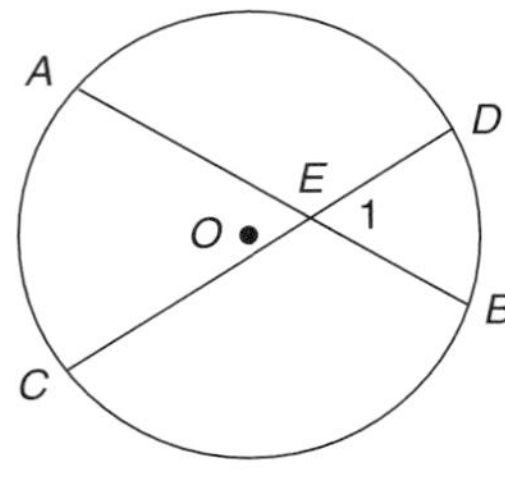

(a)

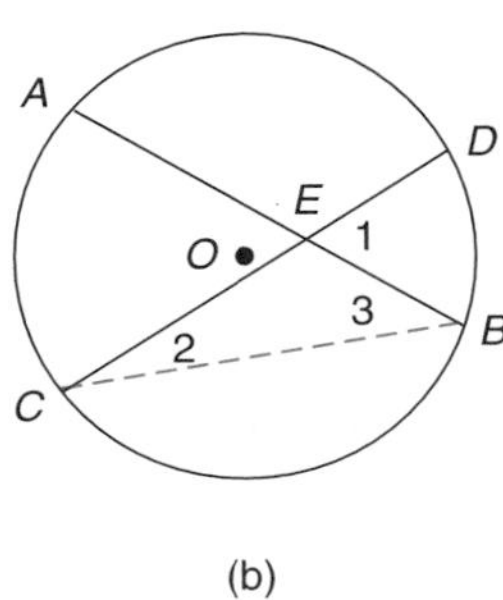

(b)

FIGURE 6.25

In Figure 6.25(a), $\angle 1$ intercepts $\overset{\frown}{DB}$ and $\angle AEC$ intercepts $\overset{\frown}{AC}$. According to Theorem 6.2.2,

$$m\angle 1 = \frac{1}{2}(m\overset{\frown}{AC} + m\overset{\frown}{DB})$$

To prove Theorem 6.2.2, we add to Figure 6.25(a) the auxiliary line segment $\overline{CB}$ [see Figure 6.25(b)].

Given: Chords $\overline{AB}$ and $\overline{CD}$ intersect at point E in $\odot O$

Prove: $m\angle 1 = \frac{1}{2}(m\overset{\frown}{AC} + m\overset{\frown}{DB})$

Proof: Draw $\overline{CB}$. Now $m\angle 1 = m\angle 2 + m\angle 3$ because $\angle 1$ is an exterior $\angle$ of $\triangle CBE$. Since $\angle 2$ and $\angle 3$ are inscribed angles of $\odot O$,

$$m\angle 2 = \frac{1}{2}m\overset{\frown}{DB} \text{ and } m\angle 3 = \frac{1}{2}m\overset{\frown}{AC}$$

Substitution leads to

$$\begin{aligned} m\angle 1 &= \frac{1}{2}m\overset{\frown}{DB} + \frac{1}{2}m\overset{\frown}{AC} \\ &= \frac{1}{2}(m\overset{\frown}{DB} + m\overset{\frown}{AC}) \end{aligned}$$

Equivalently,

$$m\angle 1 = \frac{1}{2}(m\overset{\frown}{AC} + m\overset{\frown}{DB})$$

Next, we apply Theorem 6.2.2.

EXAMPLE 1

In Figure 6.25(a), $m\overset{\frown}{AC} = 84°$ and $m\overset{\frown}{DB} = 62°$. Find $m\angle 1$.

Solution

By Theorem 6.2.2,

$$\begin{aligned} m\angle 1 &= \frac{1}{2}(m\overset{\frown}{AC} + m\overset{\frown}{DB}) \\ &= \frac{1}{2}(84° + 62°) \\ &= \frac{1}{2}(146°) = 73° \end{aligned}$$

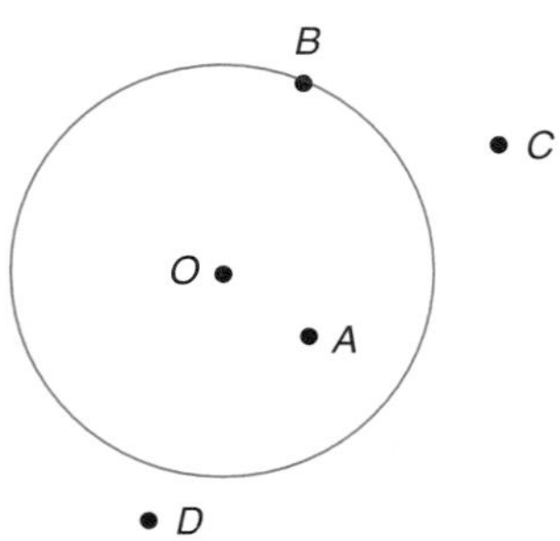

FIGURE 6.26

Recall that a circle separates points in the plane into three sets: points *in the interior* of the circle, points *on* the circle, and points *in the exterior* of the circle. In Figure 6.26, point A and center O are in the **interior** of $\odot O$ because their distances from center O are less than the length of the radius. Point B is on the circle, but points C and D are in the **exterior** of $\odot O$ because their distances from O are greater than the length of the radius. (See Exercises 36 and 37.) In the proof of Theorem 6.2.3, we use the fact that a tangent to a circle cannot contain an interior point of the circle.

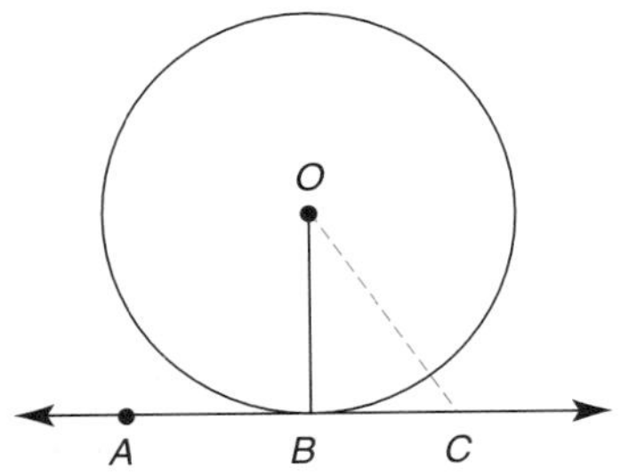

FIGURE 6.27

THEOREM 6.2.3: The radius (or any other line through the center of a circle) drawn to a tangent at the point of tangency is perpendicular to the tangent at that point.

Given: $\odot O$ with tangent $\overleftrightarrow{AB}$ and radius $\overline{OB}$ (See Figure 6.27.)

Prove: $\overline{OB} \perp \overleftrightarrow{AB}$

Proof: $\odot O$ has tangent $\overleftrightarrow{AB}$ and radius $\overline{OB}$. Let C name any point on $\overleftrightarrow{AB}$ except B. Now $OC > OB$ since C lies in the exterior of the circle. It follows that $\overline{OB} \perp \overleftrightarrow{AB}$ because the shortest distance from a point to a line is determined by the perpendicular segment from that point to the line.

The following example illustrates an application of Theorem 6.2.3.

EXAMPLE 2

A shuttle going to the moon has reached a position that is five miles above its surface. If the radius of the moon is 1080 mi, how far to the horizon can the NASA crew members see? (See Figure 6.28.)

Solution

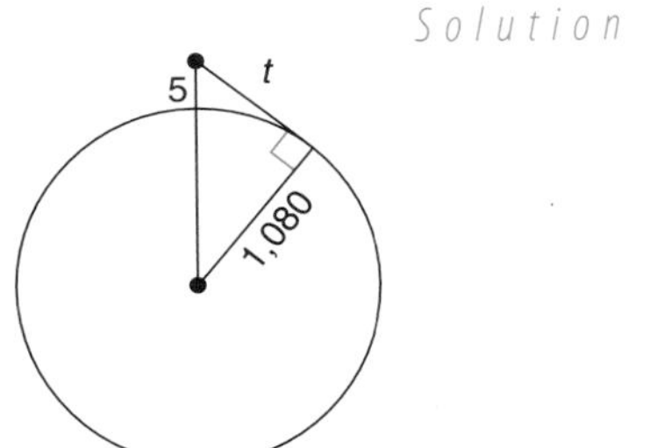

FIGURE 6.28

According to Theorem 6.2.3, the tangent determining the line of sight and the radius of the moon form a right angle. In the right triangle determined, let t represent the desired distance. Then

$$1085^2 = t^2 + 1080^2$$
$$1{,}177{,}225 = t^2 + 1{,}166{,}400$$
$$t^2 = 10{,}825 \rightarrow t = \sqrt{10{,}825} \approx 104 \text{ mi}$$

A consequence of Theorem 6.2.3 is Corollary 6.2.4, which follows. Of the three possible cases indicated in Figure 6.29, only the first is proved: the remaining two are left as exercises for the student. See Exercises 34 and 35.

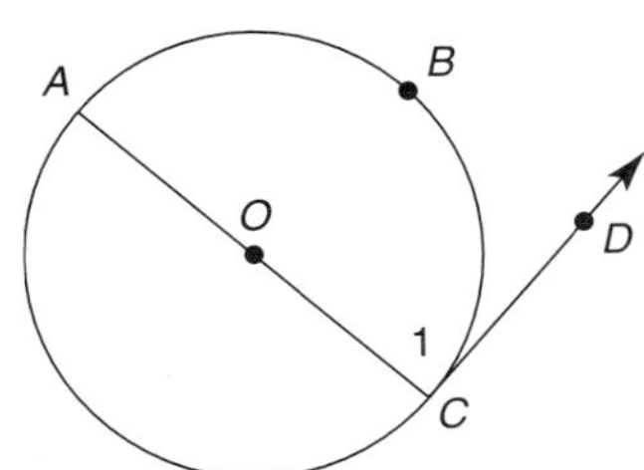

(a) Case i
The chord is a diameter

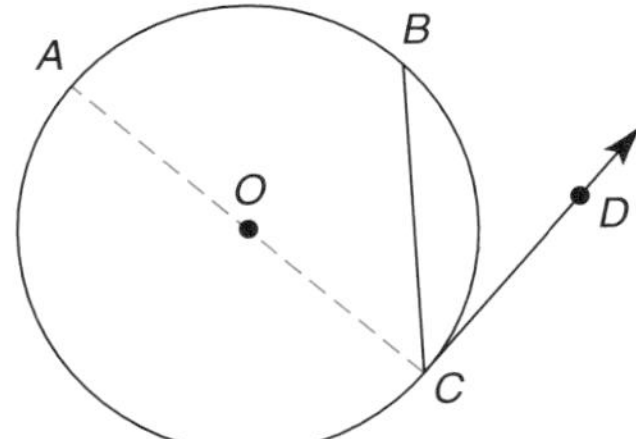

(b) Case ii
The diameter is in the exterior of the angle

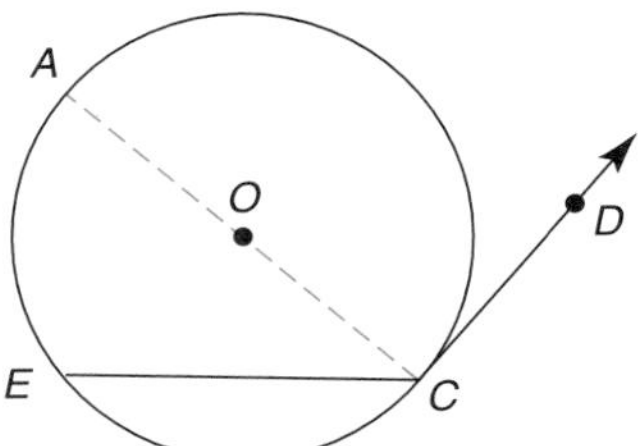

(c) Case iii
The diameter lies in the interior of the angle

FIGURE 6.29

COROLLARY 6.2.4: The measure of an angle formed by a tangent and a chord drawn to the point of tangency is one-half the measure of the intercepted arc. (See Figure 6.29.)

Given: Chord $\overline{CA}$ (which is a diameter) and tangent $\overrightarrow{CD}$ [See Figure 6.29(a).]

Prove: $m\angle 1 = \frac{1}{2}m\widehat{ABC}$.

Proof: By Theorem 6.2.3, $\overline{AC} \perp \overrightarrow{CD}$. Then $\angle 1$ is a right angle and $m\angle 1 = 90°$. Because the intercepted arc $\widehat{ABC}$ is a semicircle, $m\widehat{ABC} = 180°$. Thus it follows that $m\angle 1 = \frac{1}{2}m\widehat{ABC}$.

EXAMPLE 3

Given: In Figure 6.30, $\odot O$ with diameter $\overline{DB}$ and $m\widehat{DE} = 84°$

Find:
a) $m\angle 1$
b) $m\angle 2$
c) $m\angle ABD$
d) $m\angle ABE$

Solution

a) $\angle 1$ is an inscribed angle; $m\angle 1 = \frac{1}{2}m\widehat{DE} = 42°$

b) With $m\widehat{DE} = 84°$ and $\widehat{DEB}$ a semicircle, $m\widehat{BE} = 180° - 84° = 96°$. By Corollary 6.2.4, $m\angle 2 = \frac{1}{2}m\widehat{BE} = \frac{1}{2}(96°) = 48°$.

c) Since $\overline{DB}$ is perpendicular to $\overleftrightarrow{AB}$, $m\angle ABD = 90°$.

d) $m\angle ABE = m\angle ABD + m\angle 1 = 90° + 42° = 132°$

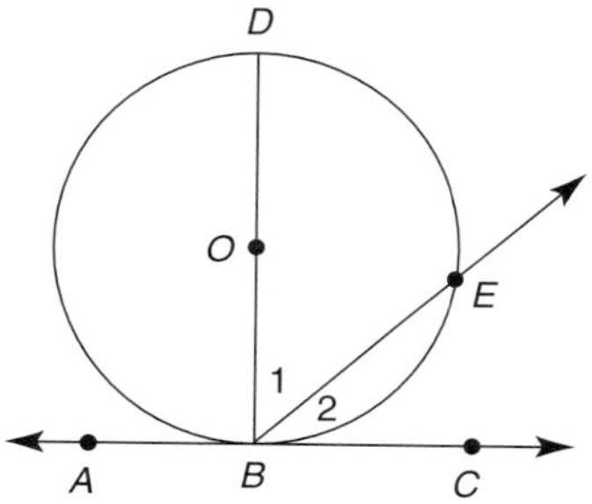

FIGURE 6.30

THEOREM 6.2.5: The measure of an angle formed when two secants intersect at a point outside the circle is one-half the difference of the measures of the two intercepted arcs.

Given: Secants $\overline{AC}$ and $\overline{DC}$ as shown in Figure 6.31

Prove: $m\angle C = \frac{1}{2}(m\widehat{AD} - m\widehat{BE})$

Proof: Draw $\overline{BD}$ to form $\triangle BCD$. Then the measure of the exterior angle of $\triangle BCD$ is given by

$$m\angle 1 = m\angle C + m\angle D$$

so

$$m\angle C = m\angle 1 - m\angle D$$

$\angle 1$ and $\angle D$ are inscribed angles, so $m\angle 1 = \frac{1}{2}m\widehat{AD}$ and $m\angle D = \frac{1}{2}m\widehat{BE}$.

Then

$$m\angle C = \frac{1}{2}m\widehat{AD} - \frac{1}{2}m\widehat{BE}$$

so

$$m\angle C = \frac{1}{2}(m\widehat{AD} - m\widehat{BE})$$

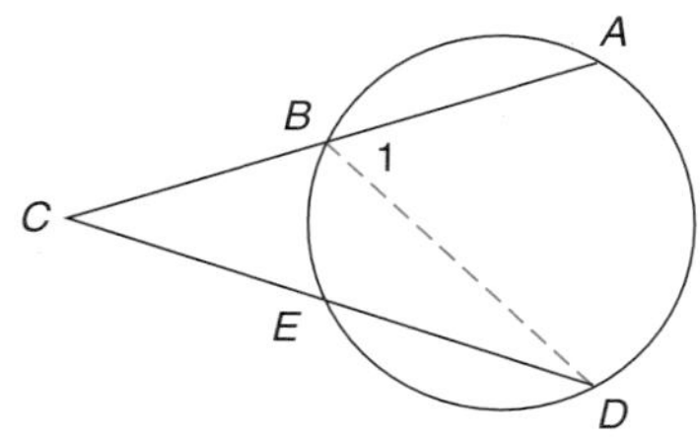

FIGURE 6.31

NOTE: In an application of Theorem 6.2.5, one subtracts the smaller arc measure from the larger arc measure.

EXAMPLE 4

Given: In $\odot O$ of Figure 6.32, $m\angle AOB = 136°$ and $m\angle DOC = 46°$

Find: $m\angle E$

Solution

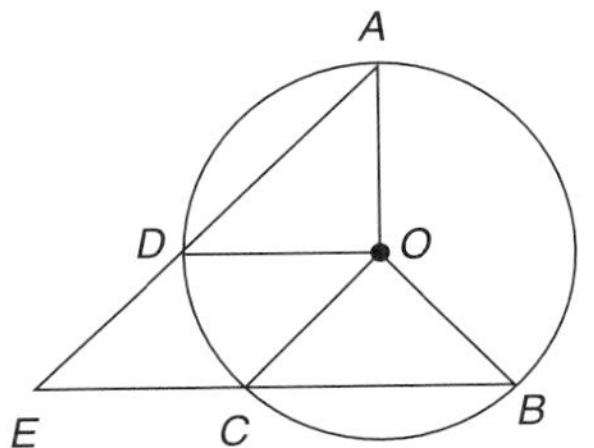

FIGURE 6.32

If $m\angle AOB = 136°$, then $m\overset{\frown}{AB} = 136°$. If $m\angle DOC = 46°$, then $m\overset{\frown}{DC} = 46°$. By Theorem 6.2.5,

$$\begin{aligned} m\angle E &= \frac{1}{2}(m\overset{\frown}{AB} - m\overset{\frown}{DC}) \\ &= \frac{1}{2}(136° - 46°) \\ &= \frac{1}{2}(90°) = 45° \end{aligned}$$

Theorems 6.2.5–6.2.7 will show that any angle formed by two lines that intersect *outside* a circle has a measure equal to one-half of the difference of the measures of the two intercepted arcs. While the next two theorems are not proved, the auxiliary lines shown will help complete the proofs.

THEOREM 6.2.6: If an angle is formed by a secant and a tangent that intersect in the exterior of a circle, then the measure of the angle is one-half the difference of the measures of its intercepted arcs.

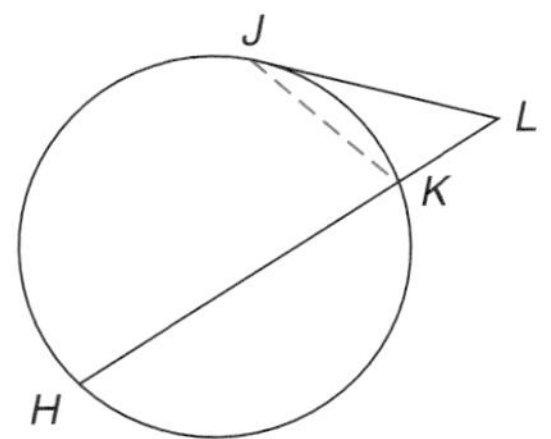

FIGURE 6.33

According to Theorem 6.2.6,

$$m\angle L = \frac{1}{2}(m\overset{\frown}{HJ} - m\overset{\frown}{JK})$$

in Figure 6.33.

A quick study of Theorems 6.2.5–6.2.7 shows that the smaller arc is "nearer" the vertex of the angle while the larger arc is "farther from" the vertex.

THEOREM 6.2.7: If an angle is formed by two intersecting tangents, then the measure of the angle is one-half the difference of the measures of the intercepted arcs.

In Figure 6.34(a), the small arc is a minor arc $(\widehat{AC})$ while the large arc is a major arc $(\widehat{ADC})$. Then

$$m\angle ABC = \frac{1}{2}(m\widehat{ADC} - m\widehat{AC})$$

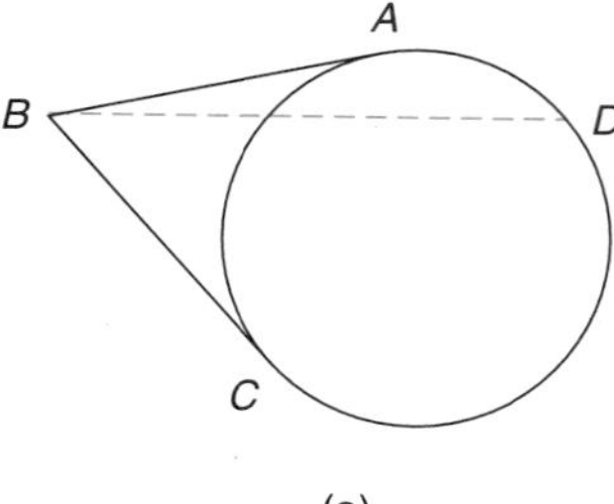

EXAMPLE 5

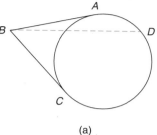

Given: In Figure 6.34(b), $m\widehat{MN} = 70°$, $m\widehat{NP} = 88°$, $m\widehat{MR} = 46°$, $m\widehat{RS} = 26°$.

Find: **a)** $m\angle MTN$
b) $m\angle NTP$
c) $m\angle MTP$

Solution

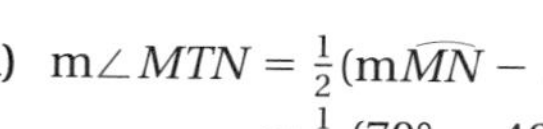

a) $m\angle MTN = \frac{1}{2}(m\widehat{MN} - m\widehat{MR})$
$= \frac{1}{2}(70° - 46°)$
$= \frac{1}{2}(24°) = 12°$

b) $m\angle NTP = \frac{1}{2}(m\widehat{NP} - m\widehat{RS})$
$= \frac{1}{2}(88° - 26°)$
$= \frac{1}{2}(62°) = 31°$

c) $m\angle MTP = m\angle MTN + m\angle NTP$.
Using results from (a) and (b),
$m\angle MTP = 12° + 31° = 43°$

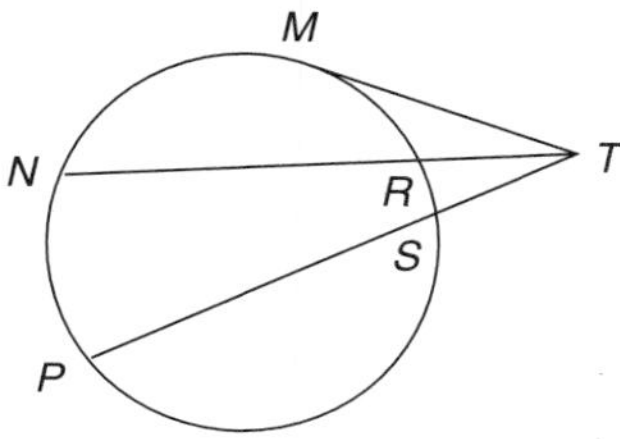

FIGURE 6.34

Before considering the final example of this section, let's review the methods used to measure different types of angles related to a circle. These are summarized in Table 6.1.

TABLE 6.1

Methods for Measuring Angles Related to a Circle

Location of the Vertex of the Angle	Rule for Measuring the Angle
In the *interior* of the circle	*One-half the sum* of the measures of the intercepted arcs
On the circle	*One-half the measure* of the intercepted arc
In the *exterior* of the circle	*One-half the difference* of the measures of the two intercepted arcs

EXAMPLE 6

Given that $m\angle 1 = 46°$ in Figure 6.35, find the measures of $\widehat{AB}$ and $\widehat{ACB}$.

Solution

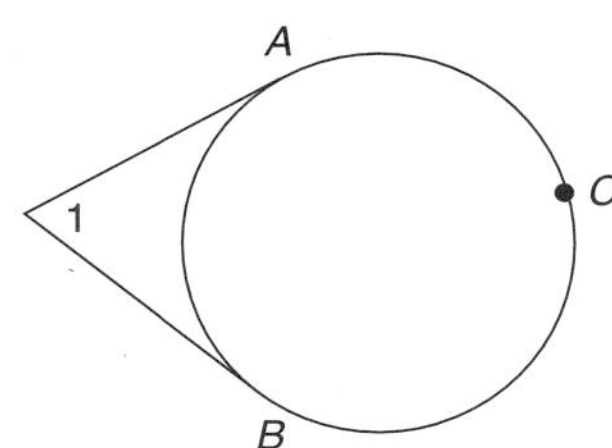

FIGURE 6.35

Let $m\widehat{AB} = x$ and $m\widehat{ACB} = y$. Now

$$m\angle 1 = \frac{1}{2}(m\widehat{ACB} - m\widehat{AB})$$

so

$$46 = \frac{1}{2}(y - x)$$

Multiplying by 2, we have $92 = y - x$.

Also, $y + x = 360$ since these two arcs form the entire circle. We add these equations as shown.

$$\begin{aligned} y + x &= 360 \\ y - x &= 92 \\ \hline 2y &= 452 \\ y &= 226 \end{aligned}$$

Since $x + y = 360$, we know that $x + 226 = 360$ and $x = 134$. Then $m\widehat{AB} = 134°$ and $m\widehat{ACB} = 226°$.

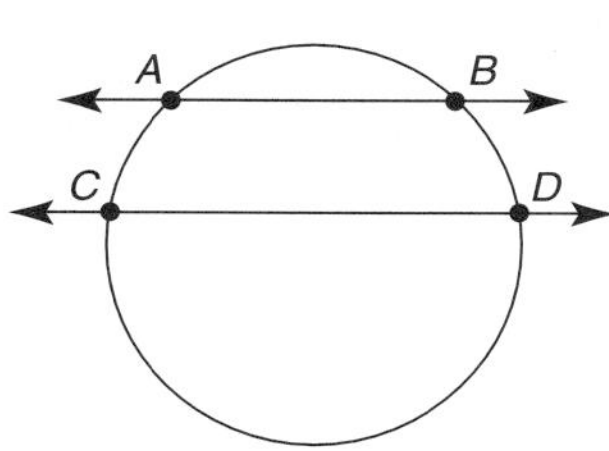

FIGURE 6.36

> **THEOREM 6.2.8:** If two parallel lines intersect a circle, the intercepted arcs between these lines are congruent. (See Figure 6.36.)

Where $\overleftrightarrow{AB} \parallel \overleftrightarrow{CD}$, it follows that $\widehat{AC} \cong \widehat{BD}$. Equivalently, $m\widehat{AC} = m\widehat{BD}$. The proof of Theorem 6.2.8 is left as an exercise for the student.

6.2 Exercises

In Exercises 1 and 2, use the drawing shown.

1. *Given:* $m\widehat{AB} = 92°$, $m\widehat{DA} = 114°$, $m\widehat{BC} = 138°$

Find:
- **a)** $m\angle 1$ ($\angle DAC$)
- **b)** $m\angle 2$ ($\angle ADB$)
- **c)** $m\angle 3$ ($\angle AFB$)
- **d)** $m\angle 4$ ($\angle DEC$)
- **e)** $m\angle 5$ ($\angle CEB$)

Exercises 1, 2

2. *Given:* $m\widehat{DC} = 30°$ and $\widehat{DABC}$ is trisected at points A and B

Find:
- **a)** $m\angle 1$
- **b)** $m\angle 2$
- **c)** $m\angle 3$
- **d)** $m\angle 4$
- **e)** $m\angle 5$

3. Is it possible for all four conditions to hold simultaneously? Explain.

i) $\overline{RS}$ is a diameter.
ii) $\overline{TS}$ is a chord.
iii) $\overrightarrow{SW}$ is a tangent.
iv) $\overrightarrow{SW} \perp \overline{TS}$

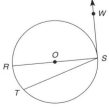

4. Is it possible for

- **a)** an inscribed rectangle in a circle to have a diameter for a side? Explain.
- **b)** a circumscribed rectangle about a circle to be a square? Explain.

5. *Given:* In $\odot Q$, $\overline{PR}$ contains Q, $\overline{MR}$ is a tangent, $m\overset{\frown}{MP} = 112°$, $m\overset{\frown}{MN} = 60°$, $m\overset{\frown}{MT} = 46°$
Find: **a)** $m\angle MRP$
b) $m\angle 1$
c) $m\angle 2$

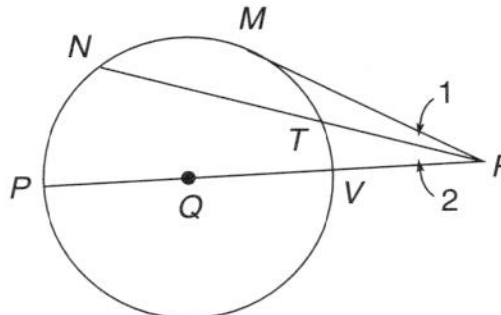

6. *Given:* $\overrightarrow{AB}$ and $\overrightarrow{AC}$ are tangent to $\odot O$, $m\overset{\frown}{BC} = 126°$
Find: **a)** $m\angle A$
b) $m\angle ABC$
c) $m\angle ACB$

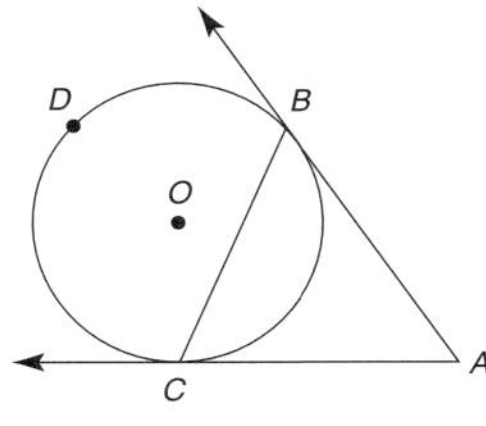

Exercises 6, 7

7. *Given:* Tangents $\overrightarrow{AB}$ and $\overrightarrow{AC}$ to $\odot O$
$m\angle ACB = 68°$
Find: **a)** $m\overset{\frown}{BC}$
b) $m\overset{\frown}{BDC}$
c) $m\angle ABC$
d) $m\angle A$

8. *Given:* $m\angle 1 = 72°$, $m\overset{\frown}{DC} = 34°$
Find: **a)** $m\overset{\frown}{AB}$
b) $m\angle 2$

9. *Given:* $m\angle 2 = 36°$
$m\overset{\frown}{AB} = 4 \cdot m\overset{\frown}{DC}$
Find: **a)** $m\overset{\frown}{AB}$
b) $m\angle 1$
(**HINT:** Let $m\overset{\frown}{DC} = x$ and $m\overset{\frown}{AB} = 4x$.)

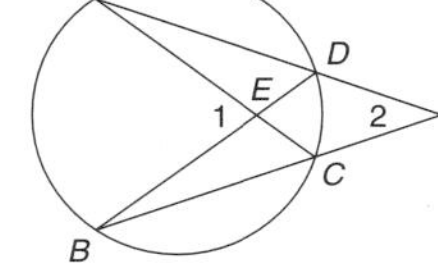

Exercises 8, 9

In Exercises 10 and 11, R and T are points of tangency.

10. *Given:* $m\angle 3 = 42°$
Find: **a)** $m\overset{\frown}{RT}$
b) $m\overset{\frown}{RST}$

11. *Given:* $\overset{\frown}{RS} \cong \overset{\frown}{ST} \cong \overset{\frown}{RT}$
Find: **a)** $m\overset{\frown}{RT}$
b) $m\overset{\frown}{RST}$
c) $m\angle 3$

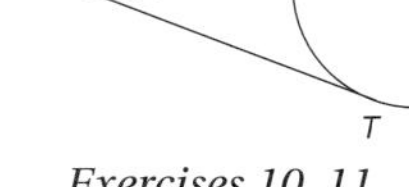

Exercises 10, 11

12. *Given:* $m\angle 1 = 63°$,
$m\overset{\frown}{RS} = 3x + 6$,
$m\overset{\frown}{VT} = x$
Find: $m\overset{\frown}{RS}$

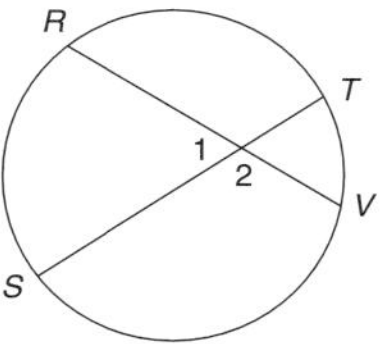

Exercises 12, 13

13. *Given:* $m\angle 2 = 124°$,
$m\overset{\frown}{TV} = x + 1$,
$m\overset{\frown}{SR} = 3(x + 1)$
Find: $m\overset{\frown}{TV}$

14. *Given:* $m\angle 1 = 71°$ and
$m\angle 2 = 33°$
Find: $m\overset{\frown}{CE}$ and $m\overset{\frown}{BD}$

15. *Given:* $m\angle 1 = 62°$,
$m\angle 2 = 26°$
Find: $m\overset{\frown}{CE}$ and $m\overset{\frown}{BD}$

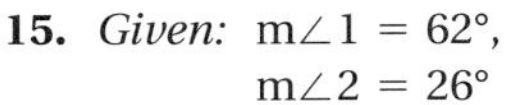

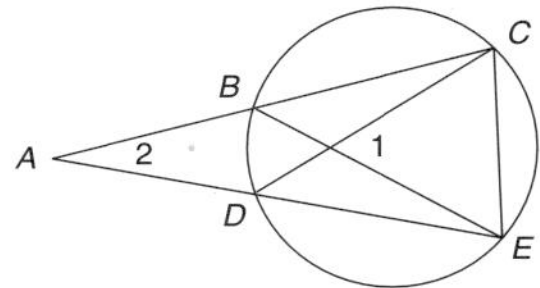

Exercises 14, 15

16. A quadrilateral $RSTV$ is circumscribed about a circle so that its tangent sides are at the endpoints of two intersecting diameters.
a) What type of quadrilateral is $RSTV$?
b) If the diameters are also perpendicular, what type of quadrilateral is $RSTV$?

In Exercises 17 and 18, complete each proof.

17. *Given:* $\overline{AB}$ and $\overline{AC}$ are tangents to $\odot O$ from point A
Prove: $\triangle ABC$ is isosceles

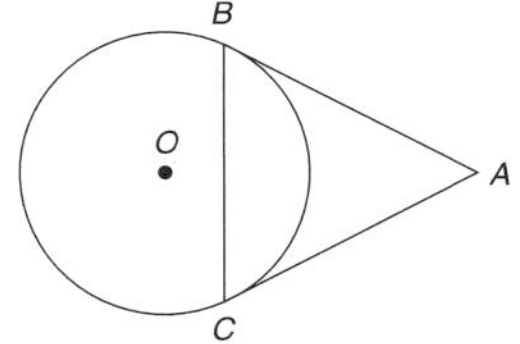

PROOF

Statements	Reasons
1. ?	1. Given
2. $m\angle B = \frac{1}{2}(m\overset{\frown}{BC})$ and $m\angle C = \frac{1}{2}(m\overset{\frown}{BC})$	2. ?
3. $m\angle B = m\angle C$	3. ?
4. $\angle B \cong \angle C$	4. ?
5. ?	5. If two $\angle$s of a $\triangle$ are $\cong$, the sides opposite the $\angle$s are $\cong$
6. ?	6. If two sides of a $\triangle$ are $\cong$, the $\triangle$ is isosceles

18. *Given:* $\overline{RS} \parallel \overline{TQ}$
Prove: $\overset{\frown}{RT} \cong \overset{\frown}{SQ}$

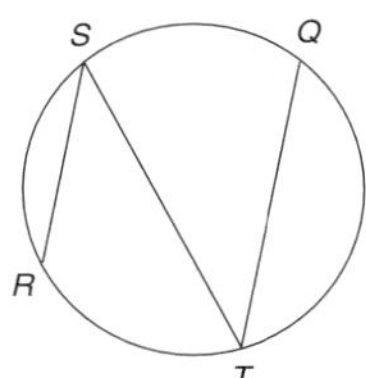

PROOF

Statements	Reasons
1. $\overline{RS} \parallel \overline{TQ}$	1. ?
2. $\angle S \cong \angle T$	2. ?
3. ?	3. If two $\angle$s are $\cong$, the $\angle$s are = in measure.
4. $m\angle S = \frac{1}{2}(m\overset{\frown}{RT})$	4. ?
5. $m\angle T = \frac{1}{2}(m\overset{\frown}{SQ})$	5. ?
6. $\frac{1}{2}(m\overset{\frown}{RT}) = \frac{1}{2}(m\overset{\frown}{SQ})$	6. ?
7. $m\overset{\frown}{RT} = m\overset{\frown}{SQ}$	7. Multiplication Property of Equality
8. ?	8. If two arcs of a $\odot$ are = in measure, the arcs are $\cong$.

In Exercises 19 to 21, complete a paragraph proof.

19. *Given:* Tangent $\overline{AB}$ to $\odot O$ at point B
$m\angle A = m\angle B$
Prove: $m\overset{\frown}{BD} = 2 \cdot m\overset{\frown}{BC}$

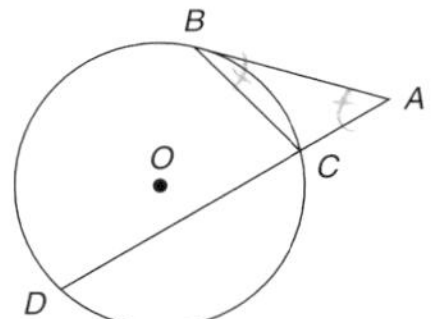

20. *Given:* Diameter $\overline{AB} \perp \overline{CE}$ at D
Prove: CD is the geometric mean of AD and DB

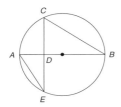

In Exercises 21 and 22, $\overline{CA}$ and $\overline{CB}$ are tangents.

21. *Given:* $m\overset{\frown}{AB} = x$
Prove: $m\angle 1 = 180° - x$

Exercises 21, 22

22. Use the result from Exercise 21 to find $m\angle 1$ if $m\overset{\frown}{AB} = 104°$.

23. An airplane reaches an altitude of 3 mi above the earth. Assuming a clear day and that a passenger has binoculars, how far can the passenger see?
(**HINT:** The radius of the earth is approximately 4000 miles.)

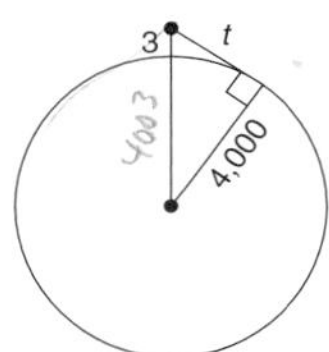

24. From the veranda of a beachfront hotel, Manny is searching the seascape through his binoculars. A ship suddenly appears on the horizon. If Manny is 80 ft above the earth, how far is the ship out at sea?
(**HINT:** See Exercise 23 and note that 1 mi = 5280 ft.)

25. For the five-pointed star (pentagram), find the measures of $\angle 1$ and $\angle 2$.

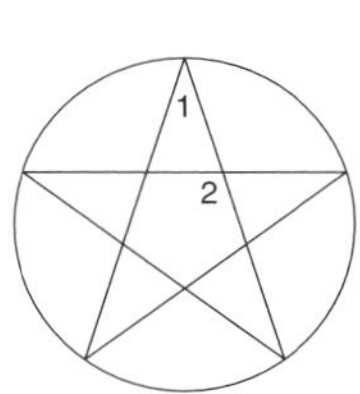

26. ➢ On a fitting for a hex wrench, the distance from the center O to a vertex is 5 mm. The length of radius $\overline{OB}$ of the circle is 10 mm. If $\overline{OC} \perp \overline{DE}$ at F, how long is $\overline{FC}$?

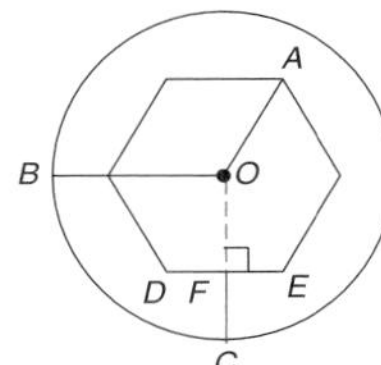

27. ➢ *Given:* $\overline{AB}$ is a diameter in $\odot O$
M is the midpoint of chord $\overline{AC}$
N is the midpoint of chord $\overline{CB}$
$MB = \sqrt{73}, AN = 2\sqrt{13}$
Find: The length of diameter $\overline{AB}$

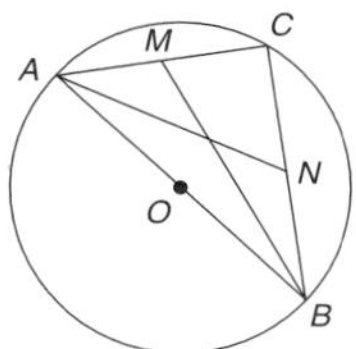

28. A surveyor sees a circular planetarium through a 60° angle. If the surveyor is 45 ft from the door, what is the diameter of the planetarium?

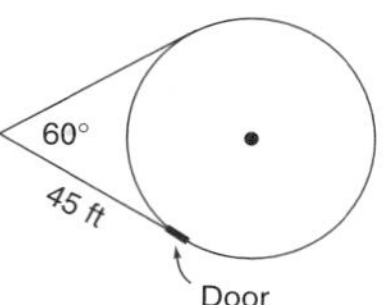

In Exercises 29 to 37, provide a paragraph proof. Be sure to provide a drawing, Given, and Prove where needed.

29. If two parallel lines intersect a circle, then the intercepted arcs between these lines are congruent.
(**HINT:** Draw an auxiliary line.)

30. The line joining the centers of two circles that intersect at two points is the perpendicular bisector of the common chord.

31. If a trapezoid is inscribed in a circle, then it is an isosceles trapezoid.

32. If a parallelogram is inscribed in a circle, then it is a rectangle.

33. If one side of an inscribed triangle is a diameter, then the triangle is a right triangle.

34. Prove case 2 of Corollary 6.2.4: The measure of an angle formed by a tangent and a chord drawn to the point of tangency is one-half the measure of the intercepted arc.

35. Prove case 3 of Corollary 6.2.4.

36. *Given:* $\odot O$ with P in its exterior; O-Y-P
Prove: $OP > OY$

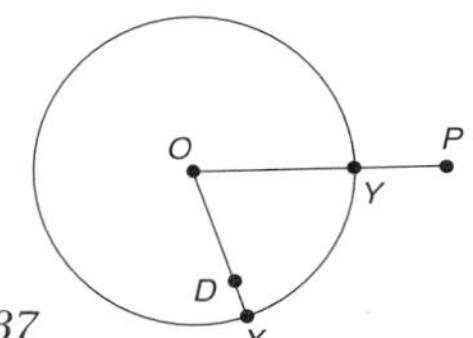

Exercises 36, 37

37. *Given:* $\odot O$ with D in its interior; O-D-X
Prove: $OD < OX$

6.3 Line and Segment Relationships in the Circle

In this section, we turn our attention to line and line segment relationships in the circle. Since some of these statements (such as Theorems 6.3.1–6.3.3) are so similar in wording, the student is strongly encouraged to make drawings and then compare the information that is given in each theorem to the conclusion of that theorem.

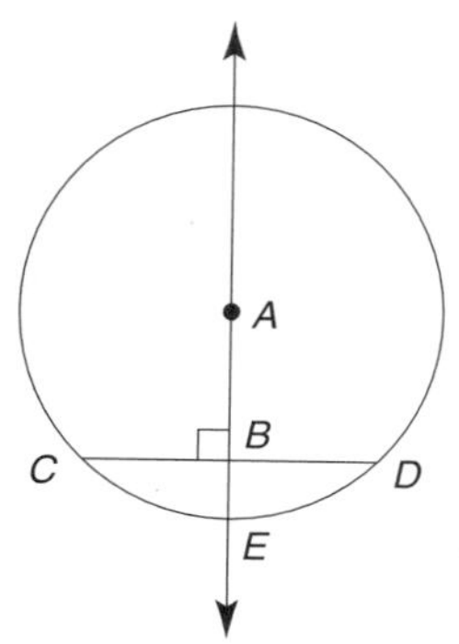

FIGURE 6.37

THEOREM 6.3.1: If a line is drawn through the center of a circle perpendicular to a chord, then it bisects the chord and its arc.

NOTE: Notice that "arc" generally refers to the minor arc, even though the major arc is also bisected.

Given: $\overleftrightarrow{AB} \perp$ chord $\overline{CD}$ in circle A (See Figure 6.37.)
Prove: $\overline{CB} \cong \overline{BD}$ and $\overset{\frown}{CE} \cong \overset{\frown}{ED}$
The proof is left as an exercise for the student.

(**HINT:** Draw $\overline{AC}$ and $\overline{AD}$.)

Even though the "Prove" statement does not match the conclusion of Theorem 6.3.1, we know that $\overline{CD}$ is bisected by $\overleftrightarrow{AB}$ if $\overline{CB} \cong \overline{BD}$ and that $\overset{\frown}{CD}$ is bisected by $\overleftrightarrow{AE}$ if $\overset{\frown}{CE} \cong \overset{\frown}{ED}$.

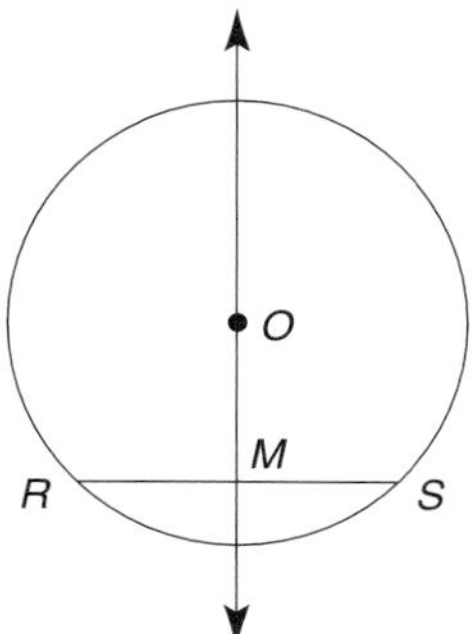

FIGURE 6.38

> **THEOREM 6.3.2:** If a line through the center of a circle bisects a chord other than a diameter, then it is perpendicular to the chord.

Given: Circle O; $\overleftrightarrow{OM}$ is the bisector of chord $\overline{RS}$ (See Figure 6.38.)
Prove: $\overleftrightarrow{OM} \perp \overline{RS}$
The proof is left as an exercise for the student.

(**HINT:** Draw radii $\overline{OR}$ and $\overline{OS}$.)

Figure 6.39(a) illustrates the following theorem. However, Figure 6.39(b) is used in the proof.

> **THEOREM 6.3.3:** The perpendicular bisector of a chord contains the center of the circle.

Given: In Figure 6.39(a), $\overleftrightarrow{QR}$ is the perpendicular bisector of chord $\overline{TV}$ in $\odot O$

Prove: $\overleftrightarrow{QR}$ contains point O

Proof (by indirect method): Suppose that O is not on $\overleftrightarrow{QR}$. Draw $\overline{OR}$ and radii $\overline{OT}$ and $\overline{OV}$. [See Figure 6.39(b).] Because $\overleftrightarrow{QR}$ is the perpendicular bisector of $\overline{TV}$, R must be the midpoint of $\overline{TV}$; then $\overline{TR} \cong \overline{RV}$. Also, $\overline{OT} \cong \overline{OV}$ (all radii of a $\odot$ are $\cong$). With $\overline{OR} \cong \overline{OR}$ by identity, we have $\triangle ORT \cong \triangle ORV$ by SSS.

Now $\angle ORT \cong \angle ORV$ by CPCTC. It follows that $\overline{OR} \perp \overline{TV}$ because these lines (segments) meet to form congruent adjacent angles.

Then $\overline{OR}$ is the perpendicular bisector of $\overline{TV}$. But $\overleftrightarrow{QR}$ is also the perpendicular bisector of $\overline{TV}$, which contradicts the uniqueness of the perpendicular bisector of a segment.

Then the supposition must be false, and it follows that center O is on $\overleftrightarrow{QR}$, the perpendicular bisector of chord $\overline{TV}$.

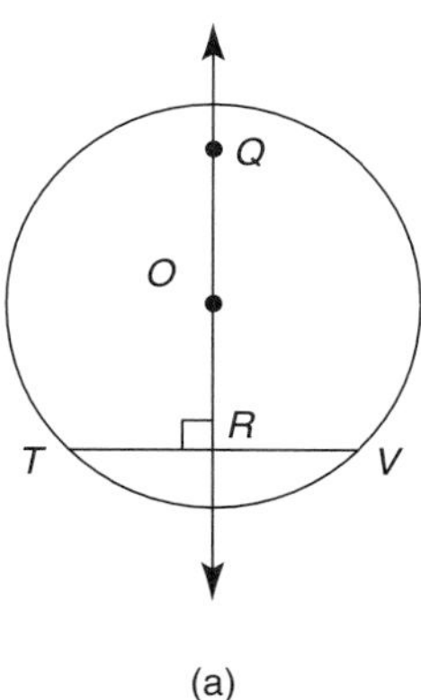

(a)

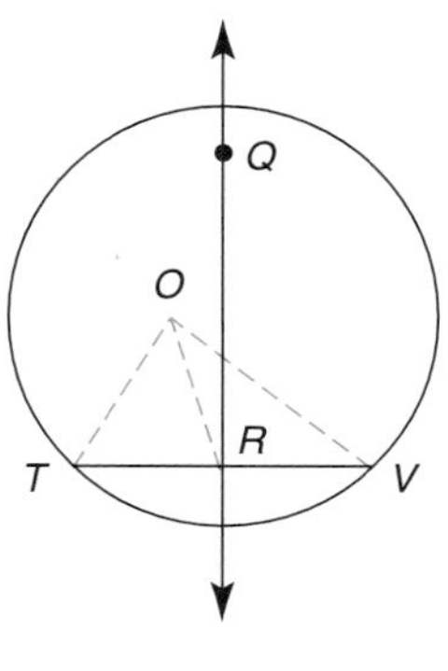

(b)

FIGURE 6.39

EXAMPLE 1

Given: In Figure 6.40, $\odot O$ has a radius of length 5
$\overline{OE} \perp \overline{CD}$ at B and $OB = 3$

Find: CD

Draw radius $\overline{OC}$. By the Pythagorean Theorem,

$$(OC)^2 = (OB)^2 + (BC)^2$$
$$5^2 = 3^2 + (BC)^2$$
$$25 = 9 + (BC)^2$$
$$(BC)^2 = 16$$
$$BC = 4$$

According to Theorem 6.3.1, we know that $CD = 2 \cdot BC$; then it follows that $CD = 2 \cdot 4 = 8$.

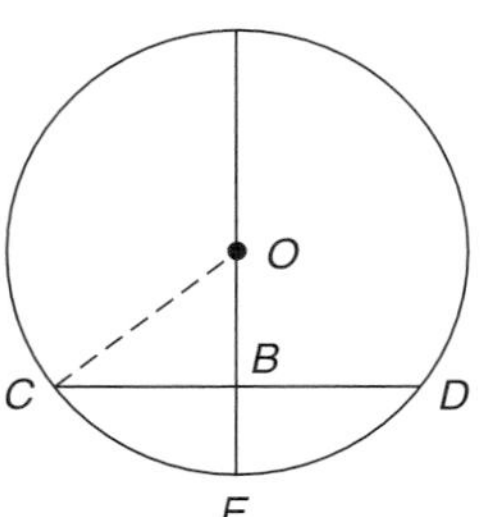

FIGURE 6.40

Circles That Are Tangent

Although concentric circles do not intersect, they do share a common center. For the concentric circles shown in Figure 6.41, the chord of the larger circle is a tangent of the smaller circle.

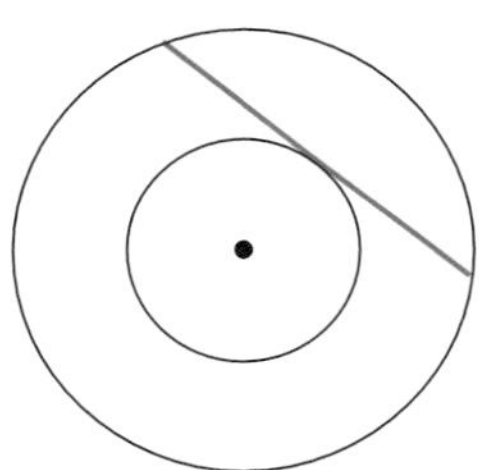

FIGURE 6.41

If two circles touch at one point, they are **tangent circles.** In Figure 6.42, circles P and Q are **internally tangent** while circles O and R are **externally tangent.**

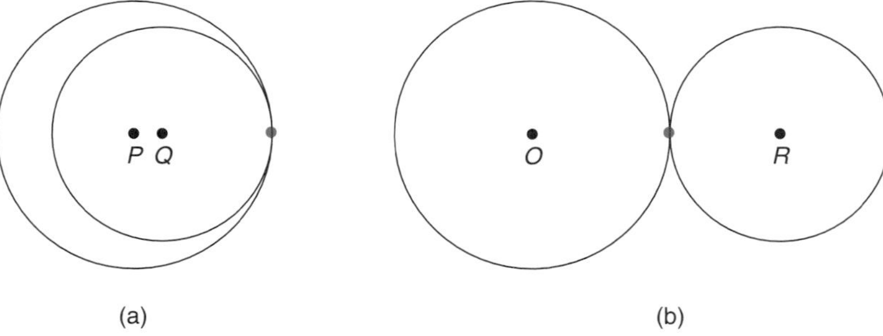

FIGURE 6.42

The line that passes through the centers of two distinct circles is known as the **line of centers** of the two circles. The line segment joining the centers of two circles is also commonly called the "line of centers" of the two circles. In Figure 6.43, $\overleftrightarrow{AB}$ or $\overline{AB}$ is the line of centers for circles A and B.

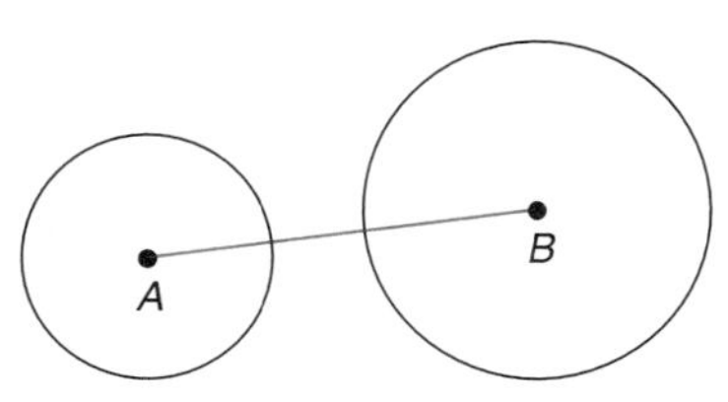

FIGURE 6.43

Common Tangent Lines to Circles

A line segment that is tangent to each of two circles is a **common tangent** for these circles. If the common tangent *does not* intersect the line of centers, it is a **common external tangent.** In Figure 6.44, circles P and Q have one common external tangent, $\overleftrightarrow{ST}$; circles A and B have two common external tangents, $\overleftrightarrow{WX}$ and $\overleftrightarrow{YZ}$.

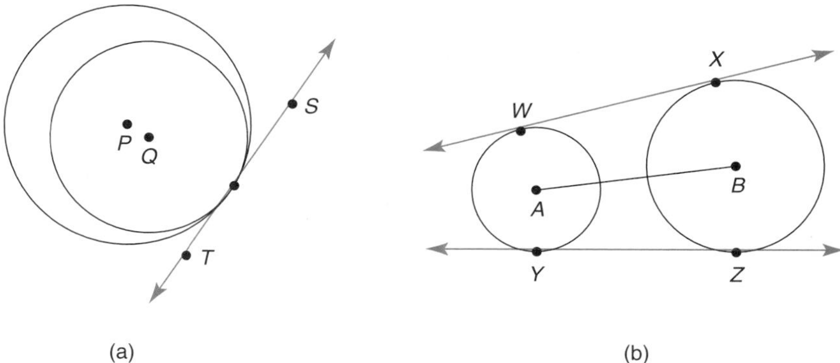

FIGURE 6.44

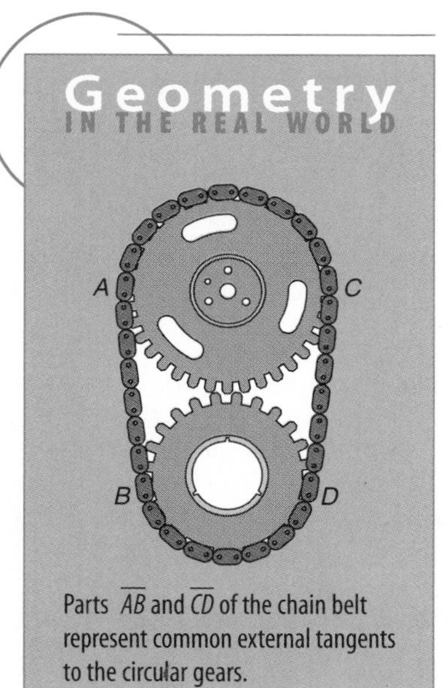

Parts $\overline{AB}$ and $\overline{CD}$ of the chain belt represent common external tangents to the circular gears.

If the common tangent *does* intersect the line of centers for two circles, it is a **common internal tangent** for the two circles. In Figure 6.45, $\overleftrightarrow{DE}$ is a common internal tangent for externally tangent circles O and R; $\overleftrightarrow{AB}$ and $\overleftrightarrow{CD}$ are common internal tangents for $\odot M$ and $\odot N$.

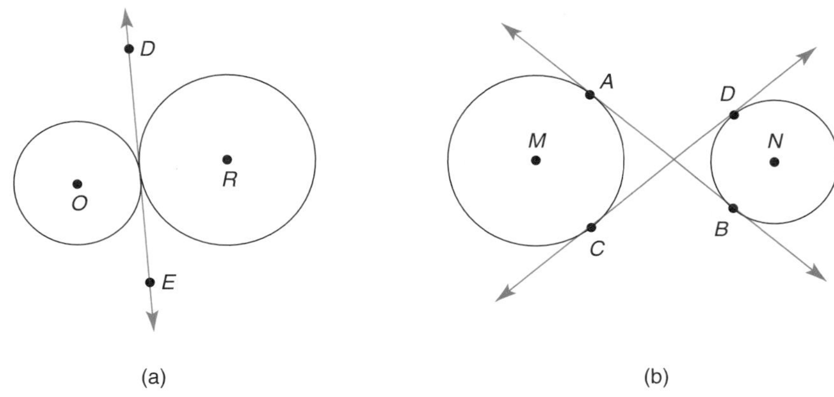

FIGURE 6.45

DISCOVER!

Measure the lengths of tangent segments $\overline{AB}$ and $\overline{AC}$ of Figure 6.46 on the next page. How do AB and AC compare?

ANSWER
They are equal.

> **THEOREM 6.3.4:** The tangent segments to a circle from an external point are congruent.

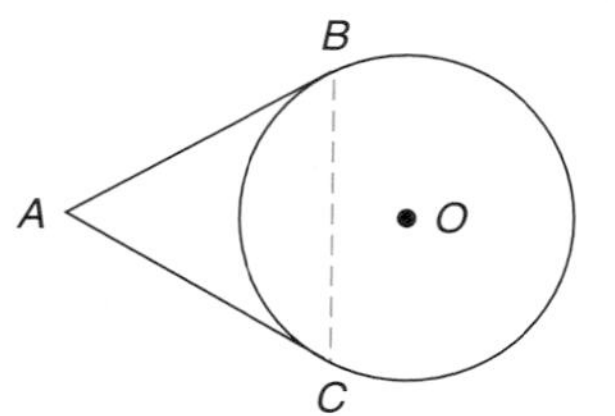

FIGURE 6.46

Given: In Figure 6.46, $\overline{AB}$ and $\overline{AC}$ are tangents to $\odot O$ from point A

Prove: $\overline{AB} \cong \overline{AC}$

Proof: Draw $\overline{BC}$. Now $m\angle B = \frac{1}{2}m\widehat{BC}$ and $m\angle C = \frac{1}{2}m\widehat{BC}$. Then $\angle B \cong \angle C$ since these angles have equal measures. In turn, the sides opposite $\angle B$ and $\angle C$ of $\triangle ABC$ are congruent. That is, $\overline{AB} \cong \overline{AC}$.

We apply Theorem 6.3.4 in Examples 2 and 3.

EXAMPLE 2

A belt used in an automobile engine wraps around two pulleys with different radius lengths. Explain why the straight pieces named $\overline{AB}$ and $\overline{CD}$ have the same length. (See Figure 6.47.)

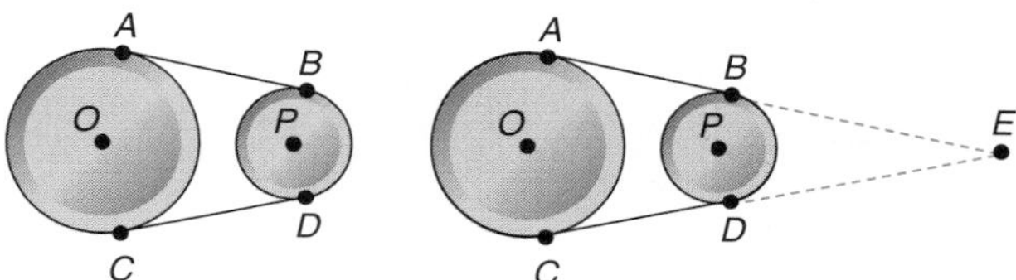

FIGURE 6.47

Solution

Since the pulley centered at O has the larger radius length, we extend $\overline{AB}$ and $\overline{CD}$ to meet at point E. Because E is an external point to both $\odot O$ and $\odot P$, we know that $EB = ED$ and $EA = EC$ by Theorem 6.3.4. By subtracting equals from equals, $EA - EB = EC - ED$. Because $EA - EB = AB$ and $EC - ED = CD$, it follows that $AB = CD$.

EXAMPLE 3

The circle shown in Figure 6.48 is inscribed in $\triangle ABC$; $AB = 9$, $BC = 8$, and $AC = 7$. Find the lengths AM, MB, and NC.

Solution

Because the tangent segments from an external point are $\cong$, we can let

$$AM = AP = x$$
$$BM = BN = y$$
$$NC = CP = z$$

Now

$$x + y = 9 \quad \text{(from } AB = 9\text{)}$$
$$y + z = 8 \quad \text{(from } BC = 8\text{)}$$
$$x + z = 7 \quad \text{(from } AC = 7\text{)}$$

Subtracting the second equation from the first, we have

$$\begin{array}{rl} x + y &= 9 \\ \underline{ y + z} &\underline{= 8} \\ x - z &= 1 \end{array}$$

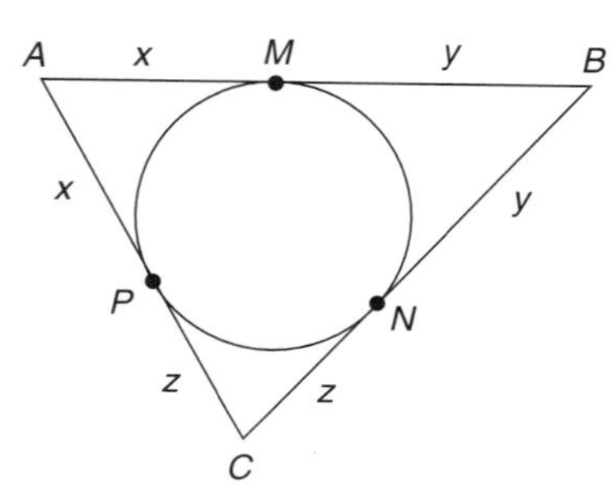

FIGURE 6.48

Now we use this new equation along with the third equation above and add:

$$\begin{aligned} x - z &= 1 \\ x + z &= 7 \\ \hline 2x &= 8 \rightarrow x = 4 \rightarrow AM = 4 \end{aligned}$$

Since $x = 4$ and $x + y = 9$, $y = 5$. Then $BM = 5$. Since $x = 4$ and $x + z = 7$, $z = 3$, so $NC = 3$. Summarizing, $AM = 4$, $BM = 5$, and $NC = 3$.

Lengths of Segments in a Circle

To complete this section, we consider three relationships involving the lengths of chords, secants, or tangents. The first theorem is proved, but the proofs of the others are left as exercises for the student.

THEOREM 6.3.5: If two chords intersect within a circle, then the product of the lengths of the segments (parts) of one chord is equal to the product of the lengths of the segments of the other.

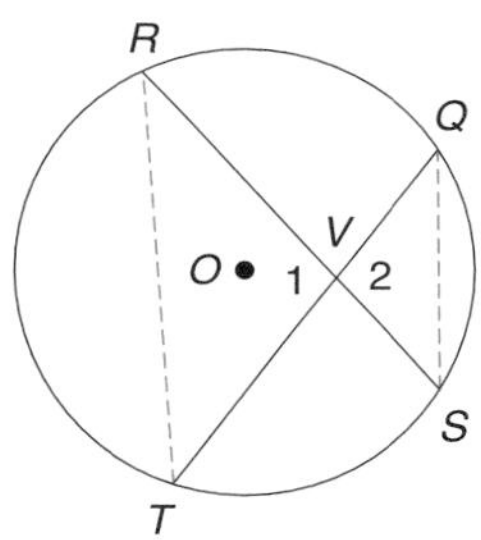

FIGURE 6.49

Given: Circle O with chords $\overline{RS}$ and $\overline{TQ}$ intersecting at point V (See Figure 6.49.)

Prove: $RV \cdot VS = TV \cdot VQ$

Proof: Draw $\overline{RT}$ and $\overline{QS}$. In $\triangle RTV$ and $\triangle QSV$, we have $\angle 1 \cong \angle 2$ (vertical $\angle$s). Also, $\angle R$ and $\angle Q$ are inscribed angles that intercept the same arc (namely $\overset{\frown}{TS}$), so $\angle R \cong \angle Q$. By AA, $\triangle RTV \sim \triangle QSV$. Using CSSTP, $\frac{RV}{VQ} = \frac{TV}{VS}$ and so $RV \cdot VS = TV \cdot VQ$.

EXAMPLE 4

In Figure 6.50, $HP = 4$, $PJ = 5$, and $LP = 8$. Find PM.

Solution Applying Theorem 6.3.5, $HP \cdot PJ = LP \cdot PM$. Then

$$\begin{aligned} 4 \cdot 5 &= 8 \cdot PM \\ 8 \cdot PM &= 20 \\ PM &= 2.5 \end{aligned}$$

EXAMPLE 5

In Figure 6.50, $HP = 6$, $PJ = 4$, and $LM = 11$. Find LP and PM.

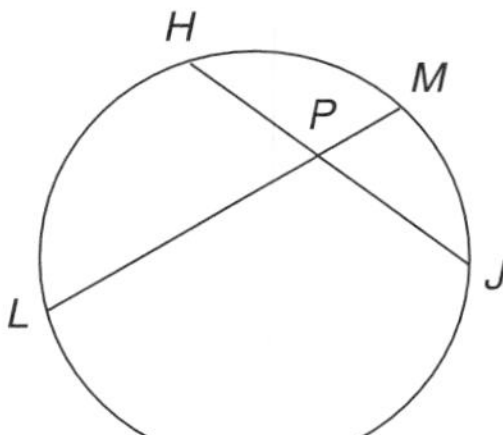

FIGURE 6.50

Solution

Since $LP + PM = LM$, it follows that $PM = LM - LP$. If $LM = 11$ and $LP = x$, then $PM = 11 - x$. Now $HP \cdot PJ = LP \cdot PM$ becomes

$$\begin{aligned} 6 \cdot 4 &= x(11 - x) \\ 24 &= 11x - x^2 \\ x^2 - 11x + 24 &= 0 \\ (x - 3)(x - 8) &= 0, \text{ so } x - 3 = 0 \text{ or } x - 8 = 0 \\ x = 3 \quad &\text{or} \quad x = 8 \end{aligned}$$

Therefore $LP = 3$ or $LP = 8$

If $LP = 3$, then $PM = 8$; conversely, if $LP = 8$, then $PM = 3$. That is, the segments of chord $\overline{LM}$ have lengths of 3 and 8.

THEOREM 6.3.6: If two secant segments are drawn to a circle from an external point, then the products of the lengths of each secant with its external segment are equal.

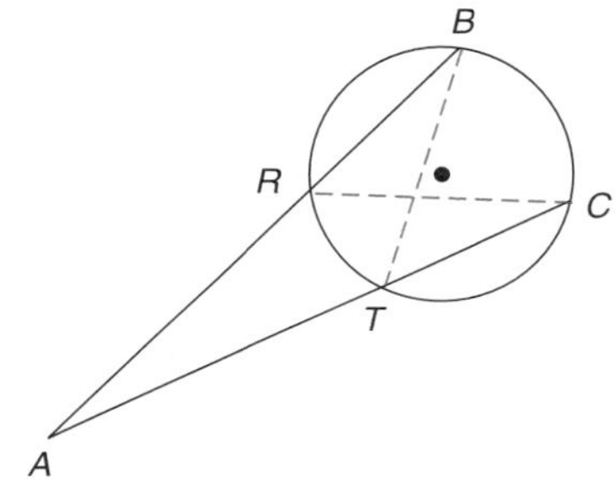

FIGURE 6.51

Given: Secants $\overline{AB}$ and $\overline{AC}$ for the circle in Figure 6.51

Prove: $AB \cdot RA = AC \cdot TA$

The proof is left as an exercise for the student.

(**HINT:** First prove that $\triangle ABT \sim \triangle ACR$.)

EXAMPLE 6

Given: In Figure 6.51, $AB = 14$, $BR = 5$, and $TC = 5$

Find: AC and TA

Solution

Let $AC = x$. Since $AT + TC = AC$, we have $AT + 5 = x$, so $TA = x - 5$. If $AB = 14$ and $BR = 5$, then $AR = 9$. The statement $AB \cdot RA = AC \cdot TA$ becomes

$$\begin{aligned} 14 \cdot 9 &= x(x - 5) \\ 126 &= x^2 - 5x \\ x^2 - 5x - 126 &= 0 \\ (x - 14)(x + 9) &= 0, \text{ so } x - 14 = 0 \text{ or } x + 9 = 0 \\ x = 14 &\text{ or } x = -9 \end{aligned}$$

($x = -9$ is discarded because the length of $\overline{AC}$ cannot be negative.)

Thus $AC = 14$, so $TA = 9$.

THEOREM 6.3.7: If a tangent segment and secant segment are drawn to a circle from an external point, then the square of the length of the tangent equals the product of the lengths of the secant with its external segment.

Given: Tangent $\overline{TV}$ and secant $\overline{TW}$ in Figure 6.52

Prove: $(TV)^2 = TW \cdot TX$

The proof is left as an exercise for the student.

(**HINT:** Use the auxiliary lines shown to prove that $\triangle TVW \sim \triangle TXV$.)

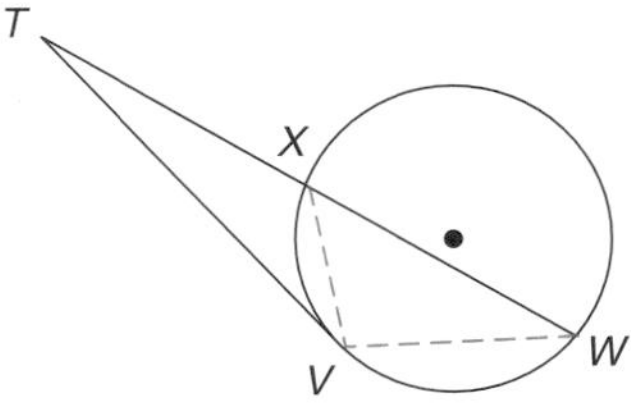

FIGURE 6.52

EXAMPLE 7

Given: In Figure 6.53, $SV = 3$ and $VR = 9$

Find: ST

Solution

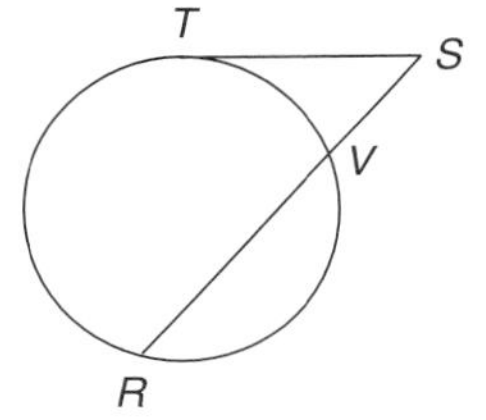

FIGURE 6.53

If $SV = 3$ and $VR = 9$, then $SR = 12$. Using Theorem 6.3.7,

$$(ST)^2 = SR \cdot SV$$
$$(ST)^2 = 12 \cdot 3$$
$$(ST)^2 = 36$$
$$ST = 6 \text{ or } -6$$

Since ST cannot be negative, $ST = 6$.

In Table 6.2 we summarize the results of Sections 6.1–6.3.

TABLE 6.2

Selected Properties of Circles

Figure	Angle Measure	Segment Relationships
Central Angle	$m\angle 1 = m\overset{\frown}{AB}$	$OA = OB$
Inscribed Angle	$m\angle 2 = \frac{1}{2}m\overset{\frown}{HJ}$	generally, $HK \neq KJ$

(continued)

TABLE 6.2 *(continued)*
Selected Properties of Circles

Figure	Angle Measure	Segment Relationships
Angle Formed by Intersecting Chords 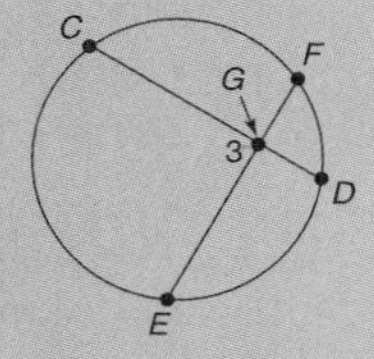	$m\angle 3 = \frac{1}{2}(m\overparen{CE} + m\overparen{FD})$	$CG \cdot GD = EG \cdot GF$
Angle Formed by Intersecting Secants 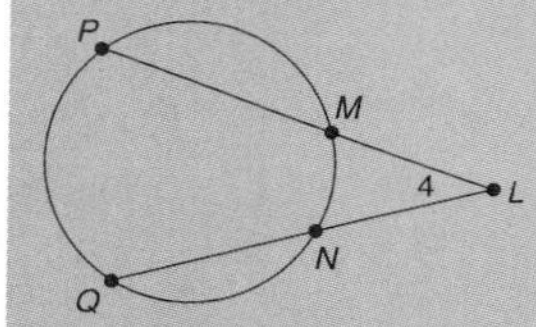	$m\angle 4 = \frac{1}{2}(m\overparen{PQ} - m\overparen{MN})$	$PL \cdot LM = QL \cdot LN$
Angle Formed by Intersecting Tangents 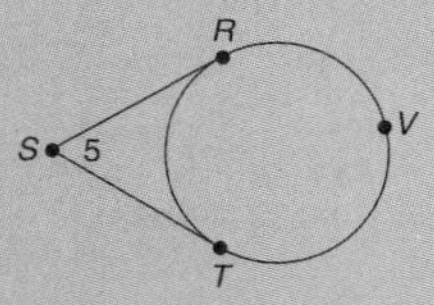	$m\angle 5 = \frac{1}{2}(m\overparen{RVT} - m\overparen{RT})$	$SR = ST$
Angle Formed by Radius Drawn to Tangent 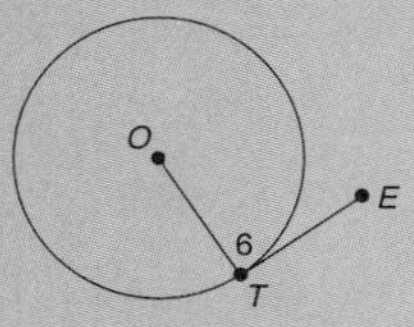	$m\angle 6 = 90°$	$\overline{OT} \perp \overline{TE}$

6.3 Exercises

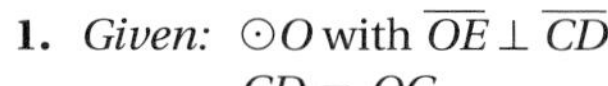

1. *Given:* $\odot O$ with $\overline{OE} \perp \overline{CD}$
   $CD = OC$
   Find: $m\overset{\frown}{CF}$
2. *Given:* $OC = 8$ and $OE = 6$
   $\overline{OE} \perp \overline{CD}$ in $\odot O$
   Find: CD

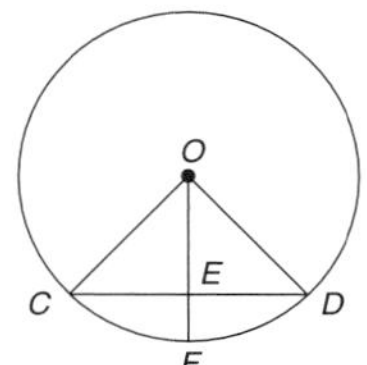

Exercises 1, 2

3. *Given:* $\overline{OV} \perp \overline{RS}$ in $\odot O$
   $OV = 9$ and $OT = 6$
   Find: RS

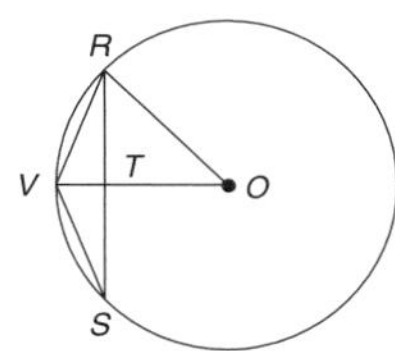

Exercises 3, 4

4. *Given:* V is the midpoint of $\overset{\frown}{RS}$ in $\odot O$
   $m\angle S = 15°$ and $OT = 6$
   Find: OR
5. Sketch two circles that have:
   a) No common tangents
   b) Exactly one common tangent
   c) Exactly two common tangents
   d) Exactly three common tangents
   e) Exactly four common tangents
6. Two congruent intersecting circles B and D (not shown) have a line (segment) of centers $\overline{BD}$ and a common chord $\overline{AC}$ that are congruent. Explain why quadrilateral $ABCD$ is a square.

In the figure for Exercises 7 to 12, O is the center of the circle.

7. *Given:* $AE = 6$, $EB = 4$, $DE = 8$
   Find: EC
8. *Given:* $DE = 12$, $EC = 5$, $AE = 8$
   Find: EB
9. *Given:* $AE = 8$, $EB = 6$, $DC = 16$
   Find: DE and EC
10. *Given:* $AE = 7$, $EB = 5$, $DC = 12$
    Find: DE and EC

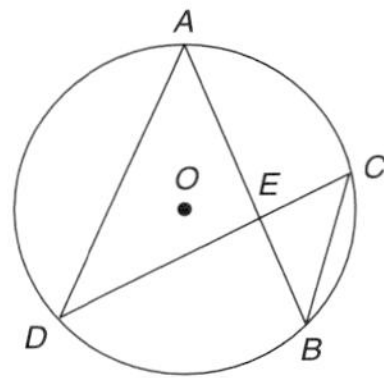

Exercises 7–12

11. *Given:* $AE = 6$, $EC = 3$, $AD = 8$
    Find: CB
12. *Given:* $AD = 10$, $BC = 4$, $AE = 7$
    Find: EC
13. *Given:* $AB = 6$, $BC = 8$, $AE = 15$
    Find: DE

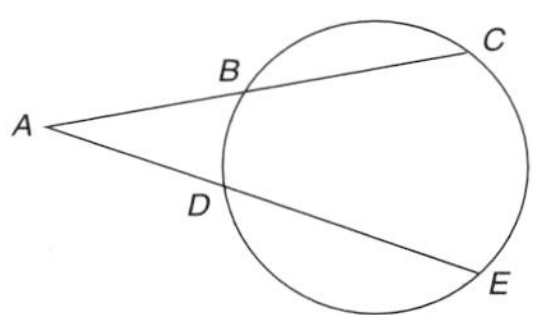

Exercises 13–16

14. *Given:* $AC = 12$, $AB = 6$, $AE = 14$
    Find: AD
15. *Given:* $AB = 4$, $BC = 5$, $AD = 3$
    Find: DE
16. *Given:* $AB = 5$, $BC = 6$, $AD = 6$
    Find: AE

In the figure for Exercises 17 to 20, $\overline{RS}$ is tangent to the circle at S.

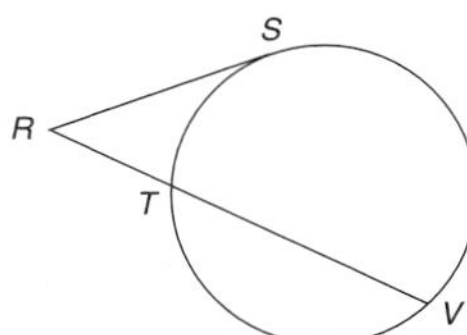

Exercises 17–20

17. *Given:* $RS = 8$ and $RV = 12$
    Find: RT
18. *Given:* $RT = 4$ and $TV = 6$
    Find: RS
19. *Given:* $\overline{RS} \cong \overline{TV}$ and $RT = 6$
    Find: RS
    (**HINT:** Use the Quadratic Formula.)
20. *Given:* $RT = \frac{1}{2} \cdot RS$ and $TV = 9$
    Find: RT

21. For the two circles in Figures (a), (b), and (c), find the total number of common tangents (internal and external).

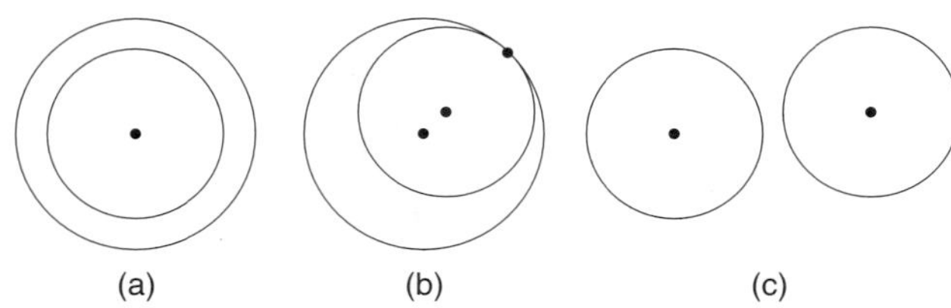
(a) (b) (c)

22. For the two circles in Figures (a), (b), and (c), find the total number of common tangents (internal and external).

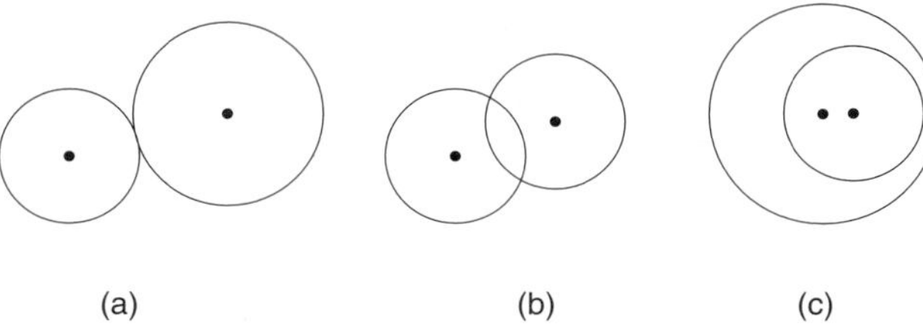
(a) (b) (c)

In Exercises 23 to 26, provide a paragraph proof.

23. *Given:* $\odot O$ and $\odot Q$ are tangent at point F
secant $\overline{AC}$ to $\odot O$
secant $\overline{AE}$ to $\odot Q$
common internal tangent $\overline{AF}$
Prove: $AC \cdot AB = AE \cdot AD$

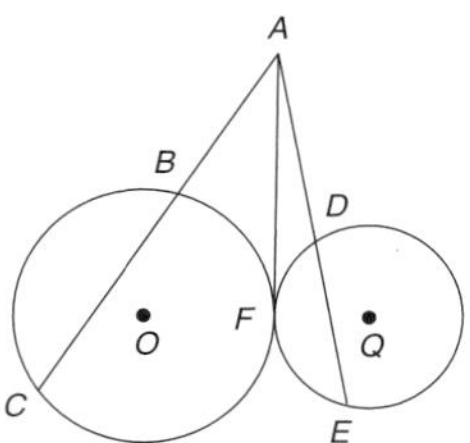

24. *Given:* $\odot O$ with $\overline{OM} \perp \overline{AB}$ and $\overline{ON} \perp \overline{BC}$
$\overline{OM} \cong \overline{ON}$
Prove: $\triangle ABC$ is isosceles

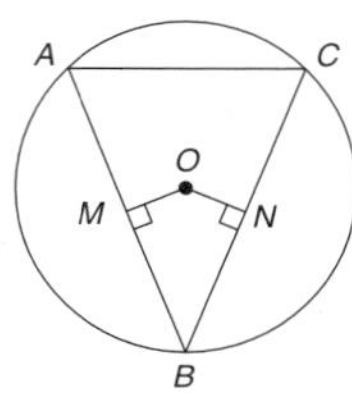

25. *Given:* $\odot Q$ with tangents $\overline{MN}$ and $\overline{MP}$ so that $\overline{MN} \perp \overline{MP}$
Prove: $MNQP$ is a square

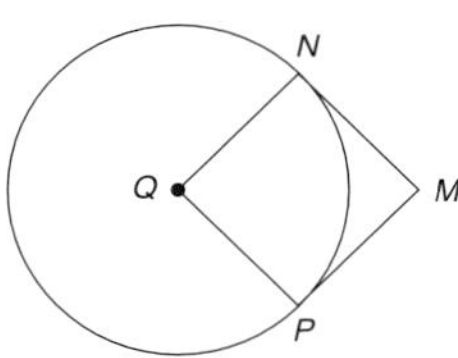

26. *Given:* $\overline{AB} \cong \overline{DC}$ in $\odot P$
Prove: $\triangle ABD \cong \triangle CDB$

27. Does it follow from Exercise 26 that $\triangle ADE$ is also congruent to $\triangle CBE$? What may you conclude about $\overline{AE}$ and $\overline{CE}$ in the drawing? What may you conclude about $\overline{DE}$ and $\overline{EB}$?

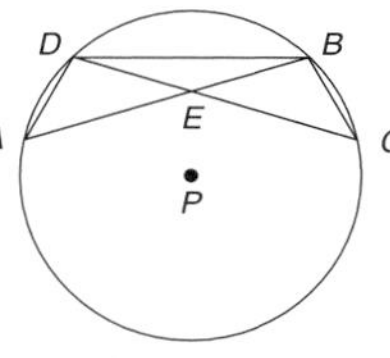

Exercises 26, 27

28. In $\odot O$ (not shown), $\overline{RS}$ is a diameter and T is the midpoint of semicircle $\overset{\frown}{RTS}$. What is the value of the ratio $\frac{RT}{RS}$? The ratio $\frac{RT}{RO}$?

29. ➢ *Given:* Tangents $\overline{AB}$, $\overline{BC}$, and $\overline{AC}$ to $\odot O$ at points M, N, and P, respectively
$AB = 14$, $BC = 16$, $AC = 12$
Find: AM, PC, and BN

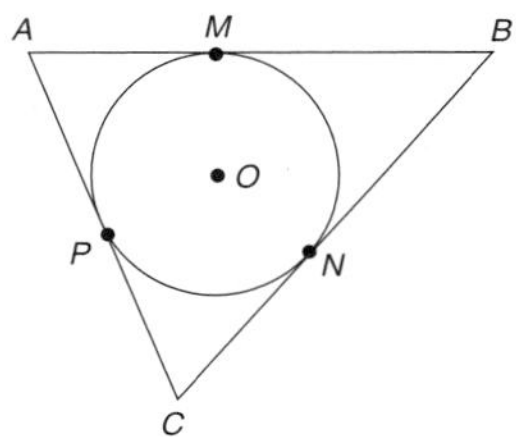

30. ➢ *Given:* $\odot Q$ is inscribed in isosceles right $\triangle RST$; the perimeter of $\triangle RST$ is $8 + 4\sqrt{2}$
Find: TM

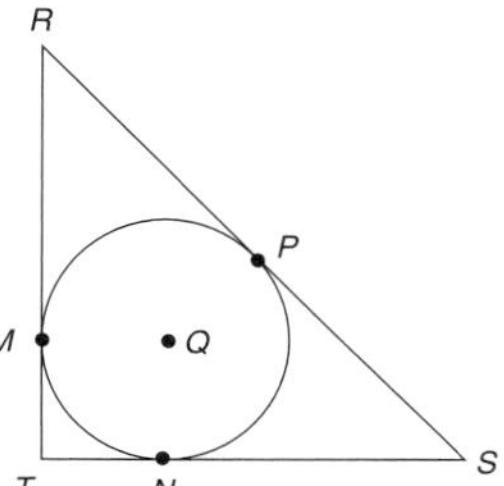

31. ➢ *Given:* $\overline{AB}$ is an external tangent to $\odot O$ and $\odot Q$ at points A and B; radii for $\odot O$ and $\odot Q$ are 4 and 9, respectively

Find: AB

(**HINT:** The line of centers $\overline{OQ}$ contains point C, the point at which $\odot O$ and $\odot Q$ are tangent.)

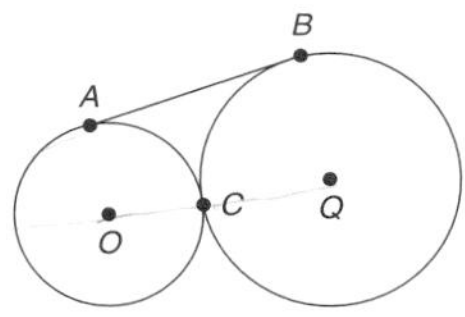

32. ➢ If the larger gear has 30 teeth and the smaller gear has 18, then the gear ratio (larger to smaller) is 5:3. When the larger gear rotates through an angle of 60°, through what angle measure does the smaller gear rotate?

Exercises 32, 33

33. ➢ In Exercise 32, suppose that the larger gear has 20 teeth and the smaller gear has 10 (the gear ratio is 2:1). If the smaller gear rotates through an angle of 90°, through what angle measure does the larger gear rotate?

In Exercises 34 to 37, prove the stated theorems.

34. If a line is drawn through the center of a circle perpendicular to a chord, then it bisects the chord and its minor arc.
(**NOTE:** The major arc is also bisected by the line.)

35. If a line is drawn through the center of a circle to the midpoint of a chord other than a diameter, then it is perpendicular to the chord.

36. If two secant segments are drawn to a circle from an external point, then the products of the lengths of each secant with its external segment are equal.

37. If a tangent segment and a secant segment are drawn to a circle from an external point, then the square of the length of the tangent equals the product of the lengths of the secant with its external segment.

6.4 Some Constructions and Inequalities for the Circle

In Section 6.3, we proved that the radius drawn to a tangent at the point of contact is perpendicular to the tangent at that point. We now show that the converse of that theorem is also true by using an indirect proof. Recall that there is only one line perpendicular to a given line at a point on that line.

THEOREM 6.4.1: The line that is perpendicular to the radius of a circle at its endpoint on the circle is a tangent to the circle.

Given: In Figure 6.54(a), $\odot O$ with radius $\overline{OT}$
$\overleftrightarrow{QT} \perp \overline{OT}$

Prove: $\overleftrightarrow{QT}$ is a tangent to $\odot O$ at point T

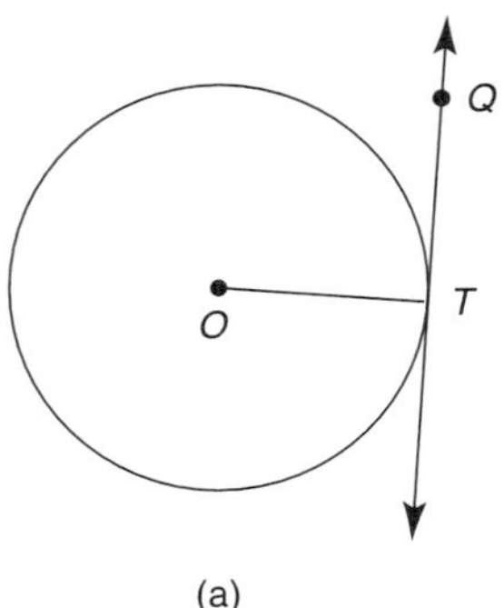

(a)

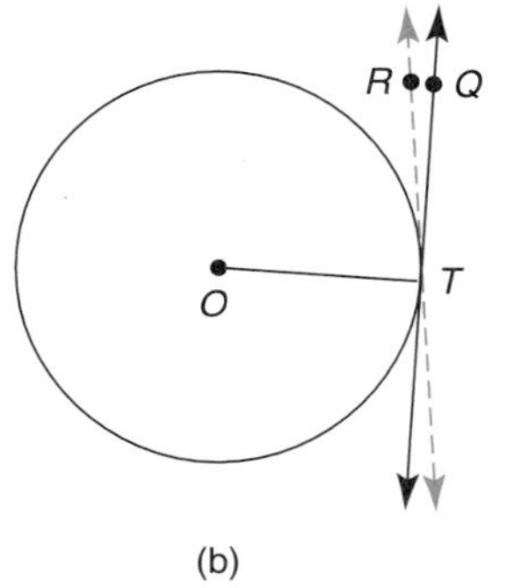

(b)

FIGURE 6.54

Proof: Suppose that $\overleftrightarrow{QT}$ is not a tangent to $\odot O$ at T. Then the tangent (call it $\overleftrightarrow{RT}$) can be drawn at T, the point of tangency. [See Figure 6.54(b).]

Now $\overline{OT}$ is the radius to tangent $\overleftrightarrow{RT}$ at T, and since a radius drawn to a tangent at the point of contact of the tangent is perpendicular to the tangent, $\overline{OT} \perp \overleftrightarrow{RT}$. But $\overline{OT} \perp \overleftrightarrow{QT}$ by hypothesis. Thus two lines are perpendicular to $\overline{OT}$ at point T, contradicting the fact that there is only one line perpendicular to a line at a point on the line. Therefore $\overleftrightarrow{QT}$ must be the tangent to $\odot O$ at point T.

Constructions of Tangents to Circles

Construction 8:
To construct a tangent to a circle at a point on the circle.

Given: $\odot P$ with point X on the circle [See Figure 6.55(a).]

Construct: A tangent $\overleftrightarrow{XW}$ to $\odot P$ at point X

Construction: First draw radius $\overline{PX}$ and extend it to form $\overrightarrow{PX}$. Using X as the center and any radius length less than XP, draw two arcs to intersect $\overrightarrow{PX}$ at points Y and Z as shown in Figure 6.55(b).

Now complete the construction of the perpendicular to $\overrightarrow{PX}$ at point P. From Y and Z, mark arcs with equal radii of length greater than XY. Calling the point of intersection W [see Figure 6.55(c)], draw $\overleftrightarrow{XW}$, the desired tangent to $\odot P$ at point X.

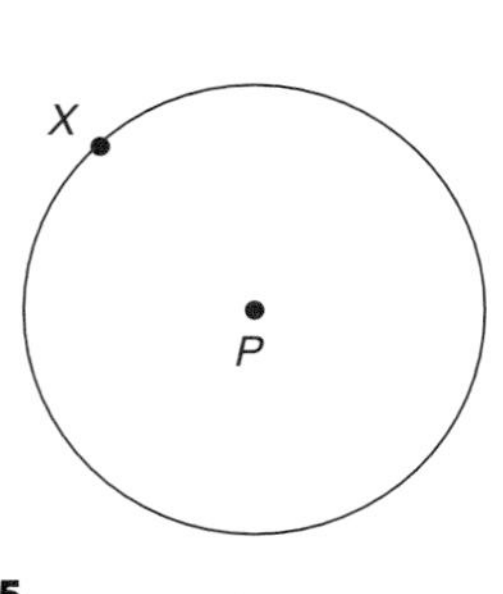

(a)

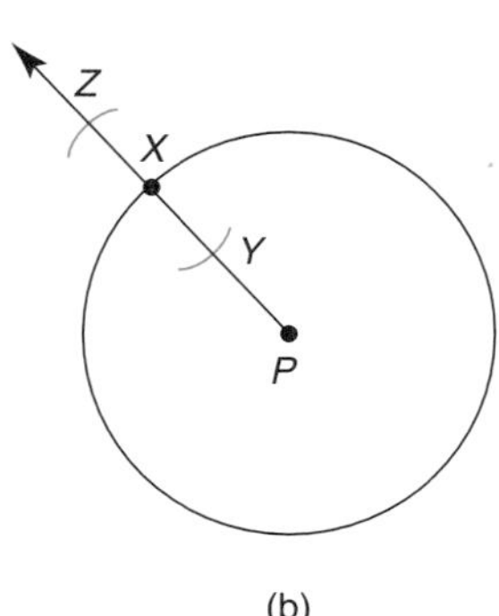

(b)

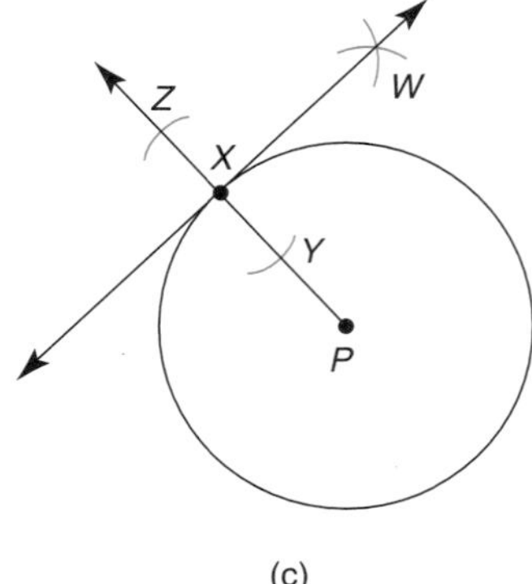

(c)

FIGURE 6.55

EXAMPLE 1

Make a drawing so that points A, B, C, and D are on $\odot O$ in that order. If tangents are constructed at points A, B, C, and D, what type of quadrilateral would be formed by the tangent segments if

a) $m\overset{\frown}{AB} = m\overset{\frown}{CD}$ and $m\overset{\frown}{BC} = m\overset{\frown}{AD}$?
b) all arcs $\overset{\frown}{AB}$, $\overset{\frown}{BC}$, $\overset{\frown}{CD}$, and $\overset{\frown}{DA}$ are congruent?

Solution

a) A rhombus (opposite $\angle$s are $\cong$; all sides $\cong$)
b) A square (all four $\angle$s are right $\angle$s; all sides $\cong$)

We now consider a more difficult construction.

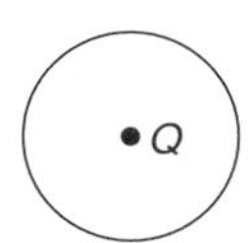

(a)

Construction 9:
To construct a tangent to a circle from an external point.

Given: $\odot Q$ and external point E [See Figure 6.56(a).]

Construct: A tangent $\overline{ET}$, with T as the point of tangency

Construction: Draw $\overline{EQ}$. Construct the perpendicular bisector of $\overline{EQ}$, to intersect $\overline{EQ}$ at its midpoint M. [See Figure 6.56(b).]
With M as center and MQ (or ME) as the length of radius, construct a circle. The points of intersection of circle M with circle Q are designated by T and V.
Draw $\overline{ET}$, the desired tangent. [See Figure 6.56(c).] (Note that $\overline{EV}$, if drawn, would also be a tangent to $\odot Q$).

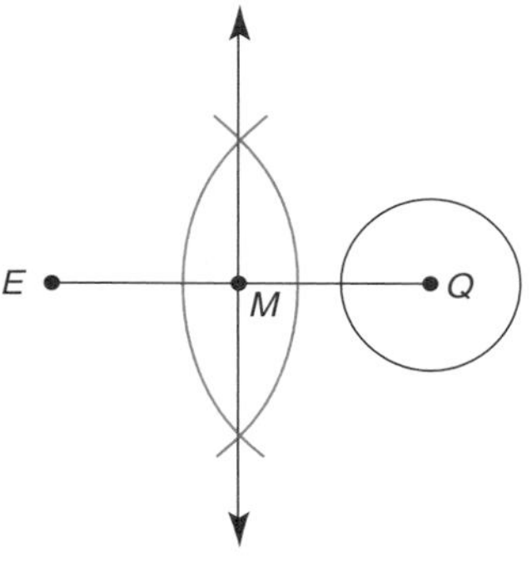

(b)

In the preceding construction, $\overline{QT}$ (not shown) is a radius of the smaller circle Q. In the larger circle M, $\angle ETQ$ is an inscribed angle that intercepts a semicircle. Thus $\angle ETQ$ is a right angle and $\overline{ET} \perp \overline{TQ}$. Since the line drawn perpendicular to the radius of a circle at its endpoint on the circle is a tangent to the circle, $\overline{ET}$ is a tangent to circle Q.

Inequalities in the Circle

The remaining theorems involve inequalities in the circle. Because some of these theorems are proven, the exercise set that follows emphasizes the applications of these theorems rather than their proofs.

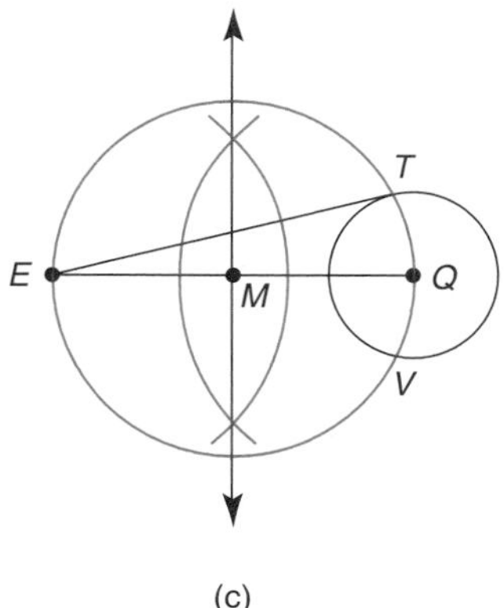

(c)

FIGURE 6.56

THEOREM 6.4.2: In a circle (or in congruent circles) containing two unequal central angles, the larger angle corresponds to the larger intercepted arc.

Given: $\odot O$ with central angles $\angle 1$ and $\angle 2$ in Figure 6.57; $m\angle 1 > m\angle 2$

Prove: $m\overset{\frown}{AB} > m\overset{\frown}{CD}$

Proof: In $\odot O$, $m\angle 1 > m\angle 2$. By the Central Angle Postulate, $m\angle 1 = m\overset{\frown}{AB}$ and $m\angle 2 = m\overset{\frown}{CD}$. By substitution, $m\overset{\frown}{AB} > m\overset{\frown}{CD}$.

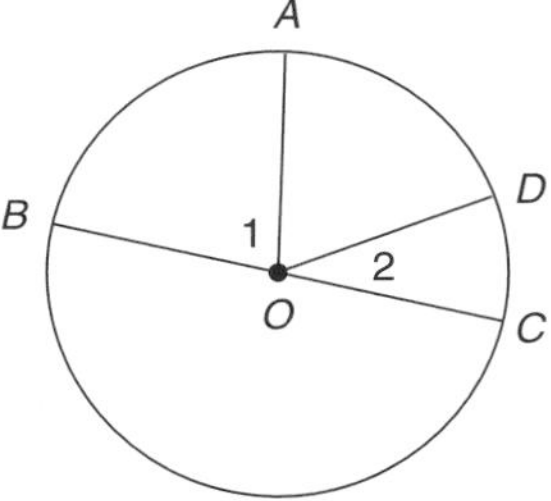

FIGURE 6.57

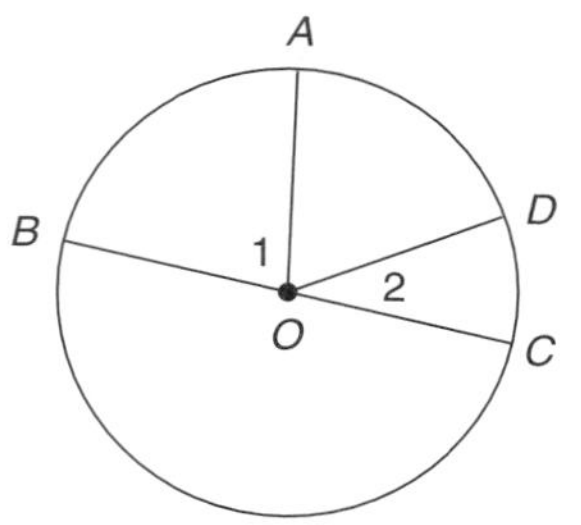

FIGURE 6.57

The converse of Theorem 6.4.2 is also quickly proved.

> **THEOREM 6.4.3:** In a circle (or in congruent circles) containing two unequal arcs, the larger arc corresponds to the larger central angle.

Given: In Figure 6.57, $\odot O$ with $\overset{\frown}{AB}$ and $\overset{\frown}{CD}$
$m\overset{\frown}{AB} > m\overset{\frown}{CD}$

Prove: $m\angle 1 > m\angle 2$

The proof is left as an exercise for the student.

EXAMPLE 2

Given: In Figure 6.58, $\odot Q$ with $m\overset{\frown}{RS} > m\overset{\frown}{TV}$.

a) Using Theorem 6.4.3, what may you conclude regarding the measures of $\angle RQS$ and $\angle TQV$?

b) What does intuition suggest regarding RS and TV?

Solution

a) $m\angle RQS > m\angle TQV$

b) $RS > TV$

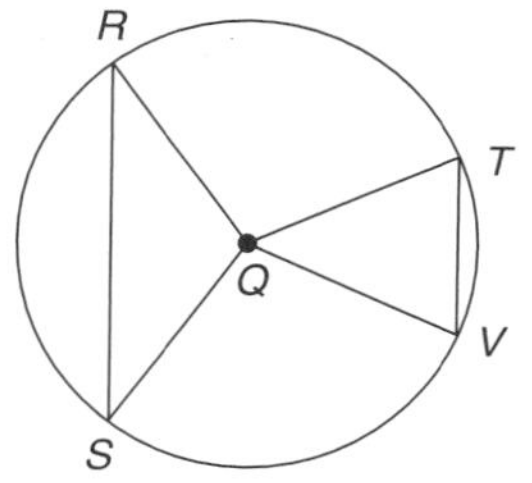

FIGURE 6.58

Before we apply Theorem 6.4.4 and prove Theorem 6.4.5, we consider the following activity. It is worth mentioning that the proof of Theorem 6.4.4 is similar to that of Theorem 6.4.5.

DISCOVER!

In Figure 6.59, $\overline{PT}$ measures the distance from center P to chord $\overline{EF}$. Likewise, $\overline{PR}$ measures the distance from P to chord $\overline{AB}$. Using a ruler, show that $PR > PT$. How do the lengths of chords $\overline{AB}$ and $\overline{EF}$ compare?

ANSWER
$AB < EF$

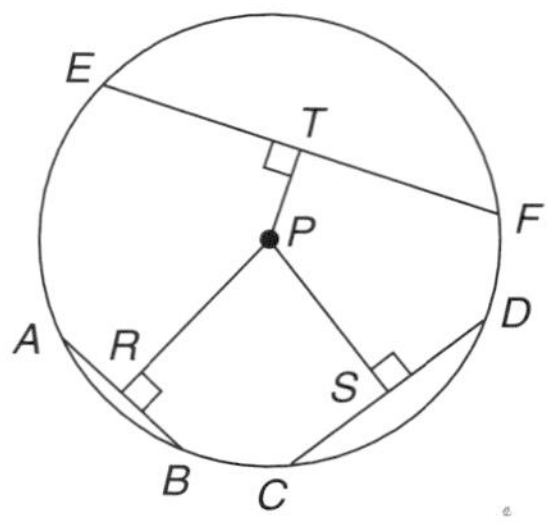

FIGURE 6.59

> **THEOREM 6.4.4:** In a circle (or in congruent circles) containing two unequal chords, the shorter chord is at the greater distance from the center of the circle.

EXAMPLE 3

In circle P of Figure 6.59 on page 268, any radius has a length of 6 cm, and the chords have lengths $AB = 4$ cm, $DC = 6$ cm, and $EF = 10$ cm. Let $\overline{PR}$, $\overline{PS}$, and $\overline{PT}$ name perpendicular segments to these chords.

a) Of $\overline{PR}$, $\overline{PS}$, and $\overline{PT}$, which is longest?
b) Which is shortest?

Solution

a) $\overline{PR}$ is longest, according to Theorem 6.4.4.
b) $\overline{PT}$ is shortest.

In the proof of Theorem 6.4.5, a and b represent the lengths of line segments. Then a and b are positive. If $a < b$, then $a^2 < b^2$; the converse is also true.

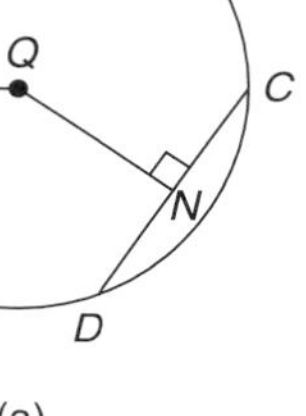

(a)

(b)

FIGURE 6.60

THEOREM 6.4.5: In a circle (or in congruent circles) containing two unequal chords, the chord nearer the center of the circle has the greater length.

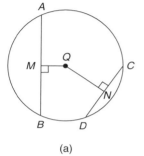

Given: In Figure 6.60(a), $\odot Q$ with chords $\overline{AB}$ and $\overline{CD}$
$\overline{QM} \perp \overline{AB}$ and $\overline{QN} \perp \overline{CD}$
$QM < QN$

Prove: $AB > CD$

Proof: In Figure 6.60(b), denote lengths of $\overline{QM}$ and $\overline{QN}$ by a and c, respectively. Draw radii $\overline{QA}$, $\overline{QB}$, $\overline{QC}$, and $\overline{QD}$, and denote all lengths by r. $\overline{QM}$ is the perpendicular bisector of $\overline{AB}$, and $\overline{QN}$ is the perpendicular bisector of $\overline{CD}$, since a radius perpendicular to a chord bisects the chord and its arc. Let $MB = b$ and $NC = d$.

With right angles at M and N, we see that $\triangle QMB$ and $\triangle QNC$ are right triangles.

According to the Pythagorean Theorem, $r^2 = a^2 + b^2$ and $r^2 = c^2 + d^2$, so $b^2 = r^2 - a^2$ and $d^2 = r^2 - c^2$. If $QM < QN$, then $a < c$ and $a^2 < c^2$. Therefore $-a^2 > -c^2$. Adding r^2, we have $r^2 - a^2 > r^2 - c^2$ or $b^2 > d^2$, which implies that $b > d$. If $b > d$, then $2b > 2d$. But $AB = 2b$ while $CD = 2d$. Therefore $AB > CD$.

It is important that the phrase "minor arc" be used in our final theorems. For the second theorem, the proof is provided because it is more involved.

THEOREM 6.4.6: In a circle (or in congruent circles) containing two unequal chords, the longer chord corresponds to the greater minor arc.

The proof of Theorem 6.4.6 is left to the student.

THEOREM 6.4.7: In a circle (or in congruent circles) containing two unequal minor arcs, the greater minor arc corresponds to the longer of the chords related to these arcs.

Given: In Figure 6.61(a), $\odot O$ with $m\widehat{AB} > m\widehat{CD}$ and chords $\overline{AB}$ and $\overline{CD}$

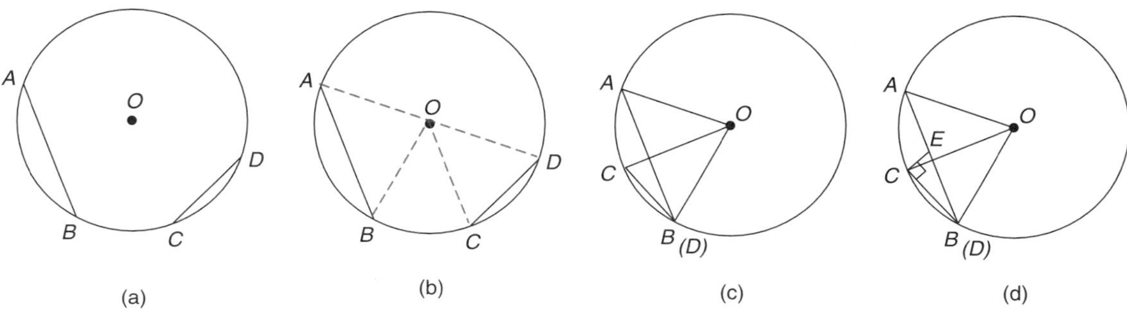

FIGURE 6.61

Prove: $AB > CD$

Proof: In circle O, draw radii $\overline{OA}$, $\overline{OB}$, $\overline{OC}$, and $\overline{OD}$. $m\widehat{AB} > m\widehat{CD}$ leads to $m\angle AOB > m\angle COD$ since the larger arc in a circle corresponds to a larger central angle. [See Figure 6.61(b).]

We now rotate $\triangle COD$ to the position on the circle for which D coincides with B, as shown in Figure 6.61(c). Because radii $\overline{OC}$ and $\overline{OB}$ are congruent, $\triangle COD$ is isosceles; also, $m\angle C = m\angle ODC$.

In $\triangle COD$, $m\angle COD + m\angle C + m\angle CDO = 180°$. Because $m\angle COD$ is positive, we have $m\angle C + m\angle CDO < 180°$ and $2 \cdot m\angle C < 180°$ by substitution. Therefore $m\angle C < 90°$.

Now construct the perpendicular segment to $\overline{CD}$ at point C, as shown in Figure 6.61(d). Denote the intersection of the perpendicular segment and $\overline{AB}$ by point E. Because $\triangle DCE$ is a right $\triangle$ with hypotenuse $\overline{EB}$, $EB > CD$ (*). Since $AB = AE + EB$ and $AE > 0$, we have $AB > EB$ (*). By the Transitive Property, the starred (*) statements reveal that $AB > CD$.

NOTE: In the preceding proof, $\overline{CE}$ must intersect $\overline{AB}$ at some point between A and B. If it were to intersect at A, the measure of inscribed $\angle BCA$ would have to be more than 90°; this follows from the facts that $\widehat{AB}$ is a minor arc and that the intercepted arc for $\angle BCA$ would have to be a major arc.

6.4 Exercises

1. Construct a circle O and choose some point D on the circle. Now construct the tangent to circle O at point D.
2. Construct a circle P and choose three points R, S, and T on the circle. Construct the triangle that has its sides tangent at R, S, and T.
3. X, Y, and Z are on circle O so that $m\widehat{XY} = 120°$, $m\widehat{YZ} = 130°$, and $m\widehat{XZ} = 110°$. Suppose that triangle XYZ is drawn and that the triangle ABC is constructed with its sides tangent to circle O at X, Y, and Z. Are $\triangle XYZ$ and $\triangle ABC$ similar triangles?

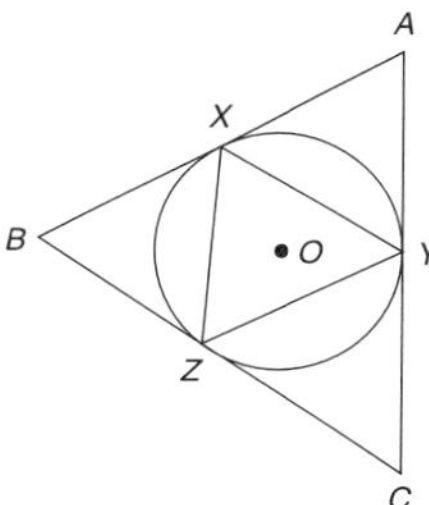

4. Construct the two tangent segments to circle P (not shown) from external point E.
5. Point V is in the exterior of circle Q (not shown) so that $\overline{VQ}$ is equal in length to the diameter of circle Q. Construct the two tangents to circle Q from point V. Then determine the measure of the angle that has vertex V and has the tangents as sides.
6. Given circle P and points R-P-T so that R and T are in the exterior of circle P, suppose that tangents are constructed from R and T to form a quadrilateral (as shown). Identify the type of quadrilateral formed
   a) when $RP > PT$.
   b) when $RP = PT$.

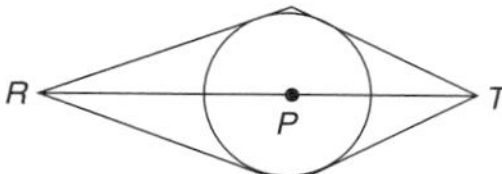

7. Given parallel chords $\overline{AB}$, $\overline{CD}$, $\overline{EF}$, and $\overline{GH}$ in circle O, which chord has the greatest length? Which has the least length? Why?

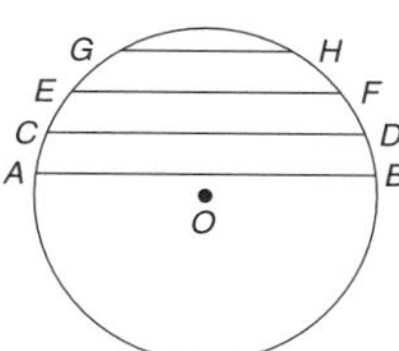

8. Given chords $\overline{MN}$, $\overline{RS}$, and $\overline{TV}$ in $\odot Q$ so that $QZ > QY > QX$, which chord has the greatest length? Which has the least length? Why?

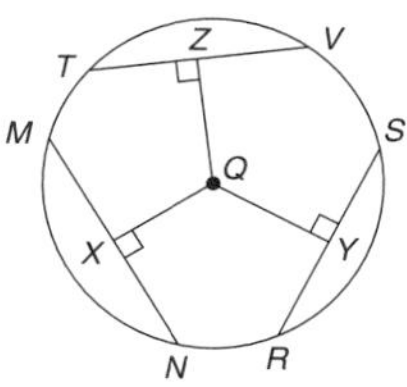

9. Given circle O with radius $\overline{OT}$, tangent $\overleftrightarrow{AD}$, and line segments $\overline{OA}$, $\overline{OB}$, $\overline{OC}$, and $\overline{OD}$:
   a) Which line segment drawn from O has the smallest length?
   b) If $m\angle 1 = 40°$, $m\angle 2 = 50°$, $m\angle 3 = 45°$, and $m\angle 4 = 30°$, which line segment from point O has the greatest length?

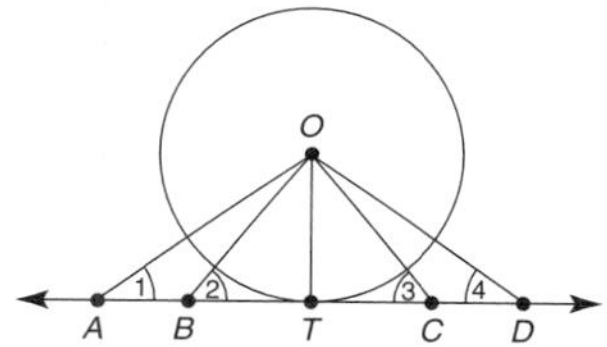

10. **a)** If $m\widehat{RS} > m\widehat{TV}$, write an inequality to compare $m\angle 1$ with $m\angle 2$.
    b) If $m\angle 1 > m\angle 2$, write an inequality to compare $m\widehat{RS}$ with $m\widehat{TV}$.

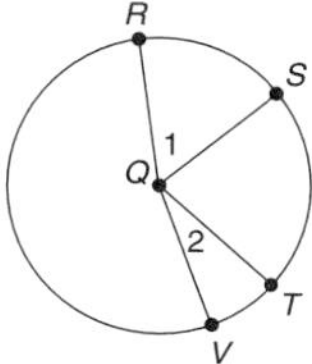

11. **a)** If $MN > PQ$, write an inequality to compare the measures of minor arcs $\widehat{MN}$ and $\widehat{PQ}$.
    b) If $MN > PQ$, write an inequality to compare the measures of major arcs $\widehat{MPN}$ and $\widehat{PMQ}$.

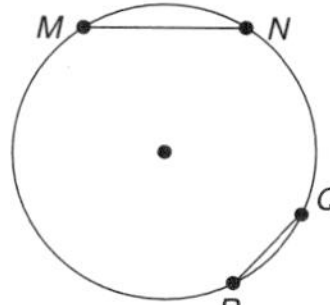

12. a) If $m\overset{\frown}{XY} > m\overset{\frown}{YZ}$, write an inequality to compare the measures of inscribed angles 1 and 2.

b) If $m\angle 1 < m\angle 2$, write an inequality to compare the measures of $\overset{\frown}{XY}$ and $\overset{\frown}{YZ}$.

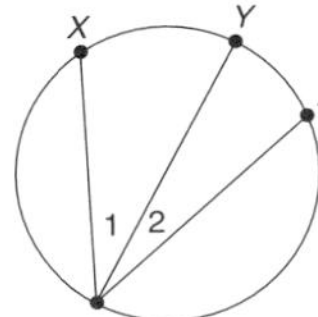

13. Quadrilateral $ABCD$ is inscribed in circle P (not shown). If $\angle A$ is an acute angle, what type of angle is $\angle C$?

14. Quadrilateral $RSTV$ is inscribed in circle Q (not shown). If arcs $\overset{\frown}{RS}$, $\overset{\frown}{ST}$, and $\overset{\frown}{TV}$ are all congruent, what type of quadrilateral is $RSTV$?

15. In circle O, points A, B, and C are on the circle so that $m\overset{\frown}{AB} = 60°$ and $m\overset{\frown}{BC} = 40°$.

a) How are $\angle AOB$ and $\angle BOC$ related?

b) How are AB and BC related?

16. In $\odot O$, $AB = 6$ cm and $BC = 4$ cm.

a) How are $m\angle AOB$ and $m\angle BOC$ related?

b) How are $m\overset{\frown}{AB}$ and $m\overset{\frown}{BC}$ related?

Exercises 15–17

17. In $\odot O$, $m\angle AOB = 70°$ and $m\angle BOC = 30°$.

a) How are $m\overset{\frown}{AB}$ and $m\overset{\frown}{BC}$ related?

b) How are AB and BC related?

18. Triangle ABC is inscribed in circle O; $AB = 5$, $BC = 6$, and $AC = 7$.

a) Which is the largest minor arc of $\odot O$ . . . $\overset{\frown}{AB}$, $\overset{\frown}{BC}$, or $\overset{\frown}{AC}$?

b) Which side of the triangle is nearest point O?

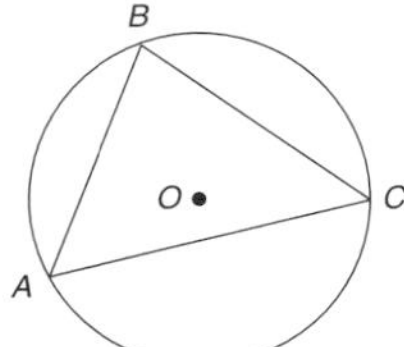

Exercises 18, 19

19. Given circle O with $m\overset{\frown}{BC} = 120°$ and $m\overset{\frown}{AC} = 130°$:

a) Which angle of triangle ABC is smallest?

b) Which side of triangle ABC is nearest point O?

20. Circle O has a diameter of length 20 cm. Chord $\overline{AB}$ has length 12 cm, and chord $\overline{CD}$ has length 10 cm. How much closer is $\overline{AB}$ than $\overline{CD}$ to point O?

21. Circle P has a radius of length 8 in. Points A, B, C, and D are on circle P so that $m\angle APB = 90°$ while $m\angle CPD = 60°$. How much closer is chord $\overline{AB}$ than $\overline{CD}$ to point P?

22. A tangent $\overline{ET}$ is constructed to circle Q from external point E. Which angle and side of triangle QTE are largest? Which angle and side are smallest?

23. Two congruent circles, $\odot O$ and $\odot P$, do not intersect. Construct a common external tangent for $\odot O$ and $\odot P$.

24. Explain why the following statement is incorrect:

"In a circle (or in congruent circles) containing two unequal chords, the longer chord corresponds to the greater major arc."

25. Prove that in a circle containing two unequal arcs, the larger arc corresponds to a larger central angle.

26. Prove that in a circle containing two unequal chords, the longer chord corresponds to a larger central angle. (**HINT:** You may use any theorems stated in this section.)

27. ➢ In $\odot O$, chord $\overline{AB} \parallel$ chord $\overline{CD}$. Radius $\overline{OE}$ is perpendicular to $\overline{AB}$ and $\overline{CD}$ at points M and N respectively. If $OE = 13$, $AB = 24$, and $CD = 10$, then the distance from O to $\overline{CD}$ is greater than the distance from O to $\overline{AB}$. Determine how much farther chord $\overline{CD}$ is from center O than chord $\overline{AB}$ is from center O; that is, find MN.

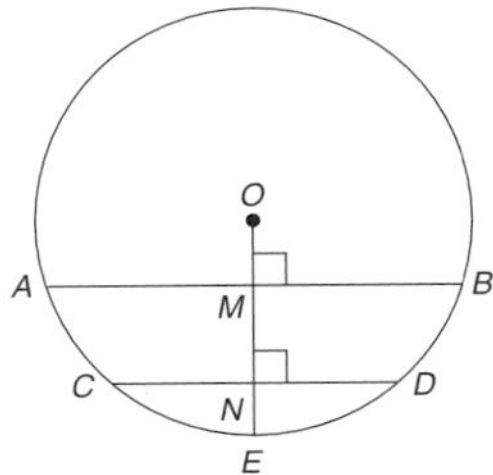

28. ➢ In $\odot P$, whose radius has length 8 in., $m\overset{\frown}{AB} = m\overset{\frown}{BC} = 60°$. Because $m\overset{\frown}{AC} = 120°$, chord $\overline{AC}$ is longer than either of the congruent chords $\overline{AB}$ or $\overline{BC}$. Determine how much longer $\overline{AC}$ is than $\overline{AB}$; that is, find the exact and the approximate value of $AC - AB$.

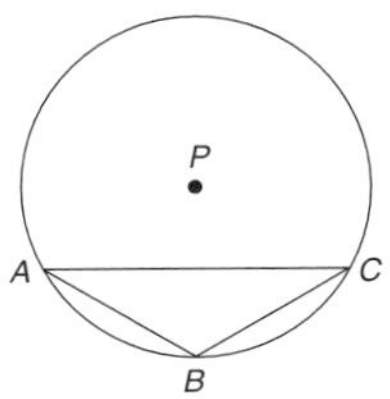

6.5 Locus of Points

At times we need to describe a set of points that satisfy a given condition or set of conditions. The term used to describe the resulting geometric figure is "locus," the plural of which is "loci" (pronounced lō-sī). The English word "location" is derived from the Latin word "locus."

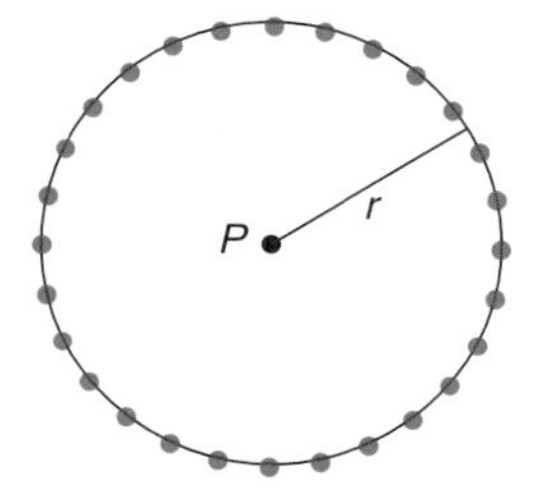

FIGURE 6.62

DEFINITION: A **locus** is the set of all points and only those points that satisfy a given condition (or set of conditions).

In this definition, the phrase "all points and only those points" has a dual meaning:

1. All points of the locus satisfy the given condition.
2. All points satisfying the given condition are included in the locus.

The set of points satisfying a given locus can be a well-known geometric figure such as a line or a circle.

EXAMPLE 1

Describe the locus of points in a plane that are at a fixed distance (r) from a given point (P).

Solution The locus is the circle with center P and radius r. (See Figure 6.62.)

EXAMPLE 2

Describe the locus of points in a plane that are equidistant from two fixed points (P and Q).

Solution The locus is a line that is the perpendicular bisector of $\overline{PQ}$. (See Figure 6.63, in which $PX = QX$ for any point X on line t.)

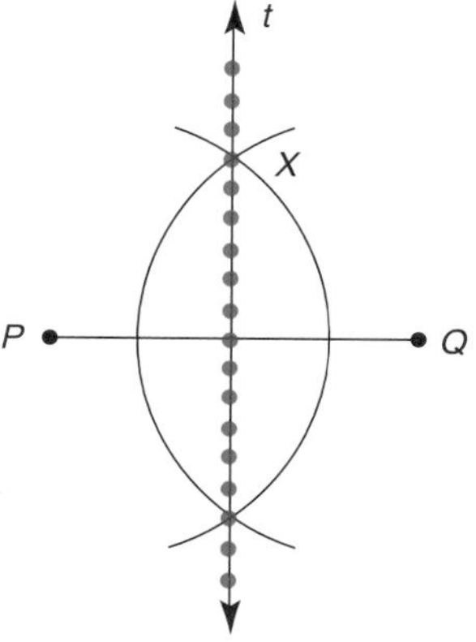

FIGURE 6.63

EXAMPLE 3

Describe the locus of points in a plane that are equidistant from the sides of an angle ($\angle ABC$) in that plane.

Solution

The locus is the ray that bisects $\angle ABC$. (See Figure 6.64.)

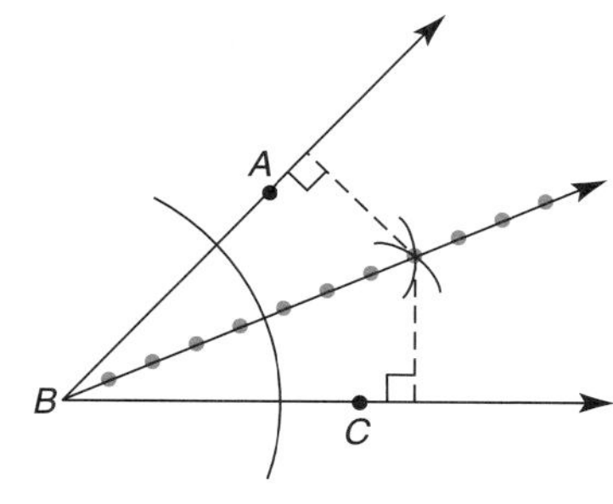

FIGURE 6.64

Some definitions are given in a locus format; for example, the following is an alternative definition of "circle."

> **DEFINITION:** A **circle** is the locus of points in a plane that are at a fixed distance from a given point.

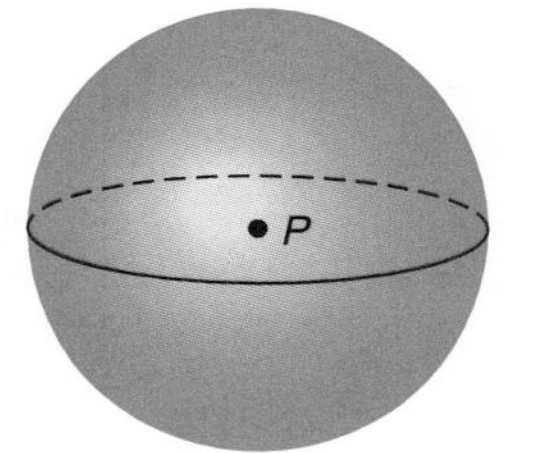

FIGURE 6.65

Each of the preceding examples includes the phrase "in a plane." If that phrase is omitted, the locus is found "in space." For instance, the locus of points that are at a fixed distance from a given point is actually a *sphere* (the three-dimensional object in Figure 6.65); the sphere has the fixed point as center, and the fixed distance determines the length of the radius. Unless otherwise stated, we will consider the locus to be restricted to a plane.

EXAMPLE 4

Describe the locus of points *in space* that are equidistant from two parallel planes (P and Q).

Solution

The locus is the plane parallel to each of the given planes and midway between them. (See Figure 6.66.)

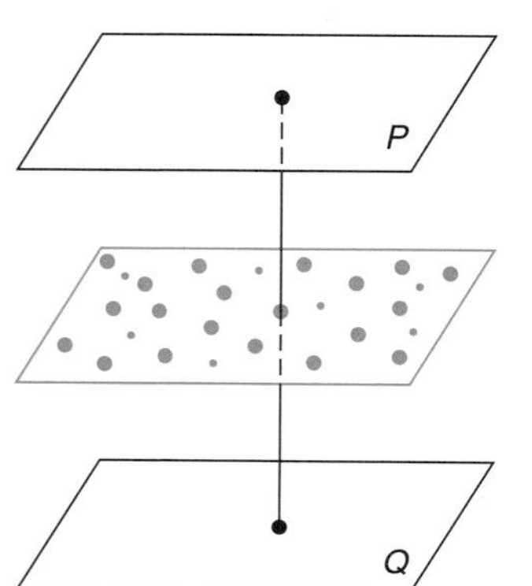

FIGURE 6.66

There are two very important theorems involving the locus concept. The results of these two theorems will be used in Section 6.6. When we verify the locus theorems, we *must* establish two results:

1. If a point is in the locus, then it satisfies the condition.
2. If a point satisfies the condition, then it is a point of the locus.

> **THEOREM 6.5.1:** The locus of points in a plane equidistant from the sides of an angle is the angle bisector.

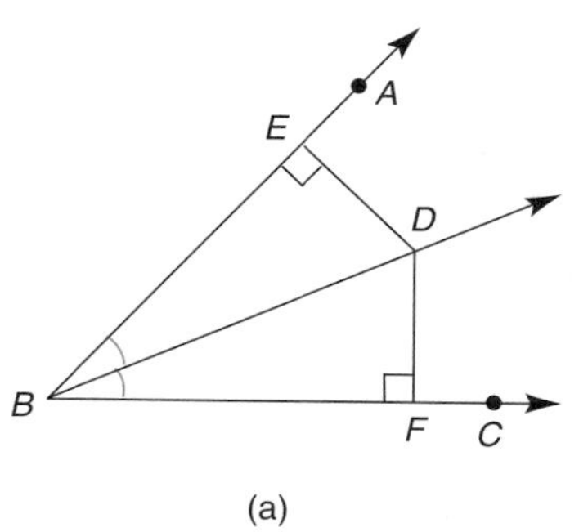

(a)

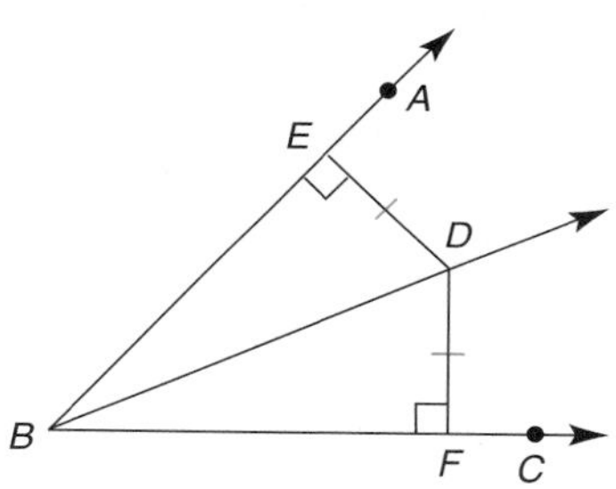

FIGURE 6.67

Proof (Notice that *both* parts i and ii are necessary):

i) If a point is on the angle bisector, then it is equidistant from the sides of the angle.

Given: $\overrightarrow{BD}$ bisects $\angle ABC$
$\overline{DE} \perp \overrightarrow{BA}$ and $\overline{DF} \perp \overrightarrow{BC}$

Prove: $\overline{DE} \cong \overline{DF}$

Proof: In Figure 6.67(a), $\overrightarrow{BD}$ bisects $\angle ABC$; thus $\angle ABD \cong \angle CBD$. $\overline{DE} \perp \overrightarrow{BA}$ and $\overline{DF} \perp \overrightarrow{BC}$, so $\angle DEB$ and $\angle DFB$ are $\cong$ right $\angle$s. $\overline{BD} \cong \overline{BD}$. By AAS, $\triangle DEB \cong \triangle DFB$. Then $\overline{DE} \cong \overline{DF}$ by CPCTC.

ii) If a point is equidistant from the sides of an angle, then it is on the angle bisector.

Given: $\angle ABC$ so that $\overline{DE} \perp \overrightarrow{BA}$ and $\overline{DF} \perp \overrightarrow{BC}$
$\overline{DE} \cong \overline{DF}$

Prove: $\overrightarrow{BD}$ bisects $\angle ABC$; that is, D is on the bisector of $\angle ABC$

Proof: In Figure 6.67(b), $\overline{DE} \perp \overrightarrow{BA}$ and $\overline{DF} \perp \overrightarrow{BC}$, so $\angle DEB$ and $\angle DFB$ are right triangles. $\overline{DE} \cong \overline{DF}$ by hypothesis. Also, $\overline{BD} \cong \overline{BD}$. $\triangle DEB \cong \triangle DFB$ by HL. Then $\angle ABD \cong \angle CBD$ by CPCTC, so $\overrightarrow{BD}$ bisects $\angle ABC$ by definition.

In both construction problems and locus problems, we can verify results. In locus problems, we must remember to demonstrate two relationships!

A second important theorem about a locus of points follows.

THEOREM 6.5.2: The locus of points in a plane that are equidistant from the endpoints of a line segment is the perpendicular bisector of that line segment.

Proof

i) If a point is equidistant from the endpoints of a line segment, then it lies on the perpendicular bisector of the line segment.

Given: $\overline{AB}$ and point X not on $\overline{AB}$, so that $AX = BX$ [See Figure 6.68(a).]

Prove: X lies on the perpendicular bisector of $\overline{AB}$

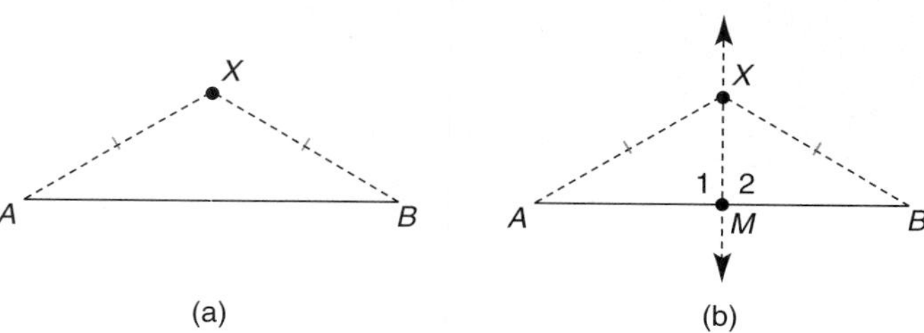

FIGURE 6.68

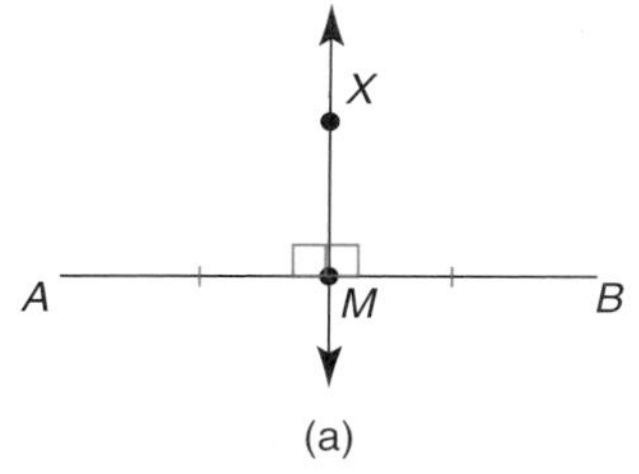

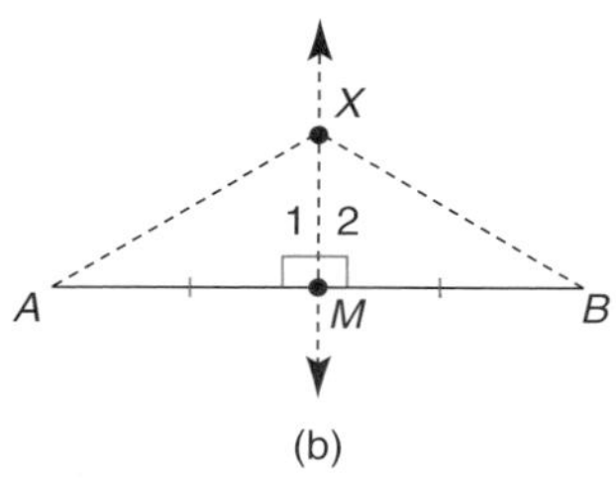

FIGURE 6.69

Proof: Let M represent the midpoint of $\overline{AB}$. [See Figure 6.68(b) on page 275.] Then $\overline{AM} \cong \overline{MB}$. Since $AX = BX$, we know that $\overline{AX} \cong \overline{BX}$. Since $\overline{XM} \cong \overline{XM}$, $\triangle AMX \cong \triangle BMX$ by SSS. By CPCTC, $\angle$s 1 and 2 are congruent and $\overline{MX} \perp \overline{AB}$. By definition, $\overline{MX}$ is the perpendicular bisector of $\overline{AB}$, so X lies on the perpendicular bisector of $\overline{AB}$.

ii) If a point is on the perpendicular bisector of a line segment, then the point is equidistant from the endpoints of the line segment.

Given: Point X lies on $\overline{MX}$, the perpendicular bisector of $\overline{AB}$ [See Figure 6.69(a).]

Prove: X is equidistant from A and B ($AX = XB$) [See Figure 6.69(b).]

Proof: X is on the perpendicular bisector of $\overline{AB}$, so $\angle$s 1 and 2 are congruent right angles and $\overline{AM} \cong \overline{MB}$. With $\overline{XM} \cong \overline{XM}$, $\triangle$s AMX and BMX are congruent by SAS; in turn, $\overline{XA} \cong \overline{XB}$ by CPCTC.

Then $XA = XB$ and X is equidistant from A and B.

We now return to further considerations of locus in a plane.

Suppose that a given line segment in a fixed location is to be used as the hypotenuse of a right triangle. How might you locate possible positions for the vertex of the right angle? One method might be to draw 30° and 60° angles at the endpoints so that the remaining angle formed must measure 90° [see Figure 6.70(a)]. This is only one possibility, but it actually provides four permissible points, due to symmetry; these points are indicated in Figure 6.70(b). This problem is completed in Example 5.

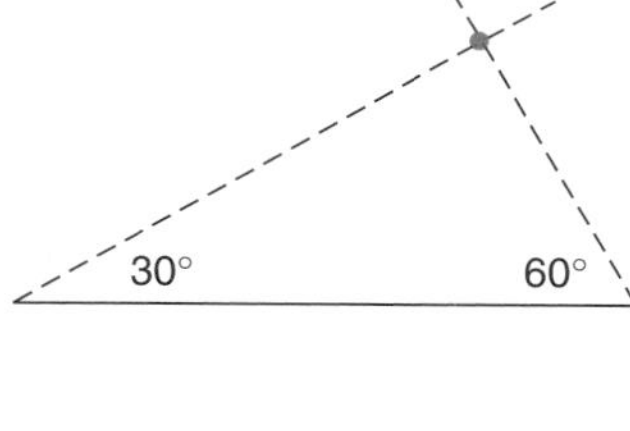

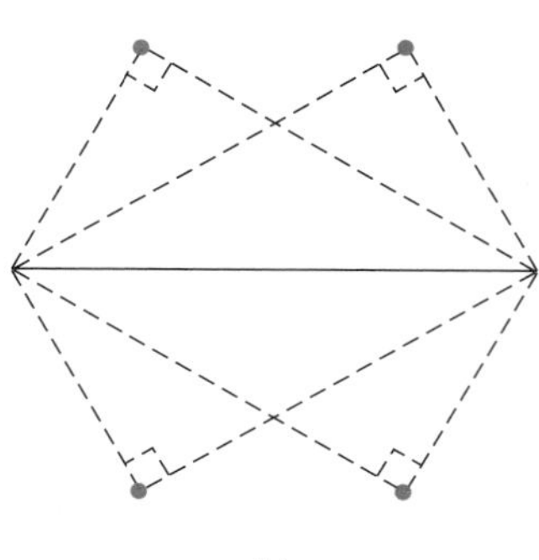

(b)

FIGURE 6.70

EXAMPLE 5

Find the locus of the vertex of the right angle of a right triangle if the hypotenuse is $\overline{AB}$ in Figure 6.71(a).

Solution

Rather than use the "hit or miss" approach for locating the possible vertices (as suggested just before this example), recall that an angle inscribed in a semicircle is a right angle.

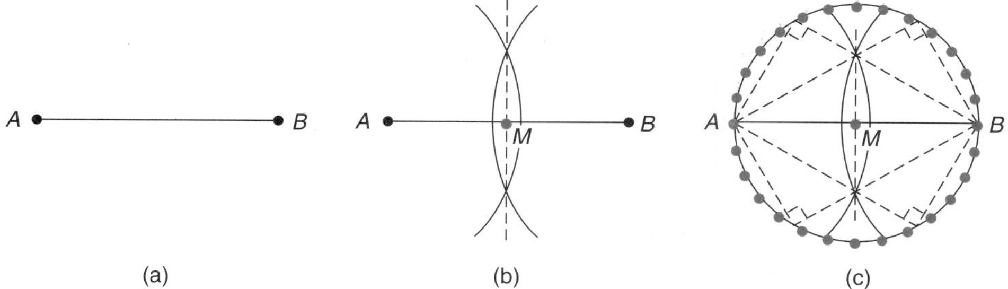

FIGURE 6.71

Thus, we construct the circle whose center is the midpoint M of the hypotenuse and whose radius equals one-half the length of the hypotenuse. First, the midpoint M of the hypotenuse $\overline{AB}$ is located. [See Figure 6.71(b)].

With the radius of the circle equal to one-half the hypotenuse, the circle with center M is drawn in Figure 6.71(c).

The locus of the vertex of a right angle whose hypotenuse is given is the circle whose center is at the midpoint of the given segment and whose radius is equal in length to half the length of the given segment. Every point (except A and B) on $\odot M$ is the vertex of a right triangle with hypotenuse $\overline{AB}$; see Theorem 6.1.9.

In Example 5, the construction involves locating the midpoint M of $\overline{AB}$ and this is found by the method for the perpendicular bisector. The compass is then opened to a radius whose length is MA or MB, and the circle is drawn. When a construction is performed, it falls into one of two categories:

1. A basic construction method
2. A construction problem that may require several steps and may involve several basic construction methods (like Example 5)

The next example illustrates another category 2 problem.

Recall that the diagonals of a rhombus are perpendicular and also bisect each other. With this information, we can locate the vertices of the rhombus whose diagonals (lengths) are known.

EXAMPLE 6

Construct rhombus $ABCD$ given its diagonals $\overline{AC}$ and $\overline{BD}$. (See Figure 6.72 on page 278.)

Solution

To begin, we construct the perpendicular bisector of $\overline{AC}$; we know that the remaining vertices B and D must lie on this line. See Figure 6.72(a) on the next page, in which M is the midpoint of $\overline{AC}$.

To locate the midpoint of $\overline{BD}$, we construct its perpendicular bisector as well. See Figure 6.72(b).

Using an arc length equal to one-half the length of $\overline{BD}$ [like MB in Figure 6.72(c)], we mark off this distance both above and below $\overline{AC}$ on the perpendicular bisector determined in Figure 6.72(a). See Figure 6.72(c).

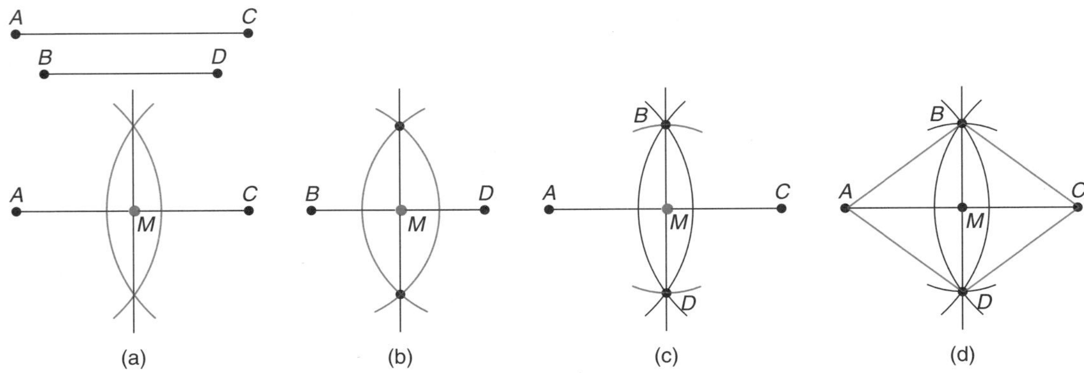

FIGURE 6.72

Using the marked arcs to locate (determine) points B and D, we join A to B, B to C, C to D, and D to A. The completed rhombus $ABCD$ is shown in Figure 6.72(d).

6.5 Exercises

In Exercises 1 to 6, use the drawing.

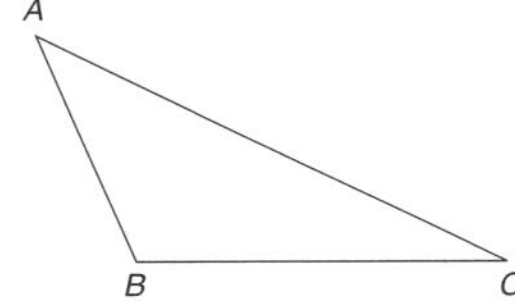

Exercises 1–6

1. *Given:* Obtuse $\triangle ABC$
   Construct: The bisector of $\angle ABC$
2. *Given:* Obtuse $\triangle ABC$
   Construct: The bisector of $\angle BAC$
3. *Given:* Obtuse $\triangle ABC$
   Construct: The perpendicular bisector of $\overline{AB}$
4. *Given:* Obtuse $\triangle ABC$
   Construct: The perpendicular bisector of $\overline{AC}$
5. *Given:* Obtuse $\triangle ABC$
   Construct: The altitude from A to $\overline{BC}$
   (**HINT:** Extend $\overline{BC}$.)
6. *Given:* Obtuse $\triangle ABC$
   Construct: The altitude from B to $\overline{AC}$

Use the figure shown for Exercises 7 and 8.

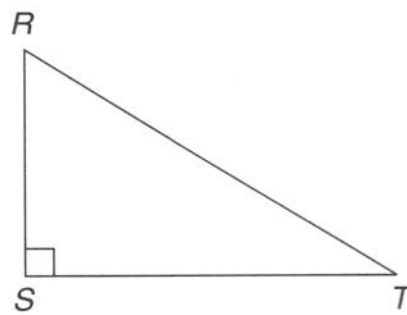

Exercises 7–8

7. *Given:* Right $\triangle RST$
   Construct: The median from S to $\overline{RT}$
8. *Given:* Right $\triangle RST$
   Construct: The median from R to $\overline{ST}$

In Exercises 9 to 20, sketch and describe each locus in the plane.

9. Find the locus of points that are at a given distance from a fixed line.
10. Find the locus of points that are equidistant from two given parallel lines.
11. Find the locus of points that are at a distance of 3 in. from a fixed point O.

12. Find the locus of points that are equidistant from two fixed points A and B.
13. Find the locus of points that are equidistant from three noncollinear points D, E, and F.
14. Find the locus of the midpoints of the radii of a circle O that has a radius of length 8 cm.
15. Find the locus of the midpoints of all chords of circle Q that are parallel to diameter $\overline{PR}$.
16. Find the locus of points in the interior of a right triangle with sides of 6 in., 8 in., and 10 in. and at a distance of 1 in. from the triangle.
17. Find the locus of points that are equidistant from two given intersecting lines.
18. ➢ Find the locus of points that are equidistant from a fixed line and a point not on that line. (**NOTE:** This figure is known as a *parabola*.)
19. Given that lines p and q intersect, find the locus of points that are at a distance of 1 cm from line p and also at a distance of 2 cm from line q.
20. Given that congruent circles O and P have radii of length 4 in. and that the line of centers has length 6 in., find the locus of points that are 1 in. from each circle.

In Exercises 21 to 28, sketch and describe the locus of points in space.

21. Find the locus of points that are at a given distance from a fixed line.
22. Find the locus of points that are equidistant from two fixed points.
23. Find the locus of points that are at a distance of 2 cm from a sphere whose radius is 5 cm.
24. Find the locus of points that are at a given distance from a given plane.
25. Find the locus of points that are the midpoints of the radii of a sphere whose center is point O and whose radius has a length of 5 m.
26. ➢ Find the locus of points that are equidistant from three noncollinear points D, E, and F.
27. In a room, find the locus of points that are equidistant from the parallel ceiling and floor, which are 8 ft apart.
28. Find the locus of points that are equidistant from all points on the surface of a sphere with center point Q.

In Exercises 29 and 30, use the method of proof of Theorem 6.5.1 to justify each construction method.

29. The perpendicular bisector method.
30. The construction of a perpendicular to a line from a point outside the line.

In Exercises 31 to 34, refer to the line segments shown.

Exercises 31–34

31. Construct an isosceles right triangle that has hypotenuse $\overline{AB}$.
32. Construct a rhombus whose sides are equal in length to AB, while one diagonal has length CD.
33. Construct an isosceles triangle in which each leg has length CD and the altitude to the base has length AB.
34. Construct an equilateral triangle in which the altitude to any side has length AB.
35. Construct the inscribed circle for obtuse $\triangle RST$.

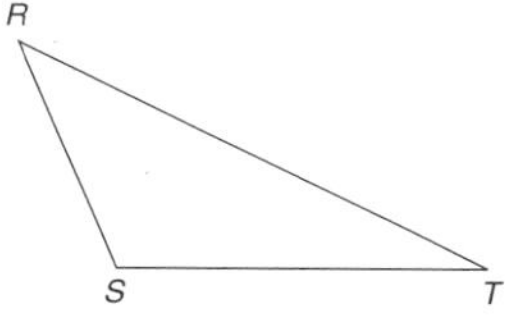

Exercises 35, 36

36. Construct the circumscribed circle for obtuse $\triangle RST$.
37. Use the following theorem to locate the center of the circle of which $\overset{\frown}{RT}$ is a part.

    Theorem: The perpendicular bisector of a chord passes through the center of a circle.

Exercise 37

38. ➢ Use the following theorem to construct the geometric mean of the numerical lengths of the segments $\overline{WX}$ and $\overline{YZ}$.

    Theorem: The length of the altitude to the hypotenuse of a right triangle is the geometric mean between the lengths of the segments of the hypotenuse.

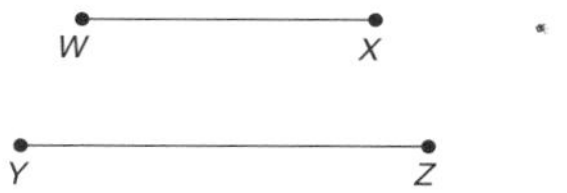

39. Use the following theorem to construct a triangle similar to the given triangle but with sides that are twice the length of those of the given triangle.

    Theorem: If the three pairs of sides for two triangles are in proportion, then those triangles are similar.

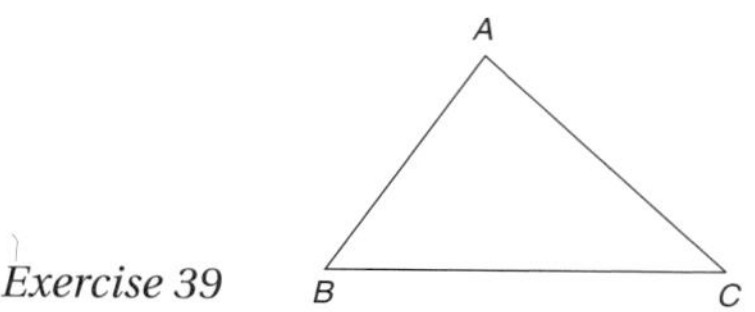

Exercise 39

40. ➢ Verify this locus theorem:

    The locus of points equidistant from two fixed points is the perpendicular bisector of the segment joining those points.

6.6 Concurrence of Lines

In this section, we consider lines that share a common point.

DEFINITION: A number of lines are **concurrent** if they have exactly one point in common.

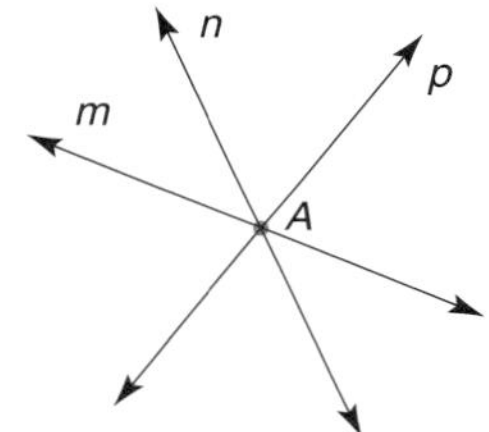

m, *n*, and *p* *are* concurrent

FIGURE 6.73

The three lines in Figure 6.73 are concurrent at point *A*. The three lines in Figure 6.74 are not concurrent even though any pair of lines (like *r* and *s*) do intersect.

Parts of lines (rays or segments) are concurrent if they are parts of concurrent lines.

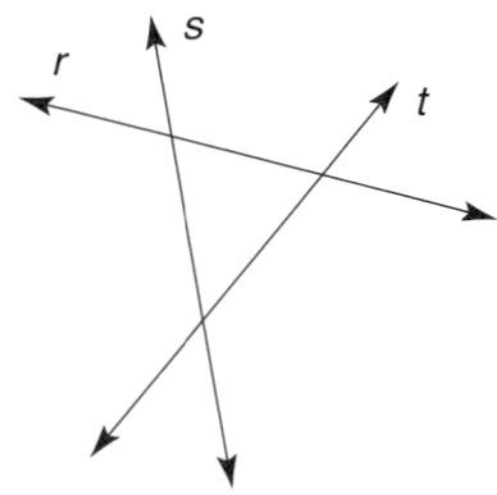

r, *s*, and *t* are *not* concurrent

FIGURE 6.74

DISCOVER!

The Geometer's Sketchpad, a computer software program, can be useful in demonstrating the concurrence of the lines described in each theorem in this section.

THEOREM 6.6.1: The three angle bisectors of the angles of a triangle are concurrent.

For the informal proofs of this section, no "Given" or "Prove" is stated. In more advanced courses, these parts of the proof are "understood."

EXAMPLE 1

Give an informal proof of Theorem 6.6.1.

Proof In Figure 6.75(a), the bisectors of $\angle BAC$ and $\angle ABC$ intersect at point *E*.

Because the angle bisector of $\angle BAC$ is the locus of points equidistant from the sides of $\angle BAC$, we know that $\overline{EM} \cong \overline{EN}$ in Figure 6.75(b). Similarly, we know that $\overline{EM} \cong \overline{EP}$ since E is on the angle bisector of $\angle ABC$.

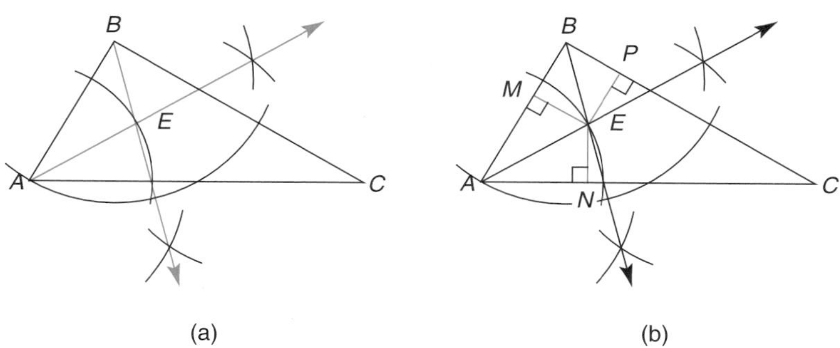

FIGURE 6.75

By the Transitive Property of Congruence, it follows that $\overline{EP} \cong \overline{EN}$.

Because the angle bisector is the locus of points equidistant from the sides of the angle, we know that E is also on the bisector of the third angle, $\angle ACB$. Thus, the angle-bisectors are concurrent.

The point E at which the angle bisectors meet in Example 1 is the **incenter** of the triangle. As the following example shows, the term "incenter" is well-deserved because this point is the *center* of the *in*scribed circle of the triangle.

EXAMPLE 2

Complete the construction of the inscribed circle for $\triangle ABC$ in Figure 6.75(b).

Solution

Having found the incenter E, we need the length of the radius. Because $\overline{EN} \perp \overline{AC}$ (as shown in Figure 6.76), the length of $\overline{EN}$ is the desired radius; thus, the circle is completed.

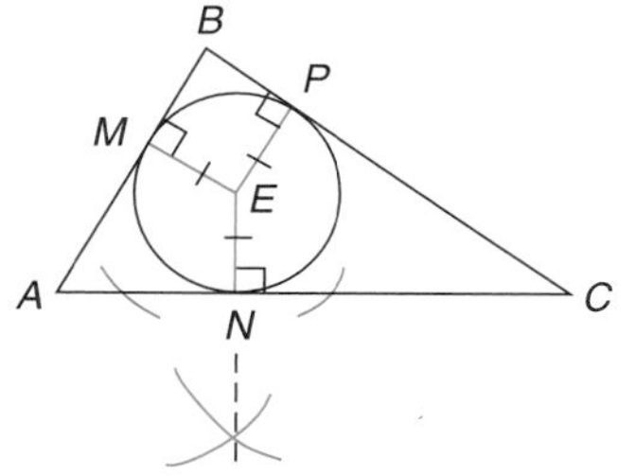

FIGURE 6.76

NOTE: The sides of the triangle are tangents for the inscribed circle, which is called the **incircle** of the triangle.

It is also possible to circumscribe a circle about a given triangle. The construction depends on the following theorem, the proof of which is sketched in Example 3.

> **THEOREM 6.6.2:** The three perpendicular bisectors of the sides of a triangle are concurrent.

EXAMPLE 3

Give an informal proof of Theorem 6.6.2.

Proof

Let $\overline{FS}$ and $\overline{FR}$ name the perpendicular bisectors of sides $\overline{BC}$ and $\overline{AC}$, respectively. See Figure 6.77(a). Using Theorem 6.5.2, the point of concurrency F is equidistant from the endpoints of $\overline{BC}$; thus, $\overline{BF} \cong \overline{FC}$. In the same manner, $\overline{AF} \cong \overline{FC}$. By the Transitive Property, it follows that $\overline{AF} \cong \overline{BF}$; again citing Theorem 6.5.2, F must be on the perpendicular bisector of $\overline{AB}$, since this point is equidistant from the endpoints of $\overline{AB}$.

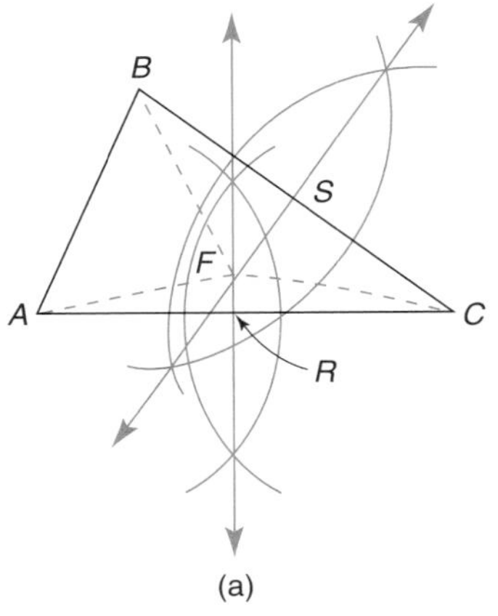

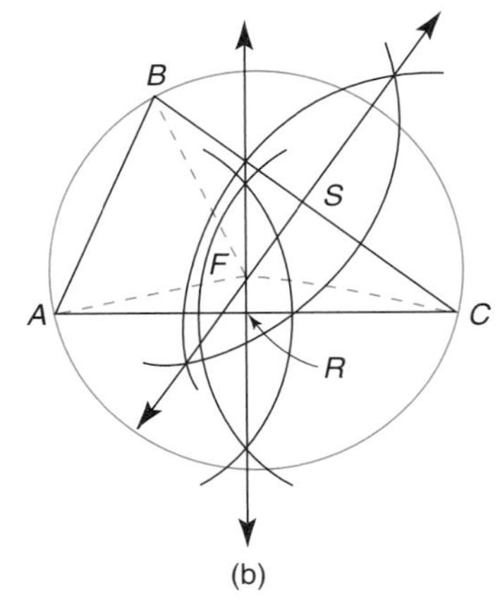

FIGURE 6.77

EXAMPLE 4

Complete the construction of the circumscribed circle for $\triangle ABC$ that was given in Figure 6.75(a).

Solution

We have already identified the center of the circle as point F. To complete the construction, we use F as the center and a radius of length equal to the distance from F to any one of the vertices A, B, or C. The circumscribed circle is shown in Figure 6.77(b).

NOTE: The sides of the triangle are chords of the circle, which is called the **circumcircle** of the triangle.

The point at which the perpendicular bisectors of the sides of a triangle meet is the **circumcenter** of the triangle. This is easily remembered as the *center* of the *circum*scribed circle.

The incenter and the circumcenter of a triangle are generally distinct points. However, it is possible for the two centers to coincide in a special type of triangle. Although the incenter of a triangle always lies in the interior of the triangle, the circumcenter of an obtuse triangle will lie in the exterior of the triangle. See Figure 6.78. The circumcenter of a right triangle is the midpoint of the hypotenuse.

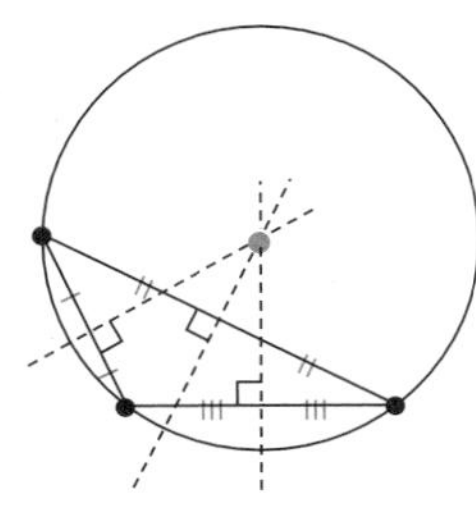

FIGURE 6.78

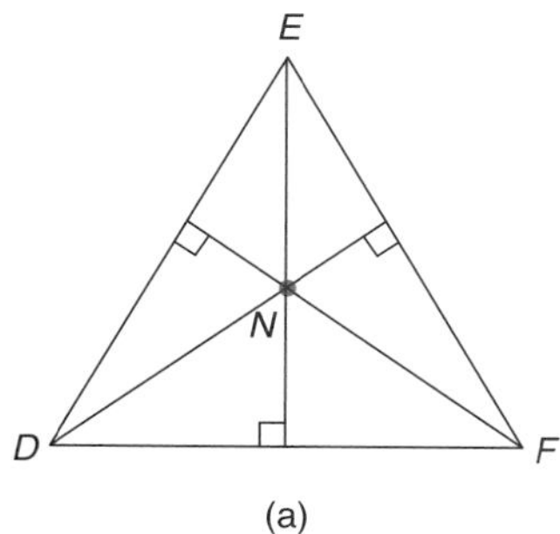

(a)

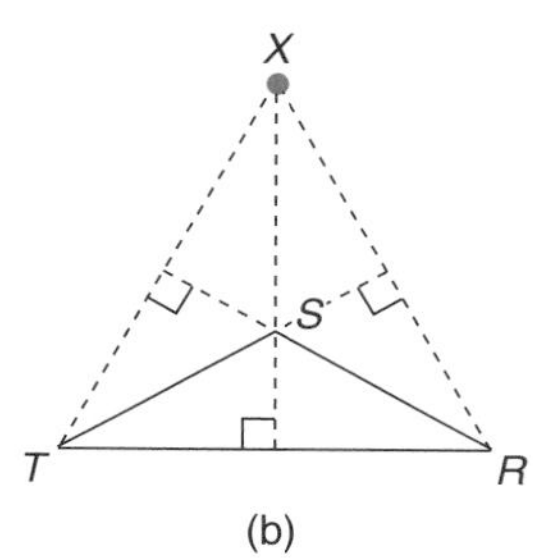

(b)

FIGURE 6.79

To complete the discussion of concurrence, we include theorems involving the altitudes and medians of a triangle.

> **THEOREM 6.6.3:** The three altitudes of a triangle are concurrent.

The point of concurrence for the three altitudes of a triangle is the **orthocenter** of the triangle. In Figure 6.79(a), point N is the orthocenter of $\triangle DEF$. For the obtuse triangle RST shown in Figure 6.79(b), we see that orthocenter X lies in the exterior.

Rather than prove Theorem 6.6.3, we sketch a part of that proof. Consider $\triangle MNP$, shown with its altitudes in Figure 6.80(a). To prove that the altitudes are concurrent requires

1. that we draw auxiliary lines through N parallel to $\overline{MP}$, through M parallel to $\overline{NP}$, and through P parallel to $\overline{NM}$. [See Figure 6.80(b).]
2. that we show that the altitudes of $\triangle MNP$ are perpendicular bisectors of the sides of the newly formed $\triangle RST$; thus, altitudes $\overline{PX}$, $\overline{MY}$, and $\overline{NZ}$ are concurrent.

Sketch of proof that $\overline{PX}$ is the $\perp$ bisector of $\overline{RS}$:

Proof

Because $\overline{PX}$ is an altitude of $\triangle MNP$, $\overline{PX} \perp \overline{MN}$. But $\overline{RS} \parallel \overline{MN}$ by construction. Because a line perpendicular to one of two parallel lines must be perpendicular to the other, we have $\overline{PX} \perp \overline{RS}$. Now we need to show that $\overline{PX}$ bisects $\overline{RS}$. Due to construction, $\overline{MR} \parallel \overline{NP}$ and $\overline{RP} \parallel \overline{MN}$, so $MRPN$ is a parallelogram. Then $\overline{MN} \cong \overline{RP}$ since the opposite sides of a parallelogram are congruent. By construction, $MPSN$ is also a parallelogram and $\overline{MN} \cong \overline{PS}$. Then $\overline{RP} \cong \overline{PS}$ since $\overline{MN}$ is congruent to each segment. Thus, $\overline{RS}$ is bisected at point P and $\overline{PX}$ is the $\perp$ bisector of $\overline{RS}$.

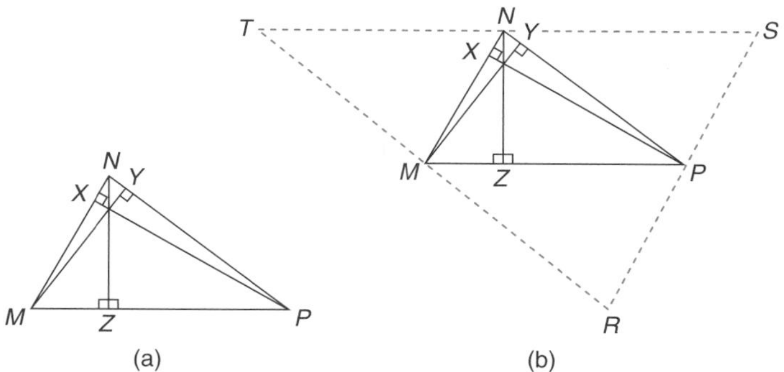

FIGURE 6.80

In Figure 6.80(b), similar arguments (leading to one long proof) could be used to show that $\overline{NZ}$ is the $\perp$ bisector of $\overline{TS}$ and that $\overline{MY}$ is the $\perp$ bisector of $\overline{TR}$. Since the concurrent perpendicular bisectors of the sides of $\triangle RST$ are also the altitudes of $\triangle MNP$, these altitudes must be concurrent.

The intersection of any two altitudes determines the orthocenter of a triangle. We use this fact in Example 5. If the third altitude were constructed, it would locate the same point of intersection.

EXAMPLE 5

Construct the orthocenter of $\triangle ABC$ in Figure 6.81.

Solution First construct the altitude from A to $\overline{BC}$; here, we draw an arc from A to intersect $\overline{BC}$ at X and Y. Now draw equal arcs from X and Y to intersect at Z. $\overline{AH}$ is the desired altitude. Repeat the process to construct altitude $\overline{CJ}$ from vertex C to side $\overline{AB}$. The point of intersection O is the orthocenter of $\triangle ABC$.

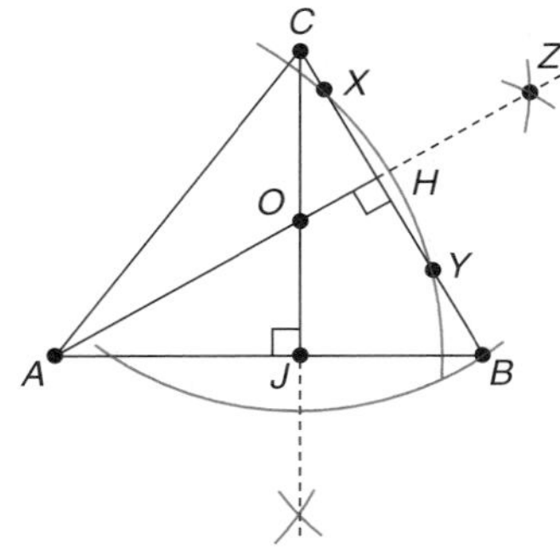

FIGURE 6.81

Recall that a median of a triangle joins a vertex to the midpoint of the opposite side of the triangle. Through construction, we can show that the three medians of a triangle are concurrent. We delay discussion of the proof of the following theorem until Chapter 9.

THEOREM 6.6.4: The three medians of a triangle are concurrent at a point that is two-thirds the distance from any vertex to the midpoint of the opposite side.

The point of concurrence for the three medians is the **centroid** of the triangle. In Figure 6.82, point C is the centroid of $\triangle RST$. According to Theorem 6.6.4, $RC = \frac{2}{3}(RM)$, $SC = \frac{2}{3}(SN)$, and $TC = \frac{2}{3}(TP)$.

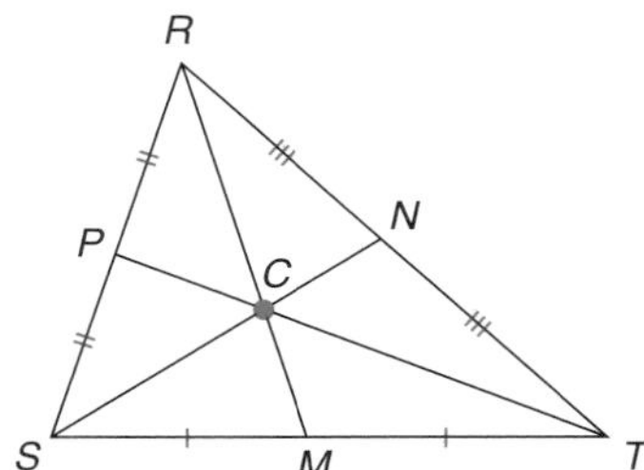

FIGURE 6.82

EXAMPLE 6

Suppose that the medians of $\triangle RST$ have the lengths $RM = 12$, $SN = 15$, and $TP = 18$. If the centroid of $\triangle RST$ is point C, find the length of:

a) RC **(b)** CM **(c)** SC

Solution

a) $RC = \frac{2}{3}(RM)$, so $RC = \frac{2}{3}(12) = 8$.
b) $CM = RM - RC$, so $CM = 12 - 8 = 4$.
c) $SC = \frac{2}{3}(SN)$, so $SC = \frac{2}{3}(15) = 10$.

EXAMPLE 7

Given: In Figure 6.83(a), isosceles $\triangle RST$ with $RS = RT = 15$, while $ST = 18$; medians $\overline{RZ}$, $\overline{TX}$, and $\overline{SY}$ meet at centroid Q

Find: RQ

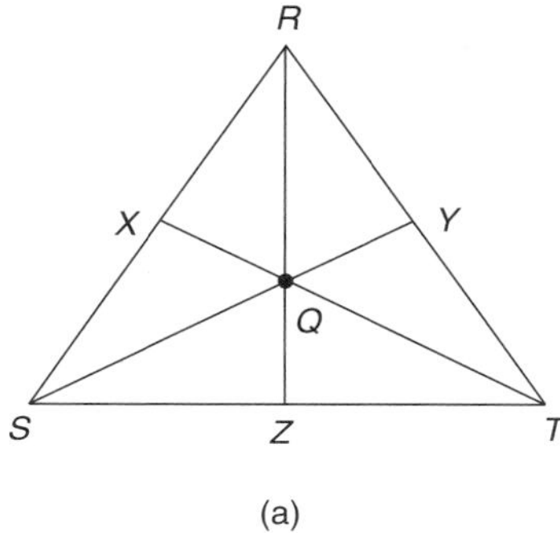

(a)

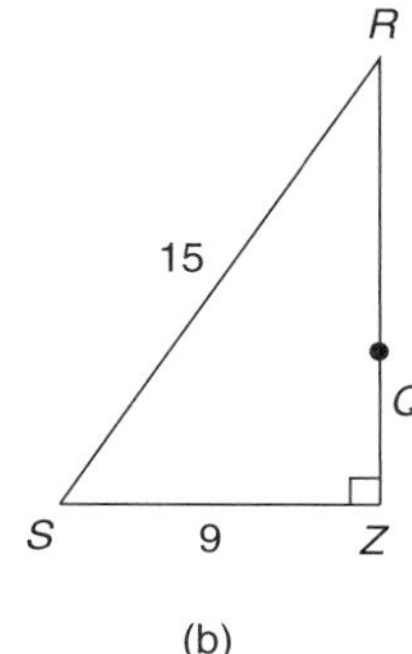

(b)

FIGURE 6.83

Solution

Median $\overline{RZ}$ separates $\triangle RST$ into two congruent right triangles, $\triangle RZS$ and $\triangle RZT$; this follows from SSS. With Z the midpoint of $\overline{ST}$, $SZ = 9$.

Using the Pythagorean Theorem with $\triangle RZS$ in Figure 6.83(b), we have

$$\begin{aligned}(RS)^2 &= (RZ)^2 + (SZ)^2\\ 15^2 &= (RZ)^2 + 9^2\\ 225 &= (RZ)^2 + 81\\ (RZ)^2 &= 144\\ RZ &= 12\end{aligned}$$

By Theorem 6.6.4,

$$RQ = \frac{2}{3}(RZ) = \frac{2}{3}(12) = 8$$

It is *possible* for the angle bisectors of certain quadrilaterals to be concurrent. Likewise, the perpendicular bisectors of the sides of a quadrilateral *can be* concurrent. Of course, there are four angle bisectors and four perpendicular bisectors of sides to consider. In Example 8, we consider this situation.

EXAMPLE 8

Use intuition and Figure 6.84 to decide which of the following are concurrent.

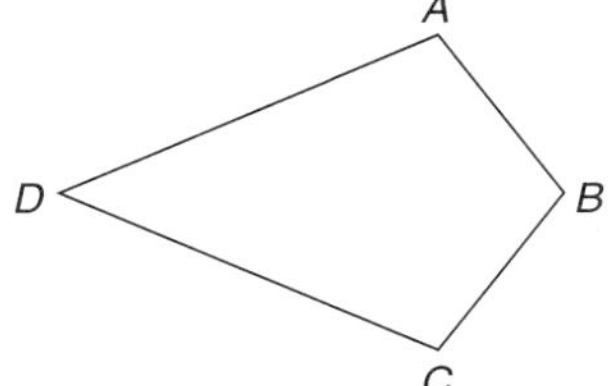

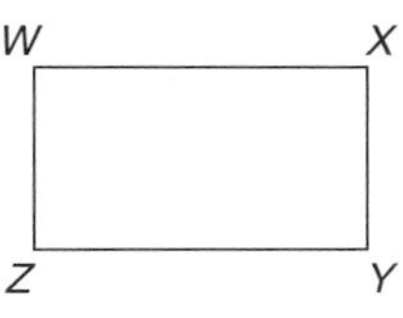

FIGURE 6.84

a) The angle bisectors of a kite

b) The perpendicular bisectors of the sides of a kite

c) The angle bisectors of a rectangle

d) The perpendicular bisectors of the sides of a rectangle

Solution

a) The angle bisectors of the kite are concurrent.
b) The $\perp$ bisectors of the sides of the kite are not concurrent (unless $\angle A$ and $\angle C$ are both right angles).
c) The angle bisectors of the rectangle are not concurrent (unless the rectangle is a square).
d) The $\perp$ bisectors of the sides of the rectangle are concurrent (circumcenter is also the point of intersection of diagonals).

NOTE: The student should make drawings to verify these results.

The centroid of a triangular region is sometimes called its *center of mass* or *center of gravity.* This is because the region of uniform thickness "balances" upon the point known as its centroid. Consider the following activity.

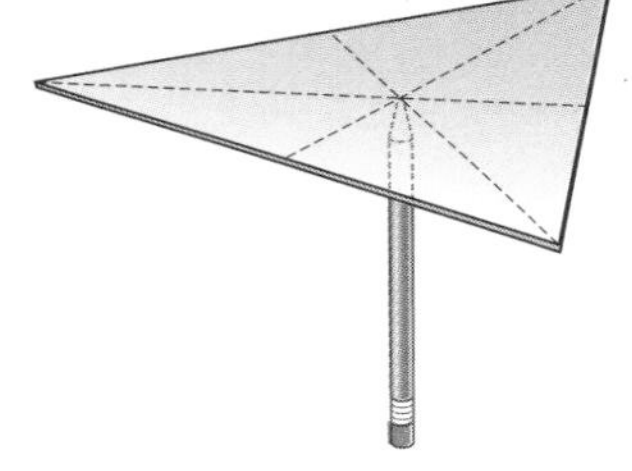

DISCOVER!

Take a piece of cardboard or heavy poster paper. Draw a triangle on the paper and cut out the triangular shape. Now use a ruler to mark the midpoints of each side and draw the medians to locate the centroid. Place the triangle on the point of a pen or pencil at the centroid and see how well you can balance the triangular region.

6.6 Exercises

1. Which lines (or line segments or rays) must be drawn (or constructed) in a triangle to locate its
   a) incenter?
   b) circumcenter?
   c) orthocenter?
   d) centroid?
2. It is really necessary to construct all three angle bisectors of the angles of a triangle to locate its incenter?
3. Is it really necessary to construct all three perpendicular bisectors of the sides of a triangle to locate its circumcenter?
4. To locate the orthocenter, is it necessary to construct all three altitudes of a right triangle?
5. For what type of triangle are the angle bisectors, the medians, the perpendicular bisectors of sides, and the altitudes all the same?
6. What point on a right triangle is the orthocenter of the right triangle?
7. What point on a right triangle is the circumcenter of the right triangle?
8. Must the centroid of an isosceles triangle lie on the altitude to the base?
9. Draw a triangle and, by construction, find its incenter.
10. Draw an acute triangle and, by construction, find its circumcenter.
11. Draw an obtuse triangle and, by construction, find its circumcenter.

12. Draw an acute triangle and, by construction, find its orthocenter.

13. Draw an obtuse triangle and, by construction, find its orthocenter. (**HINT:** You will have to extend the sides opposite the acute angles.)

14. Draw an acute triangle and, by construction, find the centroid of the triangle. (**HINT:** Begin by constructing the perpendicular bisectors of the sides.)

15. Draw an obtuse triangle and, by construction, find the centroid of the triangle. (**HINT:** Begin by constructing the perpendicular bisectors of the sides.)

16. Is the incenter always located in the interior of the triangle?

17. Is the circumcenter always located in the interior of the triangle?

18. Find the length of the radius of the inscribed circle for a right triangle whose legs measure 6 and 8.

19. Find the distance from the circumcenter to each vertex of an equilateral triangle whose sides measure 10.

20. A triangle has angles measuring 30°, 30°, and 120°. If the congruent sides measure 6 units each, find the length of the radius of the circumscribed circle.

21. *Given:* Isosceles $\triangle RST$
$RS = RT = 17$ and $ST = 16$
medians $\overline{RZ}$, $\overline{TX}$, and $\overline{SY}$ meet at centroid Q
Find: RQ and SQ

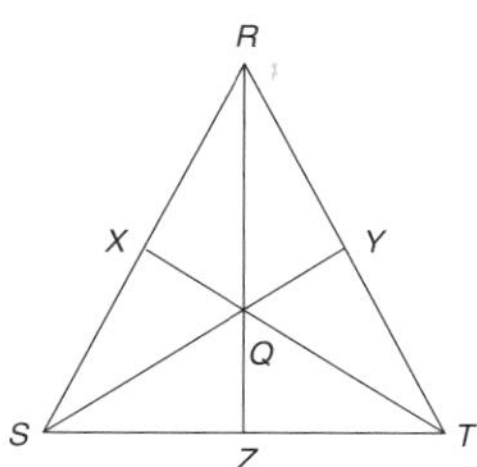

Exercises 21, 22

22. *Given:* Isosceles $\triangle RST$
$RS = RT = 10$ and $ST = 16$
medians $\overline{RZ}$, $\overline{TX}$, and $\overline{SY}$ meet at Q
Find: RQ and QT

23. In $\triangle MNP$, medians $\overline{MB}$, $\overline{NA}$, and $\overline{PC}$ intersect at centroid Q. See the figure at the top of the next column.

a) If $MQ = 8$, find QB.
b) If $QC = 3$, find PQ.
c) If $AQ = 3.5$, find AN.

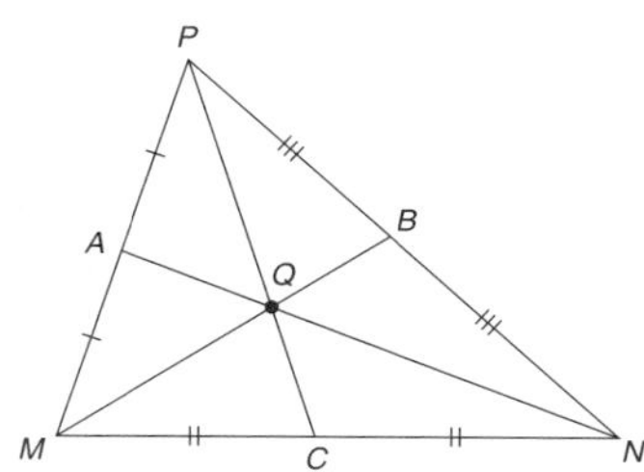

Exercises 23, 24

24. In $\triangle MNP$, medians $\overline{MB}$, $\overline{NA}$, and $\overline{PC}$ intersect at centroid Q.

a) Find QB if $MQ = 8.2$.
b) Find PQ if $QC = \frac{7}{2}$.
c) Find AN if $AQ = 4.6$.

25. Draw a triangle. Construct its inscribed circle.

26. Draw a triangle. Construct its circumscribed circle.

27. For what type of triangle will the incenter and circumcenter be the same?

28. Does a rectangle have (a) an incenter? (b) a circumcenter?

29. Does a square have (a) an incenter? (b) a circumcenter?

30. Does a regular pentagon have (a) an incenter? (b) a circumcenter?

31. Does a rhombus have (a) an incenter? (b) a circumcenter?

32. Does an isosceles trapezoid have (a) an incenter? (b) a circumcenter?

33. A distributing company plans an Illinois location that would be the same distance from each of its principal delivery sites at Chicago, St. Louis, and Indianapolis. Use a construction method to locate the approximate position of the distributing company.
(**NOTE:** Trace the outline of the two states on your own paper.)

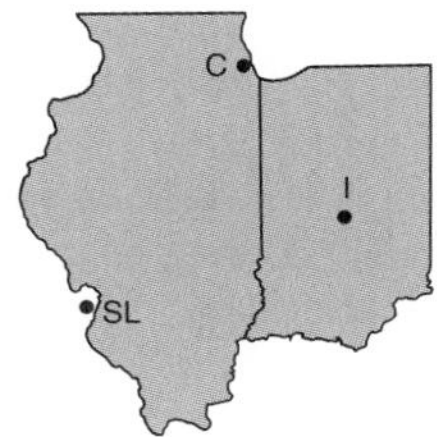

A Look Beyond: The Value of π

In geometry, any two figures that have the same shape are described as similar. Because all circles have the same shape, we say that all circles are similar to each other. Just as a proportionality exists among the corresponding sides of similar triangles, we also assume a proportionality among the circumferences (distances around) and diameters (distances across) of circles. By representing the circumferences of the circles in Figure 6.85 by C_1, C_2, and C_3 and their corresponding lengths of diameters by d_1, d_2, and d_3, we are claiming that

$$\frac{C_1}{d_1} = \frac{C_2}{d_2} = \frac{C_3}{d_3} = k$$

for some constant of proportionality k.

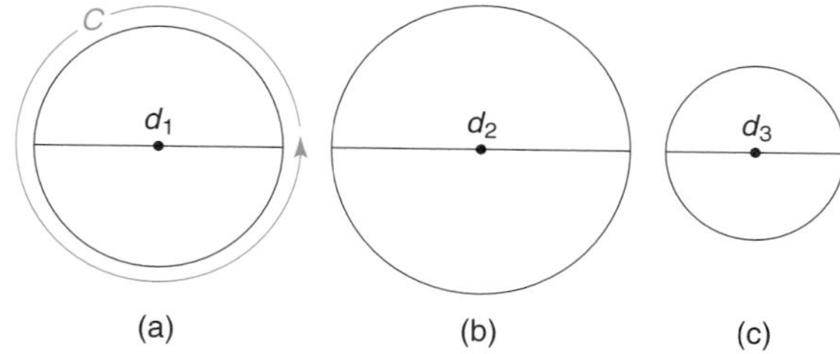

FIGURE 6.85

We denote the constant k described above by the Greek letter π. Thus $\pi = \frac{C}{d}$ in any circle. It follows that $C = \pi d$ or $C = 2\pi r$ (since $d = 2r$ in any circle). In applying these formulas for the circumference of a circle, we often wish to leave π in the answer so that the result is exact. When an approximation for the circumference (and later for the area) of a circle is needed, several common substitutions are used for π. Among these are $\pi \approx \frac{22}{7}$ and $\pi \approx 3.14$. A calculator may display the value $\pi \approx 3.1415926535$.

Because π is needed in many applications involving the circumference or area of a circle, its approximation is often necessary; but finding an accurate approximation of π was not quickly or easily done. While the formula for circumference can be expressed as $C = 2\pi r$, the formula for the area of the circle is $A = \pi r^2$. This and other area formulas will be given more attention in Chapter 7.

Several references to the value of π are made in literature. One of the earliest comes from the Bible; the passage from I Kings, verse 23 describes the distance around a vat as three times the distance across the vat (which suggests that π equals 3, a very rough approximation). Perhaps no greater accuracy was needed in applications at that time.

In the content of the Rhind papyrus (a document over 3000 years old), the Egyptian scribe Ahmes gives the formula for the area of a circle as $\left(d - \frac{1}{9}d\right)^2$. To determine the Egyptian approximation of π, we need to expand this expression as follows:

$$\left(d - \frac{1}{9}d\right)^2 = \left(\frac{8}{9}d\right)^2 = \left(\frac{8}{9} \cdot 2r\right)^2 = \left(\frac{16}{9}r\right)^2 = \frac{256}{81}r^2$$

In the formula for the area of the circle, the value of π is the multiplier (coefficient) of r^2. Because this coefficient is $\frac{256}{81}$ (which has the decimal equivalent of 3.1604), the Egyptians had a better approximation of π than was given in the book of I Kings.

Archimedes, the brilliant Greek geometer, knew that the formula for the area of a circle was $A = \frac{1}{2}Cr$ (with C the circumference and r the length of radius). His formula was equivalent to the one which we use today and is developed as follows:

$$A = \frac{1}{2}Cr = \frac{1}{2}(2\pi r)r = \pi r^2$$

The second proposition of Archimedes' work *Measure of the Circle* develops a relationship between the area of a circle and the area of the square in which it is inscribed. (See Figure 6.86.) Specifically, Archimedes claimed that the ratio of the area of the circle to that of the square was 11:14. This leads to the following set of equations and an approximation of the value of π:

$$\frac{\pi r^2}{(2r)^2} \approx \frac{11}{14}$$

$$\frac{\pi r^2}{4r^2} \approx \frac{11}{14}$$

$$\frac{\pi}{4} \approx \frac{11}{14}$$

$$\pi \approx 4 \cdot \frac{11}{14} \approx \frac{22}{7}$$

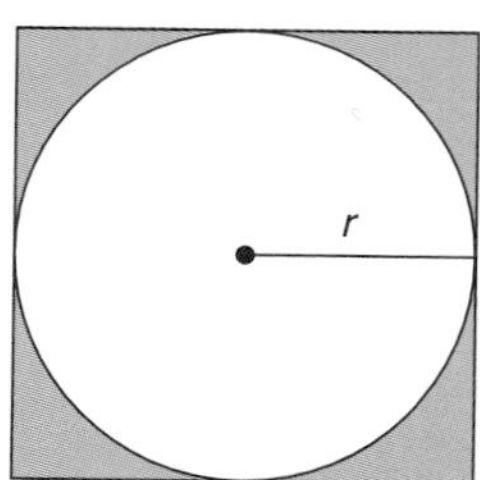

FIGURE 6.86

Archimedes later improved his approximation of π by showing that

$$3\frac{10}{71} < \pi < 3\frac{1}{7}$$

Today's calculators provide excellent approximations for the irrational number π. We should recall, however, that π is an irrational number which can be expressed exactly only by the unique symbol π.

Summary

A Look Back at Chapter 6

One goal in this chapter has been to classify angles inside, on, and outside the circle. Formulas for finding the measures of these angles were developed. Line and line segments related to a circle were defined, and some ways of finding the measures of these segments were described. Theorems involving inequalities in a circle were proved. Using the concept of locus, we justified several of the basic constructions as well as the concurrence of lines.

A Look Ahead to Chapter 7

Our goal in the next chapter is to deal with the areas of triangles, certain quadrilaterals, and regular polygons. The area of a circle and the area of a sector of a circle will be discussed. Special right triangles will play an important role in determining the areas of these plane figures.

Important Terms and Concepts of Chapter 6

6.1 Circle, Congruent Circles, Concentric Circles
Center of the Circle
Radius, Diameter, Chord
Semicircle
Arc, Major Arc, Minor Arc, Intercepted Arc, Congruent Arcs
Central Angle, Inscribed Angle

6.2 Tangent, Point of Tangency, Secant
Polygon Inscribed in a Circle, Circumscribed Circle
Polygon Circumscribed About a Circle, Inscribed Circle
Interior and Exterior of a Circle

6.3 Tangent Circles
Internally Tangent Circles
Externally Tangent Circles
Line of Centers
Common Tangent
Common External Tangents
Common Internal Tangents

6.4 Constructions of Tangents to a Circle
Inequalities in the Circle

6.5 Locus of Points

6.6 Concurrent Lines
Incircle, Circumcircle
Incenter, Circumcenter, Orthocenter, Centroid

A Look Beyond: *The Value of* π

Review Exercises

1. The radius of a circle is 15 mm. The length of a chord is 24 mm. Find the distance from the center of the circle to the chord.
2. Find the length of a chord that is 8 cm from the center of a circle that has a radius of 17 cm.
3. Two circles intersect and have a common chord 10 in. long. The radius of one circle is 13 in. and the centers of the circles are 16 in. apart. Find the radius of the other circle.
4. Two circles intersect and they have a common chord 12 cm long. The measure of the angles formed by the common chord and a radius of each circle to the points of intersection of the circles is 45°. Find the radius of each circle.

In Review Exercises 5 to 10, $\overrightarrow{BA}$ is tangent to the circle in the figure shown.

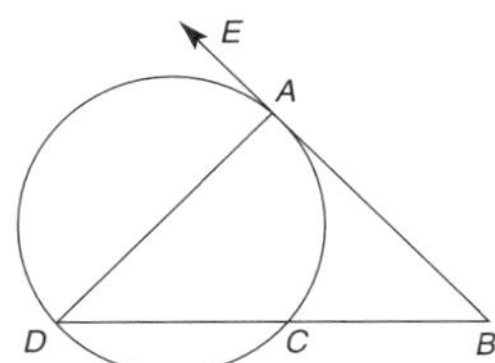

Exercises 5–10

5. $m\angle B = 25°$, $m\widehat{AD} = 140°$, $m\widehat{DC} = ?$
6. $m\widehat{DC} = 190°$, $m\widehat{AD} = 120°$, $m\angle B = ?$
7. $m\angle EAD = 70°$, $m\angle B = 30°$, $m\widehat{AC} = ?$
8. $m\angle D = 40°$, $m\widehat{DC} = 130°$, $m\angle B = ?$
9. *Given:* C is the midpoint of $\widehat{ACD}$ and $m\angle B = 40°$
   Find: $m\widehat{AD}$, $m\widehat{AC}$, $m\widehat{DC}$
10. *Given:* $m\angle B = 35°$ and $m\widehat{DC} = 70°$
    Find: $m\widehat{AD}$, $m\widehat{AC}$
11. *Given:* $\odot O$ with tangent ℓ and $m\angle 1 = 46°$
    Find: $m\angle 2$, $m\angle 3$, $m\angle 4$, $m\angle 5$

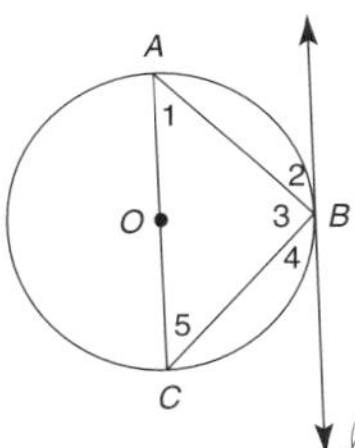

Exercises 11, 12

12. *Given:* $\odot O$ with tangent ℓ and $m\angle 5 = 40°$
    Find: $m\angle 1$, $m\angle 2$, $m\angle 3$, $m\angle 4$
13. Two circles are concentric. A chord of the larger circle is also tangent to the smaller circle. The radius of one circle is 20, and the radius of the other is 16. Find the length of the chord.
14. Two parallel chords of a circle each have length 16. The distance between them is 12. Find the radius of the circle.

In Review Exercises 15 to 22, state whether the statements are always true (A), sometimes true (S), or never true (N).

15. In a circle, congruent chords are equidistant from the center.
16. If a triangle is inscribed in a circle and one of its sides is a diameter, then the triangle is an isosceles triangle.
17. If a central angle and an inscribed angle of a circle intercept the same arc, then they are congruent.
18. A trapezoid can be inscribed in a circle.
19. If a parallelogram is inscribed in a circle, then each of its diagonals must be a diameter.
20. If two chords of a circle are not congruent, then the shorter chord is nearer the center of the circle.
21. Tangents to a circle at the endpoints of a diameter are parallel.
22. Two concentric circles have at least one point in common.
23. **a)** $m\widehat{AB} = 80°$, $m\angle AEB = 75°$, $m\widehat{CD} = ?$
    b) $m\widehat{AC} = 62°$, $m\angle DEB = 45°$, $m\widehat{BD} = ?$
    c) $m\widehat{AB} = 88°$, $m\angle P = 24°$, $m\angle CED = ?$
    d) $m\angle CED = 41°$, $m\widehat{CD} = 20°$, $m\angle P = ?$
    e) $m\angle AEB = 65°$, $m\angle P = 25°$, $m\widehat{AB} = ?$, $m\widehat{CD} = ?$
    f) $m\angle CED = 50°$, $m\widehat{AC} + m\widehat{BD} = ?$

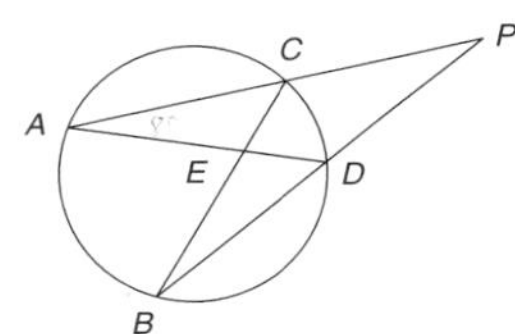

24. Given that $\overline{CF}$ is a tangent to the circle shown:
    a) $CF = 6$, $AC = 12$, $BC = ?$
    b) $AG = 3$, $BE = 10$, $BG = 4$, $DG = ?$
    c) $AC = 12$, $BC = 4$, $DC = 3$, $CE = ?$
    d) $AG = 8$, $GD = 5$, $BG = 10$, $GE = ?$
    e) $CF = 6$, $AB = 5$, $BC = ?$
    f) $EG = 2$, $GB = 4$, $AD = 9$, $GD = ?$
    g) $AC = 30$, $BC = 3$, $CD = ED$, $ED = ?$
    h) $AC = 9$, $BC = 5$, $ED = 12$, $CD = ?$
    i) $ED = 8$, $DC = 4$, $FC = ?$
    j) $FC = 6$, $ED = 9$, $CD = ?$

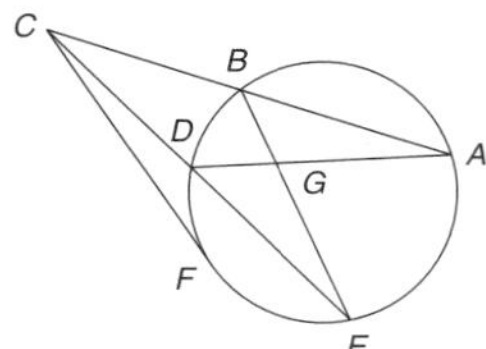

25. *Given:* $\overline{DF} \cong \overline{AC}$ in $\odot O$
$OE = 5x + 4$
$OB = 2x + 19$
Find: OE

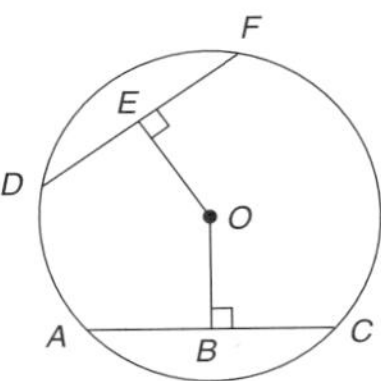

Exercises 25, 26

26. *Given:* $\overline{OE} \cong \overline{OB}$ in $\odot O$
$DF = x(x - 2)$
$AC = x + 28$
Find: DE and AC

In Review Exercises 27 to 29, give a proof for each statement.

27. *Given:* $\overline{DC}$ is tangent to circles B and A at points D and C, respectively
Prove: $AC \cdot ED = CE \cdot BD$

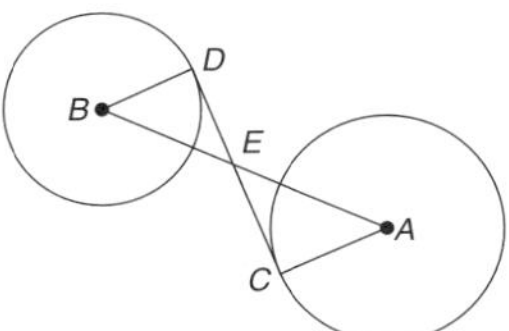

28. *Given:* $\odot O$ with $\overline{EO} \perp \overline{BC}$, $\overline{DO} \perp \overline{BA}$, $\overline{EO} \cong \overline{OD}$
Prove: $\overset{\frown}{BC} \cong \overset{\frown}{BA}$

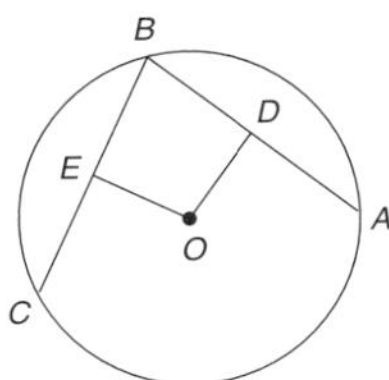

29. *Given:* $\overline{AP}$ and $\overline{BP}$ are tangent to $\odot Q$ at A and B
C is the midpoint of $\overset{\frown}{AB}$
Prove: $\overrightarrow{PC}$ bisects $\angle APB$

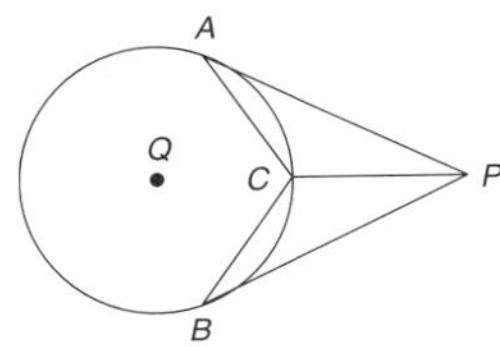

30. *Given:* $\odot O$ with diameter $\overline{AC}$ and tangent $\overleftrightarrow{DE}$
$m\overset{\frown}{AD} = 136°$ and $m\overset{\frown}{BC} = 50°$
Find: The measures of the angles, $\angle 1$ through $\angle 10$

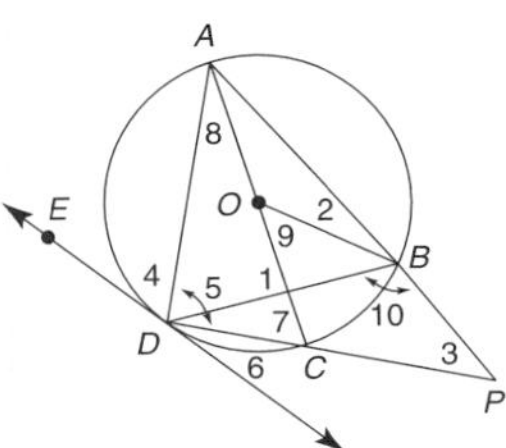

31. A square is inscribed in a circle with a radius of 6 cm. Find the perimeter of the square.

32. A 30-60-90 triangle is inscribed in a circle with a radius of 5 cm. Find the perimeter of the triangle.

33. A circle is inscribed in a right triangle. The radius of the circle is 6 cm, and the hypotenuse is 29 cm. Find the lengths of the two segments of the hypotenuse.

34. *Given:* $\odot O$ is inscribed in $\triangle ABC$
$AB = 9$, $BC = 13$, $AC = 10$
Find: AD, BE, FC

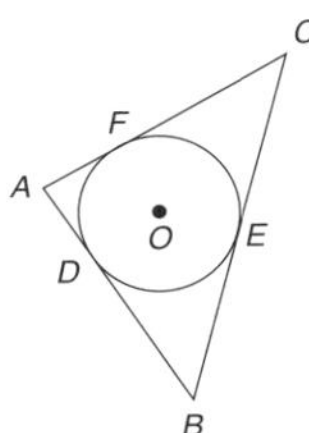

35. In $\odot Q$ with $\triangle ABQ$ and $\triangle CDQ$, $m\overset{\frown}{AB} > m\overset{\frown}{CD}$. Also, $\overline{QP} \perp \overline{AB}$ and $\overline{QR} \perp \overline{CD}$.

a) How are AB and CD related?
b) How are QP and QR related?
c) How are $\angle A$ and $\angle C$ related?

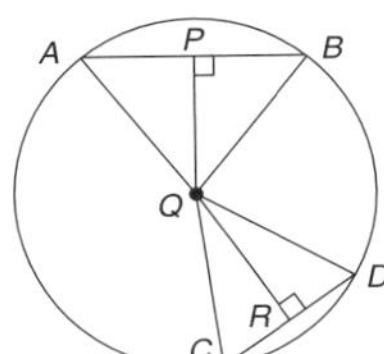

36. In $\odot O$ (not shown), secant $\overleftrightarrow{AB}$ intersects the circle at A and B; C is a point on $\overleftrightarrow{AB}$ in the exterior of the circle.

a) Construct the tangent to $\odot O$ at point B.
b) Construct the tangents to $\odot O$ from point C.

In Review Exercises 37 and 38, use the figure shown.

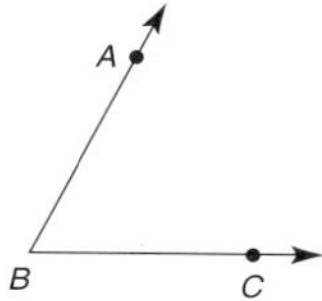

Exercises 37, 38

37. Construct a right triangle so that one leg has length AB and the other has length twice AB.

38. Construct a rhombus with side $\overline{AB}$ and $\angle ABC$.

In Review Exercises 39 to 41, sketch and describe the locus in a plane.

39. Find the locus of the midpoints of the radii of a circle.

40. Find the locus of the centers of all circles passing through two given points.

41. What is the locus of the center of a penny that rolls around a half-dollar?

In Exercises 42 and 43, sketch and describe the locus in space.

42. Find the locus of points less than 3 units from a given point.

43. Find the locus of points equidistant from two parallel planes.

In Review Exercises 44 to 49, use construction methods with the accompanying figure.

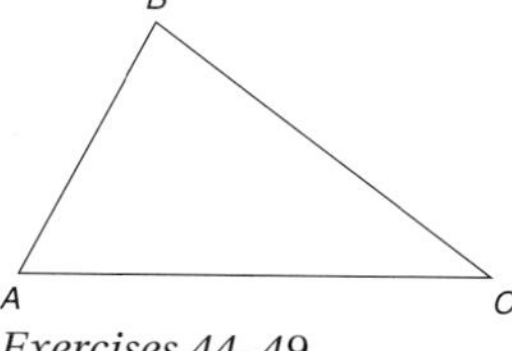

Exercises 44–49

44. *Given:* $\triangle ABC$
Find: The incenter

45. *Given:* $\triangle ABC$
Find: The circumcenter

46. *Given:* $\triangle ABC$
Find: The orthocenter

47. *Given:* $\triangle ABC$
Find: The centroid

48. Use the result from Exercise 44 to inscribe a circle in $\triangle ABC$.

49. Use the result from Exercise 45 to circumscribe the circle about the triangle.

50. *Given:* $\triangle ABC$ with medians $\overline{AE}$, $\overline{DC}$, $\overline{BF}$
Find: **a)** BG if $BF = 18$
b) GE if $AG = 4$
c) DG if $CG = 4\sqrt{3}$

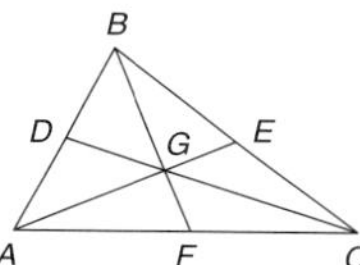

Exercises 50, 51

51. *Given:* $\triangle ABC$ with medians $\overline{AE}$, $\overline{DC}$, $\overline{BF}$
$AG = 2x + 2y$, $GE = 2x - y$
$BG = 3y + 1$, $GF = x$
Find: BF and AE

Chapter 7
Areas of Polygons and Circles

A line segment can be measured in linear units such as feet, inches, or meters. In Chapter 7, we measure the part of a plane enclosed by some plane figure such as a polygon or circle. This measure, known as the area of the region, has many practical applications. The units of measure for area include square feet, square inches, and square meters. The acre is a unit of area measure used in agriculture.

The accompanying drawing shows the floor plan of the office of a business administrator who plans to have new carpet installed. How many square yards of carpet will be needed? We will explore this and other area applications in this chapter.

CHAPTER OUTLINE

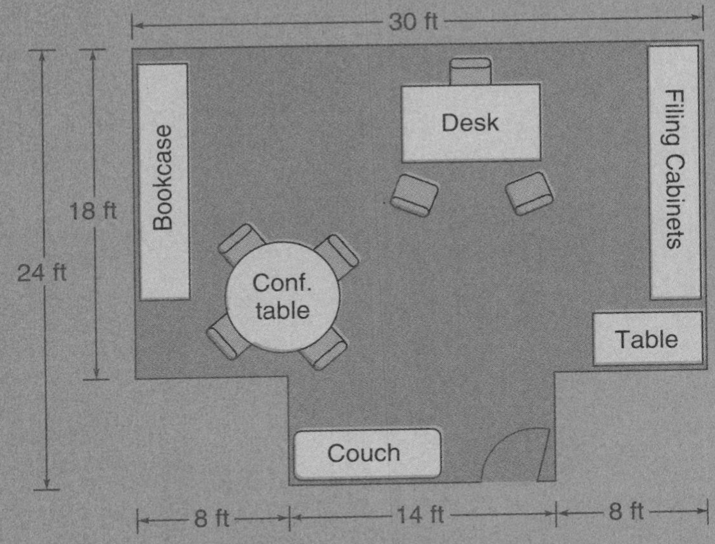

7.1 Area and Initial Postulates

A line segment is measured in linear units such as inches, centimeters, or yards. If a line segment measures 5 centimeters, we write $AB = 5$ cm or simply $AB = 5$ (if the units are apparent or are not stated). The instrument of measure is the ruler.

Lines are *one-dimensional;* that is, we speak only of the dimension "length" when measuring a line segment. On the other hand, a plane is an infinite *two-dimensional* surface. A closed or bounded portion of the plane is called a **region.**

When a region such as R in plane M [see Figure 7.1(a)] is measured, we call this measure the "area of the region." The unit used to measure area is called a **square unit** because it is a square with each side of length 1. The measure of the area of region R is the number of non-overlapping square units that can be placed in (fit into) the region.

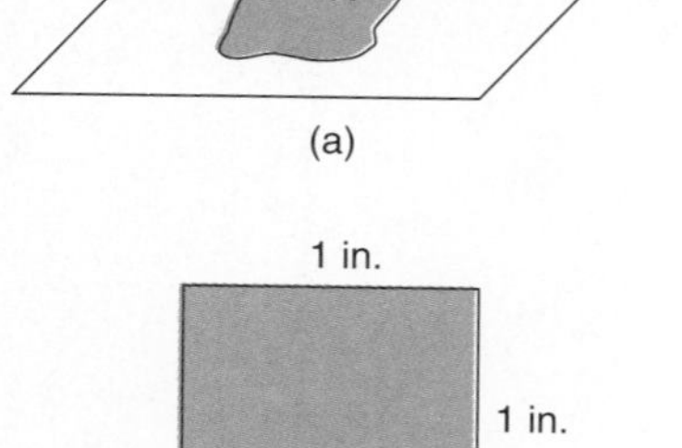

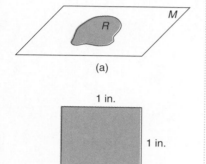

(a)

1 in.

1 in.

(b)

FIGURE 7.1

Square units (not linear units) are used to measure area. Using an exponent, square inches are written as in.^{2}. The unit represented by Figure 7.1(b) is one square inch or 1 in.^{2}.

One application of area involves measuring the floor area to be covered by carpeting, often measured in square yards (yd^2). Another involves calculating the number of squares of shingles needed to cover a roof. (A "square" refers to the number of shingles needed to cover a 100-ft^2 section of the roof.)

In Figure 7.2, the regions have measurable areas and are bounded by figures encountered in earlier chapters. A region is **bounded** if we can distinguish between its interior and exterior.

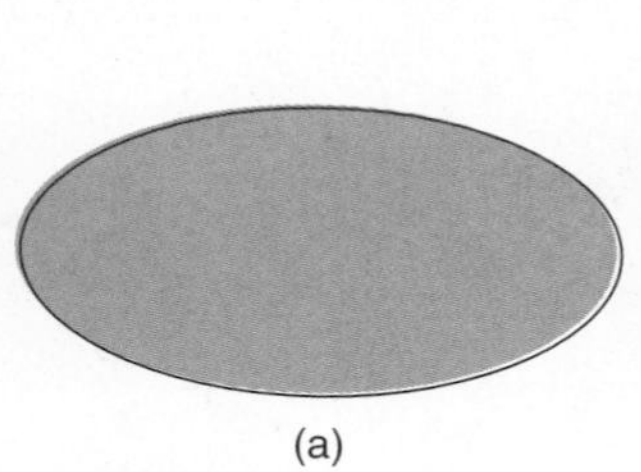
(a)

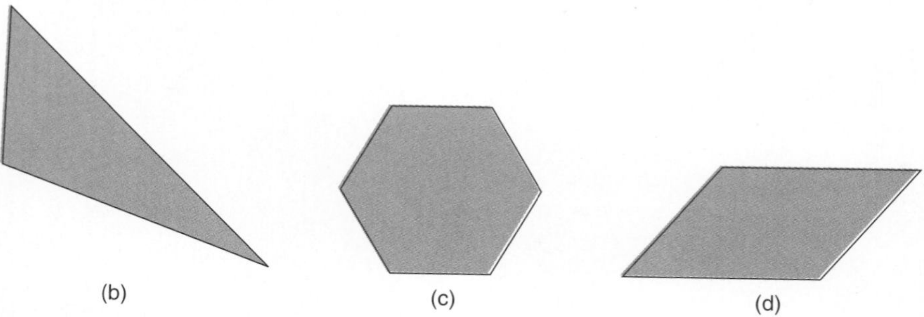
(b) (c) (d)

FIGURE 7.2

We can measure the area of the region within a triangle [see Figure 7.2(b)]. However, we cannot actually measure the area of the triangle itself (three line segments do not have area). Nonetheless, it is common to refer to the area of the region within a triangle as the *area of the triangle.*

The previous discussion does not formally define a region or its area. These are accepted as the undefined terms in the following postulate.

POSTULATE 18: (Area Postulate)
Corresponding to every bounded region is a unique positive number A, known as the area of that region.

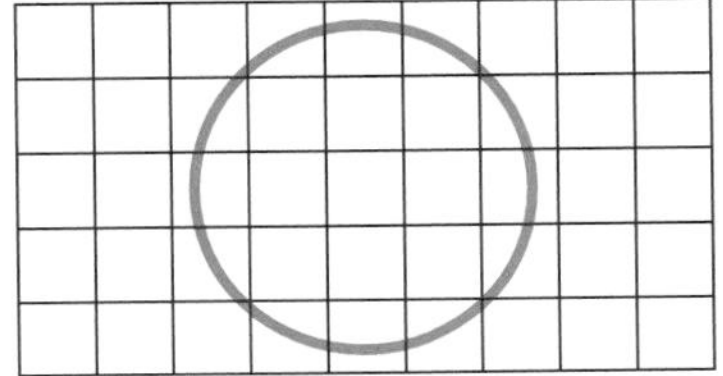

FIGURE 7.3

One way to estimate the area of a region is to place it in a grid, as shown in Figure 7.3. Counting only the number of whole squares inside the region gives an approximation that is less than the actual area. On the other hand, counting squares that are inside or partially inside provides an approximation that is greater than the actual area. A fair estimate of the area of a region is often given by the average of the smaller and larger approximations just described. If the area of the circle shown in Figure 7.3 is between 9 and 21, we might estimate its area to be $\frac{9 + 21}{2}$ or 15 square units.

To develop another property of area, we consider $\triangle ABC$ and $\triangle DEF$ (which are congruent) in Figure 7.4. One triangle can be placed over the other so that they coincide. How are the areas of the two triangles related? The answer is found in the following postulate.

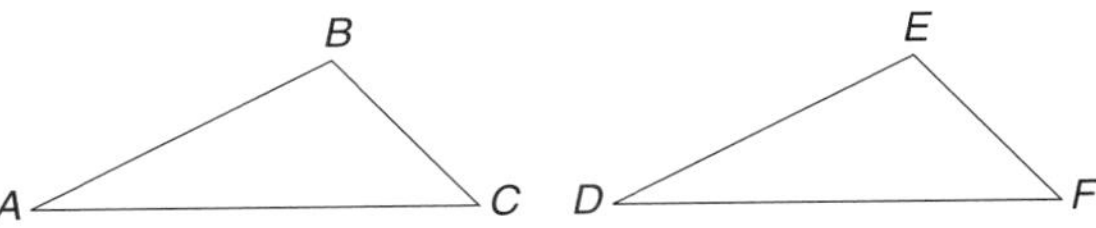

FIGURE 7.4

POSTULATE 19:
If two closed plane figures are congruent, then their areas are equal.

EXAMPLE 1

In Figure 7.5, points B and C trisect $\overline{AD}$; $\overline{EC} \perp \overline{AD}$. Name two triangles with equal areas.

Solution

$\triangle ECB \cong \triangle ECD$ by SAS. Then $\triangle ECB$ and $\triangle ECD$ have equal areas by Postulate 19.

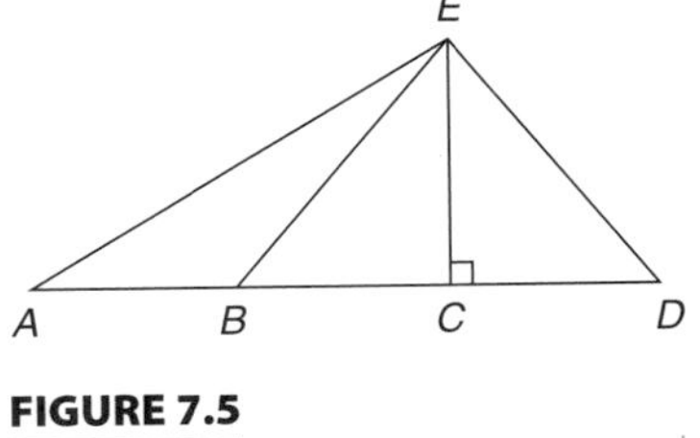

FIGURE 7.5

NOTE: $\triangle EBA$ is also equal in area to $\triangle ECB$ and $\triangle ECD$, but this cannot be established until the material in Section 7.2 is considered.

FIGURE 7.6

Consider Figure 7.6. The entire region is bounded by a curve and then subdivided by a segment into smaller regions R and S. These regions have a common boundary and do not overlap. Since a numerical area can be associated with each region R and S, the area of $R \cup S$ (read as "R union S" and meaning region R joined to region S) is equal to the sum of the areas of R and S. This leads to Postulate 20, in which A_R represents the "area of region R," A_S represents the "area of region S," and $A_{R \cup S}$ means the "area of region $R \cup S$."

POSTULATE 20: (Area-Addition Postulate)
Let R and S be two enclosed regions that do not intersect. Then

$$A_{R \cup S} = A_R + A_S$$

EXAMPLE 2

In Figure 7.7, the pentagon $ABCDE$ is formed by square $ABCD$ and $\triangle ADE$. If the area of the square is 36 in.^{2} while that of $\triangle ADE$ is 12 in.^{2}, find the area of pentagon $ABCDE$.

Solution

Because square $ABCD$ and $\triangle ADE$ do not overlap and have a common boundary $\overline{AD}$, we have Area (pentagon $ABCDE$) = area (square $ABCD$) + area ($\triangle ADE$). By the Area-Addition Postulate,

$$\text{Area (pentagon } ABCDE) = 36 \text{ in.}^2 + 12 \text{ in.}^2 = 48 \text{ in.}^2$$

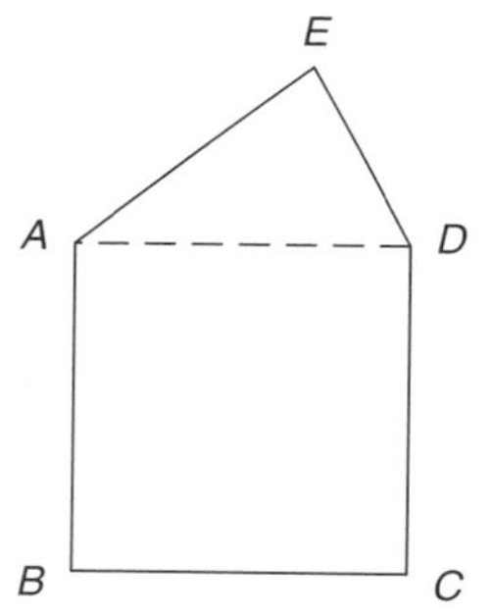

FIGURE 7.7

It is convenient to subscript A (area) with letters that name the figure whose area is indicated. The principle used in Example 2 is more conveniently and compactly stated in the form

$$A_{ABCDE} = A_{ABCD} + A_{ADE}$$

Area of a Rectangle

DISCOVER!

Study rectangle $MNPQ$ in Figure 7.8, and note that it has dimensions of 3 cm and 4 cm. The number of squares, 1 cm on a side, in the rectangle is 12. Rather than count the number of squares in the figure, how can you calculate the area?

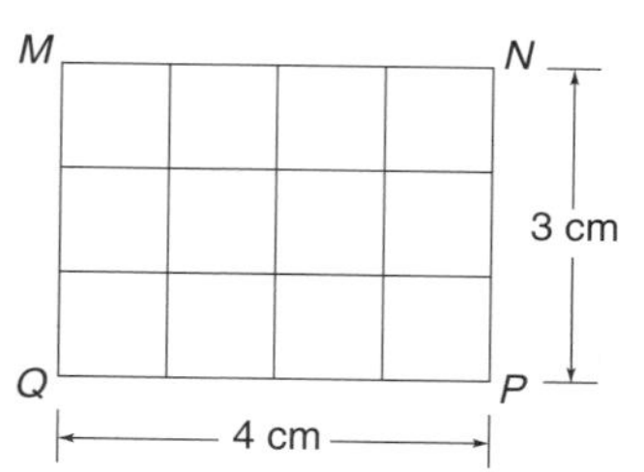

FIGURE 7.8

ANSWER
Multiply $3 \times 4 = 12$.

■ **Warning**

Although 1 ft = 12 in., 1 ft² = 144 in.²

See Figure 7.9. ■

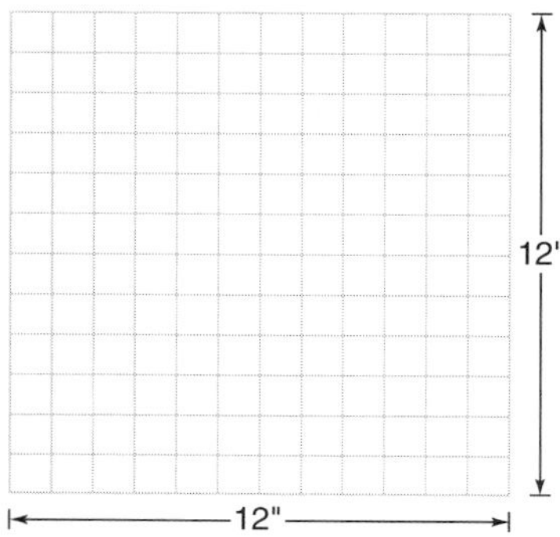

FIGURE 7.9

In the preceding *Discover!*, the unit of area is cm^2. Multiplication of dimensions is handled like algebraic multiplication. Compare:

$$3x \cdot 4x = 12x^2 \qquad 3\text{ cm} \cdot 4\text{ cm} = 12\text{ cm}^2$$

If the units used to measure the dimensions of a region are *not* the same, then they must be converted into like units in order to calculate area. For instance, if we need to multiply 2 feet by 6 inches, we note that 2 ft = 2(12 in.) = 24 in., and so $A = 2\text{ ft} \cdot 6\text{ in.} = 24\text{ in.} \cdot 6\text{ in.} = 144\text{ in.}^2$ Alternately, 6 in. = $\frac{1}{2}$ ft, so $A = 2\text{ ft} \cdot \frac{1}{2}\text{ ft} = 1\text{ ft}^2$. Because the area is unique, we know that $1\text{ ft}^2 = 144\text{ in.}^2$

Recall that one side of a rectangle is called its *base,* and either side perpendicular to the base is called the *altitude* of the rectangle.

POSTULATE 21:
The area A of a rectangle whose base has length b and whose altitude has length h is given by $A = bh$.

It is common to describe the dimensions of a rectangle as its length ℓ and width w. The area is found by $A = \ell w$.

EXAMPLE 3

Find the area of rectangle $ABCD$ in Figure 7.10 if $AB = 12$ cm and $AD = 7$ cm.

Solution

FIGURE 7.10

Because it makes little difference which dimension is chosen as base b and which as altitude h, we arbitrarily choose $AB = b = 12$ cm and $AD = h = 7$ cm. Then

$$\begin{aligned} A &= bh \\ &= 12\text{ cm} \cdot 7\text{ cm} \\ &= 84\text{ cm}^2 \end{aligned}$$

If units are not provided for the dimensions of a region, we assume that they are alike. In such a case, we simply give the area as a number of square units.

THEOREM 7.1.1: The area A of a square whose sides are each of length s is given by $A = s^2$.

No proof is given for Theorem 7.1.1, which follows immediately from Postulate 21.

Area of a Parallelogram

While a rectangle's altitude is one of its sides, that is not true of a parallelogram. An **altitude** of a parallelogram is a perpendicular segment from one side to the opposite side, known as the **base.** A side may have to be extended in order to show this altitude-base relationship in a drawing. In Figure 7.11(a), if $\overline{RS}$ is designated as the base, then any of the segments $\overline{ZR}$, $\overline{VX}$, or $\overline{YS}$ is an altitude corresponding to that base (or to base $\overline{VT}$).

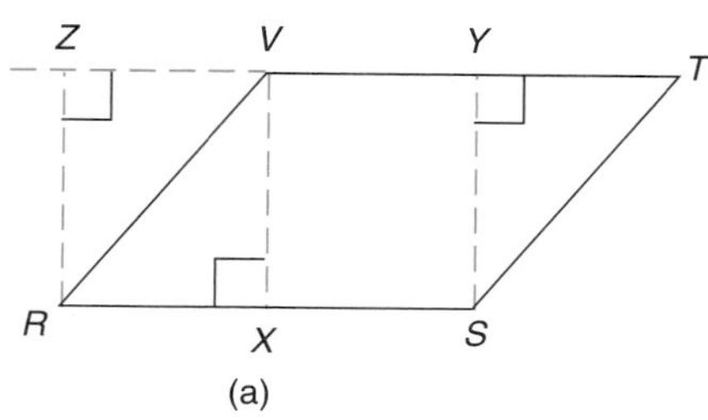

(a)

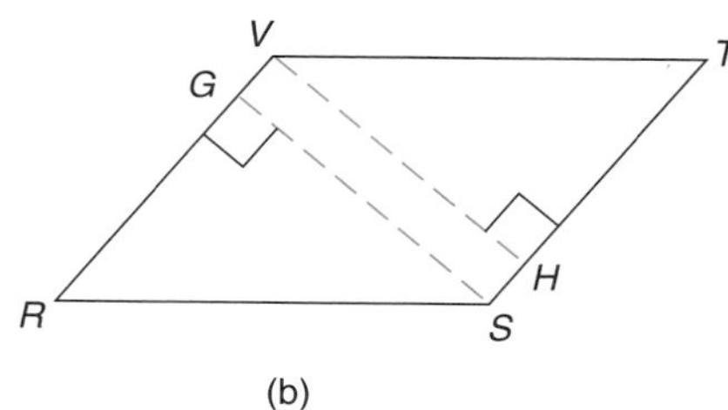

(b)

FIGURE 7.11

Another look at ▱$RSTV$ [in Figure 7.11(b)] shows that $\overline{ST}$ (or $\overline{VR}$) could just as well have been chosen as the base. Possible choices for the altitude include $\overline{VH}$ and $\overline{GS}$. In the theorem that follows, it is necessary to select a base and an altitude drawn to that base!

THEOREM 7.1.2: The area A of a parallelogram with a base of length b and with corresponding altitude of length h is given by

$$A = bh$$

Given: In Figure 7.12(a), ▱$RSTV$ with $\overline{VX} \perp \overline{RS}$
$RS = b$ and $VX = h$

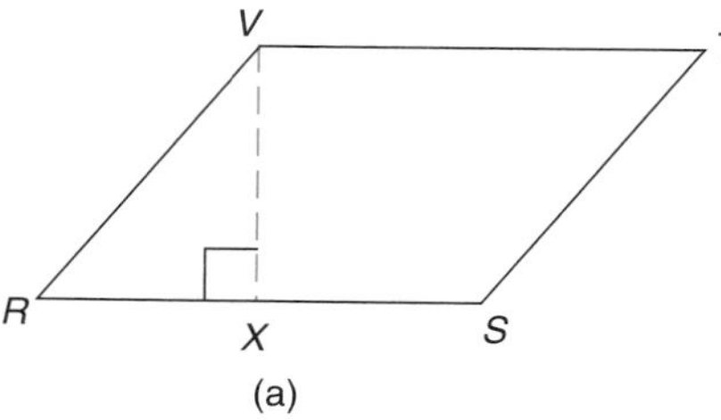

(a)

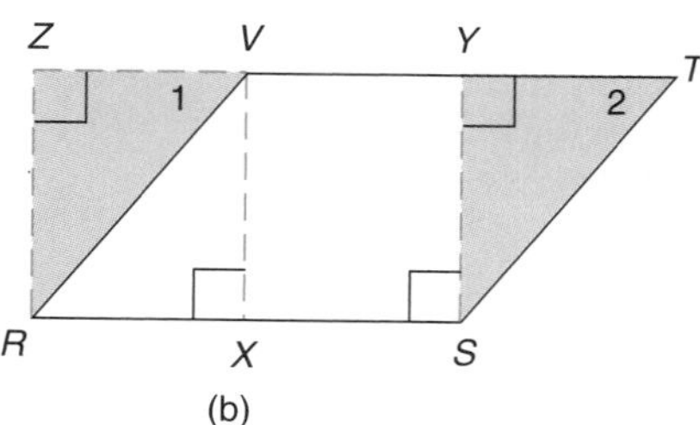

(b)

FIGURE 7.12

Prove: $A_{RSTV} = bh$

Proof: Construct $\overline{YS} \perp \overline{VT}$ and $\overline{RZ} \perp \overline{VT}$, in which Z lies on an extension of $\overline{VT}$, as shown in Figure 7.12(b). Right $\angle Z$ and right $\angle SYT$ are $\cong$. Also, $\overline{ZR} \cong \overline{SY}$ since parallel lines are everywhere equidistant.

Because $\angle 1$ and $\angle 2$ are $\cong$ corresponding angles for parallel segments $\overline{VR}$ and $\overline{TS}$, $\triangle RZV \cong \triangle SYT$ by AAS. Then $A_{RZV} = A_{SYT}$ because congruent $\triangle$s have equal areas.

Because $A_{RSTV} = A_{RSYV} + A_{SYT}$, it follows that $A_{RSTV} = A_{RSYV} + A_{RZV}$. But $RSYV \cup RZV$ is rectangle $RSYZ$, which has the area bh. Therefore $A_{RSTV} = A_{RSYZ} = bh$.

EXAMPLE 4

Given that all dimensions in Figure 7.13 are in inches, find the area of $\square MNPQ$ by using base

a) MN. **b)** PN.

Solution

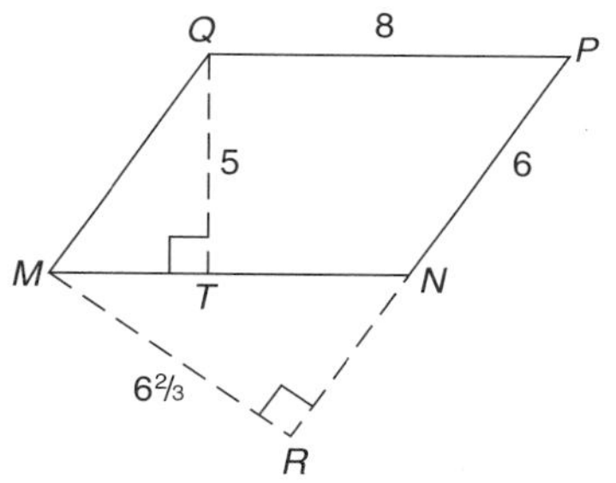

a) $MN = b = 8$, while the corresponding altitude is of length $QT = h = 5$. Then

$$A = 8 \text{ in.} \cdot 5 \text{ in.}$$
$$= 40 \text{ in.}^2$$

b) $PN = b = 6$, so the corresponding altitude length is $MR = h = 6\frac{2}{3}$. Then

$$A = 6 \cdot 6\tfrac{2}{3}$$
$$= 6 \cdot \tfrac{20}{3}$$
$$= 40 \text{ in.}^2$$

FIGURE 7.13

NOTE: The area of the $\square$ is unchanged when a different base and its corresponding altitude are used to calculate its area. See Postulate 18 on page 295.

EXAMPLE 5

Given: In Figure 7.14, $\square MNPQ$ with $PN = 8$ and $QP = 10$
Altitude $\overline{QR}$ to base $\overline{MN}$ has length $QR = 6$

Find: SN, the length of the altitude between $\overline{QM}$ and $\overline{PN}$

Solution

Choosing $MN = b = 10$ and $QR = h = 6$, we see that

$$A = bh = 10 \cdot 6 = 60$$

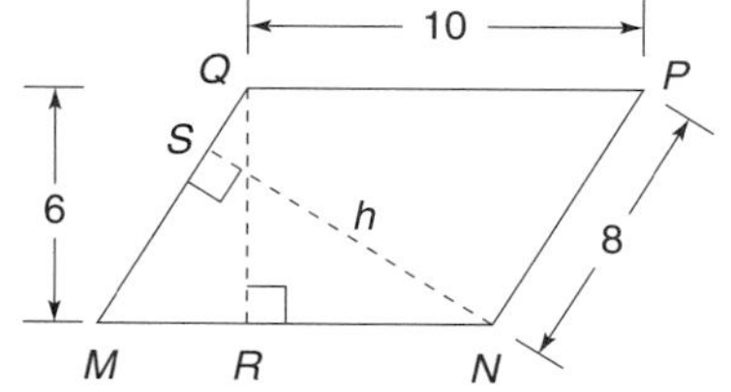

FIGURE 7.14

Now we choose $PN = b = 8$ and $SN = h$, so $A = 8h$. Because the area of the parallelogram is unique, it follows that

$$8h = 60$$

$$h = \frac{60}{8} = 7.5; \text{ that is, } SN = 7.5$$

Area of a Triangle

The last formula we consider in this section is the area of a triangle. It follows easily from the formula for the area of a parallelogram. In the formula, any side of the triangle can be chosen as its base.

THEOREM 7.1.3: The area A of a triangle whose base has length b and whose corresponding altitude has length h is given by

$$A = \frac{1}{2}bh$$

Following is a proof of Theorem 7.1.3.

Given: In Figure 7.15, $\triangle ABC$ with $\overline{CD} \perp \overline{AB}$
$AB = b$ and $CD = h$

FIGURE 7.15

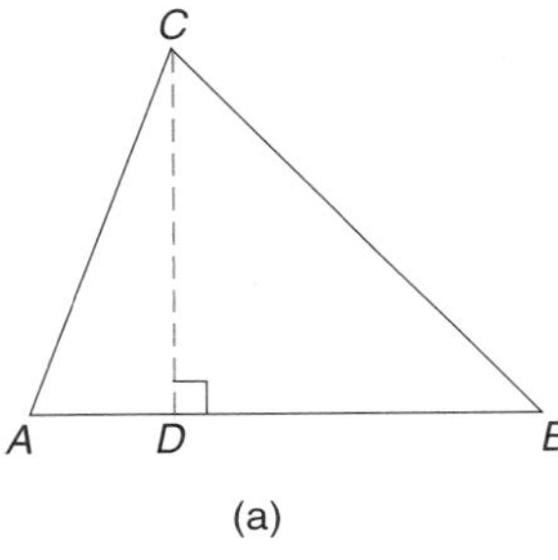

(a)

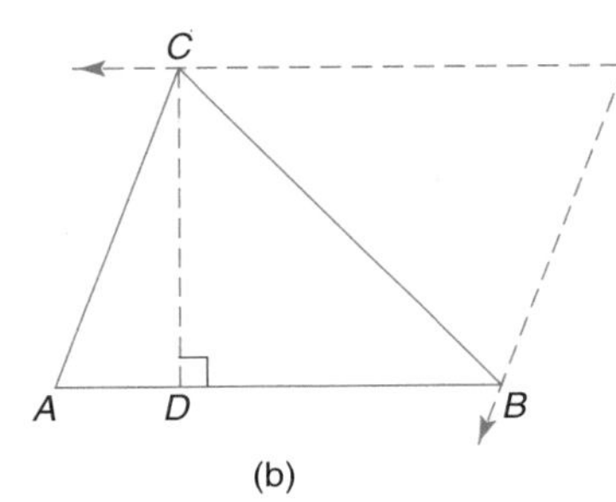

(b)

Prove: $A = \frac{1}{2}bh$

Proof: Construct a line through point C parallel to $\overline{AB}$ and a line through B parallel to $\overline{AC}$. These lines meet at a point X. [See Figure 7.15(b).] Now $ABXC$ is a parallelogram and $\overline{CB}$ is its diagonal. Then $\triangle ABC \cong \triangle XCB$ since a diagonal separates the $\square$ into two $\cong$ $\triangle$s. Thus $A_{ABC} = A_{XCB}$.

Furthermore, $A_{ABXC} = A_{ABC} + A_{XCB}$ by the Area-Addition Postulate. So

$$A_{ABXC} = A_{ABC} + A_{ABC} = 2 \cdot A_{ABC}$$

Then

$$A_{ABC} = \frac{1}{2} \cdot A_{ABXC} = \frac{1}{2}(bh)$$

Dropping subscripts, $A = \frac{1}{2}bh$.

EXAMPLE 6

Given: In Figure 7.16, right $\triangle MPN$ with $PN = 8$ and $MN = 17$

Find: A_{MNP}

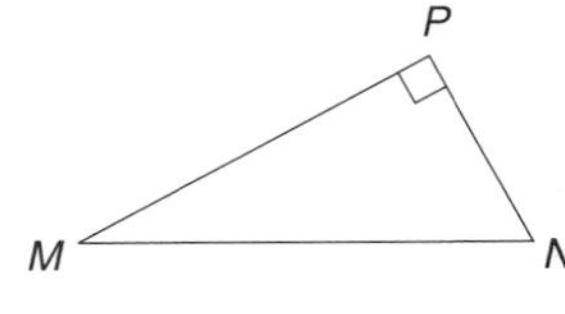

FIGURE 7.16

Solution With $\overline{PN}$ as a base, we need altitude $\overline{PM}$. By the Pythagorean Theorem,

$$17^2 = (PM)^2 + 8^2$$
$$289 = (PM)^2 + 64$$

Then $(PM)^2 = 225$, so $PM = 15$.

With $PN = b = 8$ and $PM = h = 15$, we have

$$A = \frac{1}{2} \cdot 8 \cdot 15 = 60 \text{ units}^2$$

Warning

The phrase "area of a polygon" really means the area of the region enclosed by the polygon.

7.1 Exercises

1. Suppose that two triangles have equal areas. Are the triangles congruent? Why or why not? Are two squares with equal areas necessarily congruent? Why or why not?
2. The area of the square is 12, while the area of the circle is 30. Does the shaded area equal 42? Why or why not?

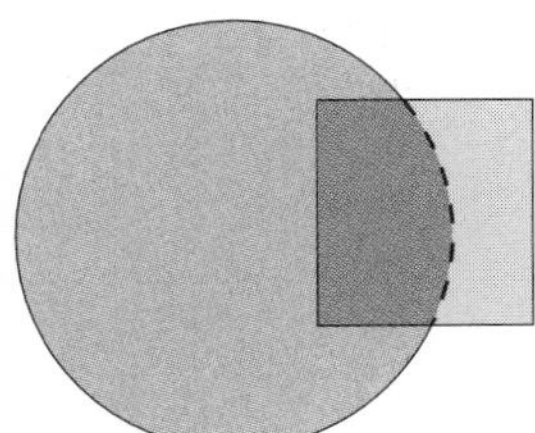

Exercises 2, 3

3. Consider the information in Exercise 2, but suppose you know that the area of the region defined by the intersection of the square and the circle measures 5. What is the area of the shaded region?
4. If $MNPQ$ is a rhombus, which formula should be used to calculate its area?

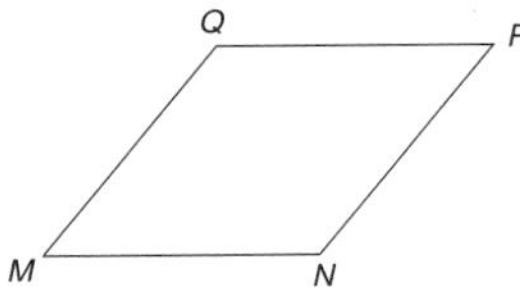

Exercises 4, 5

5. In rhombus $MNPQ$, how does the length of the altitude to $\overline{PN}$ compare to the length of the altitude to $\overline{MN}$? Explain.
6. When the diagonals of a rhombus are drawn, how do the areas of the four resulting smaller triangles compare to each other and to the given rhombus?

In Exercises 7 to 16, find areas of the figures shown or described.

7. A rectangle's length is 6 cm, and its width is 9 cm.
8. A right triangle has one leg measuring 20 in. and a hypotenuse measuring 29 in.
9. A 45-45-90 triangle has a hypotenuse measuring 6 m.
10. A triangle's altitude to the 15-in. side measures 8 in.
11. 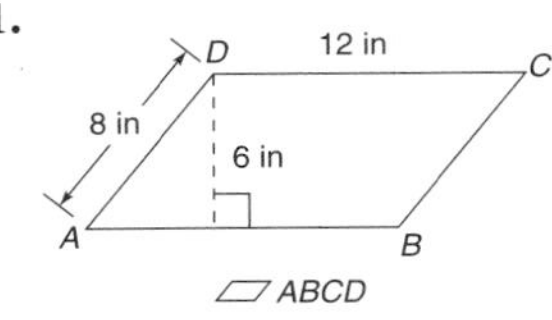

12. 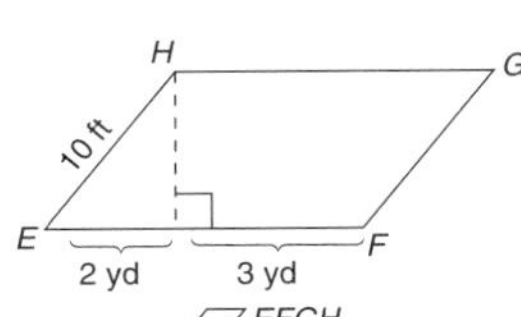

13. 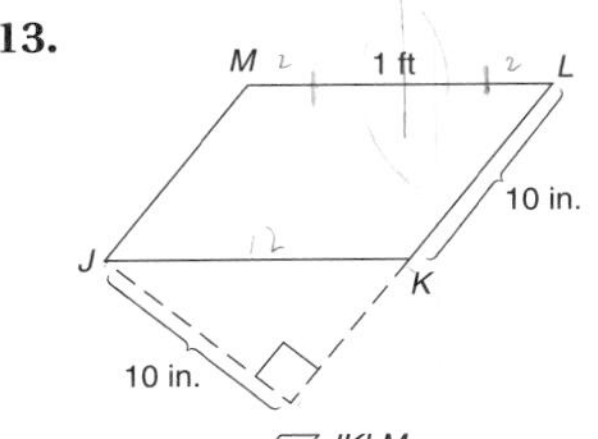

14. 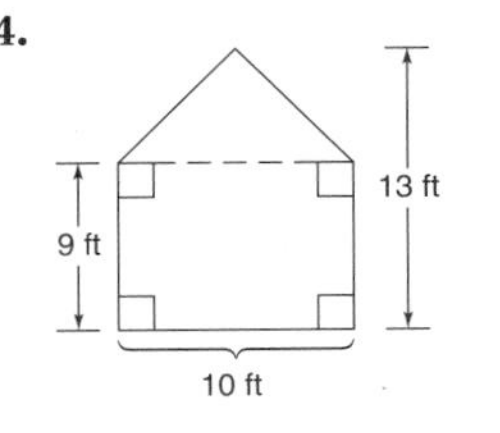

15.

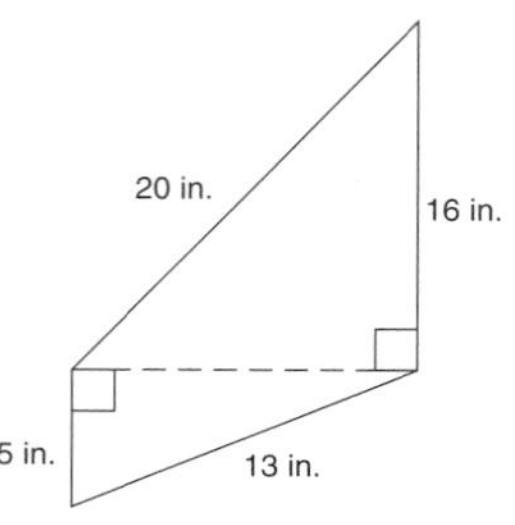

16.

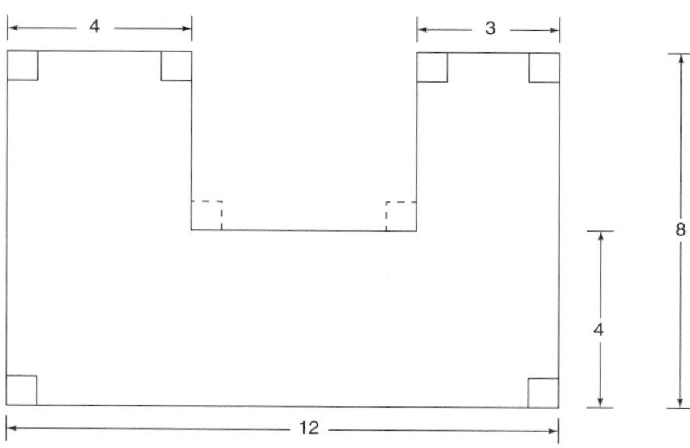

In Exercises 17 to 22, find the area of the shaded region.

17.

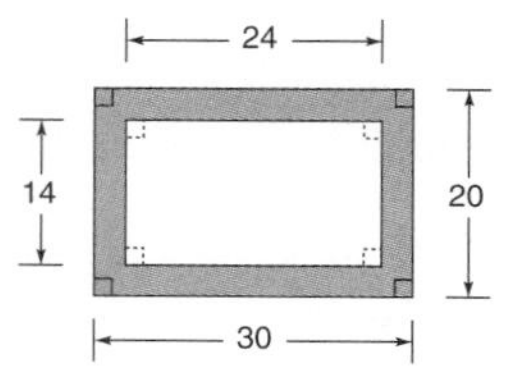

18.

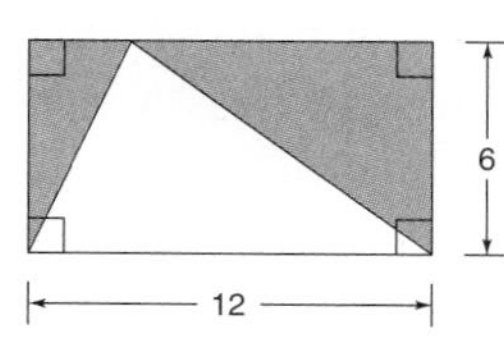

19.

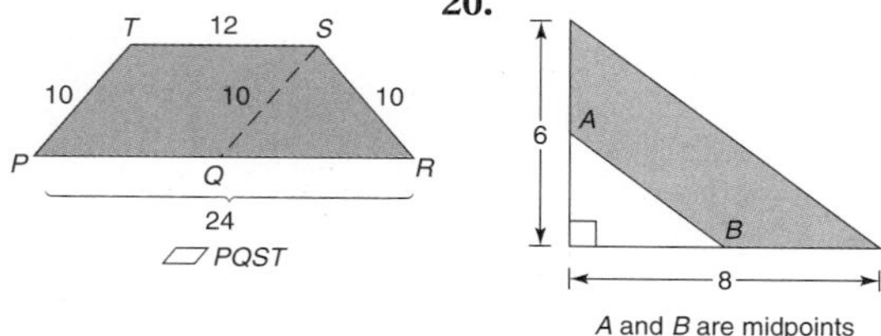

20.

A and B are midpoints

21.

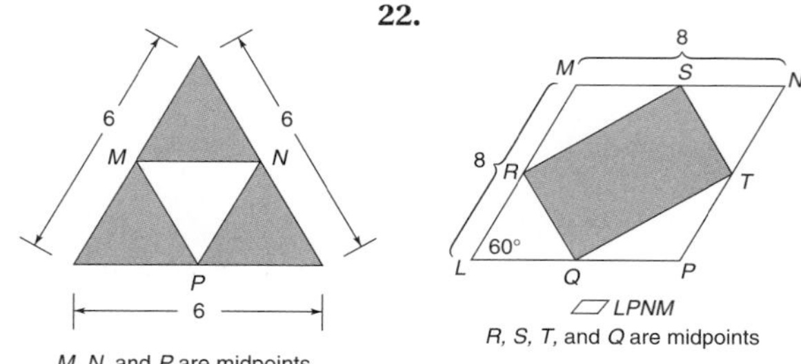

22.

23. A triangular corner of a store has been roped off to be used as an area for displaying Christmas ornaments. Find the area of the display section.

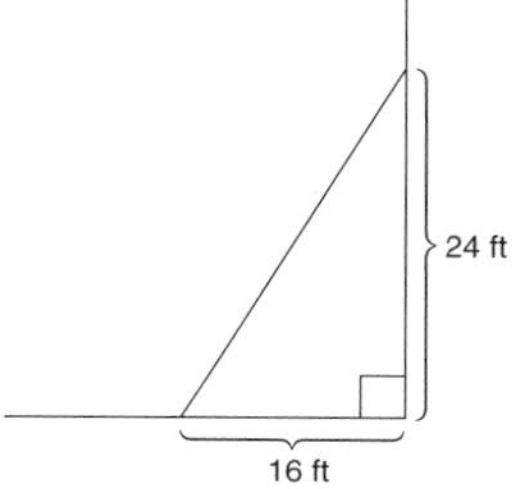

24. Carpeting is to be purchased for the family room and hallway shown. What is the area to be covered?

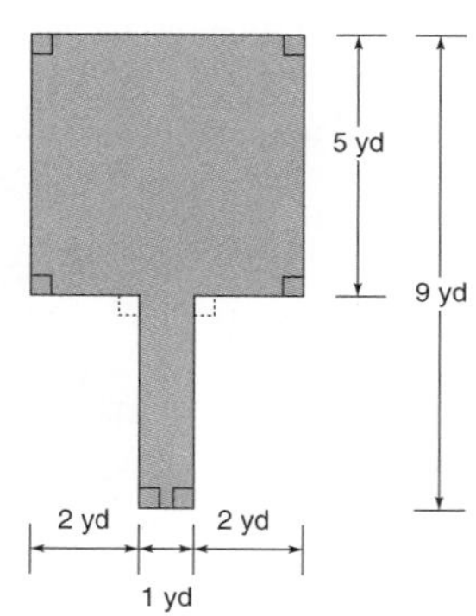

25. The exterior wall (the gabled end of the house shown) remains to be painted.

a) What is the area of the outside wall?

b) If each gallon of paint covers approximately 105 ft^2, how many gallons of paint must be purchased?

c) If each gallon of paint is on sale for \$15.50, what is the total cost of the paint?

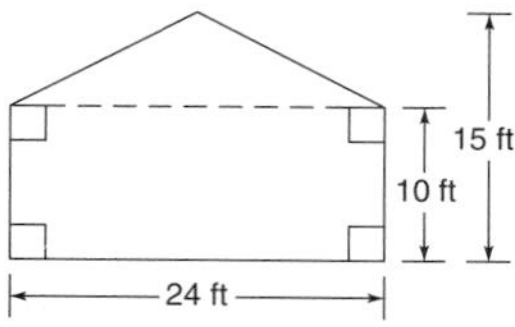

26. The roof of the house shown needs to be shingled.

a) Considering that both the front and back sections of the roof have equal areas, find the total area to be shingled.

b) If roofing is sold in squares (each covering 100 ft^2), how many squares are needed to complete the work?

c) If each square costs \$22.50 and an extra square is allowed for trimming around vents, what is the total cost of the shingles?

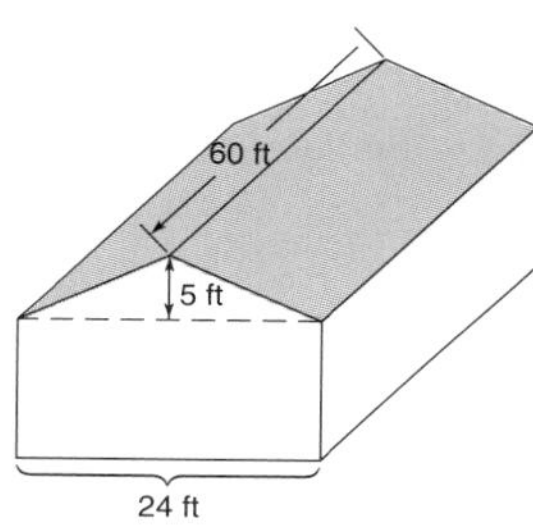

27. A beach tent is designed so that one side is open. Find the number of square feet of canvas needed to make the tent.

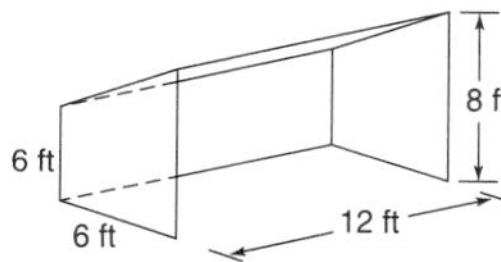

28. Gary and Carolyn plan to build the deck shown.

a) Find the total floor space (area) for the deck.

b) Find the approximate cost of building the deck if the estimated cost is \$3.20 per ft^2.

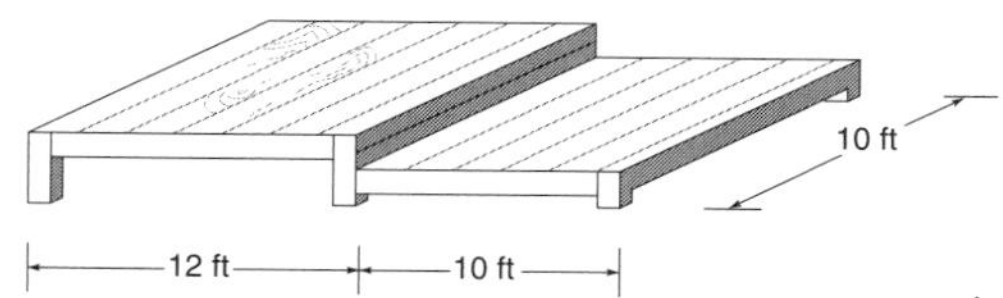

29. A *square yard* is a square with sides 1 yard in length. How many

a) *square feet* are in 1 square yard?

b) *square inches* are in 1 square yard?

30. Given $\triangle RST$ with median $\overline{RV}$, explain why $A_{RSV} = A_{RVT}$.

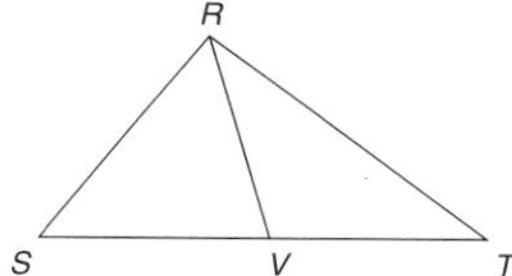

31. Given $\triangle ABC$ with midpoints M, N, and P of the sides, explain why $A_{ABC} = 4 \cdot A_{MNP}$.

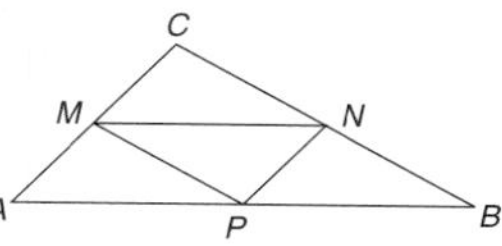

In Exercises 32 and 34, provide paragraph proofs.

32. *Given:* Right $\triangle ABC$
Prove: $h = \dfrac{ab}{c}$

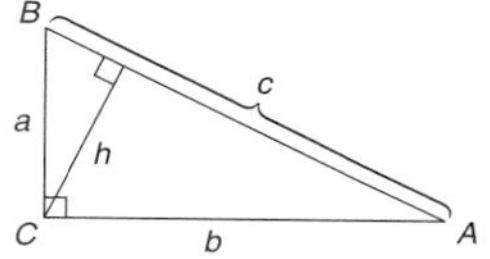

33. *Given:* Square $HJKL$ with $LJ = d$
Prove: $A_{HJKL} = \dfrac{d^2}{2}$

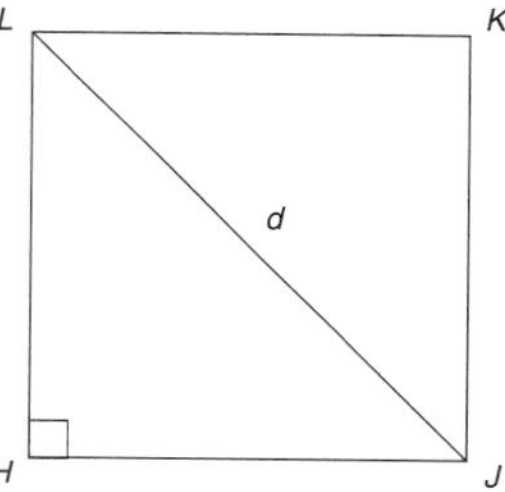

34. *Given:* $\square RSTV$ with $\overline{VW} \cong \overline{VT}$
Prove: $A_{RSTV} = (RS)^2$

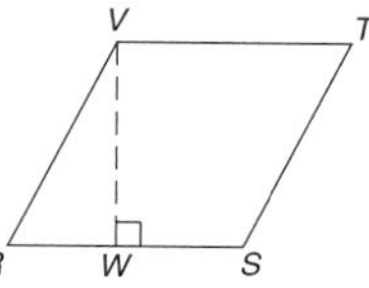

35. *Given:* The area of right $\triangle ABC$ (not shown) is 40 in.^{2}
$m\angle C = 90°$
$AC = x$
$BC = x + 2$
Find: x

36. The lengths of the legs of a right triangle are consecutive even integers. The numerical value of the area is three times that of the longer leg. Find the lengths of the legs of the triangle.

37. ➢ *Given:* △*ABC*, whose sides are 13 in., 14 in., and 15 in.

Find: **a)** *BD*, the length of the altitude to the 14-in. side. (**HINT:** Use the Pythagorean Theorem twice.)

b) The area of △*ABC*, using the result from part (a)

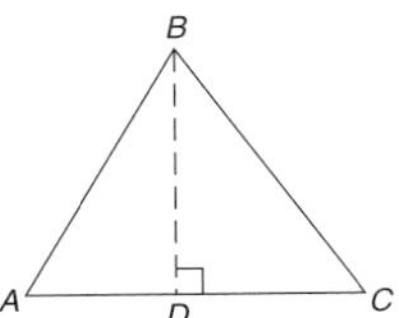

38. ➢ *Given:* △*ABC*, whose sides are 10 cm, 17 cm, and 21 cm

Find: **a)** *BD*, the length of the altitude to the 21-cm side

b) The area of △*ABC*, using the result from part (a)

39. If the base of a rectangle is increased by 20 percent and the altitude is increased by 30 percent, by what percentage is the area increased?

40. If the base of a rectangle is increased by 20 percent but the altitude is decreased by 30 percent, by what percentage is the area changed? Is this an increase or decrease in area?

41. Given region $R \cup S$, explain why $A_{R \cup S} > A_R$.

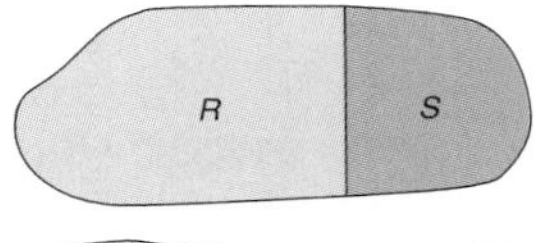

42. Given region $R \cup S \cup T$, explain why $A_{R \cup S \cup T} = A_R + A_S + A_T$.

43. The algebra method of FOIL multiplication is illustrated geometrically in the drawing. Use the drawing with rectangular regions to complete the following rule:

$(a + b)(c + d) =$ ______

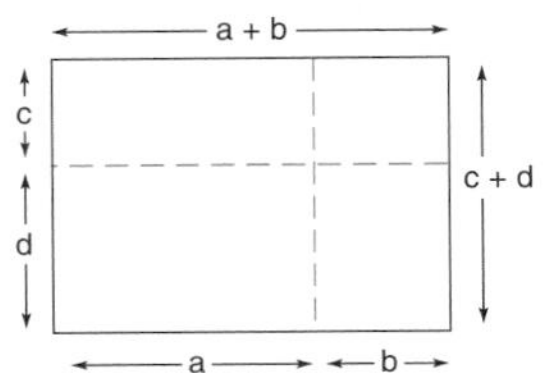

44. Use the square configuration to complete the following algebra rule:

$(a + b)^2 =$ ______

(**NOTE:** Simplify where possible.)

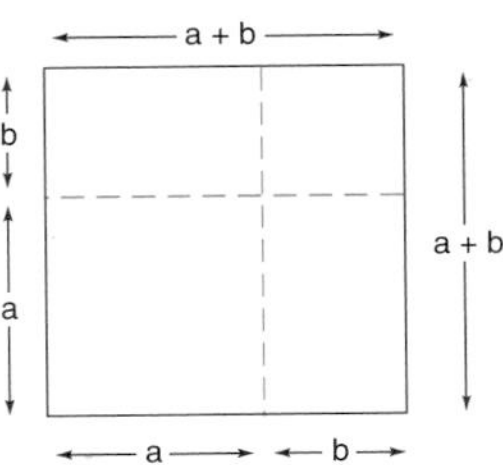

In Exercises 45 to 48, use the fact that the area of the polygon is unique.

45. In the right triangle, find the length of the altitude drawn to the hypotenuse.

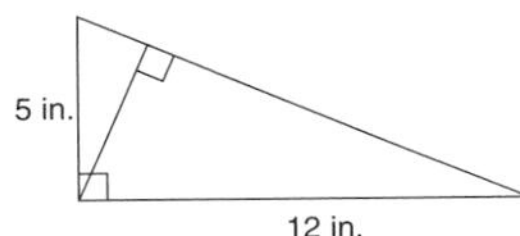

46. In the triangle whose sides are 13, 20, and 21 cm long, the length of the altitude drawn to the 21-cm side is 12 cm. Find the lengths of the remaining altitudes of the triangle.

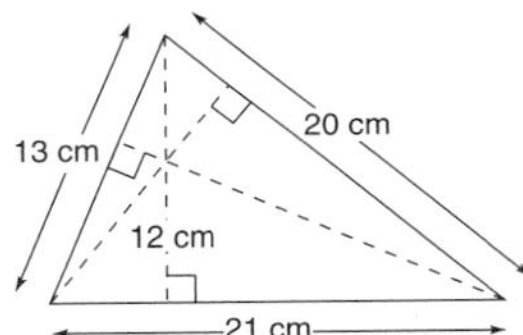

47. In ▱*MNPQ*, *QP* = 10 and *QM* = 5. The length of altitude $\overline{QR}$ (to side $\overline{MN}$) is 4. Find the length of altitude $\overline{QS}$ from *Q* to $\overline{PN}$.

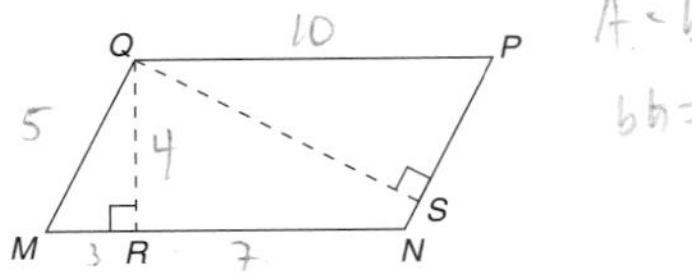

48. In ▱*ABCD*, *AB* = 7 and *BC* = 12. The length of altitude $\overline{AF}$ (to side $\overline{BC}$) is 5. Find the length of altitude $\overline{AE}$ from *A* to $\overline{DC}$.

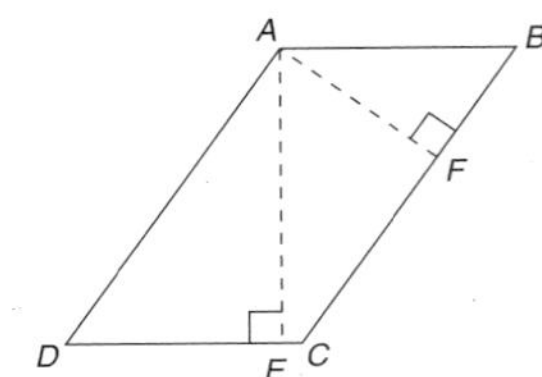

49. a) Find a lower estimate of the area of the figure by counting whole squares within the figure.
b) Find an upper estimate of the area of the figure by counting whole and partial squares within the figure.
c) Use the average of the results in parts (a) and (b) to provide a better estimate of the area of the figure.
d) Does intuition suggest that the area estimate of part (c) is the exact answer?

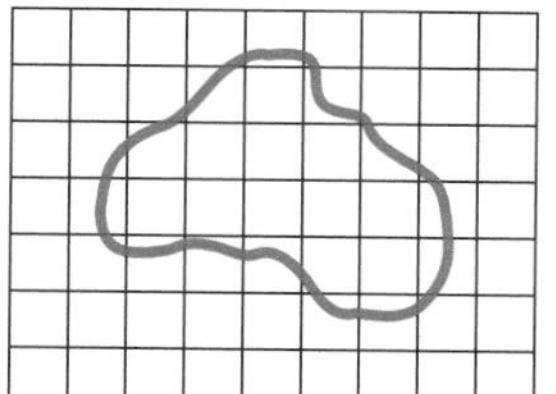

50. a) Find a lower estimate of the area of the figure by counting whole squares within the figure.
b) Find an upper estimate of the area of the figure by counting whole and partial squares within the figure.
c) Use the average of the results in parts (a) and (b) to provide a better estimate of the area of the figure.
d) Does intuition suggest that the area estimate of part (c) is the exact answer?

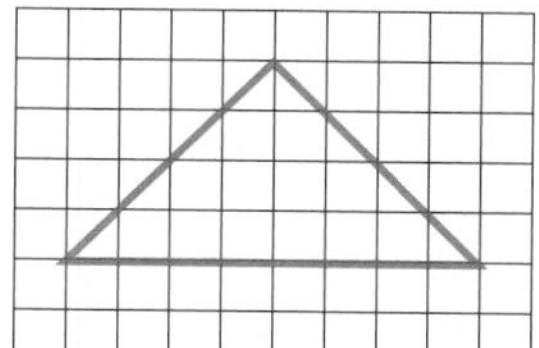

7.2 Perimeter and Area of Polygons

We begin this section with a reminder of the meaning of perimeter.

DEFINITION: The **perimeter** of a polygon is the sum of the lengths of all sides of the polygon.

Table 7.1 summarizes perimeter formulas for types of triangles, while Table 7.2 on the next page summarizes formulas for the perimeters of selected types of quadrilaterals. However, it is more important to understand the concept of perimeter than to memorize formulas. See if you can explain each formula.

TABLE 7.1
Perimeter of a Triangle

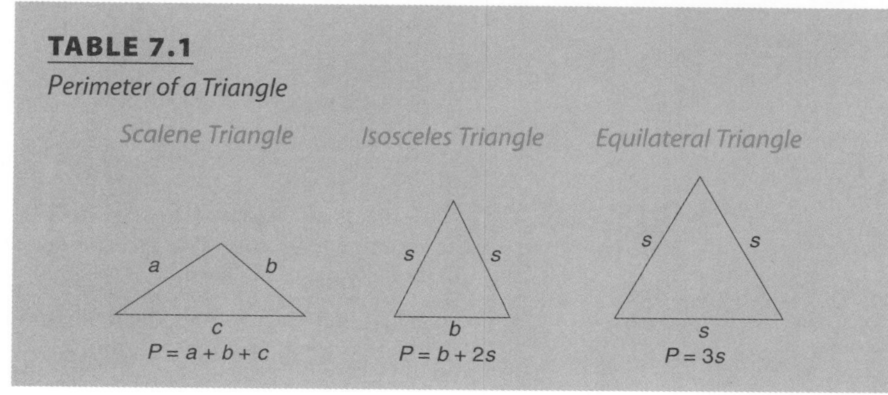

TABLE 7.2
Perimeter of a Quadrilateral

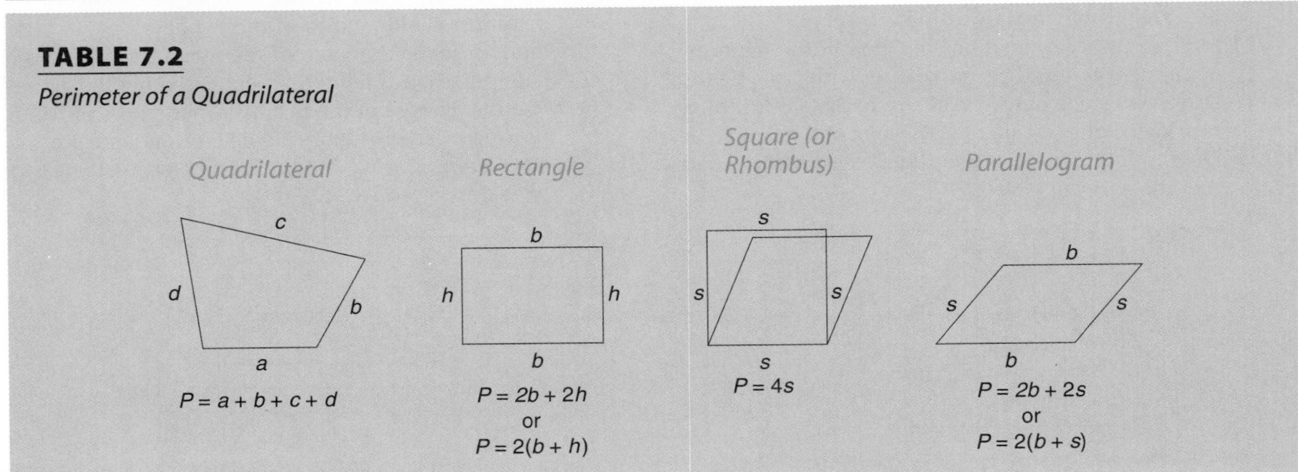

EXAMPLE 1

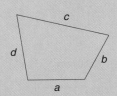

Find the perimeter of $\triangle ABC$ in Figure 7.17 if:

a) $AB = 5$ in., $AC = 6$ in., and $BC = 7$ in.
b) Altitude $AD = 8$ cm, $BC = 6$ cm, and $\overline{AB} \cong \overline{AC}$

Solution

a) $P_{ABC} = AB + AC + BC$
$= 5 + 6 + 7$
$= 18$ in.

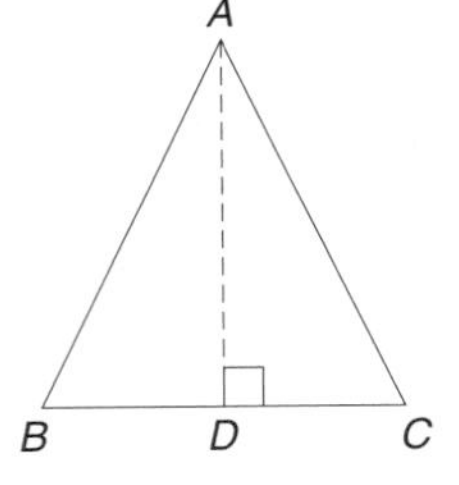

FIGURE 7.17

b) With $\overline{AB} \cong \overline{AC}$, $\triangle ABC$ is isosceles. Then $\overline{AD}$ is the $\perp$ bisector of $\overline{BC}$. If $BC = 6$, it follows that $DC = 3$. Using the Pythagorean Theorem, we have

$$(AD)^2 + (DC)^2 = (AC)^2$$
$$8^2 + 3^2 = (AC)^2$$
$$64 + 9 = (AC)^2$$
$$AC = \sqrt{73}$$

Now $P_{ABC} = 6 + \sqrt{73} + \sqrt{73} = 6 + 2\sqrt{73} \approx 23.09$ cm.

NOTE: Because $x + x = 2x$, $\sqrt{73} + \sqrt{73} = 2\sqrt{73}$.

We apply the perimeter concept in a more general manner in Example 2.

EXAMPLE 2

While remodeling, the Gibsons have decided to replace the old woodwork with Colonial-style oak woodwork.

a) Using the floor plan provided in Figure 7.18, find the amount of baseboard (in linear ft) needed for the room. Do *not* make any allowances for doors!
b) Find the cost of the baseboard if the price is \$1.32 per linear foot.

Solution

a) The perimeter or "distance around" the room is

$$12 + 6 + 8 + 12 + 20 + 18 = 76 \text{ linear feet}$$

b) The cost is $76 \cdot \$1.32 = \$100.32$.

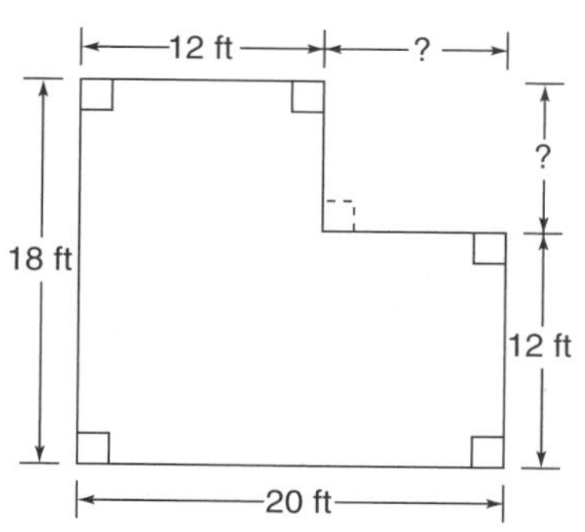

FIGURE 7.18

Heron's Formula

If the lengths of the sides of a triangle are known, the formula used to calculate the area is known as **Heron's Formula** (in honor of Heron of Alexandria, circa 75 A.D.). One of the numbers found in this formula is the *semiperimeter* of a triangle, which is defined as one-half the perimeter. For the triangle that has sides of lengths a, b, and c, the semiperimeter is $s = \frac{1}{2}(a + b + c)$. We apply Heron's Formula without proof.

THEOREM 7.2.1: **(Heron's Formula)** If the three sides of a triangle have lengths a, b, and c, then the area A of the triangle is given by

$$A = \sqrt{s(s - a)(s - b)(s - c)},$$

where the semiperimeter of the triangle is

$$s = \frac{1}{2}(a + b + c)$$

EXAMPLE 3

Find the area of a triangle which has sides of lengths 4, 13, and 15. (See Figure 7.19.)

Solution

If we designate the sides as $a = 4$, $b = 13$, and $c = 15$, the semiperimeter of the triangle is given by $s = \frac{1}{2}(4 + 13 + 15) = \frac{1}{2}(32) = 16$. Therefore

$$A = \sqrt{s(s - a)(s - b)(s - c)},$$
$$= \sqrt{16(16 - 4)(16 - 13)(16 - 15)}$$
$$= \sqrt{16(12)(3)(1)} = \sqrt{576} = 24 \text{ units}^2$$

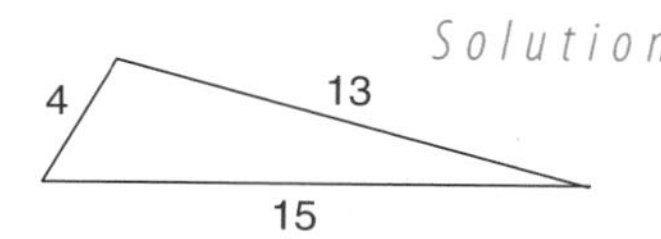

FIGURE 7.19

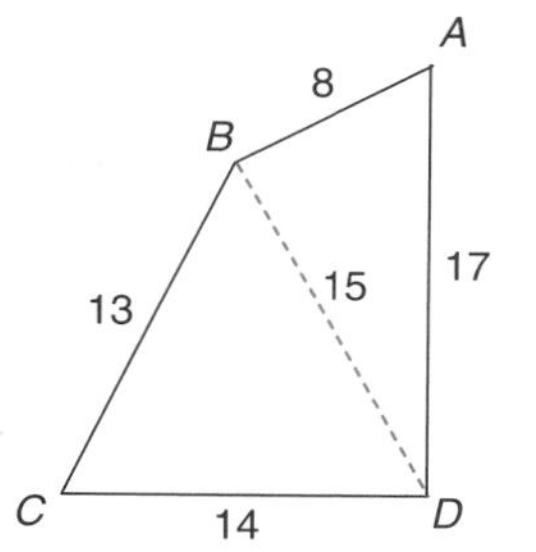

FIGURE 7.20

If the lengths of the sides of a quadrilateral are known, we can apply Heron's Formula to find the area when the length of a diagonal is also known. In quadrilateral $ABCD$ in Figure 7.20, Heron's Formula can be used to show that the area of $\triangle ABD$ is 60 while the area of $\triangle BCD$ is 84. Thus, the area of quadrilateral $ABCD$ is 144 units^2.

The remaining theorems of this section contain *numerical subscripts.* In practice, subscripts enable us to compare like quantities. For instance, the lengths of the two unequal bases of a trapezoid are written b_1 (read "b sub 1") and b_2. In particular, b_1 represents the numerical length of the first base while b_2

represents the length of the second base. The following chart illustrates the use of numerical subscripts.

Theorem	Subscripted Symbol	Meaning
Theorem 7.2.2	b_1	Length of *first* base of trapezoid
Corollary 7.2.4	d_2	Length of *second* diagonal of rhombus
Theorem 7.2.6	A_1	Area of *first* triangle

Area of a Trapezoid

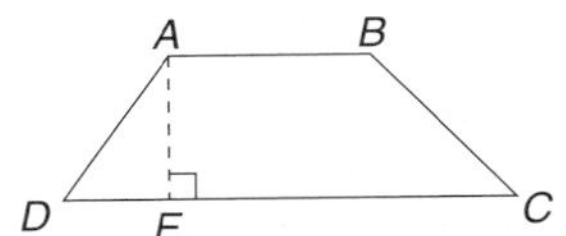

FIGURE 7.21

Recall that the two parallel sides of a trapezoid are its *bases*. The *altitude* is any line segment that is drawn perpendicular from one base to the other. In Figure 7.21, $\overline{AB}$ and $\overline{DC}$ are bases while $\overline{AE}$ is an altitude for the trapezoid.

We use the more common formula for the area of a triangle (namely, $A = \frac{1}{2}bh$) to develop our remaining theorems. In Theorem 7.2.2, b_1 and b_2 represent the lengths of the bases of the trapezoid. (In some textbooks, b represents the length of the *shorter* base while B is the length of the *longer* base.)

THEOREM 7.2.2: The area A of a trapezoid whose bases have lengths b_1 and b_2 and whose altitude has length h is given by

$$A = \frac{1}{2}h(b_1 + b_2)$$

Given: Trapezoid $ABCD$ with $\overline{AB} \parallel \overline{DC}$

Prove: $A_{ABCD} = \frac{1}{2}h(b_1 + b_2)$

Proof: Draw $\overline{AC}$ as shown in Figure 7.22(a). Now $\triangle ADC$ has an altitude of length h and a base of length b_2. As shown in Figure 7.22(b),

$$A_{ADC} = \frac{1}{2}hb_2$$

Also, $\triangle ABC$ has an altitude of length h and a base of length b_1. [See Figure 7.22(c).] Then

$$A_{ABC} = \frac{1}{2}hb_1$$

Thus
$$\begin{aligned} A_{ABCD} &= A_{ABC} + A_{ADC} \\ &= \frac{1}{2}hb_1 + \frac{1}{2}hb_2 \\ &= \frac{1}{2}h(b_1 + b_2) \end{aligned}$$

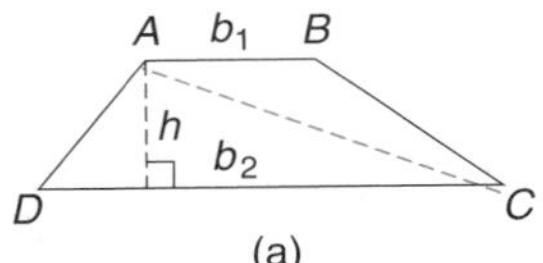

(a)

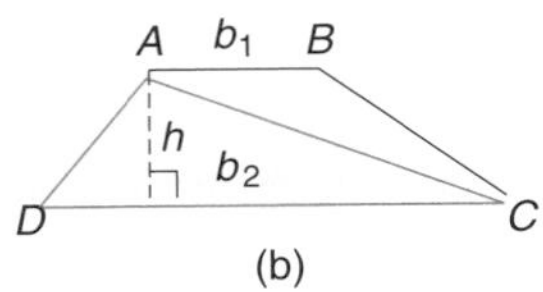

(b)

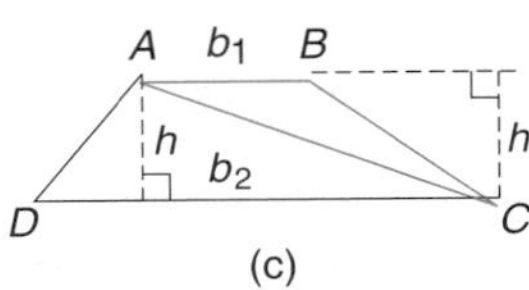

(c)

FIGURE 7.22

EXAMPLE 4

Find the area of the trapezoid in Figure 7.23 if $RS = 5$, $TV = 13$, and $RW = 6$.

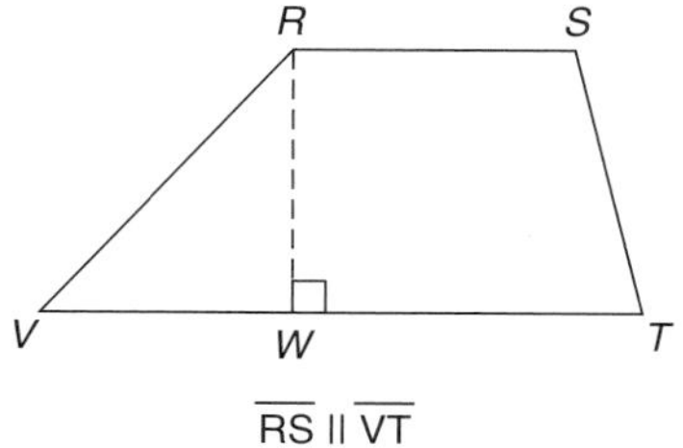

FIGURE 7.23

Solution

Let $RS = 5 = b_1$ and $TV = 13 = b_2$. Also, $RW = h = 6$. Now,

$$A = \frac{1}{2}h(b_1 + b_2)$$

becomes

$$A = \frac{1}{2} \cdot 6(5 + 13)$$

$$= \frac{1}{2} \cdot 6 \cdot 18$$

$$= 3 \cdot 18 = 54 \text{ units}^2$$

The following activity emphasizes the formula for the area of a trapezoid.

DISCOVER!

Cut out two trapezoids that are copies of each other and place one next to the other to form a parallelogram.

a) How long is the base of the parallelogram?

b) What is the area of the parallelogram?

c) What is the area of the trapezoid?

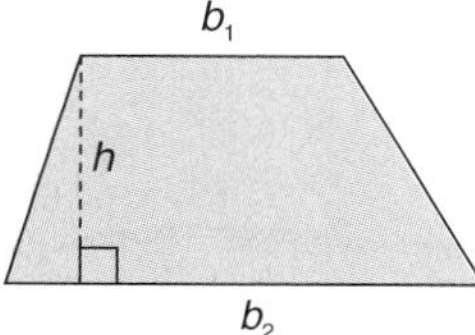

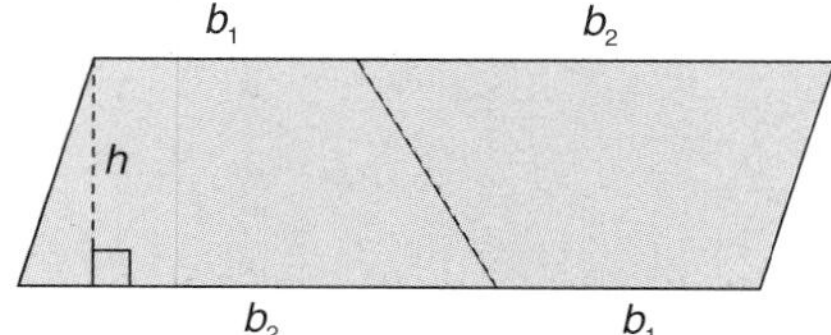

ANSWERS

a) $b_1 + b_2$ **b)** $h(b_1 + b_2)$ **c)** $\frac{1}{2}h(b_1 + b_2)$

Quadrilaterals with Perpendicular Diagonals

The following theorem leads to Corollaries 7.2.4 and 7.2.5, where the formula is also used to find the area of a rhombus and kite.

THEOREM 7.2.3: The area of any quadrilateral with perpendicular diagonals of lengths d_1 and d_2 is given by

$$A = \frac{1}{2} d_1 d_2$$

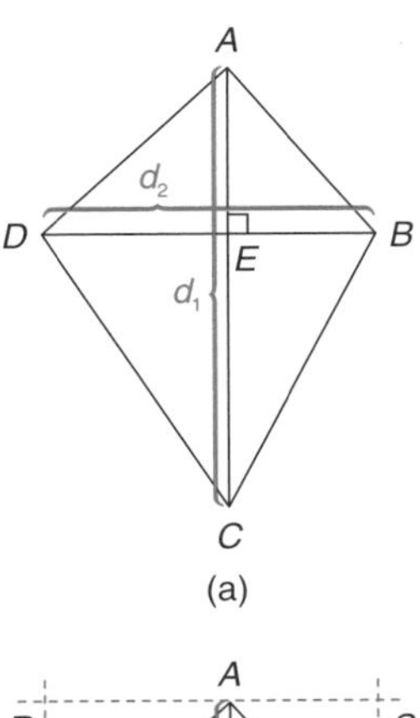

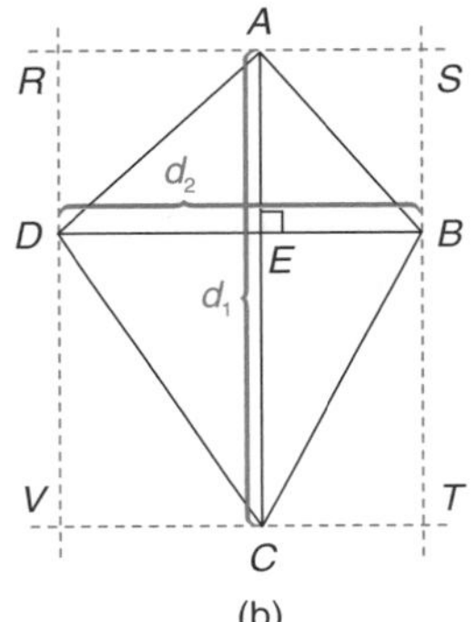

FIGURE 7.24

Given: Quadrilateral $ABCD$ with $\overline{AC} \perp \overline{BD}$ [see Figure 7.24(a).]

Prove: $A_{ABCD} = \frac{1}{2}\, d_1 d_2$

Proof: Through points A and C, draw lines parallel to $\overline{DB}$. Likewise, draw lines parallel to $\overline{AC}$ through points B and D. Let the points of intersection of these lines be R, S, T, and V, as shown in Figure 7.24(b). Because each quadrilateral $ARDE$, $ASBE$, $BECT$, and $CEDV$ is a parallelogram containing a right angle, each is a rectangle. Furthermore, $A_{\triangle ADE} = \frac{1}{2} \cdot A_{ARDE}$, $A_{\triangle ABE} = \frac{1}{2} \cdot A_{ASBE}$, $A_{\triangle BEC} = \frac{1}{2} \cdot A_{BECT}$, and $A_{\triangle DEC} = \frac{1}{2} \cdot A_{CEDV}$.

Then $A_{ABCD} = \frac{1}{2} \cdot A_{RSTV}$. But $RSTV$ is a rectangle, since it is a parallelogram containing a right angle. Because $RSTV$ has dimensions d_1 and d_2 [see Figure 7.24(b)], its area is $d_1 d_2$. By substitution, $A_{ABCD} = \frac{1}{2}\, d_1 d_2$.

Area of a Rhombus

Recall that a rhombus is a parallelogram with two congruent adjacent sides; in turn, we proved that all four sides were congruent. Because the diagonals of a rhombus are perpendicular, we have the following corollary of Theorem 7.2.3.

COROLLARY 7.2.4: The area A of a rhombus whose diagonals have lengths d_1 and d_2 is given by

$$A = \frac{1}{2}\, d_1 d_2$$

(See Figure 7.25.)

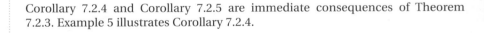

FIGURE 7.25

Corollary 7.2.4 and Corollary 7.2.5 are immediate consequences of Theorem 7.2.3. Example 5 illustrates Corollary 7.2.4.

EXAMPLE 5

Find the area of the rhombus $MNPQ$ in Figure 7.26 if $MP = 12$ and $NQ = 16$.

Solution

By Corollary 7.2.4,

$$A_{MNPQ} = \frac{1}{2}\, d_1 d_2 = \frac{1}{2} \cdot 12 \cdot 16 = 96 \text{ units}^2$$

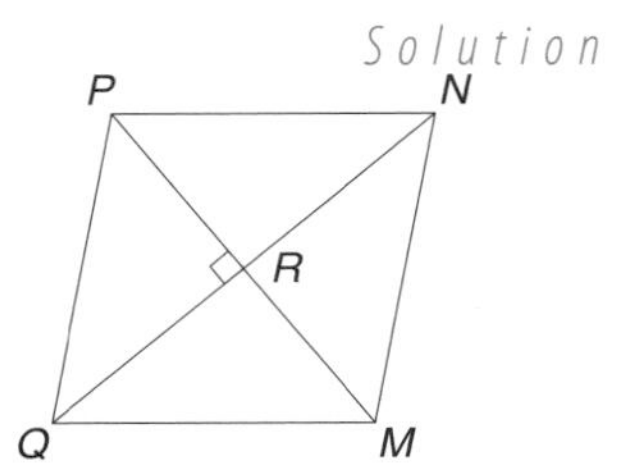

FIGURE 7.26

In problems involving the rhombus, we often use the fact that diagonals are perpendicular to find needed measures. If the length of a side and the length of either diagonal are known, the length of the other diagonal can be found by applying the Pythagorean Theorem.

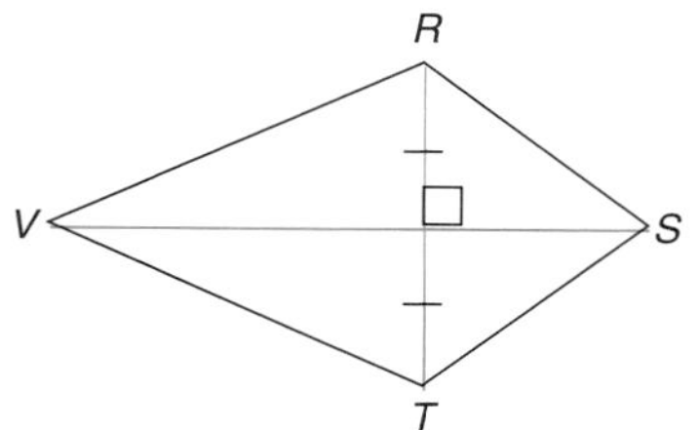

FIGURE 7.27

Area of a Kite

For a kite, we proved in Exercise 25 of page 164 that one diagonal is the perpendicular bisector of the other. (See Figure 7.27.)

COROLLARY 7.2.5: The area A of a kite whose diagonals have lengths d_1 and d_2 is given by

$$A = \frac{1}{2} d_1 d_2$$

We apply Corollary 7.2.5 in Example 6.

EXAMPLE 6

Find the length of $\overline{RT}$ in Figure 7.28 if the area of the kite $RSTV$ is 360 in.^{2} and $SV = 30$ in.

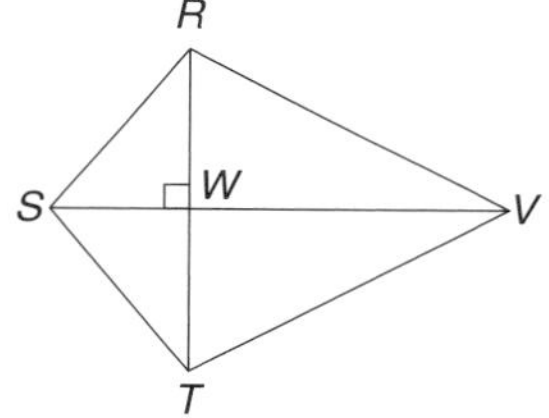

FIGURE 7.28

Solution

$A = \frac{1}{2}d_1d_2$ becomes $360 = \frac{1}{2}(30)d$, in which d is the length of the remaining diagonal $\overline{RT}$. Then $360 = 15d$, which means that $d = 24$. Then $RT = 24$ in.

Areas of Similar Polygons

The following theorem compares the areas of similar triangles. In Figure 7.29 on the next page, we refer to the areas of the similar triangles as A_1 and A_2. The triangle with area A_1 has sides of lengths a_1, b_1, and c_1 while the triangle with area A_2 has sides of lengths a_2, b_2, and c_2. Where a_1 corresponds to a_2, b_1 to b_2, and c_1 to c_2, Theorem 7.2.6 implies that

$$\frac{A_1}{A_2} = \left(\frac{a_1}{a_2}\right)^2 \text{ or } \frac{A_1}{A_2} = \left(\frac{b_1}{b_2}\right)^2 \text{ or } \frac{A_1}{A_2} = \left(\frac{c_1}{c_2}\right)^2$$

We prove only the first relationship; the other proofs are analogous.

THEOREM 7.2.6: The ratio of the areas of two similar triangles equals the square of the ratio of the lengths of any two corresponding sides; that is,

$$\frac{A_1}{A_2} = \left(\frac{a_1}{a_2}\right)^2$$

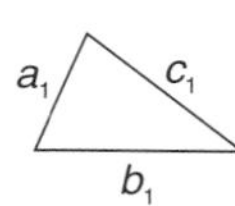

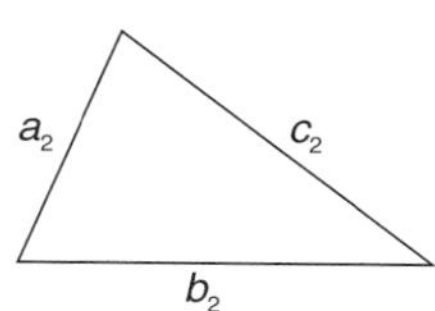

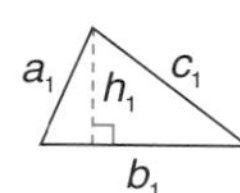

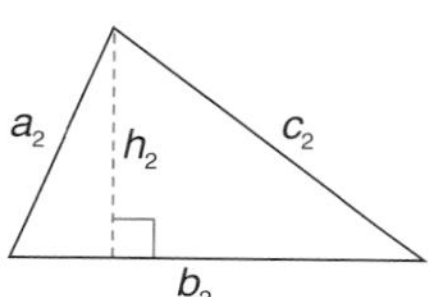

FIGURE 7.29

Given: Similar triangles as shown in Figure 7.29.

Prove: $\frac{A_1}{A_2} = \left(\frac{a_1}{a_2}\right)^2$

Proof: For the similar triangles, h_1 and h_2 are the respective lengths of altitudes to the corresponding sides of lengths b_1 and b_2. Now $A_1 = \frac{1}{2}b_1h_1$ and $A_2 = \frac{1}{2}b_2h_2$, so

$$\frac{A_1}{A_2} = \frac{\frac{1}{2}b_1h_1}{\frac{1}{2}b_2h_2} \quad \text{or} \quad \frac{A_1}{A_2} = \frac{\frac{1}{2}}{\frac{1}{2}} \cdot \frac{b_1}{b_2} \cdot \frac{h_1}{h_2}$$

Simplifying, we have

$$\frac{A_1}{A_2} = \frac{b_1}{b_2} \cdot \frac{h_1}{h_2}$$

Because the triangles are similar, we know that $\frac{b_1}{b_2} = \frac{a_1}{a_2}$. Because corresponding altitudes of similar triangles have the same ratio as a pair of corresponding sides (Theorem 5.2.2), we also know that $\frac{h_1}{h_2} = \frac{a_1}{a_2}$. Through substitution, $\frac{A_1}{A_2} = \frac{b_1}{b_2} \cdot \frac{h_1}{h_2}$ becomes $\frac{A_1}{A_2} = \frac{a_1}{a_2} \cdot \frac{a_1}{a_2}$. Then $\frac{A_1}{A_2} = \left(\frac{a_1}{a_2}\right)^2$.

Because Theorem 7.2.6 can be extended to any pair of similar polygons, we could also prove that the ratio of the areas of two squares equals the square of the ratio of the lengths of any two sides (if taken in the same order). We apply this relationship in Example 7.

EXAMPLE 7

Use the ratio $\frac{A_1}{A_2}$ to compare the areas of

a) Two similar triangles in which the sides of the first triangle are $\frac{1}{2}$ as long as the sides of the second triangle

b) Two squares in which each side of the first square is 3 times as long as each side of the second square

Solution

a) $s_1 = \frac{1}{2}s_2$, so $\frac{s_1}{s_2} = \frac{1}{2}$. (See Figure 7.30.)

Now $\frac{A_1}{A_2} = \left(\frac{s_1}{s_2}\right)^2$ becomes $\frac{A_1}{A_2} = \left(\frac{1}{2}\right)^2$ or $\frac{A_1}{A_2} = \frac{1}{4}$. That is, the area of the first triangle is $\frac{1}{4}$ the area of the second triangle.

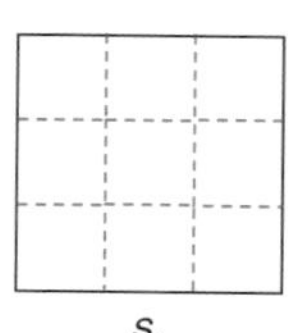

FIGURE 7.31

b) $s_1 = 3s_2$, so $\frac{s_1}{s_2} = 3$. (See Figure 7.31.)

$\frac{A_1}{A_2} = \left(\frac{s_1}{s_2}\right)^2$ becomes $\frac{A_1}{A_2} = (3)^2$ or $\frac{A_1}{A_2} = 9$. That is, the area of the first square is 9 times the area of the second square.

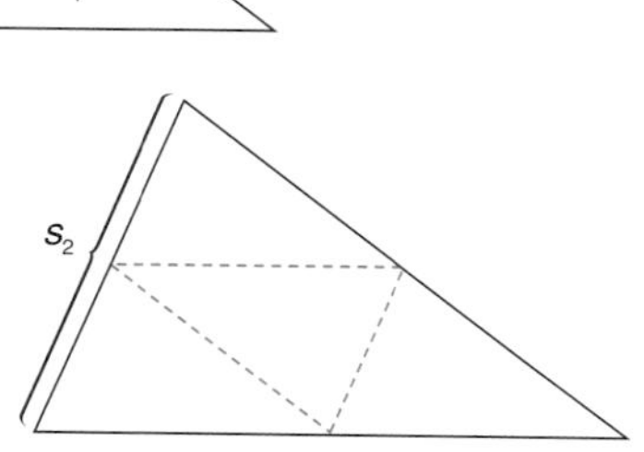

FIGURE 7.30

NOTE: Figures 7.30 and 7.31 provide visual insight into the relationship described in Theorem 7.2.6.

7.2 Exercises

In Exercises 1 to 8, find the perimeter of each figure.

1.

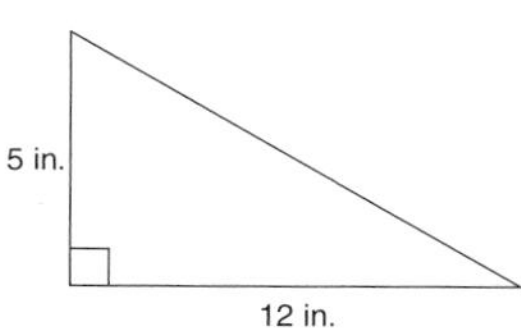

2.

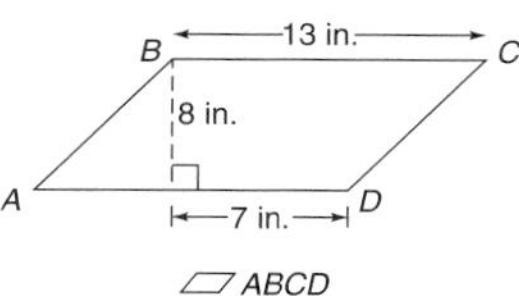

▱ $ABCD$

3.

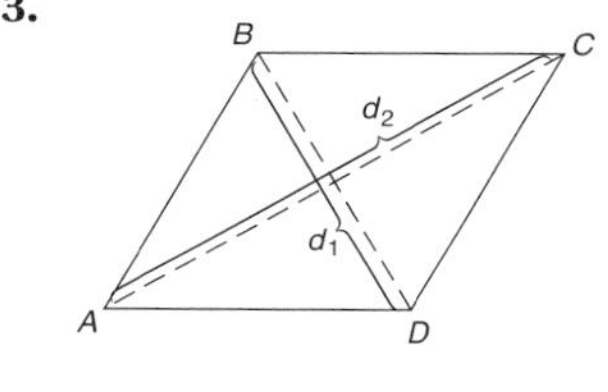

▱ $ABCD$ with $\overline{AB} \cong \overline{BC}$

$d_1 = 4$ m

$d_2 = 10$ m

4.

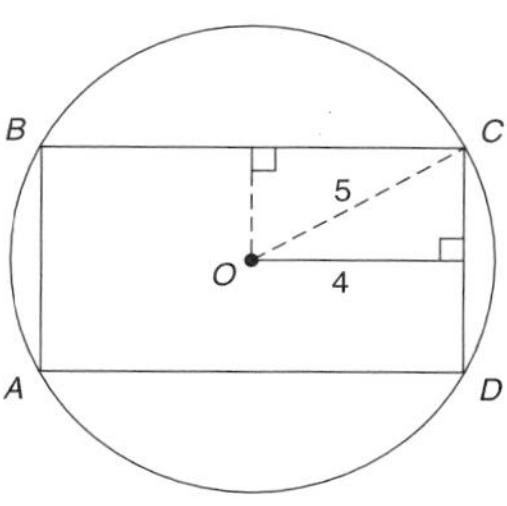

▱ $ABCD$ in ⊙ O

5.

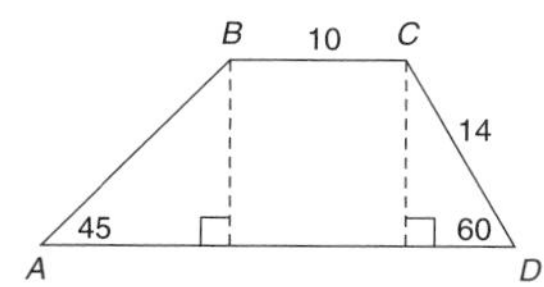

Trapezoid $ABCD$ with $\overline{BC} \parallel \overline{AD}$

$m\angle A = 45$, $m\angle D = 60$,

$BC = 10$, $CD = 14$

6.

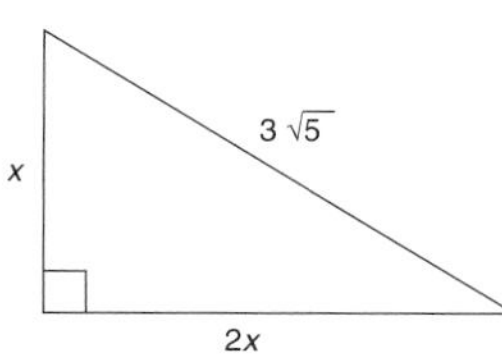

7.

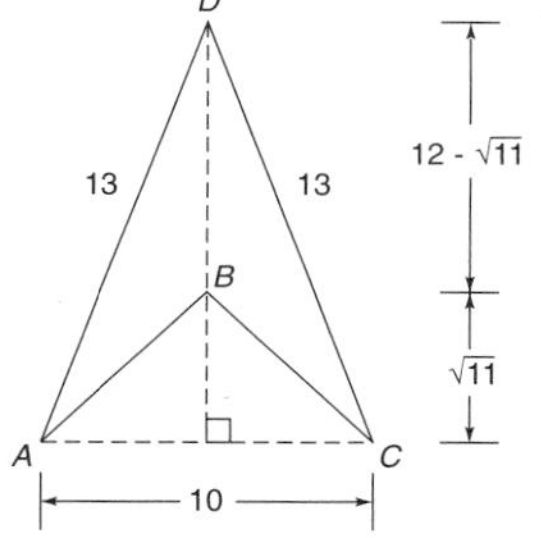

$\overline{AB} \cong \overline{BC}$ in concave quadrilateral $ABCD$

8.

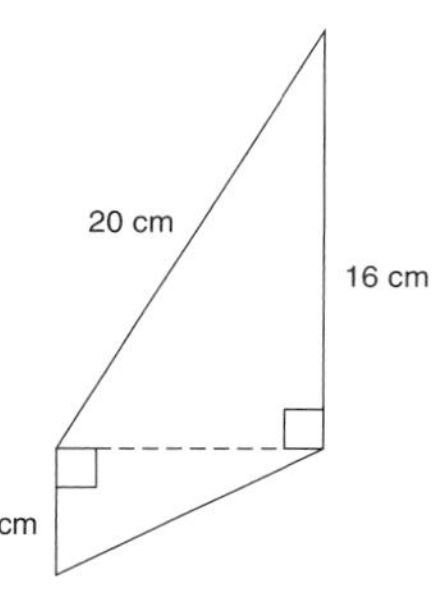

In Exercises 9 and 10, use Heron's Formula.

9. Find the area of a triangle whose sides measure 13, 14, and 15 in.

10. Find the area of a triangle whose sides measure 10, 17, and 21 cm.

In Exercises 11 to 16, find the area of the given polygon.

11.

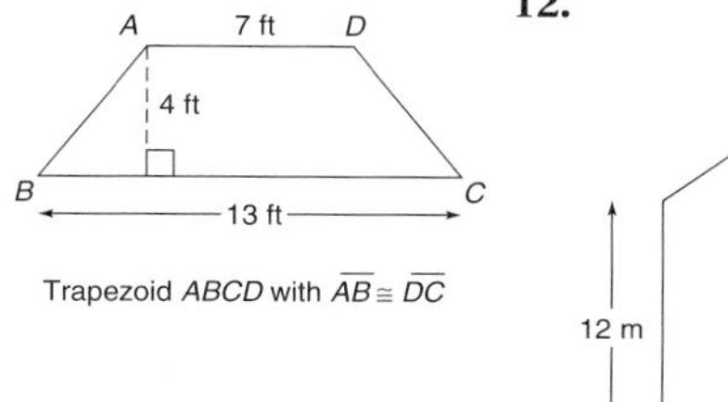

Trapezoid $ABCD$ with $\overline{AB} \cong \overline{DC}$

12.

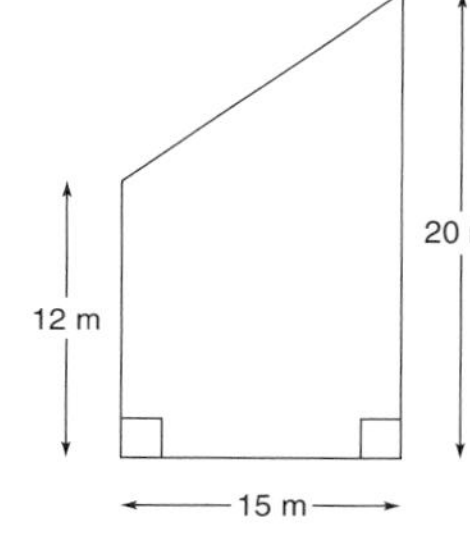

13.

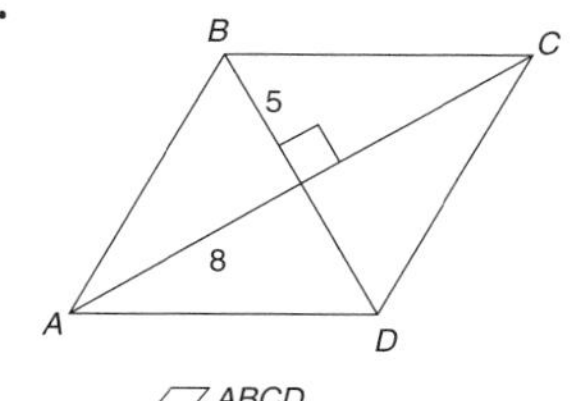

▱ $ABCD$

14.

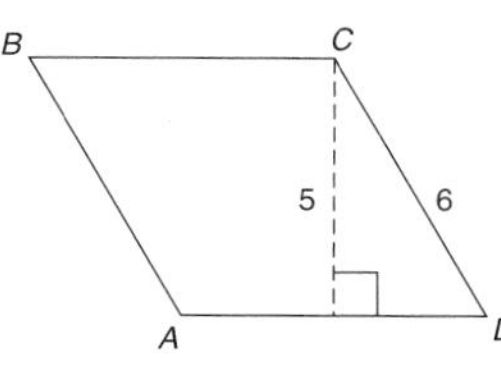

▱ $ABCD$ with $\overline{BC} \cong \overline{CD}$

15.

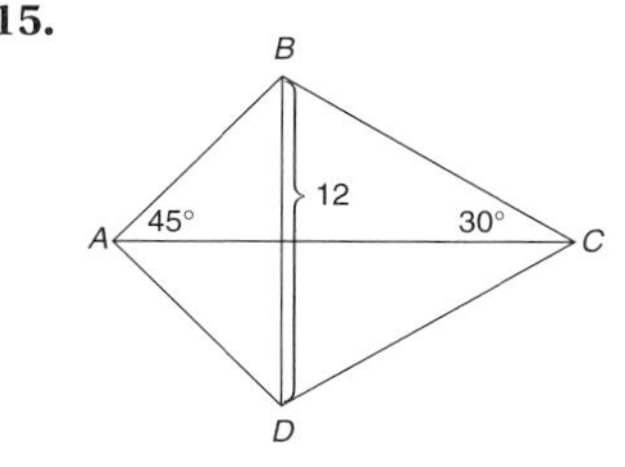

Kite $ABCD$ with $BD = 12$

$m\angle BAC = 45°$, $m\angle BCA = 30°$

16.

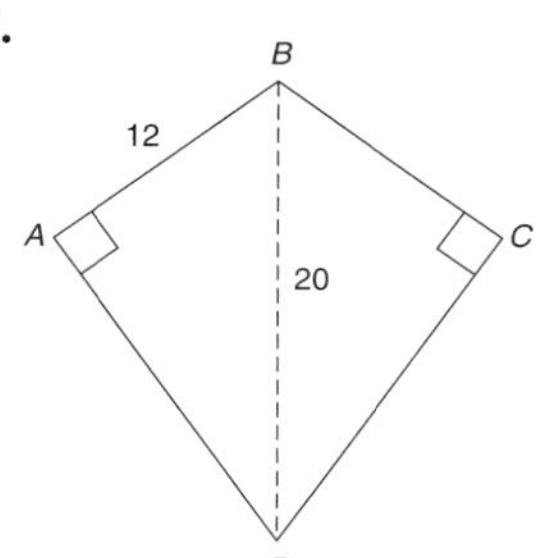

Kite $ABCD$

17. In a triangle of perimeter 76 in., the length of the first side is twice the length of the second side, and the length of the third side is 12 in. more than the length of the second side. Find the lengths of the three sides.

18. In a triangle whose area is 72 in.^{2}, the base has a length of 8 in. Find the length of the altitude.

19. A trapezoid has an area of 96 cm^2. If the altitude has a length of 8 cm and one base has a length of 9 cm, find the length of the other base.

20. The numerical difference between the area of a square and the perimeter of that square is 32. Find the length of a side of the square.

21. Find the ratio $\frac{A_1}{A_2}$ of the areas of two similar triangles if:

a) The ratio of corresponding sides is $\frac{s_1}{s_2} = \frac{3}{2}$

b) The lengths of the sides of the first triangle are 6, 8, and 10 in. while those of the second triangle are 3, 4, and 5 in.

22. Find the ratio $\frac{A_1}{A_2}$ of the areas of two similar rectangles if:

a) The ratio of corresponding sides is $\frac{s_1}{s_2} = \frac{2}{5}$

b) The length of the first rectangle is 6 m while the length of the second rectangle is 4 m.

In Exercises 23 and 24, give a paragraph form of proof. Provide drawings as needed.

23. *Given:* Equilateral $\triangle ABC$ with each side of length s

Prove: $A_{ABC} = \frac{s^2}{4}\sqrt{3}$ (**HINT:** Use Heron's Formula.)

24. *Given:* Isosceles $\triangle MNQ$ with $QM = QN = s$ and $MN = 2a$

Prove: $A_{MNQ} = a\sqrt{s^2 - a^2}$ (**NOTE:** $s > a$.)

In Exercises 25 to 28, find the area of the figure shown.

25. *Given:* In $\odot O$, $OA = 5$

$BC = 6$ and $CD = 4$

Find: A_{ABCD}

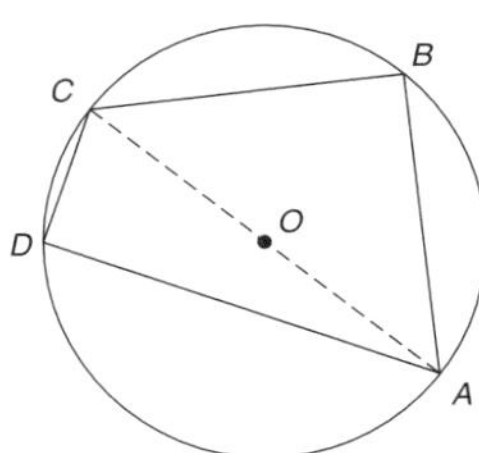

26. *Given:* Hexagon $RSTVWX$ with $\overline{WV} \parallel \overline{XT} \parallel \overline{RS}$

$RS = 10$

$ST = 8$

$TV = 5$

$WV = 16$

and $\overline{WX} \cong \overline{VT}$

Find: A_{RSTVWX}

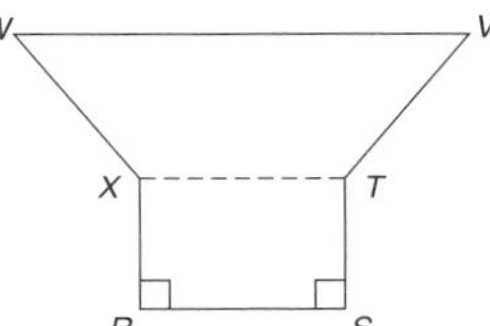

27. *Given:* Pentagon $ABCDE$ with $\overline{DC} \cong \overline{DE}$, $AE = AB = 5$, $BC = 12$

Find: A_{ABCDE}

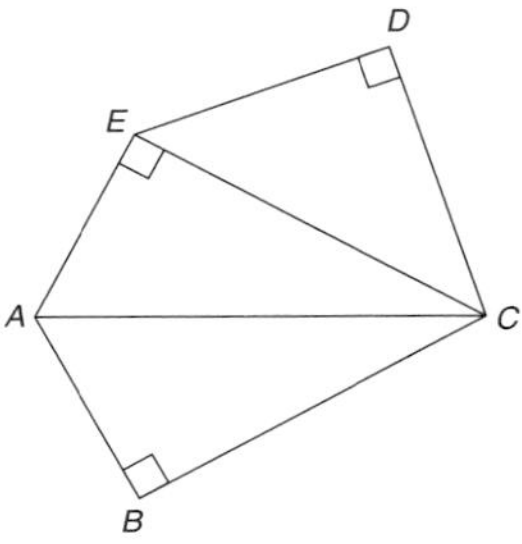

28. *Given:* Pentagon $RSTVW$ with $m\angle VRS = m\angle VSR = 60°$, $RS = 8\sqrt{2}$, $\overline{RW} \cong \overline{WV} \cong \overline{VT} \cong \overline{TS}$

Find: A_{RSTVW}

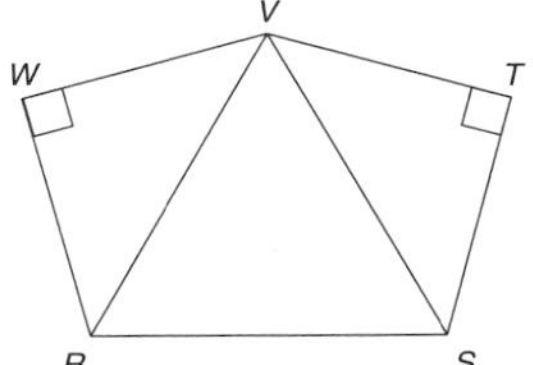

29. Mary Frances has a rectangular garden plot that encloses an area of 48 yd^2. If 28 yd of fencing are purchased to enclose the garden, what are the dimensions of the rectangular plot?

30. The perimeter of a right triangle is 12 m. If the hypotenuse has a length of 5 m, find the lengths of the two legs.

31. Farmer Watson wishes to fence a rectangular plot of ground measuring 245 ft by 140 ft.

a) What amount of fencing is needed?

b) What is the total cost of the fencing if it costs \$0.59 per ft?

32. The farmer in Exercise 31 has decided to take the fencing purchased and use it to enclose the subdivided plots shown.

a) What are the overall dimensions of the rectangular enclosure shown?

b) What is the total area of the enclosures shown?

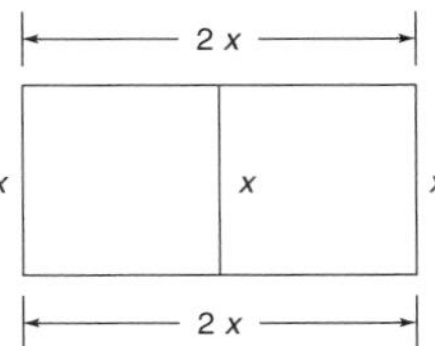

33. Find the area of the room whose floor plan is shown.

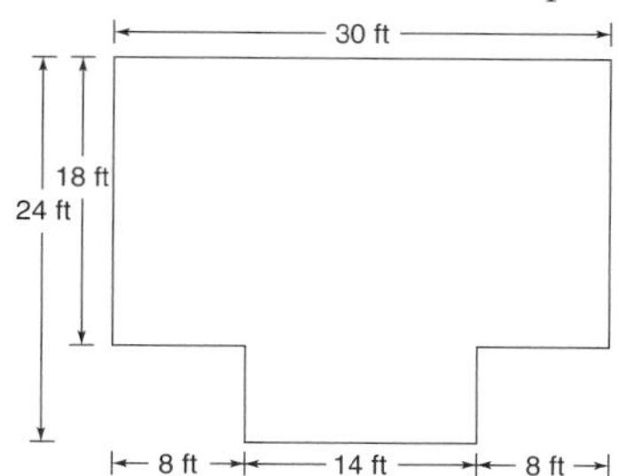

Exercises 33, 34

34. Find the perimeter of the room in Exercise 33.

35. Examine several rectangles, each with a perimeter of 40 in., and find the dimensions of the rectangle that has the largest area. What type of figure has the largest area?

36. Examine several rectangles, each with an area of 36 in.², and find the dimensions of the rectangle that has the smallest perimeter. What type of figure has the smallest perimeter?

37. Square *RSTV* is inscribed in square *WXYZ* as shown. If $ZT = 5$ and $TY = 12$, find:

a) The perimeter of *RSTV*

b) The area of *RSTV*

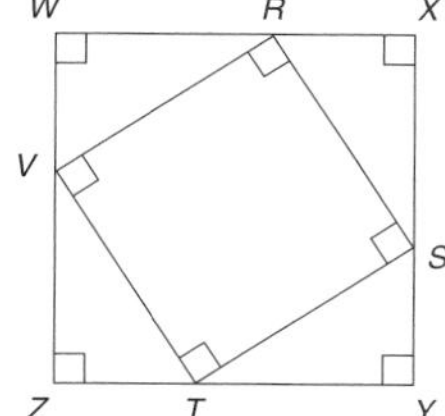

Exercises 37, 38

38. Square *RSTV* is inscribed in square *WXYZ* as shown. If $ZT = 8$ and $TY = 15$, find:

a) The perimeter of *RSTV*

b) The area of *RSTV*

For Exercises 39 and 40, use this information: Let a, b, and c be the integer lengths of sides of a triangle. If the area of the triangle is also an integer, then (a, b, c) is known as a Heron triple.

39. Which of these are Heron triples?

a) (5, 6, 7) **b)** (13, 14, 15)

40. Which of these are Heron triples?

a) (9, 10, 17) **b)** (8, 10, 12)

41. Prove that the area of a trapezoid whose altitude has length h and whose median has length m is $A = hm$.

42. Prove that the area of a square whose diagonal length is d is $A = \frac{1}{2}d^2$.

43. ➤ The shaded region is that of a trapezoid. Determine the height of the trapezoid.

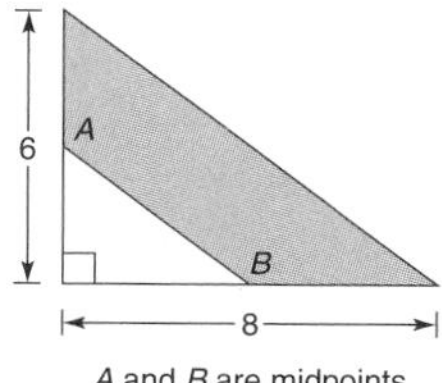

A and *B* are midpoints

7.3 Regular Polygons and Area

In this section, our main goal is to develop a formula for the area of any **regular polygon.** However, several interesting properties of regular polygons are developed as we move toward this goal. For instance, every regular polygon has both an inscribed circle and a circumscribed circle; furthermore, these two circles are concentric. In Example 1, we use angle bisectors to locate the center of the inscribed circle. The center, which is found by using the bisectors of any two consecutive angles, is equidistant from the sides of the square.

EXAMPLE 1

Given square *ABCD* in Figure 7.32(a) on the next page, construct inscribed $\odot O$.

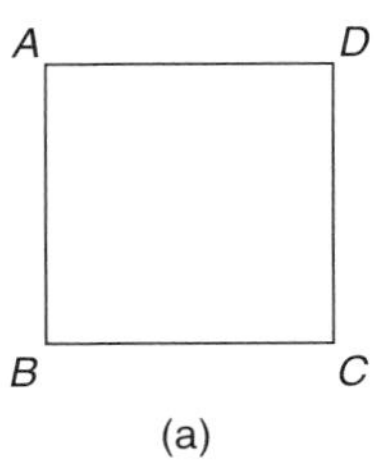

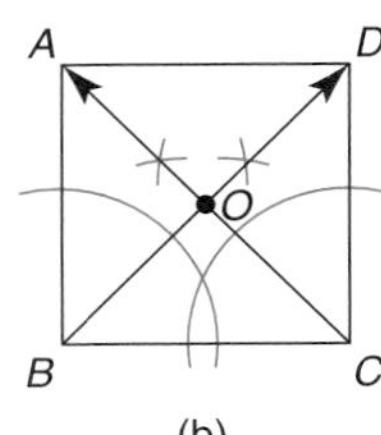

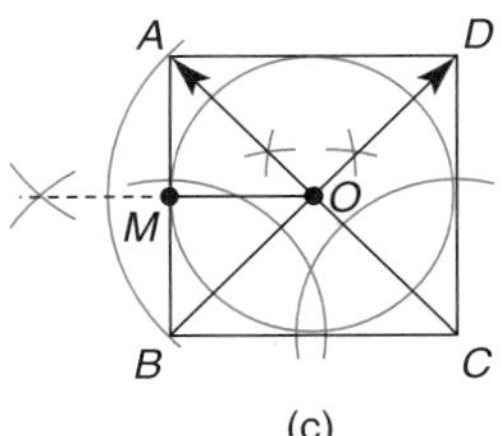

FIGURE 7.32

Solution

The center of an inscribed circle must lie at the same distance from each side. Center O is the point of concurrency of the angle bisectors of the square. In Figure 7.32(b), we construct the angle bisectors of $\angle B$ and $\angle C$ to identify point O. Constructing $\overline{OM} \perp \overline{AB}$, OM is the distance from O to $\overline{AB}$ and the length of the radius of the inscribed circle. Finally we construct inscribed $\odot O$ with radius $\overline{OM}$ in Figure 7.32(c).

In Example 2, we use the perpendicular bisectors of two consecutive sides to locate the center of the circumscribed circle. The center determines a point that is equidistant from the vertices of the hexagon.

EXAMPLE 2

Given regular hexagon $MNPQRS$ in Figure 7.33(a), construct circumscribed $\odot X$.

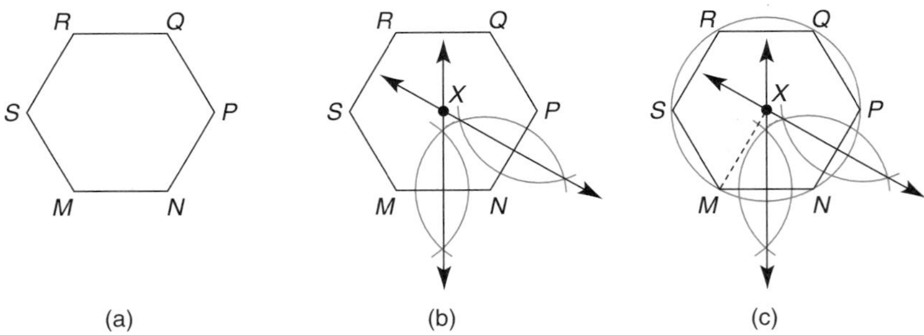

FIGURE 7.33

Solution

The center of a circumscribed circle must lie at the same distance from each vertex. Center X is the point of concurrency of the perpendicular bisectors of two consecutive sides of the hexagon. In Figure 7.33(b), we construct the perpendicular bisectors of $\overline{MN}$ and $\overline{NP}$ to locate point X. Where XM is the distance from X to vertex M, we use radius $\overline{XM}$ to construct circumscribed $\odot X$ in Figure 7.33(c).

FIGURE 7.34

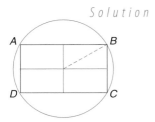

For a rectangle, which is not a regular polygon, we can only circumscribe the circle (see Figure 7.34). Why? For a rhombus (also not a regular polygon), we can only inscribe the circle (see Figure 7.35). Why?

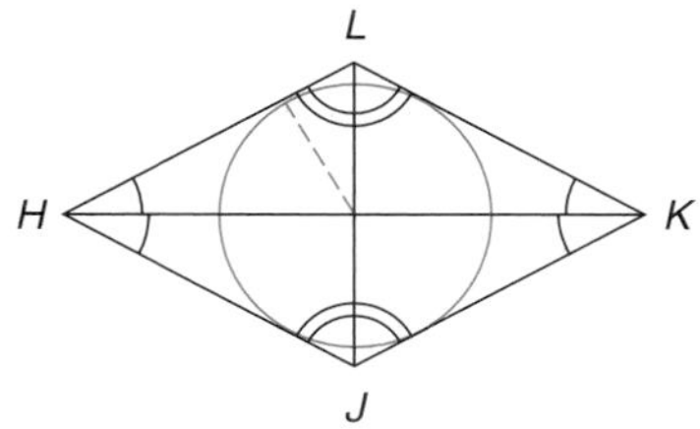

FIGURE 7.35

As we shall see, we can construct the inscribed and circumscribed circles for regular polygons because they are both equilateral and equiangular. A few of the regular polygons are shown in Figure 7.36.

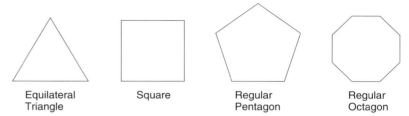

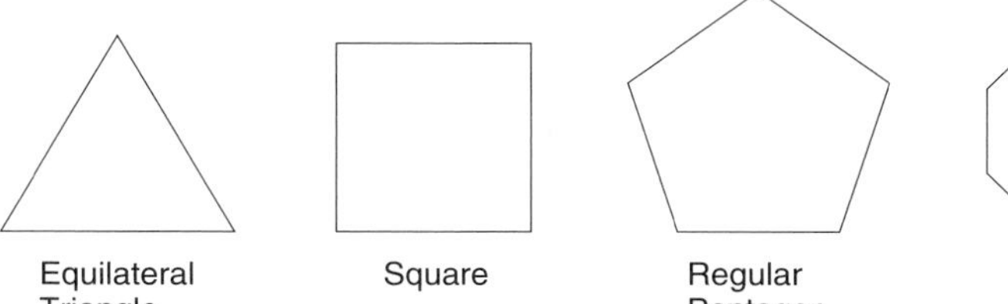

FIGURE 7.36

We saw earlier that the sum of measures of the interior angles of any polygon is given by the formula

$$S = (n - 2)180$$

where n is the number of sides of the polygon. The sum of the measures of the exterior angles is always 360°.

EXAMPLE 3

a) Find the measure of each interior angle of a regular polygon with 15 sides.
b) Find the number of sides of a regular polygon if each interior angle measures 144°.

Solution

a) Because each of the n angles have equal measures, the formula for the measure of each interior angle,

$$I = \frac{(n - 2)180}{n}$$

becomes

$$I = \frac{(15 - 2)180}{15}$$

which simplifies to 156°.

b) Since $I = 144°$, we can determine the number of sides by solving the equation

$$\frac{(n - 2)180}{n} = 144$$

Then

$$\begin{aligned} (n - 2)180 &= 144n \\ 180n - 360 &= 144n \\ 36n &= 360 \\ n &= 10 \end{aligned}$$

NOTE: In Example 3(a), we could have found the measure of each exterior angle and then used the fact that the interior angle is its supplement. In Example 3(b), the supplement of the interior angle is the exterior angle; then we could have used the formula $E = \frac{360}{n}$ to find n. Since $I = 144°$, the value of E is 36°.

Regular polygons allow us to inscribe and to circumscribe a circle. The next theorem will establish the following relationships:

1. The centers of the inscribed and circumscribed circles of a regular polygon are the same.
2. The angle bisectors of two consecutive angles or the perpendicular bisectors of two consecutive sides can be used to locate this common center.
3. The inscribed circle's radius is any line segment from the center drawn perpendicular to a side; the radius of the circumscribed circle joins the center to any vertex.

THEOREM 7.3.1: A circle can be circumscribed about (or inscribed in) any regular polygon.

Given: Regular polygon *ABCDEF*. [See Figure 7.37(a).]
Prove: A circle *O* can be circumscribed about *ABCDEF* and a circle can be inscribed in *ABCDEF*.

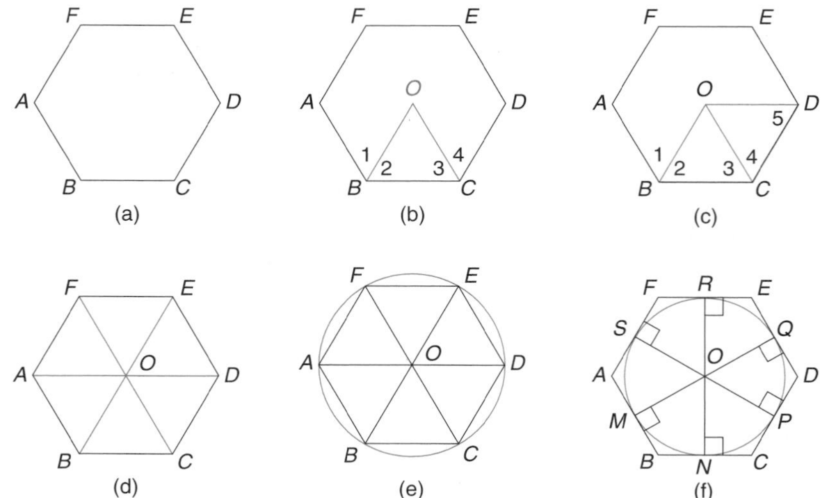

FIGURE 7.37

Proof: Let point O be the point at which the angle bisectors for $\angle ABC$ and $\angle BCD$ meet. [See Figure 7.37(b).] Then $\angle 1 \cong \angle 2$ and $\angle 3 \cong \angle 4$.

Because $\angle ABC \cong \angle BCD$ (by definition of regular polygons), it follows that

$$\frac{1}{2}m\angle ABC = \frac{1}{2}m\angle BCD$$

In turn, $m\angle 2 = m\angle 3$, so $\angle 2 \cong \angle 3$. Then $\overline{OB} \cong \overline{OC}$ (sides opposite $\cong$ $\angle$s of a $\triangle$).

From the facts that $\angle 3 \cong \angle 4$, $\overline{OC} \cong \overline{OC}$, and $\overline{BC} \cong \overline{CD}$, it follows that $\triangle OCB \cong \triangle OCD$ by SAS. [See Figure 7.37(c).] In turn, $\overline{OC} \cong \overline{OD}$ by CPCTC, so $\angle 4 \cong \angle 5$ since these lie opposite $\overline{OC}$ and $\overline{OD}$.

Because $\angle 5 \cong \angle 4$ and $m\angle 4 = \frac{1}{2}m\angle BCD$, it follows that $m\angle 5 = \frac{1}{2}m\angle BCD$. But $\angle BCD \cong \angle CDE$ since these are angles of a regular polygon. For that reason, $m\angle 5 = \frac{1}{2}m\angle CDE$, and $\overline{OD}$ bisects $\angle CDE$.

By continuing this procedure, we can show that $\overline{OE}$ bisects $\angle DEF$, $\overline{OF}$ bisects $\angle EFA$, and $\overline{OA}$ bisects $\angle FAB$. Therefore the resulting $\triangle AOB$, $\triangle BOC$, $\triangle COD$, $\triangle DOE$, $\triangle EOF$, and $\triangle FOA$ are congruent by ASA. [See Figure 7.37(d).] By CPCTC, $\overline{OA} \cong \overline{OB} \cong \overline{OC} \cong \overline{OD} \cong \overline{OE} \cong \overline{OF}$. With O as center and $\overline{OA}$ as radius, circle O can be circumscribed about $ABCDEF$, as shown in Figure 7.37(e).

Because corresponding altitudes of $\cong$ $\triangle$s are also congruent, we see that $\overline{OM} \cong \overline{ON} \cong \overline{OP} \cong \overline{OQ} \cong \overline{OR} \cong \overline{OS}$, where these are the altitudes to the bases of the triangles.

Again with O as center, but now with a radius equal in length to OM, we complete the inscribed circle in $ABCDEF$. [See Figure 7.37(f).]

In the proof of Theorem 7.3.1 a regular hexagon was drawn. The method of proof would not change, regardless of the number of sides chosen. In the proof, point O was the common center of the circumscribed and inscribed circles for $ABCDEF$.

DEFINITION: The **center of a regular polygon** is the common center for the inscribed and circumscribed circles of the polygon.

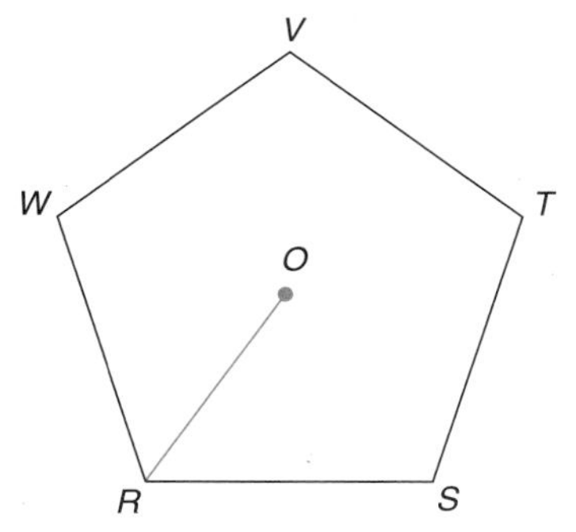

FIGURE 7.38

NOTE: The preceding definition does not tell us how to locate the center of a regular polygon. The center is the intersection of the angle bisectors of two consecutive angles; alternately, the intersection of the perpendicular bisectors of two consecutive sides can also be used to locate the center of the regular polygon. Note that a regular polygon has a center, whether or not either of the related circles are shown.

In Figure 7.38, point O is the center of the regular pentagon $RSTVW$. In this figure, $\overline{OR}$ is called a "radius" of the regular pentagon.

DEFINITION: A **radius of a regular polygon** is any line segment that joins the center of the regular polygon to one of its vertices.

In the proof of Theorem 7.3.1, we saw that "All radii of a regular polygon are congruent."

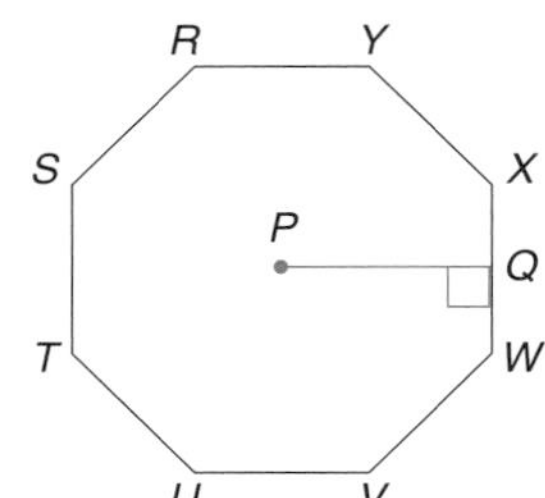

FIGURE 7.39

DEFINITION: An **apothem** of a regular polygon is any line segment drawn from the center of that polygon perpendicular to one of the sides.

In regular octagon *RSTUVWXY* in Figure 7.39, whose center is point *P*, the segment $\overline{PQ}$ is an apothem. Any regular polygon of n sides has n apothems and n radii. The proof of Theorem 7.3.1 establishes that "All apothems of a regular polygon are congruent."

DEFINITION: A **central angle of a regular polygon** is an angle formed by two consecutive radii of the regular polygon.

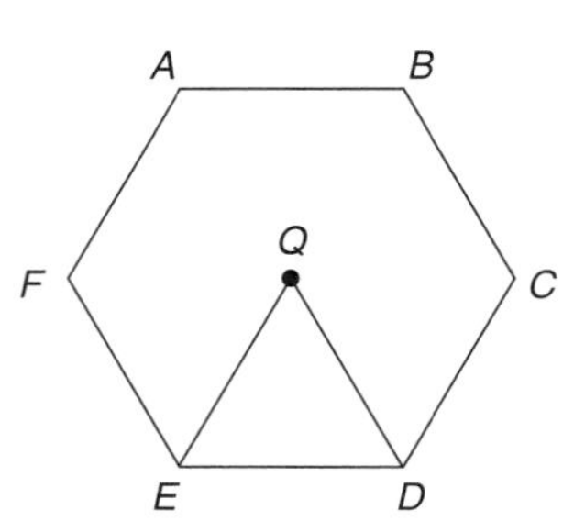

FIGURE 7.40

In regular hexagon *ABCDEF* with center *Q* (see Figure 7.40), angle *EQD* is a central angle. Because of the congruences of the triangles in the proof of Theorem 7.3.1, "All central angles of a regular polygon are congruent."

THEOREM 7.3.2 The measure of the central angle of a regular polygon of n sides is given by $c = \frac{360}{n}$.

We apply Theorem 7.3.2 in Example 4.

EXAMPLE 4

a) Find the measure of the central angle of a regular polygon of 9 sides.
b) Find the number of sides of a regular polygon whose central angle measures 72°.

Solution

a) $c = \frac{360}{9} = 40°$
b) $72 = \frac{360}{n} \rightarrow 72n = 360 \rightarrow n = 5$ sides

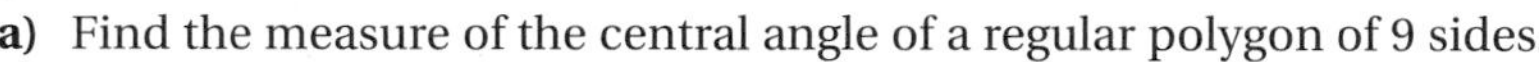

The next two theorems are stated without proof.

THEOREM 7.3.3: Any radius of a regular polygon bisects the angle at the vertex to which it is drawn.

THEOREM 7.3.4: Any apothem to a side of a regular polygon bisects the side of the polygon to which it is drawn.

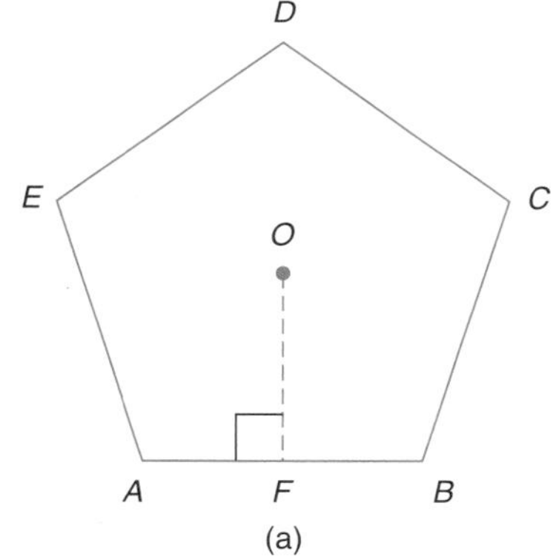

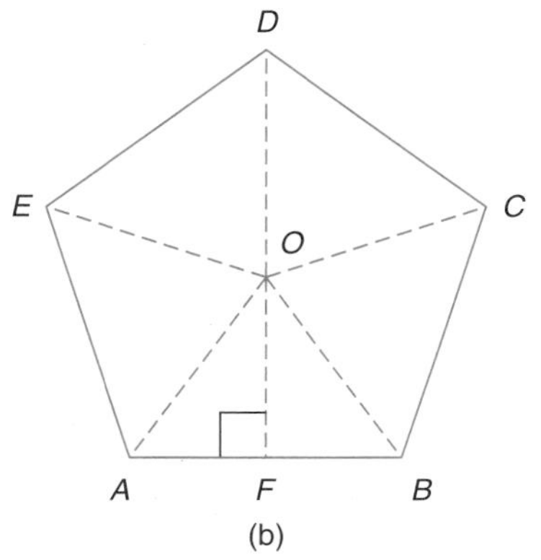

FIGURE 7.41

Area of a Regular Polygon

We have laid the groundwork for determining the area of a regular polygon. In the proof of Theorem 7.3.5, the figure chosen is a regular pentagon; however, the proof applies to regular polygons of any number of sides.

It is also worth noting that the perimeter P of a regular polygon is the sum of its equal sides. If there are n sides and each has length s, the perimeter of the regular polygon is $P = ns$.

THEOREM 7.3.5: The area A of a regular polygon whose apothem has length a and whose perimeter is P is given by

$$A = \frac{1}{2}aP$$

Given: Regular polygon $ABCDE$ in Figure 7.41(a) so that $OF = a$ and the perimeter of $ABCDE$ is P

Prove: $A_{ABCDE} = \frac{1}{2}aP$

Proof: From center O, draw radii $\overline{OA}$, $\overline{OB}$, $\overline{OC}$, $\overline{OD}$, and $\overline{OE}$. [See Figure 7.41(b).] Now $\triangle AOB$, $\triangle BOC$, $\triangle COD$, $\triangle DOE$, and $\triangle EOA$ are all $\cong$ by SSS. Where s represents the length of each of the congruent sides of the regular polygon and a is the length of an apothem, the area of each $\triangle$ is $\frac{1}{2}sa$ (from $A = \frac{1}{2}bh$). Therefore the area of the pentagon is

$$\begin{aligned} A_{ABCDE} &= \left(\frac{1}{2}sa\right) + \left(\frac{1}{2}sa\right) + \left(\frac{1}{2}sa\right) + \left(\frac{1}{2}sa\right) + \left(\frac{1}{2}sa\right) \\ &= \frac{1}{2}a(s + s + s + s + s) \end{aligned}$$

But since the sum $s + s + s + s + s$ represents the perimeter of the polygon, we have

$$A_{ABCDE} = \frac{1}{2}aP$$

EXAMPLE 5

In Figure 7.41(a) on page 321, find the area of the regular pentagon $ABCDE$ with center O if $OF = 4$ and $AB = 5.9$.

Solution

$OF = a = 4$ and $AB = 5.9$. Therefore $P = 5(5.9)$ or $P = 29.5$. Consequently,

$$\begin{aligned} A_{ABCDE} &= \frac{1}{2} \cdot 4(29.5) \\ &= 59 \text{ units}^2 \end{aligned}$$

EXAMPLE 6

Find the exact area of equilateral triangle ABC in Figure 7.42 if each side measures 12 in.

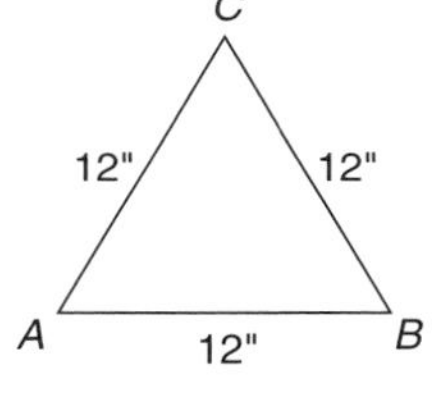

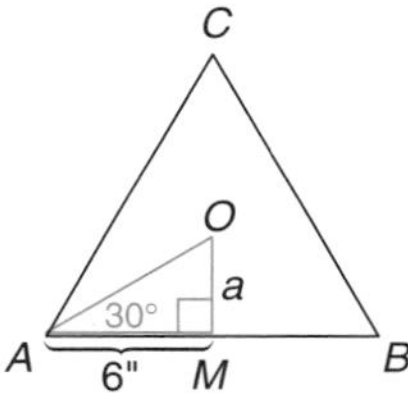

FIGURE 7.42

Solution

In $\triangle ABC$, the perimeter is $P = 3 \cdot 12$ or 36 in.

To find the length a of an apothem, we draw the radius $\overline{OA}$ from center O to point A and the apothem $\overline{OM}$ from O to side $\overline{AB}$. Because the radius bisects $\angle BAC$, $m\angle OAB = 30°$. Because apothem $\overline{OM} \perp \overline{AB}$, $m\angle OMA = 90°$. Using the 30°-60°-90° relationship in $\triangle OMA$, we see that $a\sqrt{3} = 6$. Thus

$$a = \frac{6}{\sqrt{3}} = \frac{6}{\sqrt{3}} \cdot \frac{\sqrt{3}}{\sqrt{3}} = \frac{6\sqrt{3}}{3} = 2\sqrt{3}$$

Now $A = \frac{1}{2}aP$ becomes $A = \frac{1}{2} \cdot 2\sqrt{3} \cdot 36 = 36\sqrt{3}$ in.^{2}

NOTE: Using the calculator's value for $\sqrt{3}$ leads to an approximation of the area rather than an exact area.

7.3 Exercises

1. Describe, if possible, how you would inscribe a circle within kite $ABCD$.

2. What condition must be satisfied for it to be possible to circumscribe a circle about kite $ABCD$?

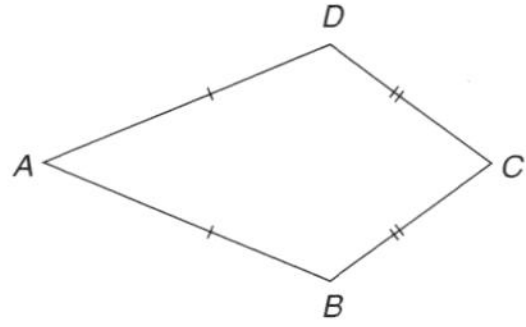

3. Inscribe a circle in rhombus *JKLM*.

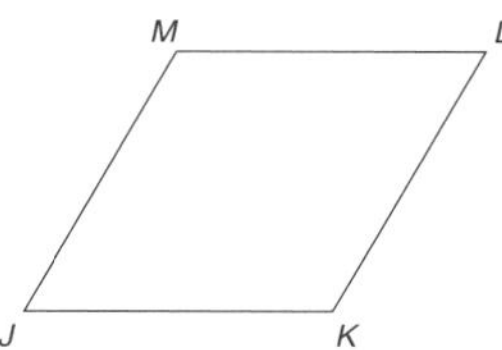

4. What condition must be satisfied for it to be possible to circumscribe a circle about trapezoid *RSTV*?

5. Inscribe a regular octagon within a circle.
6. Inscribe an equilateral triangle within a circle.
7. Circumscribe a square about a circle.
8. Circumscribe an equilateral triangle about a circle.
9. Find the lengths of the apothem and the radius of a square whose sides have length 10 in.
10. Find the lengths of the apothem and the radius of a regular hexagon whose sides have length 6 cm.
11. Find the lengths of the side and the radius of an equilateral triangle whose apothem's length is 8 ft.
12. Find the lengths of the side and the radius of a regular hexagon whose apothem's length is 10 m.
13. Find the measure of the central angle of a regular polygon of

 a) three sides. c) five sides.
 b) four sides. d) six sides.

14. Find the number of sides of a regular polygon that has a central angle measuring:

 a) 30 c) 36
 b) 72 d) 20

15. Find the area of a regular hexagon whose sides have length 6 cm.
16. Find the area of a square whose apothem measures 5 cm.
17. Find the area of an equilateral triangle whose radius measures 10 in.
18. Find the approximate area of a regular pentagon whose apothem measures 6 in. and each of whose sides measures approximately 8.9 in.
19. In a regular polygon of 12 sides, the measure of each side is 2 in. while the measure of an apothem is $(2 + \sqrt{3})$ in. Find the area.
20. In a regular octagon, the measure of each apothem is 4 cm while each side measures $8(\sqrt{2} - 1)$ cm. Find the area.
21. Find the ratio of the area of a square circumscribed about a circle to the area of a square inscribed in the circle.
22. Regular octagon *ABCDEFGH* is inscribed in a circle whose radius is $\frac{7}{2}\sqrt{2}$ cm. Considering that the area of the octagon is less than the area of the circle and greater than the area of the square *ACEG*, find the two integers between which the area of the octagon must lie.
(**NOTE:** For the circle, use $A = \pi r^2$ with $\pi \approx \frac{22}{7}$.)

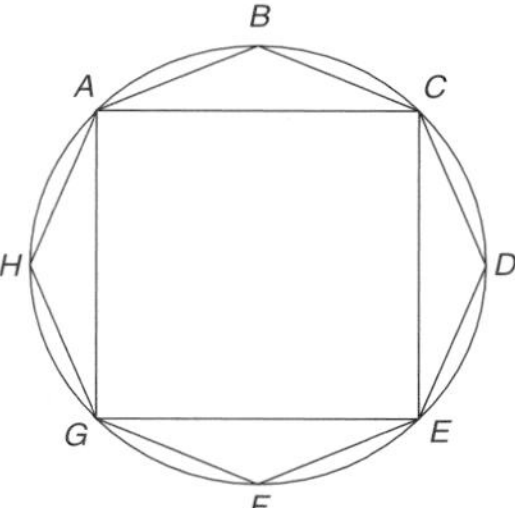

23. *Given:* Regular pentagon *RSTVQ* with equilateral $\triangle PQR$
Find: $m\angle VPS$

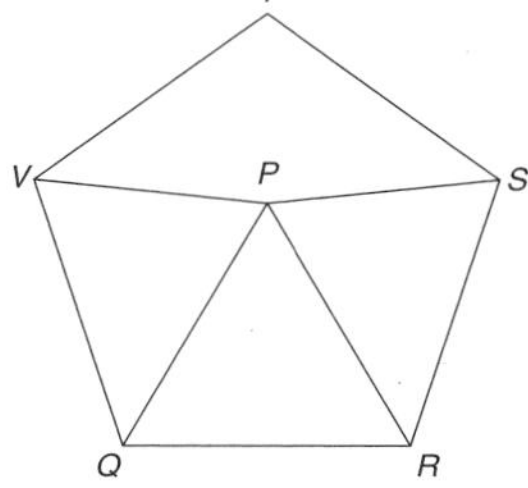

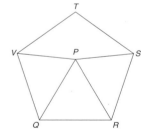

24. *Given:* Regular pentagon *JKLMN* (not shown) with diagonals $\overline{LN}$ and $\overline{KN}$
Find: $m\angle LNK$

25. *Given:* Quadrilateral *ABCD* is circumscribed about $\odot O$
Prove: $AB + CD = DA + BC$

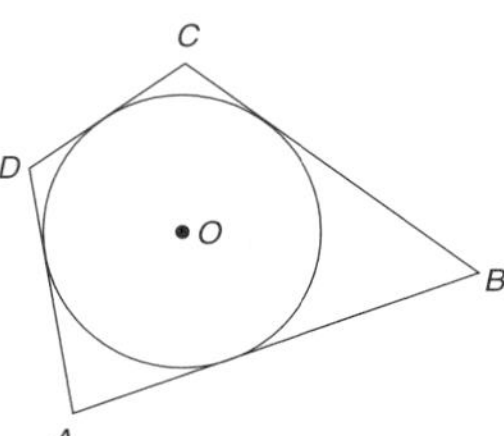

26. *Given:* Quadrilateral $RSTV$ inscribed in $\odot Q$
Prove: $m\angle R + m\angle T = m\angle V + m\angle S$

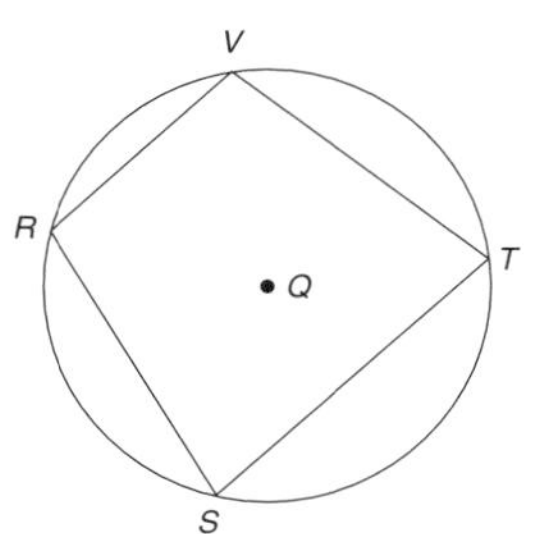

27. ➤ *Prove:* If a circle is divided into n congruent arcs ($n \geq 3$), the chords determined by joining consecutive endpoints of these arcs form a regular polygon.

28. ➤ *Prove:* If a circle is divided into n congruent arcs ($n \geq 3$), the tangents drawn at the endpoints of these arcs form a regular polygon.

7.4 Circumference and Area of a Circle

In geometry, any two figures that have the same shape are described as similar. For this reason, we say that all circles are similar to each other. Just as a proportionality exists among the sides of similar triangles, experimentation shows that there is a proportionality among the circumferences (distances around) and diameters (distances across) of circles; see the Discover! activity below. Representing the circumferences of the circles in Figure 7.43 by C_1, C_2, and C_3 and the diameters by d_1, d_2, and d_3, we claim that

$$\frac{C_1}{d_1} = \frac{C_2}{d_2} = \frac{C_3}{d_3} = k$$

where k is the constant of proportionality.

FIGURE 7.43

POSTULATE 22:
The ratio of the circumference of a circle to the length of its diameter is a unique positive constant.

DISCOVER!

Find an object of circular shape, such as the lid of a jar. Using a flexible tape measure (like a seamstress or carpenter might use), measure both the distance around (circumference) and the distance across (length of diameter) the circle. Now divide the circumference C by the diameter length d. What is your result?

ANSWER
The ratio should be slightly larger than 3.

The constant of proportionality k described in the opening paragraph of this section, in Postulate 22, and in the Discover! activity is represented by the Greek letter π (pi).

DEFINITION: π is the ratio between the circumference C and the diameter length d of any circle; thus, $\pi = \frac{C}{d}$ in any circle.

THEOREM 7.4.1: The circumference of a circle is given by the formula

$$C = \pi d \qquad \text{or} \qquad C = 2\pi r$$

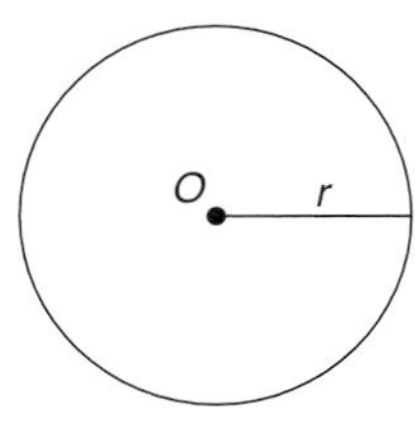

FIGURE 7.44

Given: Circle O with radius r (See Figure 7.44.)

Prove: $C = 2\pi r$

Proof: By Postulate 22, $\pi = \frac{C}{d}$. Multiplying each side of the equation by d, we have $C = \pi d$. Since $d = 2r$ (the diameter's length is twice that of the radius), the formula for the circumference can be written $C = \pi(2r)$ or $C = 2\pi r$.

Value of π

In calculating the circumference of a circle, we generally leave the symbol π in the answer in order to state an *exact* result. However, the value of π is irrational and cannot be represented exactly by a common fraction or by a terminating decimal. When an approximation is needed for π, we use a calculator. Throughout history, some commonly used approximations of π have been $\pi \approx \frac{22}{7}$, $\pi \approx 3.14$, and $\pi \approx 3.1416$.

Although these approximate values have been used for centuries, your calculator provides greater accuracy. A calculator may show that $\pi \approx 3.141592654$.

EXAMPLE 1

In $\odot O$ in Figure 7.45, $OA = 7$ cm. Using $\pi \approx \frac{22}{7}$,

a) find the approximate circumference C of $\odot O$.
b) find the approximate length of the minor arc $\widehat{AB}$.

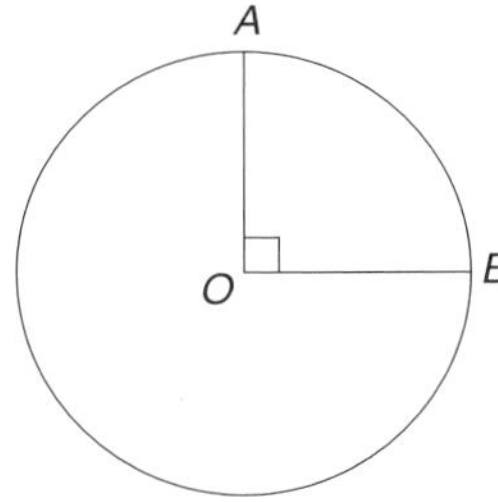

FIGURE 7.45

Solution

a) $C = 2\pi r$
$= 2 \cdot \frac{22}{7} \cdot \not{7}$
$= 44$ cm

b) Because the degree of measure of $\widehat{AB}$ is 90°, we have $\frac{90}{360}$ or $\frac{1}{4}$ of the circumference for the arc length. Then

$$\text{length of } \widehat{AB} = \frac{90}{360} \cdot 44 = \frac{1}{4} \cdot 44 = 11 \text{ cm}$$

EXAMPLE 2

The exact circumference of a circle is 17π in.

a) Find the length of the radius.
b) Find the length of the diameter.

Solution

a)

$$C = 2\pi r$$
$$17\pi = 2\pi r$$
$$\frac{17\pi}{2\pi} = \frac{2\pi r}{2\pi}$$
$$r = \frac{17}{2} = 8.5 \text{ in.}$$

b) Because $d = 2r$, $d = 2(8.5)$ or $d = 17$ in.

EXAMPLE 3

A thin circular rubber gasket is used as a seal to prevent oil from leaking from a tank (see Figure 7.46). If the gasket has a radius of 2.37 in., use the value of π provided by your calculator to find the circumference of the gasket to the nearest hundredth of an inch.

Solution

Using the calculator with $C = 2\pi r$, we have $C = 2 \cdot \pi \cdot 2.37$ or $C \approx 14.89114918$. Rounding to the nearest hundredth of an inch, $C \approx 14.89$ in.

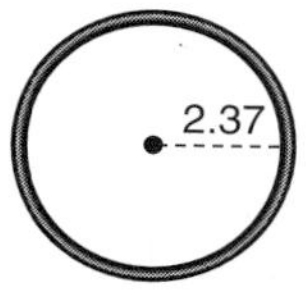

FIGURE 7.46

Length of an Arc

In Example 1(b), we used the phrase *length of arc* without a definition. Informally, the length of an arc is the distance between the endpoints of the arc as if measured along a straight line. If we measured one-third of the circumference of the rubber gasket (a 120° arc) in Example 3, we would expect the length to be slightly less than 5 in. This measurement could be accomplished by holding that part of the gasket taut in a straight line, but not so tightly that it would be stretched.

Two further observations can be made with regard to the measurement of arc length.

1. The ratio of the degree measure of the arc to 360 (the degree measure of the entire circle) is the same as the ratio of the length of the arc to the circumference; that is, $\frac{m}{360} = \frac{\ell}{C}$.
2. As $m\widehat{AB}$ denotes the degree measure of an arc, $\ell\widehat{AB}$ denotes the length of the arc. While $m\widehat{AB}$ is measured in degrees, $\ell\widehat{AB}$ is measured in linear units such as inches, feet, or centimeters.

THEOREM 7.4.2: In a circle whose circumference is C, the length ℓ of an arc whose degree measure is m is given by

$$\ell = \frac{m}{360} \cdot C$$

NOTE: For arc AB, $\ell\widehat{AB} = \frac{m\widehat{AB}}{360} \cdot C$.

EXAMPLE 4

Find the approximate length of major arc ABC in a circle of radius 7 in. if $m\widehat{AC} = 45°$. (See Figure 7.47.) Use $\pi \approx \frac{22}{7}$.

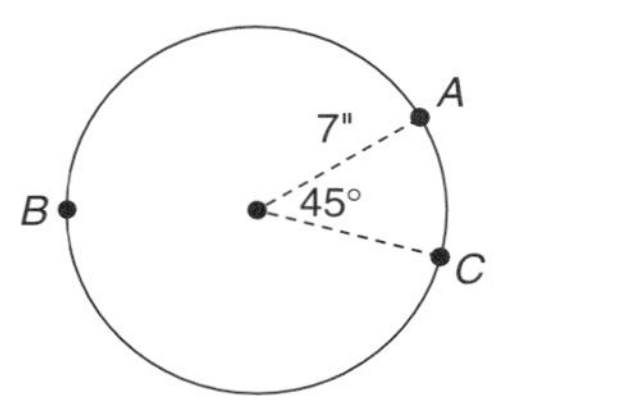

FIGURE 7.47

Solution

$m\widehat{ABC} = 360° - 45° = 315°$. Using Theorem 7.4.2, $\ell\widehat{ABC} = \frac{m\widehat{ABC}}{360} \cdot C$ or $\ell\widehat{ABC} = \frac{315}{360} \cdot 2 \cdot \frac{22}{7} \cdot 7$, which can be simplified to $\ell\widehat{ABC} = 38\frac{1}{2}$ in.

Limits

In the discussion that follows, we use the undefined term "limit;" in practice, a limit represents a numerical measure. The following example illustrates this notion.

EXAMPLE 5

Find the limit (largest possible number) for the length of a chord in a circle whose length of radius is 5 cm.

Solution

By considering several chords in the circle in Figure 7.48, we see that the greatest possible length of a chord is that of a diameter. Thus the limit of the length of a chord is 10 cm.

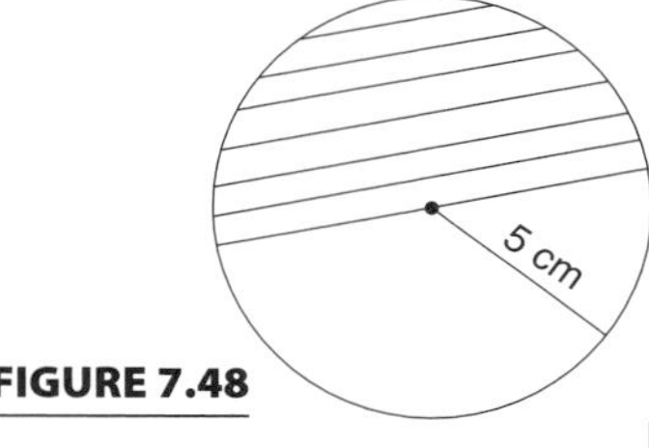

FIGURE 7.48

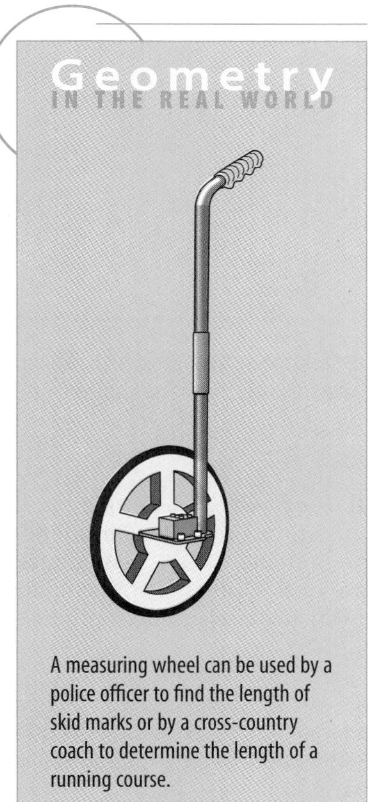

A measuring wheel can be used by a police officer to find the length of skid marks or by a cross-country coach to determine the length of a running course.

Area of a Circle

Now consider the problem of finding the area of a circle. To do so, let a regular polygon of n sides be inscribed in the circle. As we allow n to grow larger (often written as $n \to \infty$ and read as "n approaches infinity"), two observations can be made:

1. The length of an apothem of the regular polygon approaches the length of a radius of the circle as its limit ($a \to r$).
2. The perimeter of the regular polygon approaches the circumference of the circle as its limit ($P \to C$).

In Figure 7.49, the area of an inscribed regular polygon with n sides approaches the area of the circle as its limit as n increases. Using observations 1 and 2, we make the following claim. Because the formula for the area of a regular polygon is

$$A = \frac{1}{2}aP,$$

the area of the circumscribed circle is given by the limit

$$A = \frac{1}{2}rC$$

Because $C = 2\pi r$, this formula becomes

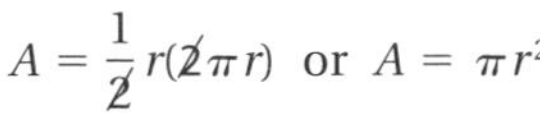

$$A = \frac{1}{\cancel{2}}r(\cancel{2}\pi r) \text{ or } A = \pi r^2$$

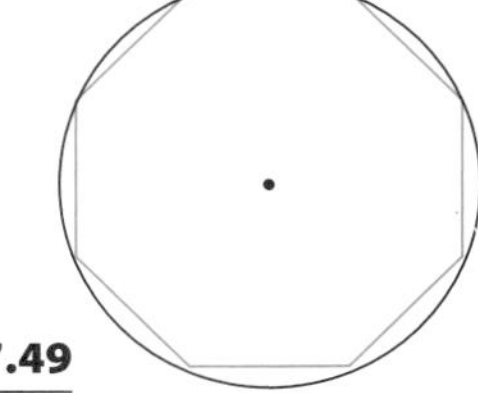

FIGURE 7.49

THEOREM 7.4.3: The area A of a circle whose radius has length r is given by $A = \pi r^2$.

EXAMPLE 6

Find the approximate area of a circle whose radius has a length of 10 in. (Use $\pi \approx 3.14$.)

Solution

$A = \pi r^2$ becomes $A = 3.14(10)^2$. Then

$$A = 3.14(100) = 314 \text{ in.}^2$$

EXAMPLE 7

The approximate area of a circle is 38.5 cm². Find the length of the radius of the circle. $\left(\text{Use } \pi \approx \frac{22}{7}\right)$.

Solution

$A = \pi r^2$ becomes $38.5 = \frac{22}{7} \cdot r^2$ or $\frac{77}{2} = \frac{22}{7} \cdot r^2$. Multiplying each side of the equation by $\frac{7}{22}$, we have

$$\frac{7}{22} \cdot \frac{77}{2} = \frac{7}{22} \cdot \frac{22}{7} \cdot r^2$$

or

$$r^2 = \frac{49}{4}$$

Taking the positive square root for the radius,

$$r = \sqrt{\frac{49}{4}} = \frac{\sqrt{49}}{\sqrt{4}} = \frac{7}{2} = 3.5 \text{ cm}$$

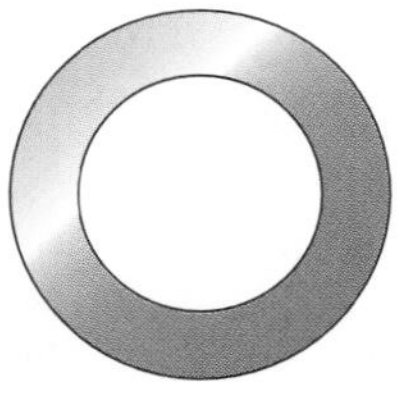

FIGURE 7.50

A plane figure bounded by concentric circles is known as a *ring* or *annulus* (see Figure 7.50). The piece of hardware known as a *washer* has the shape of an annulus.

EXAMPLE 8

A machine cuts washers from a flat piece of metal. The radius of the inside circular boundary of the washer is 0.3 in. and the radius of the outer circular boundary is 0.5 in. What is the area of the annulus? Give both an exact answer and an approximate answer rounded to tenths of a square inch. Using the approximate answer, determine the number of square inches of material used to produce 1000 of the washers.

Solution

Where R is the larger radius and r is the smaller radius, $A = \pi R^2 - \pi r^2$. Then $A = \pi(0.5)^2 - \pi(0.3)^2$ or $A = 0.16\pi$. The exact number of square inches used in producing a washer is 0.16π in.², or approximately 0.5 in.² When 1000 washers are produced, approximately 500 in.² of metal are used.

NOTE: Many people have a difficult time remembering which expression ($2\pi r$ or πr^2) is used in the formula for the circumference (or area) of a circle. This is understandable, since each expression contains a 2, a radius r, and the factor π. To remember that $C = 2\pi r$ gives the circumference while $A = \pi r^2$ gives the area, *think about the units involved.* Considering a circle of radius 3 in., $C = 2\pi r$ becomes $C = 2 \times 3.14 \times 3$ in., or Circumference equals 18.74 *inches.* (We measure the *distance around* a circle in *linear* units such as inches.)

For the circle of radius 3 in., $A = \pi r^2$ becomes $A = 3.14 \times 3$ in. $\times$ 3 in. or Area equals 28.26 in.^{2} (We measure the *area* of a circular region in *square* units.)

7.4 Exercises

1. Find the exact circumference and area of a circle whose radius has length 8 cm.

2. Find the exact circumference and area of a circle whose diameter has length 10 in.

3. Find the approximate circumference and area of a circle whose radius has length $10\frac{1}{2}$ in. Use $\pi \approx \frac{22}{7}$.

4. Find the approximate circumference and area of a circle whose diameter is 20 cm. Use $\pi \approx 3.14$.

5. Find the exact lengths of a radius and a diameter of a circle whose circumference is:

a) 44π in. **b)** 60π ft

6. Find the approximate lengths of a radius and a diameter of a circle whose circumference is:

a) 88 in. $\left(\text{Use } \pi \approx \frac{22}{7}.\right)$ **b)** 157 m (Use $\pi \approx 3.14$.)

7. Find the exact lengths of a radius and a diameter of a circle whose area is:

a) 25π in.^{2} **b)** 2.25π cm^2

8. Find the exact length of a radius and the exact circumference of a circle whose area is:

a) 36π m^2 **b)** 6.25π ft^2

9. Find the exact length $\ell\overparen{AB}$, where $\overparen{AB}$ refers to the minor arc of the circle.

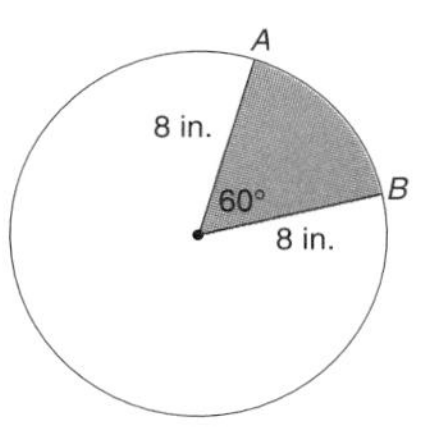

10. Find the exact length $\ell\overparen{CD}$ of the minor arc shown.

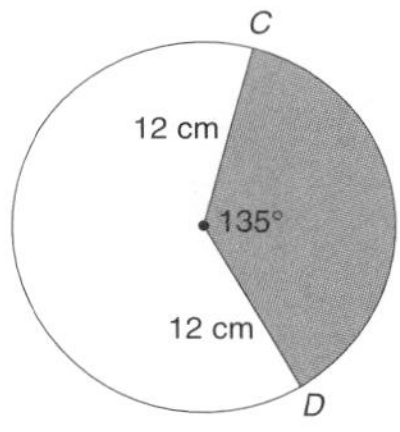

11. Use your calculator value of π to find the approximate circumference of a circle with radius 12.38 in.

12. Use your calculator value of π to find the approximate area of a circle with radius 12.38 in.

13. A metal circular disc whose area is 143 cm^2 is used as a knockout on an electrical service in a factory. Use your calculator value of π to find its radius to the nearest tenth of an inch.

14. A circular lock washer whose outside circumference measures 5.48 cm is used in an electric box to hold an electrical cable in place. Use your calculator value of π to find the outside radius to the nearest tenth of an inch.

15. The central angle corresponding to a circular brake shoe measures 60°. Approximately how long is the curved surface of the brake shoe if the circle has a radius of 7 in.?

16. Use your calculator to find the approximate radius and diameter of a circle with area 56.35 in.^{2}

17. A rectangle has a perimeter of 16 in. What is the limit (largest possible value) of the area of the rectangle?

18. Two sides of a triangle measure 5 in. and 7 in., respectively. What is the limit of the length of the third side?

19. Let N be any point on side $\overline{BC}$ of the right triangle ABC. Find the limit of the length of $\overline{AN}$.

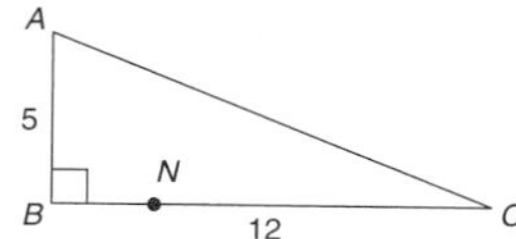

20. What is the limit of m$\angle RTS$ if T lies in the shaded region?

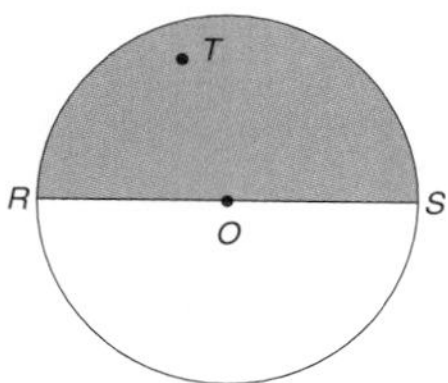

In Exercises 21 to 24, find exact areas of the shaded regions.

21.

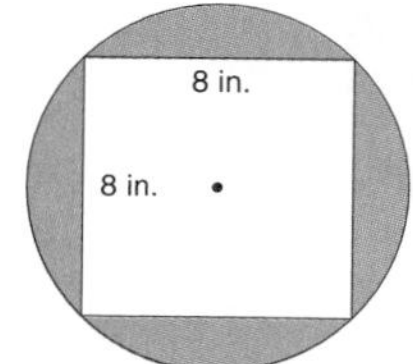

22.

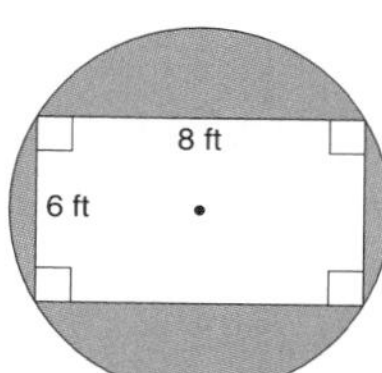

23.

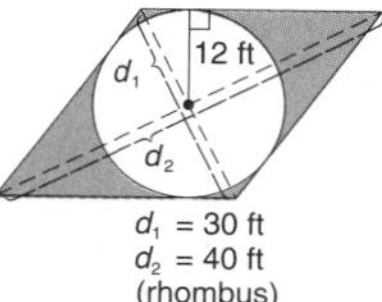

d_1 = 30 ft
d_2 = 40 ft
(rhombus)

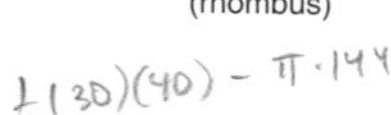

24.

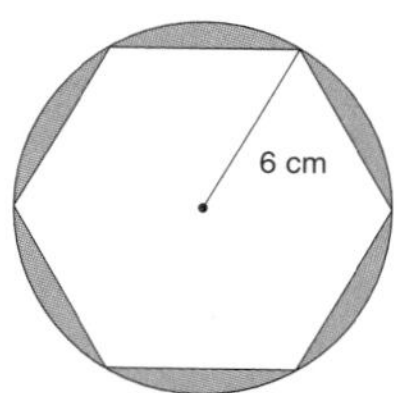

Regular hexagon inscribed in a circle

In Exercises 25 and 26, use your calculator value of π to solve each problem. Round answers to the nearest integer.

25. Find the length of the radius of a circle whose area is 154 cm^2.

26. Find the length of the diameter of a circle whose circumference is 157 in.

27. Assuming that a 90° arc has an exact length of 4π in., find the length of the radius of the circle.

28. The ratio of the circumferences of two circles is 2:1. What is the ratio of their areas?

29. Given concentric circles with radii of lengths R and r, where $R > r$, explain why $Area_{\text{ring}} = \pi(R + r)(R - r)$.

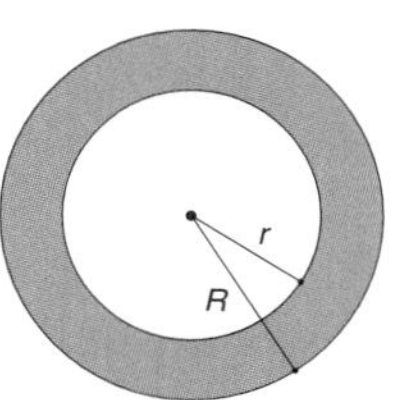

30. Given a circle with diameter of length d, explain why $A_{\text{circle}} = \frac{1}{4}\pi d^2$.

31. The radii of two concentric circles differ in length by exactly 1 in. If their areas differ by exactly 7π in.^{2}, find the lengths of the radii of the two circles.

In Exercises 32 to 42, use your calculator value of π unless otherwise stated. Round answers to two decimal places.

32. The carpet in the circular entryway of a church needs to be replaced. The diameter of the circular region to be carpeted is 18 ft.

a) What length (in feet) of a metal protective strip is needed to bind the circumference of the carpet?

b) If the metal strips are sold in lengths of 6 ft, how many will be needed? (**NOTE:** Assume that these can be bent to follow the circle *and* that they can be placed end-to-end.)

c) If the cost of the metal strip is \$1.59 per linear foot, find the cost of the metal strips needed.

33. At center court on a gymnasium floor, a large circular emblem is to be painted. The circular design has a radius of 8 ft.

a) What is the area to be painted?

b) If a pint of paint covers 70 ft^2, how many pints of paint are needed to complete the job?

c) If each pint of paint costs \$2.95, find the cost of the paint needed.

34. A track is to be constructed around the football field at a junior high school. If the straightaways are 100 yd in length, what length of radius is needed for each of the semicircles shown if the total length around the track is to be 440 yd?

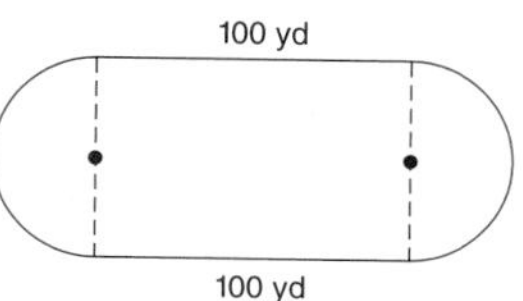

35. A circular grass courtyard at a shopping mall has a 40 ft diameter. This area needs to be reseeded.

a) What is the total area to be reseeded? (Use $\pi \approx 3.14$).
b) If 1 lb of seed is to be used to cover a 60-ft^2 region, how many pounds of seed will be needed?
c) If the cost of 1 lb of seed is \$1.65, what is the total cost of the grass seed needed?

36. Find the approximate area of a regular polygon that has 20 sides if the length of its radius is 7 cm.

37. Find the approximate perimeter of a regular polygon that has 20 sides if the length of its radius is 7 cm.

38. In a two-pulley system, the centers of the pulleys are 20 in. apart. If the radius of each pulley measures 6 in., how long is the belt used in the pulley system?

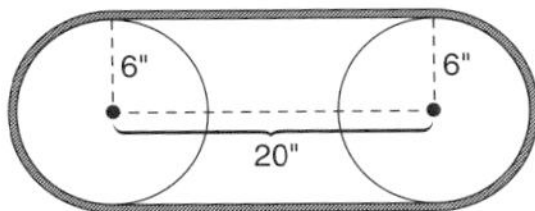

39. If two gears, each of radius 4 in., are used in a chain drive system with a chain of length 54 in., what is the distance between the centers of the gears?

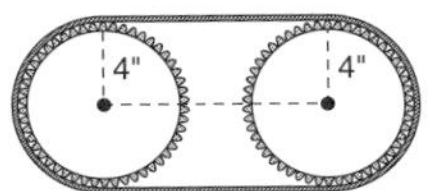

40. A pizza with a 12-in. diameter costs \$6.95. The 16-in. pizza with the same ingredients costs \$9.95. Which pizza is the better buy?

41. A communications satellite forms a circular orbit 375 mi above the earth. If the earth's radius is approximately 4000 mi, what distance is traveled by the satellite in one complete orbit?

42. The radius of the Ferris wheel's circular path is 40 ft. If a "ride" of 12 revolutions is made in 3 minutes, at what rate in *feet per second* is the passenger in a cart moving during the ride?

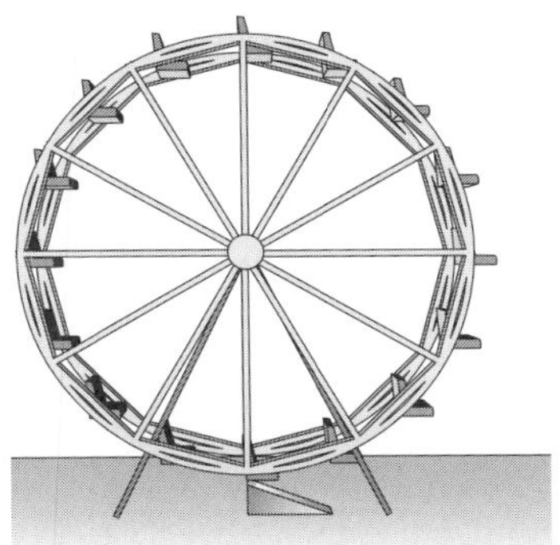

7.5 More Area Relationships in the Circle

Area of a Sector

DEFINITION: A **sector** of a circle is a region bounded by two radii of the circle and an arc intercepted by those radii. (See Figure 7.51.)

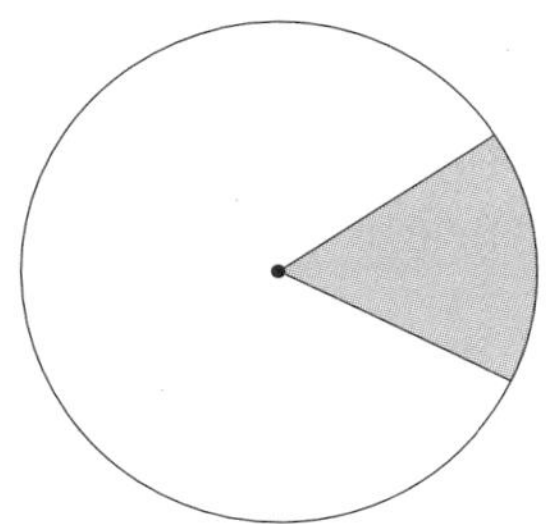

FIGURE 7.51

A sector will generally be shaded to avoid confusion about whether the arc is intended to be a major or minor arc. In simple terms, the sector of a circle generally has the shape of a piece of pie.

Just as the length of an arc is part of the circle's circumference, the area of a sector is part of its area. When fractions are illustrated using circles, $\frac{1}{4}$ is represented by shading a 90° sector while $\frac{1}{3}$ is represented by shading a 120° sector

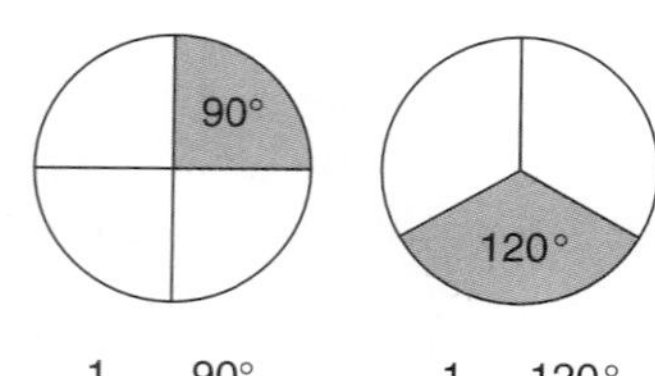

FIGURE 7.52

(see Figure 7.52). Thus, we make the following assumption about the measure of the area of a sector.

POSTULATE 23:
The ratio of the degree measure m of the central angle of a sector to 360° is the same as the ratio of the area of the sector to the area of the circle; that is, $\frac{\text{area of sector}}{\text{area of circle}} = \frac{m}{360}$.

THEOREM 7.5.1: In a circle of radius r, the area A of a sector whose arc has degree measure m is given by

$$A = \frac{m}{360}\pi r^2$$

Theorem 7.5.1 follows directly from Postulate 23.

EXAMPLE 1

If m∠O = 100°, find the area of the 100° sector shown in Figure 7.53. Use your calculator and round the answer to the nearest hundredth.

Solution

$A = \frac{m}{360}\pi r^2$ becomes

$$A = \frac{100}{360} \cdot \pi \cdot 10^2 \approx 87.27 \text{ in.}^2$$

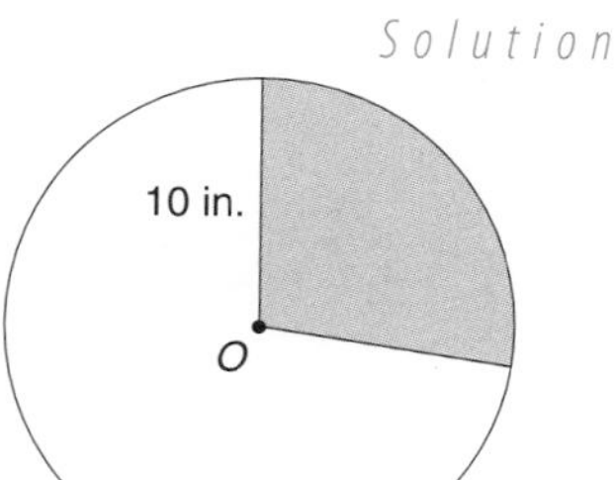

FIGURE 7.53

In applications with circles, we sometimes need exact answers for circumference and area; in such cases, we simply leave π in the result. For instance, in a circle of radius length 5 in., the exact circumference is 10π in. and the exact area is expressed as 25π in.^{2}

Because a sector is bound by two radii and an arc, the perimeter of a sector is the sum of the lengths of the two radii and the length of its arc. In Example 2, we apply this formula, $P_{\text{sector}} = 2r + \ell\widehat{AB}$.

EXAMPLE 2

Find the perimeter of the sector shown in Figure 7.53. Use the calculator value of π and round your answer to the nearest hundredth of an inch.

Solution

Since $r = 10$ and m∠O = 100°, $\ell\widehat{AB} = \frac{100}{360} \cdot 2 \cdot \pi \cdot 10$, which simplifies to 17.45. Now $P_{\text{sector}} = 2r + \ell\widehat{AB}$ becomes $P_{\text{sector}} = 2(10) + 17.45 = 37.45$ in.

Because a semicircle is one-half of a circle, a semicircular region corresponds to a central angle of 180°. As stated in the following corollary to Theorem 7.5.1, the area of the semicircle is $\frac{180}{360}$ (or one-half) the area of the entire circle.

> **COROLLARY 7.5.2:** The area of a semicircular region of radius r is $A = \frac{1}{2}\pi r^2$.

EXAMPLE 3

In Figure 7.54, a square of side 8 in. is shown with semicircles cut away. Find the exact shaded area by leaving π in the answer.

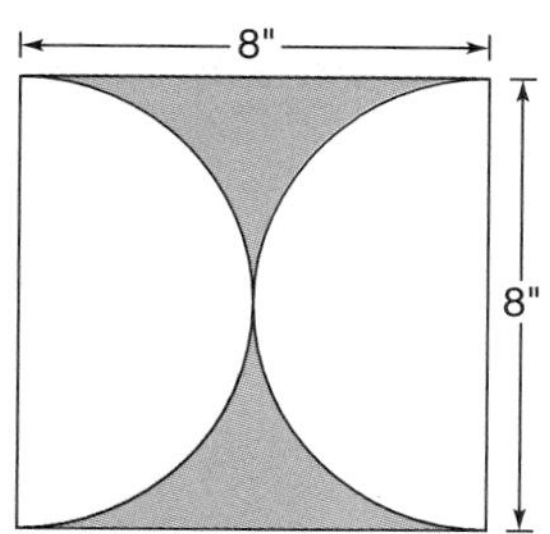

FIGURE 7.54

Solution

To find the shaded area A, we see that $A + 2 \cdot A_{\text{semicircle}} = A_{\text{square}}$. It follows that $A = A_{\text{square}} - 2 \cdot A_{\text{semicircle}}$.

If the side of the square is 8 in., then the radius of each semicircle is 4 in. Now $A = 8^2 - 2\left(\frac{1}{2}\pi \cdot 4^2\right)$ or $A = 64 - 2(8\pi)$, so $A = (64 - 16\pi)$ in.^{2}

DISCOVER!

In statistics, a pie chart can be used to represent the breakdown of a budget. In the pie chart shown, a 90° sector (one-fourth the area of the circle) is shaded to show that 25% of a person's income (one-fourth of the income) is devoted to rent payment. What degree measure of sector must be shaded if a sector indicates that 20% of the person's income is used for a car payment?

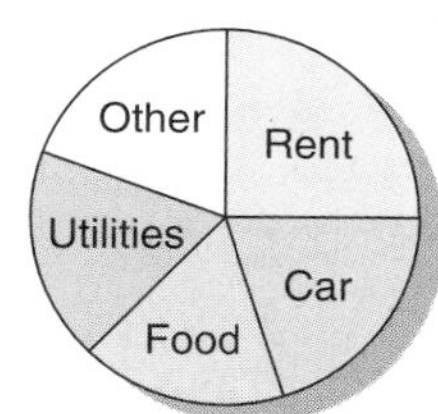

ANSWER
72° (from 20% of 360°)

Area of a Segment

> **DEFINITION:** A **segment** of a circle is a region bounded by a chord and its minor (or major) arc.

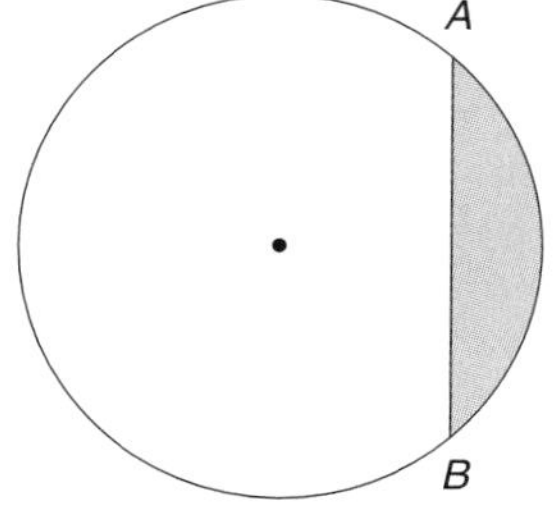

FIGURE 7.55

In Figure 7.55, the segment is bounded by chord $\overline{AB}$ and its minor arc $\widehat{AB}$. Again, we avoid confusion by shading the segment whose area we seek.

EXAMPLE 4

Find the exact area of the segment bounded by a chord and an arc whose measure is 90°. The radius has length 12 in. as shown in Figure 7.56 on page 334.

Solution Because

$$A_\triangle + A_{\text{segment}} = A_{\text{sector}}$$
$$A_{\text{segment}} = A_{\text{sector}} - A_\triangle$$
$$= \frac{90}{360} \cdot \pi \cdot 12^2 - \frac{1}{2} \cdot 12 \cdot 12$$
$$= (36\pi - 72) \text{ in.}^2$$

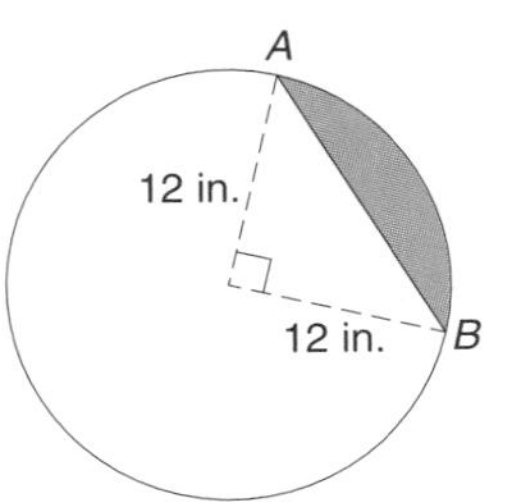

FIGURE 7.56

In Example 4, the boundaries of the segment shown are chord $\overline{AB}$ and minor arc $\widehat{AB}$. Thus, the perimeter of the segment is given by $P_{\text{segment}} = AB + \ell\widehat{AB}$. We use this formula in Example 5.

EXAMPLE 5

Find the exact perimeter of the segment described in Example 4 (see Figure 7.57). Then use your calculator to approximate this answer to the nearest hundredth of an inch.

Solution Because $\ell\widehat{AB} = \frac{90}{360} \cdot 2 \cdot \pi \cdot r$, we have $\ell\widehat{AB} = \frac{1}{4} \cdot 2 \cdot \pi \cdot 12 = 6\pi$ in.

Using either the Pythagorean Theorem or the 45°-45°-90° relationship, $AB = 12\sqrt{2}$.

Now $P_{\text{segment}} = AB + \ell\widehat{AB}$ becomes $P_{\text{segment}} = (12\sqrt{2} + 6\pi)$ in. Using a calculator, the approximate perimeter is 35.82 in.

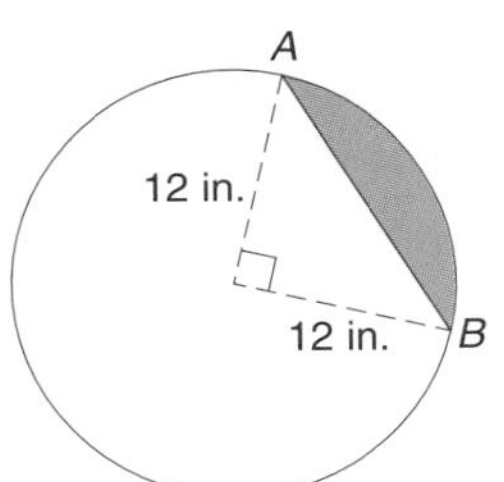

FIGURE 7.57

Area of a Triangle with an Inscribed Circle

THEOREM 7.5.3: Where P represents the perimeter of a triangle and r represents the length of the radius of its inscribed circle, the area of the triangle is given by

$$A = \frac{1}{2}rP$$

Given: A triangle [see Figure 7.58(a)] of perimeter P, whose sides measure a, b, and c; the radius of the inscribed circle measures r

Prove: $A = \frac{1}{2}rP$

Proof: In Figure 7.58(b), the triangle is separated into three smaller triangles as follows. Where O is the incenter of the triangle, we draw line segments from O to each vertex. Now r is the altitude of each of the numbered triangles. Where A_1, A_2, and A_3 are the areas of these triangles, the desired area A is the sum $A = A_1 + A_2 + A_3$. Using the formula $A = \frac{1}{2}bh$, we have $A = \frac{1}{2}r \cdot a + \frac{1}{2}r \cdot b + \frac{1}{2}r \cdot c$. Thus, $A = \frac{1}{2}r(a + b + c)$ so $A = \frac{1}{2}rP$.

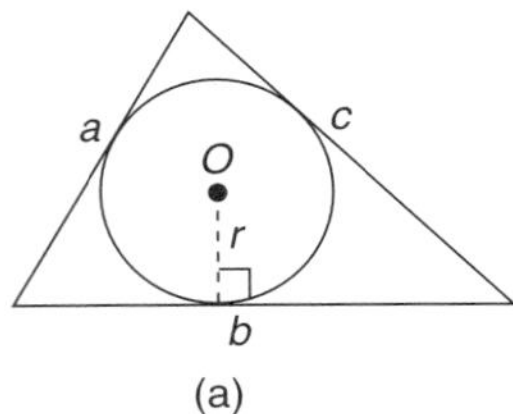

(a)

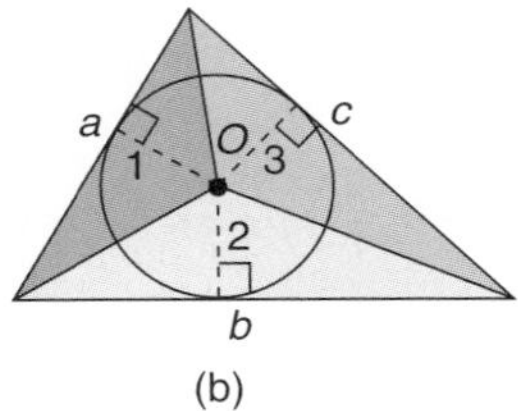

(b)

FIGURE 7.58

EXAMPLE 6

Find the area of a triangle whose sides measure 5 cm, 12 cm, and 13 cm if the radius of the inscribed circle is 2 cm. (See Figure 7.59.)

Solution

Using the given lengths of sides, the perimeter of the triangle is $P = 5 + 12 + 13 = 30$ cm. Using $A = \frac{1}{2}rP$, we have $A = \frac{1}{2} \cdot 2 \cdot 30$ or $A = 30 \text{ cm}^2$.

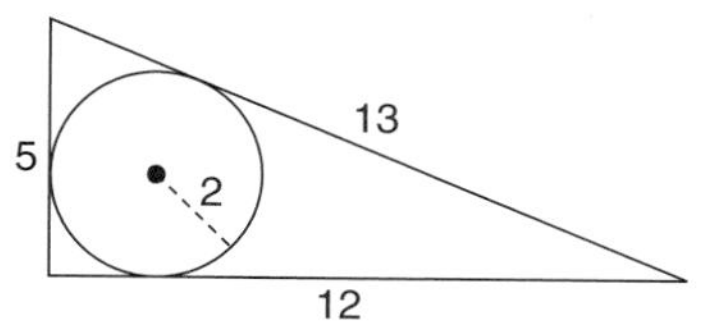

FIGURE 7.59

Because the triangle shown in Example 6 is a right triangle ($5^2 + 12^2 = 13^2$), the area of the triangle could have been determined by using either $A = \frac{1}{2}bh$ or $A = \sqrt{s(s-a)(s-b)(s-c)}$. The advantage provided by Theorem 7.5.3 lies in applications where we need to determine the length of the radius of the inscribed circle of a triangle.

EXAMPLE 7

In an attic, wooden braces supporting the roof form a triangle whose sides measure 4 ft, 6 ft, and 6 ft., see Figure 7.60. To the nearest inch, find the radius of the largest circular cold-air duct that can be run through the opening formed by the braces.

Solution

Where s is the semiperimeter of the triangle, Heron's Formula states that $A = \sqrt{s(s-a)(s-b)(s-c)}$. Since $s = \frac{1}{2}(a + b + c) = \frac{1}{2}(4 + 6 + 6) = 8$, we have $A = \sqrt{8(8-4)(8-6)(8-6)} = \sqrt{8(4)(2)(2)} = \sqrt{128}$. We can simplify the area expression to $\sqrt{64} \cdot \sqrt{2}$, so $A = 8\sqrt{2} \text{ ft}^2$.

Recalling Theorem 7.5.3, we know that $A = \frac{1}{2}rP$. Substitution leads to $8\sqrt{2} = \frac{1}{2}r(4 + 6 + 6)$ or $8\sqrt{2} = 8r$. Then $r = \sqrt{2}$. Where $r \approx 1.414$ ft, it follows that $r \approx 1.414$ (12 in.) or $r \approx 16.97$ in. ≈ 17 in.

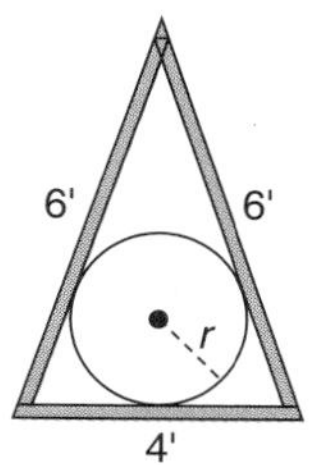

FIGURE 7.60

NOTE: If the ductwork is a flexible plastic tubing, the duct having radius 17 in. likely can be used. If the ductwork were a rigid metal or heavy plastic, the radius might need to be restricted to perhaps 16 in.

7.5 Exercises

1. In the circle, the radius length is 10 in. and the length of $\overset{\frown}{AB}$ is 14 in. What is the perimeter of the shaded sector?

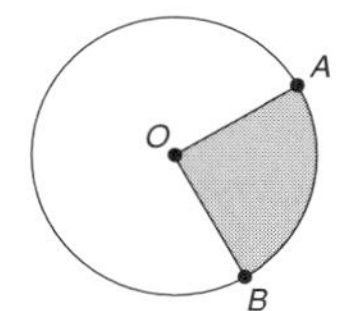

Exercises 1, 2

2. If the area of the circle is 360 in.^{2}, what is the area of the sector if its central angle measures 90°?

3. If the area of the 120° sector is 50 cm^2, what is the area of the entire circle?

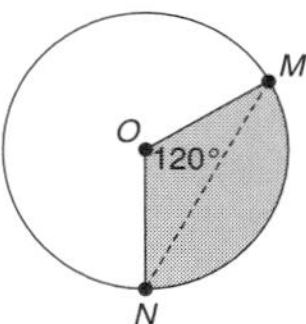

Exercises 3, 4

4. If the area of the 120° sector shown on page 335 is 40 cm^2 and the area of $\triangle MON$ is 16 cm^2, what is the area of the segment bounded by chord $\overline{MN}$ and $\overset{\frown}{MN}$?

5. Suppose that a circle of radius r is inscribed in an equilateral triangle whose sides have length s. Find an expression for the area of the triangle in terms of r and s. (**HINT:** Use Theorem 7.5.3.)

6. Suppose that a circle of radius r is inscribed in a rhombus whose sides each have length s. Find an expression for the area of the rhombus in terms of r and s.

7. Find the perimeter of a segment of a circle whose boundaries are a chord measuring 24 mm (millimeters) and an arc of length 30 mm.

8. A sector with perimeter 30 in. has a bounding arc of length 12 in. Find the length of the radius of the circle.

9. A circle is inscribed in a triangle having sides of lengths 6 in., 8 in., and 10 in. If the radius of the inscribed circle is 2 in., use $A = \frac{1}{2}rP$ to find the area of the triangle.

10. A circle is inscribed in a triangle having sides of lengths 5 in., 12 in., and 13 in. If the radius of the inscribed circle is 2 in., use $A = \frac{1}{2}rP$ to find the area of the triangle.

11. A triangle with sides of lengths 3 in., 4 in., and 5 in. has an area of 6 $in.^2$ What is the length of the radius of the inscribed circle?

12. The approximate area of a triangle with sides of lengths 3 in., 5 in., and 6 in. is 7.48 $in.^2$. What is the length of the radius of the inscribed circle?

13. Find the exact perimeter and area of the sector shown.

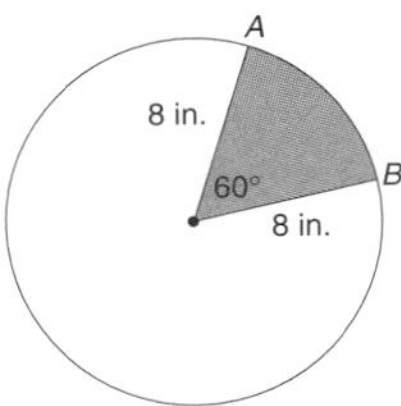

14. Find the exact perimeter and area of the sector shown.

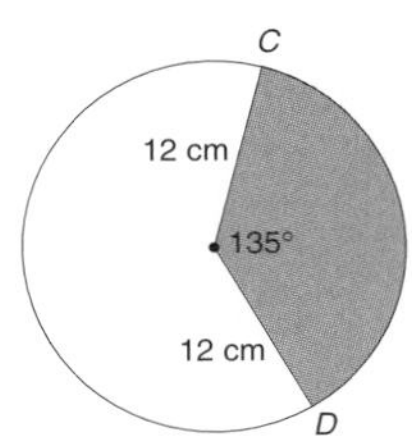

15. Find the approximate perimeter of the sector shown.

16. Find the approximate area of the sector shown.

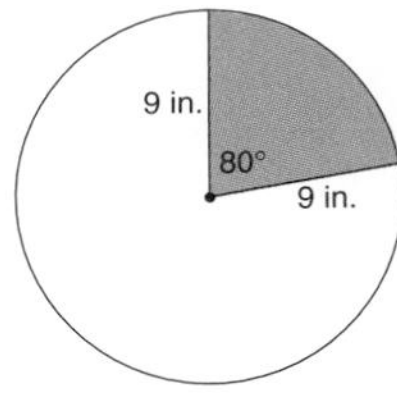

Exercises 15, 16

17. Find the exact perimeter and area of the segment shown, given that $m\angle O = 60°$ and $OA = 12$ in.

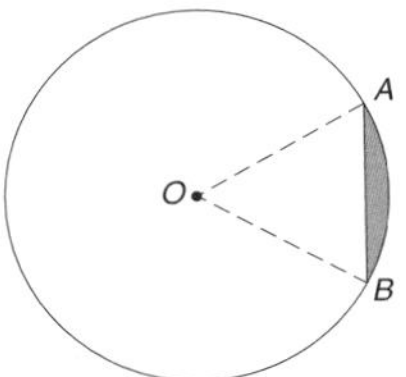

Exercises 17, 18

18. Find the exact perimeter and area of the segment shown, given that $m\angle O = 120°$ and $AB = 10$ in.

In Exercises 19 and 20, find the exact areas of the shaded regions.

19.

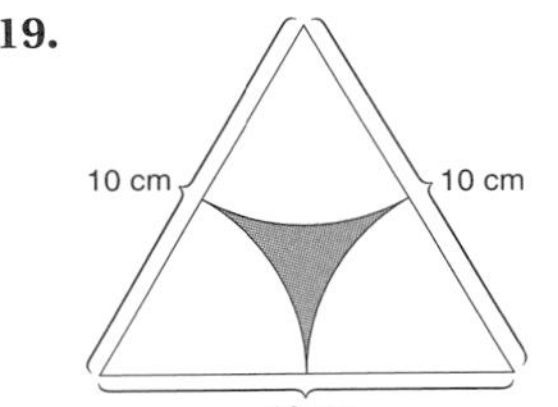

20.

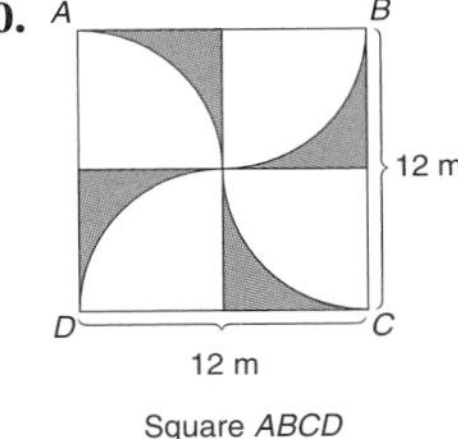

Square $ABCD$

21. Assuming that the exact area of a sector determined by a 40° arc is $\frac{9}{4}\pi$ cm^2, find the length of the radius of the circle.

22. For concentric circles with radii of lengths 3 in. and 6 in., find the area of the smaller segment determined by a chord of the larger circle that is also a tangent of the smaller circle.

23. ➢ A circle can be inscribed in the trapezoid shown. Find the area of that circle.

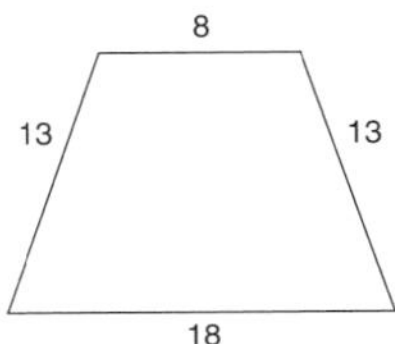

24. ➢ A circle can be inscribed in an equilateral triangle, each of whose sides has length 10 cm. Find the area of that circle.

25. In a circle whose radius has length 12 m, the length of an arc is 6π m. What is the degree measure of that arc?

26. At the Pizza Dude restaurant, a 12-in. pizza costs \$3.40 to make and the manager wants to make at least \$2.20 from the sale of each pizza. If the pizza will be sold by the slice and each pizza is cut into 6 pieces, what is the minimum charge per slice?

27. At the Pizza Dude restaurant, pizza is sold by the slice. If the pizza is cut into 6 pieces, the selling price is \$1.25 per slice. If the pizza is cut into 8 pieces, then each slice is sold for \$0.95. In which way will the Pizza Dude restaurant clear more money from sales?

28. Determine a formula for the area of the shaded region determined by the square and its inscribed circle.

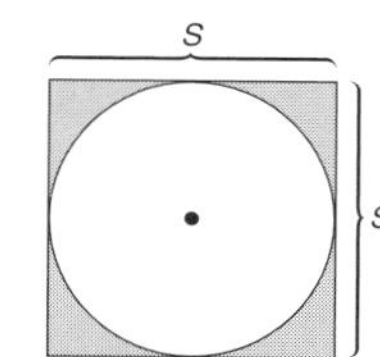

29. Determine a formula for the area of the shaded region determined by the circle and its inscribed square.

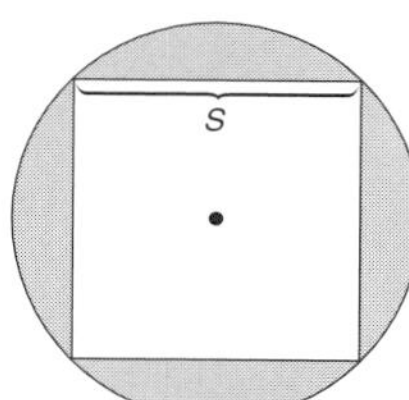

30. Find a formula for the area of the shaded region, which represents one-fourth of an annulus (ring).

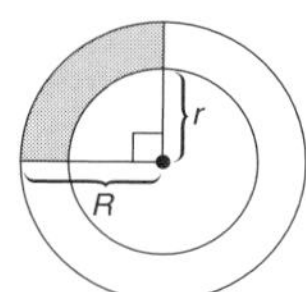

31. A company logo on the side of a building shows an isosceles triangle with an inscribed circle. If the sides of the triangle are 10 ft, 13 ft, and 13 ft, find the radius of the inscribed circle.

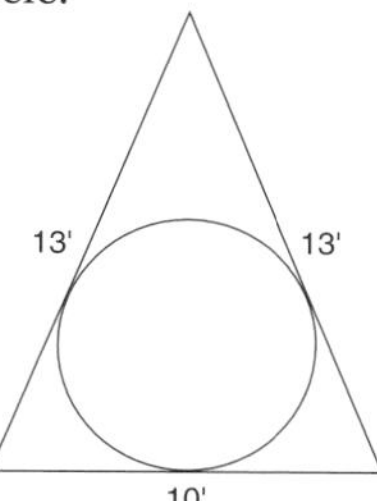

32. In a right triangle with sides of lengths a, b, and c (where c is the hypotenuse), show that the length of the radius of the inscribed circle is $r = \frac{ab}{a + b + c}$.

33. In a triangle with sides of lengths a, b, and c and semi-perimeter s, show that the length of the radius of the inscribed circle is

$$r = \frac{2\sqrt{s(s - a)(s - b)(s - c)}}{a + b + c}.$$

34. Use the results from Exercises 32 and 33 to find the length of the radius of the inscribed circle for a triangle with sides of lengths

a) 8, 15, and 17. **b)** 7, 9, and 12.

35. Use the results from Exercises 32 and 33 to find the length of the radius of the inscribed circle for a triangle with sides of lengths

a) 7, 24, and 25. **b)** 9, 10, and 17.

36. Three pipes, each of radius 4 in., are stacked as shown. What is the height of the stack?

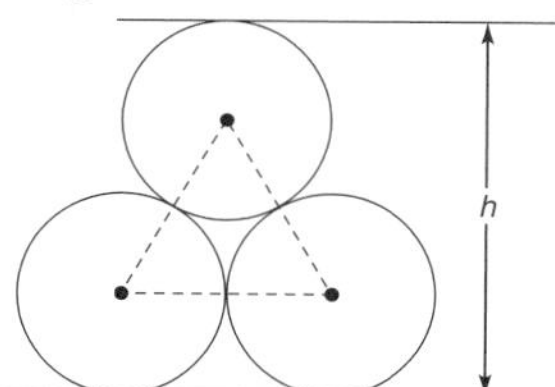

37. A windshield wiper rotates through a 120° angle as it cleans a windshield. From the point of rotation, the wiper blade begins at a distance of 4 in. and ends at a distance of 18 in. (The wiper blade is 14 in. in length.) Find the area cleaned by the wiper blade.

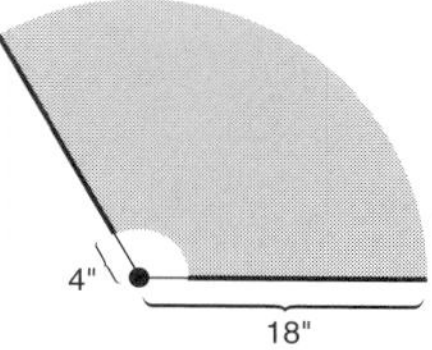

38. A goat is tethered to a barn by a 12 ft chain. If the chain is connected to the barn at a point 6 ft from one end of the barn, then what area is the goat able to graze?

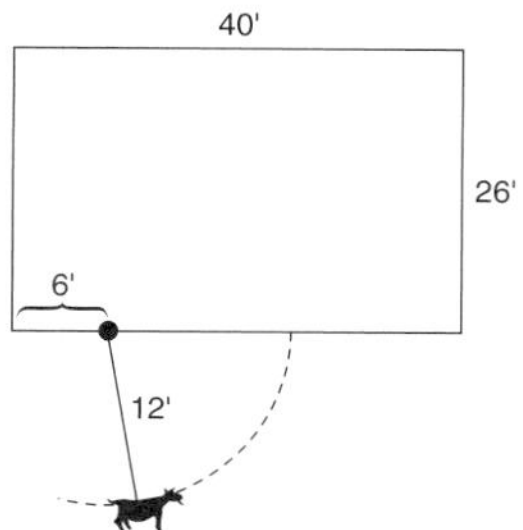

A Look Beyond: Another Look at the Pythagorean Theorem

Some of the many proofs of the Pythagorean Theorem depend on area relationships. One such proof was devised by President James A. Garfield (1831–1881), twentieth president of the United States.

In his proof, the right triangle with legs a and b and hypotenuse c is introduced into a trapezoid, as shown in Figure 7.61(b).

In Figure 7.61(b), the points A, B, and C are collinear. With $\angle 1$ and $\angle 2$ being complementary and the sum of angles about point B being 180°, it follows that $\angle 3$ is a right angle.

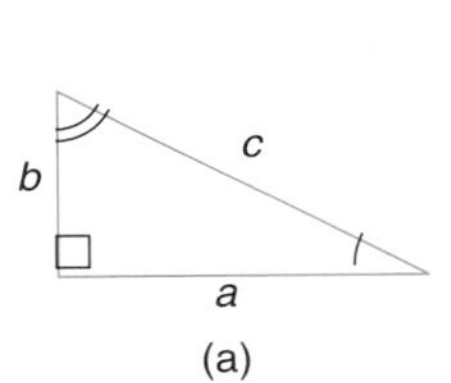

(a)

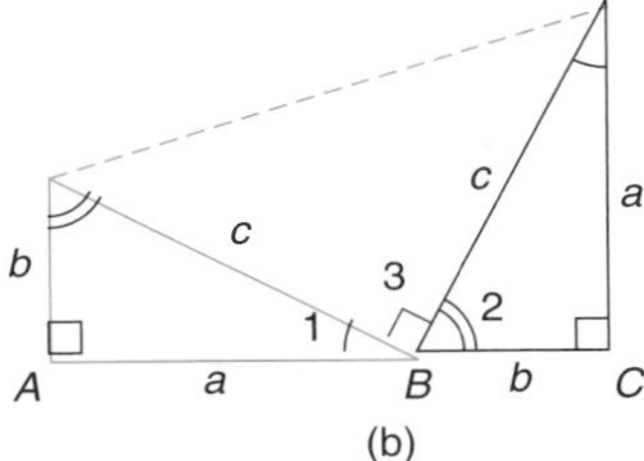

(b)

FIGURE 7.61

If the drawing is perceived as a trapezoid (as shown in Figure 7.62), the area is given by

$$A = \frac{1}{2}h(b_1 + b_2)$$
$$= \frac{1}{2}(a + b)(a + b)$$
$$= \frac{1}{2}(a + b)^2$$
$$= \frac{1}{2}(a^2 + 2ab + b^2)$$
$$= \frac{1}{2}a^2 + ab + \frac{1}{2}b^2$$

$b_1 = a$

$b_2 = b$

$h = a + b$

FIGURE 7.62

Now we treat the trapezoid as a composite of three triangles in Figure 7.63.

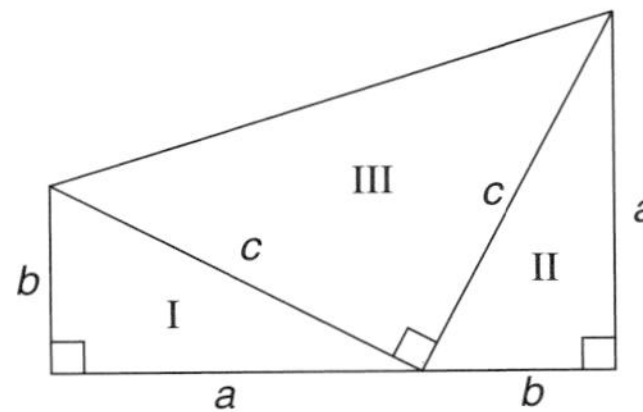

FIGURE 7.63

The total area of regions (triangles) I, II, and III is given by

$$A = A_{I} + A_{II} + A_{III}$$
$$= \frac{1}{2}ab + \frac{1}{2}ab + \left(\frac{1}{2}c \cdot c\right)$$
$$= ab + \frac{1}{2}c^2$$

Equating the areas of the trapezoid in Figure 7.62 and the composite in Figure 7.63, we find that

$$\frac{1}{2}a^2 + ab + \frac{1}{2}b^2 = ab + \frac{1}{2}c^2$$
$$\frac{1}{2}a^2 + \frac{1}{2}b^2 = \frac{1}{2}c^2$$

Multiplying by 2, we get

$$a^2 + b^2 = c^2$$

The earlier proof (over 2000 years earlier!) of this theorem by the Greek mathematician Pythagoras is found in many historical works on geometry. It is not difficult to see the relationship between the two proofs.

In the proof credited to Pythagoras, a right triangle with legs of lengths a and b and hypotenuse of length c is reproduced several times to form a square. Again, points A, B, C (and C, D, E; and so on) must be collinear. [See Figure 7.64(c).]

The area of the large square in Figure 7.65(a) is given by

$$A = (a + b)^2$$
$$= a^2 + 2ab + b^2$$

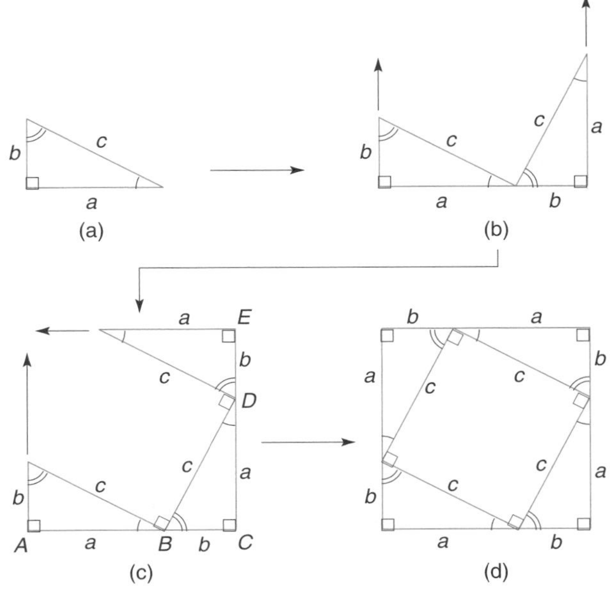

FIGURE 7.64

Considering the composite in Figure 7.65(b), we find that

$$\begin{aligned} A &= A_{I} + A_{II} + A_{III} + A_{IV} + A_{V} \\ &= 4 \cdot A_{I} + A_{V} \end{aligned}$$

since the four right triangles are congruent.

Then
$$\begin{aligned} A &= 4\left(\frac{1}{2}ab\right) + c^2 \\ &= 2ab + c^2 \end{aligned}$$

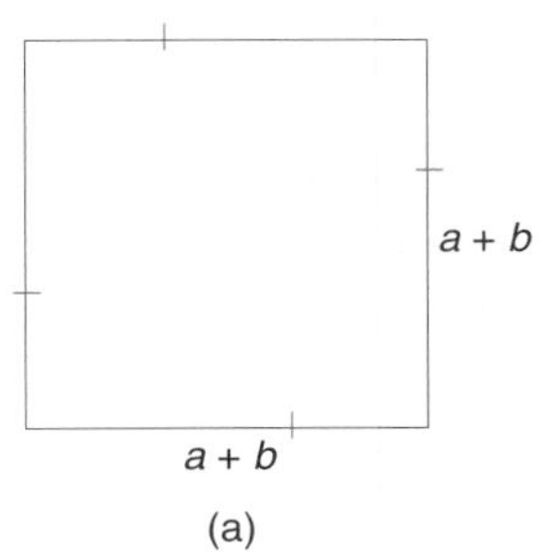

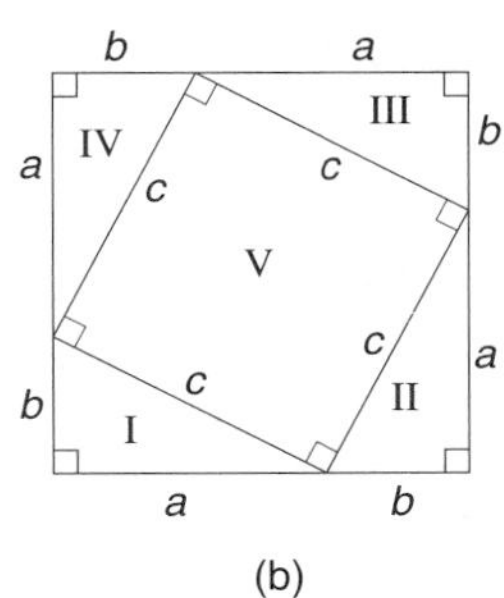

FIGURE 7.65

Again, due to the uniqueness of area, the results (area of square and area of composite) must be equal. Then

$$\begin{aligned} a^2 + 2ab + b^2 &= 2ab + c^2 \\ a^2 + b^2 &= c^2 \end{aligned}$$

Another look at the proofs by President Garfield and by Pythagoras makes it clear that the results must be consistent. In Figure 7.66, observe that Garfield's trapezoid must have one-half the area of Pythagoras' square, while maintaining the relationship that

$$c^2 = a^2 + b^2$$

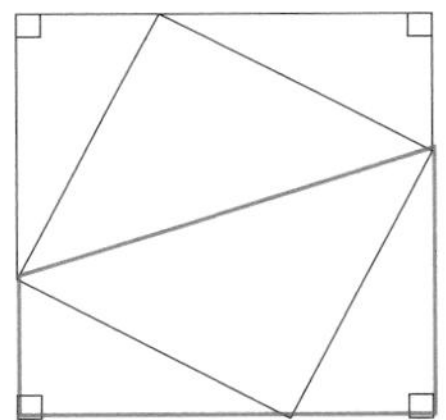

FIGURE 7.66

Summary

A Look Back at Chapter 7

Our goal in this chapter was to determine the areas of triangles, certain quadrilaterals, and regular polygons. We also explored the circumference and area of a circle and the area of a sector of a circle. The area of a circle is sometimes approximated by using $\pi \approx 3.14$ or $\pi \approx \frac{22}{7}$. At other times, the exact area is given by leaving π in the answer.

A Look Ahead to Chapter 8

Our goal in the next chapter is to deal with a type of geometry known as solid geometry. We will find the surface areas of solids with polygonal or circular bases. We will also find the volumes of these solid figures. Select polyhedra will be discussed.

Important Terms and Concepts of Chapter 7

7.1 Region
Square Unit
Area Postulates
Area of a Rectangle, Parallelogram, and Triangle
Altitude, Base of a Parallelogram

7.2 Perimeter of a Polygon
Semiperimeter of a Triangle
Heron's Formula
Area of a Trapezoid
Area of a Rhombus and Kite
Areas of Similar Polygons

7.3 Regular Polygon
Center and Central Angle of a Regular Polygon
Radius and Apothem of a Regular Polygon
Area of a Regular Polygon

7.4 Circumference of a Circle
π (Pi)
Length of an Arc
Limit
Area of a Circle

7.5 Sector
Area, Perimeter of a Sector
Segment of a Circle
Area, Perimeter of a Segment
Area of a Triangle with Inscribed Circle

A Look Beyond: Another Look at the Pythagorean Theorem

Review Exercises

In Review Exercises 1 to 3, draw a figure that allows you to solve each problem.

1. *Given:* $\square ABCD$ with $BD = 34$ and $BC = 30$
$m\angle C = 90°$
Find: A_{ABCD}

2. *Given:* $\square ABCD$ with $AB = 8$ and $AD = 10$
Find: A_{ABCD} if:

a) $m\angle A = 30°$
b) $m\angle A = 60°$
c) $m\angle A = 45°$

3. *Given:* $\square ABCD$ with $\overline{AB} \cong \overline{BD}$ and $AD = 10$
$\overline{BD} \perp \overline{DC}$
Find: A_{ABCD}

In Review Exercises 4 and 5, draw $\triangle ABC$ to solve each problem.

4. *Given:* $AB = 26$, $BC = 25$, and $AC = 17$
Find: A_{ABC}

5. *Given:* $AB = 30$, $BC = 26$, and $AC = 28$
Find: A_{ABC}

In Review Exercises 6 and 7, use the figure shown.

6. *Given:* Trapezoid $ABCD$, with $\overline{AB} \cong \overline{CD}$, $BC = 6$, $AD = 12$, and $AB = 5$
Find: A_{ABCD}

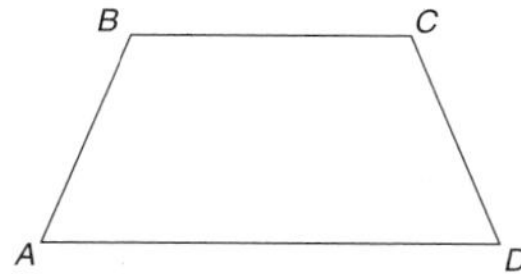

Exercises 6, 7

7. *Given:* Trapezoid $ABCD$, with $AB = 6$ and $BC = 8$,
$\overline{AB} \cong \overline{CD}$
Find: A_{ABCD} if:

a) $m\angle A = 45°$
b) $m\angle A = 30°$
c) $m\angle A = 60°$

8. Find the area and the perimeter of a rhombus whose diagonals have lengths 18 in. and 24 in.

9. Tom Morrow wants to buy some fertilizer for his yard. The lot size is 140 ft by 160 ft. The outside measurements of his house are 80 ft by 35 ft. The driveway measures 30 ft by 20 ft. All shapes are rectangular.

a) What is the square footage of his yard that needs to be fertilized?
b) If each bag of fertilizer covers 5000 ft^2, how many bags should he buy?
c) If the fertilizer costs \$18 per bag, what is his total cost?

10. Alice's mother wants to wallpaper two adjacent walls in Alice's bedroom. She also wants to put a border along the top of all four walls. The bedroom is 9 ft by 12 ft by 8 ft high.

a) If each double roll covers approximately 60 ft^2 and the wallpaper is sold in double rolls only, how many double rolls are needed?
b) If the border is sold in rolls of 5 yd each, how many rolls of the border are needed?

11. *Given:* Isosceles trapezoid $ABCD$
Equilateral $\triangle FBC$
Right $\triangle AED$
$BC = 12$, $AB = 5$, $ED = 16$

Find: **a)** A_{EAFD}
b) Perimeter of $EAFD$

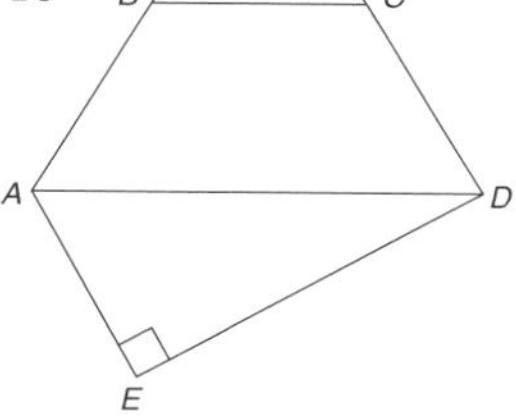

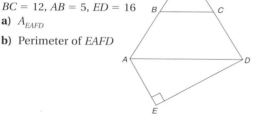

12. *Given:* Kite $ABCD$ with $AB = 10$, $BC = 17$, and $BD = 16$

Find: A_{ABCD}

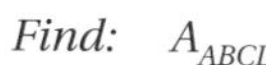

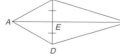

13. One side of a rectangle is 2 cm longer than the other side. If the area is 35 cm^2, find the dimensions of the rectangle.

14. One side of a triangle is 10 cm longer than a second side and the third side is 5 cm longer than the second side. The perimeter of the triangle is 60 cm.

a) Find the lengths of the three sides.
b) Find the area of the triangle.

15. Find the area of $\triangle ABD$.

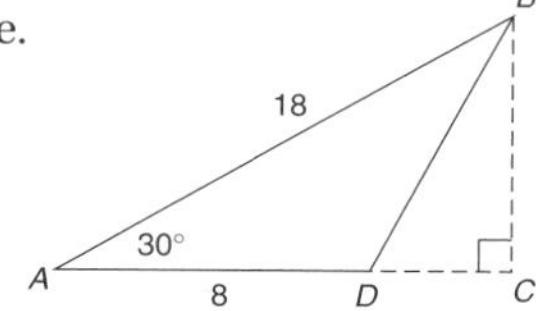

16. Find the area of an equilateral triangle, each of whose sides has length 12 cm.

17. If $\overline{AC}$ is a diameter of $\odot O$, find the area of the shaded triangle.

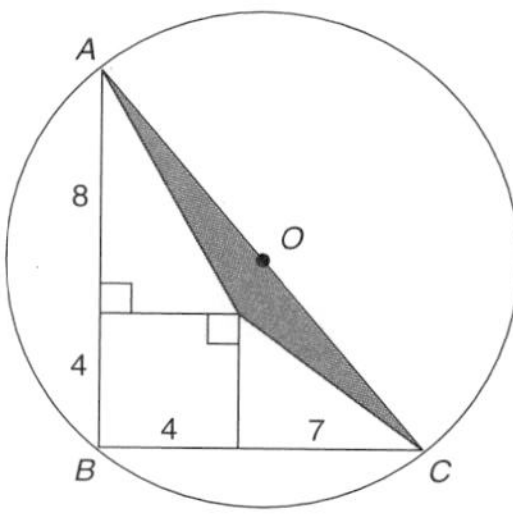

18. For a regular pentagon, find the measure of each:

a) Central angle
b) Interior angle
c) Exterior angle

19. Find the area of a regular hexagon each of whose sides has length 8 ft.

20. The area of an equilateral triangle is $108\sqrt{3}$ in.^{2} If the length of each side of the triangle is $12\sqrt{3}$ in., find the length of the apothem.

21. Find the area of a regular hexagon whose apothem has length 9 in.

22. A regular polygon has each central angle equal to 45°.

a) How many sides does the regular polygon have?
b) If each side is 5 cm and each apothem is approximately 6 cm, what is the approximate area of the polygon?

23. Can a circle be circumscribed about each of the following figures? Why or why not?

a) Parallelogram **c)** Rectangle
b) Rhombus **d)** Square

24. Can a circle be inscribed in each of the following figures? Why or why not?

a) Parallelogram **c)** Rectangle
b) Rhombus **d)** Square

25. The radius of a circle inscribed in an equilateral triangle is 7 in. Find the area of the triangle.

26. The Keiths want to carpet the cement around their rectangular pool. The dimensions for the entire area are 20 ft by 30 ft. The pool is 12 ft by 24 ft.

a) How many square feet need to be covered?
b) Since carpet is sold only by square yards, approximately how many square yards does the area in part (a) represent?
c) If the carpet costs \$9.97 per square yard, what will be the total cost of the carpet?

Find the exact area of the shaded regions in Exercises 27 to 31.

27.

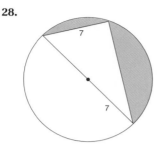

8

Square

28.

7

7

29.

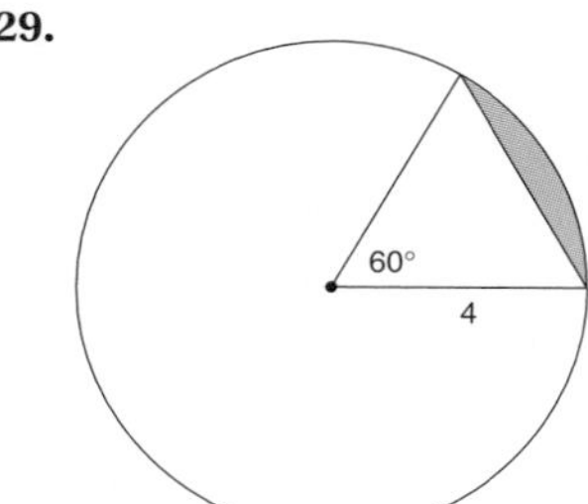

30.

6

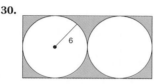

Two ≅ tangent circles, inscribed in a rectangle

31.

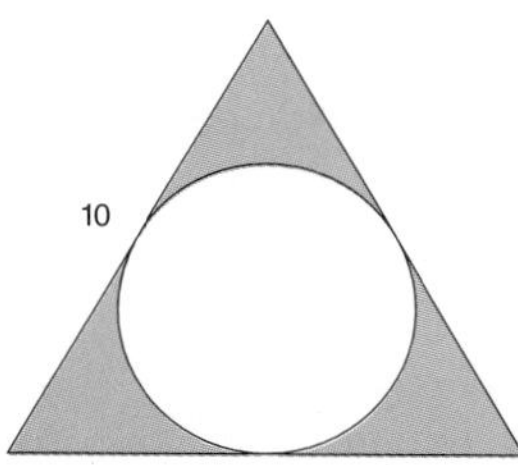

Equilateral triangle

32. An arc of a sector measures 40°. Find the exact length of the arc and the exact area of the sector if the radius is $3\sqrt{5}$ cm.

33. The circumference of a circle is 66 ft.

a) Find the diameter of the circle, using $\pi \approx \frac{22}{7}$.
b) Find the area of the circle, using $\pi \approx \frac{22}{7}$.

34. A circle has an exact area of 27π ft^2.

a) What is the area of a sector of this circle if the arc of the sector measures 80°?
b) What is the exact perimeter of the sector in part (a)?

35. An isosceles right triangle is inscribed in a circle that has a diameter of 12 in. Find the exact area between one of the legs of the triangle and its corresponding arc.

36. *Given:* Concentric circles with radii of lengths R and r, with $R > r$
Prove: $Area_{\text{ring}} = \pi(BC)^2$

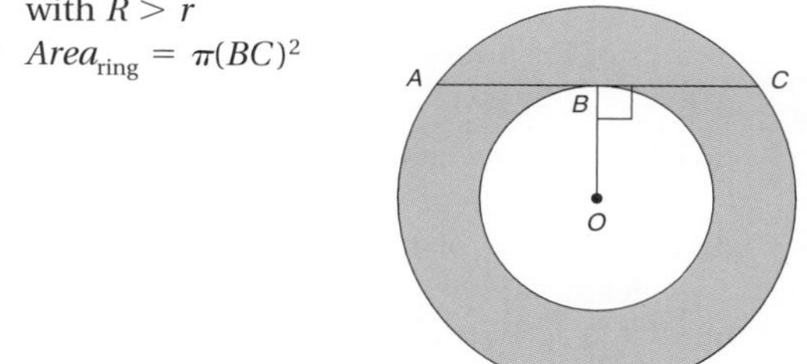

37. Prove that the area of a circle circumscribed about a square is twice the area of the circle inscribed within the square.

38. Prove that if semicircles are constructed on each of the sides of a right triangle, then the area of the semicircle on the hypotenuse is equal to the sum of the areas of the semicircles on the two legs.

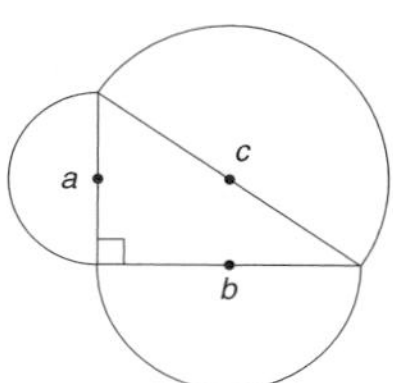

39. Jeff and Helen want to carpet their family room, except for the entranceway and the semicircle in front of the fireplace, which they want to tile.

a) How many square yards of carpeting are needed?
b) How many square feet are to be tiled?

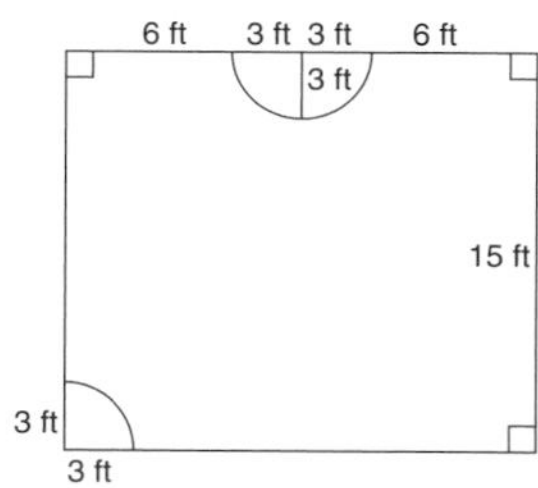

40. Sue and Dave's semicircular driveway is to be resealed, and then flowers are to be planted on either side.

a) What is the number of square feet to be resealed?
b) If the cost of resealing is $0.18 per square foot, what is the total cost?
c) If individual flowers are to be planted 1 foot from the edge of the driveway at intervals of approximately 1 foot on both sides of the driveway, how many flowers are needed?

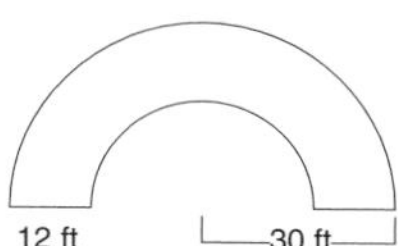

Chapter 8
Surfaces and Solids

The architectural design of the buildings shown suggests many of the solids you will encounter in this chapter. The real world is three-dimensional; that is, objects in the real world have length, width, and depth. In this chapter, formal names will be given to objects that we informally describe as "boxes" and "pop cans."

CHAPTER OUTLINE

8.1 Prisms, Area, and Volume

Prisms

Suppose that two congruent polygons lie in parallel planes in such a way that their corresponding sides are parallel. If the corresponding vertices of these polygons [such as A and A' in Figure 8.1(a)] are joined by line segments, then the "solid" that results is a **prism.** The congruent figures that lie in the parallel planes are the **bases** of the prism. The parallel planes need not be shown in the drawings of prisms.

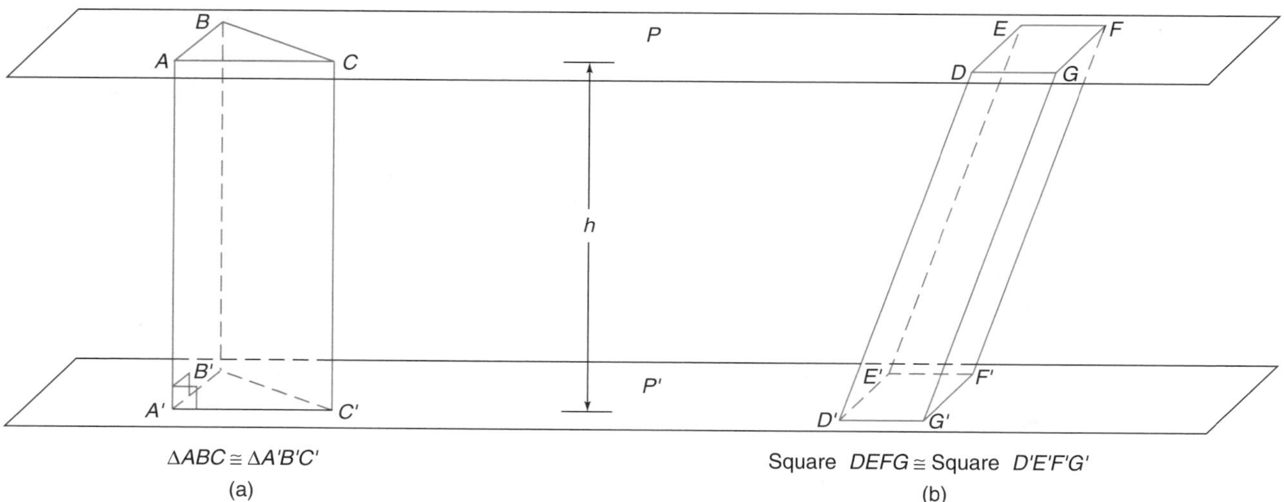

FIGURE 8.1

In Figure 8.1(a), $\overline{AB}$, $\overline{AC}$, $\overline{BC}$, $\overline{A'B'}$, $\overline{A'C'}$, and $\overline{B'C'}$ are **base edges** while $\overline{AA'}$, $\overline{BB'}$, and $\overline{CC'}$ are **lateral edges** of the prism. Because the lateral edges of this prism are perpendicular to its base edges at their points of intersection, the **lateral faces** (like quadrilateral $ACC'A'$) are rectangles. A, B, C, A', B', and C' are the **vertices** of the prism.

In Figure 8.1(b), the lateral edges of the prism are not perpendicular to its base edges. This relationship between the lateral edge and the base edge is often described as **oblique** (slanted). Considering the prisms in Figure 8.1, we are led to the following definitions.

DEFINITION: A **right prism** is a prism in which the lateral edges are perpendicular to the base edges at their points of intersection. An **oblique prism** is a prism in which the lateral edges are oblique to the base edges at their points of intersection.

Part of the description on page 344 used to name a prism depends upon its base. For instance, the prism in Figure 8.1(a) is a *right triangular prism;* in this case, the word *right* describes the prism while the word *triangular* refers to the triangular base. Similarly, the prism in Figure 8.1(b) would be an *oblique square prism,* assuming that the bases are squares. Both prisms in Figure 8.1 have an **altitude** (length of a perpendicular segment joining the planes containing the bases) of h.

EXAMPLE 1

Name each type of prism in Figure 8.2.

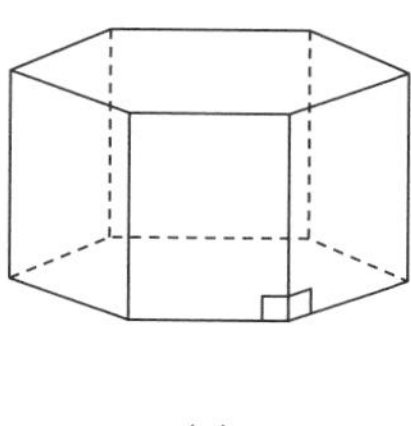

(a)

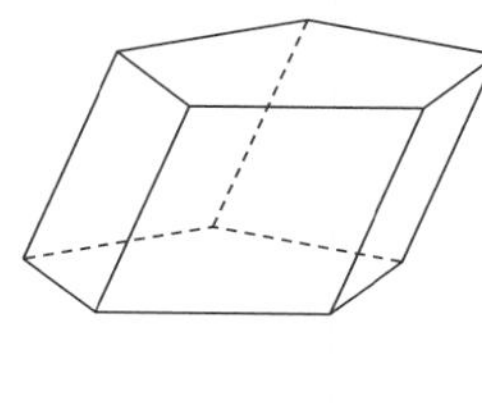

(b)

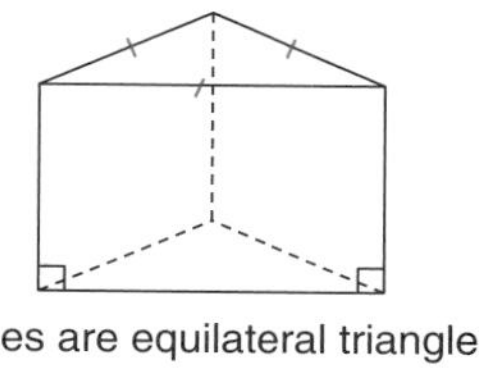

(c)

FIGURE 8.2

Solution

a) The lateral edges are perpendicular to the base edges of the hexagonal base. The prism is a *right hexagonal prism.*
b) The lateral edges are oblique to the base edges of the pentagonal base. The prism is an *oblique pentagonal prism.*
c) The lateral edges are perpendicular to the base edges of the triangular base. Because the base is equilateral, the prism is a *right equilateral triangular prism.*

Area of a Prism

DEFINITION: The **lateral area** L of a prism is the sum of the areas of all lateral faces.

In the right triangular prism of Figure 8.3, a, b, and c are the lengths of the sides of either base. These dimensions are used along with the length of the altitude (denoted by h) to calculate the lateral area, the sum of the areas of rectangles $ACC'A'$, $ABB'A'$, and $BCC'B'$. The lateral area L of the right triangular prism can be found as follows:

$$\begin{aligned} L &= ah + bh + ch \\ &= h(a + b + c) \\ &= hP \end{aligned}$$

where P is the perimeter of a base of the prism. This formula is valid for any *right* prism.

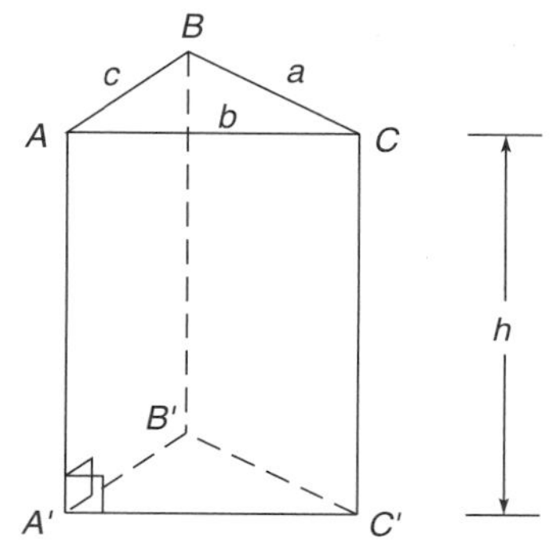

FIGURE 8.3

THEOREM 8.1.1: The lateral area L of a right prism whose altitude has measure h and whose base has perimeter P is given by $L = hP$.

Many students (and teachers) find it easier to calculate the lateral area of a prism without using the formula $L = hP$. We illustrate this in the following example.

EXAMPLE 2

The bases of the right prism shown in Figure 8.4 are equilateral pentagons with sides of length 3 in. each. If the altitude of the prism is 4 in., find the lateral area of the prism.

Solution

Each lateral face is a rectangle with dimensions 3 in. by 4 in. The area of each rectangular face is 3 in. $\times$ 4 in. = 12 in.^{2} Since there are five congruent lateral faces, the lateral area of the pentagonal prism is 5 $\times$ 12 in.^{2} = 60 in.^{2}

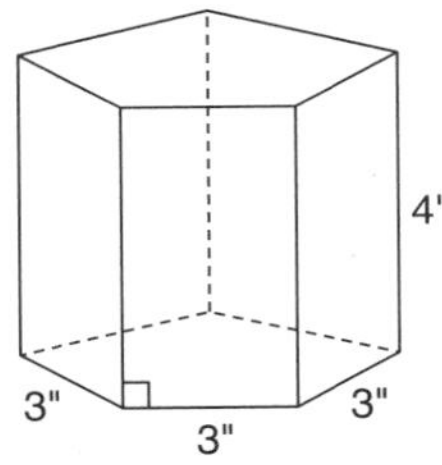

FIGURE 8.4

DEFINITION: For any prism, the **total area** T is the sum of the lateral area and the areas of the bases.

NOTE: The total area of the prism is also its **surface area.**

Both bases and lateral faces are known as *faces* of a prism. Thus, the total area T of the prism is the sum of the areas of all its faces.

Recalling Heron's Formula, we know that the base area B of the right triangular prism in Figure 8.3 can be found by the formula

$$B = \sqrt{s(s - a)(s - b)(s - c)}$$

in which s is the semiperimeter of the triangular base. We use Heron's formula in Example 3.

EXAMPLE 3

Find the total area of the right triangular prism with altitude 8 in. if the sides of the triangular bases have lengths of 13 in., 14 in., and 15 in. (See Figure 8.5.)

Solution

The lateral area is found by adding the areas of the three rectangular lateral faces. That is,

$$\begin{aligned} L &= 8 \text{ in.} \cdot 13 \text{ in.} + 8 \text{ in.} \cdot 14 \text{ in.} + 8 \text{ in.} \cdot 15 \text{ in.} \\ &= 104 \text{ in.}^2 + 112 \text{ in.}^2 + 120 \text{ in.}^2 = 336 \text{ in.}^2 \end{aligned}$$

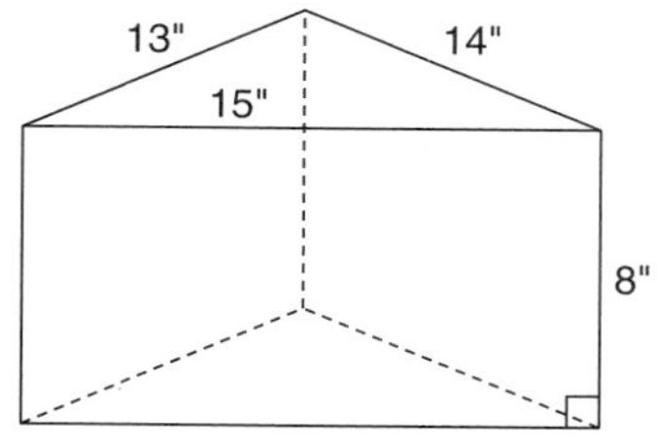

FIGURE 8.5

We use Heron's Formula to find the area of each base. With $s = \frac{1}{2}(13 + 14 + 15)$ or $s = 21$, $B = \sqrt{21(21 - 13)(21 - 14)(21 - 15)} = \sqrt{21(8)(7)(6)} = \sqrt{7056} = 84$.

Calculating the total area (or surface area) of the triangular prism,

$$T = 336 + 2(84) \qquad \text{or} \qquad T = 504 \text{ in.}^2$$

A more general formula for the total area of a prism follows.

THEOREM 8.1.2: The total area T of any prism with lateral area L and base area B is given by $T = L + 2B$.

DEFINITION: A **regular prism** is a prism whose base is a regular polygon.

In the following example, the base of the prism is a regular hexagon. Because the prism is a right prism, the lateral faces are congruent rectangles.

EXAMPLE 4

Find the lateral area L and the surface area T of the right regular hexagonal prism in Figure 8.6(a).

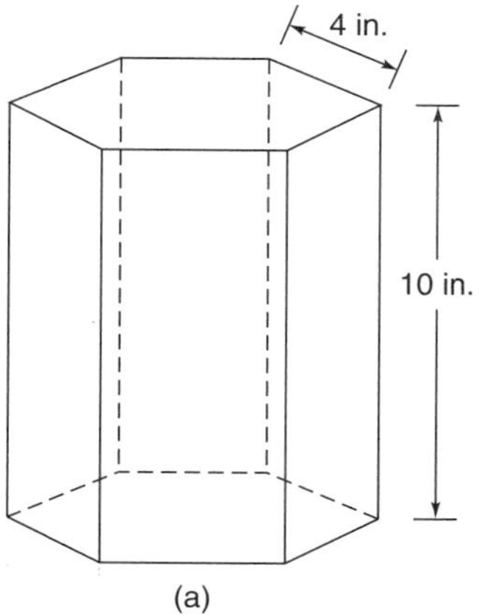

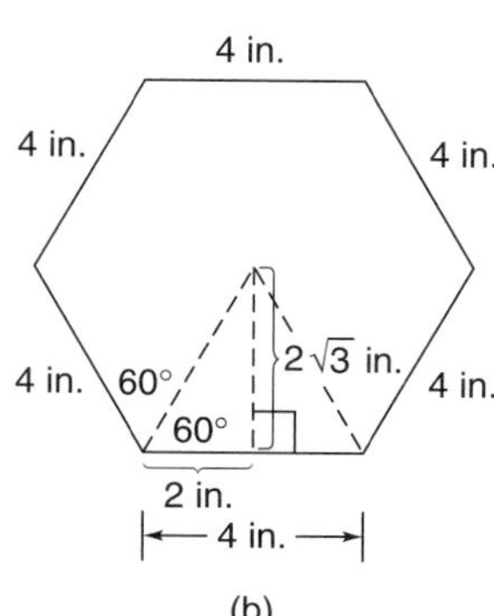

FIGURE 8.6

Solution

There are six congruent faces, each rectangular with dimensions of 4 in. by 10 in. Then

$$\begin{aligned} L &= 6(4 \cdot 10) \\ &= 240 \text{ in.}^2 \end{aligned}$$

For the regular hexagonal base [see Figure 8.6(b)], the apothem measures $a = 2\sqrt{3}$ in. while the perimeter is $P = 6 \cdot 4 = 24$ in. Then the area B of each base is given by the formula for the area of a regular polygon.

$$
\begin{aligned}
B &= \frac{1}{2}aP \\
&= \frac{1}{2} \cdot 2\sqrt{3} \cdot 24 \\
&= 24\sqrt{3}\text{ in.}^2 \approx 41.57\text{ in.}^2
\end{aligned}
$$

Now
$$
\begin{aligned}
T &= L + 2B \\
&= (240 + 48\sqrt{3})\text{ in.}^2 \approx 323.14\text{ in.}^2
\end{aligned}
$$

EXAMPLE 5

The total area of the right square prism in Figure 8.7 is 210 cm². Find the length of a side of the square base if the altitude of the prism is 8 cm.

Solution

Let x be the length of a side of the square. Then the area of the base is $B = x^2$ and the area of each of the four lateral faces is $8x$. Therefore

$$
\begin{aligned}
\underbrace{2(x^2)}_{\text{2 bases}} + \underbrace{4(8x)}_{\text{4 lateral faces}} &= 210 \\
2x^2 + 32x &= 210 \\
2x^2 + 32x - 210 &= 0 \\
x^2 + 16x - 105 &= 0 && \text{dividing by 2} \\
(x + 21)(x - 5) &= 0 && \text{factoring} \\
x + 21 = 0 \quad \text{or} \quad x - 5 &= 0 \\
x = -21 \quad \text{or} \quad x &= 5 && \text{reject } -21 \text{ as a solution}
\end{aligned}
$$

Then each side of the square base measures 5 cm.

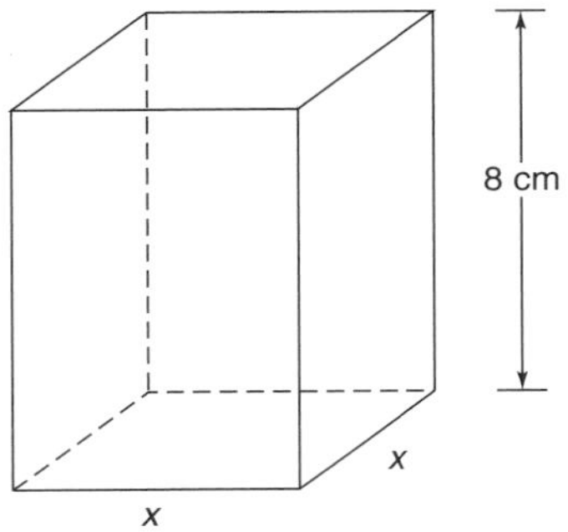

FIGURE 8.7

Volume of a Prism

To introduce the notion of *volume,* we realize that a prism encloses a portion of space. Without a formal definition, we say that **volume** is a number that measures the amount of enclosed space. To begin, we need a unit for measuring volume. Just as a meter can be used to measure length and a square yard can be used to measure area, a **cubic unit** is used to measure the amount of space enclosed within a bounded region of space. One such unit is described next.

The volume occupied by the cube (a right square prism in which lateral and base edges are congruent) shown in Figure 8.8 is 1 cubic inch or 1 in.³ The volume of a solid is the number of cubic units within the solid. Thus, we assume that the volume of any solid is a positive number of cubic units.

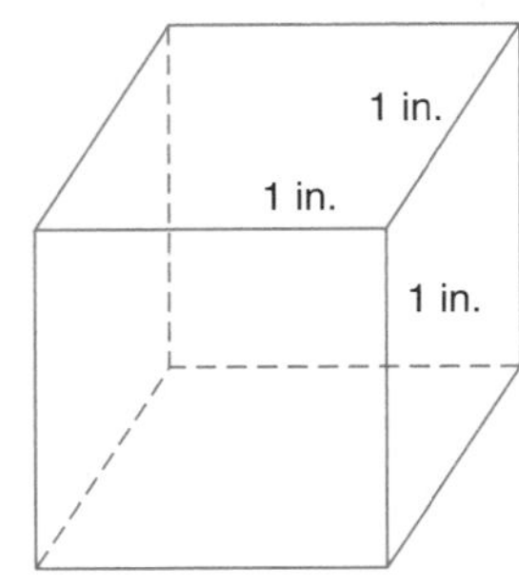

FIGURE 8.8

POSTULATE 24: (Volume Postulate)
Corresponding to every solid is a unique positive number V known as the volume of that solid.

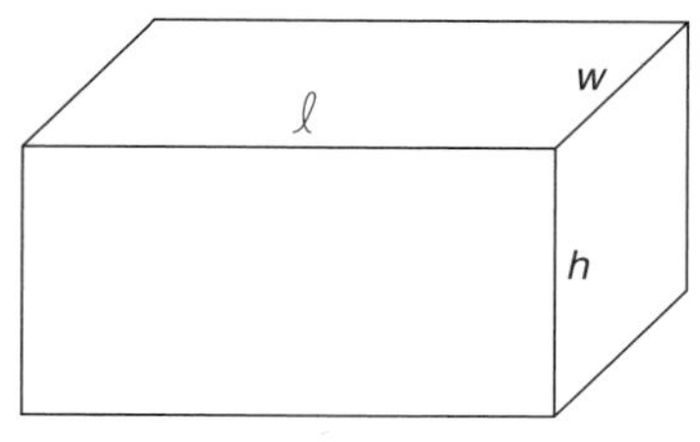

FIGURE 8.9

The simplest figure for which we can determine volume is the **right rectangular prism.** Such a solid might be described as a "box." Since boxes are used as containers for storage and shipping (like a boxcar), it is important to calculate volume as a measure of capacity. A right rectangular prism is shown in Figure 8.9; its dimensions are length ℓ, width w, and height (or altitude) h.

The volume of a right rectangular prism of length 4 in., width 3 in., and height 2 in. is easily shown to be 24 in.^{3} The volume is the product of the three dimensions of the given solid. Not only do we see that $4 \cdot 3 \cdot 2 = 24$ but also that the units are in. · in. · in. = in.^{3}

Figures 8.10(a) and (b) illustrate that the 4 by 3 by 2 box must have the volume 24 cubic units. We see that there are four layers of blocks, each of which is a 2 by 3 configuration of 6 units^3. Figure 8.10 also provides some insight into our next postulate.

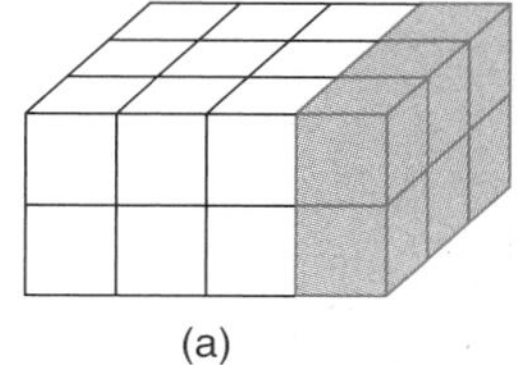

(a)

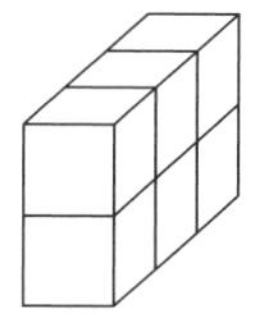
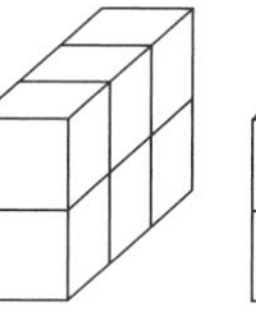
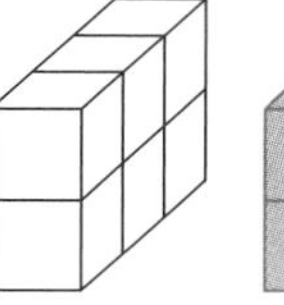
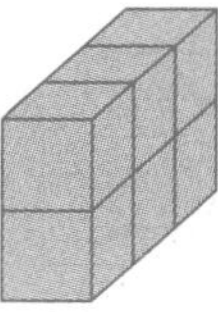

(b)

FIGURE 8.10

> **POSTULATE 25:**
> The volume of a right rectangular prism is given by
>
> $$V = \ell wh$$
>
> where ℓ measures the length, w the width, and h the altitude of the prism.

EXAMPLE 6

Find the volume of a box whose dimensions are 1 ft, 8 in., and 10 in. (See Figure 8.11.)

Solution

While it makes no difference which dimension is chosen for ℓ or w or h, it is most important that the units of measure be the same. Thus, 1 ft is replaced by 12 in. in the formula for volume:

$$\begin{aligned} V &= \ell wh \\ &= 12 \text{ in.} \cdot 8 \text{ in.} \cdot 10 \text{ in.} \\ &= 960 \text{ in.}^3 \end{aligned}$$

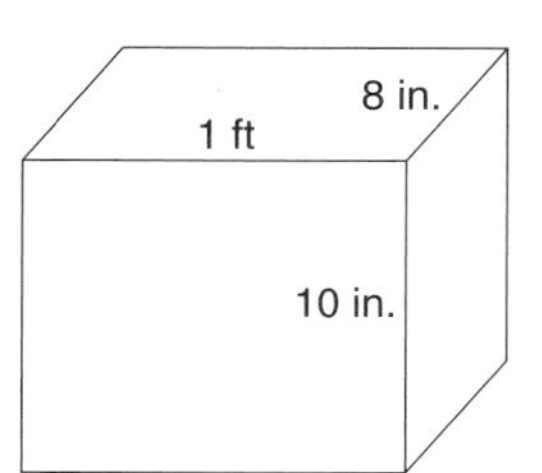

FIGURE 8.11

Notice that the formula for the volume of the right rectangular prism, $V = \ell wh$, could be replaced by the formula $V = Bh$, where B is the area of the base of the prism; that is, $B = \ell w$. As stated in the next postulate, this volume relationship is true for right prisms in general.

■ *Warning*

The upper case B found in formulas in this chapter represents the area of the base of a solid; because the base is a plane region, B is measured in square units. ■

POSTULATE 26:
The volume of a right prism is given by

$$V = Bh$$

where B is the area of a base and h is the altitude of the prism.

EXAMPLE 7

Find the volume of the right hexagonal prism in Figure 8.6 on page 347.

Solution

We found that the area of the hexagonal base was $24\sqrt{3}$ in.^{2}. Since the altitude of the hexagonal prism is 10 in., the volume is $V = Bh$ or $V = (24\sqrt{3}\text{ in.}^2)(10\text{ in.})$. Then $V = 240\sqrt{3}\text{ in.}^3 \approx 415.69\text{ in.}^3$.

NOTE: Just as $x^2 \cdot x = x^3$, the units in Example 7 are in.^{2} · in. = in.^{3}

In the final example of this section, we use the fact that 1 yd^3 = 27 ft^3. In the cube shown in Figure 8.12, each dimension measures 1 yd, or 3 ft. The cube's volume is given by 1 yd · 1 yd · 1 yd = 1 yd^3 or 3 ft · 3 ft · 3 ft = 27 ft^3.

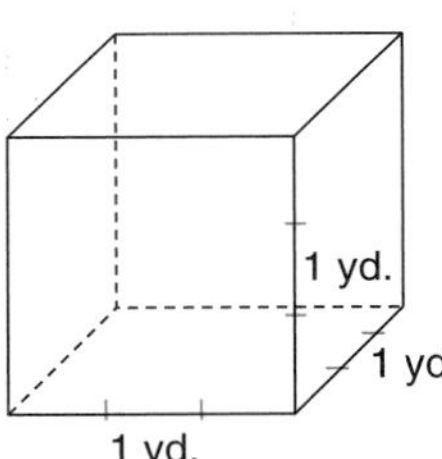

FIGURE 8.12

EXAMPLE 8

The Tarantinos are having a concrete driveway poured at their house. The section to be poured is rectangular, measuring 12 ft by 40 ft, and is 4 in. deep. How many cubic yards of concrete are needed?

Solution

Using $V = \ell wh$, we must be consistent with units. Thus, $\ell = 12$ ft, $w = 40$ ft, and $h = \frac{1}{3}$ ft (from 4 in.). Now

$$V = 12\text{ ft} \cdot 40\text{ ft} \cdot \frac{1}{3}\text{ ft}$$
$$V = 160\text{ ft}^3$$

To change 160 ft^3 to cubic yards, we divide by 27 to obtain $5\frac{25}{27}$ yd^3.

NOTE: The Tarantinos will be charged for 6 yd^3 of concrete, the result of rounding upward.

8.1 Exercises

1. Consider the solid shown.
   a) Does it appear to be a prism?
   b) Is it right or oblique?
   c) What type of base(s) does the solid have?
   d) Name the type of solid.
   e) What type of figure is each lateral face?

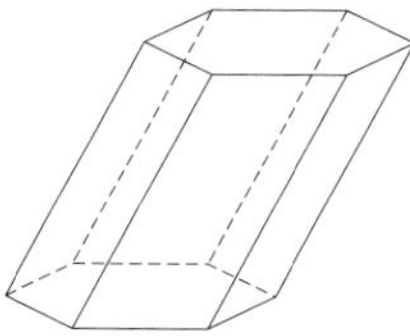

2. Consider the solid shown.
   a) Does it appear to be a prism?
   b) Is it right or oblique?
   c) What type of base(s) does the solid have?
   d) Name the type of solid.
   e) What type of figure is each lateral face?

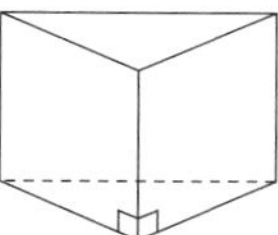

3. For the hexagonal prism shown in Exercise 1,
   a) how many vertices does it have?
   b) how many edges (lateral edges plus base edges) does it have?
   c) how many faces (lateral faces plus bases) does it have?
4. For the triangular prism shown in Exercise 2,
   a) how many vertices does it have?
   b) how many edges (lateral edges plus base edges) does it have?
   c) how many faces (lateral faces plus bases) does it have?
5. If each edge of the hexagonal prism in Exercise 1 is measured in centimeters, what unit is used to measure its (a) surface area? (b) volume?
6. If each edge of the triangular prism in Exercise 2 is measured in inches, what unit is used to measure its (a) lateral area? (b) volume?
7. Suppose that the bases of the hexagonal prism in Exercise 1 each have an area of 12 cm^2 while each lateral face has an area of 18 cm^2. Find the total (surface) area of the prism.
8. Suppose that the bases of the triangular prism in Exercise 2 each have an area of 3.4 $in.^2$ while each lateral face has an area of 4.6 in^2. Find the total (surface) area of the prism.
9. Suppose that the bases of the hexagonal prism in Exercise 1 each have an area of 12 cm^2 while the altitude of the prism measures 10 cm. Find the volume of the prism.
10. Suppose that the bases of the triangular prism in Exercise 2 each have an area of 3.4 cm^2 while the altitude of the prism measures 1.2 cm. Find the volume of the prism.
11. A solid is an octagonal prism.
    a) How many vertices does it have?
    b) How many lateral edges does it have?
    c) How many base edges are there in all?
12. A solid is a pentagonal prism.
    a) How many vertices does it have?
    b) How many lateral edges does it have?
    c) How many base edges are there in all?
13. Generalize the results found in Exercises 11 and 12 by answering each of the following questions. Assume that the number of sides in each base of the prism is n. What is the
    a) number of vertices?
    b) number of lateral edges?
    c) number of base edges?
    d) total number of edges?
    e) number of lateral faces?
    f) number of bases?
    g) total number of faces?
    (**NOTE:** Upper and lower faces = bases)
14. In the right regular pentagonal prism shown on page 352, suppose that each base edge measures 6 in. while the apothem of the base measures 4.1 in. The altitude of the prism measures 10 in.
    a) Find the lateral area of the prism.

b) Find the total area of the prism.
c) Find the volume of the prism.

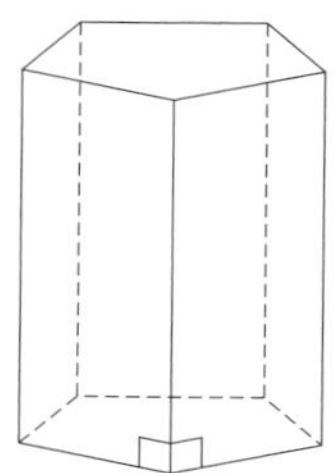

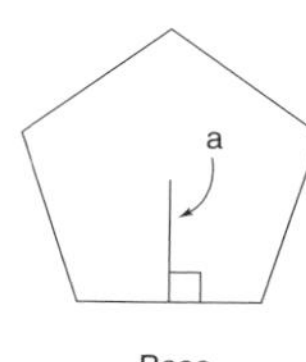

Exercises 14, 15

15. In the right regular pentagonal prism shown above, suppose that each base edge measures 9.2 cm while the apothem of the base measures 6.3 cm. The altitude of the prism measures 14.6 cm.

a) Find the lateral area of the prism.
b) Find the total area of the prism.
c) Find the volume of the prism.

16. For a right triangular prism, suppose that the sides of the triangular base measure 4 m, 5 m, and 6 m. The altitude is 7 m.

a) Find the lateral area of the prism.
b) Find the total area of the prism.
c) Find the volume of the prism.

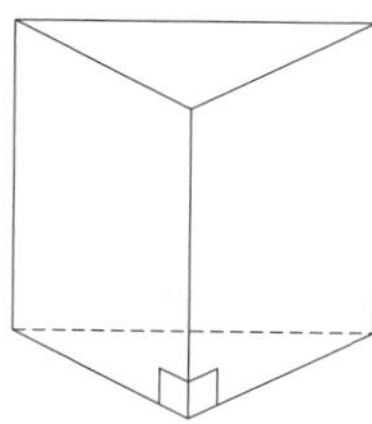

Exercises 16, 17

17. For a right triangular prism, suppose that the sides of the triangular base measure 3 ft, 4 ft, and 5 ft. The altitude is 6 ft.

a) Find the lateral area of the prism.
b) Find the total area of the prism.
c) Find the volume of the prism.

18. A cereal box measures 2 in. by 8 in. by 10 in. What is the volume of the box? How many square inches of cardboard make up its surface (disregard any hidden flaps)?

19. The measures of the sides of the square base of a box are twice the measure of the height of the box. If the volume of the box is 108 in.^{3}, find the dimensions of the box.

20. For a given box, the height measures 4 m. If the length of the rectangular base is 2 m greater than the width of the base and the lateral area L is 96 m^2, find the dimensions of the box.

21. For the box shown, the total area is 94 cm^2. Determine the value of x.

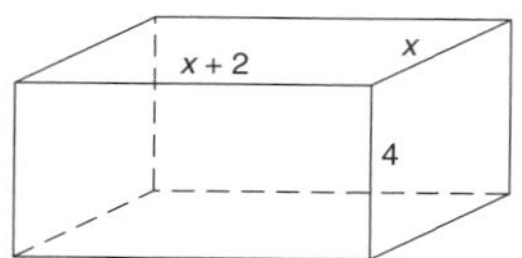

Exercises 21, 22

22. If the volume of the box is 252 in.^{3}, find the value of x.

23. The box with dimensions indicated is to be constructed of materials that cost 1 cent per square inch for the lateral surface and 2 cents per square inch for the bases. What is the total cost of constructing the box?

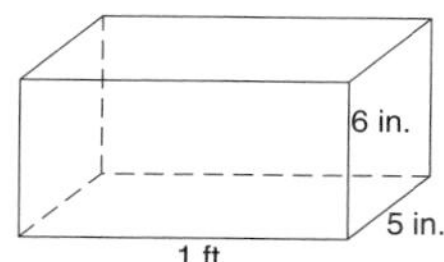

24. A steel door is 32 in. wide by 80 in. tall by $1\frac{3}{8}$ in. thick. How many cubic inches of foam insulation are needed to fill the door?

25. A storage shed is in the shape of a pentagonal prism. The front represents one of its two bases. If the shed is 10 ft deep, what is the storage capacity (volume) of its interior?

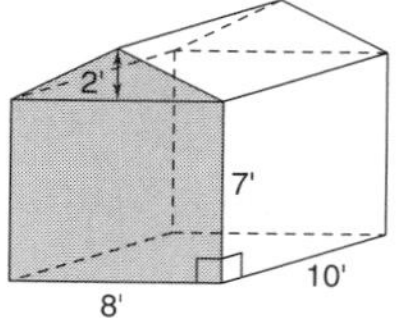

26. A storage shed is in the shape of a trapezoidal prism. Each trapezoid represents one of its bases. With dimensions as shown, what is the storage capacity (volume) of its interior?

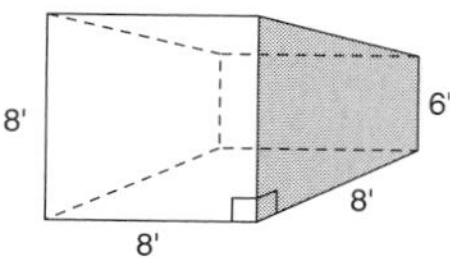

27. A cube is a right square prism in which all edges have the same length. For the cube with edge e,

a) show that the total area is $T = 6e^2$.
b) find the total area if $e = 4$ cm.
c) show that the volume is $V = e^3$.
d) find the volume if $e = 4$ cm.

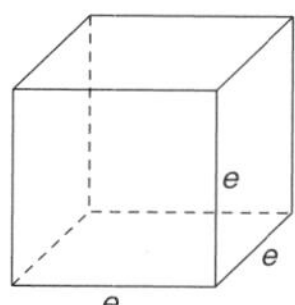

Exercises 27–29

28. Use the formulas developed in Exercise 27 to find the (a) total area T and (b) volume V of a cube with edges of length 5.3 ft each.

29. ➢ A diagonal of a cube joins two vertices so that the remaining points on the diagonal lie in the interior of the cube. Show that the diagonal of the cube having edges of length e is $e\sqrt{3}$ units long.

30. A concrete pad 4 in. thick is to have a length of 36 ft and a width of 30 ft. How many cubic yards of concrete must be poured? (**HINT:** $1 \text{ yd}^3 = 27 \text{ ft}^3$)

31. A raised flower bed is 2 ft high by 12 ft wide by 15 ft long. The mulch, soil, and peat mixture used to fill the raised bed costs \$9.60 per cubic yard. What is the total cost of the ingredients used to fill the raised garden?

32. In excavating for a new house, a contractor digs a hole in the shape of a right rectangular prism. The dimensions of the hole are 54 ft long by 36 ft wide by 9 ft deep. How many cubic yards of dirt were removed?

For Exercises 33–35, consider the oblique regular pentagonal prism shown. Each side of the base measures 12 cm and the altitude measures 12 cm.

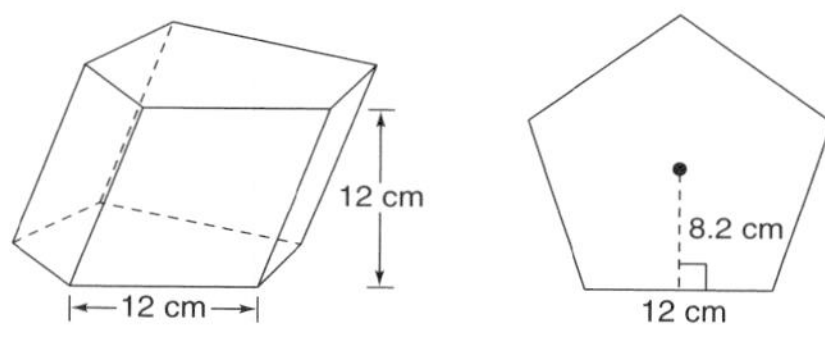

Exercises 33–35

33. Find the lateral area of the prism.
(**HINT:** Each lateral face is a parallelogram.)

34. Find the total area of the prism.

35. Find the volume of the prism.

8.2 Pyramids, Area, and Volume

The solids shown in Figure 8.13 are **pyramids.** In Figure 8.13(a), point A is noncoplanar with square $BCDE$. In Figure 8.13(b), F is noncoplanar with $\triangle GHJ$. In these pyramids, the noncoplanar point has been joined (by drawing line segments) to each vertex of the square and to each vertex of the triangle, respectively. Square $BCDE$ is the **base** of the first pyramid and $\triangle GHJ$ is the base of the second pyramid. Point A is known as the **vertex** of the **square pyramid;** likewise, point F is the vertex of the **triangular pyramid.**

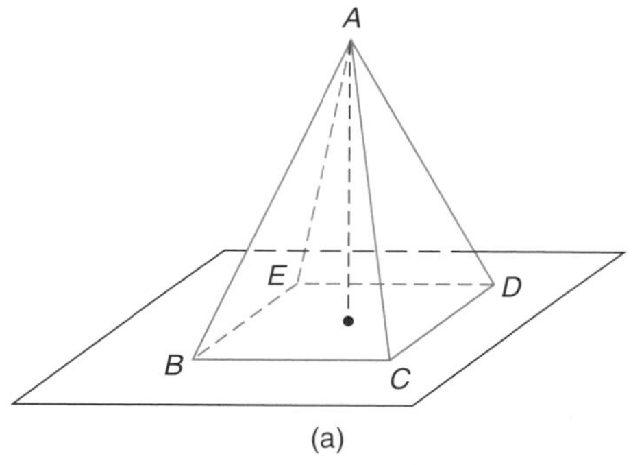

(a)

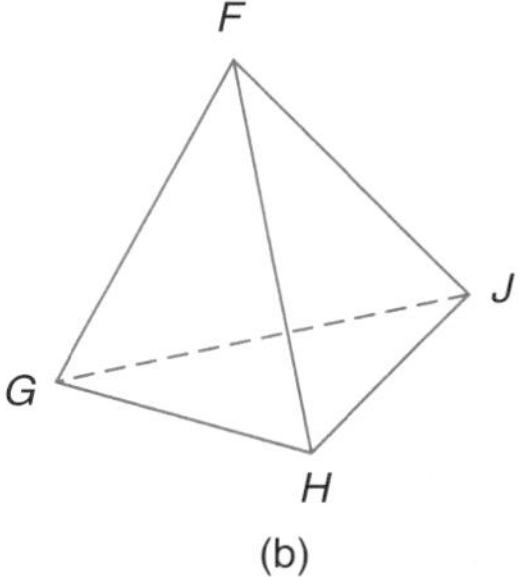

(b)

FIGURE 8.13

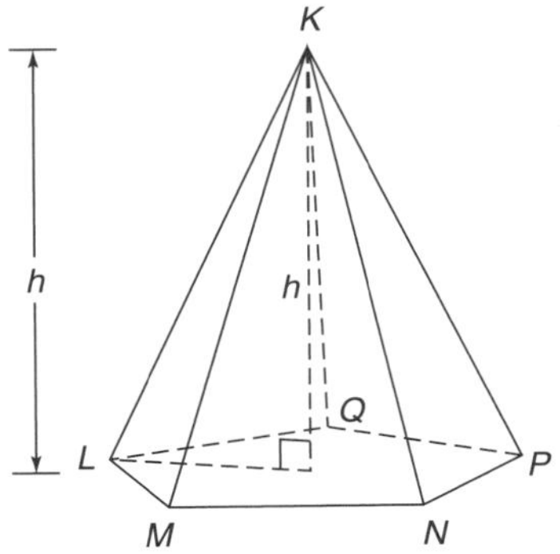

FIGURE 8.14

The pyramid in Figure 8.14 is a **pentagonal pyramid.** It has vertex K, pentagon $LMNPQ$ for a base, and **lateral edges** $\overline{KL}$, $\overline{KM}$, $\overline{KN}$, $\overline{KP}$, and $\overline{KQ}$. The sides of the base $\overline{LM}$, $\overline{MN}$, $\overline{NP}$, $\overline{PQ}$, and $\overline{QL}$ are **base edges.** All **lateral faces** of a pyramid are triangles; $\triangle KLM$ is one of the five lateral faces of the pentagonal pyramid. Including base $LMNPQ$, this pyramid has a total of six faces. The **altitude** of the pyramid, of length h, is the line segment from the vertex K perpendicular to the plane of the base.

DEFINITION: A **regular pyramid** is a pyramid whose base is a regular polygon and whose lateral edges are all congruent.

Suppose that the pyramid in Figure 8.14 is a regular pentagonal pyramid. Then the lateral faces are necessarily congruent to each other; in Figure 8.14, $\triangle KLM \cong \triangle KMN \cong \triangle KNP \cong \triangle KPQ \cong \triangle KQL$ by SSS. Each lateral face is an isosceles triangle.

DEFINITION: The **slant height** of a regular pyramid is the altitude of any of the congruent lateral faces of the regular pyramid.

NOTE: Only a regular pyramid has a slant height.

In our calculations, we use ℓ to represent the length of the slant height of a regular pyramid.

EXAMPLE 1

For a regular square pyramid with altitude 4 in. and base edges of length 6 in. each, find the length of the slant height ℓ.

Solution

In Figure 8.15, it can be shown that the apothem to any side has length 3 in. (one-half the length of the side of the square base). Also, the slant height is the hypotenuse of a right triangle with legs equal to the lengths of the altitude and the apothem. By the Pythagorean Theorem, we have

$$\ell^2 = a^2 + h^2$$
$$\ell^2 = 3^2 + 4^2$$
$$\ell^2 = 9 + 16$$
$$\ell^2 = 25 \rightarrow \ell = 5 \text{ in.}$$

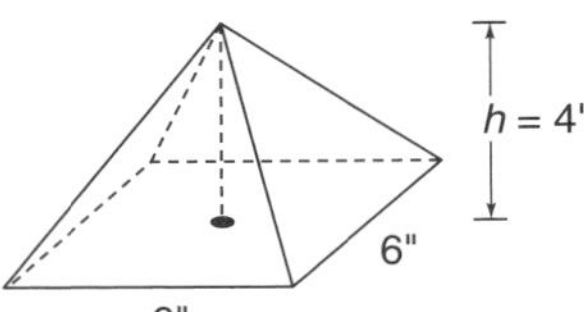

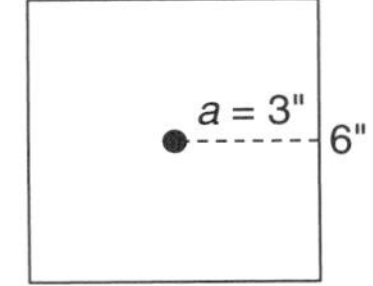

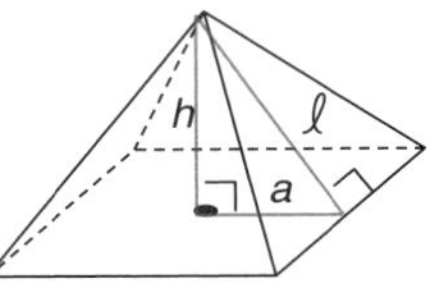

FIGURE 8.15

The following fact was used in the solution of Example 1. We accept it without further proof.

THEOREM 8.2.1: In a regular pyramid, the length a of the apothem of the base, the altitude h, and the slant height ℓ satisfy the Pythagorean Theorem; that is, $\ell^2 = a^2 + h^2$ in every regular pyramid.

Surface Area of a Pyramid

To lay the groundwork for the next theorem, we justify the result by "taking apart" one of the regular pyramids and laying it out flat. We will use a regular hexagonal pyramid for this purpose, but the argument is similar if the base is any regular polygon.

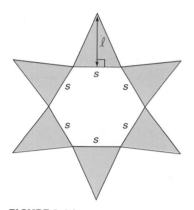

When the lateral faces of the regular pyramid are folded down into the plane, as shown in Figure 8.16, the shaded lateral area is the sum of areas of the triangular lateral faces. Using $A = \frac{1}{2}bh$, the area of each face is $\frac{1}{2} \cdot s \cdot \ell$ (each side of the base of the pyramid has length s, and the slant height is ℓ). The combined areas of the triangles give the lateral area. Because there are n triangles,

FIGURE 8.16

$$
\begin{aligned}
L &= n \cdot \frac{1}{2} \cdot s \cdot \ell \\
&= \frac{1}{2} \cdot \ell(n \cdot s) \\
&= \frac{1}{2}\ell P
\end{aligned}
$$

where P is the base perimeter.

THEOREM 8.2.2: The lateral area L of a regular pyramid with slant height of length ℓ and perimeter P of the base is given by

$$L = \frac{1}{2}\ell P$$

EXAMPLE 2

Find the lateral area of a regular pentagonal pyramid if the sides of the base measure 8 cm and the lateral edges measure 10 cm each [see Figure 8.17(a) on the next page].

Solution

For the triangular lateral face [see Figure 8.17(b)], we find the length of the slant height by applying the Pythagorean Theorem:

$$
\begin{aligned}
4^2 + \ell^2 &= 10^2, \text{ so } 16 + \ell^2 = 100 \\
\ell^2 &= 84 \\
\ell &= \sqrt{84} = \sqrt{4 \cdot 21} = \sqrt{4} \cdot \sqrt{21} = 2\sqrt{21}
\end{aligned}
$$

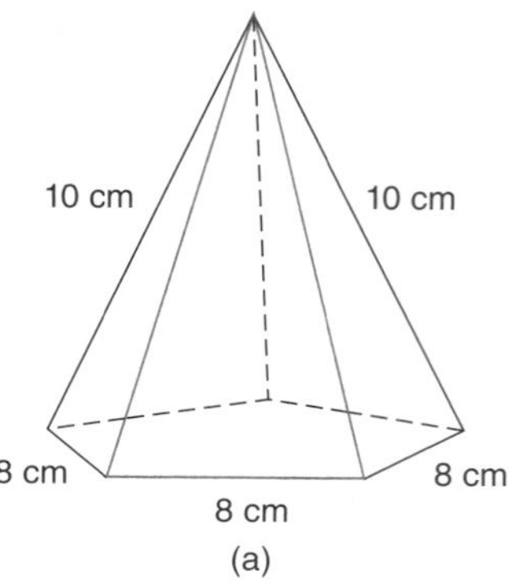

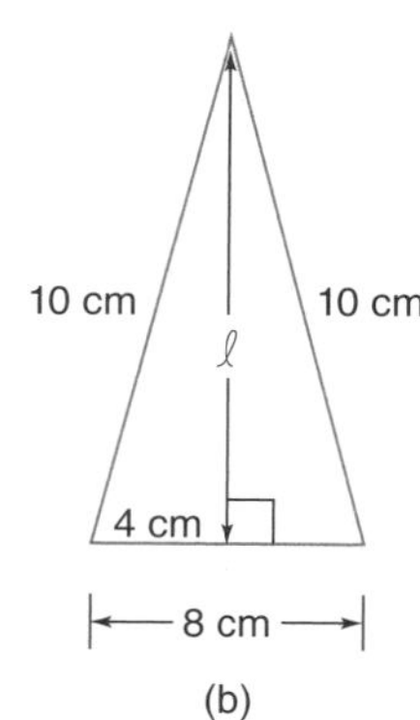

FIGURE 8.17 (a) (b)

Now $L = \frac{1}{2}\ell P$ becomes $L = \frac{1}{2} \cdot 2\sqrt{21} \cdot (5 \cdot 8) = \frac{1}{2} \cdot 2\sqrt{21} \cdot 40 = 40\sqrt{21}$ cm^2 $\approx$ 183.30 cm^2.

It may be easier to find the lateral area of a regular pyramid without using the formula of Theorem 8.2.2; simply find the area of one lateral face and multiply by the number of faces. In Example 2, for instance, the area of each triangular face is $\frac{1}{2} \cdot 8 \cdot 2\sqrt{21}$ or $8\sqrt{21}$; thus, the lateral area is $5 \cdot 8\sqrt{21} = 40\sqrt{21}$ cm^2.

THEOREM 8.2.3: The total area (surface area) T of a pyramid with lateral area L and base area B is given by $T = L + B$.

The total area T of the pyramid is the sum of the areas of all its faces.

EXAMPLE 3

Find the total area of a regular square pyramid having base edges of length 4 ft and lateral edges of length 6 ft [see Figure 8.18(a)].

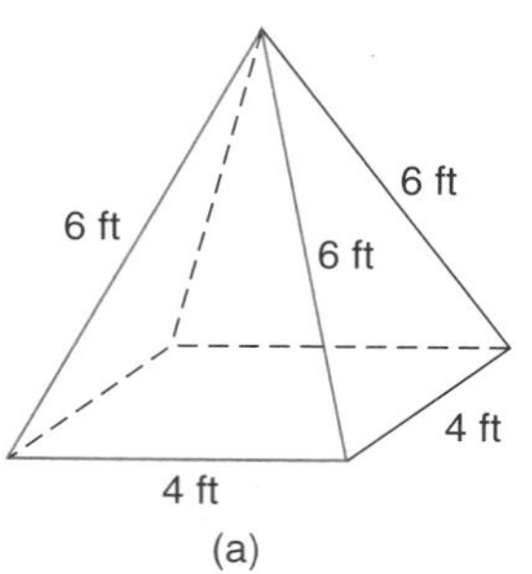

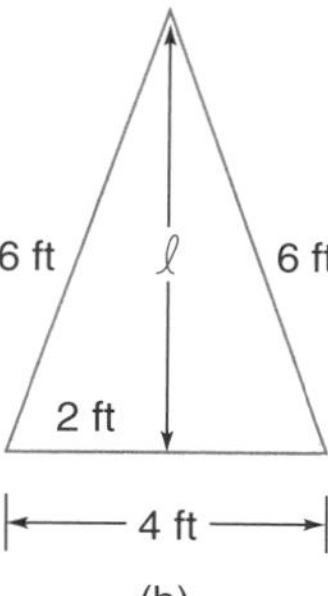

FIGURE 8.18 (a) (b)

Solution First we find the length of the slant height. [See Figure 8.18(b).]

$$\ell^2 + 2^2 = 6^2$$
$$\ell^2 + 4 = 36$$
$$\ell^2 = 32$$
$$\ell = \sqrt{32} = \sqrt{16 \cdot 2} = \sqrt{16} \cdot \sqrt{2} = 4\sqrt{2}$$

The lateral area is $L = \frac{1}{2}\ell P$. Therefore

$$L = \frac{1}{2} \cdot 4\sqrt{2}(16) = 32\sqrt{2}\text{ ft}^2$$

Because the area of the square base is 16 ft^2, the total area is

$$T = 16 + 32\sqrt{2} \approx 61.25\text{ ft}^2$$

The pyramid was described as a regular square pyramid rather than as a square pyramid. The pyramid shown in Figure 8.19(b) is oblique. It does not have congruent lateral edges or faces.

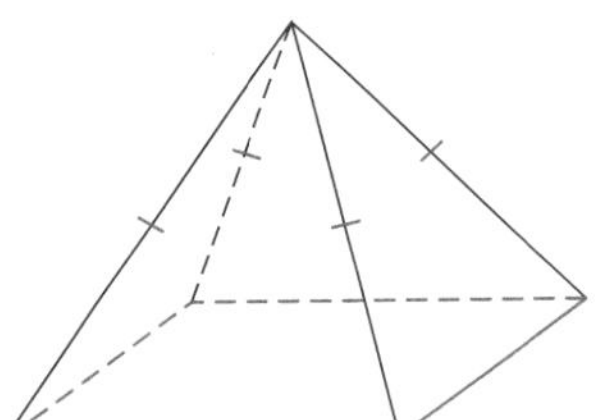

Regular square pyramid
(a)

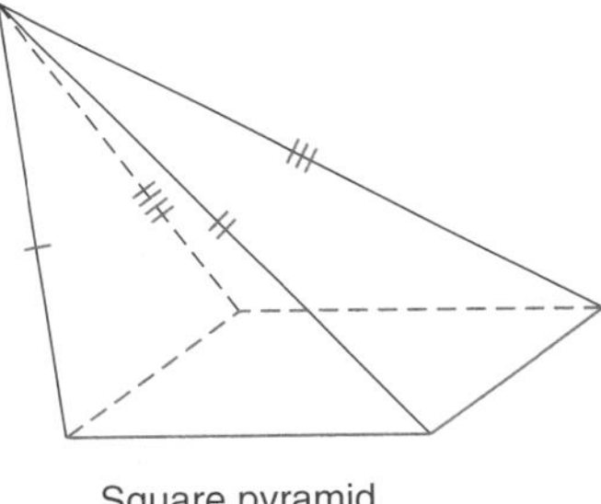

Square pyramid
(b)

FIGURE 8.19

DISCOVER!

There are kits that contain a hollow pyramid and a hollow prism that have congruent bases and the same altitude. Fill the pyramid with water and then empty the water into the prism.

a) How many times did you have to empty the pyramid in order to fill the prism?

b) As a fraction, the volume of the pyramid is what part of the volume of the prism?

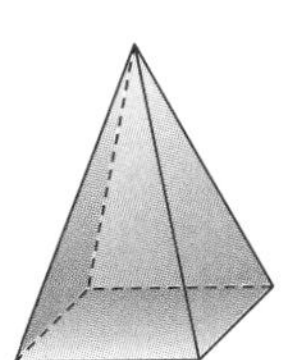

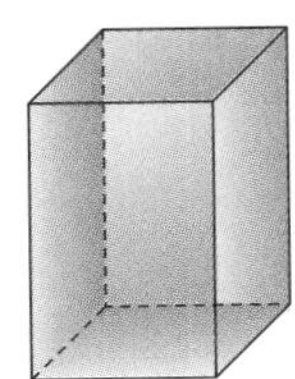

ANSWER

(a) Three times (b) $\frac{1}{3}$

Volume of a Pyramid

The final theorem in this section is presented without any attempt to construct the proof. In an advanced course such as calculus, the statement can be proved. The factor "one-third" in the formula for the volume of a pyramid provides exact results. This formula can be applied to any pyramid, even one that is not regular. See Figure 8.19(b), in which the length of altitude h is the perpendicular distance from the vertex to the plane of the square base. Read the *Discover!* before moving on to Theorem 8.2.4 and its applications.

THEOREM 8.2.4: The volume V of a pyramid having a base area B and an altitude of length h is given by

$$V = \frac{1}{3}Bh$$

EXAMPLE 4

Find the volume of the regular square pyramid with altitude $h = 4$ in. and base edges of length $s = 6$ in. (This was the pyramid of Example 1.)

Solution

The area of the square base is $B = (6 \text{ in.})^2$ or 36 in.^{2} Since $h = 4$ in., the formula $V = \frac{1}{3}Bh$ becomes

$$V = \frac{1}{3}(36 \text{ in.}^2)(4 \text{ in.}) = 48 \text{ in.}^3$$

To find the volume of a pyramid, we use the formula $V = \frac{1}{3}Bh$. In many applications, it is necessary to determine B or h from other information that has been provided. In Example 5, calculating the length of the altitude h is a challenge! In Example 6, the difficulty lies in finding the area of the base. Before we consider either problem, here is a chart that will remind us of the types of units necessary in a measurement.

TABLE 8.1

Type of Measure	*Geometric Measure*	*Type of Unit*
Linear	Length of segment, such as length of slant height	in., cm, etc.
Area	Amount of plane region enclosed, such as area of lateral face	in.^{2}, cm^2, etc.
Volume	Amount of space enclosed, such as volume of a pyramid	in.^{3}, cm^3, etc.

EXAMPLE 5

Find the volume of the regular square pyramid in Figure 8.20(a) on the next page.

Solution

The length of the altitude (of the pyramid) is represented by h, which is determined as follows.

First we see that this altitude meets the diagonals of the square base at their common midpoint [see Figure 8.20(b)]. Because each diagonal has the length $4\sqrt{2}$ ft by the 45-45-90 relationship, we have a right triangle whose legs are of lengths $2\sqrt{2}$ ft and h, while the hypotenuse has length 6 ft (the length of the lateral edge). See Figure 8.20(c).

Applying the Pythagorean Theorem in Figure 8.20(c), we have

$$\begin{aligned} h^2 + (2\sqrt{2})^2 &= 6^2 \\ h^2 + 8 &= 36 \\ h^2 &= 28 \\ h &= \sqrt{28} = \sqrt{4 \cdot 7} = \sqrt{4} \cdot \sqrt{7} = 2\sqrt{7} \end{aligned}$$

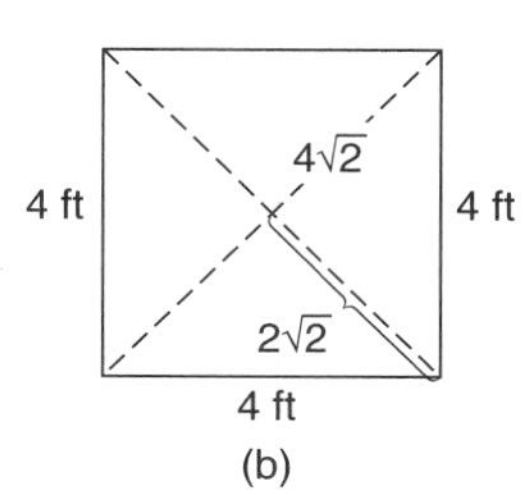

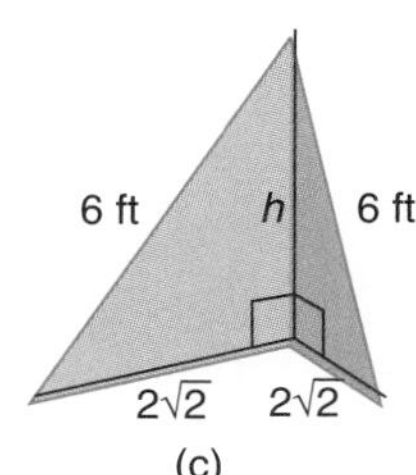

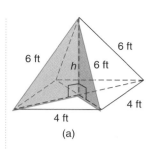

FIGURE 8.20 (a) (b) (c)

The area of the square base is $B = 4^2$ or $B = 16$ ft^2. Now we have

$$\begin{aligned} V &= \frac{1}{3}Bh \\ &= \frac{1}{3}(16)(2\sqrt{7}) \\ &= \frac{32}{3}\sqrt{7}\text{ ft}^3 \approx 28.22\text{ ft}^3 \end{aligned}$$

EXAMPLE 6

Find the volume of a regular hexagonal pyramid whose base edges have length 4 in. and whose altitude measures 12 in. [See Figure 8.21(a).]

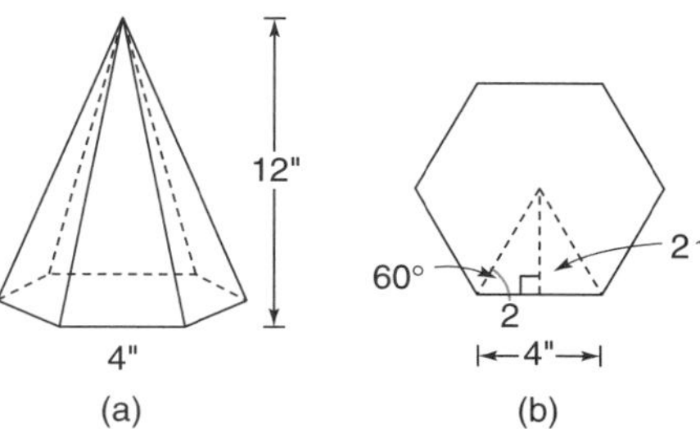

FIGURE 8.21

Solution

In the formula $V = \frac{1}{3}Bh$, the altitude is $h = 12$. To find the area of the base, we use the formula $B = \frac{1}{2}aP$ (this was written $A = \frac{1}{2}aP$ in Chapter 7). In the 30°-60°-90° triangle formed by the apothem, radius, and side of the regular hexagon, we see that

$$a = 2\sqrt{3}\text{ in.} \qquad \text{[see Figure 8.21(b)]}$$

Now $B = \frac{1}{2} \cdot 2\sqrt{3} \cdot (6 \cdot 4)$ or $B = 24\sqrt{3}$ in.^{2}

In turn, $V = \frac{1}{3}Bh$ becomes $V = \frac{1}{3}(24\sqrt{3})(12)$, so $V = 96\sqrt{3}$ in.^{3}

EXAMPLE 7

A church steeple has the shape of a regular square pyramid. Measurements taken show that the base edges measure 10 ft and that the length of a lateral edge is 13 ft. To determine the amount of roof needing to be reshingled, find the lateral area of the pyramid. (See Figure 8.22 on the next page.)

Solution The slant height ℓ of each triangular face is determined by solving the equation

$$5^2 + \ell^2 = 13^2$$
$$25 + \ell^2 = 169$$
$$\ell^2 = 44$$
$$\ell = 12$$

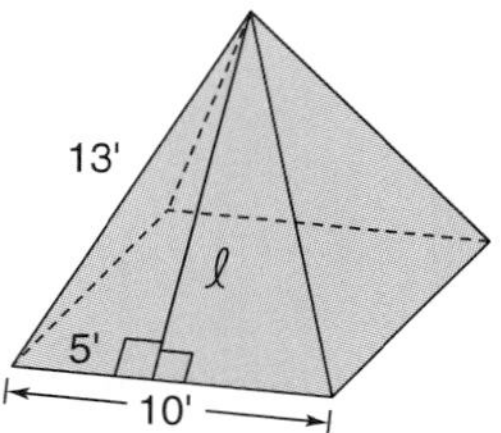

FIGURE 8.22

The formula $A = \frac{1}{2}bh$ becomes $A = \frac{1}{2} \cdot 10 \cdot 12 = 60$ ft². Considering the four lateral faces, the area to be reshingled is

$$L = 4 \cdot 60 \text{ ft}^2 \text{ or } L = 240 \text{ ft}^2$$

8.2 Exercises

In Exercises 1 to 4, name the solid that is shown. Answers are based on Sections 8.1 and 8.2.

1. a)

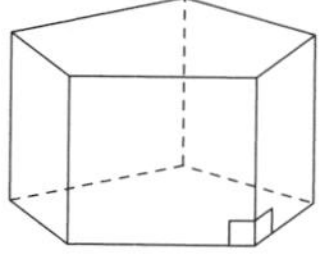

Bases not regular

b)

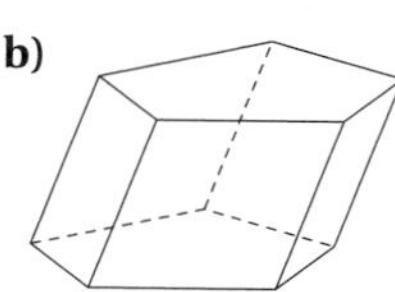

Bases not regular

2. a)

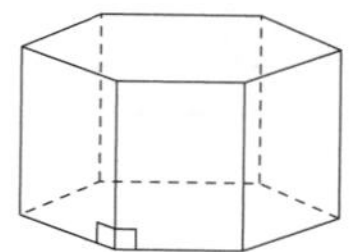

Bases are regular polygons

b)

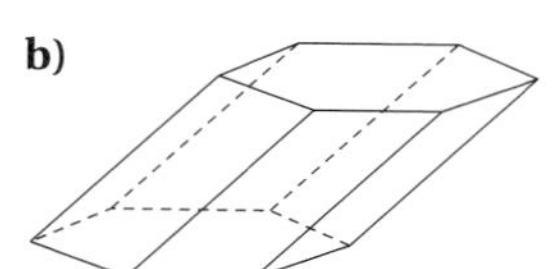

Bases not regular

3. a)

Base is a square

b)

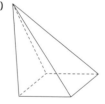

Base is a square

4. a)

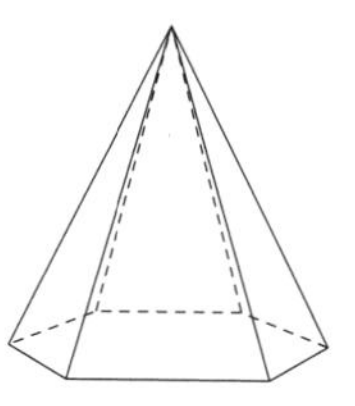

Lateral faces congruent, base is a regular polygon

b)

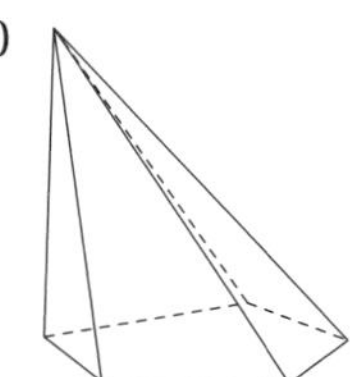

Lateral faces not congruent

5. In the solid shown, base *ABCD* is a square.
 a) Is the solid a prism or a pyramid?
 b) Name the vertex of the pyramid.
 c) Name the lateral edges.
 d) Name the lateral faces.
 e) Is the solid a regular square pyramid?

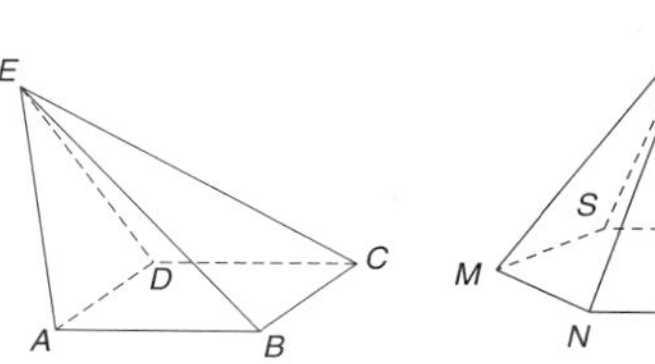

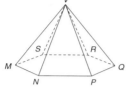

Exercises 5, 7, 9, 11 *Exercises 6, 8, 10, 12*

6. In the solid shown, the base is a regular hexagon.
 a) Name the vertex of the pyramid.
 b) Name the base edges of the pyramid.
 c) Assuming that lateral edges are congruent, are the lateral faces also congruent?
 d) Assuming that lateral edges are congruent, is the solid a regular hexagonal pyramid?

7. For the square pyramid in Exercise 5,
 a) how many vertices does it have?
 b) how many edges (lateral edges plus base edges) does it have?
 c) how many faces (lateral faces plus bases) does it have?

8. For the hexagonal pyramid in Exercise 6,
 a) how many vertices does it have?
 b) how many edges (lateral edges plus base edges) does it have?
 c) how many faces (lateral faces plus bases) does it have?

9. Suppose that the lateral faces of the pyramid in Exercise 5 have areas $A_{ABE} = 12$ in.^{2}, $A_{BCE} = 16$ in.^{2}, $A_{CED} = 12$ in.^{2}, and $A_{ADE} = 10$ in.^{2} If each side of the square base measures 4 in., find the total surface area of the pyramid.

10. Suppose that the base of the hexagonal pyramid in Exercise 6 has an area of 41.6 cm^2 while each lateral face has an area of 20 cm^2. Find the total (surface) area of the pyramid.

11. Suppose that the base of the square pyramid in Exercise 5 has an area of 16 cm^2 while the altitude of the pyramid measures 6 cm. Find the volume of the square pyramid.

12. Suppose that the base of the hexagonal pyramid in Exercise 6 has an area of 41.6 cm^2 while the altitude of the pyramid measures 3.7 cm. Find the volume of the hexagonal pyramid.

13. Assume that the number of sides in the base of a pyramid is n. Generalize the results found in earlier exercises by answering each of the following questions. What is the
 a) number of vertices?
 b) number of lateral edges?
 c) number of base edges?
 d) total number of edges?
 e) number of lateral faces?
 f) total number of faces?
 (**NOTE:** Lateral faces and base = faces)

In Exercises 14 and 15, use Theorem 8.2.1; that is, the apothem a, the altitude h, and the slant height ℓ of a regular pyramid are related by the equation $\ell^2 = a^2 + h^2$.

14. In a regular square pyramid whose base edges measure 8 in., the apothem of the base measures 4 in. If the altitude of the pyramid is 8 in., find the length of its slant height.

15. In a regular hexagonal pyramid whose base edges measure $2\sqrt{3}$ in., the apothem of the base measures 3 in. If the slant height of the pyramid is 5 in., find the length of its altitude.

16. In the regular pentagonal pyramid, each lateral edge measures 8 in. while each base edge measures 6 in. The apothem of the base measures 4.1 in.
 a) Find the lateral area of the pyramid.
 b) Find the total area of the pyramid.

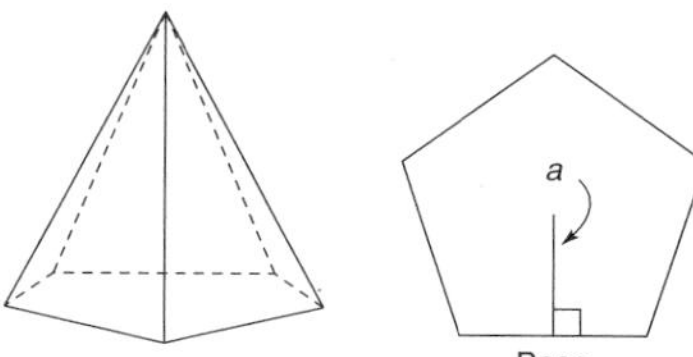

Exercises 16, 17

17. In the pentagonal pyramid, suppose that each base edge measures 9.2 cm while the apothem of the base measures 6.3 cm. The altitude of the pyramid measures 14.6 cm.
 a) Find the base area of the pyramid.
 b) Find the volume of the pyramid.

18. For the regular square pyramid shown, suppose that the sides of the square base measure 10 m each and the lateral edges measure 13 m each.

a) Find the lateral area of the pyramid.
b) Find the total area of the pyramid.
c) Find the volume of the pyramid.

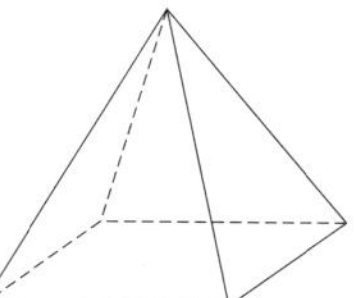

Exercises 18, 19

19. For the regular square pyramid shown, suppose that the sides of the square base measure 6 ft each and the altitude is 4 ft.

a) Find the lateral area of the pyramid.
b) Find the total area of the pyramid.
c) Find the volume of the pyramid.

20. a) Find the lateral area L of the regular hexagonal pyramid.
b) Find the total area T of the pyramid.
c) Find the volume V of the pyramid.

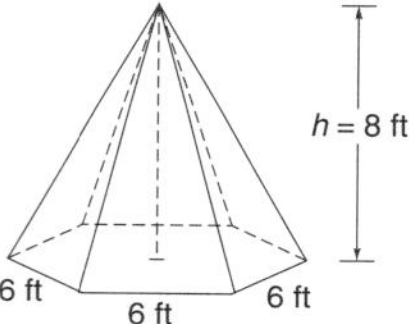

21. For a regular square pyramid, suppose that the altitude has a measure equal to that of the edges of the base. If the volume of the pyramid is 72 in.^{3}, find the total area of the pyramid.

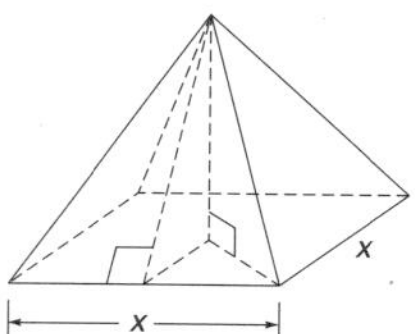

Exercises 21, 22

22. For a regular square pyramid, the slant height of each lateral face has a measure equal to that of the edges of the base. If the lateral area is 200 in.^{2}, find the volume of the pyramid.

23. A church steeple in the shape of a regular square pyramid needs to be reshingled. The part to be covered corresponds to the lateral area of the square pyramid. If lateral edges measure 17 ft and base edges measure 16 ft, how many square feet of shingles need to be replaced?

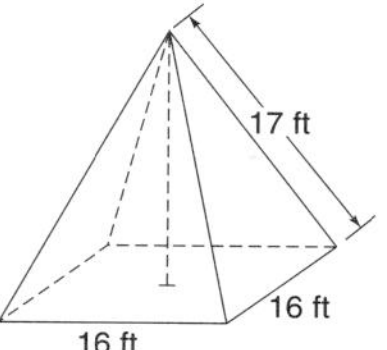

Exercises 23, 24

24. Before the shingles of the steeple (see Exercise 23) are replaced, an exhaust fan is to be installed in the steeple. To determine what size exhaust fan should be installed, it is necessary to know the volume of air in the attic (steeple). Find the volume of the regular square pyramid described in Exercise 23.

25. A teepee is constructed by using 12 poles. The construction leads to a regular pyramid with a dodecagon (12 sides) for the base. With the base as shown and knowing that the altitude of the teepee is 15 ft, what is its volume?

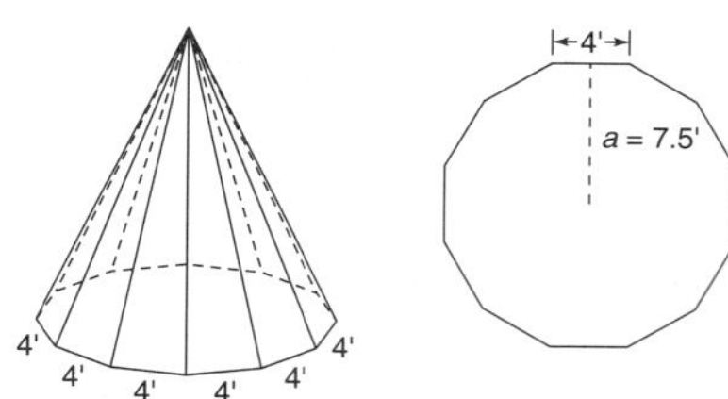

Exercises 25, 26

26. For its occupants to be protected from the elements, it was necessary that the teepee in Exercise 25 be enclosed. Find the amount of area to be covered; that is, determine the lateral area of the regular dodecagonal pyramid. Recall that its altitude measures 15 ft.

27. The street department's storage building, which is used to store the rock, gravel, and salt used on the city's roadways, is built in the shape of a regular hexagonal pyramid. The altitude of the pyramid has the same length as any side of the base. If the volume of the interior is 11,972 ft^3, find the length of the altitude and each side of the base to the nearest foot.

28. The foyer planned as an addition to an existing church is designed as a regular octagonal pyramid. Each side of the octagonal floor has a length of 10 ft and its apothem measures 12 ft. If 800 ft^2 of plywood are needed to cover the exterior of the foyer (that is, the lateral area of the pyramid is 800 ft^2), then what is the altitude of the foyer?

29. The exhaust chute on a wood chipper has a shape like the part of a pyramid known as the *frustrum of the pyramid.* With dimensions as indicated, find the volume (capacity) of the chipper's exhaust chute.

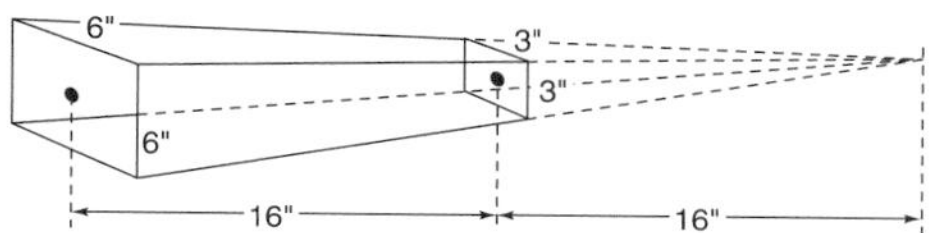

30. A popcorn container at a movie theater has the shape of a *frustrum of a pyramid* (see Exercise 29). With dimensions as indicated, find the volume (capacity) of the container.

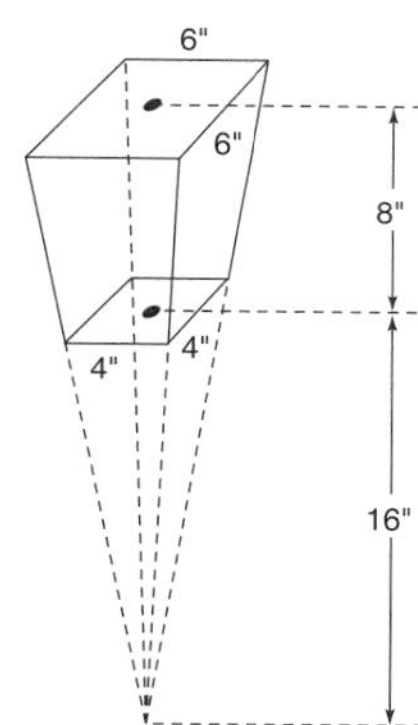

31. A regular tetrahedron is a regular triangular pyramid in which all faces (lateral faces and base) are congruent. If each edge has length e,

a) show that the area of each face is $A = \frac{e^2\sqrt{3}}{4}$.
b) show that the total area of the tetrahedron is $T = e^2\sqrt{3}$.
c) find the total area if each side measures $e = 4$ in.

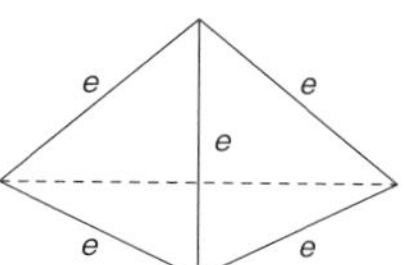

Exercises 31, 32

32. ➢ Each edge of a regular tetrahedron (see Exercise 31) has length e.

a) Show that the altitude of the tetrahedron measures $h = \frac{\sqrt{2}}{\sqrt{3}}e$.
b) Show that the volume of the tetrahedron is $V = \frac{\sqrt{2}}{12}e^3$.
c) Find the volume of the tetrahedron if each side measures $e = 4$ in.

8.3 Cylinders and Cones

Cylinders

Consider the solids in Figure 8.23 on the next page, in which congruent circles lie in parallel planes. For the circles on the left, suppose that centers O and O' are joined by $\overline{OO'}$; similarly, suppose that $\overline{QQ'}$ joins the centers of the circles on the right. Let segments such as $\overline{XX'}$ join two points of the circles on the left, so that $\overline{XX'} \parallel \overline{OO'}$. If all such segments (like $\overline{XX'}$, $\overline{YY'}$, and $\overline{ZZ'}$) are parallel to each other, then a **cylinder** is generated. Because $\overline{OO'}$ is not perpendicular to planes P and P', the solid on the left is an **oblique circular cylinder.** With $\overline{QQ'}$ perpendicular to planes P and P', the solid on the right is a **right circular cylinder.** For both cylinders, the distance between planes P and P' is the measure h, the length of the **altitude** of the cylinder. The congruent circles are known as the **bases** of each cylinder.

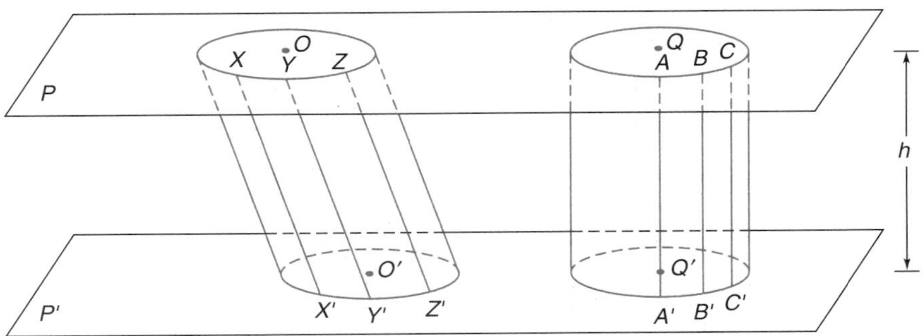

FIGURE 8.23

A right circular cylinder is shown in Figure 8.24; however, the parallel planes (like P and P' in Figure 8.23) are generally not pictured. The segment joining the centers of the two circular bases is known as the **axis** of the cylinder. For a right circular cylinder, it is necessary that the axis be perpendicular to the planes of the circular bases; in such a case, the length of the altitude h is the length of the axis.

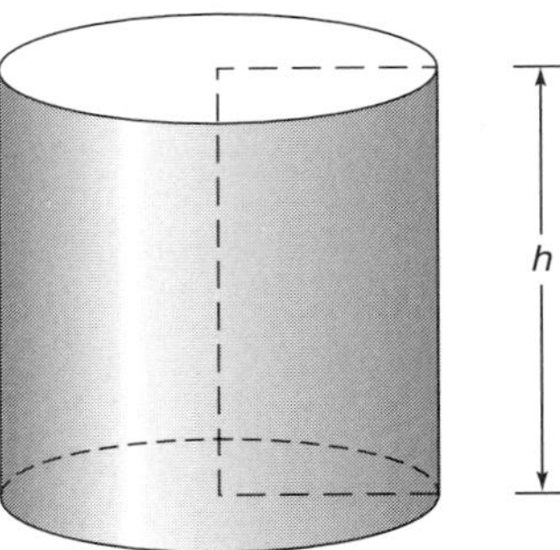

FIGURE 8.24

Surface Area of a Cylinder

The formula for the lateral area of a right circular cylinder (found in the following theorem) should be compared to the formula $L = hP$, the lateral area of a right prism whose base has perimeter P.

THEOREM 8.3.1: The lateral area L of a right circular cylinder with altitude of length h and circumference C of the base is given by $L = hC$.

Alternate Form: The lateral area of a right circular cylinder could be expressed in the form $L = 2\pi rh$, where r is the length of the radius of the circular base.

Rather than construct a formal proof of Theorem 8.3.1, consider the following activity.

DISCOVER!

Think of the aluminum can pictured in Figure (a) below as a right circular cylinder. The cylinder's circular bases are the lids of the can and the lateral surface is the "label" of the can. If the label were sliced downward by a perpendicular line between the planes, removed, and rolled out flat, it would be rectangular in shape. As shown in Figure (b), that rectangle would have a length equal to the circumference of the circular base and a width equal to the height of the cylinder. Thus the lateral area is given by $A = bh$, which becomes $L = Ch$ or $L = 2\pi rh$

(a)

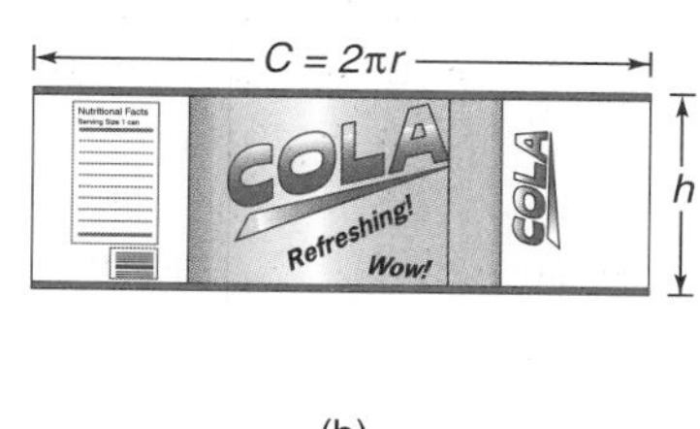

(b)

THEOREM 8.3.2: The total area T of a right circular cylinder with base area B and lateral area L is given by $T = L + 2B$.

Alternate Form: Where r is the length of the radius of the base and h is the length of the altitude of the cylinder, the total area could be expressed in the form $T = 2\pi rh + 2\pi r^2$.

EXAMPLE 1

For the right circular cylinder shown in Figure 8.25, find the

a) exact lateral area L.
b) exact surface area T.

Solution

a) $L = 2\pi rh$
$= 2 \cdot \pi \cdot 5 \cdot 12$
$= 120\pi \text{ in.}^2$

b) $T = L + 2B$
$= 2\pi rh + 2\pi r^2$
$= 2 \cdot \pi \cdot 5 \cdot 12 + 2 \cdot \pi \cdot 5^2$
$= 120\pi + 50\pi$
$= 170\pi \text{ in.}^2$

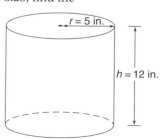

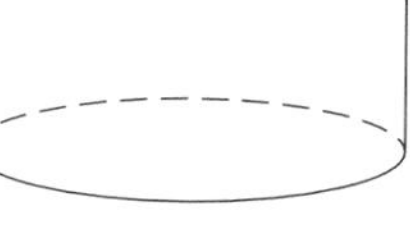

FIGURE 8.25

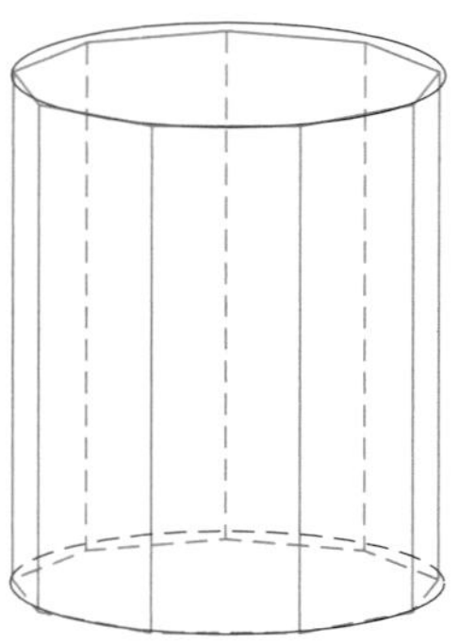

FIGURE 8.26

Volume of a Cylinder

In considering the volume of a right circular cylinder, recall that the volume of a prism is given by $V = Bh$, where B is the area of the base. In Figure 8.26, we inscribe a prism in the cylinder as shown.

Suppose that the prism is regular and that the number of sides in the inscribed polygon's base becomes larger and larger; thus, the base approaches a circle in this limiting process. The area of the polygonal base also approaches the area of the circle, while the volume of the prism approaches that of the right circular cylinder. Our conclusion is stated without proof in the following theorem.

THEOREM 8.3.3: The volume V of a right circular cylinder with base area B and altitude of length h is given by $V = Bh$.

Alternate Form: Where r is the length of the radius of the base, the volume formula for the cylinder is usually written $V = \pi r^2 h$.

EXAMPLE 2

If $d = 4$ cm and $h = 3.5$ cm, use a calculator to find the approximate volume of the right circular cylinder shown in Figure 8.27.

Solution $d = 4$, so $r = 2$. Thus $V = Bh$ or $V = \pi r^2 h$ becomes

$$V = \pi \cdot 2^2(3.5)$$
$$= \pi \cdot 4(3.5) = 14\pi \approx 43.98 \text{ cm}^3$$

EXAMPLE 3

In the right circular cylinder shown in Figure 8.27, suppose that the height equals the diameter of the circular base. If the exact volume is 128π in.^{3}, find the exact lateral area L of the cylinder.

Solution

$$h = 2r$$

so $V = \pi r^2 h$

becomes $= \pi r^2(2r)$ since $h = d = 2r$

$$= 2\pi r^3$$

Then $2\pi r^3 = 128\pi$ and dividing by 2π,

$$r^3 = 64$$
$$r = 4$$
$$h = 8 \quad \text{(from } h = 2r\text{)}$$

Now $L = 2\pi rh$

$$= 2 \cdot \pi \cdot 4 \cdot 8$$
$$= 64\pi \text{ in.}^2$$

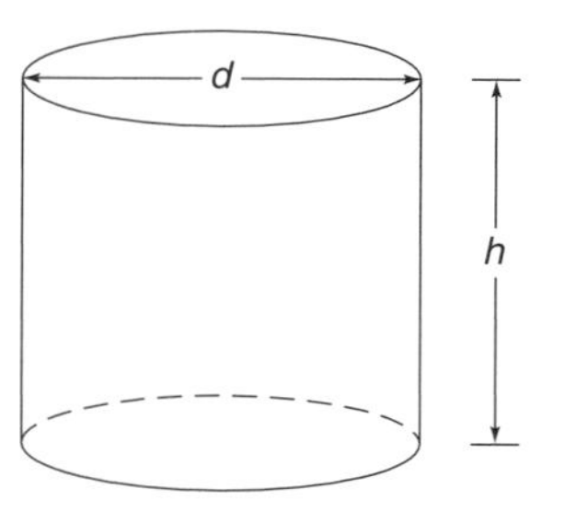

FIGURE 8.27

Table 8.2 should help us to recall and compare the area and volume formulas found in Sections 8.1 and 8.3.

TABLE 8.2

	Lateral Area	Total Area	Volume
Prism	$L = hP$	$T = L + 2B$	$V = Bh$
Cylinder	$L = hC$	$T = L + 2B$	$V = Bh$

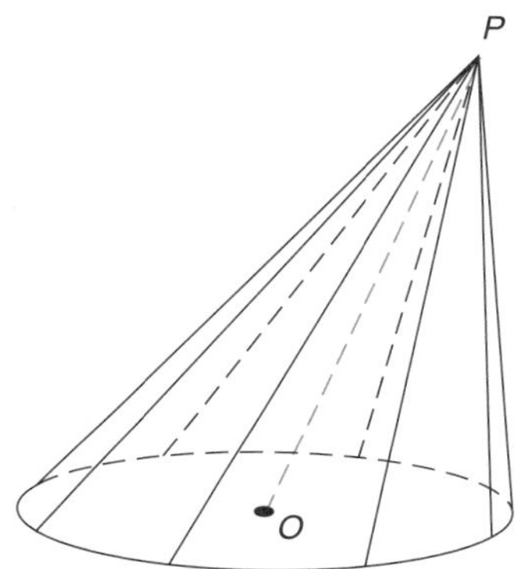

FIGURE 8.28

Cones

In Figure 8.28, consider point P, which lies outside the plane containing circle O. A surface known as a **cone** results when line segments are drawn from P to points on the circle. However, if P is joined to all possible points on the circle as well as to points in the interior of the circle, a solid is formed. If $\overline{PO}$ is not perpendicular to the plane of circle O in Figure 8.28, the cone is an **oblique circular cone.**

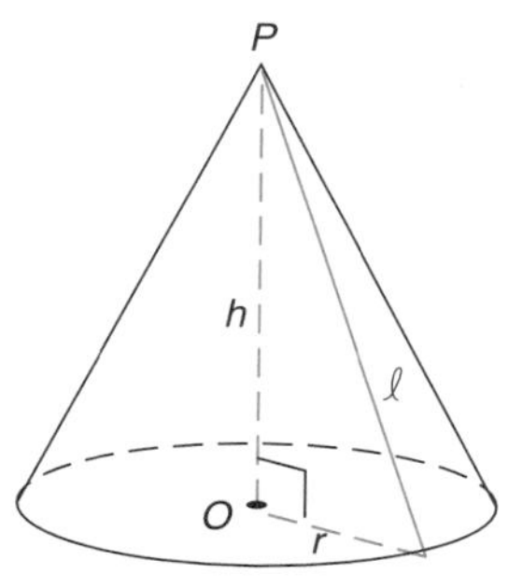

FIGURE 8.29

In Figures 8.28 and 8.29, point P is the **vertex** of the cone while circle O is the **base.** The segment $\overline{PO}$, which joins the vertex to the center of the circular base, is the **axis** of the cone. If the axis is perpendicular to the base or to the plane of the base as in Figure 8.29, the cone is a **right circular cone.** In any cone, the perpendicular segment from the vertex to the plane of the base is the **altitude** of the cone. In a right circular cone, the length h of the altitude equals the length of the axis. For a right circular cone, and only for this type of cone, any line segment that joins the vertex to a point on the circle is a **slant height** of the cone; we will denote the length of the slant height by ℓ.

Surface Area of a Cone

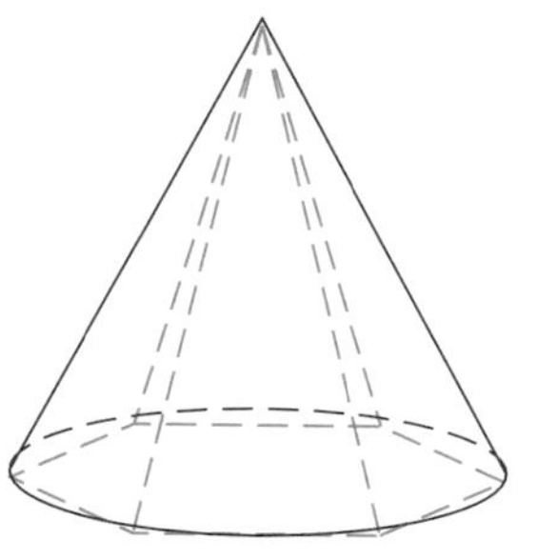
FIGURE 8.30

Recall now that the lateral area for the regular pyramid is given by $L = \frac{1}{2}\ell P$. For a right circular cone, consider an inscribed regular pyramid as in Figure 8.30. As the number of sides of the inscribed polygon's base grows larger, the perimeter of the inscribed polygon approaches the circumference of the circle as a limit. In addition, the slant height of the congruent triangular faces approaches that of the slant height of the cone. Thus the lateral area of the right circular cone is given by

$$L = \frac{1}{2}\ell C$$

in which C is the circumference of the base. The fact that $C = 2\pi r$ leads to

$$L = \frac{1}{2}\ell(2\pi r)$$

so $$L = \pi r\ell.$$

THEOREM 8.3.4: The lateral area L of a right circular cone with slant height of length ℓ and circumference C of the base is given by $L = \frac{1}{2}\ell C$.

Alternate Form: Where r is the length of the radius of the base, $L = \pi r\ell$.

The following theorem follows easily and is given without proof.

THEOREM 8.3.5: The total area T of a right circular cone with base area B and lateral area L is given by $T = B + L$.

Alternate Form: Where r is the length of the radius of the base and ℓ is the length of the slant height, $T = \pi r^2 + \pi r\ell$.

EXAMPLE 4

For the right circular cone in which $r = 3$ cm and $h = 6$ cm (see Figure 8.31), find the

a) exact and approximate lateral area L
b) exact and approximate total area T

Solution

a) We need the slant height ℓ for each part, so we apply the Pythagorean Theorem:

$$\begin{aligned} \ell^2 &= r^2 + h^2 \\ \ell^2 &= 3^2 + 6^2 \\ \ell^2 &= 9 + 36 = 45 \\ \ell &= \sqrt{45} = \sqrt{9 \cdot 5} \\ \ell &= \sqrt{9} \cdot \sqrt{5} = 3\sqrt{5} \end{aligned}$$

Using the alternate form of $L = \frac{1}{2}\ell C$, namely $L = \pi r\ell$,

we have $L = \pi \cdot 3 \cdot 3\sqrt{5}$
$= 9\pi\sqrt{5}\text{ cm}^2 \approx 63.22\text{ cm}^2$

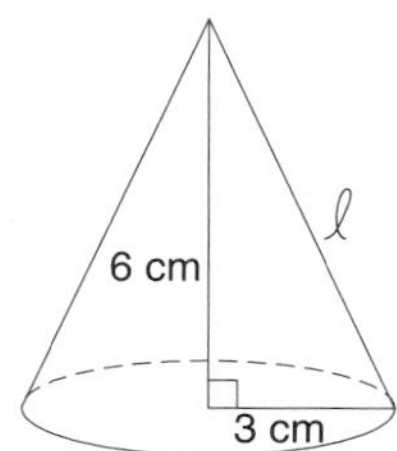

FIGURE 8.31

b) We also have

$$\begin{aligned} T &= B + L \\ &= \pi r^2 + \pi r\ell \\ &= \pi \cdot 3^2 + \pi \cdot 3 \cdot 3\sqrt{5} \\ &= (9\pi + 9\pi\sqrt{5})\text{ cm}^2 \approx 91.50\text{ cm}^2 \end{aligned}$$

The following theorem was demonstrated in the solution of Example 4.

THEOREM 8.3.6: In a right circular cone, the lengths of the radius r (of the base), the altitude h, and the slant height ℓ satisfy the Pythagorean Theorem; that is, $\ell^2 = r^2 + h^2$ in every right circular cone.

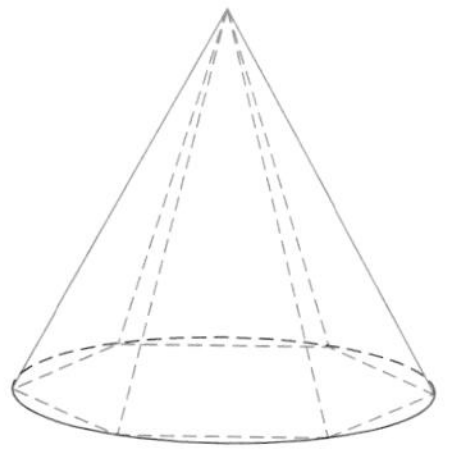

FIGURE 8.32

Volume of a Cone

Recall that the volume of a pyramid is given by the formula $V = \frac{1}{3}Bh$. Consider a regular pyramid inscribed in a right circular cone. If its number of sides increases indefinitely, the volume of the pyramid approaches that of the right circular cone (see Figure 8.32).370 Then the volume of the right circular cone is $V = \frac{1}{3}Bh$. Because the area of the base of the cone is $B = \pi r^2$, an alternate formula for the volume of the cone is

$$V = \frac{1}{3}\pi r^2 h$$

We state this result as a theorem.

THEOREM 8.3.7: The volume V of a right circular cone with base area B and altitude of length h is given by $V = \frac{1}{3}Bh$.

Alternate Form: Where r is the length of the radius of the base, the formula for the volume of the cone is usually written $V = \frac{1}{3}\pi r^2 h$.

Table 8.3 should help us to recall and compare the area and volume formulas found in Sections 8.2 and 8.3.

TABLE 8.3

	Lateral Area	*Total Area*	*Volume*	*Slant Height*
Pyramid	$L = \frac{1}{2}\ell P$	$T = B + L$	$V = \frac{1}{3}Bh$	$\ell^2 = a^2 + h^2$
Cone	$L = \frac{1}{2}\ell C$	$T = B + L$	$V = \frac{1}{3}Bh$	$\ell^2 = r^2 + h^2$

NOTE: The lateral area and slant height relationships are used only with the regular pyramid and right circular cone.

Solids of Revolution

When a plane region is revolved about one of its sides (the edge), the locus of points generated in space is called a **solid of revolution.** The complete 360° rotation moves the region about the edge until the region returns to its original position. The side (edge) used is called the **axis** of the resulting solid of revolution. Consider Example 5.

EXAMPLE 5

Describe the solid of revolution that results when

a) a rectangular region of dimensions 2 ft by 5 ft is revolved about the 5-foot side [see Figure 8.33(a)].
b) a semicircular region of radius 3 cm is revolved about the diameter shown in Figure 8.33(b).

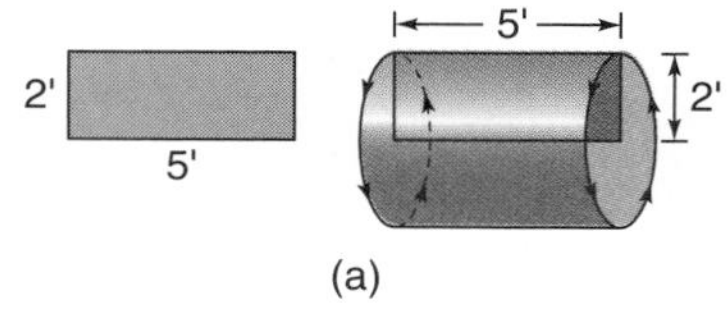

(a)

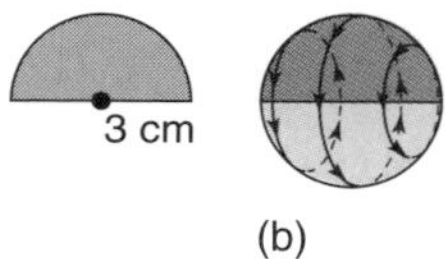

(b)

FIGURE 8.33

Solution

a) In Figure 8.33(a), the rectangle on the left is revolved about the 5-foot side to form the solid on the right. The solid of revolution generated is a right circular cylinder that has a base radius of 2 ft and an altitude of 5 ft.
b) In Figure 8.33(b), the semicircle on the left is revolved about its diameter to form the solid on the right. The solid of revolution generated is a *sphere* with a radius of length 3 cm.

NOTE: We study the sphere in greater detail in Section 8.4.

EXAMPLE 6

Determine the exact volume of the solid of revolution formed when the region bounded by a right triangle with legs of lengths 4 in. and 6 in. is revolved about the 6-in. side.

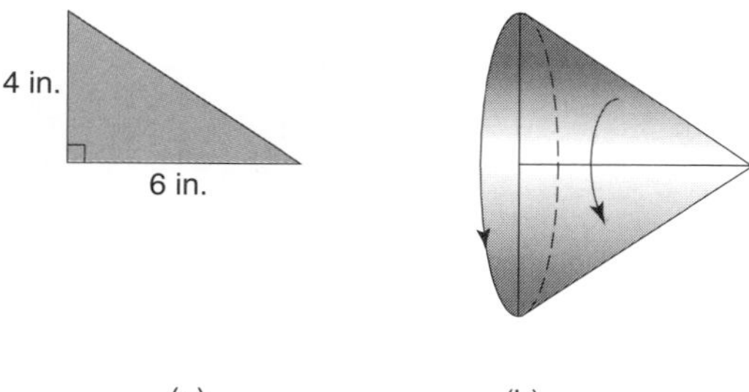

(a) (b)

FIGURE 8.34

Solution As shown in Figure 8.34, the resulting solid is a cone whose altitude measures 6 in. and whose radius of the base measures 4 in.

Using $V = \frac{1}{3}Bh$, we have

$$V = \frac{1}{3}\pi r^2 h$$
$$= \frac{1}{3} \cdot \pi \cdot 4^2 \cdot 6 = 32\pi \text{ in.}^3$$

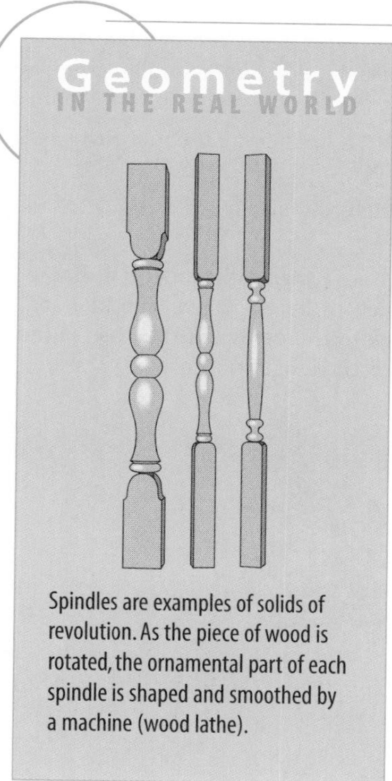

Spindles are examples of solids of revolution. As the piece of wood is rotated, the ornamental part of each spindle is shaped and smoothed by a machine (wood lathe).

It may come as a surprise that the formulas used to calculate the volumes of an oblique circular cylinder and an oblique circular cone are identical to those found earlier in this section. To see why the formula $V = Bh$ or $V = \pi r^2 h$ can be used to calculate the volume of an oblique circular cylinder, consider the stacks of pancakes shown in Figures 8.35(a) and 8.35(b). The volume is the same regardless of whether the stack is vertical or oblique.

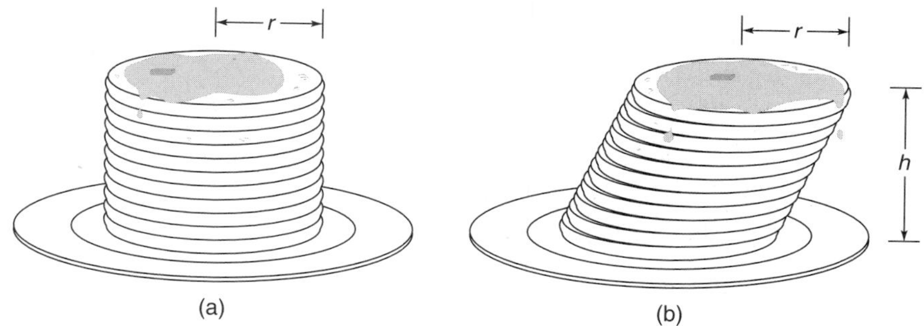

FIGURE 8.3572

It is also true that the formula for the volume of an oblique circular cone is $V = \frac{1}{3}Bh$ or $V = \frac{1}{3}\pi r^2 h$. In fact, the motivating argument preceding Theorem 8.3.7 is repeated, with the exception that the inscribed pyramid is oblique.

8.3 Exercises

1. For the right circular cylinder, suppose that $r = 5$ in. and $h = 6$ in. Find the exact and approximate:

a) Lateral area
b) Total area
c) Volume

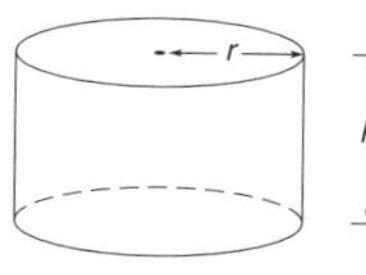

Exercises 1, 2

2. Suppose that $r = 12$ cm and $h = 15$ cm in the right circular cylinder. Find the exact and approximate:

a) Lateral area
b) Total area
c) Volume

3. The tin can shown on the next page has the indicated dimensions. Estimate the number of square inches of tin required for its construction.
(**HINT:** Include the lid and base in the result.)

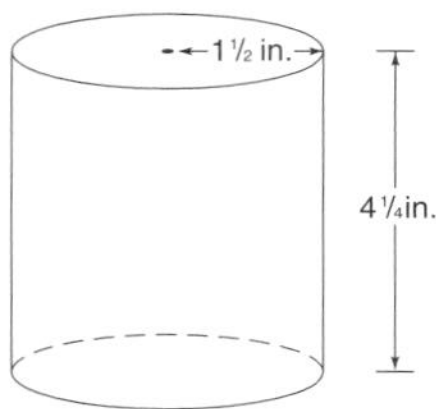

Exercises 3, 4

4. What is the volume of the tin can? If it contains 16 oz of green beans, what is the volume of the can used for 20 oz of green beans? Assume a proportionality between weight and volume.

5. If the exact volume of a right circular cylinder is 200π cm^3 and its altitude measures 8 cm, what is the measure of the radius of the circular base?

6. Suppose that the volume of an aluminum can is to be 9π in.^{3} Find the dimensions of the can if the diameter of the base is three-fourths the length of the altitude.

7. For an aluminum can, the lateral surface area is 12π in.^{2} If the length of the altitude is 1 in. greater than the length of the radius of the circular base, find the dimensions of the can.

8. Find the altitude of a storage tank in the shape of a right circular cylinder that has a circumference measuring 6π m and that has a volume measuring 81π m^3.

9. Find the volume of the oblique circular cylinder. The axis meets the plane of the base to form a 45° angle.

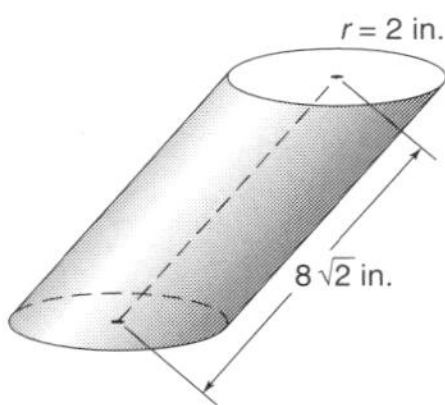

10. A cylindrical orange juice container has metal bases of radius 1 in. and a cardboard lateral surface 3 in. high. If the cost of the metal used is 0.5 cents per square inch and the cost of the cardboard is 0.2 cents per square inch, what is the approximate cost of constructing one container? Let $\pi \approx 3.14$.

In Exercises 11 to 16, use the fact that $r^2 + h^2 = \ell^2$ in a right circular cone (Theorem 8.3.6).

11. Find the slant height ℓ of a right circular cone with $r = 4$ cm and $h = 6$ cm.

12. Find the slant height ℓ of a right circular cone with $r = 5.2$ ft and $h = 3.9$ ft.

13. Find the altitude h of a right circular cone in which the diameter of the base measures $d = 9.6$ m and $\ell = 5.2$ m.

14. Find the radius of base r of a right circular cone in which $h = 6$ yd and $\ell = 8$ yd.

15. Find the slant height ℓ of a right circular cone with $r = 6$ in. and $\ell = 2h$ in.

16. Find the radius r of a right circular cone with $\ell = 12$ in. and $h = 3r$ in.

17. The oblique circular cone has an altitude and a diameter of base that are each of length 6 cm. The line segment joining the vertex to the center of the base is the **axis** of the cone. What is the length of the axis?

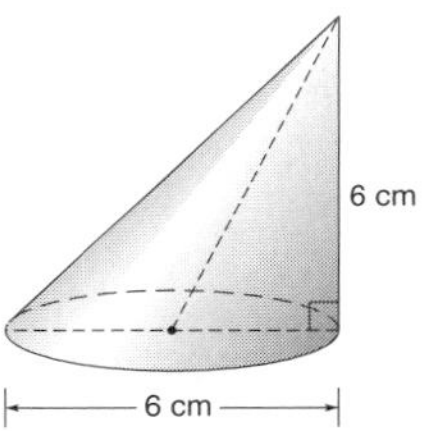

18. For the right circular cone, $h = 6$ m and $r = 4$ m. Find the exact and approximate:

   a) Lateral area
   b) Total area
   c) Volume

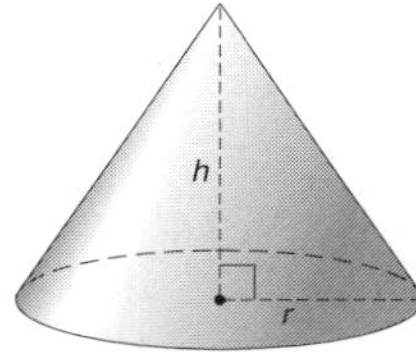

Exercises 18, 19

19. For the cone shown above, suppose that $h = 7$ in. and $r = 6$ in. Find the exact and approximate:

   a) Lateral area
   b) Total area
   c) Volume

20. The teepee has a circular floor with a radius equal to 6 ft and a height of 15 ft. Find the volume of the enclosure.

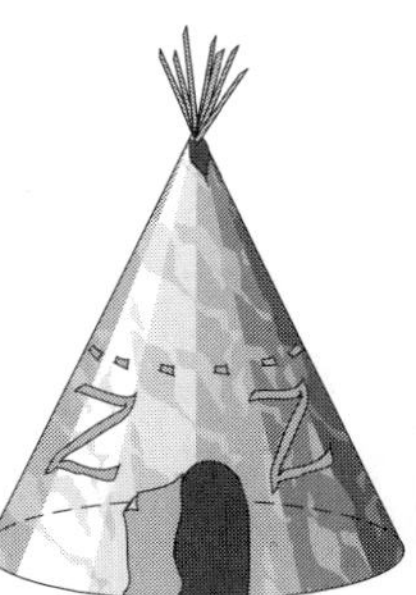

21. A rectangle has dimensions of 6 in. by 3 in. Find the exact volume of the solid of revolution formed when the rectangle is rotated about its 6-in. side.

22. A rectangle has dimensions of 6 in. by 3 in. Find the exact volume of the solid of revolution formed when the rectangle is rotated about its 3-in. side.

23. A triangle has sides that measure 15 cm, 20 cm, and 25 cm. Find the exact volume of the solid of revolution formed when the triangle is revolved about the side of length 15 cm.

24. A triangle has sides that measure 15 cm, 20 cm, and 25 cm. Find the exact volume of the solid of revolution formed when the triangle is revolved about the side of length 20 cm.

25. A triangle has sides that measure 15 cm, 20 cm, and 25 cm. Find the exact volume of the solid of revolution formed when the triangle is revolved about the side of length 25 cm.
(**HINT:** The altitude to the 25-cm side has length 12 cm.)

26. Where r is the length of the radius of a sphere, the volume of the sphere is given by $V = \frac{4}{3}\pi r^3$. Find the exact volume of the sphere that was formed in Example 5(b).

27. If a right circular cone has a circular base with a diameter of length 10 cm and a volume of 100π cm^3, find its lateral area.

28. A right circular cone has a slant height of 12 ft and a lateral area of 96π ft^2. Find its volume.

29. A solid is formed by cutting a conical section away from a right circular cylinder. If the radius measures 6 in. and the altitude measures 8 in., what is the volume of the resulting solid?

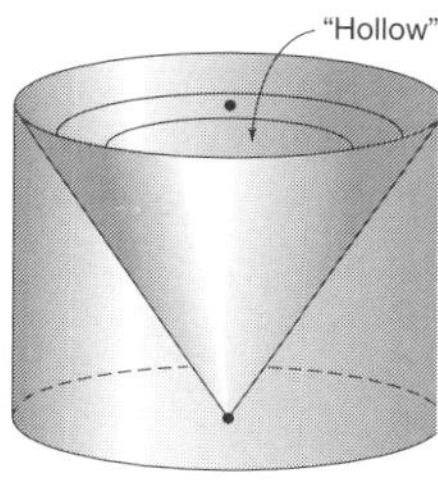

In Exercises 30 and 31, give a paragraph proof for each claim.

30. The total area T of a right circular cylinder whose altitude is of length h and whose circular base has a radius of length r is given by $T = 2\pi r(r + h)$.

31. The volume V of a washer that has an inside radius of length r, an outside radius of length R and an altitude of measure h is given by $V = \pi h(R + r)(R - r)$.

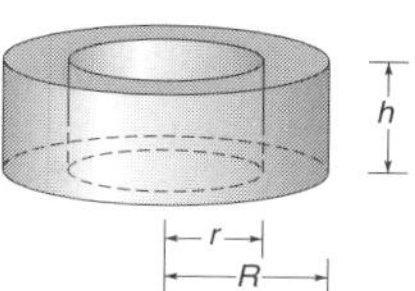

32. For a right circular cone, the slant height has a measure equal to twice that of the radius of the base. If the total area of the cone is 48π in.^{2}, what are the dimensions of the cone?

33. For a right circular cone, the ratio of the slant height to the radius is 5:3. If the volume of the cone is 96π in.^{3}, find the lateral area of the cone.

34. If the radius and height of a right circular cylinder are both doubled to form a larger cylinder, what is the ratio of the volume of the larger cylinder to the volume of the smaller cylinder?
(**NOTE:** The two cylinders are said to be "similar.")

35. For the two similar cylinders in Exercise 34, what is the ratio of the lateral area of the larger cylinder to that of the smaller cylinder?

36. For a right circular cone, the dimensions are $r = 6$ cm and $h = 8$ cm. If the radius is doubled while the height is made half as large in forming a new cone, will the volumes of the two cones be equal?

37. A cylindrical storage tank has a depth of 5 ft and a radius measuring 2 ft. If each cubic foot can hold 7.5 gal of gasoline, what is the total storage capacity of the tank (measured in gallons)?

38. If the tank in Exercise 37 needs to be painted and 1 pt of paint covers 50 ft.^{2}, how many pints are needed to paint the exterior of the storage tank?

39. A frustrum of a cone is the portion of the cone bounded between the circular base and a plane parallel to the base. With dimensions as indicated, show that the volume of the frustrum of the cone is $V = \frac{1}{3}\pi R^2 H - \frac{1}{3}\pi r^2 h$.

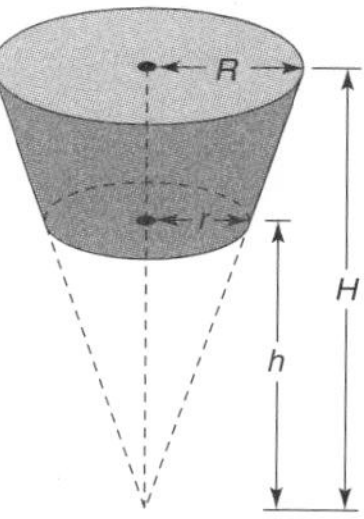

In Exercises 40 and 41, use the formula found in Exercise 39. Similar triangles were used to find h and H.

40. A margarine tub has the shape of the frustrum of a cone. With the lower base having diameter 11 cm and the upper base having diameter 14 cm, the volume of the $6\frac{2}{3}$-cm tall container can be determined by using $R = 7$ cm, $r = 5.5$ cm, $H = 32\frac{2}{3}$ cm, and $h = 26$ cm. Find its volume.

41. A container of yogurt has the shape of the frustrum of a cone. With the lower base having diameter 6 cm and the upper base having diameter 8 cm, the volume of the 7.5-cm tall container can be determined by using $R = 4$ cm, $r = 3$ cm, $H = 30$ cm, and $h = 22.5$ cm. Find its volume.

8.4 Polyhedrons and Spheres

Polyhedrons

When two planes intersect, the angle formed by two half-planes with a common edge (the line of intersection) is a **dihedral angle.** The angle shown in Figure 8.36 is such an angle. In Figure 8.36, the measure of the dihedral angle is the same as that of the angle determined by two rays that

1. have a vertex on the edge.
2. lie in the planes so that they are perpendicular to the edge.

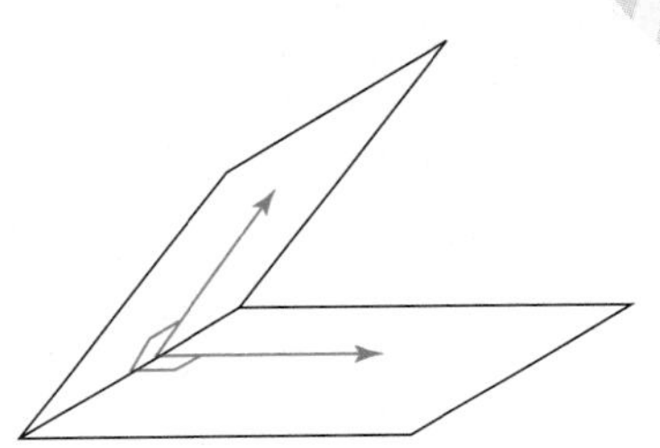

FIGURE 8.36

A **polyhedron** (plural *polyhedrons* or *polyhedra*) is a solid bounded by plane regions. Polygons form the **faces** of the solid and the segments common to these polygons are the **edges** of the polyhedron. Endpoints of the edges are the **vertices** of the polyhedron. When a polyhedron is **convex,** each face determines a plane for which all remaining faces lie on the same side of that plane. Figure 8.37(a) illustrates a convex polyhedron while Figure 8.37(b) illustrates a **concave** polyhedron.

The prisms and pyramids discussed in Sections 8.1 and 8.2 were special types of polyhedrons. For instance, a pentagonal pyramid can be described as a hexahedron because it has six faces. Because some of their surfaces do not lie in planes, the cylinders and cones of Section 8.3 are not polyhedrons.

Leonhard Euler (*Swiss:* 1707 A.D.–1763 A.D.) found that the number of vertices, the number of edges, and the number of faces of a polyhedron (not necessarily a regular polyhedron) are related by **Euler's equation.** This equation is given in the following theorem.

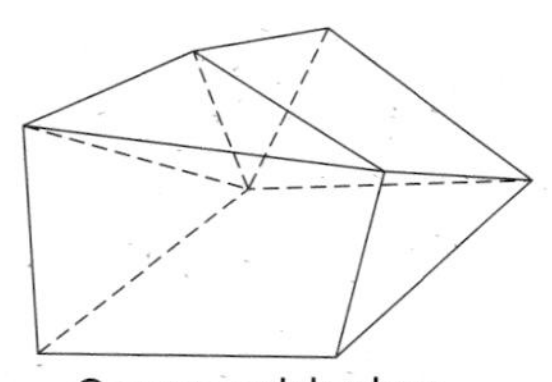

Convex polyhedron

(a)

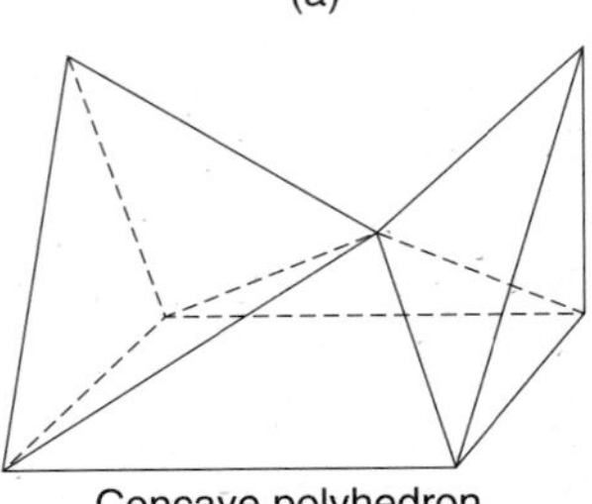

Concave polyhedron

(b)

FIGURE 8.37

THEOREM 8.4.1: (Euler's equation): The number of vertices V, the number of edges E, and the number of faces F of a polyhedron are related by the equation

$$V + F = E + 2$$

EXAMPLE 1

Verify Euler's equation for the (a) tetrahedron and (b) square pyramid shown in Figure 8.38.

Solution

a) The tetrahedron has four vertices ($V = 4$), six edges ($E = 6$), and four faces ($F = 4$), so the equation becomes $4 + 4 = 6 + 2$, which is true.

b) The pyramid has five vertices ("vertex" + 4 vertices from the base), eight edges (4 base edges + 4 lateral edges), and five faces (4 triangular regions + 1 square base). Now $V + F = E + 2$ becomes $5 + 5 = 8 + 2$, which is also true.

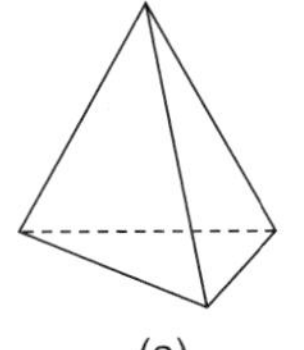

(a)

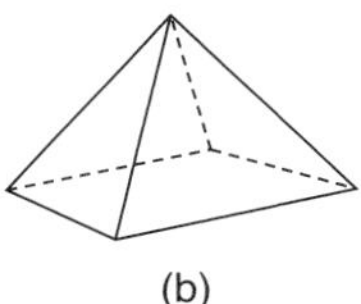

(b)

FIGURE 8.38

Regular Polyhedra

DEFINITION: A **regular polyhedron** is a convex polyhedron whose faces are congruent regular polygons arranged in such a way that adjacent faces form congruent dihedral angles.

The five regular polyhedrons are as follows:

1. Regular **tetrahedron,** which has 4 faces (congruent equilateral triangles)
2. Regular **hexahedron** (or **cube**), which has 6 faces (congruent squares)
3. Regular **octahedron,** which has 8 faces (congruent equilateral triangles)
4. Regular **dodecahedron,** which has 12 faces (congruent regular pentagons)
5. Regular **icosahedron,** which has 20 faces (congruent equilateral triangles)

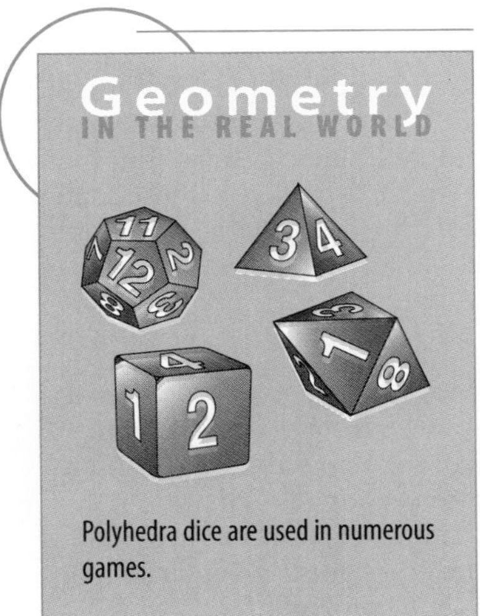

Polyhedra dice are used in numerous games.

Four of the regular polyhedrons are shown in Figure 8.39.

Regular Polyhedrons

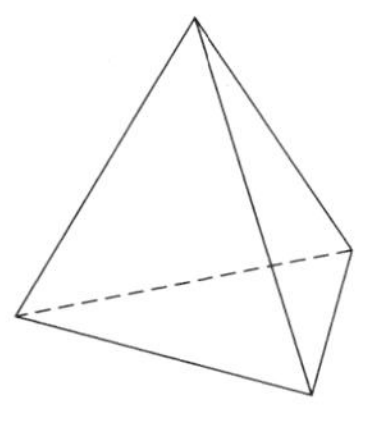

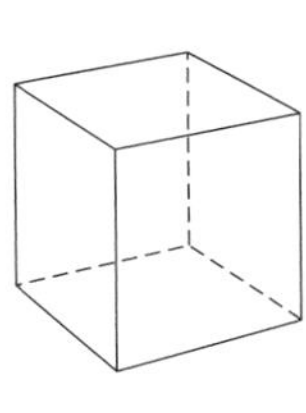

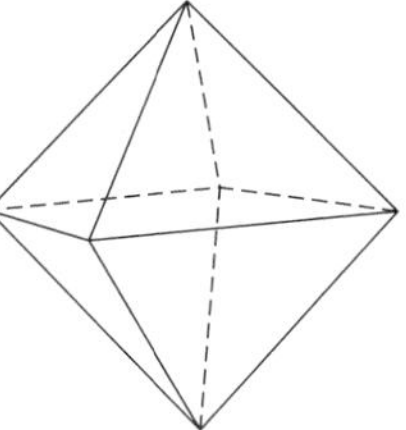

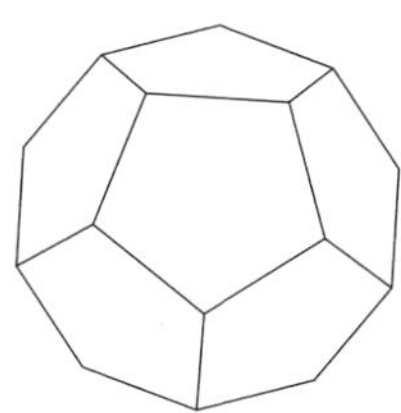

Tetrahedron Hexahedron Octahedron Dodecahedron

FIGURE 8.39

EXAMPLE 2

Consider a die that is a regular tetrahedron. Assuming that each face has an equal chance of being rolled, what is the likelihood (probability) that one roll produces (a) a "1"? (b) a result larger than "1"?

Solution

a) With four equally likely results (1, 2, 3, and 4), the probability of a "1" is $\frac{1}{4}$.
b) With four equally likely results (1, 2, 3, and 4) and three "favorable" outcomes (2, 3, and 4), the probability of rolling a number larger than a "1" is $\frac{3}{4}$.

Spheres

Another type of solid with which you are familiar is the sphere. While the surface of a basketball correctly depicts the sphere, we often use the term *sphere* to refer to a solid like a baseball as well.

In space, the sphere is characterized in three ways:

1. A **sphere** is the set of all points at a fixed distance r from a given point O. Point O is known as the **center** of the sphere, even though it is not a part of the spherical surface.
2. A **sphere** is the surface determined when a circle (or semicircle) is rotated about any of its diameters.
3. A **sphere** is the surface that represents the theoretical limit of an "inscribed" regular polyhedron whose number of faces increases without limit.

NOTE: In characterization 3, suppose that the number of faces of the regular polyhedron could grow without limit. In theory, the resulting regular polyhedra would appear rather "spherical" as the number of faces increases without limit. In reality, a regular polyhedron can have no more than 20 faces (the icosahedron). It will be convenient and appropriate to use this third characterization of the sphere when we determine the formula for its volume.

Each characterization of the sphere has its advantages.

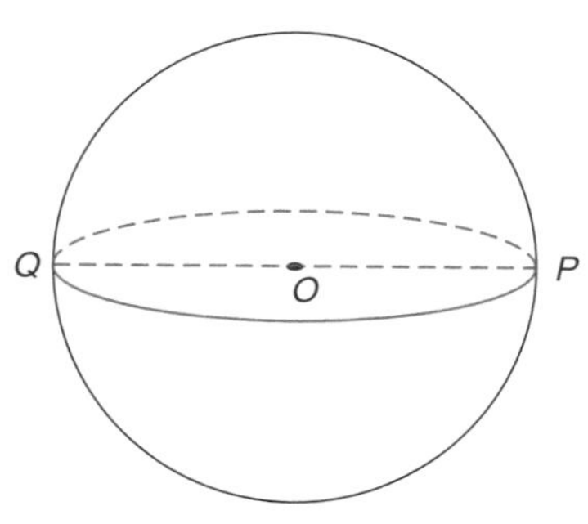

FIGURE 8.40

Characterization 1: In Figure 8.40, a sphere was generated as the locus of points in space at a distance r from point O. The line segment $\overline{OP}$ is a **radius** of sphere O, and $\overline{QP}$ is a **diameter** of the sphere. The intersection of a sphere and a plane that contains its center is a **great circle** of the sphere. For the earth, the equator is a great circle that separates the earth into two **hemispheres.**

Surface Area of a Sphere

Characterization 2: The following theorem claims that the surface area of a sphere equals four times the area of a great circle of that sphere. This theorem, which is proven in calculus, treats the sphere as a surface of revolution.

THEOREM 8.4.2: The surface area S of a sphere whose radius has length r is given by $S = 4\pi r^2$.

EXAMPLE 3

Find the surface area of a sphere whose radius is $r = 7$ in. Use your calculator to approximate the result.

Solution

$$S = 4\pi r^2 \rightarrow S = 4\pi \cdot 7^2 = 196\pi$$

Then $S \approx 615.75$ in.^{2}

DISCOVER!

A farmer's silo is a composite shape. That is, it is actually composed of two solids. What are they?

ANSWER
cylinder and hemisphere

Volume of a Sphere

Characterization 3: The third description of the sphere enables us to find its volume. To accomplish this, we treat the sphere as the theoretical limit of an inscribed regular polyhedron whose number of faces n increases without limit. The polyhedron can be separated into n pyramids; the center of the sphere is the vertex of each pyramid. As n increases, the altitude of each pyramid approaches the radius of the sphere in length. Next we find the sum of the volumes of these pyramids, the limit of which is the volume of the sphere.

In Figure 8.41, one of the pyramids described in the preceding paragraph is shown. Where the areas of the bases of the pyramids are written B_1, B_2, B_3, and so on, the sum of the volumes of the n pyramids forming the polyhedron is

$$\frac{1}{3}B_1 h + \frac{1}{3}B_2 h + \frac{1}{3}B_3 h + \ldots + \frac{1}{3}B_n h$$

Because the pyramids are congruent, they all have the same altitude h.

Next we write the volume of the polyhedron in the form

$$\frac{1}{3}h(B_1 + B_2 + B_3 + \ldots + B_n)$$

As n increases, $h \rightarrow r$ and $B_1 + B_2 + B_3 + \ldots + B_n \rightarrow S$, the surface area of the sphere. Because the surface area of the sphere is $S = 4\pi r^2$, the sum approaches the following limit as the volume of the sphere:

$$\frac{1}{3}h(B_1 + B_2 + B_3 + \ldots + B_n) \rightarrow \frac{1}{3}rS \text{ or } \frac{1}{3}r \cdot 4\pi r^2 = \frac{4}{3}\pi r^3$$

This leads us to the following theorem.

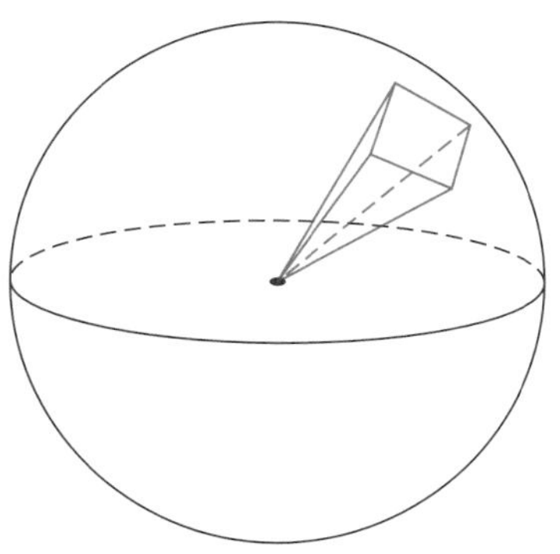

FIGURE 8.41

THEOREM 8.4.3: The volume V of a sphere with a radius of length r is given by $V = \frac{4}{3}\pi r^3$.

EXAMPLE 4

Find the exact volume of a sphere whose length of radius is 1.5 in.

Solution This calculation can be done more easily if we replace 1.5 by $\frac{3}{2}$.

$$\begin{aligned} V &= \frac{4}{3}\pi r^3 \\ &= \frac{4}{3} \cdot \pi \cdot \frac{3}{2} \cdot \frac{3}{2} \cdot \frac{3}{2} \\ &= \frac{9\pi}{2} \text{ in.}^3 \end{aligned}$$

EXAMPLE 5

A spherical propane gas storage tank has a volume of $\frac{792}{7}$ ft^3. Using $\pi \approx \frac{22}{7}$, find the radius of the sphere.

Solution $V = \frac{4}{3}\pi r^3$, which becomes $\frac{792}{7} = \frac{4}{3} \cdot \frac{22}{7} \cdot r^3$. Then $\frac{88}{21}r^3 = \frac{792}{7}$.

In turn, $\frac{21}{88} \cdot \frac{88}{21}r^3 = \frac{21}{88} \cdot \frac{792}{7} \rightarrow r^3 = 27 \rightarrow r = 3$

The radius of the tank is 3 ft.

Just as two concentric circles have the same center but different lengths of radii, so also can two spheres be concentric. This fact is the basis for the following example.

EXAMPLE 6

A child's hollow plastic ball has an inside diameter of 10 in. and is approximately $\frac{1}{8}$ in. thick (see Figure 8.42). Approximately how many cubic inches of plastic were needed to construct the ball?

Solution The volume of plastic used is the difference between the inside volume and outside volume. Where R denotes the length of the outside radius and r denotes the length of the inside radius, $R \approx 5.125$ and $r = 5$.

$$V = \frac{4}{3}\pi R^3 - \frac{4}{3}\pi r^3, \text{ so } V = \frac{4}{3}\pi(5.125)^3 - \frac{4}{3}\pi \cdot 5^3$$

Then $V \approx 563.86 - 523.60 \approx 40.26$

The volume of plastic used was approximately 40.26 in.^{3}

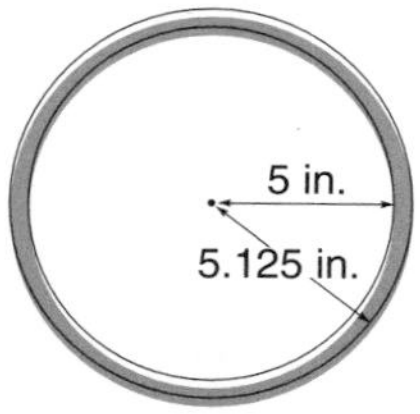

5 in.
5.125 in.

FIGURE 8.42

More Solids of Revolution

In Section 8.3, each solid of revolution was generated by revolving a plane region about a horizontal line segment. It is also possible to form a solid of revolution by rotating a region about a vertical line segment. See Examples 7 and 8.

EXAMPLE 7

Describe the solid of revolution formed when the rectangular region having dimensions of 2 in. by 5 in. [see Figure 8.43(a)] is rotated about a vertical side of length 5 in. Then find the exact volume of the solid formed [see Figure 8.43(b)].

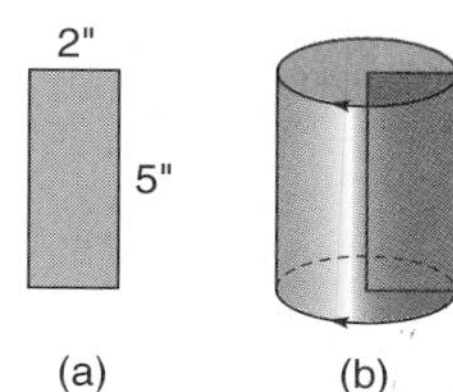

FIGURE 8.43

Solution The solid formed is a right circular cylinder with radius of base $r = 2$ and altitude $h = 5$. To find the volume, $V = \pi r^2 h$ becomes

$$V = \pi \cdot 2^2 \cdot 5, \text{ so } V = 20\pi \text{ in.}^3$$

EXAMPLE 8

Describe the solid of revolution formed when a semicircular region having a vertical diameter of length 12 cm [see Figure 8.44(a)] is revolved about that diameter. Then find the exact volume of the solid formed [see Figure 8.44(b)].

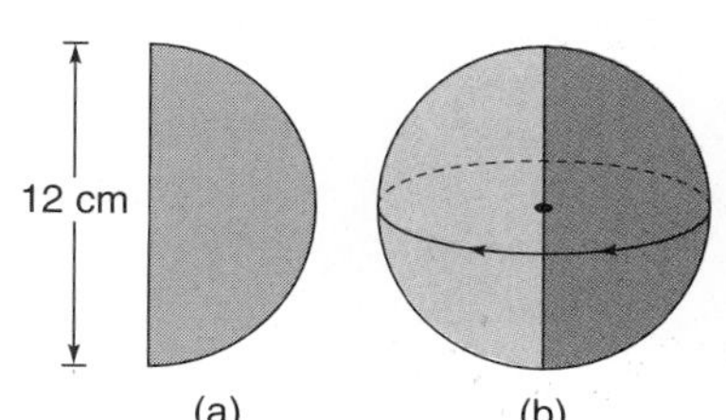

FIGURE 8.44

Solution The solid formed is a sphere with radius of base $r = 6$ cm. To find the volume, $V = \frac{4}{3}\pi r^3$ becomes $V = \frac{4}{3}\pi \cdot 6^3$, which simplifies to $V = 288\pi \text{ cm}^3$.

When a circular region is revolved about a line in the circle's exterior, a doughnut-shaped solid results. The formal name of the resulting solid of revolution, shown in Figure 8.45, is the *torus*. Calculus is necessary to calculate either the surface area or the volume of the torus.

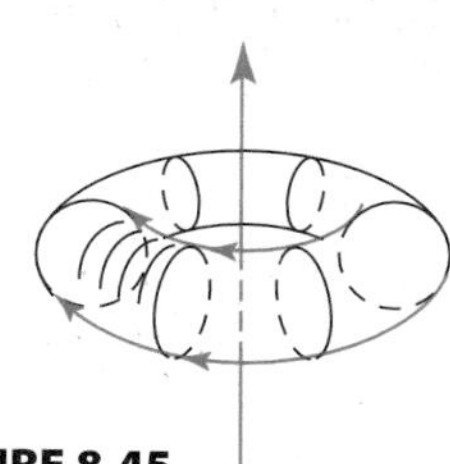

FIGURE 8.45

8.4 Exercises

1. Which of the two polyhedrons is concave? Notice that the interior dihedral angle formed by the planes containing $\triangle EJF$ and $\triangle KJF$ is larger than 180°.

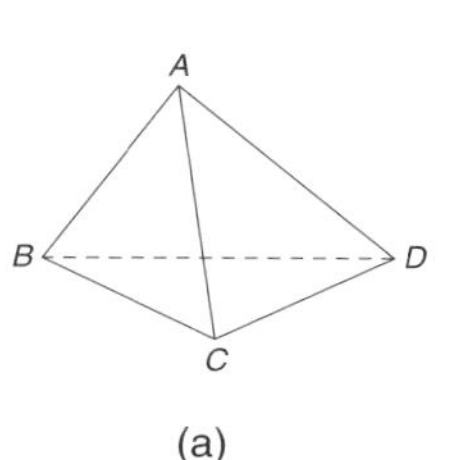

(a)

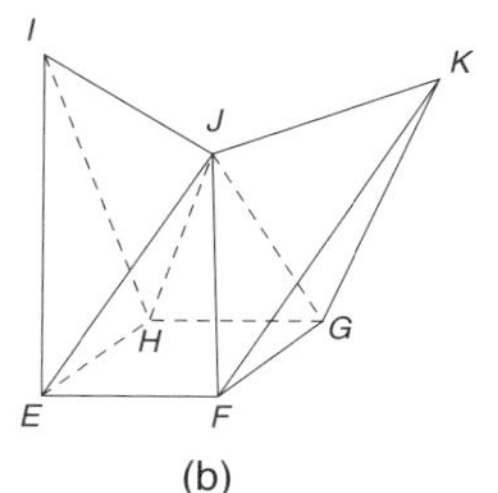

(b)

2. For Figure (a) of Exercise 1, find the number of faces, vertices, and edges in the polyhedron. Then verify Euler's equation for that polyhedron.
3. For Figure (b) of Exercise 1, find the number of faces, vertices, and edges in the polyhedron. Then verify Euler's equation for that polyhedron.
4. For a regular tetrahedron, find the number of faces, vertices, and edges in the polyhedron. Then verify Euler's equation for that polyhedron.
5. For a regular hexahedron, find the number of faces, vertices, and edges in the polyhedron. Then verify Euler's equation for that polyhedron.

In Exercises 6 to 8, the probability is the ratio $\frac{\text{number of favorable outcomes}}{\text{number of possible outcomes}}$. Use Example 2 of this section as a guide.

6. Assume that the most commonly-shaped die, a hexahedron, is rolled. What is the likelihood that
   a) a "2" results?
   b) an even number results?
   c) the result is larger than 2?
7. Assume that a die in the shape of a dodecahedron is rolled. What is the probability that
   a) an even number results?
   b) a prime number (2,3,5,7, or 11) results?
   c) the result is larger than 2?
8. Assume that a die in the shape of an icosahedron is rolled. What is the likelihood that
   a) an odd number results?
   b) a prime number (2, 3, 5, 7, 11, 13, 17, or 19) results?
   c) the result is larger than 2?
9. In sphere O, the length of radius $\overline{OP}$ is 6 in. Find the length of the chord:
   a) $\overline{QR}$ if $m\angle QOR = 90°$
   b) $\overline{QS}$ if $m\angle SOP = 60°$

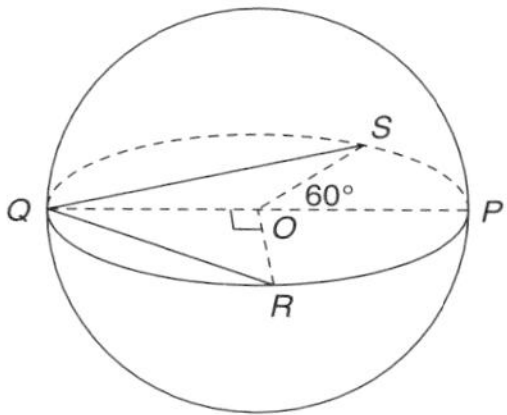

Exercises 9, 10

10. Find the approximate surface area and volume of the sphere if $OP = 6$ in. Use your calculator.
11. A sphere is inscribed within a right circular cylinder whose altitude and diameter have equal measures.
    a) Find the ratio of the surface area of the cylinder to that of the sphere.
    b) Find the ratio of the volume of the cylinder to that of the sphere.

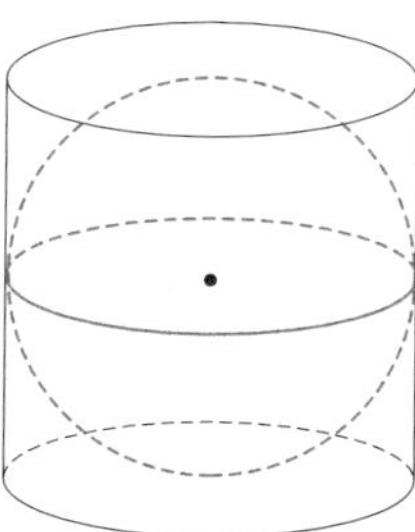

12. Given that a right circular cylinder is inscribed within a sphere, what is the least possible volume of the cylinder? (**HINT:** Consider various lengths for radius and altitude.)

13. In calculus, it can be shown that the largest possible volume for the inscribed right circular cylinder in Exercise 12 occurs when its altitude has a length equal to the diameter of the circular base. Find the length of the radius and the altitude of the cylinder of greatest volume if the radius of the sphere is 6 in.

14. Given that a *regular* polyhedron of n faces is inscribed in a sphere of radius 6 in., find the maximum (largest) possible volume for the polyhedron.

15. A right circular cone is inscribed in a sphere. If the slant height of the cone has a length equal to that of its diameter, find the length of the

a) radius of the base of the cone.
b) altitude of the cone.

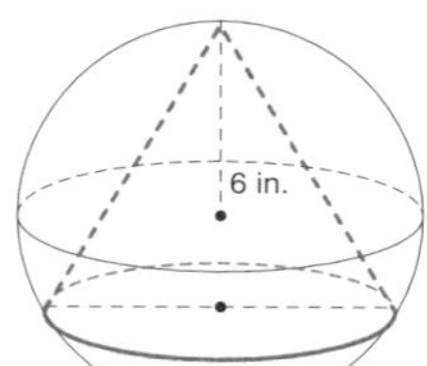

The radius of the sphere has a length of 6 in.

16. A sphere is inscribed in a right circular cone whose slant height has a length equal to the diameter of its base. What is the length of the radius of the sphere if the slant height and the diameter of the cone both measure 12 cm?

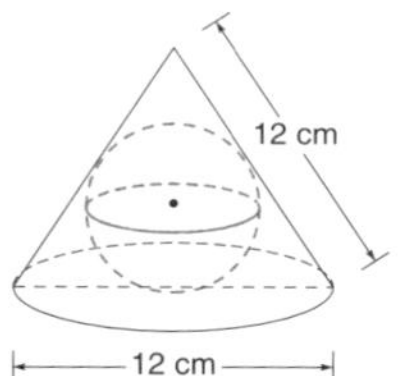

In Exercises 17 and 18, use the calculator value of π.

17. For a sphere whose radius has length 3 m, find the

a) surface area.
b) volume.

18. For a sphere whose radius has length 7 cm, find the

a) surface area.
b) volume.

19. A sphere has a volume equal to $\frac{99}{7}$ in.^{3} Determine the length of the radius of the sphere. (Let $\pi \approx \frac{22}{7}$.)

20. A sphere has a surface area equal to 154 in.^{2} Determine the length of the radius of the sphere. (Let $\pi \approx \frac{22}{7}$.)

21. The spherical storage tank described in Example 5 had a length of radius of 3 ft. Since it needs to be painted, find its surface area. Also determine the number of pints of rust-proofing paint needed to paint the tank if 1 pt covers approximately 40 ft^2. Use your calculator.

22. An observatory has the shape of a right circular cylinder surmounted by a hemisphere. If the radius of the cylinder is 14 ft and its altitude measures 30 ft, what is the surface area of its observatory? If 1 gal of paint covers 300 ft^2, how many gallons are needed to paint the surface if it needs two coats? Use your calculator.

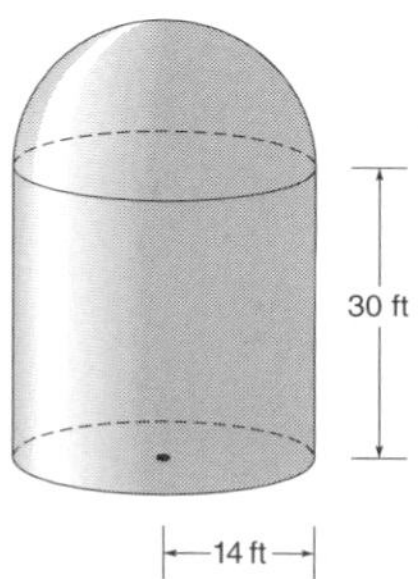

23. A leather soccer ball has an inside diameter of 8.5 in. and a thickness of 0.1 in. Find the volume of leather needed for its construction. Use your calculator.

24. An ice cream cone is completely filled with ice cream. What is the volume of the ice cream? Use your calculator.

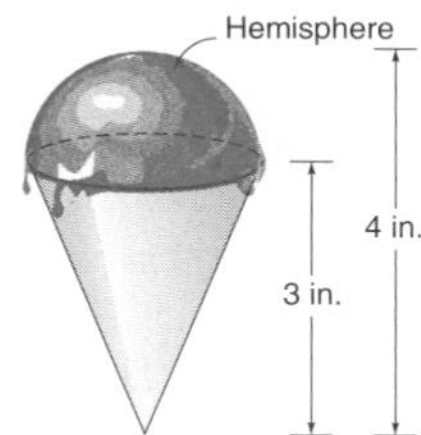

25. Suppose that a semicircular region with a vertical diameter of length 6 is rotated about that diameter. Determine the exact surface area and the exact volume of the resulting solid of revolution.

26. Suppose that a semicircular region with a vertical diameter of length 4 is rotated about that diameter. Determine the exact surface area and the exact volume of the resulting solid of revolution.

27. Sketch the torus that results when the given circle of radius 1 is revolved about the horizontal line that lies 4 units below the center of that circle.

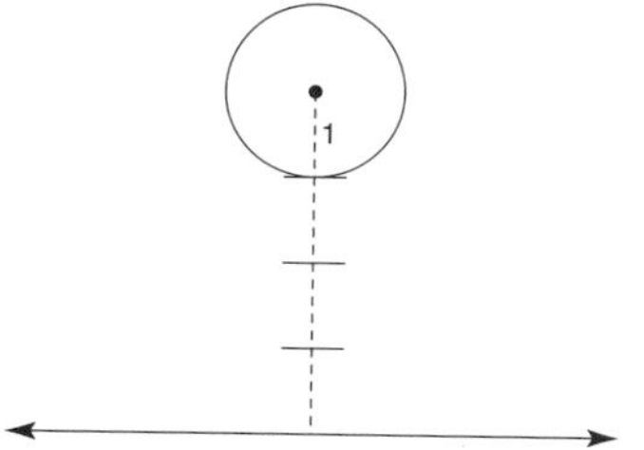

28. Sketch the solid that results when the given circle of radius 1 is revolved about the horizontal line that lies 1 unit below the center of that circle.

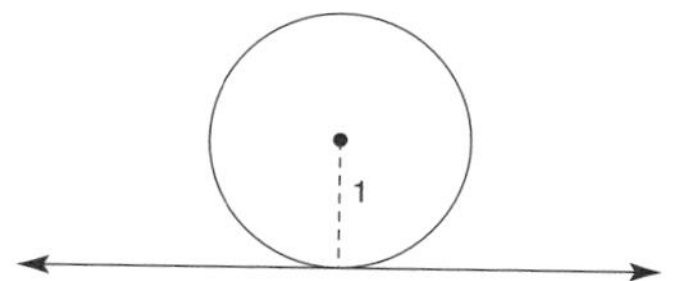

29. Explain how the following formula used in Example 6 was obtained:

$$V = \frac{4}{3}\pi R^3 - \frac{4}{3}\pi r^3$$

30. Derive a formula for the total surface area of the hollow-core sphere. (**NOTE:** Include both interior and exterior surface areas.)

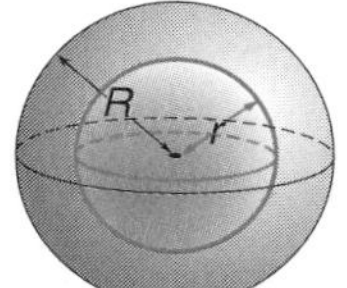

A Look Beyond: Historical Sketch of René Descartes

René Descartes was born in Tours, France, on March 31, 1596, and died in Stockholm, Sweden, on February 11, 1650. He was a contemporary of Galileo, the Italian scientist responsible for many discoveries in the science of dynamics. Descartes was also a friend of the French mathematicians Marin Mersenne (Mersenne Numbers) and Blaise Pascal (Pascal's Triangle).

As a small child, René Descartes was in poor health much of the time. Because he spent so much time reading in bed during his illnesses, he became known as a very intelligent young man. When Descartes was older and stronger, he joined the French army. It was during his time as a soldier that Descartes had three dreams that shaped much of his future. The dreams, dated to November 10, 1619, shaped his philosophy and laid the framework for his discoveries in mathematics.

Descartes resigned his commission with the army in 1621 so that he could devote his life to studies of philosophy, science, and mathematics. In the ensuing years, Descartes came to be highly regarded as a philosopher and mathematician and was invited to the learning centers of France, Holland, and Sweden.

Descartes' chief contribution to mathematics was the development of analytical geometry, using the rectangular (Cartesian) coordinate system as a means of representing points. This convention led, in turn, to the algebraic description (equations) of various geometric figures; subsequently, many conjectured properties concerning those figures could be established through proof. Using the Cartesian system, it was also possible to locate the points of intersection of certain figures such as circles, parabolas, and lines, as well as some combinations of these.

Generally, the phrase "conic sections" refers to four geometric figures: the **circle,** the **parabola,** the **ellipse,** and the **hyperbola.** These figures are shown in Figure 8.46 individually and also in relation to the upper and lower nappes of a cone. The conic sections are formed when a plane intersects the nappes of a cone.

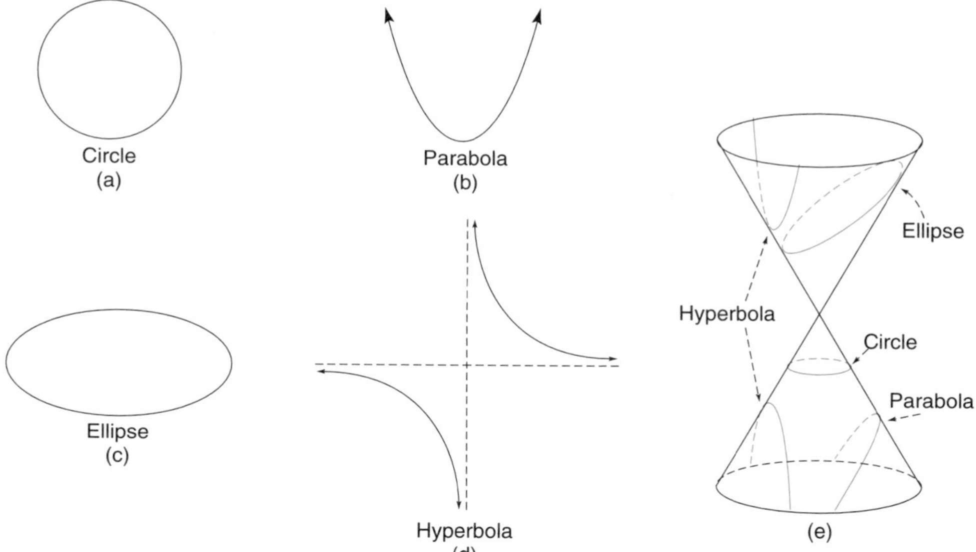

FIGURE 8.46

Other mathematical works of Descartes were devoted to the study of tangent lines to curves. The notion of a tangent to a curve is illustrated in Figure 8.47; this concept is the basis for the branch of mathematics known as **differential calculus.**

Descartes' final contributions to mathematics involved his standardizing the use of many symbols. To mention a few of these, Descartes used (1) a^2 rather than *aa* and a^3 rather than *aaa*; (2) *ab* to indicate multiplication; and (3) *a*, *b*, and *c* as constants and *x*, *y*, and *z* as variables.

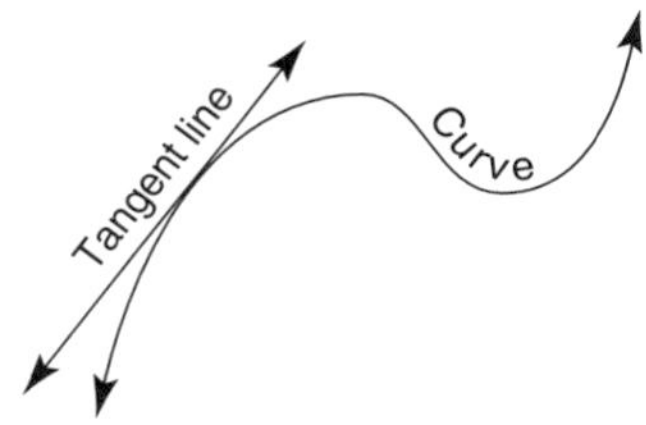

FIGURE 8.47

Summary

■ *A Look Back at Chapter 8*

Our goal in this chapter was to deal with a type of geometry known as solid geometry. We found formulas for the lateral area, the total area (surface area), and the volume of prisms, pyramids, cylinders, cones, and spheres. Some of the formulas used in this chapter were developed using the concept of "limit."

A Look Ahead to Chapter 9

Our focus in the next chapter is analytic (or coordinate) geometry. This type of geometry relates algebra and geometry. Formulas for the midpoint of a line segment, the length of a line segment, and the slope of a line will be developed. Equations of lines and circles will be graphed. Concrete examples will lead to more general proofs using analytic geometry.

Important Terms and Concepts of Chapter 8

8.1 Prisms (Right and Oblique)
Base and Altitude
Vertices, Edges, and Faces
Lateral Area, Total Area (Surface Area), and Volume
Regular Prism
Cube, Cubic Unit
Right Prism

8.2 Pyramids
Base and Altitude
Vertices, Edges, and Faces
Vertex of a Pyramid
Regular Pyramid
Slant Height of a Regular Pyramid
Lateral Area, Total Area (Surface Area), and Volume

8.3 Cylinders (Right and Oblique)
Bases and Altitude of a Cylinder
Axis of a Cylinder
Cones (Right and Oblique)
Base and Altitude of a Cone
Vertex and Slant Height of a Cone
Axis of a Cone
Lateral Area, Total Area, and Volume of Cylinders and Cones
Solid of Revolution
Axis of a Solid of Revolution

8.4 Dihedral Angle
Polyhedron (Convex and Concave)
Vertices, Edges, and Faces
Euler's Equation
Regular Polyhedrons (Tetrahedron, Hexahedron, Octahedron, Dodecahedron, and Icosahedron)
Sphere (Center, Radius, Diameter, Great Circle, Hemispheres)
Surface Area and Volume of a Sphere

A Look Beyond: *Historical Sketch of René Descartes*

Review Exercises

1. Each side of the base of a right octagonal prism is 7 in. long. The altitude of the prism is 12 in. Find the lateral area.
2. The base of a right prism is a triangle whose sides measure 7 cm, 8 cm, and 12 cm, respectively. The altitude of the prism is 11 cm. Calculate the lateral area of the right prism.
3. The height of a square box is 2 inches more than 3 times the length of a side of the base. If the lateral area is 480 in.^{2}, find the dimensions of the box and the volume of the box.
4. The base of a right prism is a rectangle whose length is 3 cm more than its width. If the altitude of the prism is 12 cm and the lateral area is 360 cm^2, find the total area and the volume.
5. The base of a right prism is a triangle whose sides are 9 in., 15 in., and 12 in. The height of the prism is 10 in. Find the

   a) lateral area. **b)** total area. **c)** volume.
6. The base of a right prism is a regular hexagon whose sides are 8 cm. The altitude of the prism is 13 cm. Find the

   a) lateral area. **b)** total area. **c)** volume.

7. A regular square pyramid has a base whose sides are 10 cm each. The altitude of the pyramid measures 8 cm. Find the length of the slant height.

8. A regular hexagonal pyramid has a base whose sides are $6\sqrt{3}$ in. each. If the slant height is 12 in., find the length of the altitude of the pyramid.

9. The radius of the base of a right circular cone measures 5 in. If the altitude of the cone measures 7 in., what is the length of the slant height?

10. The diameter of the base of a right circular cone is equal in length to the slant height. If the altitude of the cone is 6 cm, find the length of the radius of the base.

11. The slant height of a regular square pyramid is 15 in. One side of the base is 18 in. Find the

    a) lateral area.
    b) total area.
    c) volume.

12. The base of a regular pyramid is an equilateral triangle each of whose sides is 12 cm. The altitude of the pyramid is 8 cm. Find the

    a) lateral area.
    b) total area.
    c) volume.

13. The radius of the base of a right circular cylinder is 6 in. The height of the cylinder is 10 in. Find the exact

    a) lateral area.
    b) total area.
    c) volume.

14. **a)** For the trough in the shape of a half-cylinder, find the volume of water it will hold. (Use $\pi \approx 3.14$ and disregard the thickness.)
    b) If the trough is to be painted inside and out, find the number of square feet to be painted. (Use $\pi \approx 3.14$.)

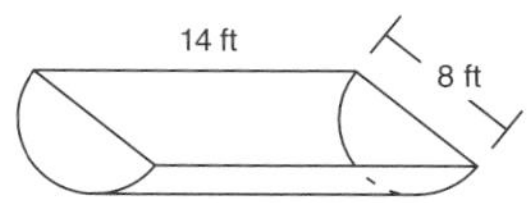

15. The slant height of a right circular cone is 12 cm. The angle formed by the slant height and the altitude is 30°. Find the exact and approximate

    a) lateral area.
    b) total area.
    c) volume.

16. The volume of a right circular cone is 96π in.^{3} If the radius of the base is 6 in., find the length of the slant height.

17. Find the surface area of a sphere if the radius is 7 in. Use $\pi \approx \frac{22}{7}$.

18. Find the volume of a sphere if the diameter is 12 cm. Use $\pi \approx 3.14$.

19. The solid shown consists of a hemisphere (half of a sphere), a cylinder, and a cone. Find the exact volume of the solid.

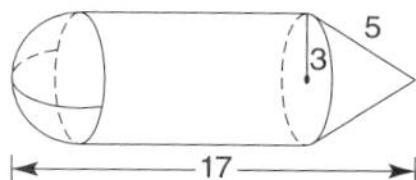

20. If the radius of one sphere is three times as long as the radius of another sphere, how do the surface areas of the spheres compare? How do the volumes compare?

21. Find the volume of the solid of revolution that results when the right triangle with legs of lengths 5 in. and 7 in. is rotated about the 7 in. leg. Use $\pi \approx \frac{22}{7}$.

22. Find the exact volume of the solid of revolution that results when the rectangular region with dimensions of 6 cm and 8 cm is rotated about a side of length 8 cm.

23. Find the exact volume of the solid of revolution that results when a semicircular region with diameter of length 4 in. is rotated about that diameter.

24. A plastic pipe is 3 ft long and has an inside radius of 4 in. and an outside radius of 5 in. How many cubic inches of plastic are in the pipe? (Use $\pi \approx 3.14$.)

25. A sphere with a diameter of 14 in. is inscribed in a hexahedron. Find the exact volume of the space inside the hexahedron but outside the sphere.

26. **a)** An octahedron has ________ faces that are ________.
    b) A tetrahedron has ________ faces that are ________.
    c) A dodecahedron has ________ faces that are ________.

27. A drug manufacturing company wants to manufacture a capsule that contains a spherical pill inside. The diameter of the pill is 4 mm and the capsule is cylindrical, with hemispheres on either end. The length of the capsule between the two hemispheres is 10 mm. What is the exact volume the capsule will hold, excluding the volume of the pill?

28. For each of the following solids, verify Euler's equation by determining *V*, the number of vertices, *E*, the number of edges, and *F*, the number of faces.

a) Right octagonal prism
b) Regular equilateral triangular pyramid
c) Octahedron

29. Find the volume of cement used in the block shown.

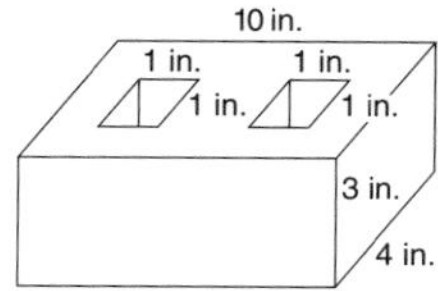

Chapter 9
Analytic Geometry

The French mathematician René Descartes is considered the father of analytic geometry. His ingenious device for relating algebra and geometry, the Cartesian coordinate system, was a major breakthrough in the development of much of mathematics. Some have written that Descartes was inspired while watching a spider weave its web in the corner of his room. Others say that the brilliant idea came to Descartes as he was dreaming. In any event, the branch of mathematics known as calculus would not have grown as rapidly without analytic geometry.

The periscope reminds us of the coordinate system for which Descartes is well known.

CHAPTER OUTLINE

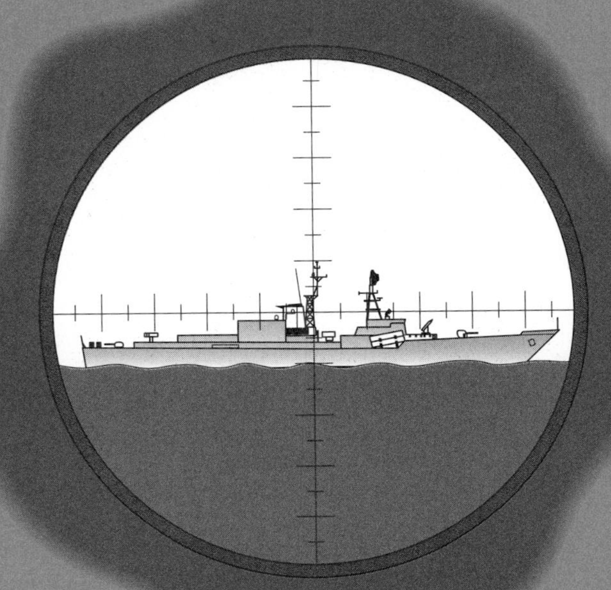

9.1 The Rectangular Coordinate System

Graphing the solution sets for $3x - 2 = 7$ and $3x - 2 > 7$ required a single number line to indicate the value of x. In this chapter, we deal with equations containing two variables; to relate algebra to plane geometry, we will need two number lines.

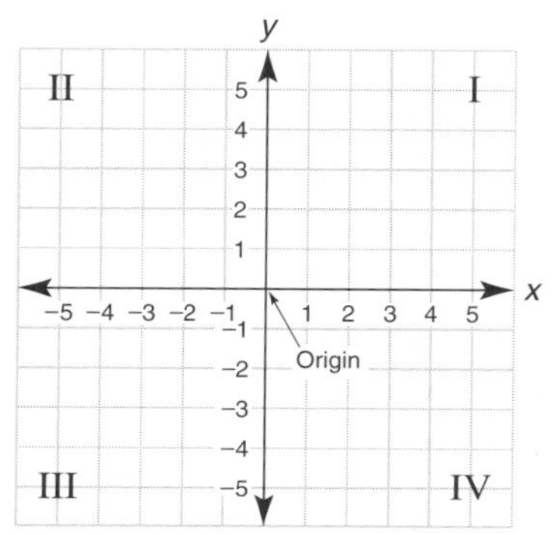

FIGURE 9.1

The study of the relationships between number pairs and points is usually referred to as **analytic geometry.** The **Cartesian coordinate system** or **rectangular coordinate system** is the plane that results when two number lines intersect perpendicularly at the origin (the point corresponding to the number 0) of each line. The horizontal number line is known as the ***x* axis,** and its numerical coordinates increase from left to right. On the vertical number line, the ***y* axis,** values increase from bottom to top; see Figure 9.1. The two axes separate the plane into four sections known as **quadrants;** the quadrants are numbered I, II, III, and IV, as shown. The point that marks the common origin of the two number lines is the **origin** of the rectangular coordinate system. It is convenient to identify the origin as (0, 0); this notation indicates that the ***x* coordinate** (listed first) is 0 and also that the ***y* coordinate** (listed second) is 0.

In the coordinate system in Figure 9.2, the point $(3, -2)$ is shown. For each point, we have the order (x, y); these pairs are referred to as **ordered pairs** since we require x before y. To plot (or locate) this point, we see that $x = 3$ and that $y = -2$. Thus the point is located by moving 3 units to the right of the origin and then 2 units down from the x axis. The dashed lines shown are used to emphasize the reason for the name "rectangular" coordinate system. Notice that this point $(3, -2)$ could also have been located by first moving down 2 units and then moving 3 units to the right of the y axis. This point is located in Quadrant IV. In Figure 9.2, ordered pairs of plus and minus signs characterize the signs of the coordinates of a point in each quadrant.

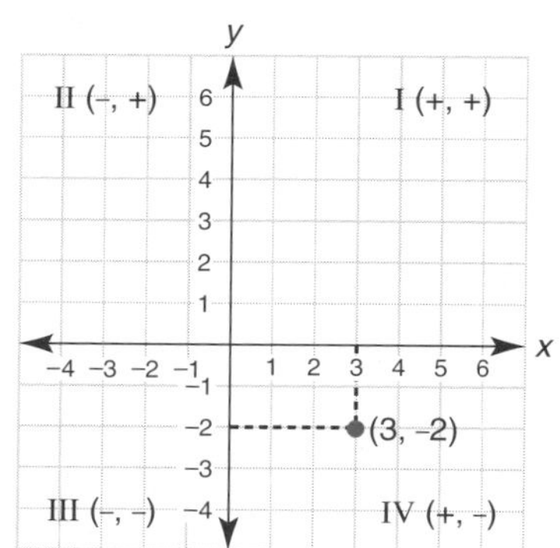

FIGURE 9.2

EXAMPLE 1

Plot points $A\,(-3, 4)$ and $B\,(2, 4)$, and find the distance between them.

Solution

Point A is located by moving 3 units to the left of the origin and then 4 units up from the x axis. Point B is located by moving 2 units to the right of the origin and then 4 units up from the x axis. In Figure 9.3, $\overline{AB}$ is a horizontal segment.

In the rectangular coordinate system, $ABCD$ is a rectangle in which $DC = 5$; $\overline{DC}$ is easily measured since it lies on the x axis. Because the opposite sides of a rectangle are congruent, it follows that $AB = 5$.

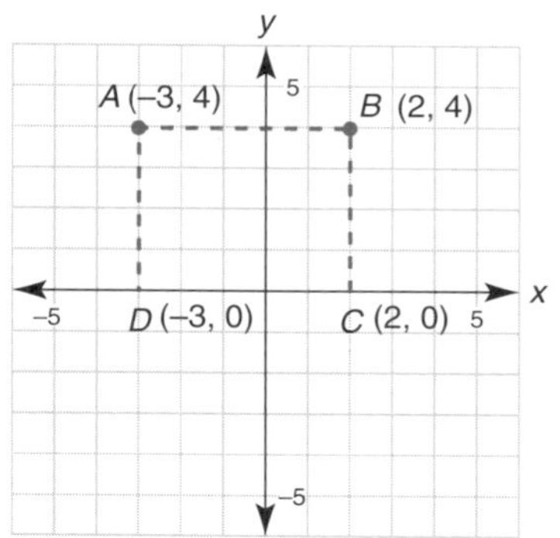

FIGURE 9.3

In Example 1, the points $(-3, 4)$ and $(2, 4)$ have the same y coordinates. In this case, the distance between the points on a horizontal line is merely the difference in the x coordinates; thus, the distance is $2 - (-3)$ or 5. It is also easy to find the distance between two points on a vertical line. When the x coordinates are the same, the distance between points is the difference in the y coordinates. In Figure 9.3, where C is $(2, 0)$ and B is $(2, 4)$, the distance between the points is $4 - 0$ or 4.

DEFINITION: Given points $A(x_1, y_1)$ and $B(x_2, y_1)$ on a horizontal line segment $\overline{AB}$, the **distance** between these points is

$$AB = x_2 - x_1 \text{ if } x_2 > x_1 \quad \text{or} \quad AB = x_1 - x_2 \text{ if } x_1 > x_2$$

In the preceding definition, repeated y coordinates characterize a horizontal line segment. In the following definition, repeated x coordinates determine a vertical line segment. In each definition, the distance is found by subtracting the smaller from the larger of the two unequal coordinates.

DEFINITION: Given points $C(x_1, y_1)$ and $D(x_1, y_2)$ on a vertical line segment $\overline{CD}$, the **distance** between these points is

$$CD = y_2 - y_1 \text{ if } y_2 > y_1 \quad \text{or} \quad CD = y_1 - y_2 \text{ if } y_1 > y_2$$

EXAMPLE 2

In Figure 9.4, name the coordinates of points C and D, and find the distance between them.

Solution

C is the point $(0, 1)$ since C is 1 unit above the origin; similarly, D is the point $(0, 5)$. We designate the coordinates of point C by $x_1 = 0$ and $y_1 = 1$ and the coordinates of point D by $x_1 = 0$ and $y_2 = 5$. Using the preceding definition,

$$CD = y_2 - y_1 = 5 - 1 = 4$$

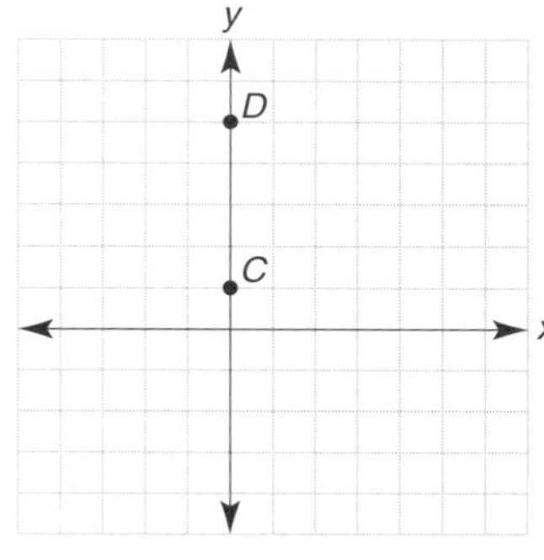

FIGURE 9.4

We now turn our attention to the more general problem of finding the distance between any two points.

The Distance Formula

THEOREM 9.1.1: (Distance Formula) The distance between two points (x_1, y_1) and (x_2, y_2) is given by the formula

$$d = \sqrt{(x_2 - x_1)^2 + (y_2 - y_1)^2}$$

Proof In the coordinate system in Figure 9.5 are points $P_1\ (x_1, y_1)$ and $P_2\ (x_2, y_2)$. In addition to drawing the segment joining these points, we draw an auxiliary horizontal segment through P_1 and an auxiliary vertical segment through P_2; these meet at point C in Figure 9.5(a). Using Figure 9.5(b) and the definitions for lengths of horizontal and vertical segments,

$$P_1C = x_2 - x_1 \quad \text{and} \quad P_2C = y_2 - y_1$$

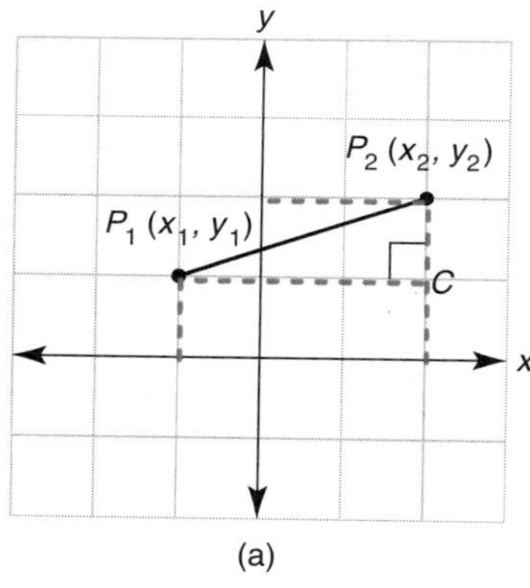

(a)

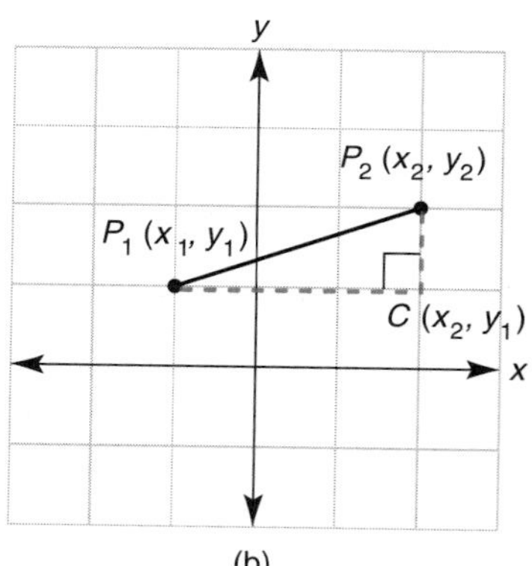

(b)

FIGURE 9.5

In right triangle P_1P_2C in Figure 9.5(b), let $d = P_1P_2$. By the Pythagorean Theorem,

$$d^2 = (x_2 - x_1)^2 + (y_2 - y_1)^2$$

Taking the positive square root for length,

$$d = \sqrt{(x_2 - x_1)^2 + (y_2 - y_1)^2}$$

EXAMPLE 3

In Figure 9.6, find the distance between points $A\ (5, -1)$ and $B\ (-1, 7)$.

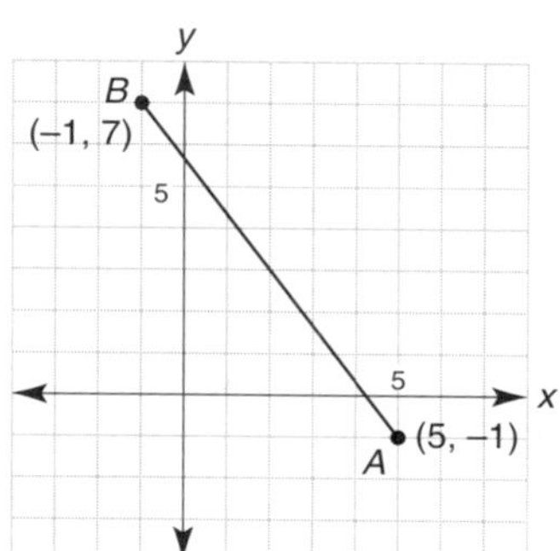

FIGURE 9.6

Solution

Using the Distance Formula and choosing $x_1 = 5$ and $y_1 = -1$ (from point A) and $x_2 = -1$ and $y_2 = 7$ (from point B), we obtain

$$\begin{aligned} d &= \sqrt{(-1-5)^2 + [7-(-1)]^2} \\ &= \sqrt{(-6)^2 + (8)^2} \\ &= \sqrt{100} = 10 \end{aligned}$$

NOTE: If the coordinates of point A were designated as $x_2 = 5$ and $y_2 = -1$ and those of point B were designated as $x_1 = -1$ and $y_1 = 7$, the distance would remain the same.

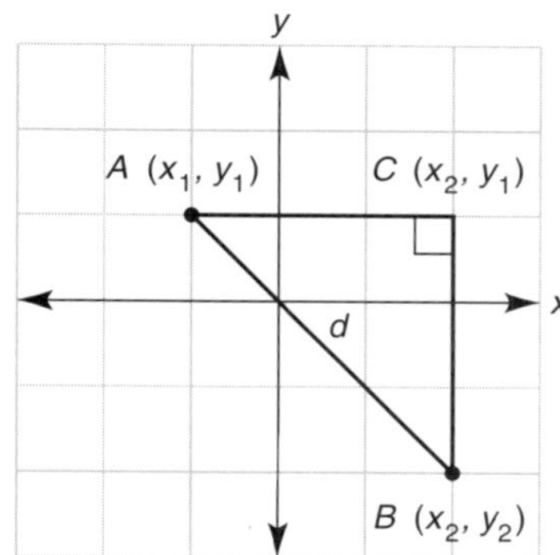

FIGURE 9.7

Looking back at the proof of the Distance Formula, Figure 9.5 shows only one of several possible placements of points. If the placement had been as in Figure 9.7, then $AC = x_2 - x_1$ since $x_2 > x_1$, while $BC = y_1 - y_2$ since $y_1 > y_2$. The Pythagorean Theorem leads to a different result:

$$d^2 = (x_2 - x_1)^2 + (y_1 - y_2)^2$$

but this can be converted to the earlier formula by using the fact that

$$(y_1 - y_2)^2 = (y_2 - y_1)^2$$

This follows from the fact that $(-a)^2 = a^2$ for any real number a.

The following example reminds us of the form of a **linear equation**, an equation whose graph is a straight line. In general, this form is $Ax + By = C$ for constants A, B, and C (where A and B do not both equal 0). We will consider the graphing of linear equations in Section 9.2, and we will determine the equation of a line when its graph is provided (or described) in Section 9.3.

EXAMPLE 4

Find the equation that describes all points (x, y) that are equidistant from $A\,(5, -1)$ and $B\,(-1, 7)$.

Solution

In Chapter 6, we saw that the locus of points equidistant from two fixed points was a line. This line, $\overleftrightarrow{MX}$ in Figure 9.8, is the perpendicular bisector of $\overline{AB}$.

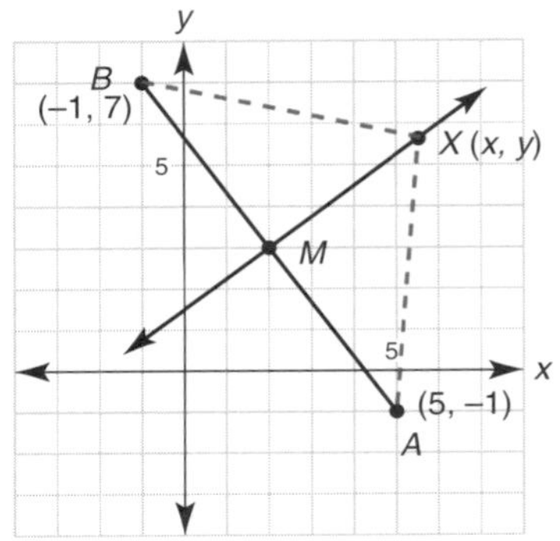

FIGURE 9.8

If X is on the locus, then $AX = BX$. By the Distance Formula, we have

$$\sqrt{(x-5)^2 + [y-(-1)]^2} = \sqrt{[x-(-1)]^2 + (y-7)^2}$$

or

$$(x-5)^2 + (y+1)^2 = (x+1)^2 + (y-7)^2$$

after simplifying and squaring. Then

$$x^2 - 10x + 25 + y^2 + 2y + 1 = x^2 + 2x + 1 + y^2 - 14y + 49$$

Eliminating x^2 and y^2 terms by subtraction leads to the equation

$$-12x + 16y = 24$$

Dividing by 4, the equation of the line becomes

$$-3x + 4y = 6$$

If we divide the equation $-12x + 16y = 24$ by -4, an equivalent solution is

$$3x - 4y = -6$$

In Figure 9.8, point M was the midpoint of $\overline{AB}$. It will be shown in Example 5(a) that M is the point (2, 3).

The Midpoint Formula

A generalized midpoint formula is given in Theorem 9.1.2. The result shows that the coordinates of the midpoint of a line segment are the averages of the coordinates of the endpoints.

To prove the Midpoint Formula, we need to establish two things:

1. $BM + MA = BA$, which establishes that the three points A, M, and B are collinear; and
2. $BM = MA$, which establishes that point M is midway between A and B.

THEOREM 9.1.2: **(Midpoint Formula)** The midpoint M of the line segment joining $A(x_1, y_1)$ and $B(x_2, y_2)$ has coordinates x_M and y_M, where

$$(x_M, y_M) = \left(\frac{x_1 + x_2}{2}, \frac{y_1 + y_2}{2}\right)$$

that is,

$$M = \left(\frac{x_1 + x_2}{2}, \frac{y_1 + y_2}{2}\right)$$

EXAMPLE 5

Use the Midpoint Formula to find the midpoint of the segment joining:

a) $(5, -1)$ and $(-1, 7)$ **b)** (a, b) and (c, d)

Solution

a) Using the Midpoint Formula and setting $x_1 = 5$, $y_1 = -1$, $x_2 = -1$, and $y_2 = 7$, we have

$$M = \left(\frac{5 + (-1)}{2}, \frac{-1 + 7}{2}\right) \qquad \text{or} \qquad M = (2, 3)$$

b) Using the Midpoint Formula and setting $x_1 = a$, $y_1 = b$, $x_2 = c$, and $y_2 = d$, we have

$$M = \left(\frac{a + c}{2}, \frac{b + d}{2}\right)$$

In part (a) of Example 5, it may be helpful to make a sketch of the segment; this will allow you to test whether your solution is reasonable! In part (b), we are generalizing the coordinates in preparation for the analytic geometry proofs that appear later in the chapter. In those sections, we will choose the x and y values of each point in such a way as to be as general as possible. When the Midpoint Formula is used, it is a good idea to select coordinates such as $(2a, 2b)$ for a point so that division by 2 will not introduce fractions.

Proof of the Midpoint Formula (Optional)

For the segment joining P_1 and P_2, we designate the midpoint by M, as shown in Figure 9.9. Let the coordinates of M be designated by (x_M, y_M). Now construct horizontal segments through P_1 and M and vertical segments through M and P_2 to intersect at points A and B, as shown in Figure 9.9(b). Because $\angle A$ and $\angle B$ are right angles, $\angle A \cong \angle B$.

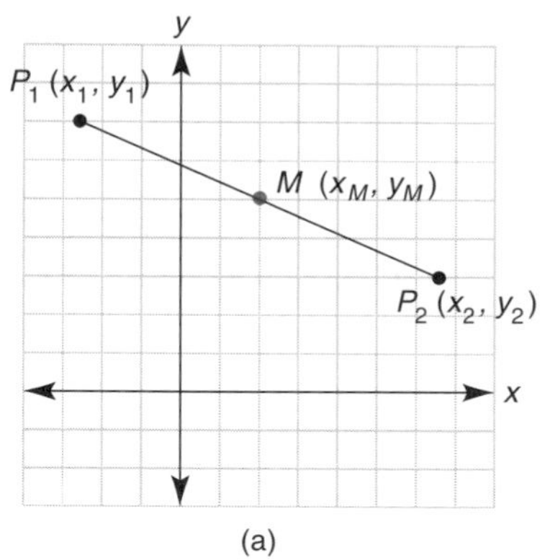

(a)

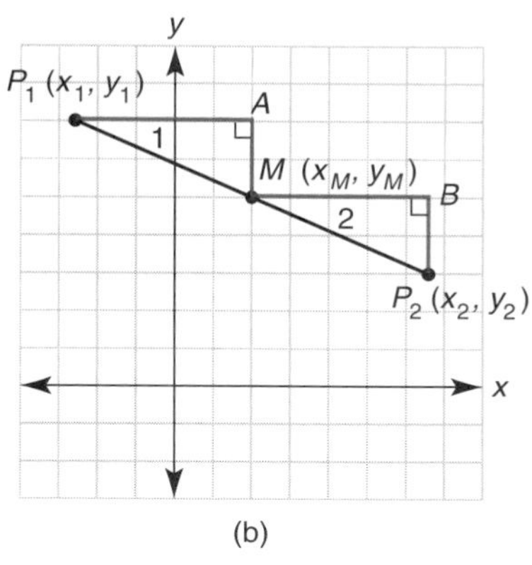

(b)

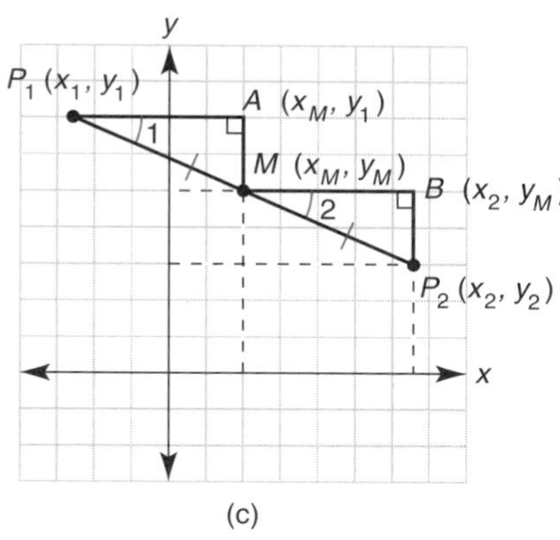

(c)

FIGURE 9.9

Since $\overline{P_1A}$ and $\overline{MB}$ are both horizontal, these segments are parallel. Then $\angle 1 \cong \angle 2$. With $\overline{P_1M} \cong \overline{MP_2}$ by the definition of a midpoint, we have $\triangle P_1AM \cong \triangle MBP_2$ by AAS. Because A is the point (x_M, y_1), we have $P_1A = x_M - x_1$. Likewise, the coordinates of B are (x_2, y_M), so $MB = x_2 - x_M$. Because $\overline{P_1A} \cong \overline{MB}$ by CPCTC, we represent the common length of the segments $\overline{P_1A}$ and $\overline{MB}$ by a. From the first equation, $x_M - x_1 = a$, so $x_M = x_1 + a$. From the second equation,

$x_2 - x_M = a$, so $x_2 = x_M + a$. Substituting $x_1 + a$ for x_M, we have

$$(x_1 + a) + a = x_2$$
$$x_1 + 2a = x_2$$

Then $$2a = x_2 - x_1$$

so $$a = \frac{x_2 - x_1}{2}$$

It follows that

$$\begin{aligned} x_M &= x_1 + a \\ &= x_1 + \frac{x_2 - x_1}{2} \\ &= \frac{2x_1}{2} + \frac{x_2 - x_1}{2} \\ &= \frac{x_1 + x_2}{2} \end{aligned}$$

The y coordinate of the midpoint can be determined in a similar manner, to show that $y_M = \frac{y_1 + y_2}{2}$

Then $$M = \left(\frac{x_1 + x_2}{2}, \frac{y_1 + y_2}{2}\right)$$

9.1 Exercises

1. Plot and then label the points $A(0, -3)$, $B(3, -4)$, $C(5, 6)$, $D(-2, -5)$, and $E(-3, 5)$.
2. Give the coordinates of each point A, B, C, D, and E. Also name the quadrant in which each point lies.

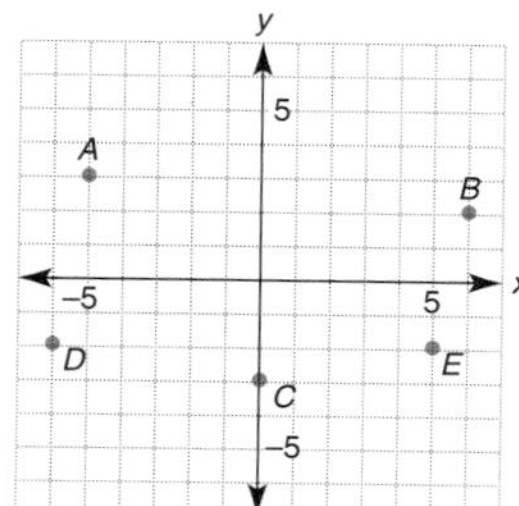

3. Find the distance between each pair of points:
   a) (5, −3) and (5, 1)
   b) (−3, 4) and (5, 4)
   c) (0, 2) and (0, −3)
   d) (−2, 0) and (7, 0)
4. If the distance between (−2, 3) and (−2, a) is 5 units, find all possible values of a.
5. If the distance between (b, 3) and (7, 3) is 3.5 units, find all possible values of b.
6. Find an expression for the distance between (a, b) and (a, c) if $b > c$.
7. Find the distance between each pair of points:
   a) (0, −3) and (4, 0)
   b) (−2, 5) and (4, −3)
   c) (3, 2) and (5, −2)
   d) (a, 0) and (0, b)
8. Find the distance between each pair of points:
   a) (−3, −7) and (2, 5)
   b) (0, 0) and (−2, 6)
   c) ($-a$, $-b$) and (a, b)
   d) ($2a$, $2b$) and ($2c$, $2d$)
9. Find the midpoint of the line segment that joins each pair of points:
   a) (0, −3) and (4, 0)
   b) (−2, 5) and (4, −3)
   c) (3, 2) and (5, −2)
   d) (a, 0) and (0, b)

10. Find the midpoint of the line segment that joins each pair of points:

a) $(-3, -7)$ and $(2, 5)$ c) $(-a, -b)$ and (a, b)
b) $(0, 0)$ and $(-2, 6)$ d) $(2a, 2b)$ and $(2c, 2d)$

11. The origin $(0, 0)$ is the midpoint of $\overline{AB}$. Find the coordinates of B if A is the point:

a) $(3, -4)$ c) $(a, 0)$
b) $(0, 2)$ d) (b, c)

12. The x axis is the perpendicular bisector of $\overline{AB}$. Find the coordinates of B if A is the point:

a) $(3, -4)$ c) $(0, a)$
b) $(0, 2)$ d) (b, c)

13. The y axis is the perpendicular bisector of $\overline{AB}$. Find the coordinates of B if A is the point:

a) $(3, -4)$ c) $(a, 0)$
b) $(2, 0)$ d) (b, c)

14. $M(3, -4)$ is the midpoint of $\overline{AB}$, in which A is the point $(-5, 7)$. Find the coordinates of B.
(**NOTE:** Algebra can be used!)

15. $M(2.1, -5.7)$ is the midpoint of $\overline{AB}$, in which A is the point $(1.7, 2.3)$. Find the coordinates of B.

16. A circle has its center at the point $(-2, 3)$. If one endpoint of the diameter is at $(3, -5)$, find the other endpoint.

17. A rectangle $ABCD$ has three of its vertices at $A(2, -1)$, $B(6, -1)$, and $C(6, 3)$. Find the fourth vertex D and the area of rectangle $ABCD$.

18. A rectangle $MNPQ$ has three of its vertices at $M(0, 0)$, $N(a, 0)$, and $Q(0, b)$. Find the fourth vertex P and the area of the rectangle $MNPQ$.

19. Use the Distance Formula to determine the type of triangle that has these vertices:

a) $A(0, 0)$, $B(4, 0)$, and $C(2, 5)$
b) $D(0, 0)$, $E(4, 0)$, and $F(2, 2\sqrt{3})$
c) $G(-5, 2)$, $H(-2, 6)$, and $K(2, 3)$

20. Use the method of Example 4 to find the equation of the line that describes all points equidistant from the points $(-3, 4)$ and $(3, 2)$.

21. Use the method of Example 4 to find the equation of the line that describes all points equidistant from the points $(1, 2)$ and $(4, 5)$.

22. For coplanar points A, B, and C, suppose that you have used the Distance Formula to show that $AB = 5$, $BC = 10$, and $AC = 15$. What may you conclude regarding points A, B, and C?

23. If two vertices of an equilateral triangle are at $(0, 0)$ and $(2a, 0)$, what is the third vertex?

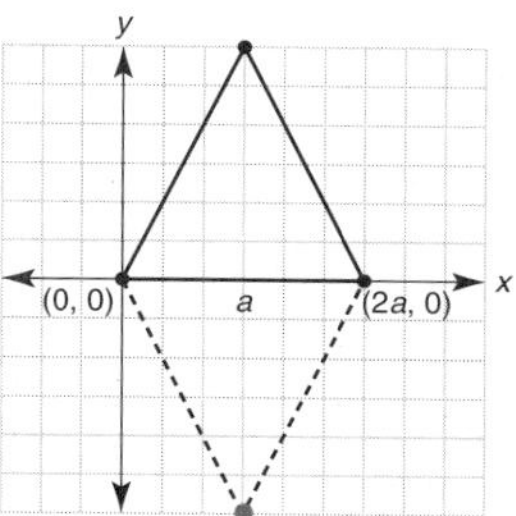

24. The rectangle whose vertices are $A(0,0)$, $B(a,0)$, $C(a, b)$, and $D(0, b)$ is shown. Use the Distance Formula to draw a conclusion concerning the lengths of the diagonals $\overline{AC}$ and $\overline{BD}$.

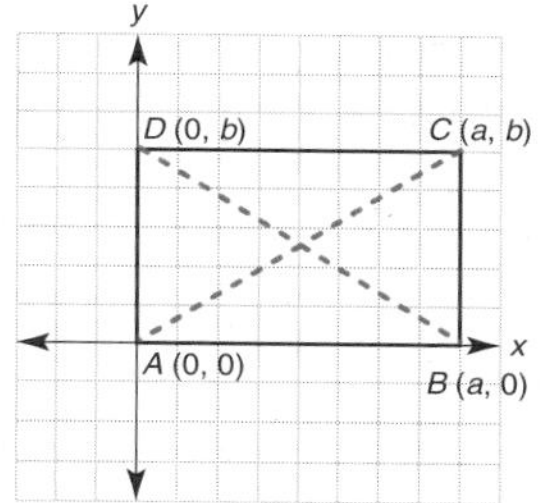

25. ➤ There are two points on the y axis at a distance of 6 units from the point $(3, 1)$. Determine the coordinates of each point.

26. ➤ There are two points on the x axis at a distance of 6 units from the point $(3, 1)$. Determine the coordinates of each point.

27. The triangle that has vertices at $M(-4, 0)$, $N(3, -1)$, and $Q(2, 4)$ has been boxed in as shown. Find the area of $\triangle MNQ$.

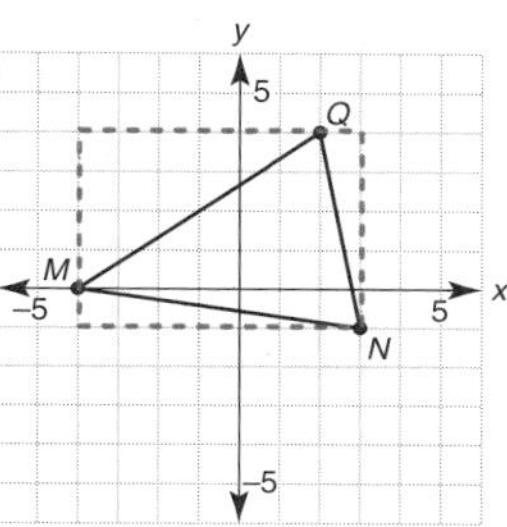

Exercises 27, 28

28. Use the method suggested in Exercise 27 to find the area of $\triangle RST$, with $R(-2, 4)$, $S(-1, -2)$, and $T(6, 5)$.

29. Determine the area of $\triangle ABC$ if $A = (2, 1)$, $B = (5, 3)$, and C is the "reflection" of B across the x axis.

30. Find the area of $\triangle ABC$ in Exercise 29, but assume C is the "reflection" of B across the y axis.

For Exercises 31 to 36, refer to Example 5 and 6 of Section 8.3.

31. Find the exact volume of the right circular cone that results when the triangular region with vertices at (0, 0), (5, 0), and (0, 9) is rotated about the
   a) x axis.
   b) y axis.

32. Find the exact volume of the solid that results when the triangular region with vertices at (0, 0), (6, 0), and (6, 4) is rotated about the
   a) x axis.
   b) y axis.

33. Find the exact volume of the solid formed when the rectangular region with vertices at (0, 0), (6, 0), (6, 4), and (0, 4) is revolved about the
   a) x axis. b) y axis.

34. Find the exact volume of the solid formed when the region bounded in Quadrant I by the lines $x = 9$ and $y = 5$ is revolved about the
   a) x axis. b) y axis.

35. Find the exact lateral area of each solid in Exercise 34.

36. ➢ Find the volume of the solid formed when the triangular region having vertices at (2, 0), (4, 0), and (2, 4) is rotated about the y axis.

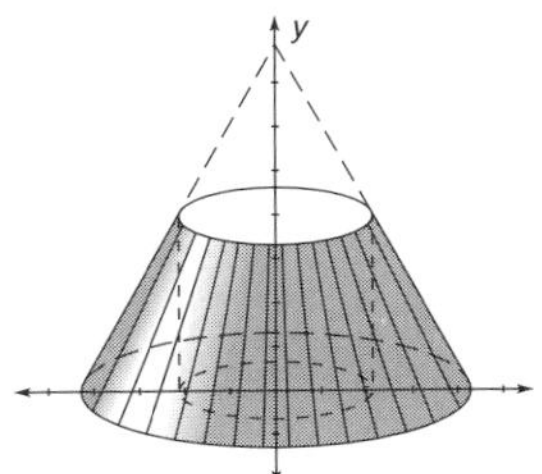

9.2 Graphs of Linear Equations and Slope

In Section 9.1, we were reminded that the general form of the equation of a line is $Ax + By = C$ (where A and B do not both equal 0). Some examples of linear equations are $2x + 3y = 12$, $3x - 4y = 12$, and $3x = -6$; as we shall see, the graph of each of these equations is a line.

The Graph of an Equation

DEFINITION: The **graph of an equation** is the set of all points (x, y) in the rectangular coordinate system whose ordered pairs satisfy the equation.

EXAMPLE 1

Draw the graph of the equation $2x + 3y = 12$.

Solution We begin by completing a table. It is convenient to use one point for which $x = 0$, a second point for which $y = 0$, and a third point as a check:

$$x = 0 \to 2(0) + 3y = 12 \to y = 4$$
$$y = 0 \to 2x + 3(0) = 12 \to x = 6$$
$$x = 3 \to 2(3) + 3y = 12 \to y = 2$$

x	y	(x, y)
0	4	(0, 4)
6	0	(6, 0)
3	2	(3, 2)

Upon plotting the third point, we see that the three points are collinear; the graph of a linear equation must be the straight line shown in Figure 9.10.

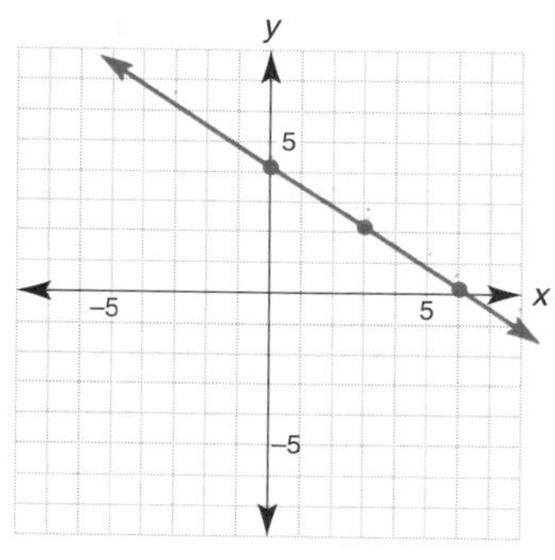

FIGURE 9.10

Because the graph in Example 1 is a locus, every point on the line must also satisfy the given equation. It is easy to see that the point $(-3, 6)$ lies on the line shown in Figure 9.10. Notice that this ordered pair also satisfies the equation $2x + 3y = 12$; that is, $2(-3) + 3(6) = 12$ or $-6 + 18 = 12$.

For the equation in Example 1, the number 6 is known as the ***x* intercept** because (6, 0) is the point at which the graph crosses the x axis; similarly, the number 4 is known as the ***y* intercept.** Most linear equations have two intercepts; these are generally represented by a (the x intercept) and b (the y intercept). To determine the x intercept, let $y = 0$ in the given equation and solve for x. The y intercept can be found similarly by choosing $x = 0$ and solving for y.

EXAMPLE 2

Find the x and y intercepts of the equation $3x - 4y = -12$, and use them to graph the equation.

Solution The x intercept is found when $y = 0$: $3x - 4(0) = -12$, so $x = -4$. The x intercept is -4, so $(-4, 0)$ is on the graph. The y intercept results when $x = 0$: $3(0) - 4y = -12$ so $y = 3$. The y intercept is 3, so (0, 3) is on the graph. Once the points $(-4, 0)$ and (0, 3) are plotted, the graph can be completed by drawing the line through these points (see Figure 9.11).

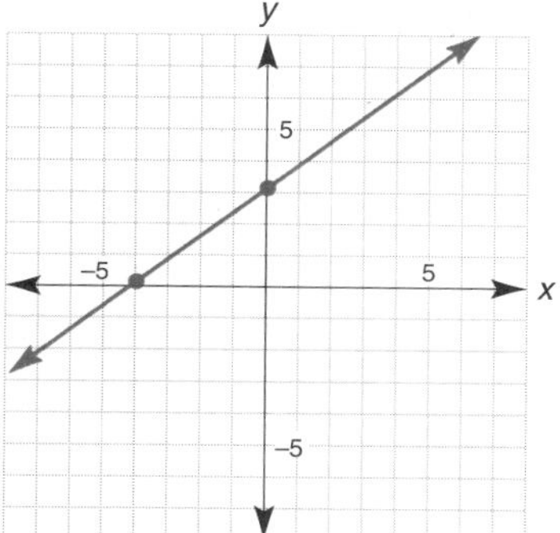

FIGURE 9.11

As we shall see in Example 3, a linear equation may have only one intercept. Is it possible for a linear equation to have no intercepts at all?

EXAMPLE 3

Draw the graphs of the following equations:

a) $x = -2$ **b)** $y = 3$

Solution

First note that each equation is a linear equation:

$$x = -2 \text{ is equivalent to } (1 \cdot x) + (0 \cdot y) = -2$$
$$y = 3 \text{ is equivalent to } (0 \cdot x) + (1 \cdot y) = 3$$

a) The equation $x = -2$ claims that the value of x is -2 regardless of the value of y; this leads to the following table:

x	y	$\rightarrow$	(x, y)
-2	-2	$\rightarrow$	$(-2, -2)$
-2	0	$\rightarrow$	$(-2, 0)$
-2	5	$\rightarrow$	$(-2, 5)$

b) The equation $y = 3$ claims that the value of y is 3 regardless of the value of x; this leads to the following table:

x	y	$\rightarrow$	(x, y)
-4	3	$\rightarrow$	$(-4, 3)$
0	3	$\rightarrow$	$(0, 3)$
5	3	$\rightarrow$	$(5, 3)$

The graphs of the equations are shown in Figure 9.12.

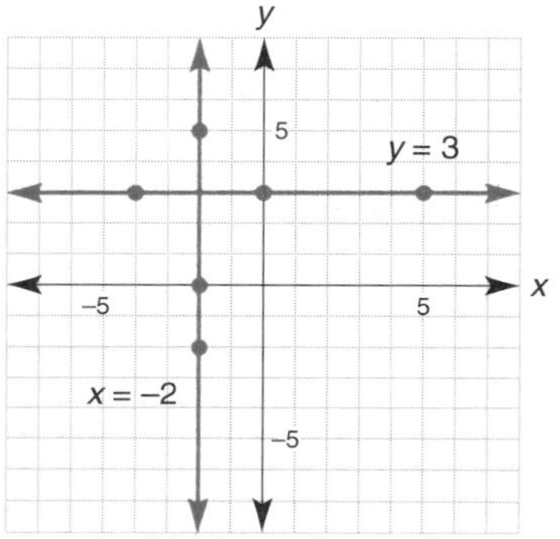

FIGURE 9.12

NOTE: When an equation can be written in the form $x = a$ (for constant a), its graph is the vertical line containing the point $(a, 0)$. When an equation can be written in the form $y = b$ (for constant b), its graph is the horizontal line containing the point $(0, b)$. Most lines are oblique (or slanted); in such cases, it is convenient to describe the slope of the line.

The Slope of a Line

DEFINITION: **(Slope Formula)** The slope of the line that contains the points (x_1, y_1) and (x_2, y_2) is given by

$$m = \frac{y_2 - y_1}{x_2 - x_1} \quad \text{for } x_1 \neq x_2$$

NOTE: When $x_1 = x_2$, the line in question is vertical and we say that the slope is **undefined.**

While the uppercase italic *M* means midpoint, we use the lowercase italic *m* for slope. Other terms used to describe the slope of a line include "pitch" and "grade." A carpenter may say that a roofline has a $\frac{5}{12}$ pitch. [See Figure 9.13(a).] In constructing a stretch of roadway, an engineer may say that there is a grade of $\frac{3}{100}$ or 3 percent. [See Figure 9.13(b).]

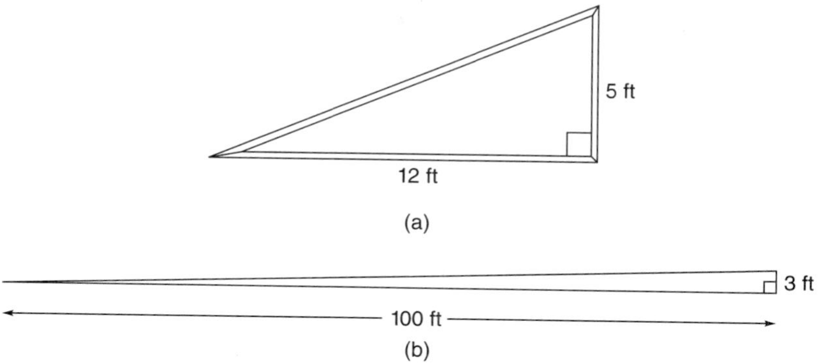

FIGURE 9.13

Whether in geometry, carpentry, or engineering, the term **slope** of line refers to the ratio of the change along the vertical to the change along the horizontal, for any two points on the line in question. A line that "rises" from left to right has a *positive* slope and a line that "falls" from left to right has a *negative* slope. (See Figure 9.14.)

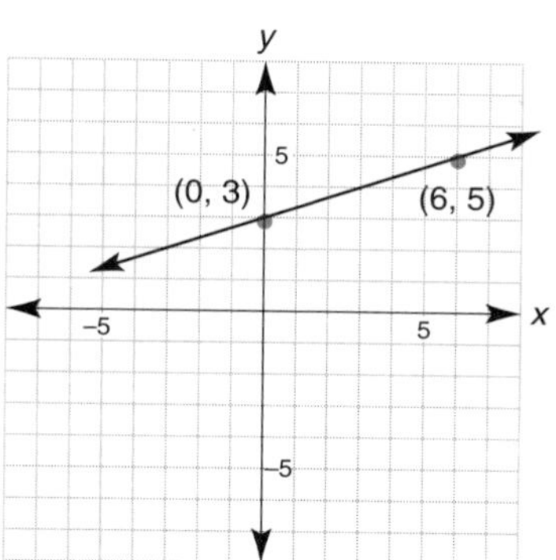

$$m = \frac{y_2 - y_1}{x_2 - x_1}$$

$$m = \frac{5 - 3}{6 - 0} = \frac{2}{6} = \frac{1}{3}$$

(a)

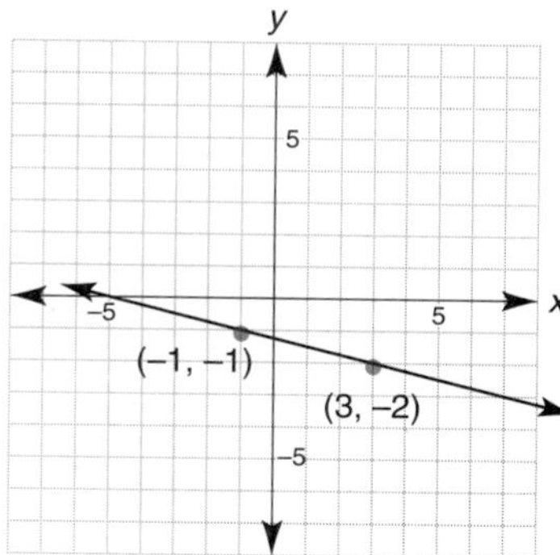

$$m = \frac{y_2 - y_1}{x_2 - x_1}$$

$$m = \frac{-2 - (-1)}{3 - (-1)} = -\frac{1}{4}$$

(b)

FIGURE 9.14

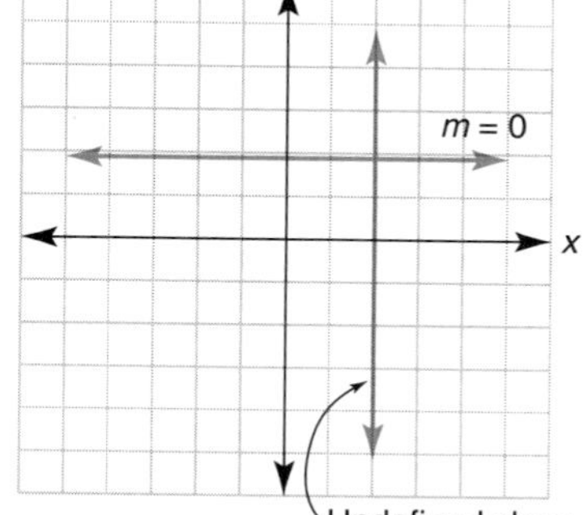

FIGURE 9.15

Any horizontal line has slope 0; any vertical line has an undefined slope. Figure 9.15 shows an example of each of these types of lines.

EXAMPLE 4

Without graphing, find the slope of the line that contains:

a) (2, 2) and (5, 3) **b)** (1, −1) and (1, 3)

Solution

a) Using the Slope Formula and choosing $x_1 = 2$, $y_1 = 2$, $x_2 = 5$, and $y_2 = 3$, we have

$$m = \frac{3 - 2}{5 - 2} = \frac{1}{3}$$

NOTE: If drawn, this line will slant upward from left to right.

b) Let $x_1 = 1$, $y_1 = -1$, $x_2 = 1$, and $y_2 = 3$. Then we calculate

$$m = \frac{3 - (-1)}{1 - 1} = \frac{4}{0}$$

which is undefined.

NOTE: If drawn, this line will be vertical because the x-coordinates are the same.

The slope of a line is unique; that is, the slope does not change when the following changes occur:

1. The order of the two points is reversed.
2. Different points on the line are selected.

The first situation is clear from the fact that $\frac{-a}{-b} = \frac{a}{b}$. The second situation is more difficult but depends on similar triangles.

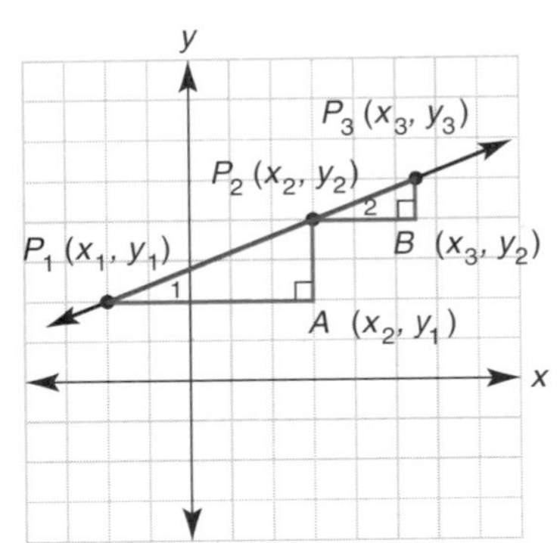

FIGURE 9.16

For an explanation of point 2, consider Figure 9.16, in which points P_1, P_2, and P_3 are collinear. What we wish to show is that the slope of line ℓ is the same whether P_1 and P_2, or P_2 and P_3, are used in the Slope Formula. If horizontal and vertical segments are drawn as shown in Figure 9.16, we can show that triangles P_1P_2A and P_2P_3B are similar.

The similarity follows from the facts that $\angle 1 \cong \angle 2$ (since $\overline{P_1A} \parallel \overline{P_2B}$) and that $\angle A$ and $\angle B$ are right angles. Then $\frac{P_2A}{P_3B} = \frac{P_1A}{P_2B}$ since these are corresponding sides of similar triangles. By interchanging the means, we have $\frac{P_2A}{P_1A} = \frac{P_3B}{P_2B}$. But

$$\frac{P_2A}{P_1A} = \frac{y_2 - y_1}{x_2 - x_1} \quad \text{and} \quad \frac{P_3B}{P_2B} = \frac{y_3 - y_2}{x_3 - x_2}$$

so

$$\frac{y_2 - y_1}{x_2 - x_1} = \frac{y_3 - y_2}{x_3 - x_2}$$

Thus the slope is not changed by having used either pair of points. For that matter, a fourth point could have been shown, in which case the slope of the line would not be changed if P_1 and P_2, or P_3 and P_4, were used; in every case, the slopes agree because of similar triangles.

In summary, if points P_1, P_2, and P_3 are collinear, then the slopes of $\overline{P_1P_2}$, $\overline{P_1P_3}$, and $\overline{P_2P_3}$ are the same. The converse of this statement is also true.

EXAMPLE 5

Are the points $A\ (2, -3)$, $B\ (5, 1)$, and $C\ (-4, -11)$ collinear?

Solution

Let $m_{\overline{AB}}$ and $m_{\overline{BC}}$ represent the slopes of $\overline{AB}$ and $\overline{BC}$, respectively. By the Slope Formula, we have

$$m_{\overline{AB}} = \frac{1 - (-3)}{5 - 2} = \frac{4}{3} \quad \text{and} \quad m_{\overline{BC}} = \frac{-11 - 1}{-4 - 5} = \frac{-12}{-9} = \frac{4}{3}$$

Because $m_{\overline{AB}} = m_{\overline{BC}}$, it follows that A, B, and C are collinear.

As we trace a line from one point to a second point, the Slope Formula tells us that

$$m = \frac{\text{change in } y}{\text{change in } x} \quad \text{or} \quad m = \frac{\text{vertical change}}{\text{horizontal change}}$$

This interpretation is used in Example 6.

EXAMPLE 6

Draw the line through $(-1, 5)$ with slope $m = -\frac{2}{3}$.

Solution

First we plot the point $(-1, 5)$. The slope can be written as $m = \frac{-2}{3}$. Thus we let the change in y from the first to the second point be -2 while the change in x is 3. From the first point $(-1, 5)$, we locate the second point by moving 2 units down and 3 units to the right. The line is then drawn as shown in Figure 9.17.

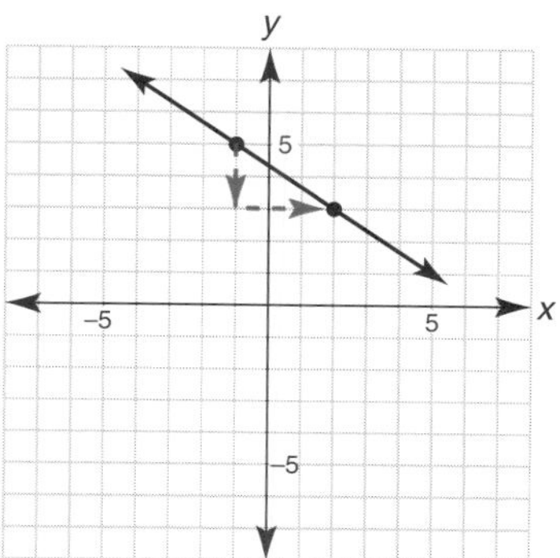

FIGURE 9.17

Two theorems are now stated without proof. However, drawings are provided, and these are to be used in the following exercises. Each proof depends on the fact that similar triangles are created through the use of the auxiliary segments included in the drawings.

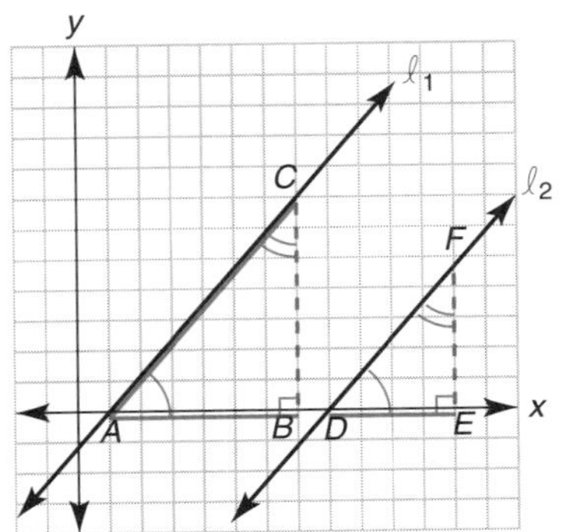

FIGURE 9.18

THEOREM 9.2.1: If two nonvertical lines are parallel, then their slopes are equal.

NOTE: If $\ell_1 \parallel \ell_2$, then $m_1 = m_2$.

In Figure 9.18, notice that $\overline{AC} \parallel \overline{DF}$. Also, $\overline{AB}$ and $\overline{DE}$ are horizontal, while $\overline{BC}$ and $\overline{EF}$ are auxiliary vertical segments. In the proof of Theorem 9.2.1, the goal is to show that $m_{\overline{AC}} = m_{\overline{DF}}$. The converse of Theorem 9.2.1 is also true; that is, if $m_1 = m_2$, then $\ell_1 \parallel \ell_2$.

THEOREM 9.2.2: If two lines (neither horizontal nor vertical) are perpendicular, then the product of their slope is -1.

NOTE: If $\ell_1 \perp \ell_2$, then $m_1 \cdot m_2 = -1$.

In Figure 9.19, auxiliary segments have again been added. The goal in proving Theorem 9.2.2 is to show that $m_{\overline{AC}} \cdot m_{\overline{CE}} = -1$. When the product of the slopes is -1, we say that the slopes are **negative reciprocals.** The converse of Theorem 9.2.2 is also true; if $m_1 \cdot m_2 = -1$, then $\ell_1 \perp \ell_2$.

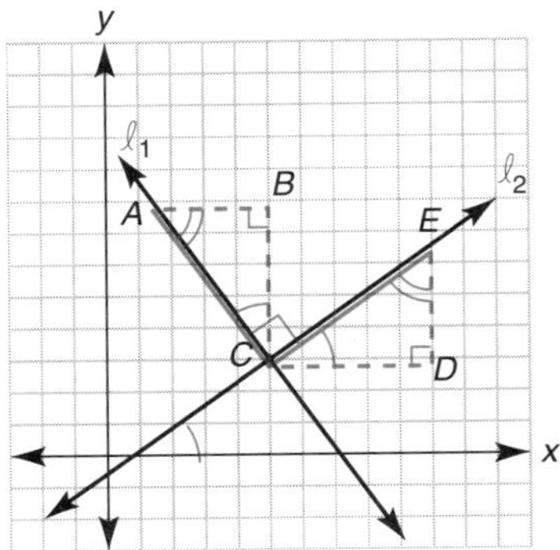

FIGURE 9.19

EXAMPLE 7

Given the points $A(-2, 3)$, $B(2, 1)$, $C(-1, 8)$, and $D(7, 3)$, are $\overline{AB}$ and $\overline{CD}$ parallel, perpendicular, or neither? (See Figure 9.20.)

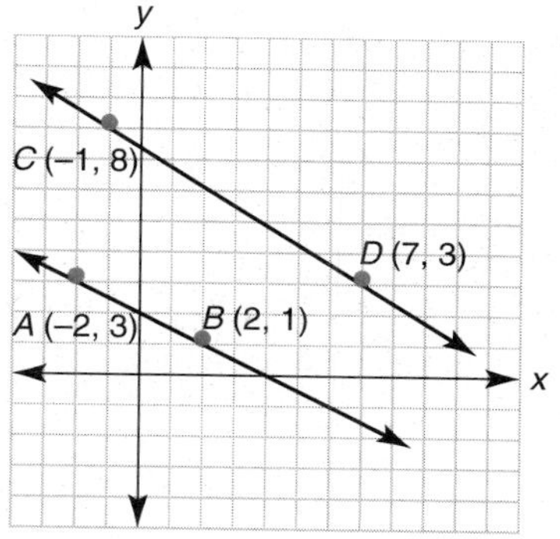

FIGURE 9.20

Solution

$$m_{\overline{AB}} = \frac{1 - 3}{2 - (-2)} = \frac{-2}{4} = -\frac{1}{2}$$
$$m_{\overline{CD}} = \frac{3 - 8}{7 - (-1)} = \frac{-5}{8} \text{ or } -\frac{5}{8}$$

Since $m_{\overline{AB}} \neq m_{\overline{CD}}$, $\overline{AB} \not\parallel \overline{CD}$. The slopes are not negative reciprocals, so $\overline{AB}$ is not perpendicular to $\overline{CD}$. Neither relationship holds for $\overline{AB}$ and $\overline{CD}$.

In Example 7, it would have been worthwhile to sketch the lines described. It is apparent from Figure 9.20 that no special relationship exists between the lines. While a sketch may help show that a relationship does *not* exist, sketching is not a dependable method for showing that lines *are* parallel or perpendicular.

EXAMPLE 8

Are the lines that are the graphs of $2x + 3y = 6$ and $3x - 2y = 12$ parallel, perpendicular, or neither?

Solution

Because $2x + 3y = 6$ contains the points (3, 0) and (0, 2), its slope is $\frac{2 - 0}{0 - 3} = -\frac{2}{3}$. The line $3x - 2y = 12$ contains (0, −6) and (4, 0); thus its slope is equal to $\frac{0 - (-6)}{4 - 0} = \frac{6}{4}$ or $\frac{3}{2}$. Because the product of the slopes is $-\frac{2}{3} \cdot \frac{3}{2}$ or -1, the lines described are perpendicular.

EXAMPLE 9

Determine the value of a for which the line through (2, −3) and (5, a) is perpendicular to the line $3x + 4y = 12$.

Solution

The line $3x + 4y = 12$ contains the points (4, 0) and (0, 3); this line has the slope

$$m = \frac{3 - 0}{0 - 4} = -\frac{3}{4}$$

For the two lines to be perpendicular, the second line must have slope $\frac{4}{3}$. Using the Slope Formula, the second line has the slope

$$\frac{a - (-3)}{5 - 2}$$

so $$\frac{a + 3}{3} = \frac{4}{3}$$

Multiplying by 3, we obtain $a + 3 = 4$, so $a = 1$.

EXAMPLE 10

In Figure 9.21, show that the quadrilateral $ABCD$, with vertices $A\,(0, 0)$, $B\,(a, 0)$, $C\,(a, b)$, and $D\,(0, b)$, is a rectangle.

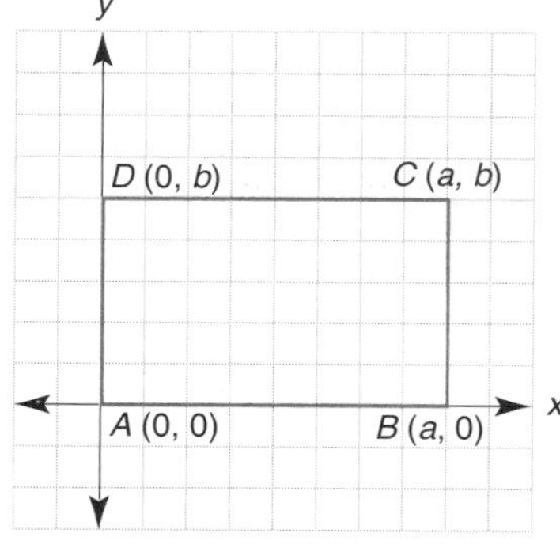

FIGURE 9.21

Solution By applying the Slope Formula, we see that

$$m_{\overline{AB}} = \frac{0 - 0}{a - 0} = 0$$

and $$m_{\overline{DC}} = \frac{b - b}{a - 0} = 0$$

Then $\overline{AB}$ and $\overline{DC}$ are horizontal and therefore are parallel to each other.

For $\overline{DA}$ and $\overline{CB}$, the slopes are undefined since the denominators in the Slope Formula each equal 0. Then $\overline{DA}$ and $\overline{CB}$ are vertical and parallel to each other.

Thus $ABCD$ is a parallelogram. With $\overline{AB}$ being horizontal and $\overline{DA}$ vertical, it follows that $\overline{DA} \perp \overline{AB}$. Therefore $ABCD$ is a rectangle by definition.

9.2 Exercises

In Exercises 1 to 8, draw the graph of each equation. Name any intercepts.

1. $3x + 4y = 12$

2. $3x + 5y = 15$

3. $x - 2y = 5$

4. $x - 3y = 4$

5. $2x + 6 = 0$

6. $3y - 9 = 0$

7. $\frac{1}{2}x + y = 3$
(**HINT:** Multiply each term by 2; then graph the resulting equation.)

8. $\frac{2}{3}x - y = 1$

9. Find the slopes of the lines containing:

a) $(2, -3)$ and $(4, 5)$
b) $(3, -2)$ and $(3, 7)$
c) $(1, -1)$ and $(2, -2)$
d) $(-2.7, 5)$ and $(-1.3, 5)$
e) (a, b) and (c, d)
f) $(a, 0)$ and $(0, b)$

10. Find the slopes of the lines containing

a) $(3, -5)$ and $(-1, 2)$
b) $(-2, -3)$ and $(-5, -7)$

c) $(2\sqrt{2}, -3\sqrt{6})$ and $(3\sqrt{2}, 5\sqrt{6})$
d) $(\sqrt{2}, \sqrt{7})$ and $(\sqrt{2}, \sqrt{3})$
e) $(a, 0)$ and $(a + b, c)$
f) (a, b) and $(-b, -a)$

11. Find x so that $\overline{AB}$ has slope m, where:

a) A is $(2, -3)$, B is $(x, 5)$, and $m = 1$
b) A is $(x, -1)$, B is $(3, 5)$, and $m = -0.5$

12. Find y so that $\overline{CD}$ has slope m, where:

a) C is $(2, -3)$, D is $(4, y)$, and $m = \frac{3}{2}$
b) C is $(-1, -4)$, D is $(3, y)$, and $m = -\frac{2}{3}$

13. Are these points collinear?

a) $A(-2, 5)$, $B(0, 2)$, and $C(4, -4)$
b) $D(-1, -1)$, $E(2, -2)$, and $F(5, -5)$

14. Are these points collinear?

a) $A(-1, -2)$, $B(3, 2)$, and $C(5, 5)$
b) $D(a, c - d)$, $E(b, c)$, and $F(2b - a, c + d)$

15. Parallel lines ℓ_1 and ℓ_2 have slopes m_1 and m_2, respectively. Find m_2 if m_1 equals:

a) $\frac{3}{4}$ b) $-\frac{5}{3}$ c) -2 d) $\frac{a - b}{c}$

16. Perpendicular lines ℓ_1 and ℓ_2 have slopes m_1 and m_2, respectively. Find m_2 if m_1 equals:

a) 5 b) $-\frac{5}{3}$ c) $-\frac{1}{2}$ d) $\frac{a - b}{c}$

In Exercises 17 to 20, state whether the lines are parallel, perpendicular, the same, or none of these.

17. $2x + 3y = 6$ and $2x - 3y = 12$

18. $2x + 3y = 6$ and $4x + 6y = -12$

19. $2x + 3y = 6$ and $3x - 2y = 12$

20. $2x + 3y = 6$ and $4x + 6y = 12$

21. Find x such that the points $A(x, 5)$, $B(2, 3)$, and $C(4, -5)$ are collinear.

22. Find a such that the points $A(1, 3)$, $B(4, 5)$, and $C(a, a)$ are collinear.

23. Find x such that the line through $(2, -3)$ and $(3, 2)$ is perpendicular to the line through $(-2, 4)$ and $(x, -1)$.

24. Find x such that the line through $(2, -3)$ and $(3, 2)$ is parallel to the line through $(-2, 4)$ and $(x, -1)$.

In Exercises 25 to 30, draw the line described.

25. Through $(3, -2)$ and with $m = 2$

26. Through $(-2, -5)$ and with $m = \frac{5}{7}$

27. With y intercept 5 and with $m = -\frac{3}{4}$

28. With x intercept -3 and with $m = 0.25$

29. Through $(-2, 1)$ and parallel to the line $2x - y = 6$

30. Through $(-2, 1)$ and perpendicular to the line that has intercepts $a = -2$ and $b = 3$

31. Use slopes to decide whether the triangle with vertices at $(6, 5)$, $(-3, 0)$, and $(4, -2)$ is a right triangle.

32. If $A(2, 2)$, $B(7, 3)$, and $C(4, x)$ are the vertices of a right triangle with right angle C, find the value of x.

33. ➢ If $(2, 3)$, $(5, -2)$, and $(7, 2)$ are three vertices (not necessarily consecutive) of a parallelogram, find the possible locations of the fourth vertex.

34. Three vertices of rectangle $ABCD$ are $A(-5, 1)$, $B(-2, -3)$, and $C(6, y)$. Find the value of y and also the fourth vertex.

35. Show that quadrilateral $RSTV$ is an isosceles trapezoid.

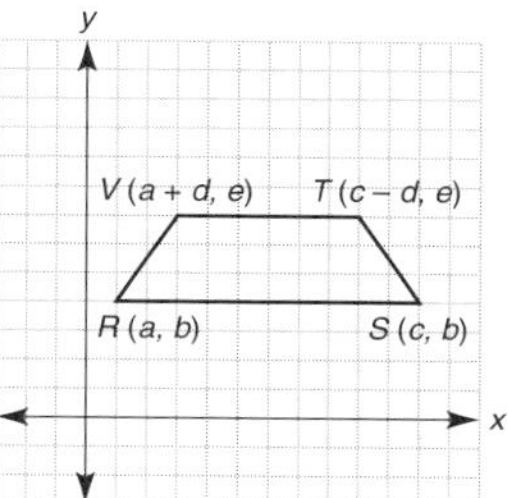

36. Show that quadrilateral $ABCD$ is a parallelogram.

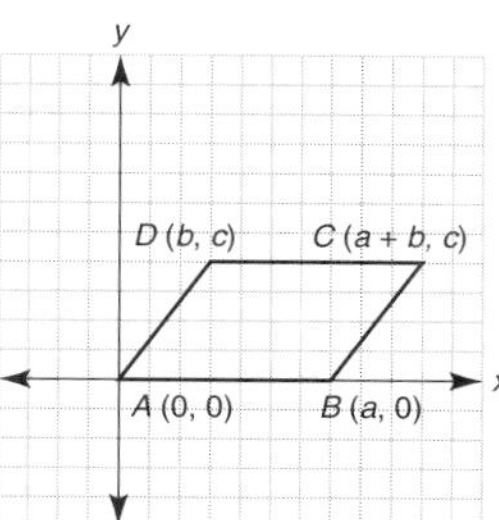

37. Quadrilateral $EFGH$ has the vertices $E(0, 0)$, $F(a, 0)$, $G(a + b, c)$, and $H(2b, 2c)$. Verify that $EFGH$ is a trapezoid by showing that the slopes of two sides are equal.

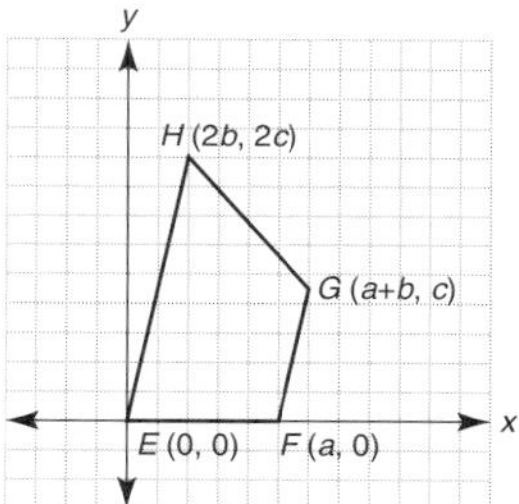

38. Find an equation involving a, b, c, d, and e if $\overleftrightarrow{AC} \perp \overleftrightarrow{BC}$.
(**HINT:** Use slopes.)

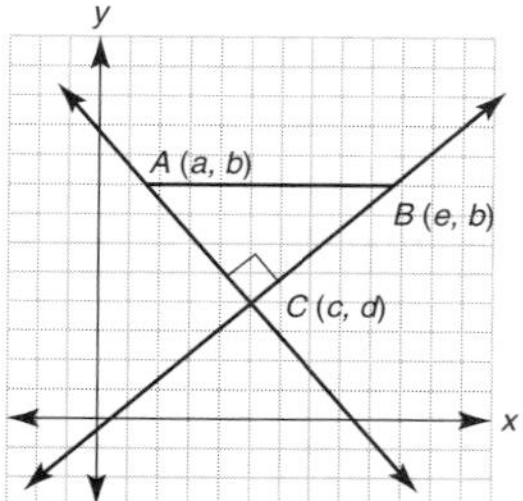

39. Prove that if two nonvertical lines are parallel, then their slopes are equal.
(**HINT:** See Figure 9.18.)

40. ➢ Prove that if two lines (neither horizontal nor vertical) are perpendicular, then the product of their slopes is -1.
(**HINT:** See Figure 9.19. You need to show and use the fact that $\triangle ABC \sim \triangle EDC$.)

9.3 Equations of Lines

The graph of $2x + 3y = 6$ and the graph of $4x + 6y = 12$ are both lines. Because the intercepts are $a = 3$ and $b = 2$ for both lines, the lines are the same and are described as **coincident lines.** In general, the graphs of $Ax + By = C$ and $kAx + kBy = kC$ (in which k is a nonzero multiplier of the first equation) are the same. For that reason, we may replace the equation $\frac{1}{2}x + y = 3$ with the equation $x + 2y = 6$ when graphing. (See Exercise 7, Section 9.2.) Just as there is a graph for every linear equation, there is also an equation for every line drawn in the rectangular coordinate system. Because $4x + 6y = 12$ and $2x + 3y = 6$ represent the same line, we will designate the equation with reduced coefficients, namely $2x + 3y = 6$, as the one we seek.

EXAMPLE 1

Write an equation whose graph is the same as that of $4x - 12y = 60$.

Solution

Dividing by 4, we have $x - 3y = 15$. If we divide by -4, the result is $-x + 3y = -15$. Either is correct; these equations are described as "equivalent equations" because their graphs are identical.

> Many exercises in this section ask for an equation of a line in the form $Ax + By = C$. Your answer is correct if either:
>
> 1. It matches perfectly the solution equation provided.
> 2. It is a nonzero multiple of the solution equation.
>
> **NOTE:** In general, an answer expressed in integers is preferable to one that uses fractions for A, B, and C. When integer choices are possible for A, B, and C, the smallest choices are generally used; thus, $12x + 16y = 36$ should be replaced by $3x + 4y = 9$.

Slope-Intercept Form of a Line

We now turn our attention to methods for finding the equation of a line. In the first technique, the equation can be found if the slope and the y intercept of the line are known.

THEOREM 9.3.1: (Slope-Intercept Form of a Line) The line whose slope is m and whose y intercept is b has the equation $y = mx + b$.

Proof Consider the line whose slope is m (see Figure 9.22). Using the Slope Formula

$$m = \frac{y_2 - y_1}{x_2 - x_1},$$

we designate (x, y) as P_2 and $(0, b)$ as P_1. Then

$$m = \frac{y - b}{x - 0} \quad \text{or} \quad m = \frac{y - b}{x}$$

Multiplying by x, we have $mx = y - b$. Then $mx + b = y$ or $y = mx + b$.

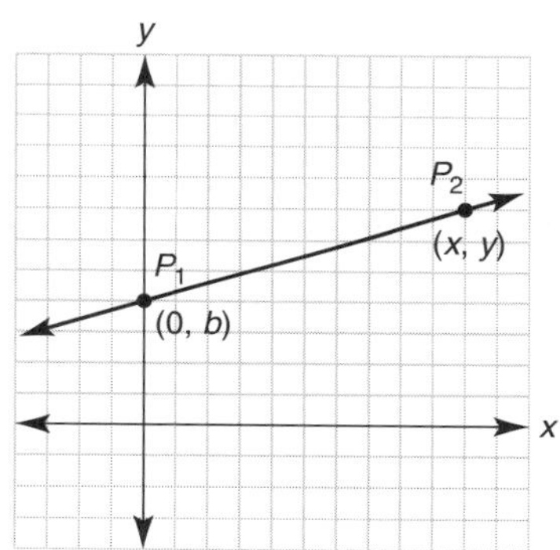

FIGURE 9.22

EXAMPLE 2

Find the general equation $Ax + By = C$ for the line with slope $m = -\frac{2}{3}$ and y intercept -2.

Solution With $y = mx + b$, we have

$$y = -\frac{2}{3}x - 2$$

Multiplying by 3, we obtain

$$3y = -2x - 6 \quad \text{or} \quad 2x + 3y = -6$$

NOTE: An equivalent and correct solution is $-2x - 3y = 6$.

It is often easier to graph an equation if it is in the form $y = mx + b$. Sometimes the equation can quickly be changed to this form, at which time we know that its graph is a line with slope m and containing $(0, b)$.

EXAMPLE 3

Draw the graph of $\frac{1}{2}x + y = 3$.

Solution

Solving for y, we have $y = -\frac{1}{2}x + 3$. Then $m = -\frac{1}{2}$, and the y intercept is 3. We first plot the point (0, 3). Because $m = -\frac{1}{2}$ or $\frac{-1}{2}$, the vertical change -1 corresponds to a horizontal change of $+2$. Thus the second point is located 1 unit down from and 2 units to the right of the first point. The line is drawn in Figure 9.23.

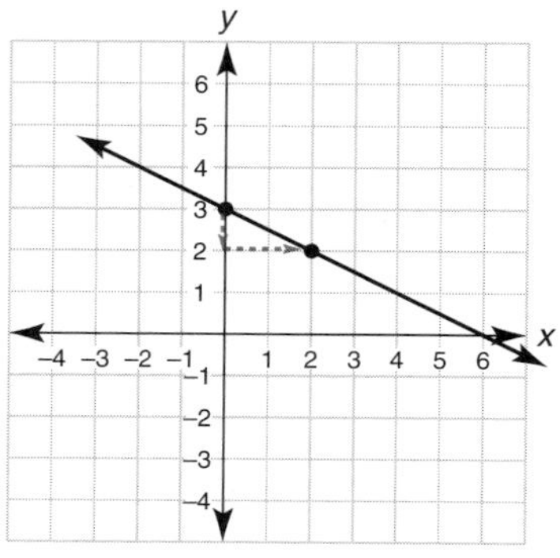

FIGURE 9.23

Point-Slope Form of a Line

If a point other than the y intercept of the line is known, we generally do not use Slope-Intercept Form to find the equation of the line. Instead, the Point-Slope Form of the equation of a line is used. This form is also used when the coordinates of two points of the line are known; in that case, the value of m is found by the Slope Formula.

THEOREM 9.3.2: **(Point-Slope Form of a Line)** The line with slope m and containing the point (x_1, y_1) has the equation

$$y - y_1 = m(x - x_1)$$

Proof

Let P_1 be the given point (x_1, y_1) on the line, and let P_2 be (x, y), which represents any other point on the line. (See Figure 9.24.) Using the Slope Formula, we have

$$m = \frac{y - y_1}{x - x_1}$$

Multiplying the equation by $(x - x_1)$, we get

$$m(x - x_1) = y - y_1$$

or

$$y - y_1 = m(x - x_1)$$

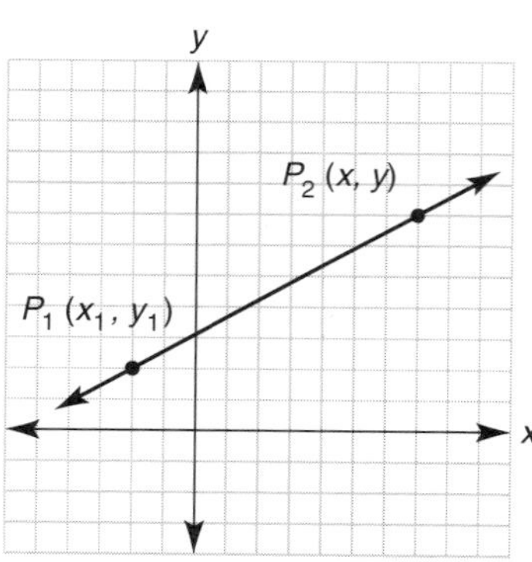

FIGURE 9.24

EXAMPLE 4

Find the general equation, $Ax + By = C$, for the line with $m = 2$ and containing $(-1, 3)$.

Solution

Applying the Point-Slope Form, the line in Figure 9.25 has the equation

$$y - 3 = 2[x - (-1)]$$
$$y - 3 = 2(x + 1)$$
$$y - 3 = 2x + 2$$
$$-2x + y = 5$$

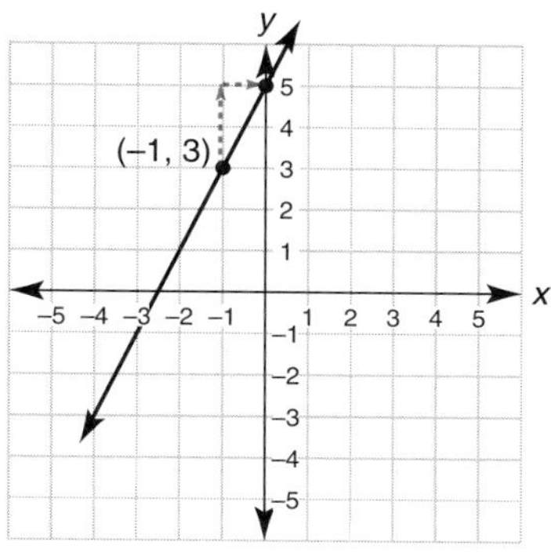

FIGURE 9.25

NOTE: An equivalent answer for Example 4 is the equation $2x - y = -5$.

EXAMPLE 5

Find an equation for the line containing the points $(-1, 2)$ and $(4, 1)$.

Solution

To use the Point-Slope Form, we need to know the slope of the line (see Figure 9.26). Choosing P_1 $(-1, 2)$ and P_2 $(4, 1)$, we have

$$m = \frac{1 - 2}{4 - (-1)} = \frac{-1}{5} = -\frac{1}{5}$$

Therefore $\quad y - 2 = -\frac{1}{5}[x - (-1)]$

$$y - 2 = -\frac{1}{5}x - \frac{1}{5}$$

Multiplying the equation by 5, we get

$$5y - 10 = -1x - 1 \text{ so that } x + 5y = 9$$

FIGURE 9.26

NOTE: Other correct forms of the answer are $-x - 5y = -9$ and $y = -\frac{1}{5}x + \frac{9}{5}$. In any correct form, the given points must satisfy the equation.

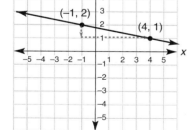

We now use the Point-Slope Form of the equation of a line as a means of drawing the graph of a linear equation. Note that the form of the equation in Example 6 makes it easy for us to recognize the slope of the line and a point on that line.

EXAMPLE 6

Draw the graph of the equation $y + 4 = 2(x - 1)$.

Solution

Comparing this equation to the Point-Slope Form, we can write it as

$$y - (-4) = 2(x - 1)$$

Then it follows that $y_1 = -4$ and $x_1 = 1$, while $m = 2$. Therefore the line contains $(1, -4)$, which we plot first.

Because $m = 2 = \frac{2}{1}$, the vertical change of 2 corresponds to a horizontal change of 1. Starting at $(1, -4)$, a second point is plotted by moving 2 units up and 1 unit to the right. The line is then drawn through the two points as shown in Figure 9.27.

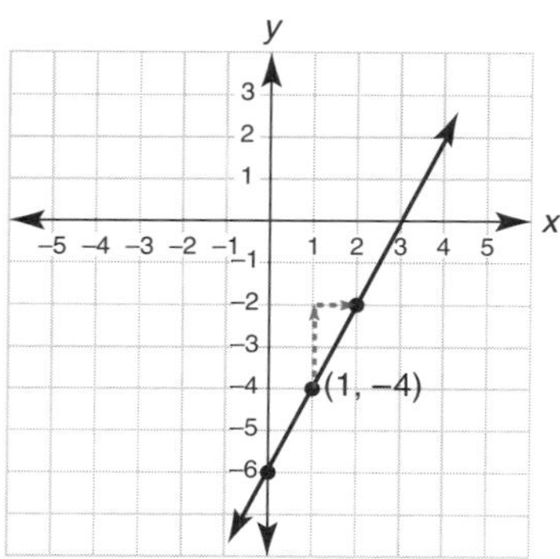

FIGURE 9.27

Solving Systems of Equations

In earlier chapters, we solved systems of equations such as

$$\begin{aligned} x + 2y &= 6 \\ 2x - y &= 7 \end{aligned}$$

by using the Addition Property or the Subtraction Property of Equality. We review the method in Example 7.

EXAMPLE 7

Solve the following system by using algebra:

$$\begin{cases} x + 2y = 6 \\ 2x - y = 7 \end{cases}$$

Solution

Multiplying the second equation by 2, the system becomes

$$\begin{cases} x + 2y = 6 \\ 4x - 2y = 14 \end{cases}$$

Adding these equations yields $5x = 20$ so that $x = 4$. Substituting $x = 4$ into the first equation, we get $4 + 2y = 6$, so $2y = 2$. Then $y = 1$. The solution is the ordered pair $(4, 1)$.

Another method for solving a system of equations is geometric and requires graphing. Solving by graphing amounts to searching for the point of intersection of the linear graphs. That point is the ordered pair that constitutes the common solution (when one exists) for the two equations. Notice that Example 8 repeats the system of Example 7.

EXAMPLE 8

Solve the following system by graphing:

$$\begin{cases} x + 2y = 6 \\ 2x - y = 7 \end{cases}$$

Solution Each equation is changed to the form $y = mx + b$ so that the slope and the y intercept are used in graphing:

$$x + 2y = 6 \rightarrow 2y = -1x + 6 \rightarrow y = -\frac{1}{2}x + 3$$
$$2x - y = 7 \rightarrow -y = -2x + 7 \rightarrow y = 2x - 7$$

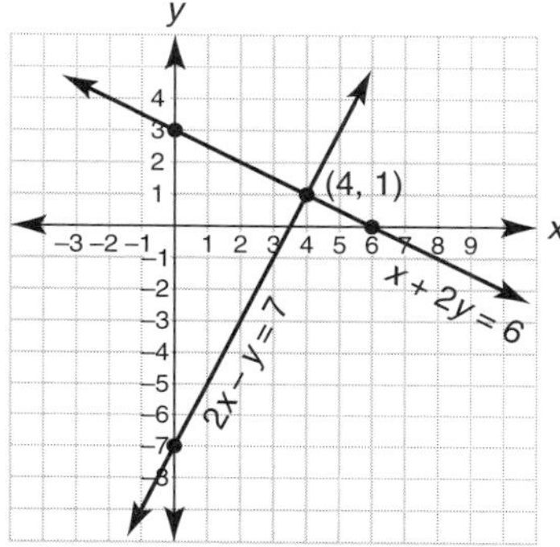

FIGURE 9.28

The graph of $y = -\frac{1}{2}x + 3$ is a line with y intercept 3 and slope $m = -\frac{1}{2}$. The graph of $y = 2x - 7$ is a line with y intercept -7 and slope $m = 2$.

The graphs are drawn in Figure 9.28 in the same coordinate system. The point of intersection (4, 1) is the common solution for each of the given equations and thus is the solution of the system.

NOTE: To check the result of Examples 7 and 8, we show that (4, 1) satisfies both of the given equations:

$$x + 2y = 6 \rightarrow 4 + 2(1) = 6 \text{ is } \textit{true}$$
$$2x - y = 7 \rightarrow 2(4) - 1 = 7 \text{ is } \textit{true}$$

The check of the solution requires that both statements resulting from substitution into the given equations be true. If either or both statements are false, we do not have the solution of the system.

The graphing method of solving a system has both advantages and disadvantages. We begin with the disadvantages:

1. Graphing the two equations can be very time-consuming.
2. Sometimes it is difficult to read the solution from the graphs, particularly when solutions involve coordinates with fractions or decimals.

Advantages of the method of solving a system of equations by graphing include the following:

1. It is easy to understand why a system such as

$$\begin{cases} x + 2y = 6 \\ 2x - y = 7 \end{cases} \quad \text{can be replaced by} \quad \begin{cases} x + 2y = 6 \\ 4x - 2y = 14 \end{cases}$$

when solving by addition or subtraction. The graphs of $2x - y = 7$ and $4x - 2y = 14$ are the same line.

2. It is easy to understand why a system such as

$$\begin{cases} x + 2y = 6 \\ 2x + 4y = -4 \end{cases}$$

has no solution. In Figure 9.29, the graphs are parallel lines.

3. If the graph of one (or both) of the equations is not a straight line, the method of graphing can still be used.

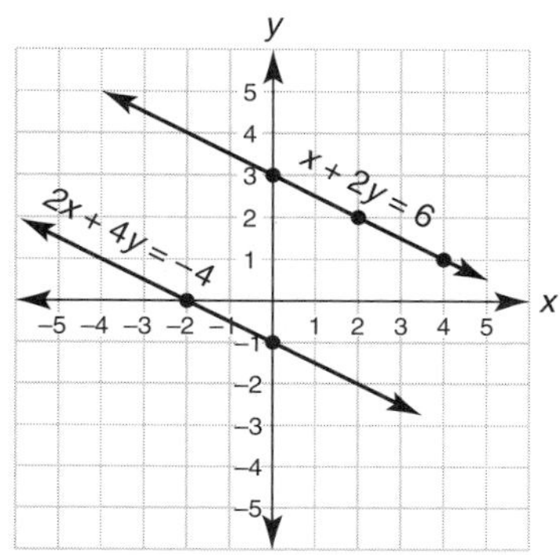

FIGURE 9.29

In Section 9.6, we consider the equation of a circle and its graph. A circle and a line could intersect once (the line is tangent to the circle), twice (the line is a secant of the circle), or not intersect at all!

9.3 Exercises

In Exercises 1 to 4, use division to write an equation of the form $Ax + By = C$ that is equivalent to the one provided. Then write the given equation in the form $y = mx + b$.

1. $8x + 16y = 48$
2. $15x - 35y = 105$
3. $-6x + 18y = -240$
4. $27x - 36y = 108$

In Exercises 5 to 10, draw the graph of each equation by using the methods of Examples 3 and 6.

5. $y = 2x - 3$
6. $y = -2x + 5$
7. $\frac{2}{5}x + y = 6$
8. $3x - 2y = 12$
9. $y - 3 = \frac{3}{4}(x - 1)$
10. $y + 3 = -2(x + 4)$

In Exercises 11 to 24, find the equation of the line described. Leave the solution in the form $Ax + By = C$.

11. The line has slope $m = -\frac{2}{3}$ and contains $(0, 5)$.
12. The line has slope $m = -3$ and contains $(0, -2)$.
13. The line contains $(2, 4)$ and $(0, 6)$.
14. The line contains $(-2, 5)$ and $(2, -1)$.
15. The line contains $(0, -1)$ and $(3, 1)$.
16. The line contains $(-2, 0)$ and $(4, 3)$.
17. The line has intercepts $a = 2$ and $b = -2$.
18. The line has intercepts $a = -3$ and $b = 5$.
19. The line goes through $(-1,5)$ and is parallel to the line $5x + 2y = 10$.
20. The line goes through $(0,3)$ and is parallel to the line $3x + y = 7$.
21. The line goes through $(0,-4)$ and is perpendicular to the line $y = \frac{3}{4}x - 5$.
22. The line goes through $(2,-3)$ and is perpendicular to the line $2x - 3y = 6$.
23. The line is the perpendicular bisector of the line segment that joins $(3, 5)$ and $(5,-1)$.
24. The line is the perpendicular bisector of the line segment that joins $(-4, 5)$ and $(1, 1)$.

In Exercises 25 to 30, use graphing to find the point of intersection of the two lines.

25. $y = \frac{1}{2}x - 3$ and $y = \frac{1}{3}x - 2$
26. $y = 2x + 3$ and $y = 3x$
27. $2x + y = 6$ and $3x - y = 19$
28. $\frac{1}{2}x + y = -3$ and $\frac{3}{4}x - y = 8$
29. $4x + 3y = 18$ and $x - 2y = 10$
30. $2x + 3y = 3$ and $3x - 2y = 24$

In Exercises 31 to 36, use algebra to find the point of intersection of the two lines whose equations are provided.

31. $2x + y = 8$ and $3x - y = 7$
32. $2x + 3y = 7$ and $x + 3y = 2$
33. $2x + y = 11$ and $3x + 2y = 16$
34. $x + y = 1$ and $4x - 2y = 1$
35. $2x + 3y = 4$ and $3x - 4y = 23$
36. $5x - 2y = -13$ and $3x + 5y = 17$
37. ➢ For constants a and b, the graphs of $ax + by = 7$ and $ax - by = 13$ intersect at $(5,-1)$. Using algebra, find the values of a and b.
38. ➢ For constants a and b, the graphs of $ax + by = 13$ and $ax - by = 7$ intersect at $(2,-1)$. Using algebra, find the values of a and b.
39. ➢ The graphs of the lines $2x - 3y = 12$, $x + 2y = -1$, and $kx + y = 13$ all contain the same point. Find that point and also the value of k.
40. ➢ The graphs of the lines $3x + 4y = 2$, $5x - y = 11$, and $x + ky = -2$ all contain the same point. Find that point and also the value of k.

For Exercises 41 and 42, assume that $A \neq 0$ and $B \neq 0$.

41. *Prove:* For $C \neq D$, the lines $Ax + By = C$ and $Ax + By = D$ are parallel.
42. *Prove:* The lines $Ax + By = C$ and $Bx - Ay = D$ are perpendicular.

9.4 Preparing to Do Analytic Proofs

In this section, our goal is to lay the groundwork for analytic proofs of geometric theorems. An analytic proof requires the use of the coordinate system and the application of the formulas found in earlier sections of this chapter. Because of the need for these formulas, a summary follows. Be sure that you have these formulas memorized and know when and how to use them.

FORMULAS OF ANALYTIC GEOMETRY

Distance	$d = \sqrt{(x_2 - x_1)^2 + (y_2 - y_1)^2}$
Midpoint	$M = \left(\frac{x_1 + x_2}{2}, \frac{y_1 + y_2}{2}\right)$
Slope	$m = \frac{y_2 - y_1}{x_2 - x_1}$
Equations of lines	$y = mx + b$ $y - y_1 = m(x - x_1)$ $Ax + By = C$
Special relationships for lines	$\ell_1 \parallel \ell_2 \leftrightarrow m_1 = m_2$ $\ell_1 \perp \ell_2 \leftrightarrow m_1 \cdot m_2 = -1$

NOTE: Neither ℓ_1 nor ℓ_2 is a vertical line in the preceding claims.

To see how the preceding list might be used in this and the next section, consider the following examples.

EXAMPLE 1

Identify the graph of each equation:

a) $y = 2x - 3$ **b)** $2x + 3y = 6$

Solution

a) $y = 2x - 3$ has the form $y = mx + b$. It is the line whose slope is $m = 2$ and whose y intercept is -3.

b) $2x + 3y = 6$ can be written in the form $y = mx + b$. If $2x + 3y = 6$, then $3y = -2x + 6$, so $y = -\frac{2}{3}x + 2$. The graph is a line with $m = -\frac{2}{3}$ and $b = 2$.

EXAMPLE 2

Suppose that you are to prove the following relationships:

a) Two lines are parallel. **b)** Two lines are perpendicular.
c) Two line segments are congruent.

Which formula(s) would you need to use? How would you complete your proof?

Solution

a) Use the Slope Formula first, to find the slope of each line. Then show that the slopes are equal.

b) Use the Slope Formula first, to find the slope of each line. Then show that $m_1 \cdot m_2 = -1$.

c) Use the Distance Formula first, to find the length of each segment. Then show that the resulting lengths are equal.

The following example has a proof that is subtle. A drawing is provided to help you understand the concept.

EXAMPLE 3

How can the Midpoint Formula be used to show that two line segments bisect each other?

Solution

If $\overline{AB}$ bisects $\overline{CD}$, and conversely, then M is the common midpoint of the two segments. (See Figure 9.30.) The Midpoint Formula is used to find the midpoint of each segment, and the results are then shown to be the same point. This establishes that each segment has been bisected by a point that is on the other segment.

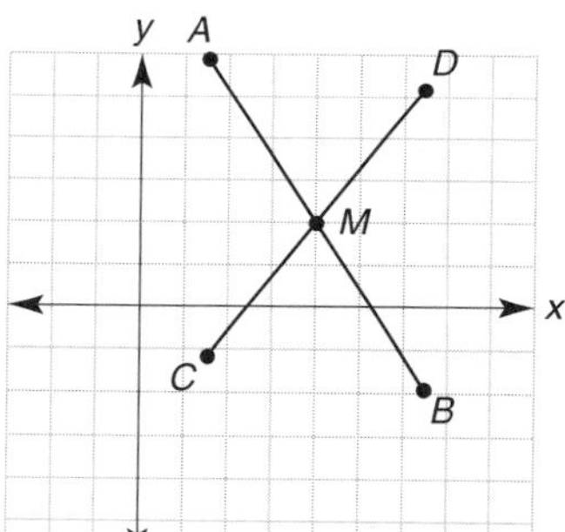

FIGURE 9.30

EXAMPLE 4

Suppose that line ℓ_1 has slope $\frac{c}{d}$. Use this fact to identify the slopes of the following lines:

a) ℓ_2 if $\ell_1 \parallel \ell_2$ **b)** ℓ_3 if $\ell_1 \perp \ell_3$

Solution

a) $m_2 = \frac{c}{d}$ because $m_1 = m_2$ when $\ell_1 \parallel \ell_2$.

b) $m_3 = -\frac{d}{c}$ because $m_1 \cdot m_3 = -1$ when $\ell_1 \perp \ell_3$.

EXAMPLE 5

What may you conclude if you know that the point (p, q) lies on the line $y = mx + b$?

Solution Since (p, q) is on the line, it is also a solution for the equation $y = mx + b$. Therefore, $q = mp + b$.

To construct proofs of geometric theorems by analytic methods, we must use the hypothesis to determine the drawing. Unlike the drawings in Chapters 1–8, the figure must be placed in the coordinate system. Making the drawing requires careful placement of the figure and proper naming of the vertices, using coordinates of the rectangular system. The following guidelines should prove helpful in positioning the figure and in naming its vertices.

MAKING THE DRAWING FOR AN ANALYTIC PROOF

Some considerations for preparing the drawing:

1. Coordinates of the vertices must be general; for instance, you may use (a, b) as a vertex, but do *not* use (2, 3).
2. Make the drawing satisfy the hypothesis without providing any additional qualities; if the theorem describes a rectangle, draw and label a rectangle but *not* a square.
3. For simplicity in your calculations, drop the figure into the rectangular coordinate system in such a manner that
   - **a)** As many 0 coordinates are used as possible.
   - **b)** The remaining coordinates represent positive numbers, due to your positioning of the figure in Quadrant I.

NOTE: In some cases, it is convenient to place a figure so that it has symmetry with respect to the y axis, in which case some negative coordinates are present.

4. When possible, use horizontal and vertical segments since you know their parallel and perpendicular relationships.
5. Use as few variable names in the coordinates as possible.

Now we consider Example 6, which clarifies the preceding list of suggestions. As you observe the drawing in each part of the example, imagine that $\triangle ABC$ has been cut out of a piece of cardboard and dropped into the coordinate system in the position indicated. Since we have freedom of placement, we want to choose the positioning that allows us the simplest possible solution.

EXAMPLE 6

Suppose that you are asked to make a drawing for the following theorem, which is to be proved analytically: “The midpoint of the hypotenuse of a right triangle is equidistant from the three vertices of the triangle.” Explain why the placement of right $\triangle ABC$ in each part of Figure 9.31 on the next page is poor.

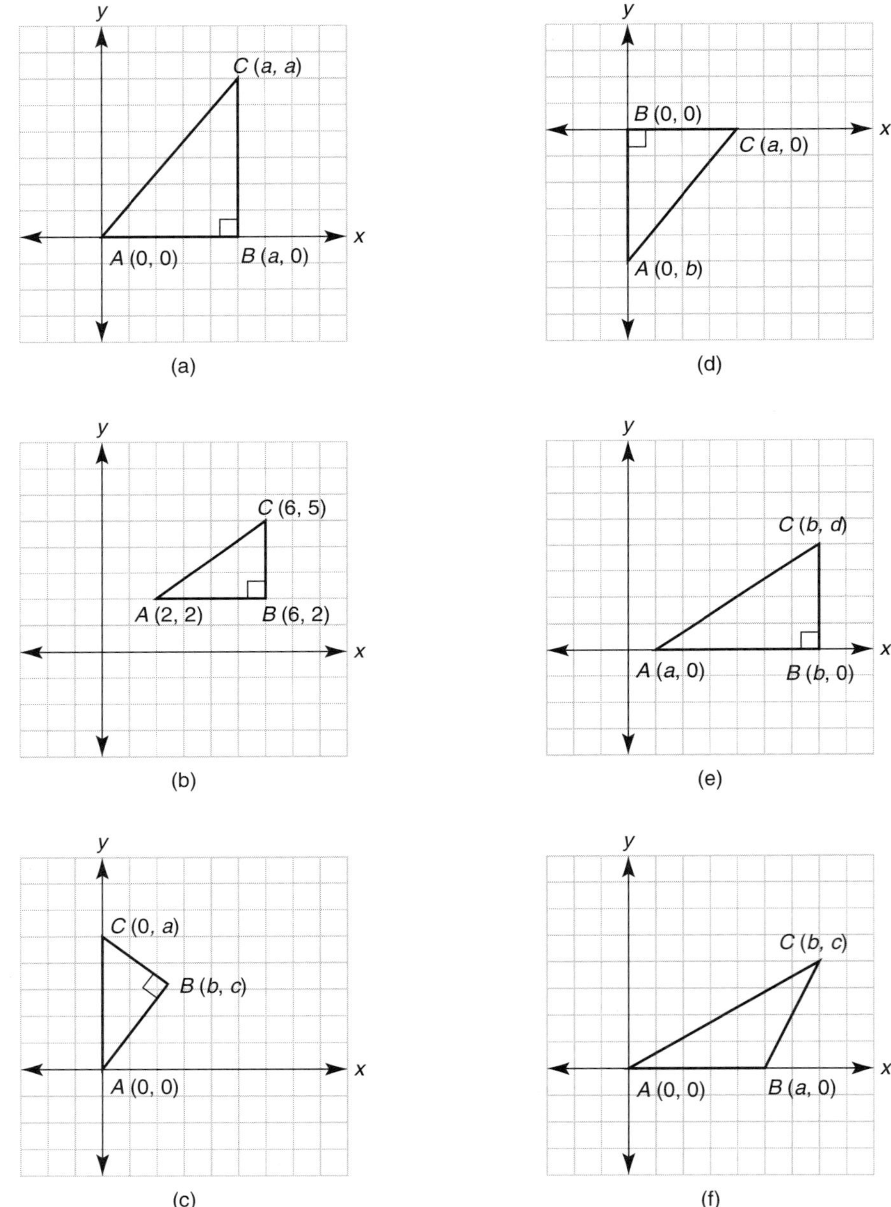

FIGURE 9.31

Solution

a) The choice of vertices causes $AB = BC$, so the triangle is also an isosceles triangle. This contradicts point 2 of the list of suggestions.

b) Coordinates are too specific! This contradicts point 1 of the list. A proof with these coordinates would *not* establish the general case.

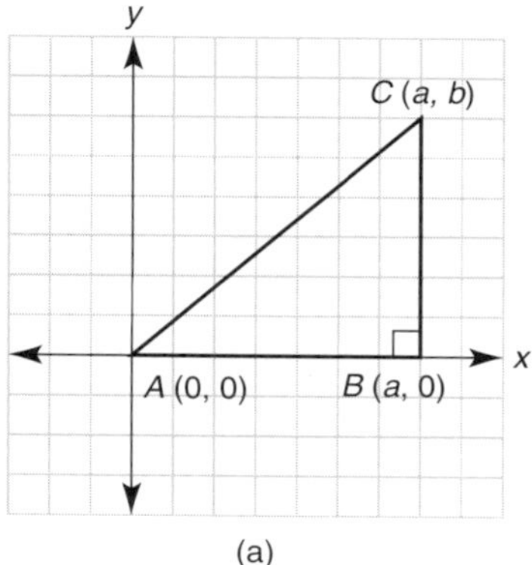

(a)

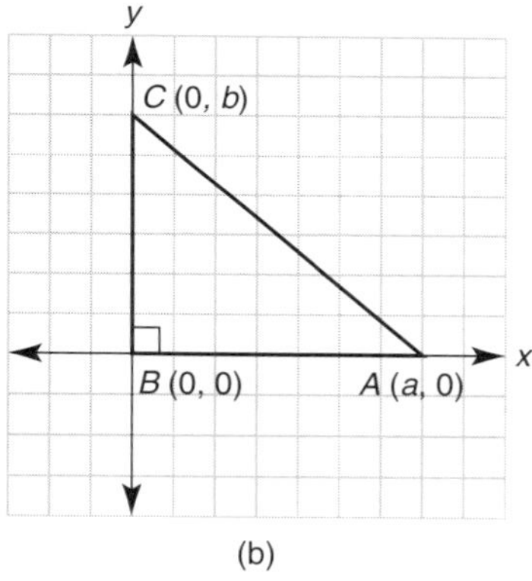

(b)

FIGURE 9.32

c) The drawing does *not* make use of horizontal and vertical lines to obtain the right angle. This violates point 4 of the list.
d) This placement fails point 3 of the list, because b is a negative number. The length of $\overline{AB}$ would be $-b$, which could be confusing.
e) This placement fails point 3 because we have not used as many 0 coordinates as we could have used. As we shall see, it also fails point 5.
f) This placement fails point 2. The triangle is not a right triangle unless $a = b$.

In Example 6, we wanted to place $\triangle ABC$ so that we met as many of the conditions listed on page 415 as possible. Two convenient placements are given in Figure 9.32. The triangle in Figure 9.32(b) is slightly better than the one in 9.32(a) in that it uses four 0 coordinates rather than three. Another advantage of Figure 9.32(b) is that the placement forces angle B to be a right angle, since the x and y axes are perpendicular.

We now turn our attention to the conclusion of the theorem. A second list examines some considerations for proving statements analytically.

USING THE CONCLUSION TO DO AN ANALYTIC PROOF

Three considerations for using the conclusion as a guide:

1. If the conclusion is a conjunction "P and Q," be sure to verify both parts of the conclusion.
2. The following pairings indicate how to prove statements of the type shown in the left column.

To prove the conclusion:	*Use the:*
a) Segments have equal lengths (like $AB = CD$)	Distance Formula
b) Segments are parallel (like $\overline{AB} \parallel \overline{CD}$)	Slope Formula (need $m_{\overline{AB}} = m_{\overline{CD}}$)
c) Segments are perpendicular (like $\overline{AB} \perp \overline{CD}$)	Slope Formula (need $m_{\overline{AB}} \cdot m_{\overline{CD}} = -1$)
d) A segment is bisected	Distance Formula
e) Segments bisect each other	Midpoint Formula

3. Anticipate the proof by thinking of the steps of the proof in reverse order; that is, reason "back" from the conclusion.

EXAMPLE 7

a) Provide an ideal drawing for the following theorem: The midpoint of the hypotenuse of a right triangle is equidistant from the three vertices of the triangle.
b) By studying the theorem, name at least two of the formulas that will be used to complete the proof.

Solution

a) We improve Figure 9.32(b) by giving the value $2a$ to the x coordinate of A and the value $2b$ to the y coordinate of C. (A factor of 2 makes it easier to calculate the midpoint M of $\overline{AC}$. See Figure 9.33.)

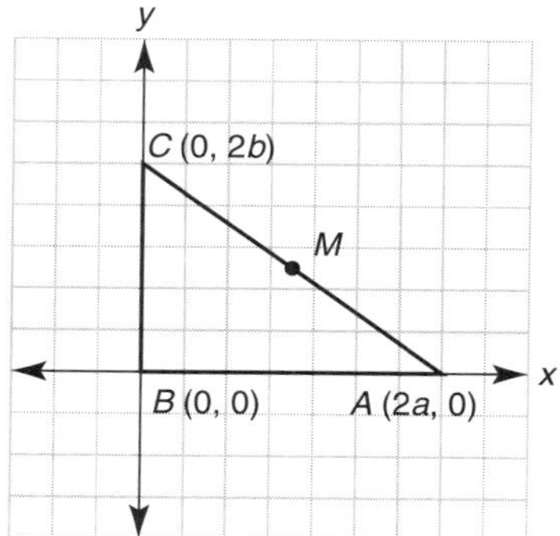

FIGURE 9.33

b) The Midpoint Formula is applied to describe the midpoint of $\overline{AC}$. Using the formula,

$$M = \left(\frac{x_1 + x_2}{2}, \frac{y_1 + y_2}{2}\right) = \left(\frac{2a + 0}{2}, \frac{0 + 2b}{2}\right)$$

so the midpoint is (a, b). The Distance Formula will also be needed since the theorem states that the distances from M to A, from M to B, and from M to C should all be equal.

The purpose of our next example is to demonstrate efficiency in the labeling of vertices.

EXAMPLE 8

If $MNPQ$ is a parallelogram in Figure 9.34, provide the coordinates of point P.

Solution

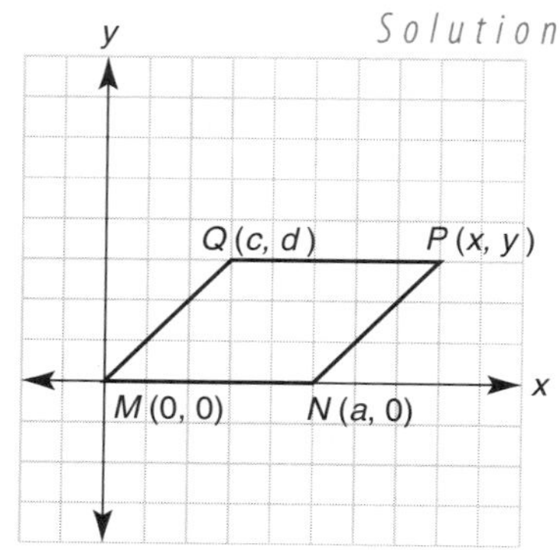

FIGURE 9.34

Consider $\square MNPQ$ in Figure 9.34. For the moment, we refer to point P as (x, y).

Because $\overline{MN} \parallel \overline{QP}$, we have $m_{\overline{MN}} = m_{\overline{QP}}$. But $m_{\overline{MN}} = \frac{0-0}{a-0} = 0$, while $m_{\overline{QP}} = \frac{y-d}{x-c}$, so we are led to the equation

$$\frac{y - d}{x - c} = 0 \rightarrow y - d = 0 \rightarrow y = d$$

Now P is described by (x, d). Because $\overline{MQ} \parallel \overline{NP}$, we are also led to the equation equating slopes of these segments. But

$$m_{\overline{MQ}} = \frac{d - 0}{c - 0} = \frac{d}{c} \quad \text{while} \quad m_{\overline{NP}} = \frac{d - 0}{x - a} = \frac{d}{x - a}$$

so

$$\frac{d}{c} = \frac{d}{x - a}$$

By using the Means-Extremes Property, we have

$$d(x - a) = d \cdot c \quad \text{with } d \neq 0$$
$$x - a = c \quad \text{dividing by } d$$
$$x = a + c \quad \text{adding } a$$

Therefore P is the point $(a + c, d)$.

In Example 8, we named the vertices of a parallelogram with the fewest possible letters. We now extend our result in Example 8 to allow for a rhombus—a parallelogram with two congruent adjacent sides.

EXAMPLE 9

In Figure 9.35, find an equation that relates a, b, c, and d if $ABCD$ is a rhombus.

Solution

As we saw in Example 8, the coordinates of the vertices of $ABCD$ define a parallelogram. For emphasis, we note that $AB \parallel DC$ and $AD \parallel BC$ because

$$m_{\overline{AB}} = m_{\overline{DC}} = 0 \qquad \text{and} \qquad m_{\overline{AD}} = m_{\overline{BC}} = \frac{d}{c}$$

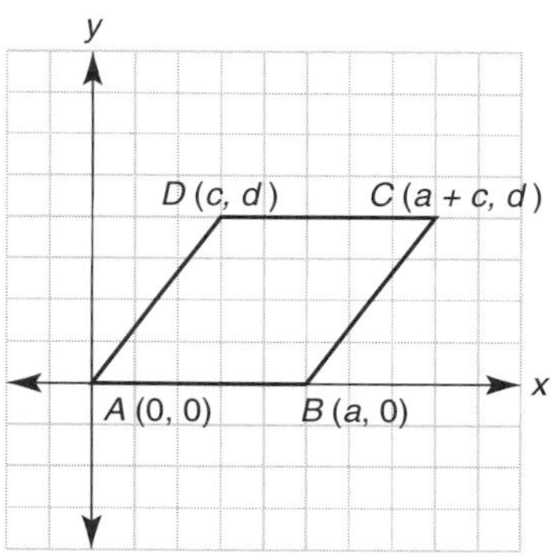

FIGURE 9.35

For Figure 9.35 to represent a rhombus, it is necessary that $AB = AD$. Now $AB = a - 0 = a$ because $\overline{AB}$ is a horizontal segment. To find an expression for the length of $\overline{AD}$, we need to use the Distance Formula:

$$d = \sqrt{(x_2 - x_1)^2 + (y_2 - y_1)^2}$$
$$AD = \sqrt{(c - 0)^2 + (d - 0)^2}$$
$$= \sqrt{c^2 + d^2}$$

Because $AB = AD$, we are led to $a = \sqrt{c^2 + d^2}$. Squaring, we have the desired relationship $a^2 = c^2 + d^2$.

EXAMPLE 10

If $\ell_1 \perp \ell_2$ in Figure 9.36, find a relationship among the variables a, b, c, and d.

Solution

First we find the slopes of lines ℓ_1 and ℓ_2. For ℓ_1, we have

$$m_1 = \frac{0 - d}{a - 0} = -\frac{d}{a}$$

For ℓ_2, we have

$$m_2 = \frac{c - 0}{b - 0} = \frac{c}{b}$$

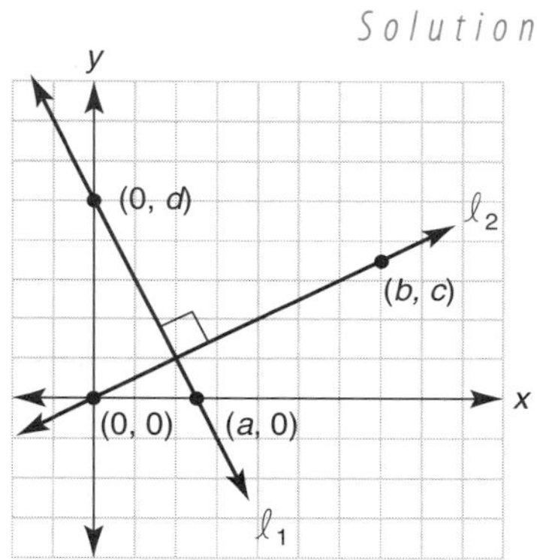

FIGURE 9.36

With $\ell_1 \perp \ell_2$, it follows that $m_1 \cdot m_2 = -1$. Substituting the slopes found above, we have

$$-\frac{d}{a} \cdot \frac{c}{b} = -1 \qquad \text{so} \qquad -\frac{dc}{ab} = -1$$

Equivalently, $\frac{dc}{ab} = 1$ and $dc = ab$.

9.4 Exercises

1. Find an expression for:
   a) The distance between $(a, 0)$ and $(0, a)$
   b) The slope of the segment joining (a, b) and (c, d)
2. Find the coordinates of the midpoint of the segment that joins the points
   a) $(a, 0)$ and $(0, b)$. b) $(2a, 0)$ and $(0, 2b)$.
3. Find the slope-intercept form of the equation of the line containing the points
   a) $(a, 0)$ and $(0, a)$. b) $(a, 0)$ and $(0, b)$.
4. Find the slope of the line that is:
   a) Parallel to the line containing $(a, 0)$ and $(0, b)$
   b) Perpendicular to the line through $(a, 0)$ and $(0, b)$

In Exercises 5 to 10, the real numbers a, b, c, and d are positive.

5. Consider the triangle with vertices at $A(0, 0)$, $B(a, 0)$, and $C(a, b)$. Explain why $\triangle ABC$ is a right triangle.

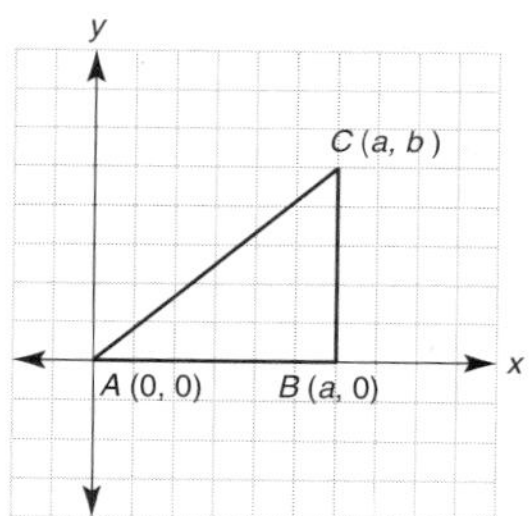

6. Consider the triangle with vertices at $R(-a, 0)$, $S(a, 0)$, and $T(0, b)$. Explain why $\triangle RST$ is an isosceles triangle.

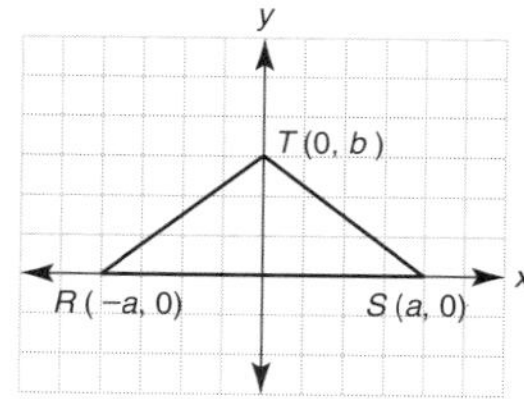

7. Consider the quadrilateral with vertices at $M(0, 0)$, $N(a, 0)$, $P(a + b, c)$, and $Q(b, c)$. Explain why $MNPQ$ is a parallelogram.

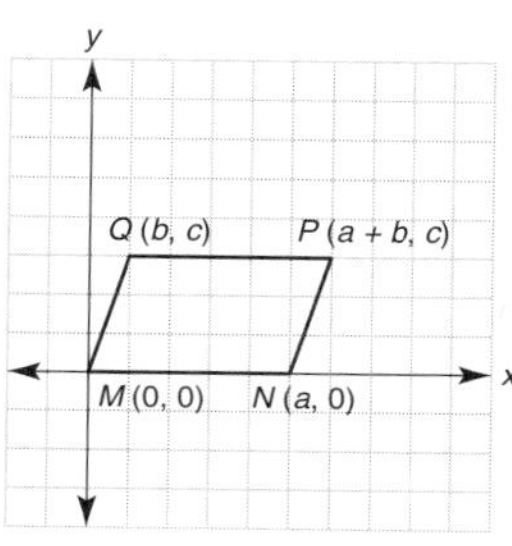

8. Consider the quadrilateral with vertices at $A(0, 0)$, $B(a, 0)$, $C(b, c)$, and $D(d, c)$. Explain why $ABCD$ is a trapezoid.

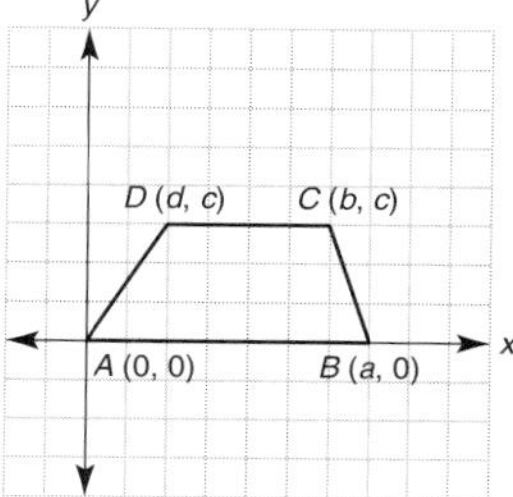

9. Consider the quadrilateral with vertices at $M(0, 0)$, $N(a, 0)$, $P(a, b)$, and $Q(0, b)$. Explain why $MNPQ$ is a rectangle.

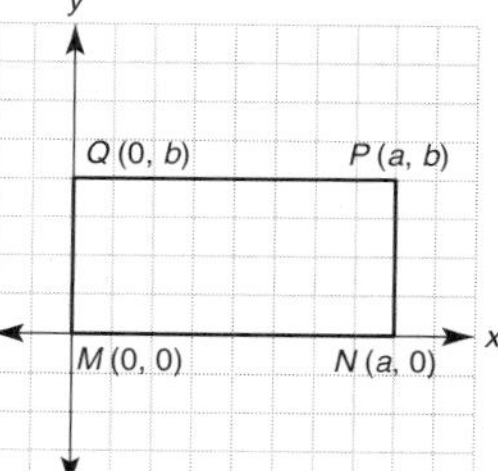

10. Consider the quadrilateral with vertices at $R(0, 0)$, $S(a, 0)$, $T(a, a)$, and $V(0, a)$. Explain why $RSTV$ is a square.

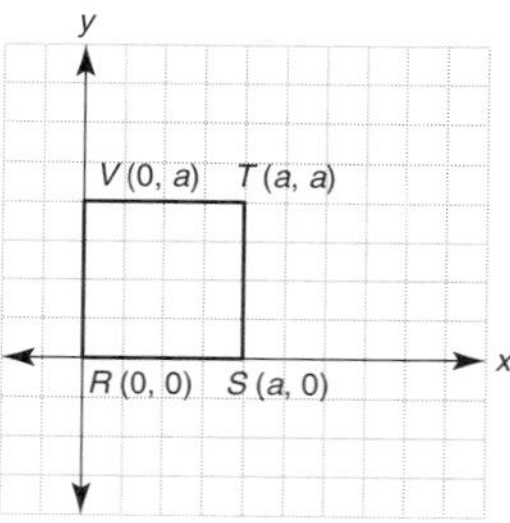

In Exercises 11 to 16, supply the missing coordinates for the vertices, using as few variables as possible.

11.

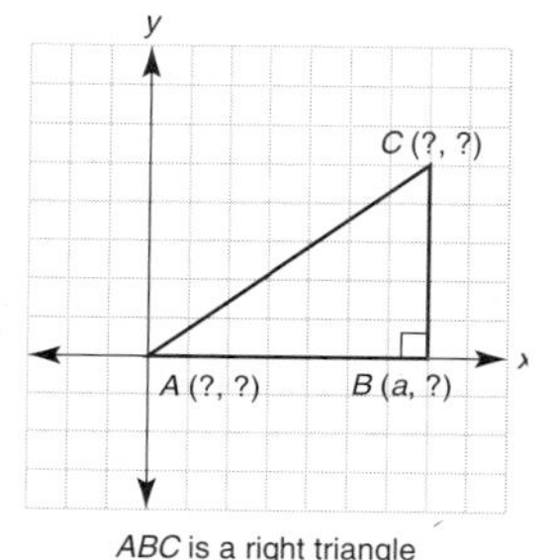

ABC is a right triangle

12.

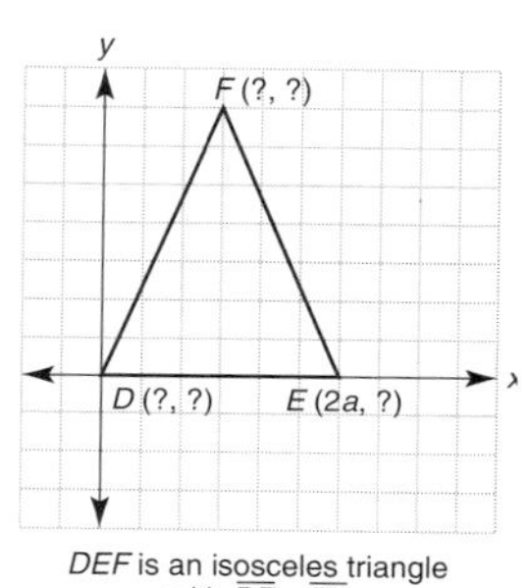

DEF is an isosceles triangle with $\overline{DF} \cong \overline{FE}$

13.

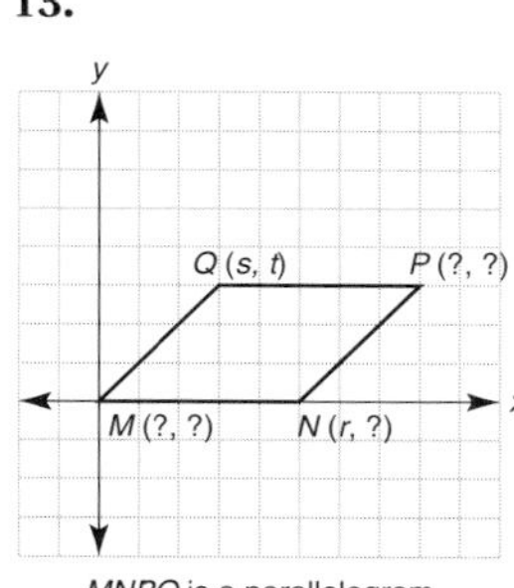

MNPQ is a parallelogram

14.

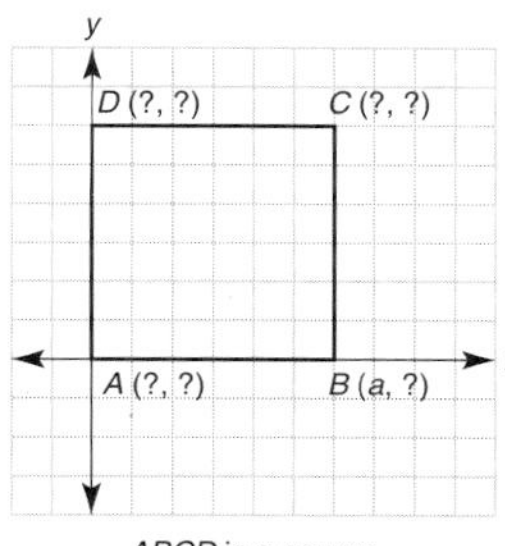

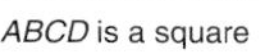

ABCD is a square

15.

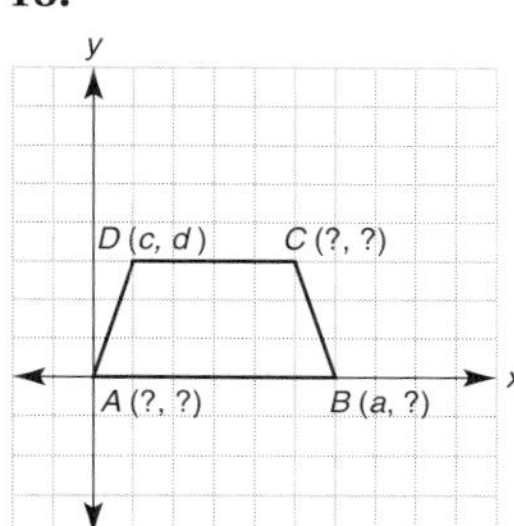

ABCD is an isosceles trapezoid; $\overline{AB} \parallel \overline{DC}$ and $\overline{AD} \cong \overline{BC}$

16.

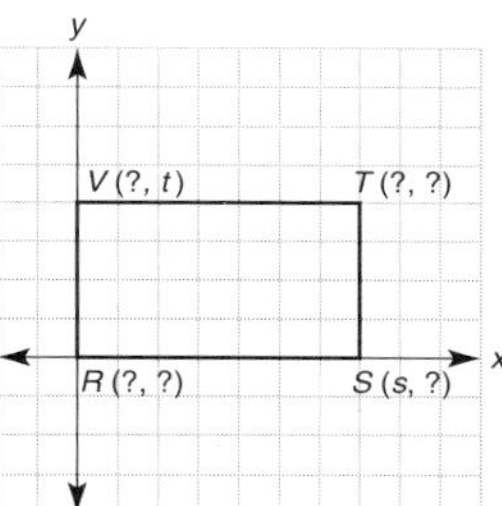

RSTV is a rectangle

In Exercises 17 to 22, draw an ideally placed figure in the coordinate system; then name the coordinates of each vertex of the figure.

17. a) A square
b) A square (midpoints of sides are needed)

18. a) A rectangle
b) A rectangle (midpoints of sides are needed)

19. a) A parallelogram
b) A parallelogram (midpoints of sides are needed)

20. a) A triangle
b) A triangle (midpoints of sides are needed)

21. a) An isosceles triangle
b) An isosceles triangle (midpoints of sides are needed)

22. a) A trapezoid
b) A trapezoid (midpoints of sides are needed)

In Exercises 23 to 28, find the equation (relationship) requested, and eliminate fractions or radicals from the equation.

23. If ▱*MNPQ* is a rhombus, state an equation that relates *r*, *s*, and *t*.

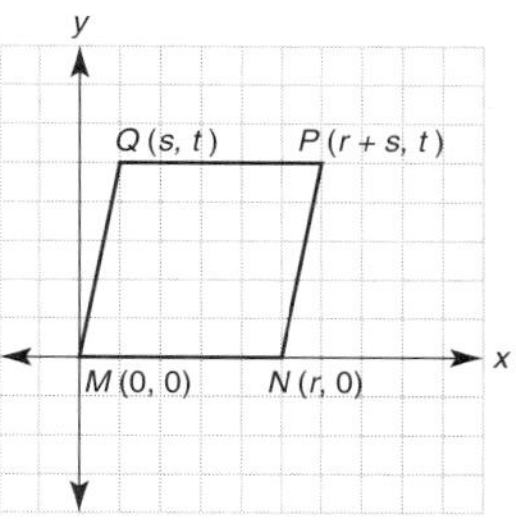

24. For ▱*RSTV*, suppose that $RT = VS$. State an equation that relates *s*, *t*, and *v*.

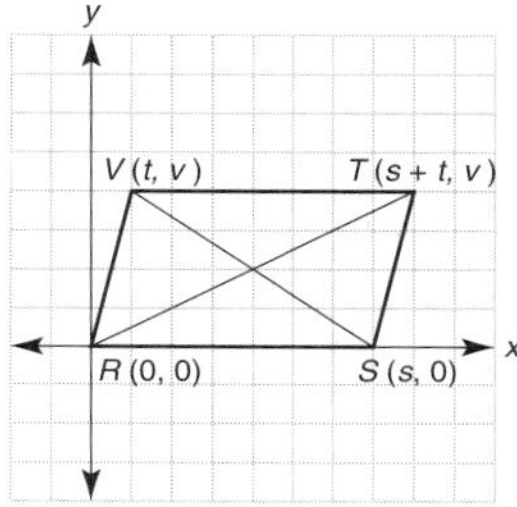

25. For ▱*ABCD*, suppose that diagonals $\overline{AC}$ and $\overline{DB}$ are ⊥. State an equation that relates *a*, *b*, and *c*.

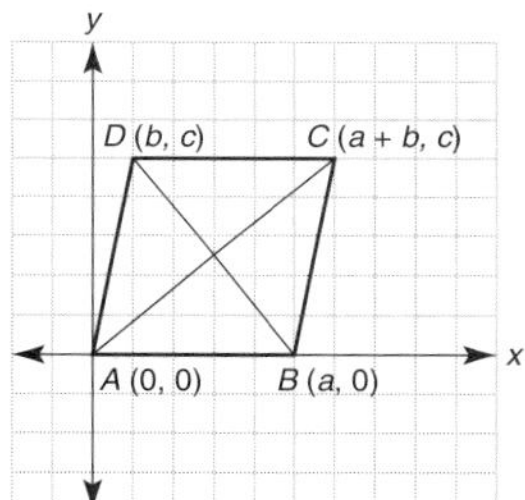

26. For quadrilateral *RSTV*, suppose that $\overline{RV} \parallel \overline{ST}$. State an equation that relates *m*, *n*, *p*, *q*, and *r*.

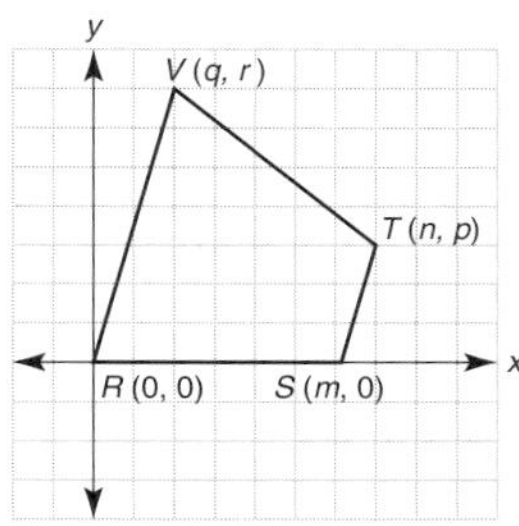

27. Suppose that △*ABC* is an equilateral triangle. State an equation that relates variables *a* and *b*.

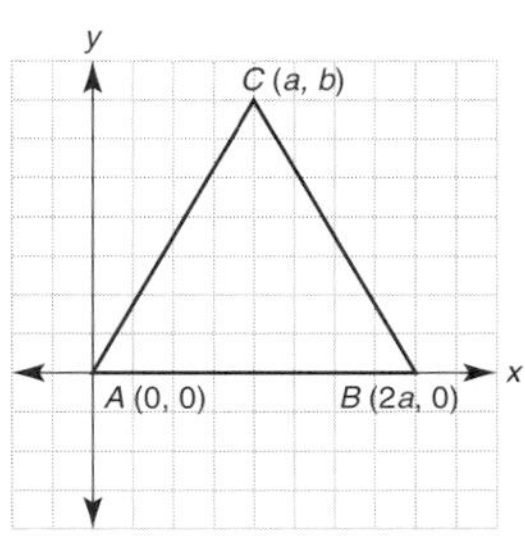

28. Suppose that $\triangle RST$ is an isosceles, with $\overline{RS} \cong \overline{RT}$. State an equation that relates s, t, and v.

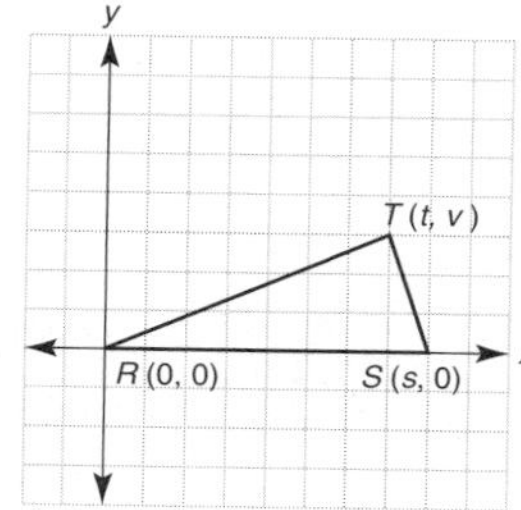

29. The drawing shows isosceles $\triangle ABC$ with $\overline{AC} \cong \overline{BC}$.

a) What type of number is a?

b) What type of number is $-a$?

c) Find an expression for the length of $\overline{AB}$.

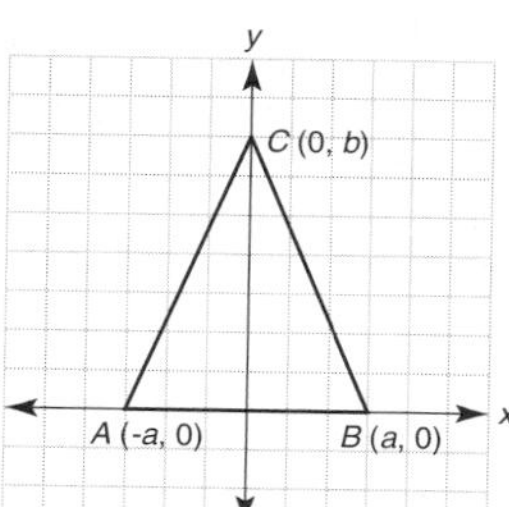

30. The drawing shows parallelogram $RSTV$.

a) What type of number is r?

b) Find an expression for RS.

c) Describe the coordinate t in terms of the other variables.

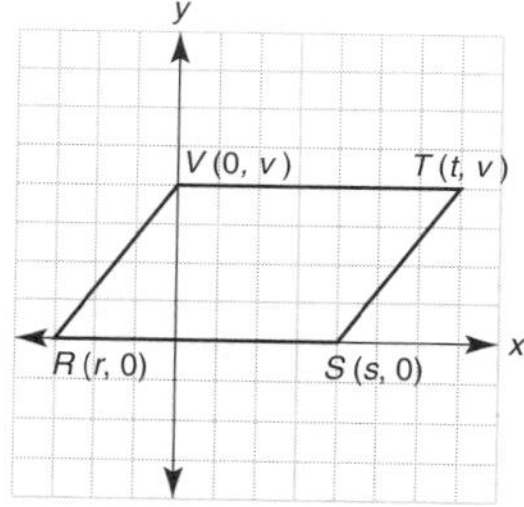

31. Which formula would you use to establish each of the following claims?

a) $\overline{AC} \perp \overline{DB}$

b) $AC = DB$

c) $\overline{DB}$ and $\overline{AC}$ bisect each other

d) $\overline{AD} \parallel \overline{BC}$

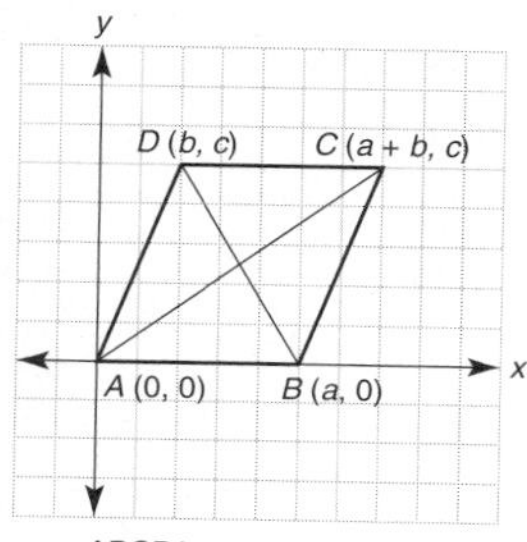

ABCD is a parallelogram

32. Which formula would you use to establish each of the following claims?

a) The coordinates of X are (d, c)

b) $m_{\overline{VT}} = 0$

c) $\overline{VT} \parallel \overline{RS}$

d) The length of $\overline{RV}$ is $2\sqrt{d^2 + c^2}$

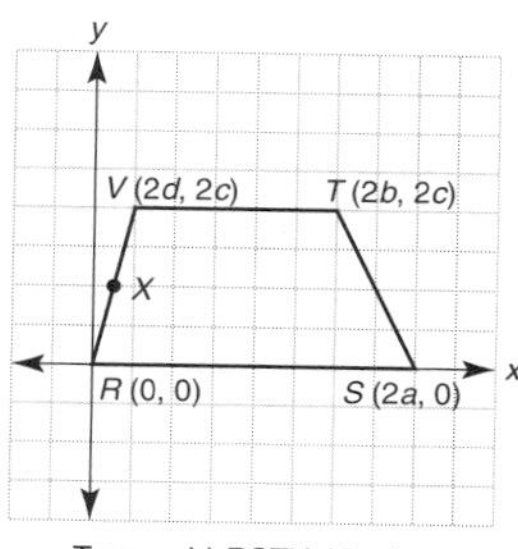

Trapezoid *RSTV*; *X* is the midpoint of $\overline{RV}$

In Exercises 33 to 36, draw and label a well-placed figure in the coordinate system for each theorem. Do not attempt to prove the theorem!

33. The line segment joining the midpoints of the two nonparallel sides of a trapezoid is parallel to the bases of the trapezoid.

34. If the midpoints of the sides of a quadrilateral are joined in order, the resulting quadrilateral is a parallelogram.

35. The diagonals of a rectangle are equal in length.

36. The diagonals of a rhombus are perpendicular to each other.

37. Find the point of intersection of the line $y = \frac{a}{b}x$ and the vertical line $x = c$.

38. At the point on the graph of $y = x^2$ where $x = a$, the slope of the tangent line is $m = 2a$. Show that the equation of the tangent line is $y = 2ax - a^2$.

39. ➢ Let $\triangle DEF$ have its vertices at $D\,(0, 0)$, $E\,(a, b)$, and $F\,(c, 0)$.

a) Show that an equation for the altitude from D to $\overline{EF}$ is $y = \frac{c-a}{b}x$.

b) Show that the general equation for the altitude from F to $\overline{DE}$ is $ax + by = ac$.

40. ➢ Show that the general equation of the perpendicular bisector of the segment joining $A(0, 0)$ to $B(2c, 2d)$ is $cx + dy = c^2 + d^2$.

9.5 Analytic Proofs

When we use algebra along with the rectangular coordinate system to prove a geometric theorem, the proof is termed **analytic.** The analytic (algebraic) approach relies heavily on the placement of the figure in the coordinate system and the application of the Distance Formula, the Midpoint Formula, or the Slope Formula (at the appropriate time). In order to contrast analytic proof with **synthetic** proof (the two-column or paragraph proofs used in earlier chapters), we repeat in this section some of our earlier theorems.

In Section 9.4, we saw how to place triangles having special qualities in the coordinate system. We review this information in Table 9.1 and then, in Example 1, we consider the proof of a theorem involving triangles. In Table 9.1, you will find that the figure determined by any positive choices of a, b, and c matches the type of triangle described.

TABLE 9.1

Analytic Proof: Suggestions for Placement of the Triangle

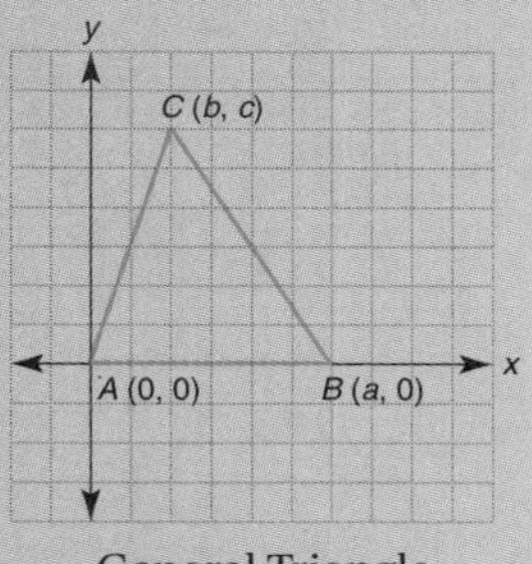

General Triangle

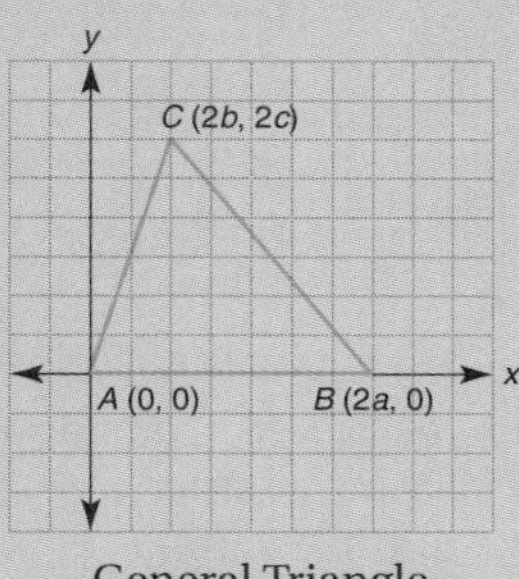

General Triangle (Midpoints)

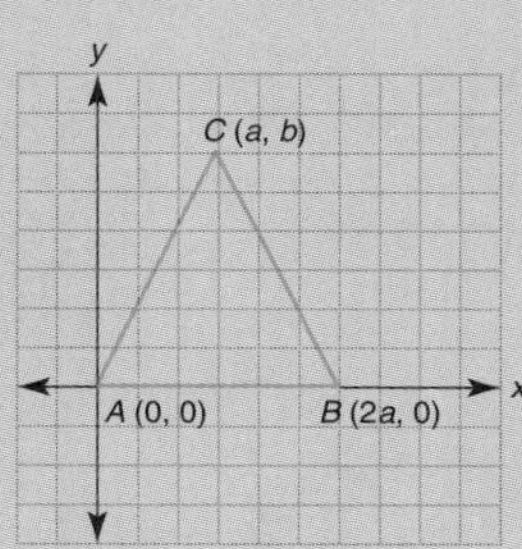

Isosceles Triangle

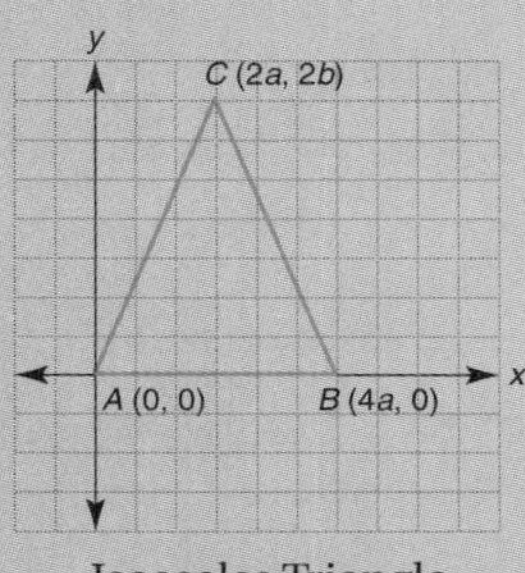

Isosceles Triangle (Midpoints)

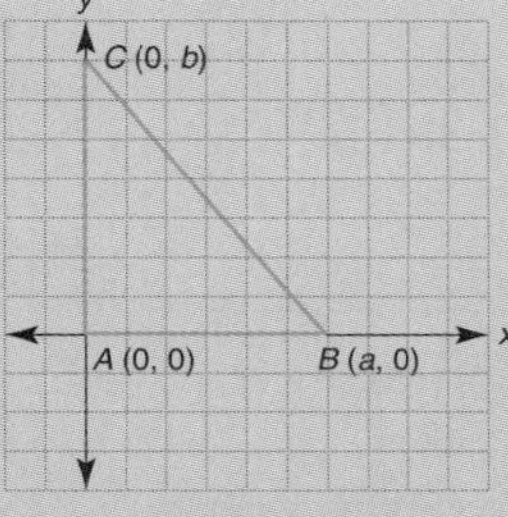

Right Triangle

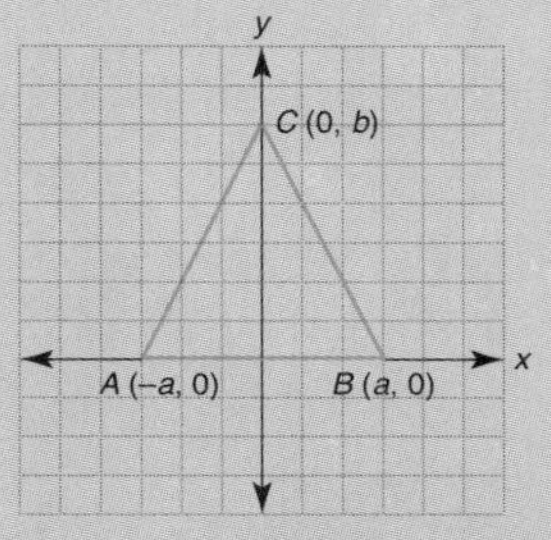

Equilateral Triangle (where $2a = \sqrt{a^2 + b^2}$, so $3a^2 = b^2$)

EXAMPLE 1

Prove the following theorem by the analytic method.

> **THEOREM 9.5.1:** The line segment determined by the midpoints of two sides of a triangle is parallel to the third side.

Plan: Use the Slope Formula; if $m_{\overline{MN}} = m_{\overline{AC}}$, then $\overline{MN} \parallel \overline{AC}$.

Proof

As shown in Figure 9.37, $\triangle ABC$ has vertices at $A(0, 0)$, $B(2a, 0)$, and $C(2b, 2c)$. With M the midpoint of $\overline{BC}$ and N the midpoint of $\overline{AB}$,

$$M = \left(\frac{2a + 2b}{2}, \frac{0 + 2c}{2}\right) \text{ which simplifies to } (a + b, c).$$

Also,

$$N = \left(\frac{0 + 2a}{2}, \frac{0 + 0}{2}\right) \text{ which simplifies to } (a, 0).$$

Next we apply the Slope Formula to determine $m_{\overline{MN}}$ and $m_{\overline{AC}}$. Now $m_{\overline{MN}} = \frac{c - 0}{(a + b) - a} = \frac{c}{b}$; also, $m_{\overline{AC}} = \frac{2c - 0}{2b - 0} = \frac{2c}{2b} = \frac{c}{b}$. Because $m_{\overline{MN}} = m_{\overline{AC}}$, we see that $\overline{MN} \parallel \overline{AC}$.

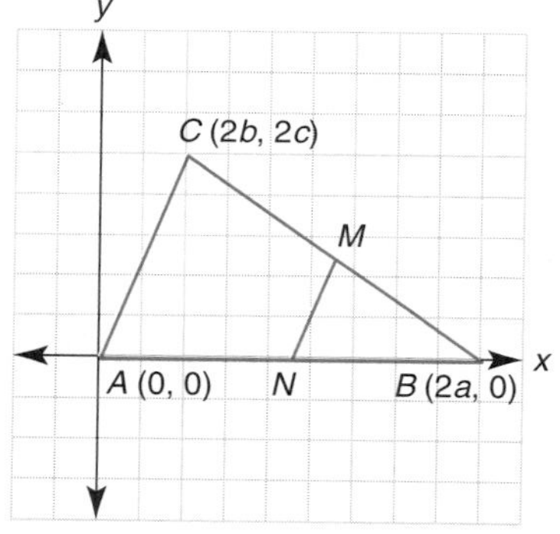

FIGURE 9.37

In Table 9.2 which follows Example 2 on the next page, we review possible placements of various quadrilaterals in the rectangular coordinate system. Again, the variables named represent positive numbers.

As we did in Example 1, we include a "plan" for Example 2. Although no plan is shown for Example 3 or Example 4, one is necessary before the proof can be written. In Example 5, we focus only on the plan.

EXAMPLE 2

Prove the following theorem by the analytic method:

> **THEOREM 9.5.2:** The diagonals of a parallelogram bisect each other.

Plan: Use the Midpoint Formula to show that the two diagonals have a common midpoint. Use a factor of 2 in nonzero coordinates.

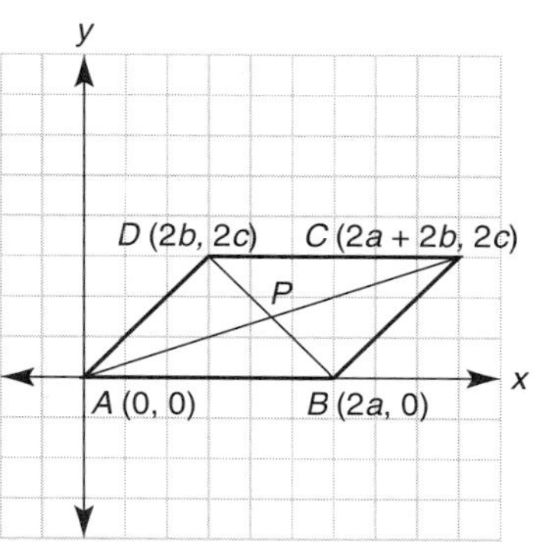

FIGURE 9.38

Proof In Figure 9.38, quadrilateral $ABCD$ is a parallelogram. The diagonals intersect at point P. By the Midpoint Formula, we have

$$\begin{aligned} M_{\overline{AC}} &= \left(\frac{0 + (2a + 2b)}{2}, \frac{0 + 2c}{2}\right) \\ &= (a + b, c) \end{aligned}$$

Also, the midpoint of $\overline{DB}$ is

$$\begin{aligned} M_{\overline{DB}} &= \left(\frac{2a + 2b}{2}, \frac{0 + 2c}{2}\right) \\ &= (a + b, c) \end{aligned}$$

Thus $(a + b, c)$ is the common midpoint of the two diagonals and must be the point of intersection of $\overline{AC}$ and $\overline{DB}$. Then $\overline{AC}$ and $\overline{DB}$ must bisect each other at point P.

TABLE 9.2

Analytic Proof: Suggestions for Placement of the Quadrilateral

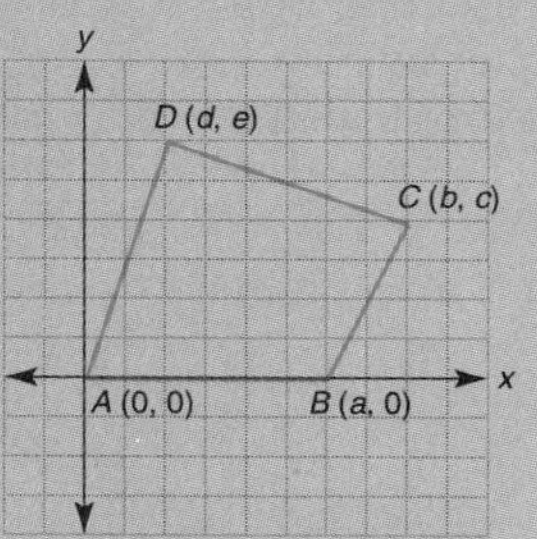

General Quadrilateral

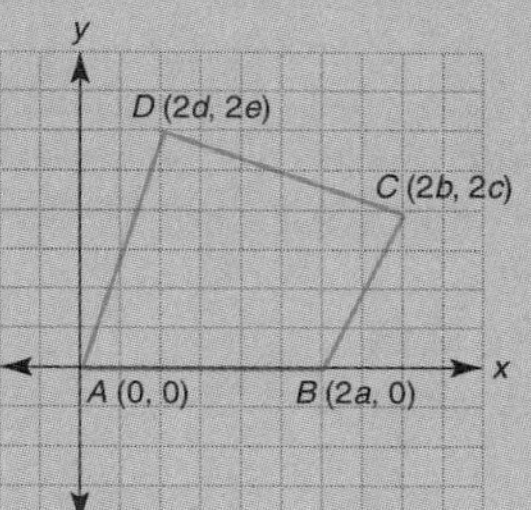

General Quadrilateral (Midpoints)

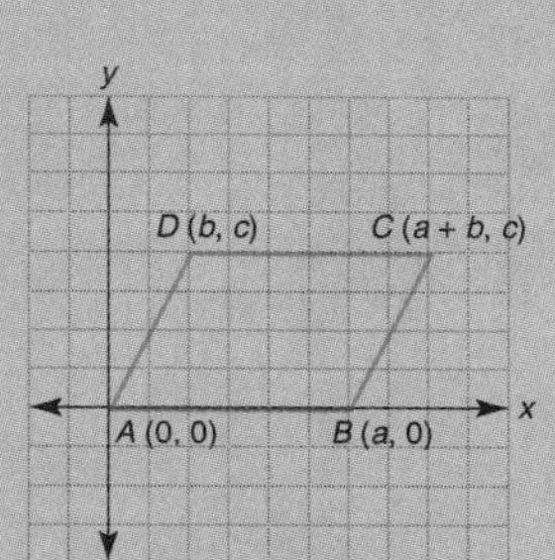

Parallelogram

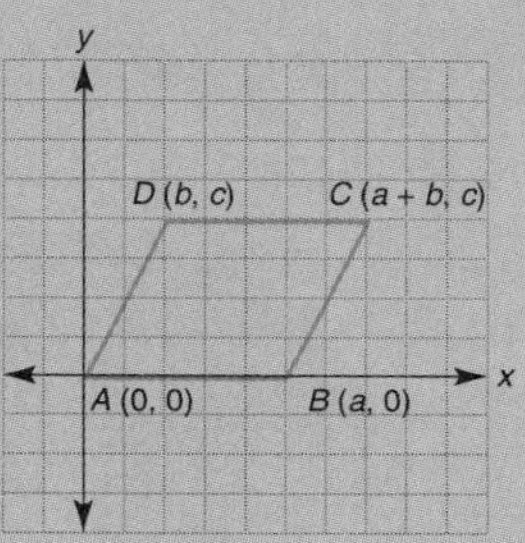

Rhombus (where $a = \sqrt{b^2 + c^2}$ so $a^2 = b^2 + c^2$)

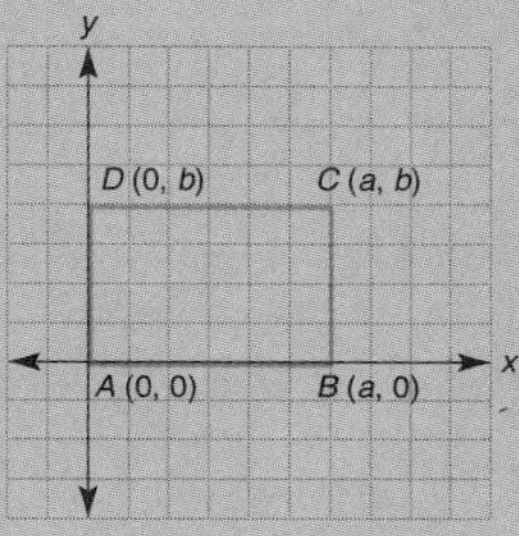

Rectangle

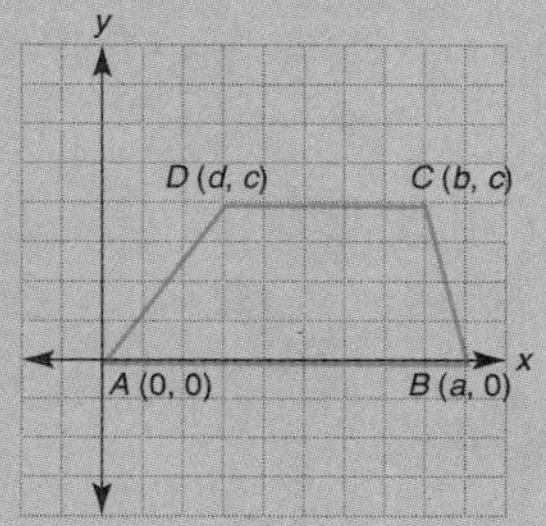

Trapezoid

The proof of Theorem 9.5.2 is not unique. In some cases, an alternative approach may be better! In Example 2, we could have used a three-step proof:

1. Find the equations of the two lines.
2. Determine the point of intersection of these lines.
3. Show that this point of intersection is the common midpoint.

But the word "bisect" in the theorem implied the use of the Midpoint Formula. Our approach to Example 2 was far easier and just as correct. The use of the Midpoint Formula is generally the best approach when the word "bisect" appears in the statement of a theorem.

We now outline the method of analytic proof.

COMPLETING AN ANALYTIC PROOF

1. Read the theorem carefully to distinguish the hypothesis and the conclusion. The hypothesis characterizes the figure to use.
2. Use the hypothesis (and nothing more) to determine a convenient placement of the figure in the rectangular coordinate system. Then label the figure. See Tables 9.1 and 9.2.
3. If any special quality is provided by the hypothesis, be sure to state this early in the proof. (For example, a rhombus should be described as a parallelogram that has two congruent adjacent sides.)
4. Study the conclusion, and devise a plan to prove this claim; this may involve reasoning back from the conclusion step by step until the hypothesis is reached.
5. Write the proof, being careful to properly order and justify each statement.

EXAMPLE 3

Prove Theorem 9.5.3 by the analytic method.

THEOREM 9.5.3: The diagonals of a rhombus are perpendicular.

Proof

In Figure 9.39, $ABCD$ has the coordinates of a parallelogram. Because $\square ABCD$ is a rhombus, $AB = AD$. Then $a = \sqrt{b^2 + c^2}$ by the Distance Formula, and squaring gives $a^2 = b^2 + c^2$. The Slope Formula leads to

$$m_{\overline{AC}} = \frac{c - 0}{(a + b) - 0} \quad \text{and} \quad m_{\overline{DB}} = \frac{0 - c}{a - b}$$

so

$$m_{\overline{AC}} = \frac{c}{a + b} \quad \text{and} \quad m_{\overline{DB}} = \frac{-c}{a - b}$$

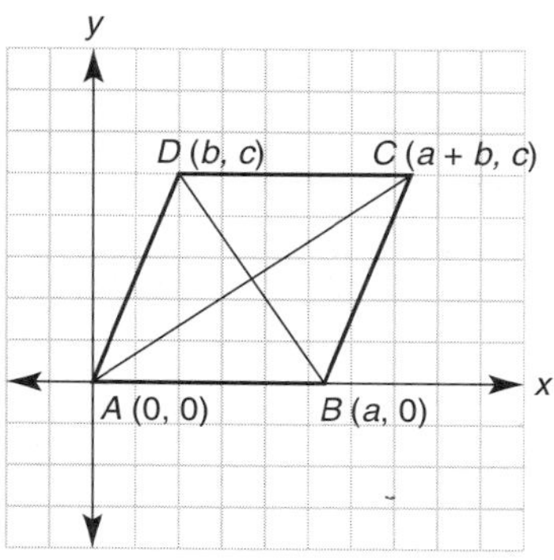

FIGURE 9.39

Then the product of the slopes of the diagonals is

$$\begin{aligned} m_{\overline{AC}} \cdot m_{\overline{DB}} &= \frac{c}{a+b} \cdot \frac{-c}{a-b} \\ &= \frac{-c^2}{a^2 - b^2} \\ &= \frac{-c^2}{(b^2 + c^2) - b^2} \qquad \text{(replaced } a^2 \text{ by } b^2 + c^2\text{)} \\ &= \frac{-c^2}{c^2} = -1 \end{aligned}$$

Then $\overline{AC} \perp \overline{DB}$ since the product of their slopes equals -1.

In Example 3, we had to use the condition that two adjacent sides of the rhombus were congruent in order to complete the proof. Had that condition been omitted, the product of slopes could not have been shown to equal -1. In general, the diagonals of a parallelogram are *not* perpendicular.

In our next example, we consider the proof of the converse of an earlier theorem. While it is easy to complete an analytic proof of the statement, "The diagonals of a rectangle are equal in length," the proof of the converse is not as straightforward.

EXAMPLE 4

Prove Theorem 9.5.4 by the analytic method.

> **THEOREM 9.5.4:** If the diagonals of a parallelogram are equal in length, then the parallelogram is a rectangle.

Proof

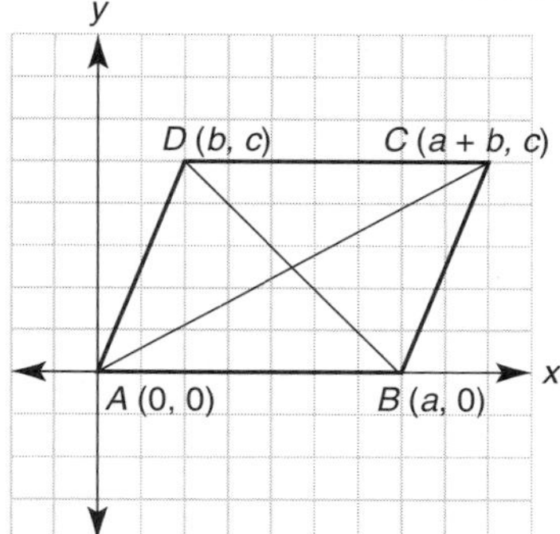

FIGURE 9.40

In parallelogram $ABCD$ in Figure 9.40, $AC = DB$. Applying the Distance Formula,

$$AC = \sqrt{[(a+b) - 0]^2 + (c-0)^2}$$

and

$$DB = \sqrt{(a-b)^2 + (0-c)^2}$$

Then it follows that

$$\begin{aligned} \sqrt{(a+b)^2 + c^2} &= \sqrt{(a-b)^2 + (-c)^2} \\ (a+b)^2 + c^2 &= (a-b)^2 + (-c)^2 \qquad \text{squaring} \\ a^2 + 2ab + b^2 + c^2 &= a^2 - 2ab + b^2 + c^2 \qquad \text{simplifying} \\ 4ab &= 0 \\ a \cdot b &= 0 \qquad \text{dividing by 4} \end{aligned}$$

Thus $a = 0$ or $b = 0$

Because $a \neq 0$ (otherwise points A and B would coincide), it is necessary that $b = 0$, so point D is on the y axis. The resulting coordinates of the figure are $A(0, 0)$, $B(a, 0)$, $C(a, c)$, and $D(0, c)$. Then $ABCD$ must be a rectangle because $\overline{AB}$ is horizontal while $\overline{AD}$ is vertical.

Our final example illustrates the use of an analytic proof to prove a theorem that was not proven by the synthetic method.

EXAMPLE 5

Formulate a plan to complete the proof of Theorem 9.5.5.

> **THEOREM 9.5.5:** The three medians of a triangle are concurrent at a point that is two-thirds the distance from any vertex to the midpoint of the opposite side.

Solution

The proof can be completed as follows:

1. Find the coordinates of the two midpoints X and Y. See Figures 9.41(a) and 9.41(b). Notice that

$$X = (a + b, c) \qquad \text{and} \qquad Y = (b, c)$$

2. Find the equations of the lines containing $\overline{AX}$ and $\overline{BY}$. The equations for $\overline{AX}$ and $\overline{BY}$ are $y = \dfrac{c}{a+b}x$ and $y = \dfrac{-c}{2a-b}x + \dfrac{2ac}{2a-b}$, respectively.

3. Find the point of intersection Z of $\overline{AX}$ and $\overline{BY}$. Solving the system provides the solution

$$Z = \left(\frac{2}{3}(a+b), \frac{2}{3}c\right)$$

4. It can now be shown that $AZ = \frac{2}{3} \cdot AX$ and $BZ = \frac{2}{3} \cdot BY$. See Figure 9.41(b), in which we can show that

$$AZ = \frac{2}{3}\sqrt{(a+b)^2 + c^2} \qquad \text{and} \qquad AX = \sqrt{(a+b)^2 + c^2}$$

5. It can also be shown that point Z lies on the third median $\overline{CW}$, whose equation is $y = \frac{2c}{2b-a}(x - a)$. See Figure 9.41(c).

6. We can also show that $CZ = \frac{2}{3} \cdot CW$, which completes the proof.

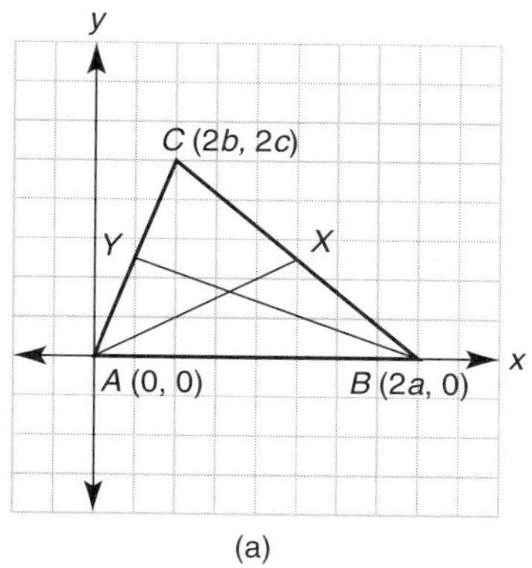

(a)

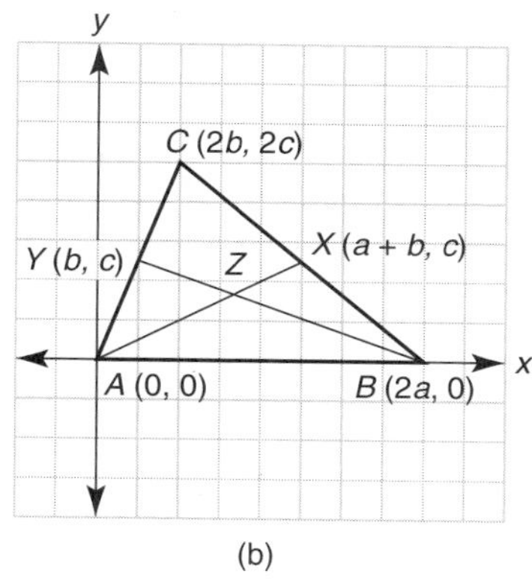

(b)

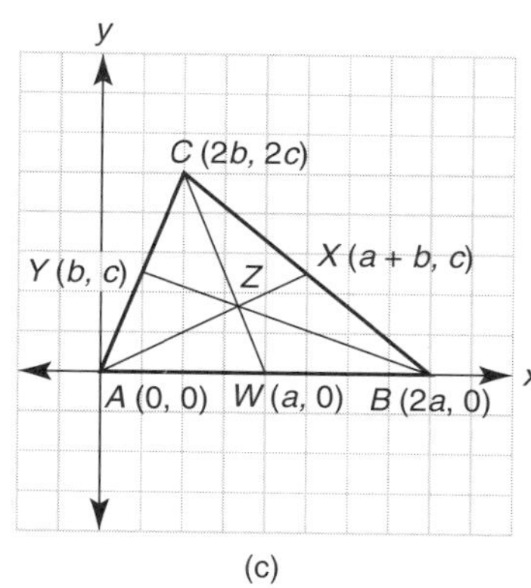

(c)

FIGURE 9.41

9.5 Exercises

In Exercises 1 to 19, complete an analytic proof for each theorem.

1. The diagonals of a rectangle are equal in length.
2. The opposite sides of a parallelogram are equal in length.
3. The diagonals of a square are perpendicular bisectors of each other.
4. The diagonals of an isosceles trapezoid are equal in length.
5. The median from the vertex of an isosceles triangle to the base is perpendicular to the base.
6. The medians to the congruent sides of an isosceles triangle are equal in length.
7. The segments that join the midpoints of the consecutive sides of a quadrilateral form a parallelogram.
8. The segments that join the midpoints of the opposite sides of a quadrilateral bisect each other.
9. The segments that join the midpoints of the consecutive sides of a rectangle form a rhombus.
10. The segments that join the midpoints of the consecutive sides of a rhombus form a rectangle.
11. The midpoint of the hypotenuse of a right triangle is equidistant from the three vertices of the triangle.
12. The median of a trapezoid is parallel to the bases of the trapezoid and has a length equal to one-half the sum of the lengths of the two bases.
13. The segment that joins the midpoints of two sides of a triangle is parallel to the third side and has a length equal to one-half the length of the third side.
14. The perpendicular bisector of the base of an isosceles triangle contains the vertex of the triangle.
15. If the midpoint of one side of a rectangle is joined to the endpoints of the opposite side, an isosceles triangle is formed.
16. ➢ If the median to one side of a triangle is also an altitude of the triangle, then the triangle is isosceles.
17. ➢ If the diagonals of a parallelogram are perpendicular, then the parallelogram is a rhombus.
18. ➢ The perpendicular bisectors of the sides of a triangle are concurrent.
19. ➢ The altitudes of a triangle are concurrent.
20. Use the analytic method to decide what type of quadrilateral is formed when the midpoints of the consecutive sides of a parallelogram are joined by line segments.
21. Use the analytic method to decide what type of triangle is formed when the midpoints of the sides of an isosceles triangle are joined by line segments.

22. Describe the steps of the procedure that allows us to find the distance from a point $P(a, b)$ to the line $Ax + By = C$.

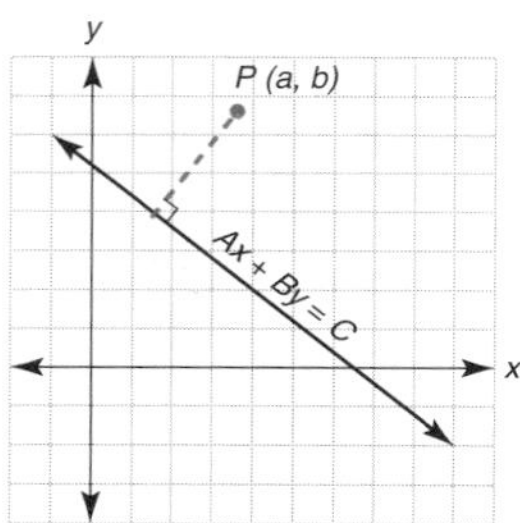

23. Would the theorem of Exercise 7 remain true for a concave quadrilateral like the one shown?

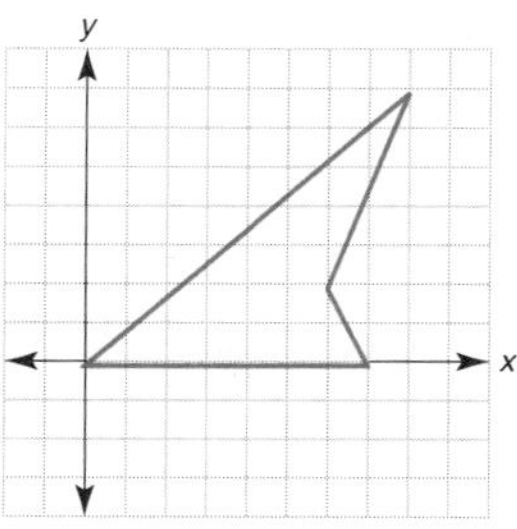

Exercise 23

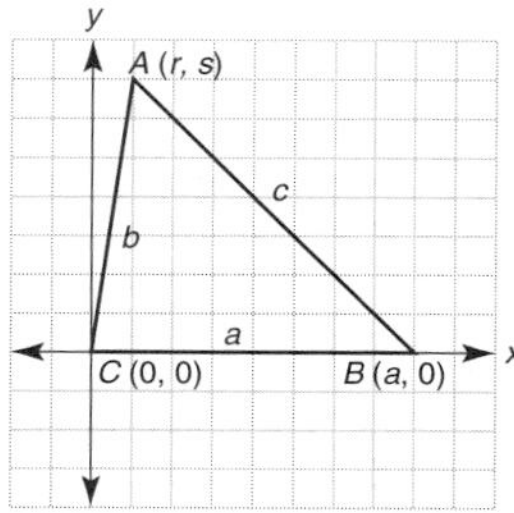

Exercise 24

24. ➢ Complete an analytic proof of the following theorem: In a triangle that has sides of lengths a, b, and c, if $c^2 = a^2 + b^2$, then the triangle is a right triangle.

9.6 The Equation of a Circle

The general form of the equation of a line $Ax + By = C$ has variable terms of degree 1; that is, the understood exponent for each variable term is 1 since $x^1 = x$ and $y^1 = y$. The equation $Ax + By = C$ is said to have degree 1.

The **degree** of each term of a variable expression is the sum of the exponents of its variable factors. For example, the term y^2z^3 has degree 5 (five variable factors) and $5n^4$ has degree 4. The **degree of an equation** equals the degree of its highest-degree term. The equations $x^2 + y^2 = 9$ and $y = x^2$ both have degree 2. Equations containing terms of degree 2 and larger are nonlinear; generally, their graphs possess the quality of curvature. For instance, the graphs of $x^2 + y^2 = 9$ (a circle) and $y = x^2$ (a parabola) are curved as shown in Figure 9.42(a) and (b) respectively.

FIGURE 9.42

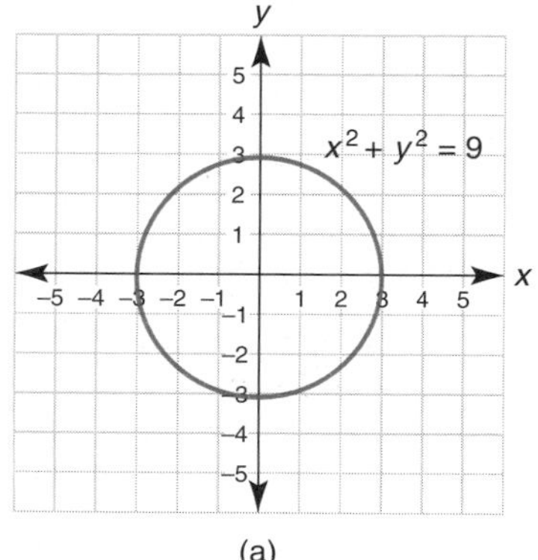

(a)

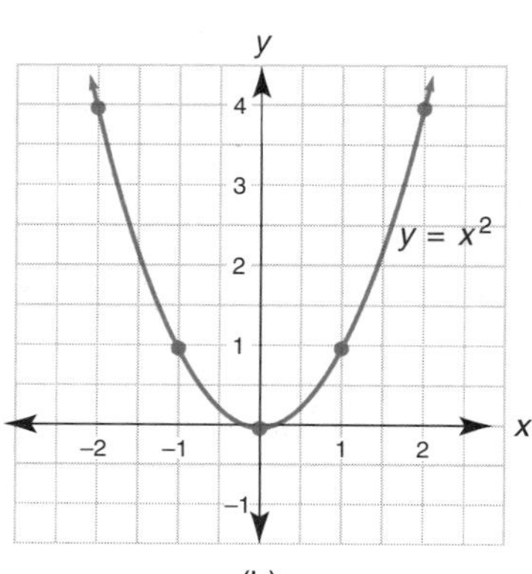

(b)

Standard Form for the Equation of a Circle

THEOREM 9.6.1: The circle whose center is (h, k) and whose radius has length r, where $r > 0$, has the equation

$$(x - h)^2 + (y - k)^2 = r^2$$

Proof

In Figure 9.43, let (x, y) represent the general point on the circle. Then the distance from center (h, k) to (x, y) is r. The Distance Formula,

$$d = \sqrt{(x_2 - x_1)^2 + (y_2 - y_1)^2}$$

becomes

$$r = \sqrt{(x - h)^2 + (y - k)^2}$$

when we designate (x, y) as P_2 and (h, k) as P_1. Squaring each side of this equation leads to the form

$$r^2 = (x - h)^2 + (y - k)^2$$

or

$$(x - h)^2 + (y - k)^2 = r^2$$

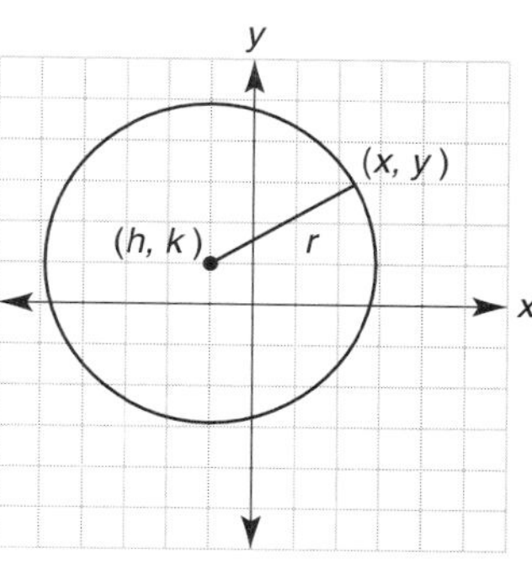

FIGURE 9.43

EXAMPLE 1

Find the equation for the circle whose center is (0, 0) and whose radius has length 5. (See Figure 9.44.)

Solution

Where $h = 0$, $k = 0$, and $r = 5$,

$$(x - h)^2 + (y - k)^2 = r^2 \quad \text{becomes}$$
$$(x - 0)^2 + (y - 0)^2 = 5^2$$
$$x^2 + y^2 = 25$$

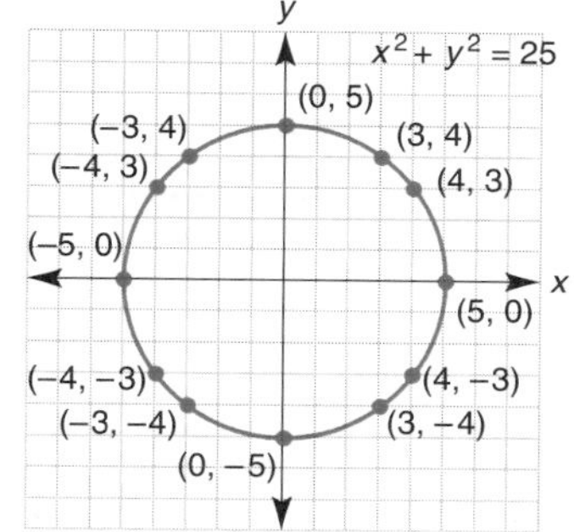

FIGURE 9.44

Notice that Figure 9.44 indicates several solutions for the equation obtained. Each ordered pair satisfies the equation $x^2 + y^2 = 25$.

A corollary of Theorem 9.6.1 follows. It deals with the circle centered at the origin. Example 1 is the inspiration for this corollary.

COROLLARY 9.6.2: The equation of the circle with center (0, 0) and radius r is $x^2 + y^2 = r^2$.

Every point at distance r from (0, 0) lies on the circle described in Corollary 9.6.2. Also, each point on the circle is a solution of the equation $x^2 + y^2 = r^2$. Consider Example 1 and the following more general example.

EXAMPLE 2

What conclusion can you draw if you know that the point (a, b) lies on the circle $x^2 + y^2 = r^2$?

Solution

Since (a, b) is on the circle, it is also a solution for the equation $x^2 + y^2 = r^2$. Therefore $a^2 + b^2 = r^2$.

When a circle is not centered at the origin, the standard form $(x - h)^2 + (y - k)^2 = r^2$ is often changed to an expanded form. Using FOIL and combining like terms, we can change the standard form of the equation to a more general form.

General Form for the Equation of a Circle

The *general form* of the equation of a circle is

$$x^2 + y^2 + Dx + Ey + F = 0$$

where D, E, and F are constants. This form shows that the equation of a circle always contains the sum $x^2 + y^2$. In the following example, the standard equation of a circle is changed to its general form.

EXAMPLE 3

Find the general form of the equation of the circle with center (2, −3) and radius of length 4. (See Figure 9.45.)

Solution

Using $(x - h)^2 + (y - k)^2 = r^2$, we have

$$(x - 2)^2 + [y - (-3)]^2 = 4^2$$

$$(x - 2)^2 + (y + 3)^2 = 16$$

$$x^2 - 4x + 4 + y^2 + 6y + 9 = 16$$

(using FOIL twice)

$$x^2 + y^2 - 4x + 6y - 3 = 0$$

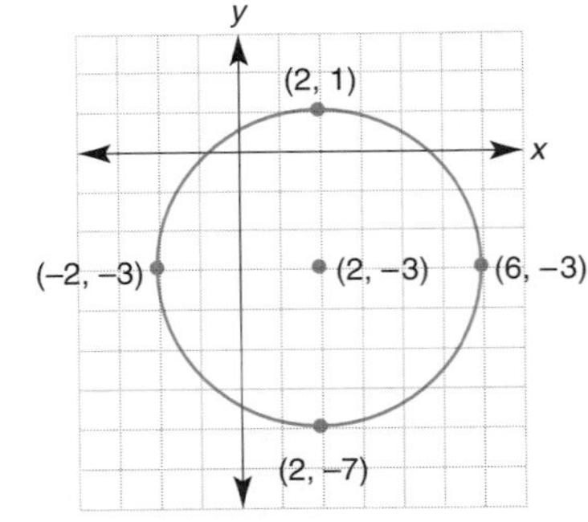

FIGURE 9.45

Graphing Circles

Before graphing the circle whose equation is in general form, we must complete the squares of the expressions $x^2 + Dx$ and $y^2 + Ey$. The goal is to find a number to add to $x^2 + Dx$ so that the result can be factored as the square of a binomial; that is, we want to find a number G so that $x^2 + Dx + G = (x + N)^2$. The procedure follows.

COMPLETING THE SQUARE

Given the expression $x^2 + Dx$, we complete the square as follows:

1. Take one-half of the linear coefficient D to obtain $\frac{1}{2}D$.
2. Square the result found in step 1 so that $\left(\frac{1}{2}D\right)^2 = \frac{1}{4}D^2$.
3. Add the result in step 2 to each side of the equation.

EXAMPLE 4

Find the number that completes the square of the expression $x^2 - 8x$.

Solution The linear coefficient (number that x is multiplied by) is $D = -8$. Now $\frac{1}{2}D = \frac{1}{2}(-8) = -4$, so $\left(\frac{1}{2}D\right)^2 = (-4)^2 = 16$. The number that completes the square is 16.

NOTE: It can be shown that $x^2 - 8x + 16$ is the *square of* $(x - 4)$; that is, $(x - 4)^2 = x^2 - 8x + 16$.

EXAMPLE 5

Draw the graph of $x^2 + y^2 - 4x + 6y - 3 = 0$.

Solution We write the equation in a convenient form for completing the squares:

$$(x^2 - 4x \qquad) + (y^2 + 6y \qquad) = 3$$

Now $\left(\frac{-4}{2}\right)^2 = (-2)^2 = 4$, and $\left(\frac{6}{2}\right)^2 = 3^2 = 9$; therefore, 4 and 9 are added to each side of the equation as follows:

$$(x^2 - 4x + \mathbf{4}) + (y^2 + 6y + \mathbf{9}) = 3 + \mathbf{4} + \mathbf{9}$$
$$(x - 2)^2 + (y + 3)^2 = 16$$
$$(x - 2)^2 + [y - (-3)]^2 = 4^2$$

The graph is a circle whose center is $(2, -3)$ and whose radius has length 4. See Example 3 for the graph.

Usually, the graph of an equation of the form $x^2 + y^2 + Dx + Ey + F = 0$ is a circle; however, two counterexamples follow.

Suppose that the equation given in Example 5 had read

$$x^2 + y^2 - 4x + 6y + 13 = 0$$

When completing the squares, we would obtain

$$(x^2 - 4x + 4) + (y^2 + 6y + 9) = -13 + 4 + 9$$

which becomes

$$(x - 2)^2 + (y + 3)^2 = 0$$

It appears that we have a circle centered at (2, −3) and with radius of length 0. The graph is actually only one point, (2, −3).

If the equation in Examples 3 and 5 had read

$$x^2 + y^2 - 4x + 6y + 14 = 0$$

we would obtain

$$(x - 2)^2 + (y + 3)^2 = -1$$

Since the square of the radius cannot possibly equal −1, no graph is possible.

Summarizing the above results, we see that the graph of the equation $x^2 + y^2 + Dx + Ey + F = 0$ could be a circle, a point, or the empty set.

Table 9.3 compares the general forms of the equation of the line and the equation of the circle with the forms that are convenient for graphing.

■ *Warning*

An equation of the form $x^2 + y^2 + Dx + Ey + F = 0$ is not always a circle. ■

TABLE 9.3

Figure	General Form	Form for Graphing
Line	$Ax + By = C$	$y = mx + b$ or $y - y_1 = m(x - x_1)$
Circle	$x^2 + y^2 + Dx + Ey + F = 0$	$(x - h)^2 + (y - k)^2 = r^2$

Finding Points of Intersection

EXAMPLE 6

Use geometry to determine the points of intersection of the circle $x^2 + y^2 = 25$ and the line $y = 3x - 5$.

Solution

The equation of the circle $x^2 + y^2 = 25$ can be expressed in the form $(x - 0)^2 + (y - 0)^2 = 5^2$; thus, the circle has center (0, 0) and a radius of length 5. The graph of $y = 3x - 5$ is a line with y intercept (0, −5) and slope m = 3. In Figure 9.46, the graphs intersect at (0, −5) and (3, 4).

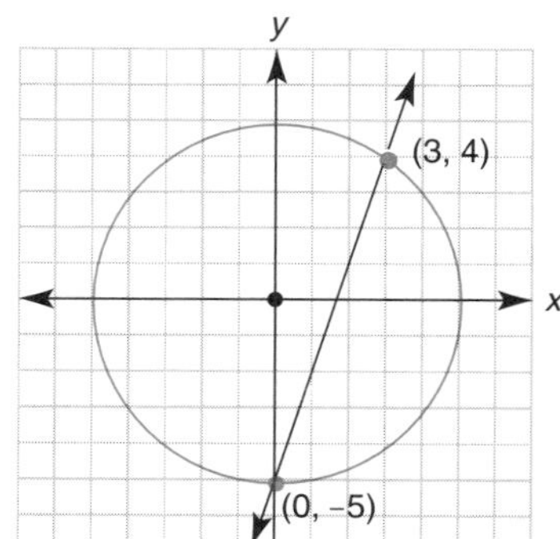

FIGURE 9.46

EXAMPLE 7

Use algebra to determine the points of intersection of the circle $x^2 + y^2 = 25$ and the line $y = 3x - 5$.

Solution

Using $3x - 5$ for y (since $y = 3x - 5$), we substitute into the equation of the circle $x^2 + y^2 = 25$ to obtain

$$\begin{aligned} x^2 + (3x - 5)^2 &= 25 \\ x^2 + 9x^2 - 30x + 25 &= 25 \\ 10x^2 - 30x &= 0 \\ 10x(x - 3) &= 0 \end{aligned}$$

Then $x = 0$ or $x = 3$ at the points of intersection. If $x = 0$, then $y = 3(0) - 5$ or $y = -5$; that is, $(0, -5)$ is one point of intersection. If $x = 3$, then $y = 3(3) - 5$ or $y = 4$; that is, $(3,4)$ is a second point of intersection. As in Example 6, the points of intersection are $(0,-5)$ and $(3,4)$.

EXAMPLE 8

Using geometry, find the intersection of the circle $x^2 + y^2 - 4x + 6y - 3 = 0$ and the line $y = x + 2$.

Solution

In Example 5, we found the graph of the first equation in this system to be a circle with center at $(2, -3)$ and radius of length $r = 4$. The second graph is the line with y intercept 2 and slope $m = 1$. (See Figure 9.47.) The graphs do not intersect, so the solution is the empty set.

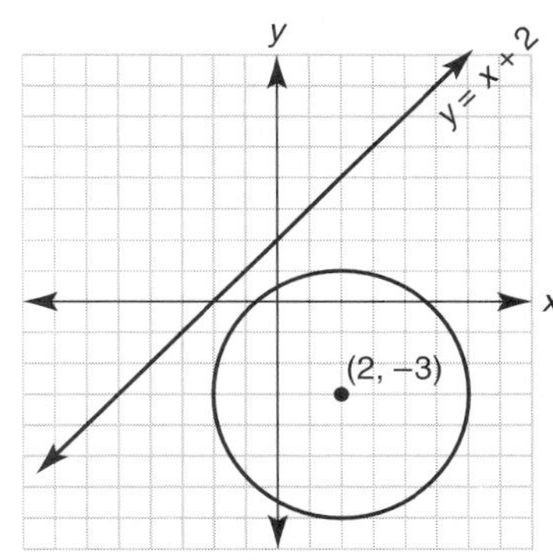

FIGURE 9.47

One difficulty when solving by the graphing method is that it is often difficult to read the graph. For this reason, we repeat Example 8, this time solving by the algebraic method. The result should underscore the need to understand both the algebraic method and the geometric method and how they are related.

EXAMPLE 9

Using algebra, find the intersection of the circle $x^2 + y^2 - 4x + 6y - 3 = 0$ and the line $y = x + 2$.

Solution

Using $x + 2$ for y (since $y = x + 2$), we substitute into the first equation to obtain

$$\begin{aligned} x^2 + (x + 2)^2 - 4x + 6(x + 2) - 3 &= 0 \\ x^2 + (x^2 + 4x + 4) - 4x + 6x + 12 - 3 &= 0 \\ 2x^2 + 6x + 13 &= 0 \end{aligned}$$

But $2x^2 + 6x + 13$ cannot be factored. To use the Quadratic Formula*, let $a = 2$, $b = 6$, and $c = 13$. Then we find

$$\begin{aligned} x &= \frac{-b \pm \sqrt{b^2 - 4ac}}{2a} \\ &= \frac{-6 \pm \sqrt{6^2 - 4(2)(13)}}{2(2)} \\ &= \frac{-6 \pm \sqrt{-68}}{4} \end{aligned}$$

Because $\sqrt{-68}$ is not a real number, there are no points of intersection for the graphs of the equations in the given system. This verifies our claim in Example 8. The solution is the empty set.

* Refer to Appendix A.

9.6 Exercises

In Exercises 1 to 6, write the equation of the circle in the form $(x - h)^2 + (y - k)^2 = r^2$. Then name the center (h, k), give the length of radius r, and sketch the graph.

1. $x^2 + y^2 = 25$
2. $x^2 + y^2 = 16$
3. $(x - 2)^2 + (y + 3)^2 = 9$
4. $(x + 1)^2 + (y - 2)^2 = 19$ (Approximate r to tenths.)
5. $x^2 + y^2 - 2x + 4y + 1 = 0$
   (**HINT:** See Example 5.)
6. $x^2 + y^2 - 4x + 6y - 12 = 0$
   (**HINT:** See Example 5.)
7. Explain why the graph of the equation $x^2 + y^2 = 0$ is a single point. Name the point.
8. Explain why the graph of the equation $(x - 3)^2 + (y + 6)^2 = 0$ is a single point. Name the point.
9. Explain why the equation $x^2 + y^2 = -4$ has no solution.
10. Explain why the equation $x^2 + y^2 + 1 = 0$ has no solution.

In Exercises 11 to 14, use geometry (draw graphs) to find the points of intersection.

11. $x^2 + y^2 = 25$ and $x + 3y = 5$
12. $x^2 + y^2 = 16$ and $x + y = 4$
13. $x^2 + y^2 = 4$ and $y = x - 4$
14. $x^2 + y^2 = 25$ and $(x - 8)^2 + y^2 = 25$

In Exercises 15 to 18, use algebra to find the points of intersection.

15. $x^2 + y^2 = 25$ and $x + 3y = 5$
16. $x^2 + y^2 = 4$ and $y = x - 4$
17. $y = x^2$ and $y = -1x + 2$
18. $y = x^2$ and $x - y = 3$

In Exercises 19 to 26, find the general equation for the circle described. The length of the radius is r.

19. Center is $(2, -5)$; $r = 4$.
20. Center is $(3, -1)$; $r = 2$.

21. Center is $(-3, -1)$; $r = 5$.

22. Center is $(-2, 6)$; $r = 3$.

23. Endpoints of the diameter are $(-3, 2)$ and $(5, -4)$.

24. Endpoints of the diameter are $(8, 5)$ and $(-2, -1)$.

25. $r = 5$; circle lies in Quadrant II and is tangent to both axes.

26. $r = 4$; circle lies in Quadrant I and is tangent to both axes.

27. ➢ The tangent line to the graph of the circle $x^2 + y^2 = 25$ at the point $(3, 4)$ is shown. Write the equation of the tangent line at $(3, 4)$ in the form $Ax + By = C$.
(**HINT:** The tangent line and the radius to it are $\perp$.)

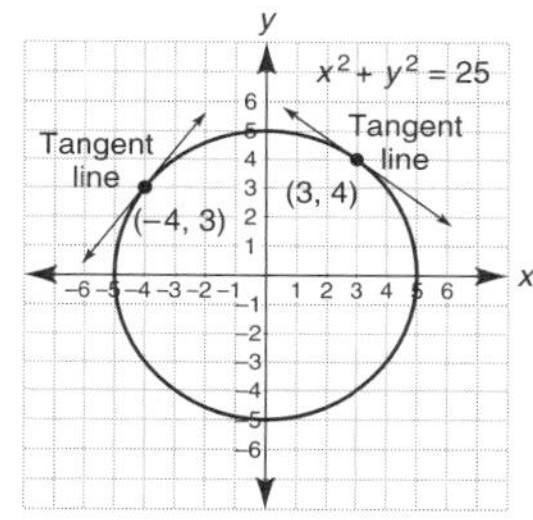

28. ➢ In the form $Ax + By = C$, write the equation of the tangent line to the circle $x^2 + y^2 = 25$ at the point $(-4, 3)$.

29. ➢ How are the two tangent lines to the circle in Exercises 27 and 28 related? Explain.

30. In the circle $x^2 + y^2 = 25$, the points $A(5, 0)$ and $B(-5, 0)$ are the endpoints of diameter $\overline{AB}$. Let C be the point $(3, 4)$. Draw line segments $\overline{AC}$ and $\overline{BC}$. How are $\overline{AC}$ and $\overline{BC}$ related?

31. The parabola with equation $y = x^2$ was shown in Figure 9.42(b) on page 430. How many points of intersection does the circle $x^2 + y^2 = 25$ and the parabola $y = x^2$ have?

32. The parabola with equation $y = x^2$ was shown in Figure 9.42(b) on page 430. How many points of intersection does the circle $x^2 + (y + 4)^2 = 9$ and the parabola $y = x^2$ have?

33. ➢ A circle passes through $(-3, 3)$, $(-1, 7)$, and $(7, 3)$. What point (h, k) is the center of the circle?

34. ➢ The circle $x^2 + y^2 = 100$ has chords $\overline{AB}$ and $\overline{CD}$. If A is $(-10, 0)$, B is $(6, 8)$, C is $(0, -10)$, and D is $(0, 10)$, find X, the point of intersection of the two chords. Then show that $AX \cdot XB = CX \cdot XD$.

35. ➢ Show that an angle inscribed in a semicircle is a right angle.
(**HINT:** Place the center of the circle at the origin.)

36. ➢ Show that the equation for the circle whose center is (h, k) and whose radius has length r is

$$x^2 + y^2 + Dx + Ey + F = 0$$

in which $D = -2h$, $E = -2k$, and $F = h^2 + k^2 - r^2$.

37. ➢ Show that the Point-Slope Form of the tangent to the circle with center (h, k) and radius length r at point of tangency (a, b) is

$$y - b = \frac{a - h}{k - b}(x - a)$$

A Look Beyond: The Banach-Tarski Paradox

In the 1920s, two Polish mathematicians proposed a mathematical dilemma to their colleagues. Known as the Banach-Tarski paradox, their proposal has puzzled students of geometry for decades. What was most baffling was that the proposal indicated that matter could be created through rearrangement of the pieces of a figure. The following steps outline the Banach-Tarski paradox.

First consider the square whose sides are each of length 8. [See Figure 9.48(a) on page 438.] By counting squares or by applying a formula, it is clearly the case that the 8-by-8 square must have an area of 64 square units. We now subdivide the square (as shown) to form two right triangles and two trapezoids. Notice the dimensions indicated on each piece of the square in Figure 9.48(b).

(a)

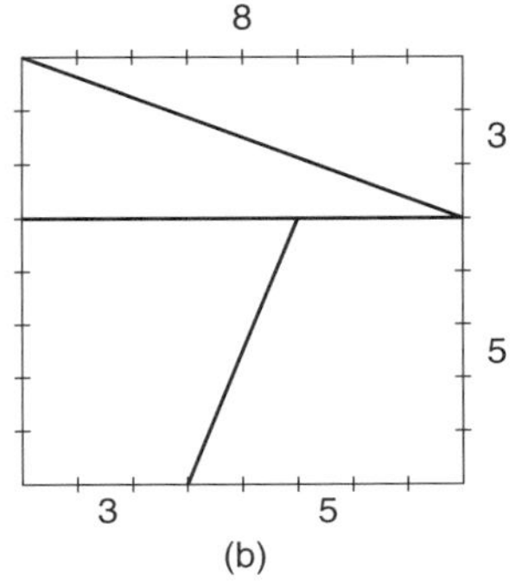

(b)

FIGURE 9.48

The parts of the square are now rearranged to form a rectangle (See Figure 9.49) whose dimensions are 13 and 5. This rectangle clearly has an area that measures 65 square units, 1 square unit more than the given square! How is it possible that the second figure has an area greater than the first?

While the puzzle is real, you may also realize that something is wrong. This paradox can be explained by considering the slopes of lines. The triangles, which have legs of lengths 3 and 8, determine a hypotenuse whose slope is $-\frac{3}{8}$. Although the side of the trapezoid appears to be collinear with the hypotenuse, it actually has a slope of $-\frac{2}{5}$. It was easy to accept that the segments were collinear because the slopes are nearly equal; in fact, $-\frac{3}{8} = -0.375$, while $-\frac{2}{5} = -0.400$. In Figure 9.50, (which is somewhat exaggerated), a very thin parallelogram appears in the space between the original segments of the cut-up square. One may quickly conclude that the area of that parallelogram is 1 square unit, and the paradox has been resolved once more!

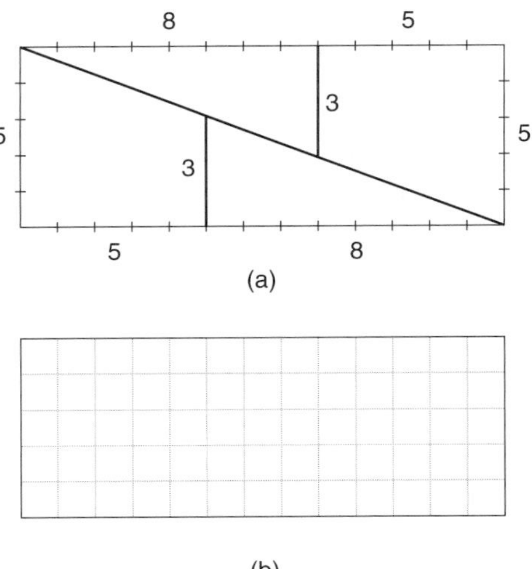

(a)

(b)

FIGURE 9.49

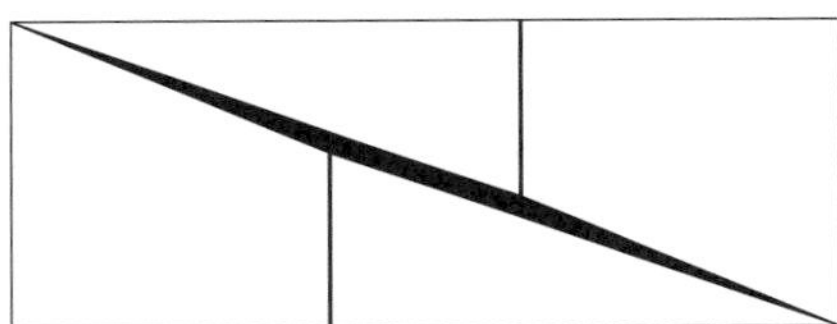

FIGURE 9.50

Summary

A Look Back at Chapter 9

Our goal in this chapter was to relate algebra and geometry. This relationship is called analytic geometry or coordinate geometry. Formulas for the length of a line segment, the midpoint of a line segment, and the slope of a line were developed. We found the general equation for a line and a circle and we used these forms for graphing. Analytic proofs were provided for a number of theorems of geometry.

A Look Ahead to Chapter 10

In the next chapter, we will again deal with the right triangle. Three trigonometric ratios (sine, cosine, and tangent) will be defined for an acute angle of the right triangle in terms of its sides. An area formula for triangles will be derived using the sine ratio. We will also prove the Law of Sines and the Law of Cosines for acute triangles. Another unit for measuring an angle, called the "radian," will be introduced.

Important Terms and Concepts of Chapter 9

9.1 Analytic Geometry
Cartesian Coordinate System
Rectangular Coordinate System
x Axis, y Axis, Quadrants, Origin
x Coordinate, y Coordinate
Ordered Pair
Distance Formula
Linear Equation
Midpoint Formula

9.2 Graphs of Equations
x Intercept, y Intercept
Slope, Slope Formula
Negative Reciprocal

9.3 Equations of Lines
Coincident Lines
Slope-Intercept Form of a Line
Point-Slope Form of a Line

9.4 Preparation for Analytic Proofs

9.5 Analytic Proofs
Synthetic Proofs

9.6 Degree of a Term
Degree of an Equation
Standard Form for the Equation of a Circle
General Form for the Equation of a Circle
Graphing Circles
Completing the Square
Finding Points of Intersection

A Look Beyond: *The Banach-Tarski Paradox*

Review Exercises

1. Find the distance between each pair of points:

a) $(6, 4)$ and $(6, -3)$
b) $(1, 4)$ and $(-5, 4)$
c) $(-5, 2)$ and $(7, -3)$
d) $(x - 3, y + 2)$ and $(x, y - 2)$

2. Find the distance between each pair of points:

a) $(2, -3)$ and $(2, 5)$
b) $(3, -2)$ and $(-7, -2)$
c) $(-4, 1)$ and $(4, 5)$
d) $(x - 2, y - 3)$ and $(x + 4, y + 5)$

3. Find the midpoint of the line segment that joins each pair of points in Exercise 1.

4. Find the midpoint of the line segment that joins each pair of points in Exercise 2.

5. Find the slope of the line joining each pair of points in Exercise 1.

6. Find the slope of the line joining each pair of points in Exercise 2.

7. $(2, 1)$ is the midpoint of $\overline{AB}$, in which A has coordinates $(8, 10)$. Find the coordinates of B.

8. The y axis is the perpendicular bisector of $\overline{RS}$. Find the coordinates of R if S is the point $(-3, 7)$.

9. If A has coordinates $(2, 1)$ and B has coordinates $(x, 3)$, find x so that the slope of $\overleftrightarrow{AB}$ is -3.

10. If R has coordinates $(-5, 2)$ and S has coordinates $(2, y)$, find y so that the slope of $\overleftrightarrow{RS}$ is $\frac{-6}{7}$.

11. Without graphing, determine whether the pairs of lines are parallel, perpendicular, the same, or none of these:

a) $x + 3y = 6$ and $3x - y = -7$
b) $2x - y = -3$ and $y = 2x - 14$
c) $y + 2 = -3(x - 5)$ and $2y = 6x + 11$
d) $0.5x + y = 0$ and $2x - y = 10$

12. Determine whether the points $(-6, 5)$, $(1, 7)$, and $(16, 10)$ are collinear.

13. Find x so that $(-2, 3)$, $(x, 6)$, and $(8, 8)$ are collinear.

14. Draw the graph of $3x + 7y = 21$, and name the x and y intercepts.

15. Draw the graph of $4x - 3y = 9$ by changing the equation to Slope-Intercept Form.

16. Draw the graph of $y + 2 = \frac{-2}{3}(x - 1)$.

17. Write the equation of the figure described:

a) The line through $(2, 3)$ and $(-3, 6)$
b) The line through $(-2, -1)$ and parallel to the line through $(6, -3)$ and $(8, -9)$
c) The line through $(3, -2)$ and perpendicular to $x + 2y = 4$

d) The line through $(-3, 5)$ and parallel to the x axis
e) The circle with center $(3, -1)$ and radius 4
f) The circle with center $(2, -3)$ and tangent to the y axis

18. Show that the triangle whose vertices are $A(-2, -3)$, $B(4, 5)$, and $C(-4, 1)$ is a right triangle.

19. Show that the triangle whose vertices are $A(3, 6)$, $B(-6, 4)$, and $C(1, -2)$ is an isosceles triangle.

20. Show that the quadrilateral whose vertices are $R(-5, -3)$, $S(1, -11)$, $T(7, -6)$, and $V(1, 2)$ is a parallelogram.

Draw the graph of each equation in Exercises 21 and 22.

21. a) $x^2 + y^2 = 49$
b) $(x - 1)^2 + (y + 3)^2 = 25$
c) $x^2 + y^2 - 4x + 10y + 20 = 0$

22. a) $x^2 + y^2 = 18$
b) $(x + 2)^2 + (y - 1)^2 = 16$
c) $x^2 + y^2 + 10x + 2y - 23 = 0$

In Exercises 23 to 26, find the intersection of the graphs of the two equations by graphing.

23. $4x - 3y = -3$
$x + 2y = 13$

24. $x^2 + y^2 = 40$
$3x - y = 20$

25. $y = x^2 + 3$
$y = 4x$
(**HINT:** The parabola of Figure 9.42 is shifted up 3 units.)

26. $x^2 + y^2 = 36$
$x^2 + (y + 3)^2 = 9$

In Exercises 27 to 30, solve the system of equations in Exercises 23 to 26 by using algebraic methods.

27. Refer to Exercise 23.

28. Refer to Exercise 24.

29. Refer to Exercise 25.

30. Refer to Exercise 26.

31. The center of circle O is $(5, 3)$. Point A, with coordinates $(9, 6)$, lies on this circle. Find the equation of circle O.

32. Find the equation of the line tangent to circle O through point A, as described in Exercise 31.

33. Three of four vertices of a parallelogram are $(0, -2)$, $(6, 8)$, and $(10, 1)$. Find the possibilities for the coordinates of the remaining vertex.

34. Find the center and radius of a circle whose equation is $x^2 + y^2 = 8x - 14y + 35$.

35. Find the center and radius of a circle whose equation is $x^2 + y^2 = 12y - 10x - 10$.

36. A $(3, 1)$, B $(5, 9)$, and C $(11, 3)$ are the vertices of $\triangle ABC$.

a) Find the length of the median from B to $\overline{AC}$.
b) Find the slope of the altitude from B to $\overline{AC}$.
c) Find the slope of a line through B parallel to $\overline{AC}$.

In Exercises 37 to 40, supply the missing coordinates for the vertices, using as few variables as possible.

37.

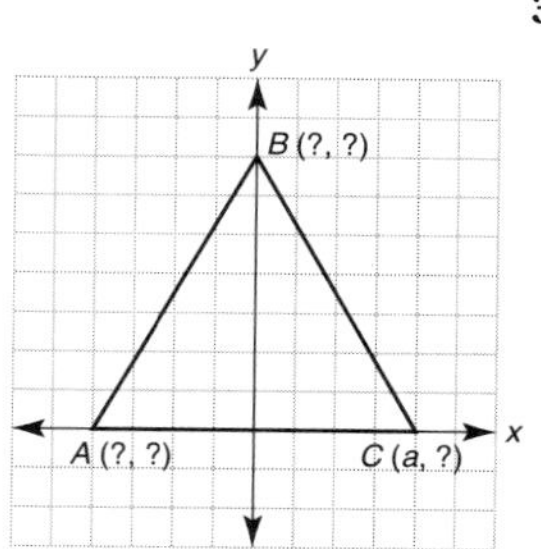

$\triangle ABC$ is isosceles, with base $\overline{AC}$

38.

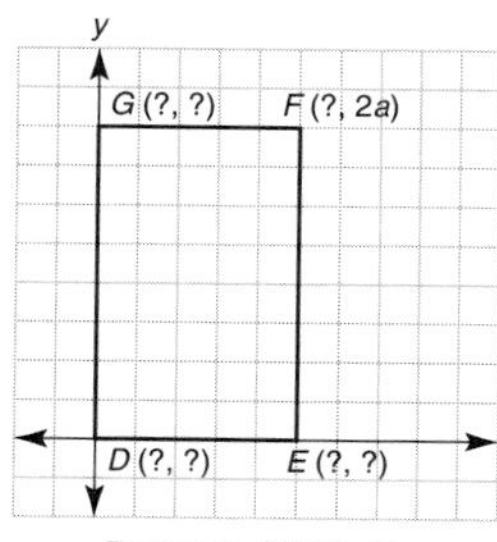

Rectangle $DEFG$ with $\overline{DG} = 2 \cdot \overline{DE}$

39.

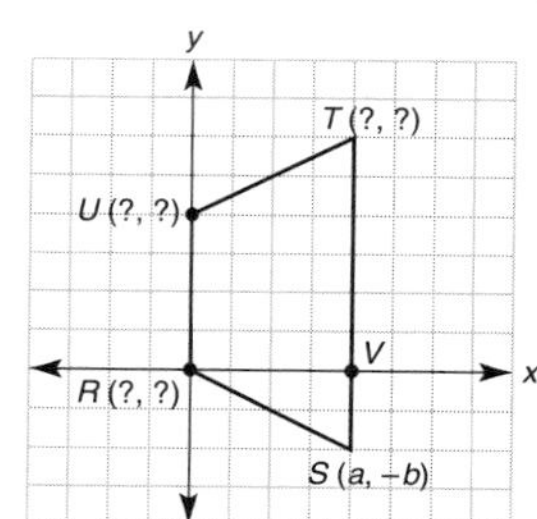

Isosceles trapezoid $RSTU$ with $\overline{RV} \cong \overline{RU}$

40.

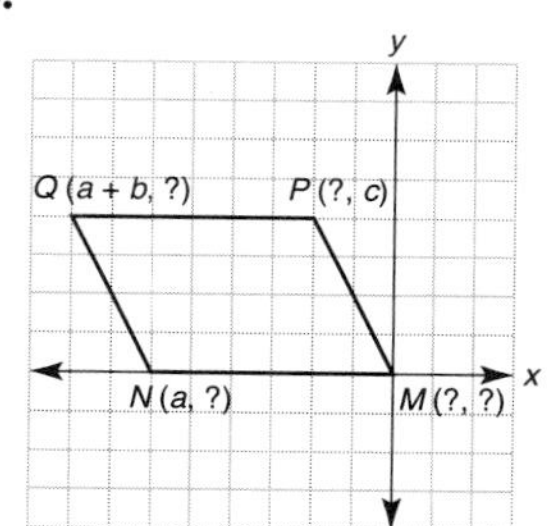

Parallelogram $MPQN$

41. A $(2a, 2b)$, B $(2c, 2d)$, and C $(0, 2e)$ are the vertices of $\triangle ABC$.

a) Find the length of the median from C to $\overline{AB}$.
b) Find the slope of the altitude from B to $\overline{AC}$.
c) Find the equation of the altitude from B to $\overline{AC}$.

Prove the statements in Exercises 42 to 46 by using analytic geometry.

42. The segments that join the midpoints of consecutive sides of a parallelogram form another parallelogram.

43. If the diagonals of a rectangle are perpendicular, then the rectangle is a square.

44. If the diagonals of a trapezoid are equal in length, then the trapezoid is an isosceles trapezoid.

45. If two medians of a triangle are equal in length, then the triangle is isosceles.

46. The segments joining the midpoints of consecutive sides of an isosceles trapezoid form a rhombus.

Chapter 10
Introduction to Trigonometry

A surveyor uses a transit to obtain accurate angle measurements. The surveyor can then use a trigonometric ratio to find an unknown length. The word "trigonometry" refers to the measurement of triangles. In this chapter, you will discover methods for measuring the angles as well as lengths of sides of right triangles when the measures of other parts of the triangle are known. These techniques can be extended to other types of triangles.

For this chapter, you will need to use a scientific or graphics calculator.

CHAPTER OUTLINE

10.1 The Sine Ratio and Applications

In this section, we will deal strictly with similar right triangles. In Figure 10.1, $\triangle ABC \sim \triangle DEF$ and $\angle C$ and $\angle F$ are right angles. Consider corresponding angles A and D. If we compare the length of the side opposite each angle to the length of the hypotenuse of each triangle, we see that

$$\frac{BC}{AB} = \frac{EF}{DE} \quad \text{or} \quad \frac{3}{5} = \frac{6}{10}$$

In the two similar right triangles, the ratio of this pair of corresponding sides depends on the measure of acute $\angle A$ (or $\angle D$ since $m\angle A = m\angle D$); for this angle, the numerical value of the ratio

$$\frac{\text{length of side opposite the acute angle}}{\text{length of hypotenuse}}$$

is unique. This ratio becomes smaller for smaller measures of $\angle A$ and larger for larger measures of $\angle A$. This ratio is unique for each measure of an acute angle even though the lengths of the sides of the two similar right triangles containing the angle are different.

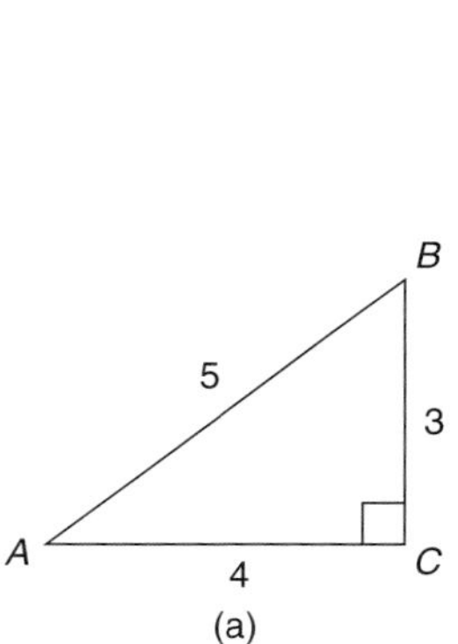

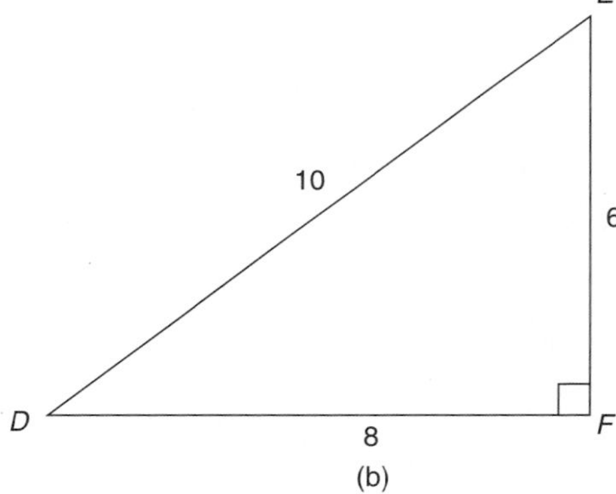

FIGURE 10.1

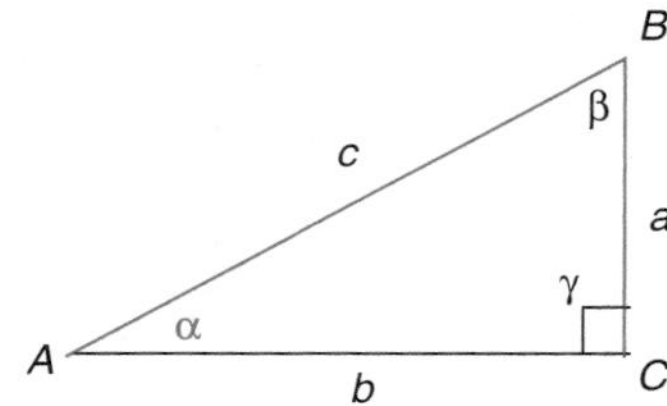

FIGURE 10.2

In Figure 10.2, we name the measures of the angles of the right triangle by the Greek letters α (alpha) at vertex A, β (beta) at vertex B, and γ (gamma) at vertex C. The lengths of the sides opposite vertices A, B, and C are a, b, and c, respectively. Relative to the acute angle, the lengths of the sides of the right triangle in the following definition are described as "opposite" and "hypotenuse." The word **opposite** is used to mean the length of the side opposite the angle named; the word **hypotenuse** is used to mean the length of the hypotenuse.

DEFINITION: In a right triangle, the **sine ratio** for an acute angle is the ratio $\frac{\text{opposite}}{\text{hypotenuse}}$.

NOTE: In right $\triangle ABC$ in Figure 10.2, we say that $\sin \alpha = \frac{a}{c}$ and $\sin \beta = \frac{b}{c}$, in which "sin" is an abbreviation of the word "sine" (pronounced like "sign").

EXAMPLE 1

In Figure 10.3, find $\sin \alpha$ and $\sin \beta$ for right $\triangle ABC$.

Solution

$a = 3$, $b = 4$, and $c = 5$. Therefore

$$\sin \alpha = \frac{a}{c} = \frac{3}{5}$$

and

$$\sin \beta = \frac{b}{c} = \frac{4}{5}$$

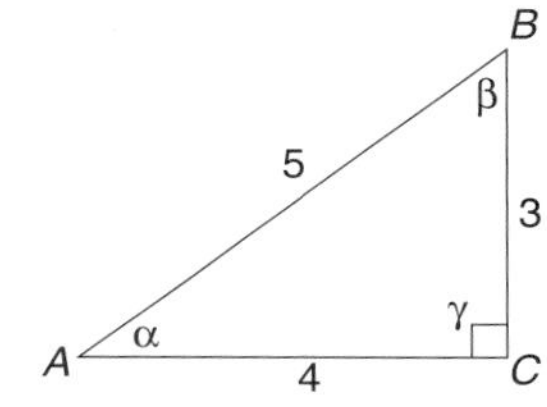

FIGURE 10.3

NOTE: Some textbooks use $\sin \alpha$ and $\sin A$ interchangeably. In Example 1, it is possible to state that $\sin A = \frac{3}{5}$ and $\sin B = \frac{4}{5}$.

EXAMPLE 2

In Figure 10.4, find $\sin \alpha$ and $\sin \beta$ for right $\triangle ABC$.

Solution

Where $a = 5$ and $c = 13$, we know that $b = 12$, based on the Pythagorean triple (5, 12, 13). We verify this result using the Pythagorean Theorem

$$\begin{aligned} c^2 &= a^2 + b^2 \\ 13^2 &= 5^2 + b^2 \\ 169 &= 25 + b^2 \\ b^2 &= 144 \\ b &= 12 \end{aligned}$$

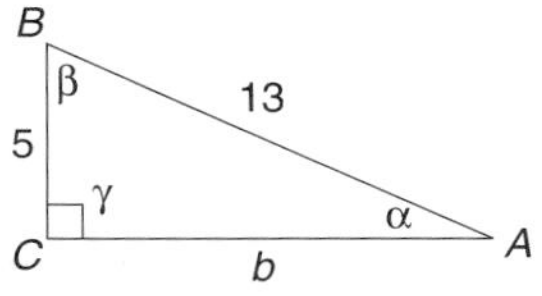

Therefore $\quad \sin \alpha = \frac{a}{c} = \frac{5}{13} \quad$ and $\quad \sin \beta = \frac{b}{c} = \frac{12}{13}$

FIGURE 10.4

The following activity is designed to give you a stronger comprehension of the meaning of expressions such as sin 53°.

DISCOVER!

Given that an acute angle of a right triangle measures 53°, find the approximate value of sin 53°. We can estimate the value of sin 53° as follows (refer to the triangle at the left):

1. Draw right $\triangle ABC$ so that $\alpha = 53°$ and $\gamma = 90°$.
2. For convenience, mark off the length of the hypotenuse as 4 cm.
3. Using a ruler, measure the length of the leg opposite the angle measuring 53°. It is approximately 3.2 cm long.
4. Now divide $\frac{\text{opposite}}{\text{hypotenuse}}$ or $\frac{3.2}{4}$ to find that $\sin 53° \approx 0.8$.

NOTE: A calculator provides greater accuracy than this geometric approach by giving the result $\sin 53° \approx 0.7986$.

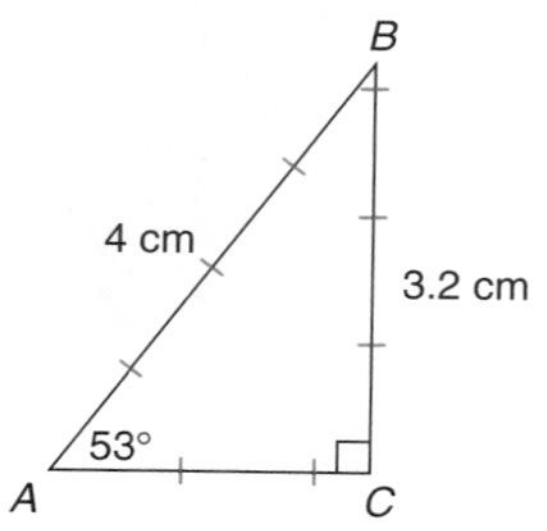

Warning

Be sure to write $\sin \alpha = \frac{5}{13}$ or $\sin 54° \approx 0.8090$. It is incorrect to write "sin" in a claim without naming the angle or its measure; for example, $\sin = \frac{5}{13}$ and $\sin \approx 0.8090$ are both absolutely meaningless.

Repeat the procedure in the preceding "Discover!" and use it to find an approximation for sin 37°. You will need to use the Pythagorean Theorem to find *AC*. You should find that $\sin 37° \approx 0.6$.

Although the sine ratios for angle measures are readily available on a calculator, we can justify several of the calculator's results by using special triangles. In fact, we can find *exact* results where the calculator provides approximations.

Recall the 30-60-90 relationship, in which the side opposite the 30° angle has a length equal to one-half that of the hypotenuse; the remaining leg has a length equal to the product of the length of the shorter leg and $\sqrt{3}$. In this case, we write $\sin 30° = \frac{x}{2x} = \frac{1}{2}$. (See Figure 10.5.)

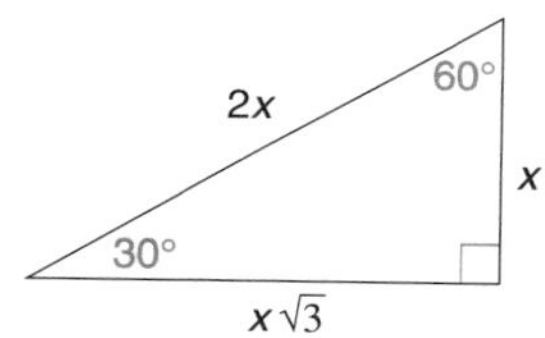

FIGURE 10.5

It is also true that $\sin 60° = \frac{x\sqrt{3}}{2x} = \frac{\sqrt{3}}{2}$. Although the exact value of sin 30° is 0.5 and the exact value of sin 60° is $\frac{\sqrt{3}}{2}$, a calculator would give an approximate value for sin 60°, such as 0.8660254. If we round the ratio for sin 60° to four decimal places, then $\sin 60° \approx 0.8660$. Use your calculator to show that $\frac{\sqrt{3}}{2} \approx 0.8660$.

EXAMPLE 3

Find exact and approximate values for sin 45°.

Solution

Using the 45-45-90 triangle in Figure 10.6, we see that $\sin 45° = \frac{x}{x\sqrt{2}} = \frac{1}{\sqrt{2}}$. Equivalently, $\sin 45° = \frac{\sqrt{2}}{2}$. A calculator approximation is $\sin 45° \approx 0.7071$.

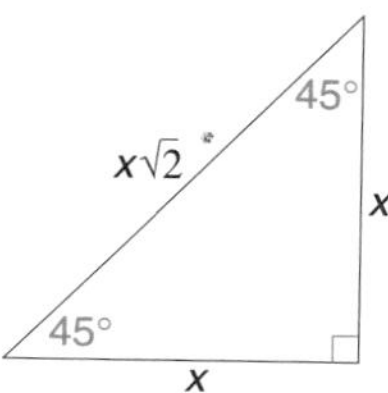

FIGURE 10.6

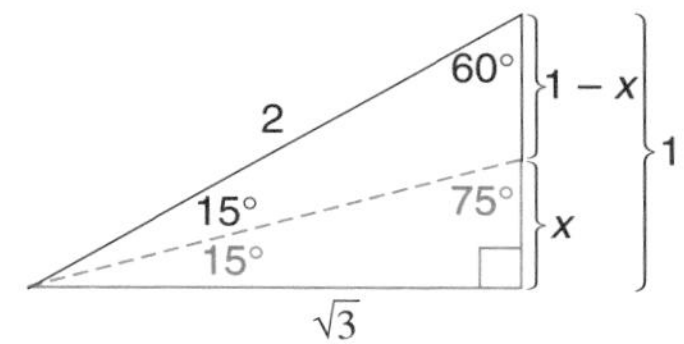

FIGURE 10.7

We will now use an earlier result (from Section 5.5) to determine the sine ratios for angles that measure 15° and 75°. Recall that an angle bisector of one angle of a triangle divides the opposite side into two segments that are proportional to the sides forming the bisected angle. Using this fact in the 30-60-90 triangle in Figure 10.7, we are led to the proportion

$$\frac{x}{1 - x} = \frac{\sqrt{3}}{2}$$

Applying the Means-Extremes Property,

$$\begin{aligned} 2x &= \sqrt{3} - \sqrt{3}x \\ 2x + \sqrt{3}x &= \sqrt{3} \\ (2 + \sqrt{3})x &= \sqrt{3} \\ x &= \frac{\sqrt{3}}{2 + \sqrt{3}} \approx 0.4641 \end{aligned}$$

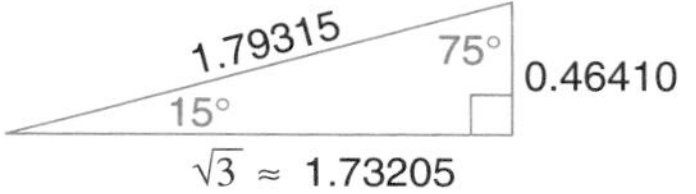

FIGURE 10.8

The number 0.4641 is the length of the side that is opposite the 15° angle of the 15-75-90 triangle (see Figure 10.8). Using the Pythagorean Theorem, we can show that the length of the hypotenuse is approximately 1.79315. Then $\sin 15° = \frac{0.46410}{1.79315} \approx 0.2588$. Using the same triangle, $\sin 75° = \frac{1.73205}{1.79315} \approx 0.9659$.

We now begin to formulate a small table of values of sine ratios. In Table 10.1, the Greek letter θ (theta) designates the angle measure in degrees. The second column has the heading $\sin\theta$ and provides the ratio for the corresponding angle; this ratio is generally given to four decimal places of accuracy. Notice that the values of $\sin\theta$ increase as θ increases in measure.

■ *Warning*

Notice that $\sin\left(\frac{1}{2}\theta\right) \neq \frac{1}{2}\sin\theta$ *in Table 10.1. When* $\theta = 60°$; $\sin 30° \neq \frac{1}{2}\sin 60°$ *since* $0.5000 \neq \frac{1}{2}(0.8660)$. ■

TABLE 10.1

Sine Ratios

θ	$\sin\theta$
15°	0.2588
30°	0.5000
45°	0.7071
60°	0.8660
75°	0.9659

NOTE: Most values provided in tables or given by a calculator are approximations. Although we use the equality symbol (=) when reading values from a table (or calculator), the solutions to the problems that follow are generally approximations.

In Figure 10.9 on the next page, let $\angle\theta$ be the acute angle whose measure increases as shown. In the figure, notice that the length of the hypotenuse is constant—it is always equal to the length of the radius of the circle. However, the side opposite $\angle\theta$ gets larger as θ increases. In fact, as θ approaches 90° ($\theta \to 90°$), the length of the leg opposite $\angle\theta$ approaches the length of the hypotenuse. As $\theta \to 90°$, $\sin\theta \to 1$. As θ decreases, $\sin\theta$ also decreases. As θ decreases ($\theta \to 0°$), the length of the side opposite $\angle\theta$ approaches 0. As $\theta \to 0°$, $\sin\theta \to 0$. These observations lead to the following definition.

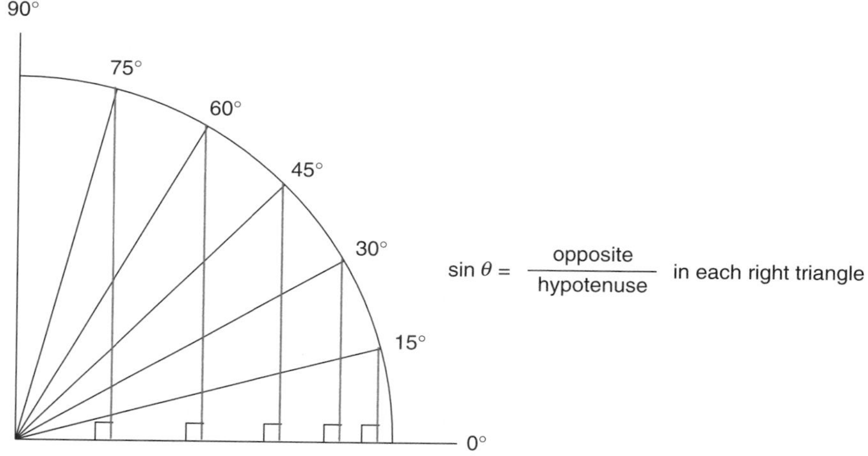

FIGURE 10.9

DEFINITION: $\sin 0° = 0$ and $\sin 90° = 1$.

NOTE: A calculator will verify these results.

EXAMPLE 4

Using Table 10.1 on page 445, find the length of a in Figure 10.10 to the nearest tenth.

Solution

$$\sin 15° = \frac{\text{opposite}}{\text{hypotenuse}} = \frac{a}{10}$$

From the table, we have $\sin 15° = 0.2588$.

$$\frac{a}{10} = 0.2588 \qquad \text{by substitution}$$
$$a = 2.588$$

Therefore $a = 2.6$ in. when rounded to tenths.

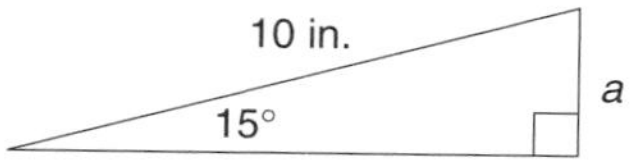

10 in.
15°
a

FIGURE 10.10

In an application problem, the sine ratio can be used to find the measure of either a side or an angle of a triangle. To find the sine ratio of the angle involved, you may use a table of ratios or a calculator. Table 10.2 provides ratios for many more angle measures than does Table 10.1. Like calculators, tables generally provide only approximations of the sine ratios indicated.

TABLE 10.2

Sine Ratios

θ	$\sin \theta$	θ	$\sin \theta$	θ	$\sin \theta$	θ	$\sin \theta$
0°	0.0000	23°	0.3907	46°	0.7193	69°	0.9336
1°	0.0175	24°	0.4067	47°	0.7314	70°	0.9397
2°	0.0349	25°	0.4226	48°	0.7431	71°	0.9455
3°	0.0523	26°	0.4384	49°	0.7547	72°	0.9511
4°	0.0698	27°	0.4540	50°	0.7660	73°	0.9563
5°	0.0872	28°	0.4695	51°	0.7771	74°	0.9613
6°	0.1045	29°	0.4848	52°	0.7880	75°	0.9659
7°	0.1219	30°	0.5000	53°	0.7986	76°	0.9703
8°	0.1392	31°	0.5150	54°	0.8090	77°	0.9744
9°	0.1564	32°	0.5299	55°	0.8192	78°	0.9781
10°	0.1736	33°	0.5446	56°	0.8290	79°	0.9816
11°	0.1908	34°	0.5592	57°	0.8387	80°	0.9848
12°	0.2079	35°	0.5736	58°	0.8480	81°	0.9877
13°	0.2250	36°	0.5878	59°	0.8572	82°	0.9903
14°	0.2419	37°	0.6018	60°	0.8660	83°	0.9925
15°	0.2588	38°	0.6157	61°	0.8746	84°	0.9945
16°	0.2756	39°	0.6293	62°	0.8829	85°	0.9962
17°	0.2924	40°	0.6428	63°	0.8910	86°	0.9976
18°	0.3090	41°	0.6561	64°	0.8988	87°	0.9986
19°	0.3256	42°	0.6691	65°	0.9063	88°	0.9994
20°	0.3420	43°	0.6820	66°	0.9135	89°	0.9998
21°	0.3584	44°	0.6947	67°	0.9205	90°	1.0000
22°	0.3746	45°	0.7071	68°	0.9272		

NOTE: In later sections, we will use the calculator (rather than tables) to find values of trigonometric ratios such as sin 36°.

EXAMPLE 5

Find sin 36°, using

a) Table 10.2.

b) a scientific or graphics calculator.

Solution

a) Find 36° under the heading θ. Now read the number under the sin θ heading:

$$\sin 36° = 0.5878$$

b) On a scientific calculator that is *in degree mode,* use the following key sequence:

[3] → [6] → [sin] → [**0.5878**]

The result is sin 36° = 0.5878, correct to four decimal places.

NOTE 1: The boldfaced number in the box represents the final answer.

NOTE 2: The key sequence for a graphics calculator follows. Here, the calculator is in degree mode and the answer is rounded to four decimal places.

[sin] → [3] → [6] → [ENTER] → [**0.5878**]

The table or a calculator can also be used to find the measure of an angle. This is possible when the sine of the angle is known.

EXAMPLE 6

If sin $\theta = 0.7986$, find θ to the nearest degree by using

a) Table 10.2. **b)** a calculator.

Solution

a) Find 0.7986 under the heading sin θ. Now look to the left to find the degree measure of the angle in the θ column:

$$\sin \theta = 0.7986 \rightarrow \theta = 53°$$

b) On some scientific calculators, you can use the following key sequence (while in the degree mode) to find θ:

[.] → [7] → [9] → [8] → [6] → [inv] → [sin] → [**53**]

The combination "inv" and "sin" yields the angle whose sine ratio is known, so $\theta = 53°$.

NOTE: On a graphics calculator that is in degree mode, use this sequence:

[$\sin^{-1}$] → [.] → [7] → [9] → [8] → [6] → [ENTER] → [**53**]

The expression $\sin^{-1}.7986$ means "the angle whose sine is 0.7986." The calculator function [$\sin^{-1}$] is found by pressing [2nd] followed by pressing [sin].

In most application problems, a drawing provides a good deal of information and allows some insight into the method of solution. For some drawings and applications, the phrases "angle of elevation" and "angle of depression" are used. These angles are measured from the horizontal as illustrated in Figures 10.11(a) and (b). In Figure 10.11(a), the angle α measured upward from the horizontal ray is the **angle of elevation.** In Figure 10.11(b), the angle β measured downward from the horizontal ray is the **angle of depression.**

FIGURE 10.11

EXAMPLE 7

The tower for a radio station stands 200 ft tall. A guy wire 250 ft long supports the antenna, as shown in Figure 10.12. Find the measure of the angle of elevation α to the nearest degree.

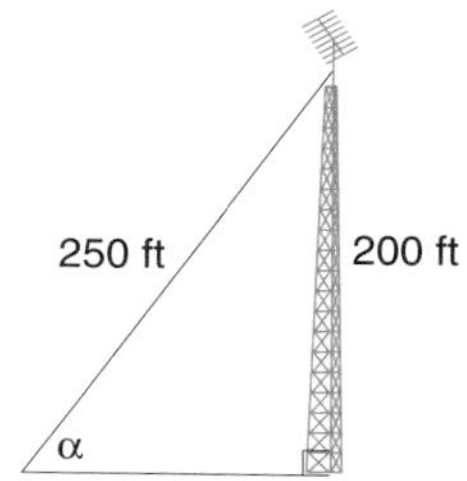

FIGURE 10.12

Solution

$$\sin \alpha = \frac{\text{opposite}}{\text{hypotenuse}} = \frac{200}{250} = 0.8$$

From Table 10.2 (or from a calculator), we find that the angle whose sine ratio is 0.8 is $\alpha = 53°$.

10.1 Exercises

In Exercises 1 to 6, find $\sin \alpha$ and $\sin \beta$ for the triangle shown.

1.

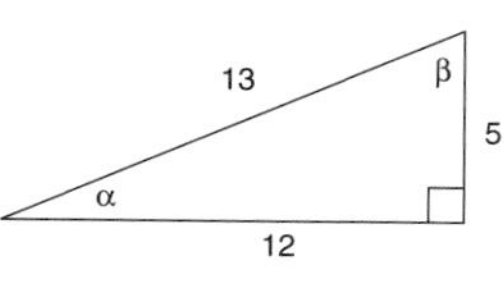

2.

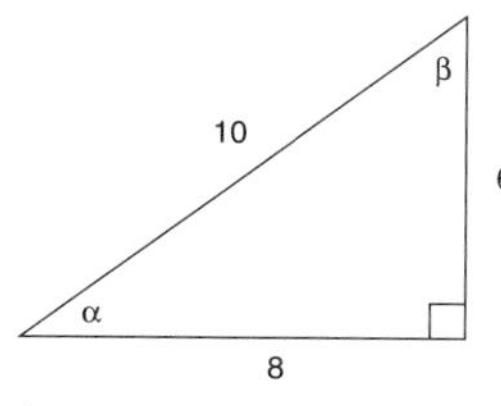

3.

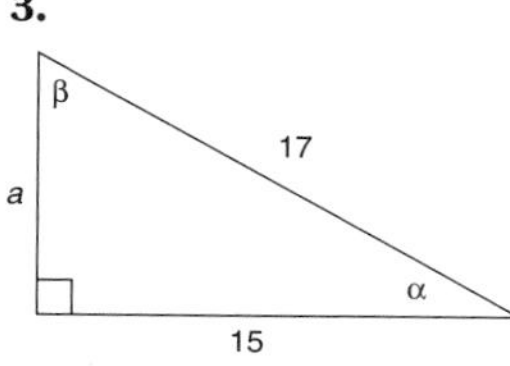

4.

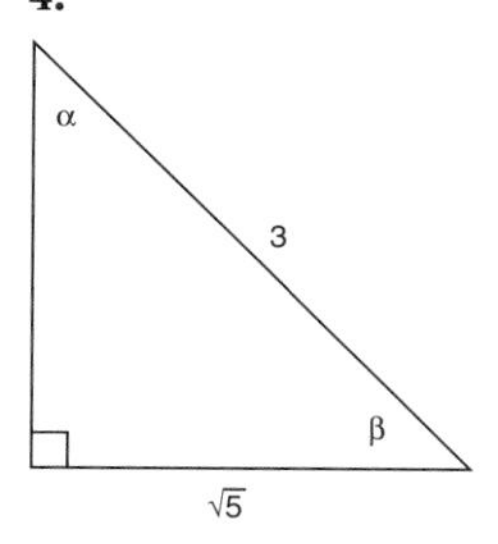

5.

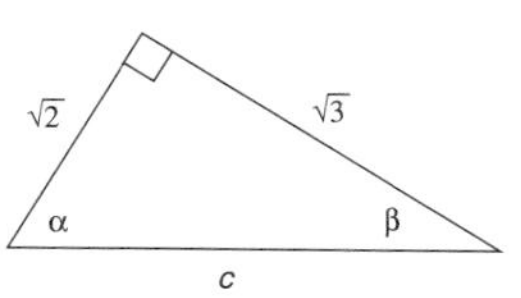

6.

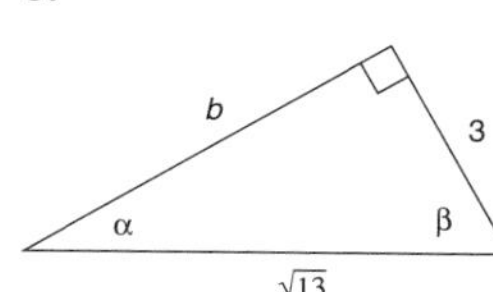

In Exercises 7 to 14, use either Table 10.2 or a calculator to find the sine of the indicated angle to four decimal places.

7. $\sin 90°$

8. $\sin 0°$

9. $\sin 17°$

10. $\sin 23°$

11. $\sin 82°$

12. $\sin 46°$

13. $\sin 72°$

14. $\sin 57°$

In Exercises 15 to 20, find the lengths of the sides indicated by the variables. Use either Table 10.2 or a calculator and round answers to the nearest tenth of a unit.

15.

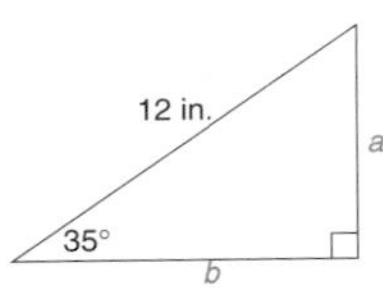

16.

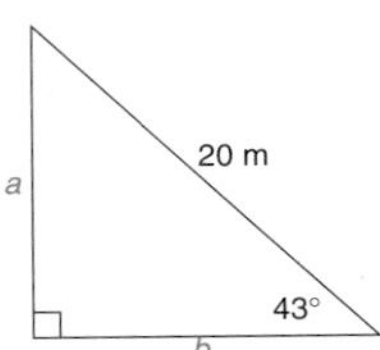

17.

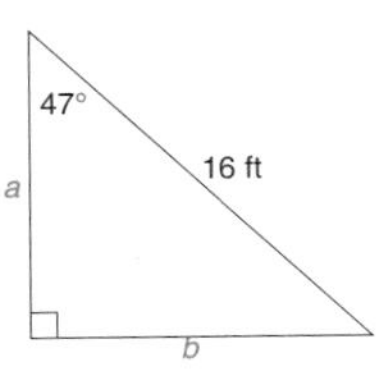

18.

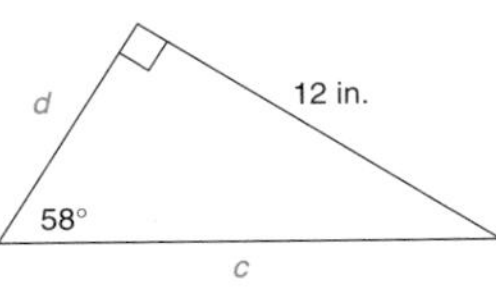

19.

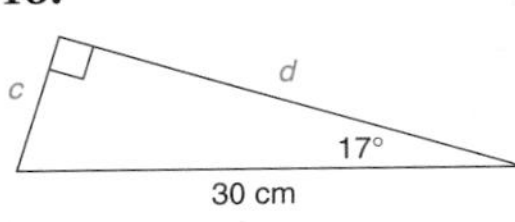

20.

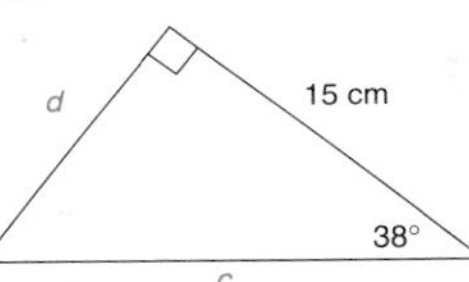

In Exercises 21 to 26, find the measures of the angles named to the nearest degree.

21.

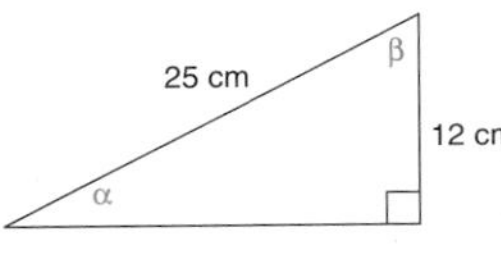

22.

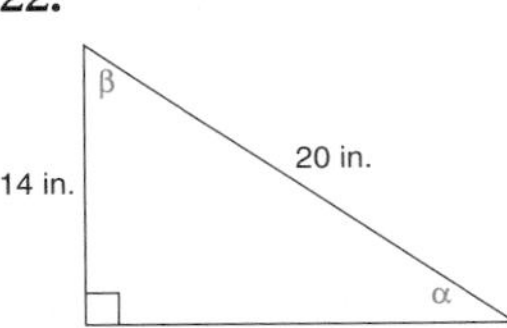

23.

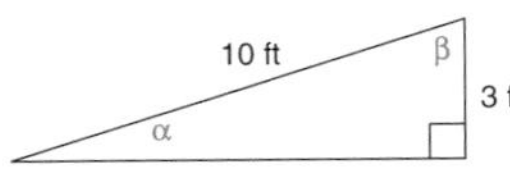

24.

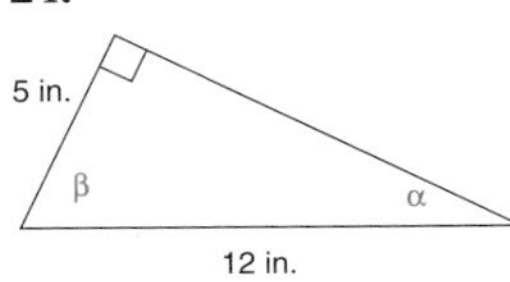

25.

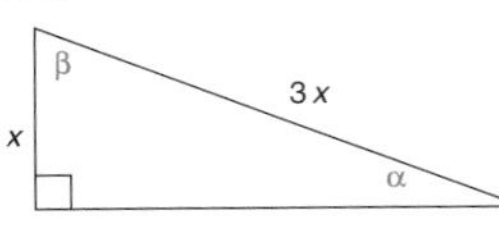

26.

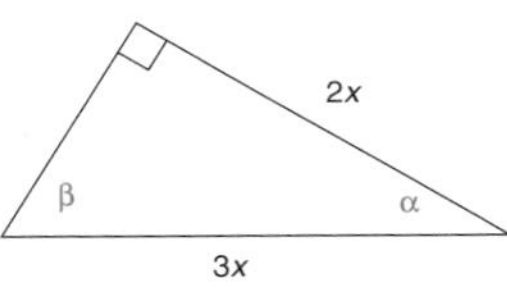

In Exercises 27 to 34, use the drawings provided to solve each problem. Angle measures should be given to the nearest degree; distances should be given to the nearest tenth of a unit.

27. The pitch or slope of a roofline is 5 to 12. Find the measure of angle α.

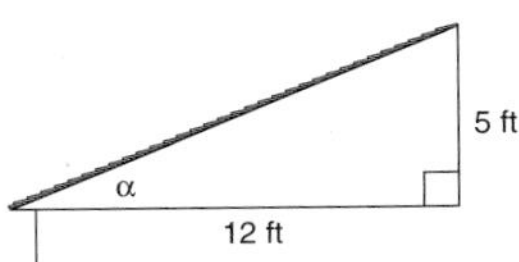

28. A kite is flying at an angle of elevation of 67° from a point on the ground. If 100 ft of kite string is out, how far is the kite above the ground?

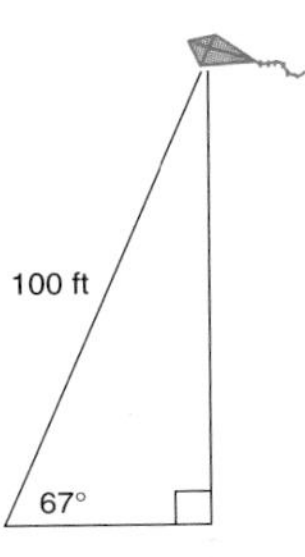

29. Danny sees a balloon that is 100 ft above the ground. If the angle of elevation from Danny to the balloon is 75°, how far from Danny is the balloon?

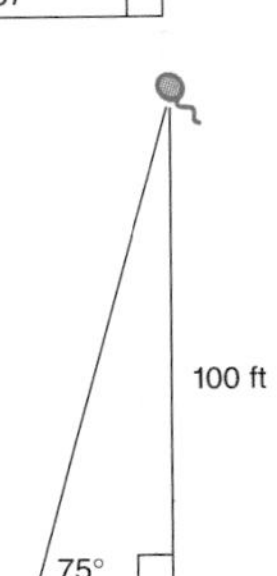

30. Over a 2000-ft span of highway through a hillside, there is a 100-ft rise in the roadway. What is the measure of the angle formed by the road and the horizontal?

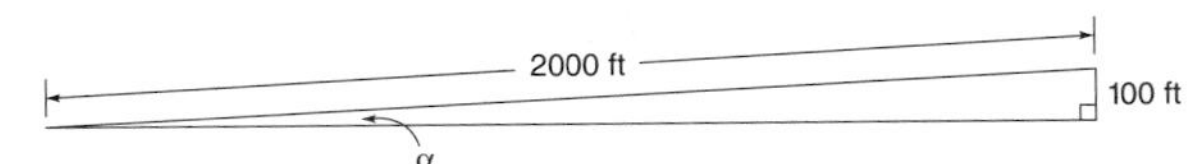

31. From a cliff, a person observes an automobile through an angle of depression of 23°. If the cliff is 50 ft high, how far is the automobile from the person?

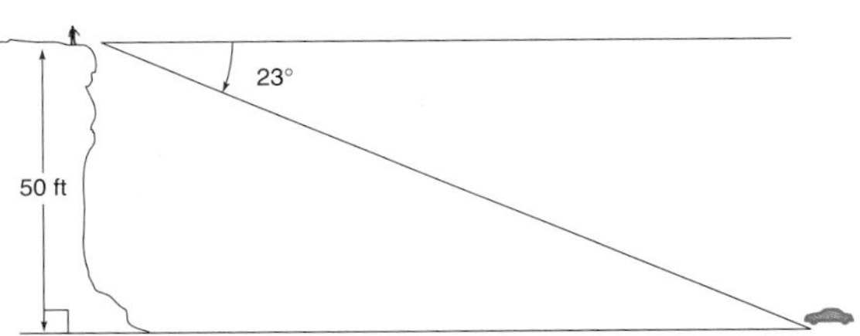

32. A 12-ft rope secures a rowboat to a pier that is 4 ft above the water. What is the angle formed by the rope and the water?

33. A 10-ft ladder is leaning against a vertical wall so that the bottom of the ladder is 4 ft away from the base of the wall. How large is the angle formed by the ladder and the wall?

34. An airplane flying at the rate of 350 ft per sec begins to climb at an angle of 10°. What is the increase in altitude over the next 15 sec?

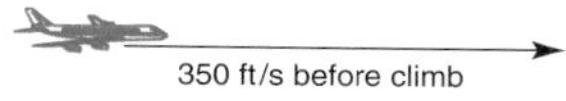

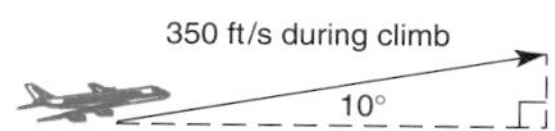

For Exercises 35–38, make drawings as needed.

35. In parallelogram $ABCD$, $AB = 6$ ft and $AD = 10$ ft. If $m\angle A = 65°$ and $\overline{BE}$ is the altitude to $\overline{AD}$, find:

a) BE correct to tenths
b) The area of $\square ABCD$

36. In right $\triangle ABC$, $\gamma = 90°$ and $\beta = 55°$. If $AB = 20$ in., find:

a) a (the length of $\overline{BC}$) correct to tenths
b) b (the length of $\overline{AC}$) correct to tenths
c) The area of right $\triangle ABC$

37. In a right circular cone, the slant height is 13 cm and the altitude is 10 cm. To the nearest degree, find the measure of the angle θ that is formed by the radius and slant height.

38. In a right circular cone, the slant height is 13 cm. Where θ is the angle formed by the radius and the slant height, $\theta = 48°$. Find the length of the altitude of the cone, correct to tenths.

10.2 The Cosine Ratio and Applications

Again we deal strictly with similar right triangles, as shown in Figure 10.13. While $\overline{BC}$ is the leg opposite angle A, we say that $\overline{AC}$ is the leg **adjacent** to angle A. In the two triangles, the ratios of the form

$$\frac{\text{length of adjacent leg}}{\text{length of hypotenuse}}$$

are equal; that is,

$$\frac{AC}{AB} = \frac{DF}{DE} \text{ or } \frac{4}{5} = \frac{8}{10}$$

because corresponding sides of ~ △s are proportional.

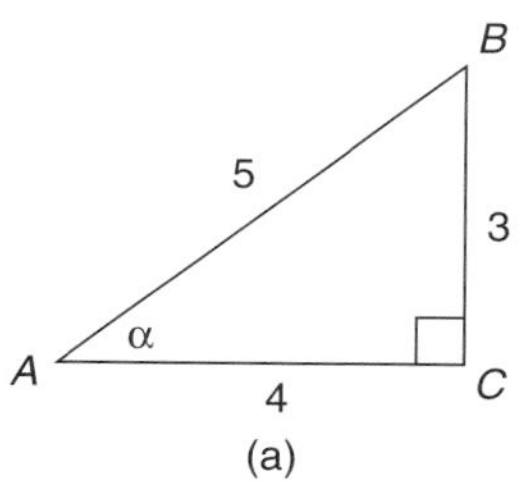

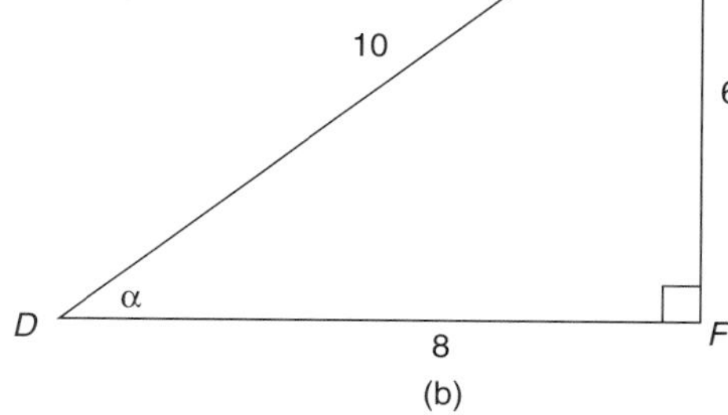

FIGURE 10.13

As with the sine ratio, the *cosine ratio* depends on the measure of acute angle A (or D). In the following definition, "adjacent" refers to the length of the leg that is adjacent to the angle named.

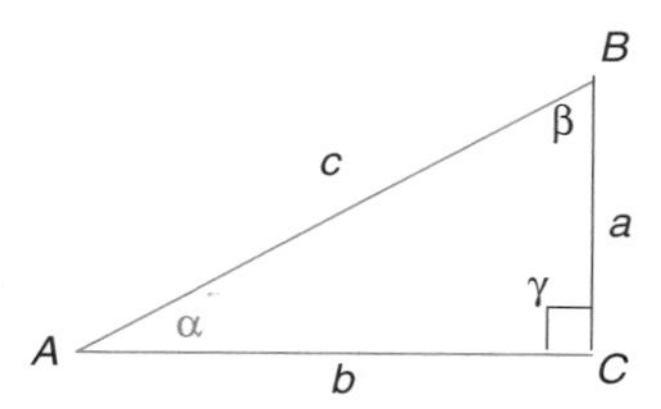

FIGURE 10.14

DEFINITION: In a right triangle, the **cosine ratio** for an acute angle is the ratio $\frac{\text{adjacent}}{\text{hypotenuse}}$.

NOTE: In right $\triangle ABC$ in Figure 10.14, we have $\cos\alpha = \frac{b}{c}$ and $\cos\beta = \frac{a}{c}$, in which "cos" is an abbreviated form of the word "cosine."

EXAMPLE 1

Find $\cos\alpha$ and $\cos\beta$ for right $\triangle ABC$ in Figure 10.15.

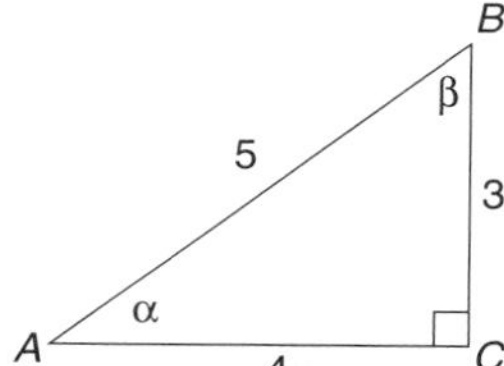

FIGURE 10.15

Solution $a = 3$, $b = 4$, and $c = 5$ for the triangle shown in Figure 10.15. Because b is the length of the leg adjacent to α and a is the length of the leg adjacent to β,

$$\cos\alpha = \frac{b}{c} = \frac{4}{5} \quad \text{and} \quad \cos\beta = \frac{a}{c} = \frac{3}{5}$$

EXAMPLE 2

Find $\cos\alpha$ and $\cos\beta$ for right $\triangle ABC$ in Figure 10.16.

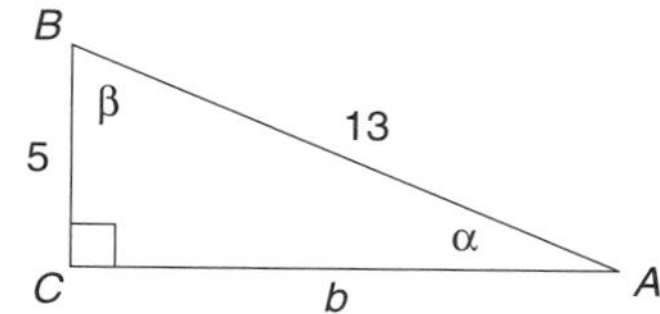

FIGURE 10.16

Solution $a = 5$ and $c = 13$. Then $b = 12$ from the Pythagorean triple (5, 12, 13).

Consequently,

$$\cos\alpha = \frac{b}{c} = \frac{12}{13} \quad \text{and} \quad \cos\beta = \frac{a}{c} = \frac{5}{13}$$

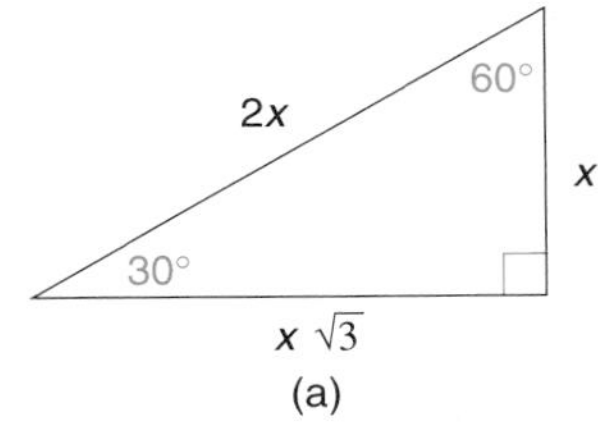

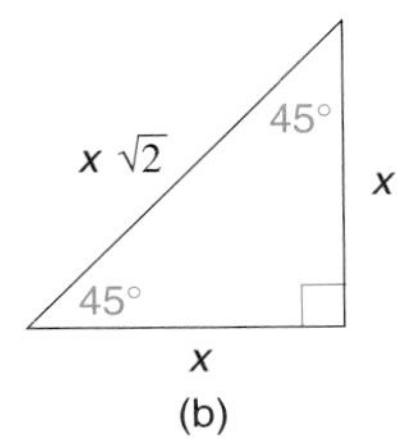

FIGURE 10.17

Just as the sine ratio of any angle is unique, the cosine ratio of any angle is also unique. Using the 30-60-90 and 45-45-90 triangles of Figure 10.17, we see that

$$\cos 30° = \frac{x\sqrt{3}}{2x} = \frac{\sqrt{3}}{2} \approx 0.8660$$

$$\cos 45° = \frac{x}{x\sqrt{2}} = \frac{1}{\sqrt{2}} = \frac{\sqrt{2}}{2} \approx 0.7071$$

$$\cos 60° = \frac{x}{2x} = \frac{1}{2} = 0.5$$

Now we use the 15-75-90 triangle shown in Figure 10.18 to find cos 75° and cos 15°. From Section 10.1, $\sin 15° = \frac{a}{c}$ and $\sin 15° = 0.2588$. But $\cos 75° = \frac{a}{c}$, so $\cos 75° = 0.2588$. Similarly, because $\sin 75° = \frac{b}{c} = 0.9659$, we see that $\cos 15° = \frac{b}{c} = 0.9659$.

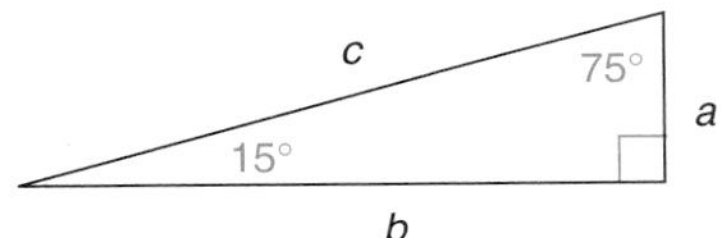

FIGURE 10.18

In Figure 10.19, we see that the cosine ratios become larger as θ decreases and become smaller as θ increases. To understand why, consider the definition

$$\cos \theta = \frac{\text{length of adjacent leg}}{\text{length of hypotenuse}} \text{ and Figure 10.19}$$

Recall that the symbol $\to$ is read "approaches." As $\theta \to 0°$, length of adjacent leg $\to$ length of hypotenuse, and therefore $\cos 0° \to 1$. Similarly, $\cos 90° \to 0$ because the adjacent leg grows smaller. Consequently, we have the following definition.

DEFINITION: $\cos 0° = 1$ and $\cos 90° = 0$.

NOTE: A calculator will verify these results.

We summarize cosine ratios in Table 10.3.

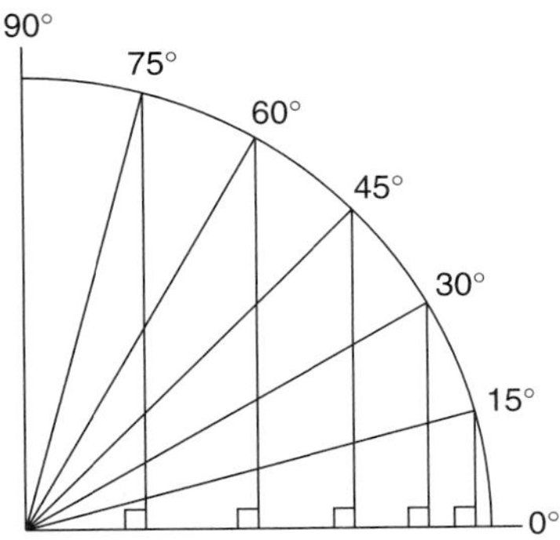

FIGURE 10.19

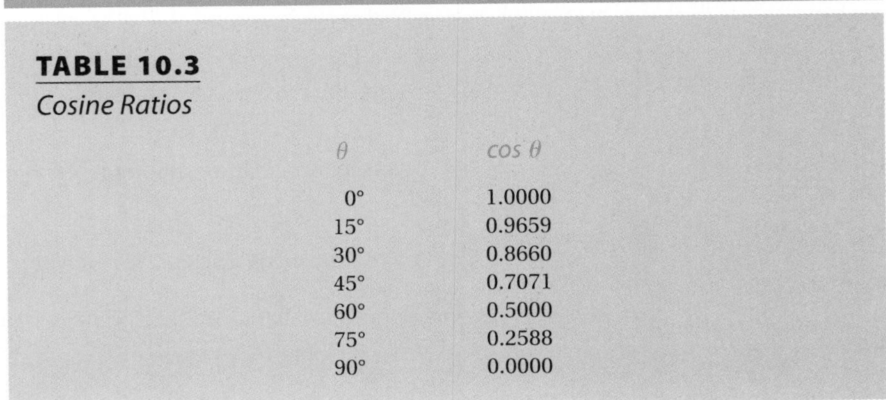

TABLE 10.3
Cosine Ratios

θ	$\cos \theta$
0°	1.0000
15°	0.9659
30°	0.8660
45°	0.7071
60°	0.5000
75°	0.2588
90°	0.0000

Some textbooks provide an expanded table of cosine ratios comparable to Table 10.2 for sine ratios. Although this text does not provide an expanded table of cosine ratios, we illustrate the application of such a table in Example 3.

EXAMPLE 3

Using Table 10.3, find the length of b in Figure 10.20 correct to the nearest tenth.

Solution

$\cos 15° = \frac{\text{adjacent}}{\text{hypotenuse}} = \frac{b}{10}$ from the triangle. Also, $\cos 15° = 0.9659$ from the table. Then

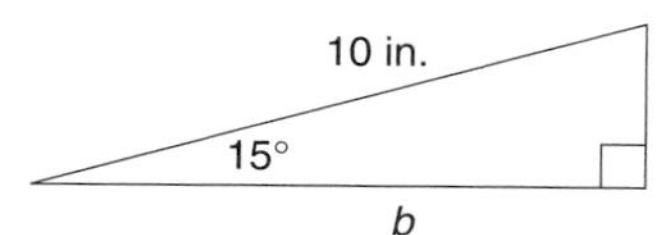

FIGURE 10.20

$$\frac{b}{10} = 0.9659 \quad \text{(since both equal } \cos 15°\text{)}$$

$$b = 9.659$$

Therefore $b = 9.7$ in.

when rounded to the nearest tenth of an inch.

In a right triangle, the cosine ratio can often be used to find an unknown length or an unknown angle measure. While the sine ratio requires that we use "opposite" and "hypotenuse," the cosine ratio requires that we use "adjacent" and "hypotenuse."

An equation of the form $\sin \alpha = \frac{a}{c}$ or $\cos \alpha = \frac{b}{c}$ contains three variables: for $\cos \alpha = \frac{b}{c}$, the variables are α, b, and c. When the values of two of the variables are known, the value of the third variable can be determined. However, we must decide which trigonometric ratio is needed to solve the problem.

EXAMPLE 4

In Figure 10.21, which trigonometric ratio would you use to find

a) α, if a and c are known?
b) b, if α and c are known?
c) c, if a and α are known?
d) β, if a and c are known?

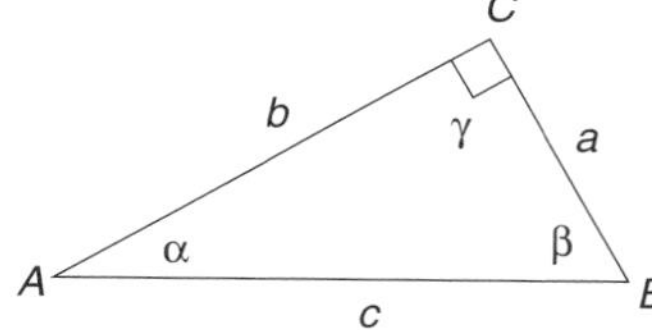

FIGURE 10.21

Solution

a) sine, since $\sin \alpha = \frac{a}{c}$ and a and c are known
b) cosine, since $\cos \alpha = \frac{b}{c}$ and α and c are known
c) sine, since $\sin \alpha = \frac{a}{c}$ and a and α are known
d) cosine, since $\cos \beta = \frac{a}{c}$ and a and c are known

To solve application problems, you generally use a calculator.

EXAMPLE 5

Find cos 67° by using a scientific calculator.

Solution

On a scientific calculator that is in degree mode, use the following key sequence:

6 → 7 → cos → 0.3907

EXAMPLE 6

Use a calculator to find the measure of angle θ if $\cos \theta = 0.5878$.

Solution

Using a scientific calculator (in degree mode), follow this key sequence:

[.] → [5] → [8] → [7] → [8] → [inv] → [cos] → [54]

Using a graphics calculator (in degree mode), follow this key sequence:

[$\cos^{-1}$] → [.] → [5] → [8] → [7] → [8] → [ENTER] → [54]

Thus, $\theta = 54°$.

EXAMPLE 7

For a regular pentagon, the length of the apothem is 12 in. Find the length of the pentagon's radius to the nearest tenth.

Solution

The central angle of the regular pentagon measures $\frac{360}{5}$ or 72°. An apothem bisects this angle, so the angle formed by the apothem and the radius measures 36°.

In Figure 10.22,

$$\cos 36° = \frac{\text{adjacent}}{\text{hypotenuse}} = \frac{12}{r}$$

Using a calculator, $\cos 36° = 0.8090$. Then $\frac{12}{r} = 0.8090$ and $0.8090r = 12$, so $r = 14.8$ in.

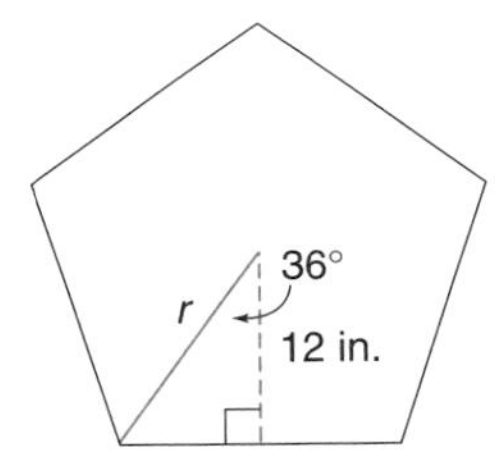

FIGURE 10.22

NOTE: The solution in Example 7 can be calculated as $r = \frac{12}{\cos 36°}$.

We now consider the proof of a statement called an **identity** because it is true for all angles; we refer to this statement as a theorem. As you will see, the statement is based entirely upon the Pythagorean Theorem.

> **THEOREM 10.2.1:** In any right triangle in which α is the measure of an acute angle,
>
> $$\sin^2 \alpha + \cos^2 \alpha = 1$$
>
> **NOTE:** $\sin^2 \alpha$ means $(\sin \alpha)^2$ and $\cos^2 \alpha$ means $(\cos \alpha)^2$.

Proof

In Figure 10.23 on the next page, $\sin \alpha = \frac{a}{c}$ and $\cos \alpha = \frac{b}{c}$. Then

$$\sin^2 \alpha + \cos^2 \alpha = \left(\frac{a}{c}\right)^2 + \left(\frac{b}{c}\right)^2 = \frac{a^2}{c^2} + \frac{b^2}{c^2} = \frac{a^2 + b^2}{c^2}$$

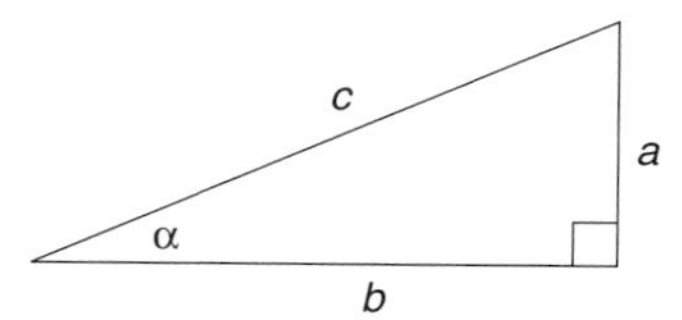

FIGURE 10.23

In the right triangle in Figure 10.23, $a^2 + b^2 = c^2$ by the Pythagorean Theorem. Substituting c^2 for $a^2 + b^2$, we have

$$\sin^2 \alpha + \cos^2 \alpha = \frac{c^2}{c^2} = 1$$

It follows that $\sin^2 \alpha + \cos^2 \alpha = 1$ for any angle α.

NOTE: Use your calculator to show that $(\sin 67°)^2 + (\cos 67°)^2 = 1$.

EXAMPLE 8

In right triangle ABC (not shown), $\sin \alpha = \frac{2}{3}$. Find $\cos \alpha$.

Solution

$$\sin^2 \alpha + \cos^2 \alpha = 1$$
$$\left(\frac{2}{3}\right)^2 + \cos^2 \alpha = 1$$
$$\frac{4}{9} + \cos^2 \alpha = 1$$
$$\cos^2 \alpha = \frac{5}{9}. \text{ Therefore, } \cos \alpha = \sqrt{\frac{5}{9}} = \frac{\sqrt{5}}{\sqrt{9}} = \frac{\sqrt{5}}{3}$$

NOTE: Because $\cos \alpha > 0$, $\cos \alpha = \frac{\sqrt{5}}{3}$ rather than $-\frac{\sqrt{5}}{3}$.

10.2 Exercises

In Exercises 1 to 6, find $\cos \alpha$ and $\cos \beta$.

1.

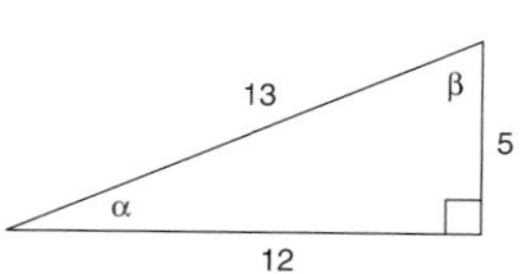

2.

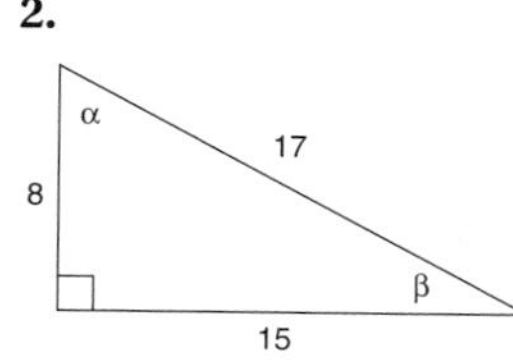

3.

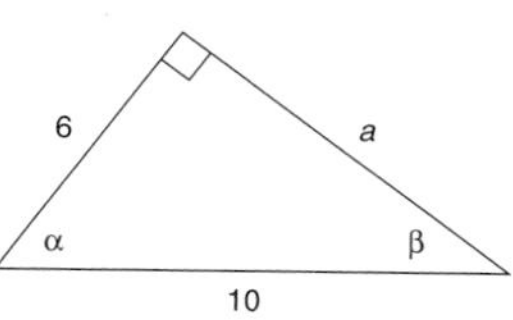

4.

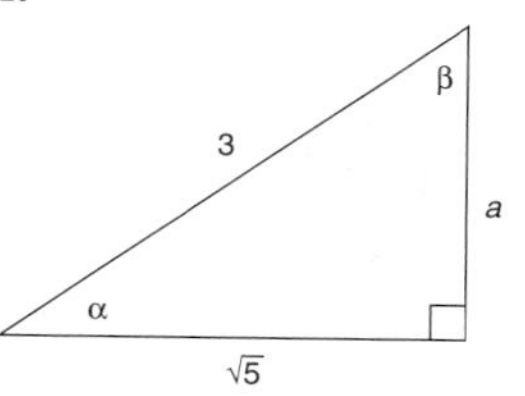

5.

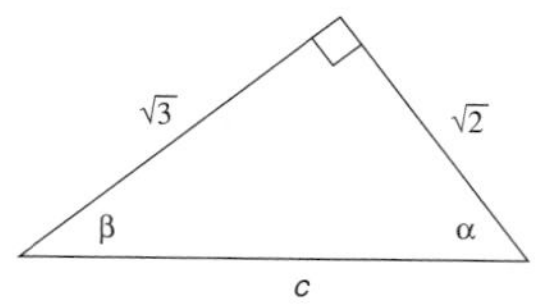

6.

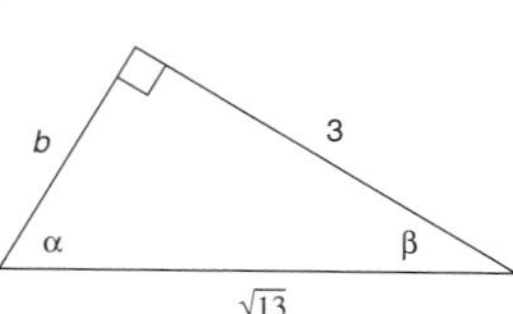

7. In Exercises 1 to 6:

a) Why does $\sin \alpha = \cos \beta$?

b) Why does $\cos \alpha = \sin \beta$?

8. Using the right triangle from Exercise 1, show that $\sin^2 \alpha + \cos^2 \alpha = 1$.

In Exercises 9 to 16, use a scientific calculator to find the indicated cosine ratio to four decimal places.

9. cos 23° **10.** cos 0° **11.** cos 17° **12.** cos 73°
13. cos 90° **14.** cos 42° **15.** cos 82° **16.** cos 7°

In Exercises 17 to 22, use either the sine ratio or the cosine ratio to find the lengths of the indicated sides of the triangle, correct to the nearest tenth of a unit.

17.

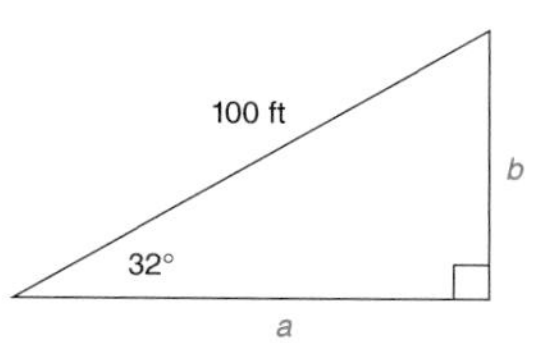

18.

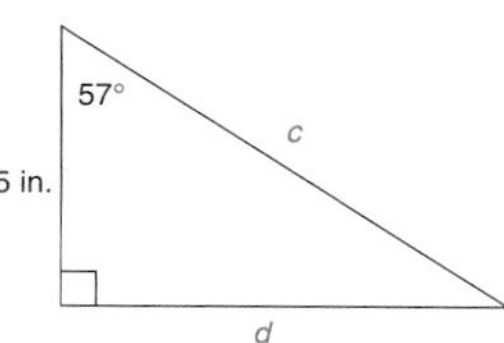

19.

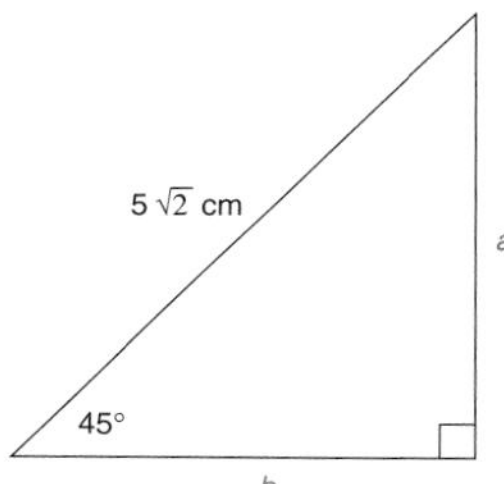

20.

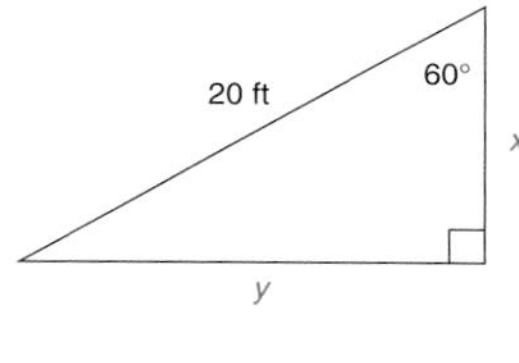

21.

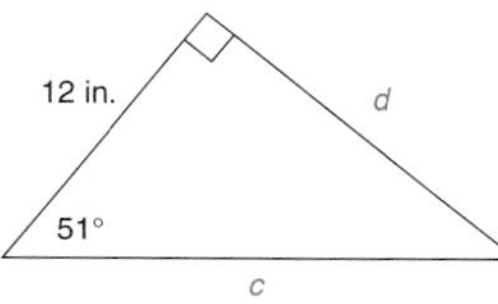

22.

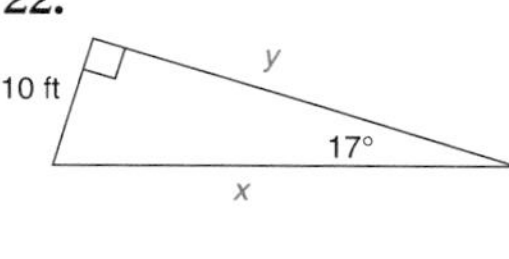

In Exercises 23 to 28, use the sine ratio or the cosine ratio as needed to find the measure of each indicated angle to the nearest degree.

23.

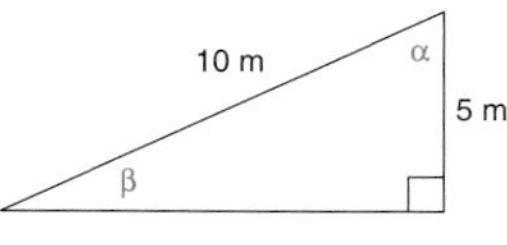

24.

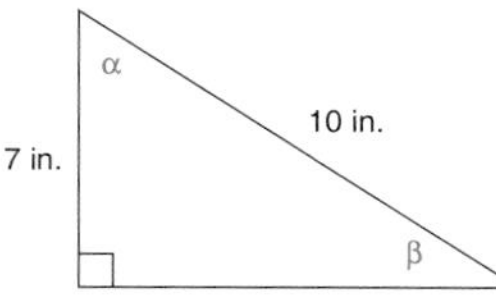

25.

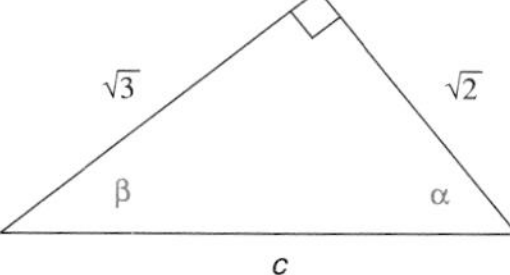

26.

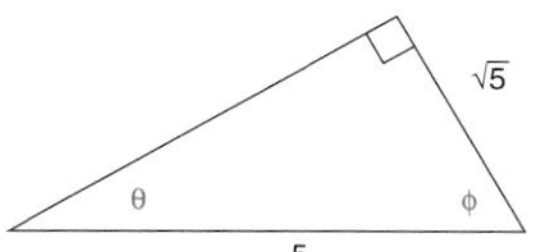

27.

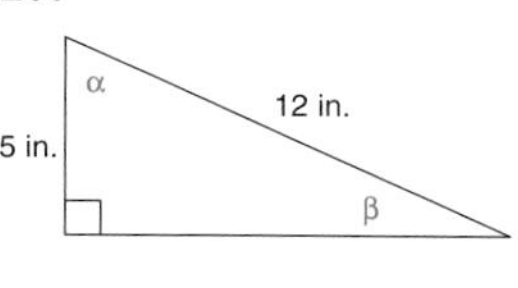

28.

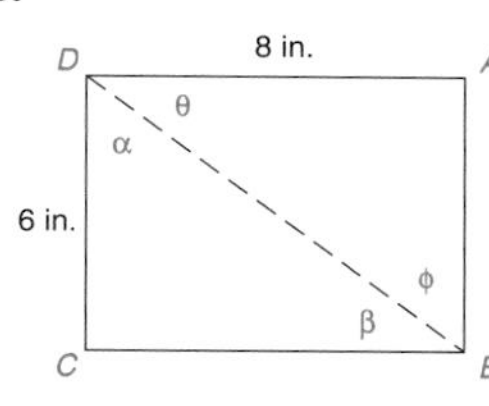

Rectangle *ABCD*

In Exercises 29 to 37, angle measures should be given to the nearest degree; distances should be given to the nearest tenth of a unit.

29. In building a garage onto his house, Gene wants to use a sloped 12-ft roof to cover an expanse that is 10 ft wide. Find the measure of angle θ.

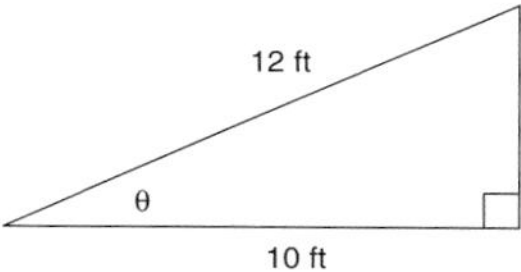

30. Gene redesigned the garage from Exercise 29 so that the 12-ft roof would rise 2 ft as shown. Find the measure of angle θ.

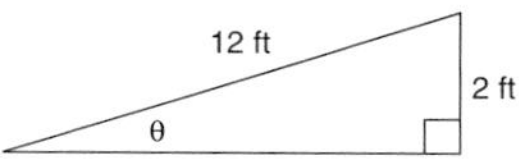

31. When an airplane is descending to land, the angle of depression is 5°. When the plane has a reading of 100 ft on the altimeter, what is its distance x from touchdown?

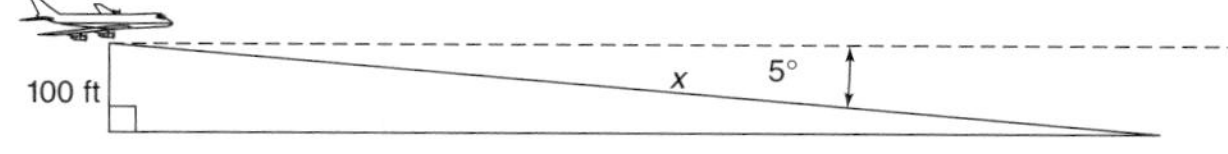

32. At a point 200 ft from the base of a cliff, the top of the cliff is seen through an angle of elevation of 37°. How tall is the cliff?

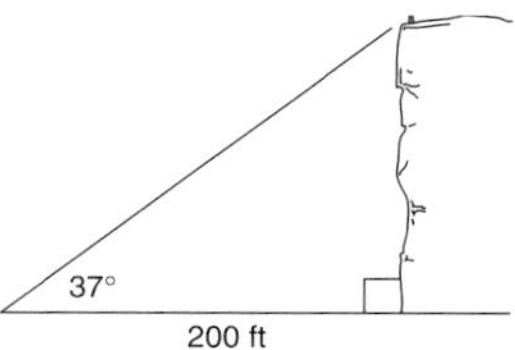

33. Find the length of each apothem in a regular pentagon whose radii measure 10 inches each.

34. Dale looks up to see his friend Lisa waving from her apartment window 30 ft from him. If Dale is standing 10 ft from the building, what is the angle of elevation as Dale looks up at Lisa?

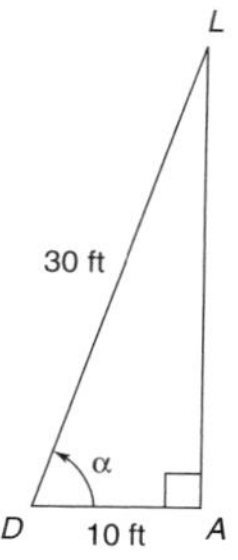

35. Find the length of the radius in a regular decagon for which each apothem has a length of 12.5 cm.

36. In searching for survivors of a boating accident, a helicopter moves horizontally across the ocean at an altitude of 200 ft above the water. If a man clinging to a life raft is seen through an angle of depression of 12°, what is the distance from the helicopter to the man in the water?

37. ➢ What is the size of the angle α formed by a diagonal of a cube and one of its edges?

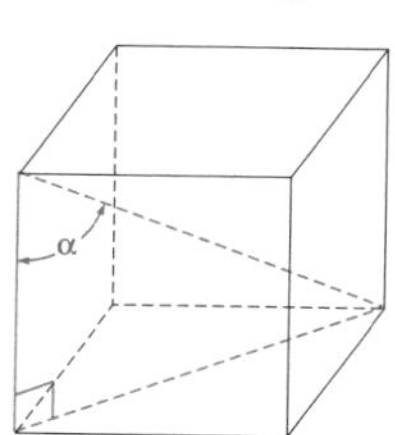

38. In the right circular cone,

a) find r correct to tenths.
b) use $L = \pi r\ell$ to find the lateral area of the cone.

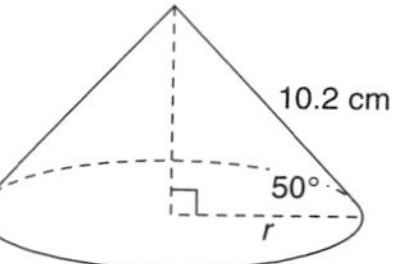

39. In parallelogram $ABCD$, find to the nearest degree:

a) $m\angle A$
b) $m\angle B$

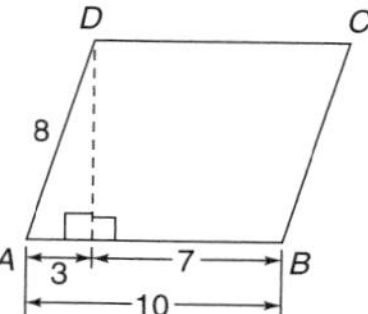

40. A ladder is carried horizontally through an L-shaped turn in a hallway. Show that the ladder has the length $L = \frac{6}{\sin\theta} + \frac{6}{\cos\theta}$.

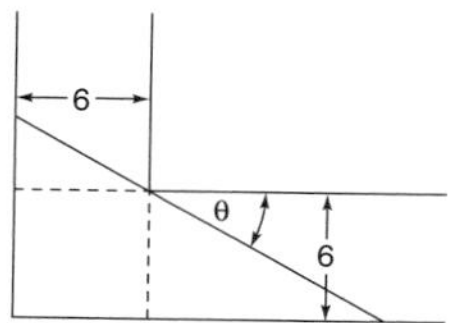

41. Use the drawing provided to show that the area of the isosceles triangle is $A = s^2 \sin\theta \cos\theta$.

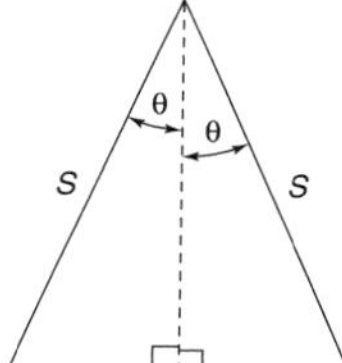

10.3 The Tangent Ratio and Other Ratios

As in Sections 10.1 and 10.2, we deal strictly with right triangles in Section 10.3. The third trigonometric ratio is the **tangent** ratio, which is defined for an acute angle of the right triangle by

$$\frac{\text{length of leg opposite acute angle}}{\text{length of leg adjacent to acute angle}}$$

Like the sine ratio, the tangent ratio increases as the measure of the acute angle increases. Unlike the sine and cosine ratios, whose values range from 0 to 1, the value of the tangent ratio is from 0 upward; that is, there is no greatest value for the tangent.

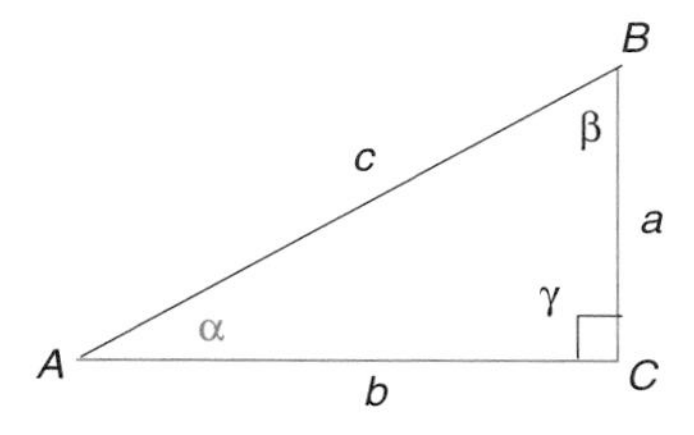

FIGURE 10.24

DEFINITION: In a right triangle, the **tangent ratio** for an acute angle is the ratio $\frac{\text{opposite}}{\text{adjacent}}$.

NOTE: In right $\triangle ABC$ in Figure 10.24, $\tan \alpha = \frac{a}{b}$ and $\tan \beta = \frac{b}{a}$, in which "tan" is an abbreviated form of the word "tangent."

EXAMPLE 1

Find the values of $\tan \alpha$ and $\tan \beta$ for the triangle in Figure 10.25.

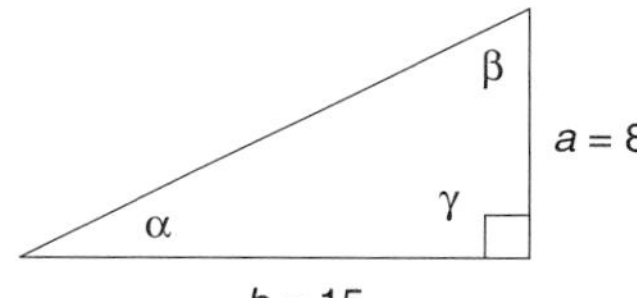

FIGURE 10.25

Solution Using the fact that the tangent ratio is $\frac{\text{opposite}}{\text{adjacent}}$,

$$\tan \alpha = \frac{a}{b} = \frac{8}{15} \quad \text{and}$$
$$\tan \beta = \frac{b}{a} = \frac{15}{8}$$

The value of $\tan \theta$ changes from 0 for a 0° angle to an immeasurably large value as the measure of the acute angle draws near 90°. That the tangent ratio $\frac{\text{opposite}}{\text{adjacent}}$ becomes infinitely large as $\theta \to 90°$ follows from the fact that the denominator becomes smaller and approaches 0 as the numerator increases.

Study Figure 10.26 to see why the value of the tangent of an angle grows immeasurably large as the angle approaches 90° in size. We often express this relationship by writing: As $\theta \to 90°$, $\tan \theta \to \infty$. The symbol ∞ is read "infinity" and implies that tan 90° is not measurable or *undefined.*

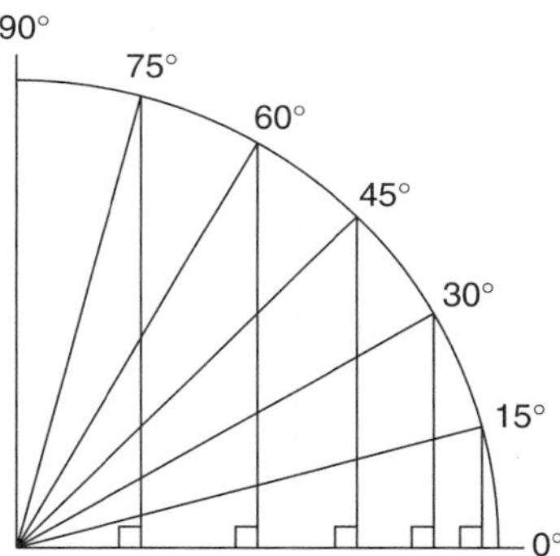

FIGURE 10.26

DEFINITION: $\tan 0° = 0$ and $\tan 90°$ is undefined.

NOTE: Use your calculator to verify that $\tan 0° = 0$. What happens when you use your calculator to find $\tan 90°$?

Certain tangent ratios are found through special triangles. By observing the triangles in Figure 10.27 and using the fact that $\tan \theta = \frac{\text{opposite}}{\text{adjacent}}$, we have

$$\tan 30° = \frac{x}{x\sqrt{3}} = \frac{1}{\sqrt{3}} \quad \text{or} \quad \frac{\sqrt{3}}{3} \approx 0.5774$$
$$\tan 45° = \frac{x}{x} = 1$$
$$\tan 60° = \frac{x\sqrt{3}}{x} = \sqrt{3} \approx 1.7321$$

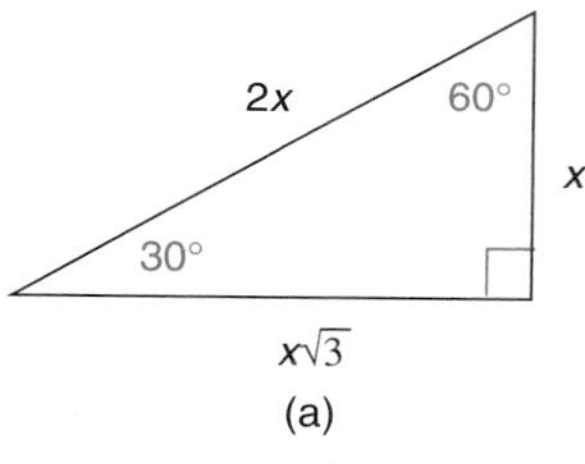

(a)

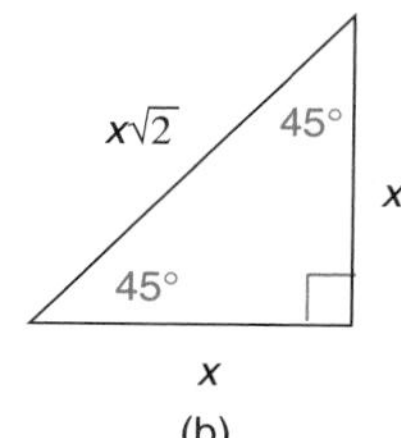

(b)

FIGURE 10.27

We apply the tangent ratio in Example 2.

EXAMPLE 2

A ski lift moves each chair through an angle of 25° as shown in Figure 10.28. What vertical change (rise) accompanies a horizontal change (run) of 845 ft?

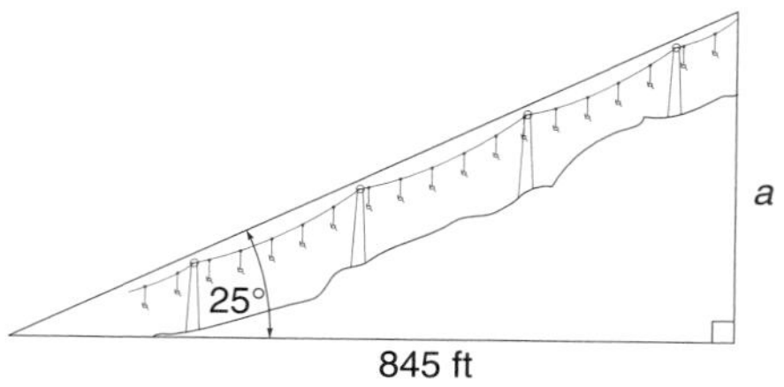

FIGURE 10.28

Solution

In the triangle, $\tan 25° = \frac{\text{opposite}}{\text{adjacent}} = \frac{a}{845}$. From $\tan 25° = \frac{a}{845}$, we multiply by 845 to obtain $a = 845 \cdot \tan 25°$.
Using a calculator, $a \approx 394$ ft.

The tangent ratio can also be used to find the measure of an angle if the lengths of the legs of a right triangle are known. This is illustrated in Example 3.

EXAMPLE 3

An airplane is seen flying just over Mission Rock, which is 1 mi away. If Mission Rock is known to be 135 ft high and the airplane is 50 ft above it, then what is the angle of elevation through which the plane is seen?

Solution From Figure 10.29 and the fact that 1 mi = 5280 ft,

$$\tan \theta = \frac{\text{opposite}}{\text{adjacent}} = \frac{185}{5280}$$

Then $\tan \theta = 0.0350$, so $\theta = 2°$ to the nearest degree.

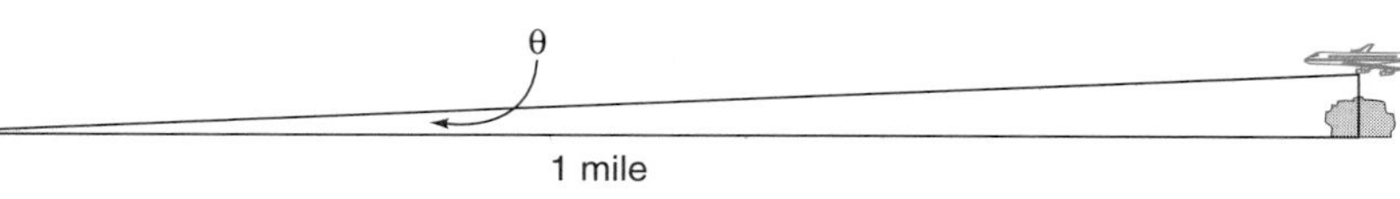

FIGURE 10.29

NOTE: Using a scientific calculator in degree mode, the typical key sequence is

0 → . → 0 → 3 → 5 → inv → tan → 2

Using a graphics calculator in degree mode, the typical key sequence is

$\tan^{-1}$ → . → 0 → 3 → 5 → 0 → ENTER → 2

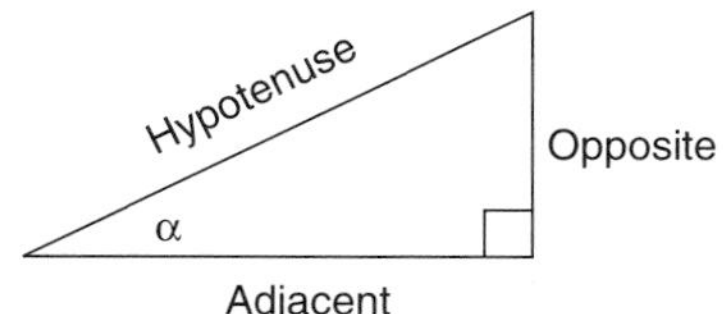

FIGURE 10.30

For the right triangle in Figure 10.30, we now have three ratios that can be used in problem solving. These are summarized as follows.

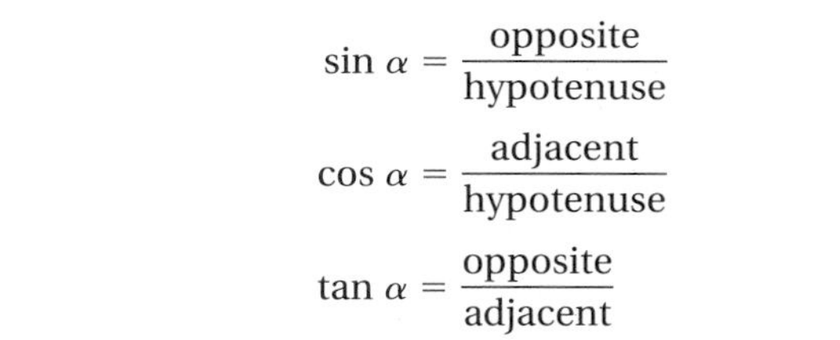

$$\sin \alpha = \frac{\text{opposite}}{\text{hypotenuse}}$$

$$\cos \alpha = \frac{\text{adjacent}}{\text{hypotenuse}}$$

$$\tan \alpha = \frac{\text{opposite}}{\text{adjacent}}$$

The equation $\tan \alpha = \frac{a}{b}$ contains three variables: α, a, and b. If the values of two of the variables are known, the value of the third variable can be found.

EXAMPLE 4

In Figure 10.31 on the next page, name the ratio that should be used to find:

a) a, if α and c are known

b) α, if a and b are known

c) β, if a and c are known

d) b, if a and β are known

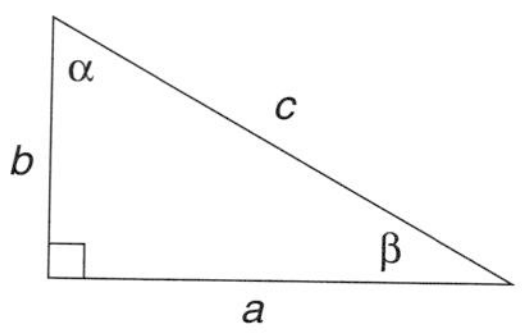

FIGURE 10.31

Solution

a) sine, since $\sin \alpha = \frac{a}{c}$
b) tangent, since $\tan \alpha = \frac{a}{b}$
c) cosine, since $\cos \beta = \frac{a}{c}$
d) tangent, since $\tan \beta = \frac{b}{a}$

EXAMPLE 5

Two apartment buildings are 40 feet apart. From a window in her apartment, Vicki can see the top of the other apartment building through an angle of elevation of 47°. She can also see the base of the other building through an angle of depression of 33°. Approximately how tall is the other building?

Solution

In Figure 10.32, the height of the building is the sum $x + y$. Using the upper and lower right triangles,

$$\tan 47° = \frac{x}{40} \quad \text{and} \quad \tan 33° = \frac{y}{40}$$

Now $$x = 40 \cdot \tan 47° \quad \text{and} \quad y = 40 \cdot \tan 33°$$

Then $x \approx 43$ and $y \approx 26$, so $x + y = 43 + 26 = 69$. The building is *approximately* 69 ft tall.

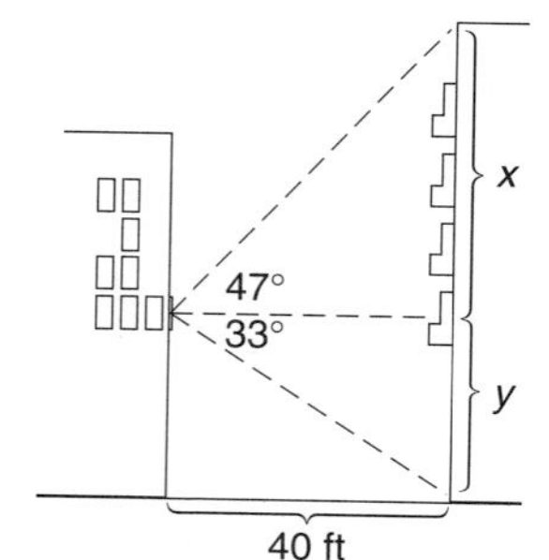

FIGURE 10.32

NOTE: In Example 5, you can determine the height of the building $(x + y)$ by entering the expression $40 \cdot \tan 47° + 40 \cdot \tan 33°$ on your calculator.

There are a total of six trigonometric ratios. We define the remaining three for completeness; however, we will be able to solve all application problems in this chapter by using only the sine, cosine, and tangent ratios. The remaining ratios are the **cotangent** (abbreviated "cot"), **secant** (abbreviated "sec"), and **cosecant** (abbreviated "csc"). These are defined in terms of the right triangle shown in Figure 10.33.

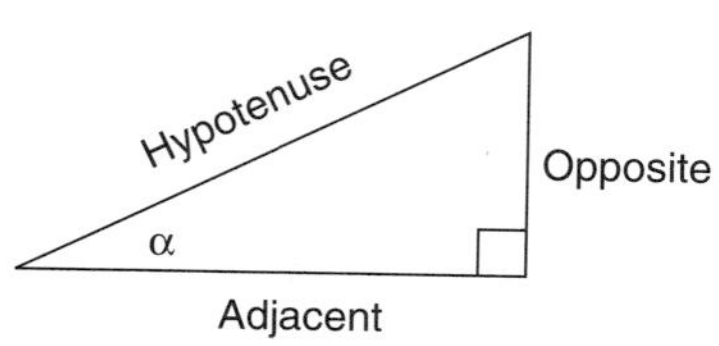

FIGURE 10.33

$$\cot \alpha = \frac{\text{adjacent}}{\text{opposite}}$$

$$\sec \alpha = \frac{\text{hypotenuse}}{\text{adjacent}}$$

$$\csc \alpha = \frac{\text{hypotenuse}}{\text{opposite}}$$

It is easy to see that cot α is the reciprocal of tan α; sec α is the reciprocal of cos α; and csc α is the reciprocal of sin α. In the following chart, we invert the trigonometric ratio on the left to obtain the reciprocal ratio named on the right.

Trigonometric Ratio	*Reciprocal Ratio*
$\text{sine} = \dfrac{\text{opposite}}{\text{hypotenuse}}$	$\text{cosecant} = \dfrac{\text{hypotenuse}}{\text{opposite}}$
$\text{cosine} = \dfrac{\text{adjacent}}{\text{hypotenuse}}$	$\text{secant} = \dfrac{\text{hypotenuse}}{\text{adjacent}}$
$\text{tangent} = \dfrac{\text{opposite}}{\text{adjacent}}$	$\text{cotangent} = \dfrac{\text{adjacent}}{\text{opposite}}$

Calculators display only the sine, cosine, and tangent ratios. By using the reciprocal key, [1/x] or [x^{-1}], you can obtain the values for the remaining ratios.

EXAMPLE 6

Use a calculator to evaluate csc 37°.

Solution First use the calculator to find sin 37° ≈ 0.6081. Now press [1/x] or [x^{-1}] to show that csc 37° ≈ 1.6616.

In Example 7, we are reminded that a calculator is not necessary for all calculations.

EXAMPLE 7

For the triangle in Figure 10.34, find the exact values of all six trigonometric ratios for angle θ.

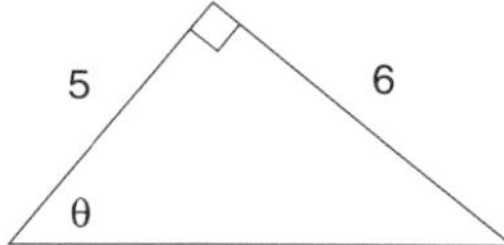

FIGURE 10.34

Solution We will need the length of the hypotenuse, which we find by the Pythagorean Theorem. With c the length of the hypotenuse,

$$\begin{aligned} c^2 &= 5^2 + 6^2 \\ c^2 &= 25 + 36 \\ c^2 &= 61 \\ c &= \sqrt{61} \end{aligned}$$

Therefore

$$\sin\theta = \frac{\text{opposite}}{\text{hypotenuse}} = \frac{6}{\sqrt{61}} = \frac{6\sqrt{61}}{61}$$

$$\cos\theta = \frac{\text{adjacent}}{\text{hypotenuse}} = \frac{5}{\sqrt{61}} = \frac{5\sqrt{61}}{61}$$

$$\tan\theta = \frac{\text{opposite}}{\text{adjacent}} = \frac{6}{5}$$

$$\cot\theta = \frac{\text{adjacent}}{\text{opposite}} = \frac{5}{6}$$

$$\sec\theta = \frac{\text{hypotenuse}}{\text{adjacent}} = \frac{\sqrt{61}}{5}$$

$$\csc\theta = \frac{\text{hypotenuse}}{\text{opposite}} = \frac{\sqrt{61}}{6}$$

NOTE: The arrows in Example 7 remind us which ratios are reciprocals of each other.

EXAMPLE 8

Evaluate the ratio named by using the given ratio:

a) $\tan\theta$, if $\cot\theta = \frac{2}{3}$
b) $\sin\alpha$, if $\csc\alpha = 1.25$
c) $\sec\beta$, if $\cos\beta = \frac{\sqrt{3}}{2}$
d) $\csc\gamma$, if $\sin\gamma = 1$

Solution

a) If $\cot\theta = \frac{2}{3}$, then $\tan\theta = \frac{3}{2}$ (the reciprocal of $\cot\theta$)
b) If $\csc\alpha = 1.25$ or $\frac{5}{4}$, then $\sin\alpha = \frac{4}{5}$ (the reciprocal of $\csc\alpha$)
c) If $\cos\beta = \frac{\sqrt{3}}{2}$, then $\sec\beta = \frac{2}{\sqrt{3}}$ or $\frac{2\sqrt{3}}{3}$ (the reciprocal of $\cos\beta$)
d) If $\sin\gamma = 1$, then $\csc\gamma = 1$ (the reciprocal of $\sin\gamma$)

EXAMPLE 9

To the nearest degree, how large is θ in the triangle in Figure 10.35?

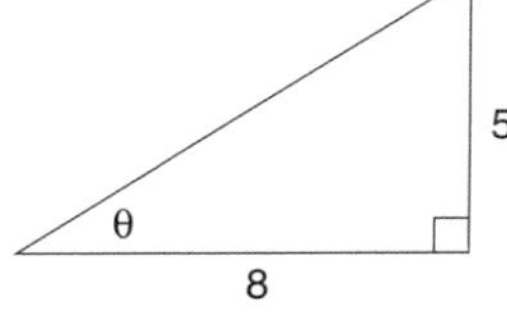

FIGURE 10.35

Solution

Because the lengths "opposite" and "adjacent" are known, we can use the tangent ratio to find θ.

$$\tan\theta = \frac{5}{8}$$

Using a scientific calculator, we determine θ by using the key sequence

[5] → [÷] → [8] → [=] → [inv] → [tan] → [32].

Using a graphics calculator, the key sequence is

$\boxed{\tan^{-1}} \rightarrow \boxed{5} \rightarrow \boxed{\div} \rightarrow \boxed{8} \rightarrow \boxed{\text{ENTER}} \rightarrow \boxed{32}$. Thus $\theta \approx 32°$.

In the application exercises that follow this section, you will have to decide which trigonometric ratio allows you to solve the problem. The Pythagorean Theorem can be used as well.

EXAMPLE 10

As his fishing vessel moves into the bay, the captain notes that the angle of elevation to the top of the lighthouse is 11°. If the lighthouse is 200 ft tall, how far is the vessel from the lighthouse?

Solution

Again we use the tangent ratio; in Figure 10.36,

$$\tan 11° = \frac{200}{x}$$
$$x \cdot \tan 11° = 200$$
$$x = \frac{200}{\tan 11°} \approx 1028.91$$

FIGURE 10.36

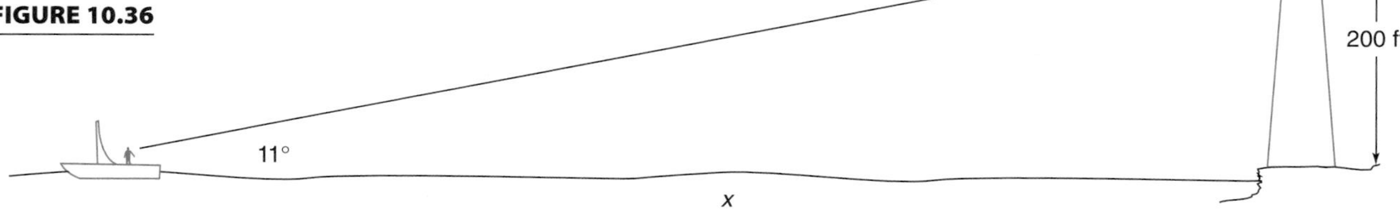

The vessel is approximately 1029 ft from the lighthouse.

10.3 Exercises

In Exercises 1 to 4, find tan α and tan β for each triangle.

1.

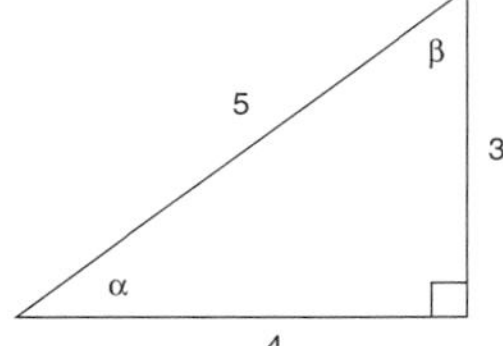

2.

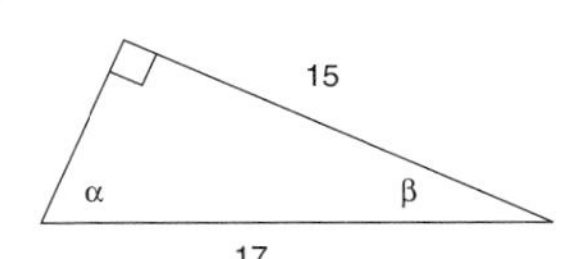

3.

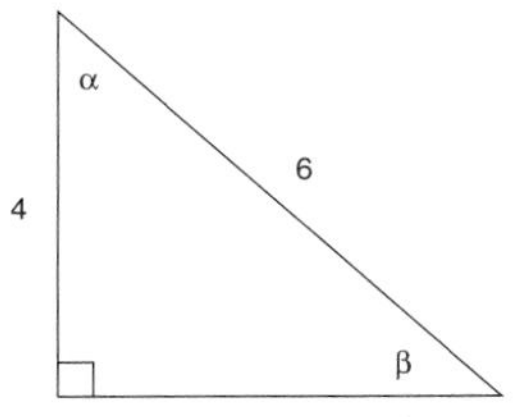

4.

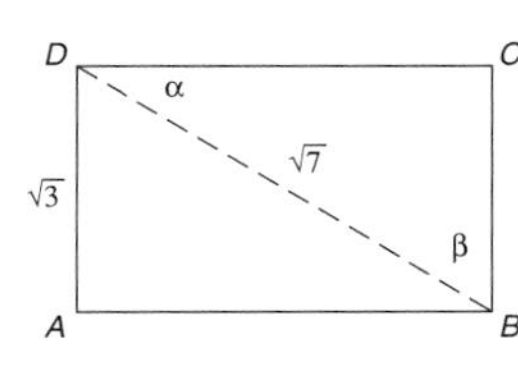

In Exercises 5 to 10, find the value (or expression) for each of the six trigonometric ratios of angle α. Use the Pythagorean Theorem as needed.

5.

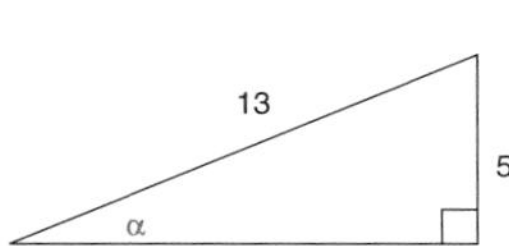

6.

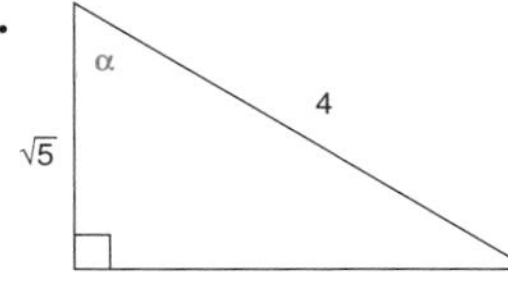

7.

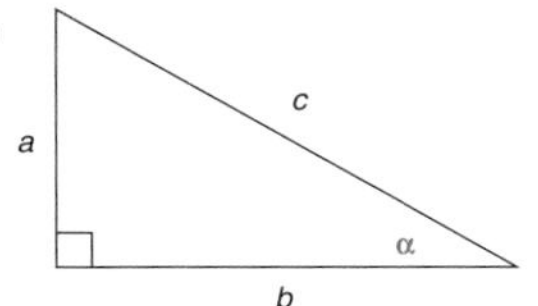

8.

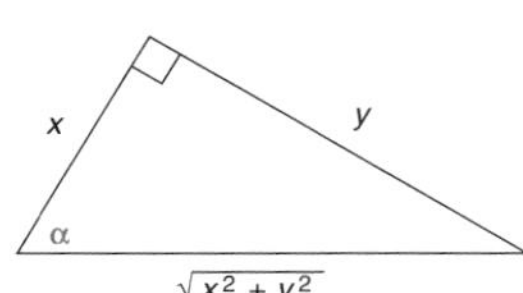

9.

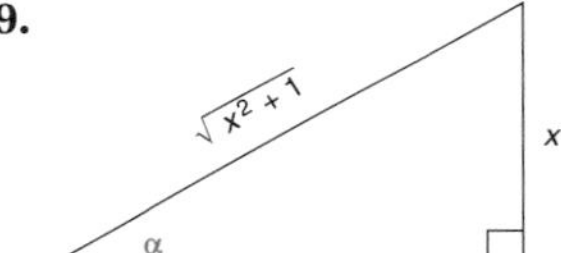

10.

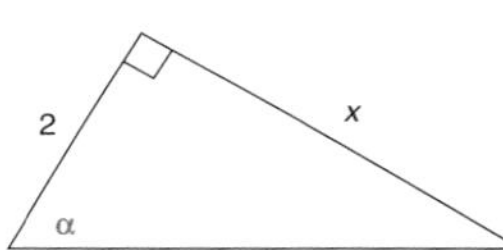

In Exercises 11 to 14, use a calculator to find the indicated tangent ratio correct to four decimal places.

11. tan 15°

12. tan 45°

13. tan 57°

14. tan 78°

In Exercises 15 to 20, use the sine, cosine, or tangent ratio to find the lengths of the indicated sides to the nearest tenth of a unit.

15.

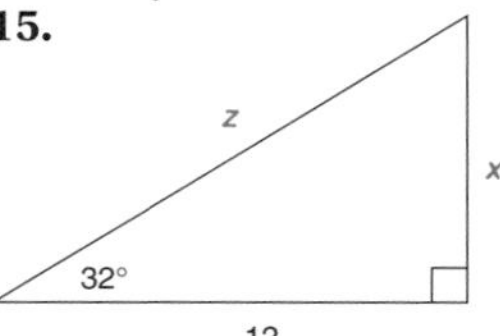

16.

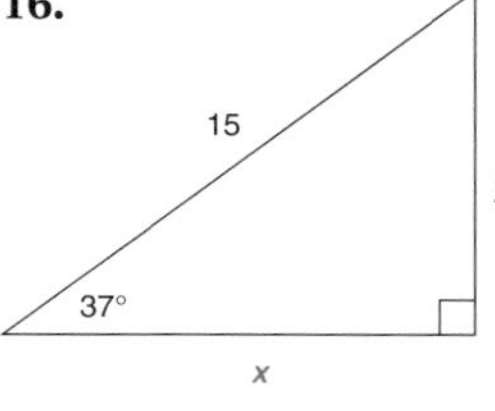

17.

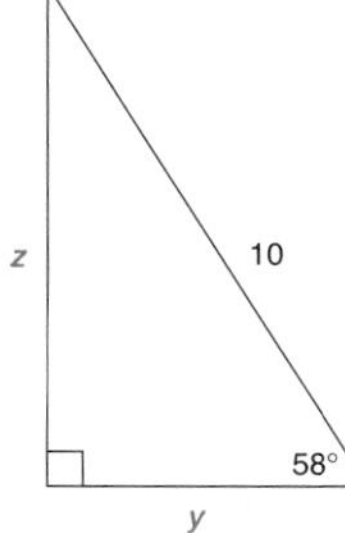

18.

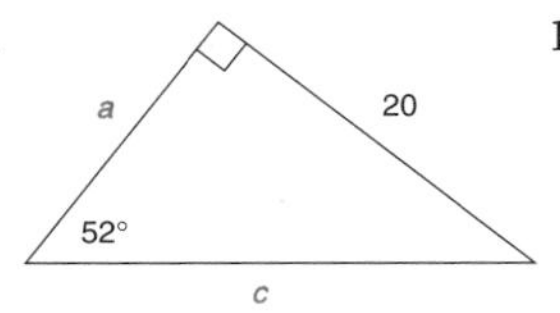

19.

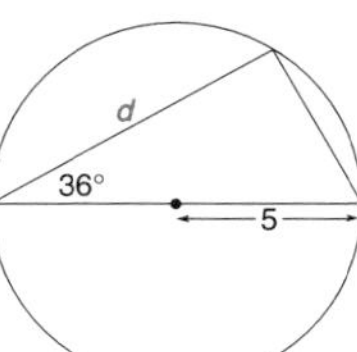

20.

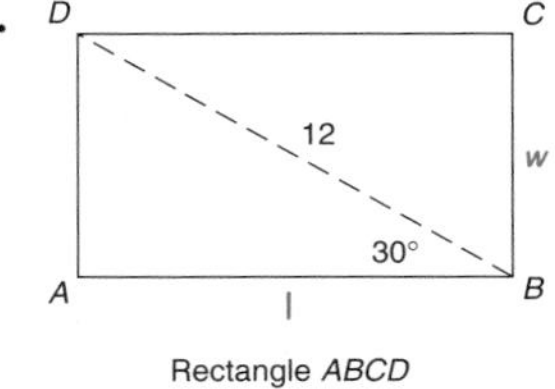

In Exercises 21 to 26, use the sine, cosine, or tangent ratio to find the indicated angle measures to the nearest degree.

21.

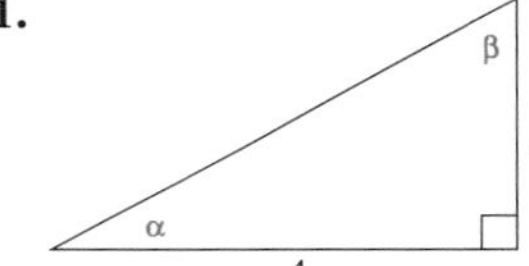

22.

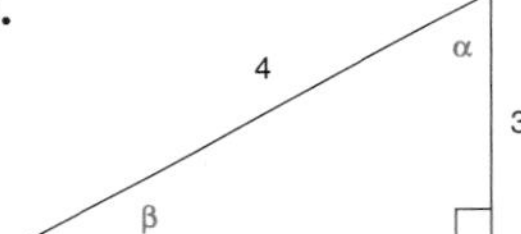

23.

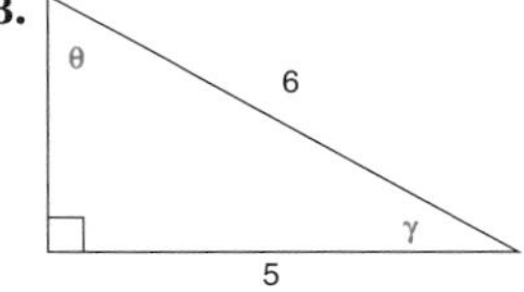

24.

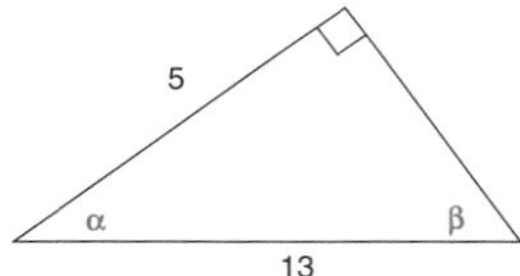

25.

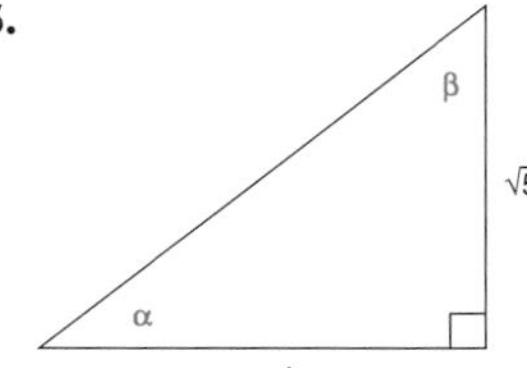

26.

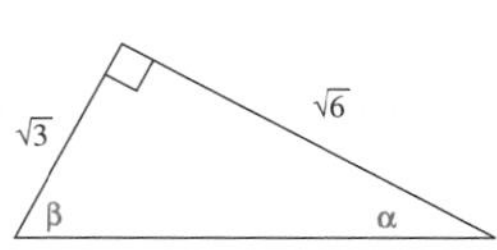

In Exercises 27 to 32, use a calculator and reciprocal relationships to find each ratio correct to four decimal places.

27. cot 34°

28. sec 15°

29. csc 30°

30. cot 67°

31. sec 42°

32. csc 72°

In Exercises 33 to 39, angle measures should be given to the nearest degree; distances should be given to the nearest tenth of a unit.

33. When her airplane is descending to land, the pilot notes an angle of depression of 5°. If the altimeter shows an altitude reading of 120 ft, what is the distance x of the plane from touchdown?

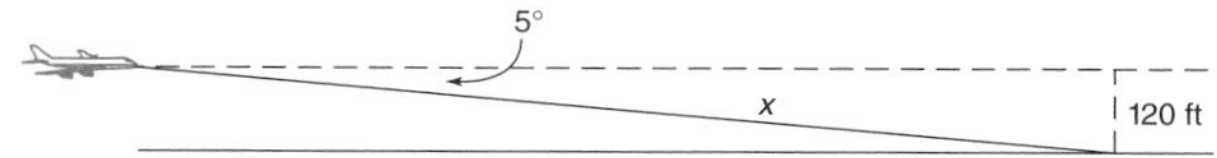

34. The top of a lookout tower is seen from a point 270 ft from its base. If the angle of elevation is 37°, how tall is the tower?

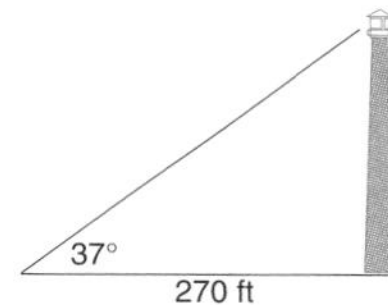

35. Find the length of the apothem to each of the 6-in. sides of a regular pentagon.

36. ➢ What is the measure of the angle between the diagonal of a cube and the diagonal of the face of the cube?

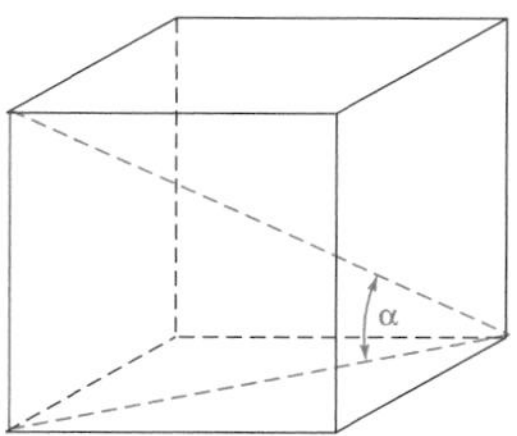

37. Upon approaching a house, Liz hears Lynette shout to her. Liz, who is standing 10 ft from the house, looks up to see Lynette in the third-story window approximately 32 ft away. What is the measure of the angle of elevation as Liz looks up at Lynette?

38. ➢ While a helicopter hovers 1000 ft above the water, its pilot spies a man in a lifeboat through an angle of depression of 28°. Along a straight line, a rescue boat can also be seen through an angle of depression of 14°. How far is the rescue boat from the lifeboat?

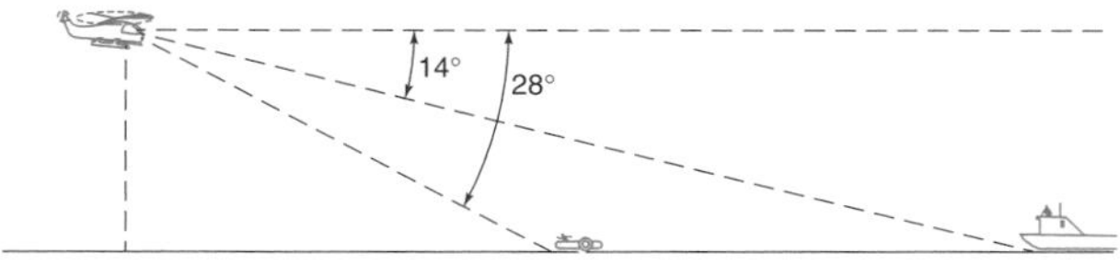

39. ➢ From atop a 200-ft lookout tower, a fire is spotted due north through an angle of depression of 12°. Firefighters located 1000 ft due east of the tower must work their way through heavy foliage to the fire. By their compasses, through what angle (measured from the north toward the west) must the firefighters travel?

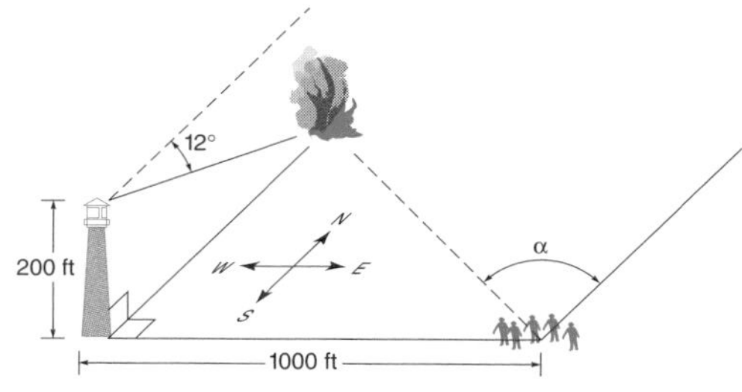

40. In the triangle shown, find each measure to the nearest tenth of a unit.

a) x

b) y

c) A, the area of the triangle

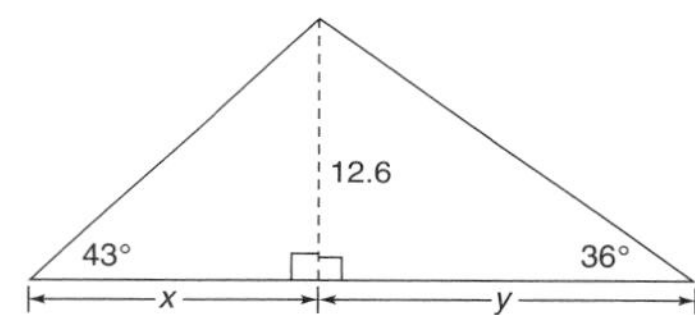

41. At an altitude of 12,000 ft, a pilot sees two towns through angles of depression of 37° and 48° as shown. To the nearest ten feet, how far apart are the towns?

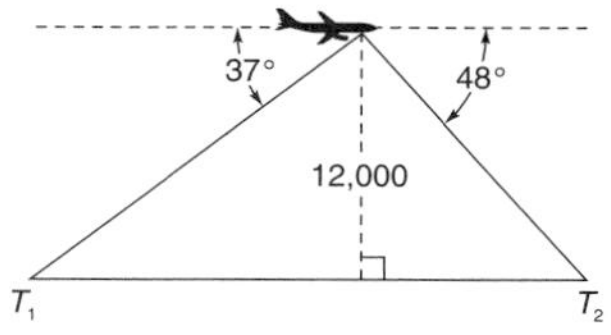

42. In the regular square pyramid shown, find

a) the slant height ℓ correct to tenths. (**HINT:** Apothem $a = 3$.)

b) Use ℓ from part (a) to find the lateral area L of the pyramid.

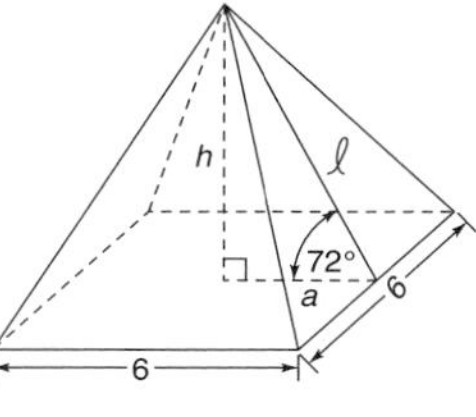

Exercises 42, 43

43. In the regular square pyramid shown,

a) find the altitude h correct to tenths. (**HINT:** Apothem $a = 3$.)

b) Use h from part (a) to find the volume of the pyramid.

10.4 More Trigonometric Relationships

Trigonometric Identities

Many of the equations that relate trigonometric ratios are known as **trigonometric identities.** These relationships are true for all angles for which the ratios involved are defined; for instance, the ratio tan 90° is not defined because it involves division by 0.

The relationships that follow are known as **reciprocal identities** because they involve ratios that are reciprocals of each other. Because

$$\sin\theta = \frac{\text{opposite}}{\text{hypotenuse}} \quad \text{and} \quad \csc\theta = \frac{\text{hypotenuse}}{\text{opposite}}$$

in a right triangle, we say that the sine and cosecant of an acute angle are reciprocals. For instance, $\sin 30° = \frac{1}{2}$ and $\csc 30° = 2$ (the inverted form of $\frac{1}{2}$).

RECIPROCAL IDENTITIES

$$\cot\theta = \frac{1}{\tan\theta} \text{ when } \tan\theta \neq 0$$

$$\sec\theta = \frac{1}{\cos\theta} \text{ when } \cos\theta \neq 0$$

$$\csc\theta = \frac{1}{\sin\theta} \text{ when } \sin\theta \neq 0$$

Some equivalent forms of the reciprocal identities are found by multiplication. If $\cot\theta = \frac{1}{\tan\theta}$, then $\tan\theta \cdot \cot\theta = 1$ when each side of the equation is multiplied by tangent θ. In turn, $\tan\theta = \frac{1}{\cot\theta}$ by division.

RECIPROCAL IDENTITIES

$$\tan\theta = \frac{1}{\cot\theta} \quad \text{and} \quad \tan\theta \cdot \cot\theta = 1$$

$$\cos\theta = \frac{1}{\sec\theta} \quad \text{and} \quad \cos\theta \cdot \sec\theta = 1$$

$$\sin\theta = \frac{1}{\csc\theta} \quad \text{and} \quad \sin\theta \cdot \csc\theta = 1$$

EXAMPLE 1

Use the given value to find the desired ratio.

a) Find $\sin\theta$ if $\csc\theta = \frac{5}{3}$. **b)** Find $\cos\alpha$ if $\sec\alpha = \sqrt{2}$.

c) Find $\tan\beta$ if $\cot\beta = \frac{a}{b}$.

Solution

a) $\sin\theta = \frac{3}{5}$, the reciprocal of $\csc\theta$.

b) $\cos\alpha = \frac{1}{\sqrt{2}} = \frac{\sqrt{2}}{2}$, the reciprocal of $\sec\alpha$.

c) $\tan\beta = \frac{b}{a}$, the reciprocal of $\cot\beta$.

In a **quotient identity,** one trigonometric ratio is the quotient of two other trigonometric ratios.

The quotient identities that follow are true for any angle θ for which the denominator does not equal 0. Recall that $\cos 90° = 0$ and $\sin 0° = 0$.

QUOTIENT IDENTITIES

$$\tan\theta = \frac{\sin\theta}{\cos\theta} \quad \text{for } \theta \neq 90°$$

$$\cot\theta = \frac{\cos\theta}{\sin\theta} \quad \text{for } \theta \neq 0°$$

Proof

We will prove only that $\tan\theta = \dfrac{\sin\theta}{\cos\theta}$. In the right triangle in Figure 10.37,

$$\sin\theta = \frac{a}{c} \text{ and } \cos\theta = \frac{b}{c}$$

Then $$\frac{\sin\theta}{\cos\theta} = \frac{\frac{a}{c}}{\frac{b}{c}} = \frac{a}{c} \div \frac{b}{c} = \frac{a}{\not{c}} \cdot \frac{\not{c}}{b} = \frac{a}{b} = \tan\theta$$

Therefore $$\tan\theta = \frac{\sin\theta}{\cos\theta}$$

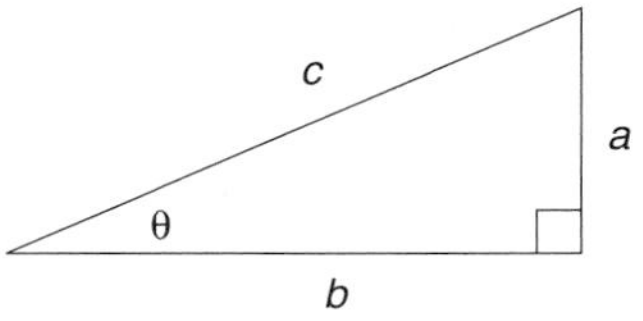

c a θ b

FIGURE 10.37

In Section 10.2, you saw that $\sin^2\theta + \cos^2\theta = 1$ is a consequence of the Pythagorean Theorem. Through division, we can add two other **Pythagorean Identities** to our list.

PYTHAGOREAN IDENTITIES

$\sin^2\theta + \cos^2\theta = 1$

$\tan^2\theta + 1 = \sec^2\theta$ for $\theta \neq 90°$

$\cot^2\theta + 1 = \csc^2\theta$ for $\theta \neq 0°$

Proof We will prove only that $\tan^2\theta + 1 = \sec^2\theta$. Using $\sin^2\theta + \cos^2\theta = 1$, division by $\cos^2\theta$ leads to

$$\frac{\sin^2\theta}{\cos^2\theta} + \frac{\cos^2\theta}{\cos^2\theta} = \frac{1}{\cos^2\theta} \qquad (\cos\theta \neq 0\text{, so } \theta \neq 90°)$$
$$\left(\frac{\sin\theta}{\cos\theta}\right)^2 + 1 = \left(\frac{1}{\cos\theta}\right)^2$$
$$(\tan\theta)^2 + 1 = (\sec\theta)^2$$
$$\tan^2\theta + 1 = \sec^2\theta$$

NOTE 1: $\tan^2\theta$ means the square of $\tan\theta$; that is, $\tan^2\theta = \tan\theta \cdot \tan\theta$.

NOTE 2: We assume that $\theta \neq 90°$ since $\tan 90°$ and $\sec 90°$ are meaningless!

EXAMPLE 2

Given that $\tan\theta = \frac{3}{4}$, find $\sec\theta$.

Solution The identity that relates $\tan\theta$ and $\sec\theta$ is $\tan^2\theta + 1 = \sec^2\theta$, so

$$\left(\frac{3}{4}\right)^2 + 1 = \sec^2\theta$$
$$\frac{9}{16} + 1 = \sec^2\theta$$
$$\frac{25}{16} = \sec^2\theta$$
$$\sec\theta = \sqrt{\frac{25}{16}} = \frac{\sqrt{25}}{\sqrt{16}} = \frac{5}{4}$$

We now turn our attention to some relationships that we will prove for *acute* triangles. The first of these is an area application.

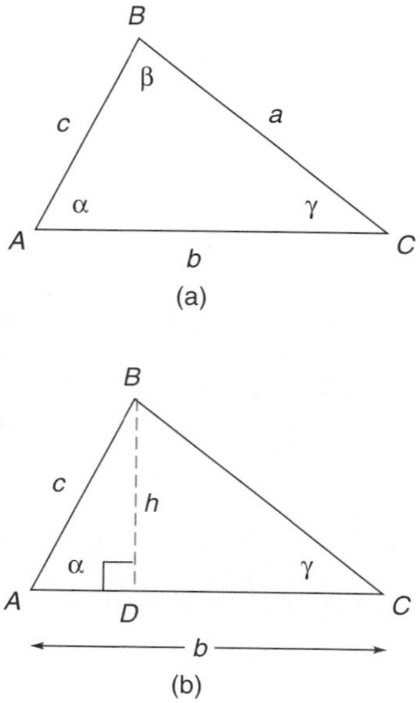

Area of a Triangle

THEOREM 10.4.1: The area of a triangle equals one-half the product of the lengths of two sides and the sine of the included angle.

Given: Acute $\triangle ABC$, as shown in Figure 10.38(a)
Prove: $A = \frac{1}{2}bc\sin\alpha$
Proof: The area of the triangle is given by $A = \frac{1}{2}bh$. With the altitude $\overline{BD}$ [see Figure 10.38(b)], we see that $\sin\alpha = \frac{h}{c}$ in right $\triangle ABD$. Then $h = c\sin\alpha$. Consequently, $A = \frac{1}{2}bh$ becomes

$$A = \frac{1}{2}b(c\sin\alpha)\text{, so } A = \frac{1}{2}bc\sin\alpha$$

(a)

(b)

FIGURE 10.38

AREA OF A TRIANGLE

$$A = \frac{1}{2}bc \sin \alpha$$

Equivalently, we can prove that

$$A = \frac{1}{2}ac \sin \beta$$

$$A = \frac{1}{2}ab \sin \gamma$$

In the more advanced course called trigonometry, this area formula can also be proved for obtuse triangles. If the triangle is a right triangle with $\gamma = 90°$, then $A = \frac{1}{2}ab \sin \gamma$ becomes $A = \frac{1}{2}ab$ (equivalent to $A = \frac{1}{2}bh$).

EXAMPLE 3

In Figure 10.39, find the area of $\triangle ABC$.

Solution

We use $A = \frac{1}{2}bc \sin \alpha$ since α, b, and c are known.

$$A = \frac{1}{2} \cdot 6 \cdot 10 \cdot \sin 33°$$
$$A = 30 \cdot \sin 33°$$
$$A \approx 16.3 \text{ in.}^2$$

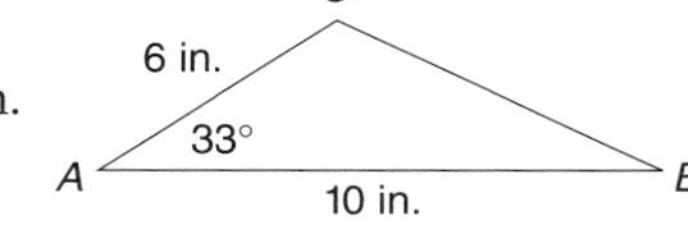

FIGURE 10.39

Law of Sines

Because the area of a triangle is unique, we can equate the three area expressions characterized by Theorem 10.4.1 as follows:

$$\frac{1}{2}bc \sin \alpha = \frac{1}{2}ac \sin \beta = \frac{1}{2}ab \sin \gamma$$

Dividing each part of this equality by $\frac{1}{2}abc$, we find

$$\frac{\frac{1}{2}bc \sin \alpha}{\frac{1}{2}bca} = \frac{\frac{1}{2}ac \sin \beta}{\frac{1}{2}acb} = \frac{\frac{1}{2}ab \sin \gamma}{\frac{1}{2}abc}$$

$$\frac{\sin \alpha}{a} = \frac{\sin \beta}{b} = \frac{\sin \gamma}{c}$$

This relationship between the lengths of sides of an acute triangle and the sines of their opposite angles is known as the Law of Sines. In trigonometry, it is shown that the Law of Sines is true for right triangles and obtuse triangles.

THEOREM 10.4.2: **(Law of Sines)** In any triangle, the three ratios between the sines of the angles and the lengths of the opposite sides are equal. That is,

$$\frac{\sin \alpha}{a} = \frac{\sin \beta}{b} = \frac{\sin \gamma}{c} \quad \text{or} \quad \frac{a}{\sin \alpha} = \frac{b}{\sin \beta} = \frac{c}{\sin \gamma}$$

Generally, only two of the equal ratios described in Theorem 10.4.2 are equated when problem solving. For instance, we can use

$$\frac{\sin \alpha}{a} = \frac{\sin \beta}{b} \quad \text{or} \quad \frac{\sin \alpha}{a} = \frac{\sin \gamma}{c} \quad \text{or} \quad \frac{\sin \beta}{b} = \frac{\sin \gamma}{c}$$

EXAMPLE 4

Use the Law of Sines to find the exact value of ST in Figure 10.40.

Solution

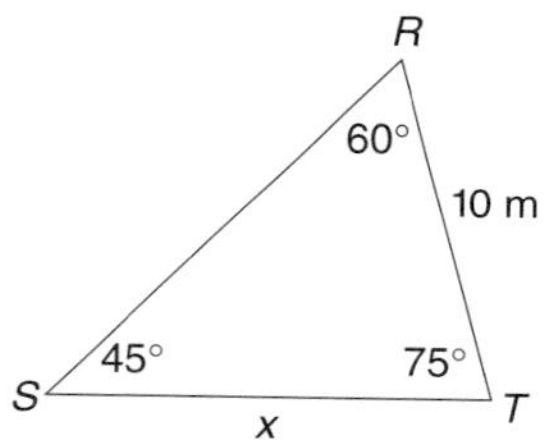

FIGURE 10.40

Since we know RT and the measure of angles S and R, we use $\frac{\sin S}{RT} = \frac{\sin R}{ST}$.

$$\frac{\sin 45°}{10} = \frac{\sin 60°}{x}$$

Because $\sin 45° = \frac{\sqrt{2}}{2}$ and $\sin 60° = \frac{\sqrt{3}}{2}$,

we have $$\frac{\frac{\sqrt{2}}{2}}{10} = \frac{\frac{\sqrt{3}}{2}}{x}$$

By the Means-Extremes Property,

$$\frac{\sqrt{2}}{2} \cdot x = \frac{\sqrt{3}}{2} \cdot 10$$

Multiplying by $\frac{2}{\sqrt{2}}$, we have

$$\frac{2}{\sqrt{2}} \cdot \frac{\sqrt{2}}{2} \cdot x = \frac{2}{\sqrt{2}} \cdot \frac{\sqrt{3}}{2} \cdot 10$$

$$x = \frac{10\sqrt{3}}{\sqrt{2}} = \frac{10\sqrt{3}}{\sqrt{2}} \cdot \frac{\sqrt{2}}{\sqrt{2}} = \frac{10\sqrt{6}}{2} = 5\sqrt{6}$$

Then $ST = 5\sqrt{6}$ m.

Law of Cosines

The final relationship that we consider is again proved only for an acute triangle. Like the Law of Sines, this relationship (known as the Law of Cosines) can be used to find unknown measures in a triangle. The Law of Cosines (which can also be established for obtuse triangles in a more advanced course) can be stated in words, "The square of one side of a triangle equals the sum of squares of the two remaining sides decreased by twice the product of those two sides and the cosine of their included angle." See Figure 10.41 as you read Theorem 10.4.3.

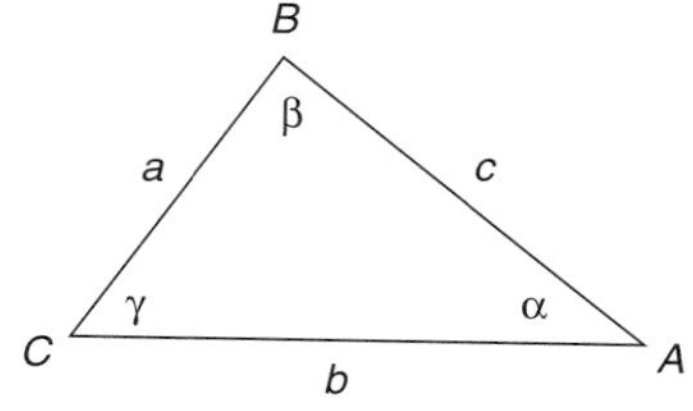

FIGURE 10.41

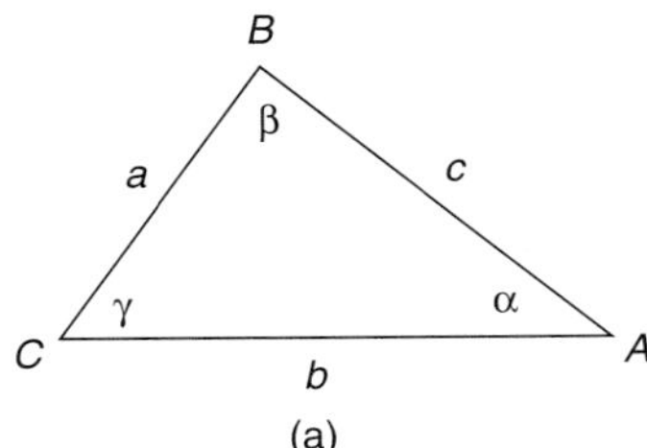

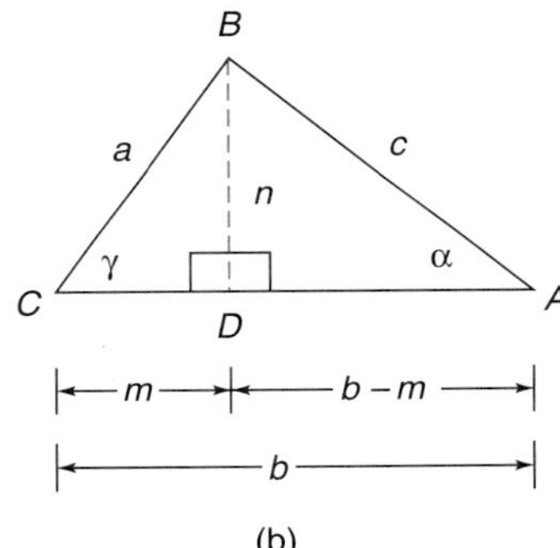

FIGURE 10.42

THEOREM 10.4.3: (Law of Cosines) In acute $\triangle ABC$,

$$c^2 = a^2 + b^2 - 2ab \cos \gamma$$
$$b^2 = a^2 + c^2 - 2ac \cos \beta$$
$$a^2 = b^2 + c^2 - 2bc \cos \alpha$$

The proof of the first form of the Law of Cosines follows:

Given: Acute $\triangle ABC$

Prove: $c^2 = a^2 + b^2 - 2ab \cos \gamma$

Proof: In Figure 10.42(a), draw the altitude $\overline{BD}$ from B to $\overline{AC}$. We designate lengths as shown in Figure 10.42(b). Now

$$(b - m)^2 + n^2 = c^2 \qquad \text{and} \qquad m^2 + n^2 = a^2$$

by the Pythagorean Theorem.

The second statement is equivalent to $m^2 = a^2 - n^2$. After we expand $(b - m)^2$, the first equation becomes

$$b^2 - 2bm + m^2 + n^2 = c^2$$

Then we replace m^2 by $(a^2 - n^2)$ to obtain

$$b^2 - 2bm + (a^2 - n^2) + n^2 = c^2$$

Simplifying yields

$$c^2 = a^2 + b^2 - 2bm$$

In right $\triangle CDB$,

$$\cos \gamma = \frac{m}{a} \qquad \text{so} \qquad m = a \cos \gamma$$

Hence we write

$$c^2 = a^2 + b^2 - 2bm$$
$$c^2 = a^2 + b^2 - 2b(a \cos \gamma)$$
$$c^2 = a^2 + b^2 - 2ab \cos \gamma$$

Similar proofs can be constructed for both remaining forms of the Law of Cosines. Although the Law of Cosines holds true for right triangles, the statement $c^2 = a^2 + b^2 - 2ab \cos \gamma$ reduces to the Pythagorean Theorem when $\gamma = 90°$.

EXAMPLE 5

Find the length of $\overline{AB}$ in the triangle in Figure 10.43.

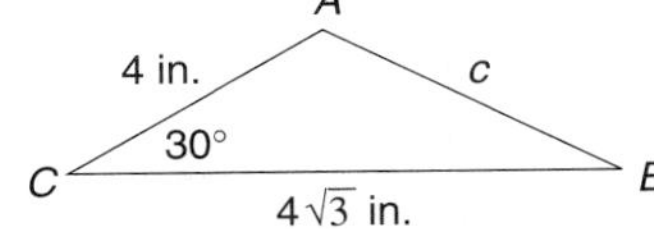

FIGURE 10.43

Solution Referring to the 30° angle as γ, we use the form

$$\begin{aligned} c^2 &= a^2 + b^2 - 2ab\cos\gamma \\ c^2 &= (4\sqrt{3})^2 + 4^2 - 2\cdot 4\sqrt{3}\cdot 4\cdot \cos 30^\circ \\ c^2 &= 48 + 16 - 2\cdot 4\sqrt{3}\cdot 4\cdot \frac{\sqrt{3}}{2} \\ c^2 &= 48 + 16 - 48 \\ c^2 &= 16 \\ c &= 4 \end{aligned}$$

Therefore $AB = 4$ in.

The Law of Cosines can also be used to find an angle of a triangle when its sides are known.

EXAMPLE 6

In acute $\triangle ABC$ in Figure 10.44, find β to the nearest degree.

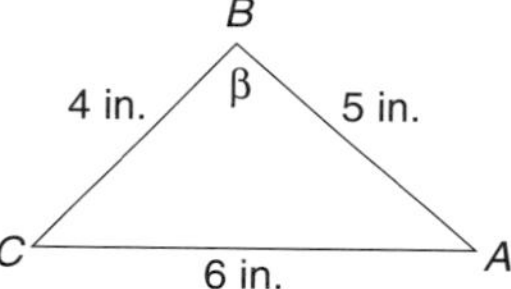

FIGURE 10.44

Solution The form of the Law of Cosines involving β is $b^2 = a^2 + c^2 - 2ac\cos\beta$. Since $b = 6$ is the length of the side opposite β, we have

$$\begin{aligned} 6^2 &= 4^2 + 5^2 - 2\cdot 4\cdot 5\cdot \cos\beta \\ 36 &= 16 + 25 - 40\cos\beta \\ 36 &= 41 - 40\cos\beta \end{aligned}$$

Therefore

$$\begin{aligned} 40\cos\beta &= 5 \\ \cos\beta &= \frac{5}{40} = \frac{1}{8} = 0.1250 \end{aligned}$$

To find β, we use a calculator to find $\beta \approx 83^\circ$.

In finding the measure of a side or an angle of an acute triangle, a decision must be made as to which form of the Law of Sines or of the Law of Cosines should be applied. Table 10.4 deals with that question and is based on the acute triangle shown in the accompanying drawing. Note that a, b, and c represent the lengths of the sides, while α, β, and γ represent the measures of the opposite angles, respectively (see Figure 10.45).

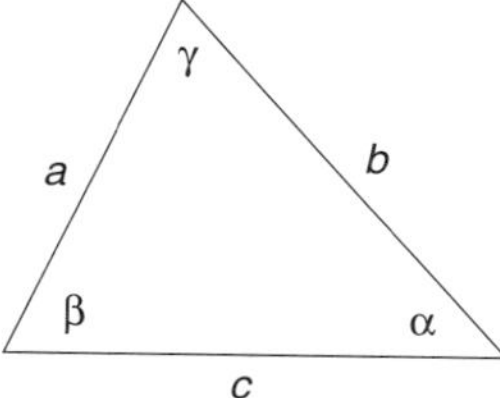

FIGURE 10.45

■ *Warning*

If we know only the measures of the three angles of the triangle, then no length of side can be determined. ■

TABLE 10.4

When to Use the Law of Sines/Law of Cosines

1. *Three sides are known:* Use the Law of Cosines to find *any* angle.
   Known measures: a, b, and *c*
   Desired measure: α

   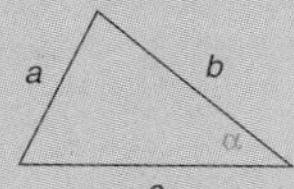

   $\therefore$ *Use* $a^2 = b^2 + c^2 - 2bc \cos \alpha$

2. *Two sides and a nonincluded angle are known:* Use the Law of Sines to find the remaining nonincluded angle.
   Known measures: a, b, and α
   Desired measure: β

   $\therefore$ *Use* $\dfrac{\sin \alpha}{a} = \dfrac{\sin \beta}{b}$

3. *Two sides and an included angle are known:* Use the Law of Cosines to find the remaining side.
   Known measures: a, b, and γ
   Desired measure: c

   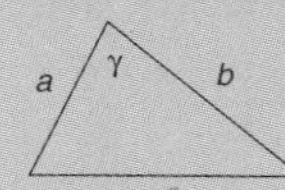

   $\therefore$ *Use* $c^2 = a^2 + b^2 - 2ab \cos \gamma$

4. *Two angles and a nonincluded side are known:* Use the Law of Sines to find the other nonincluded side.
   Known measures: a, α, and β
   Desired measure: b

   $\therefore$ *Use* $\dfrac{\sin \alpha}{a} = \dfrac{\sin \beta}{b}$

EXAMPLE 7

In the design of a child's swing set, the two metal posts that support the top bar each measure 8 ft. At ground level, the posts are to be 6 ft apart (see Figure 10.46). At what angle should the two metal posts be secured? Give the answer to the nearest degree.

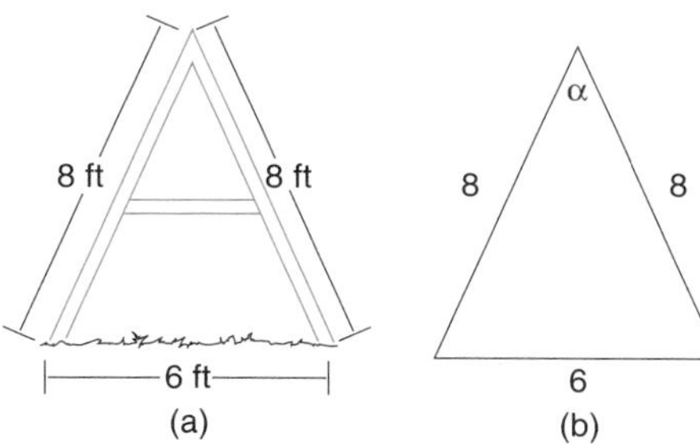

FIGURE 10.46

Solution Call the desired angle measure α. Because the three sides of the triangle are known, we use the Law of Cosines of the form $a^2 = b^2 + c^2 - 2bc\cos\alpha$.

Because a represents the length of the side opposite the angle α, $a = 6$ while $b = 8$ and $c = 8$. Consequently, we have

$$\begin{aligned}6^2 &= 8^2 + 8^2 - 2\cdot 8\cdot 8\cdot \cos\alpha\\ 36 &= 64 + 64 - 128\cos\alpha\\ 36 &= 128 - 128\cos\alpha\\ -92 &= -128\cos\alpha\\ \cos\alpha &= \frac{-92}{-128}\\ \cos\alpha &= 0.7188\end{aligned}$$

Using a calculator, $\alpha \approx 44°$.

10.4 Exercises

1. Show that $\frac{\cos\theta}{\sin\theta} = \cot\theta$.
2. Show that $\sec\theta\cdot\sin\theta = \tan\theta$.

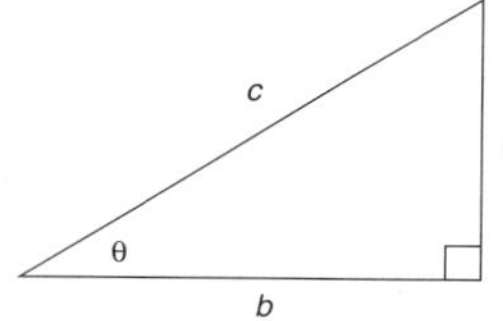

Exercises 1, 2

In Exercises 3 and 4, use the quotient relationships.

3. If $\sin\theta = 0.6$ and $\cos\theta = 0.8$, find $\tan\theta$ and $\cot\theta$.
4. If $\sin\theta = \frac{8}{17}$ and $\cos\theta = \frac{15}{17}$, find $\tan\theta$ and $\cot\theta$.
5. If $\sin\theta = \frac{3}{5}$, find $\cos\theta$.
   (**HINT:** Use the fact that $\sin^2\theta + \cos^2\theta = 1$.)
6. If $\cos\theta = \frac{\sqrt{3}}{2}$, find $\sin\theta$.
7. If $\tan\theta = \frac{5}{12}$, find $\sec\theta$.
8. If $\sec\theta = \frac{17}{15}$, find $\tan\theta$.
9. Show that $\cot^2\theta + 1 = \csc^2\theta$.
10. State whether each equation is true or false:
    a) $\cos^2\theta - \sin^2\theta = 1$
    b) $\csc^2\theta - \cot^2\theta = 1$
    c) $\tan\theta\cdot\cot\theta = 1$
    d) $\sin\theta\cdot\cos\theta = 1$

In Exercises 11 to 14, find the area of each triangle shown. Give the answer to the nearest tenth of a square unit.

11.

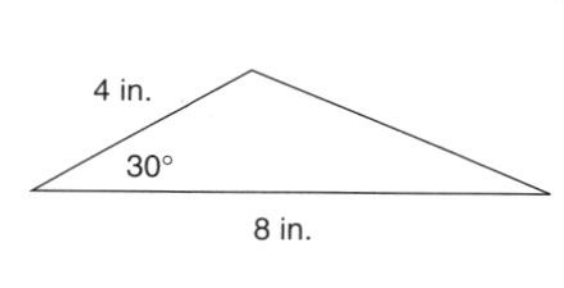

12.

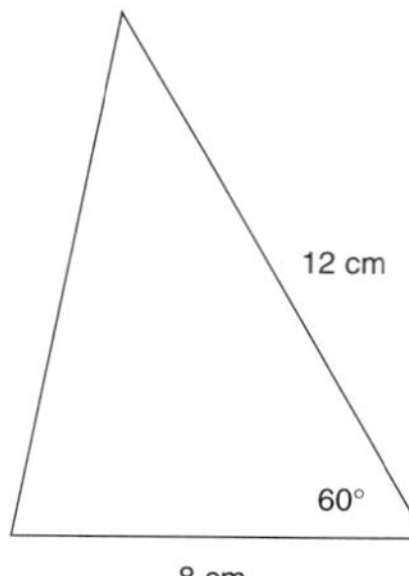

13.

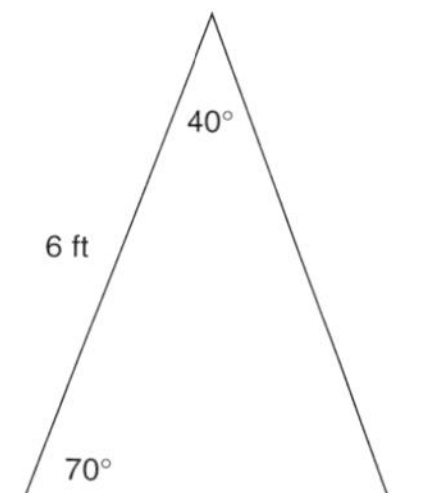

14.

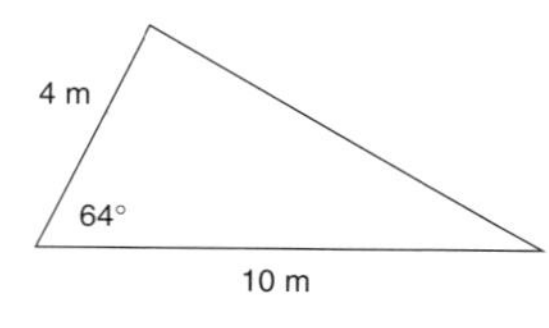

In Exercises 15 and 16, find the area of the given figure. Give the answer to the nearest tenth of a square unit.

15.

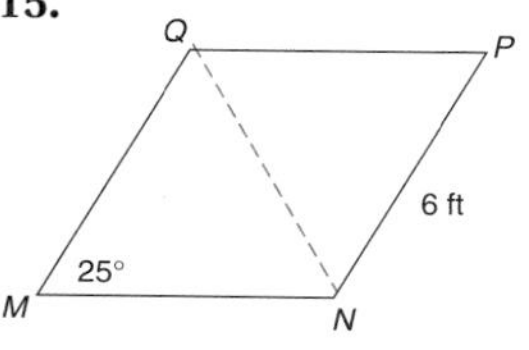

Rhombus *MNPQ*

16.

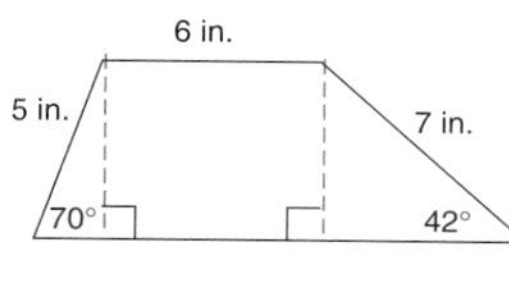

In Exercises 17 to 22, use a form of the Law of Sines to find the measure of the indicated side or angle. Angle measures should be found to the nearest degree and lengths to the nearest tenth of a unit.

17.

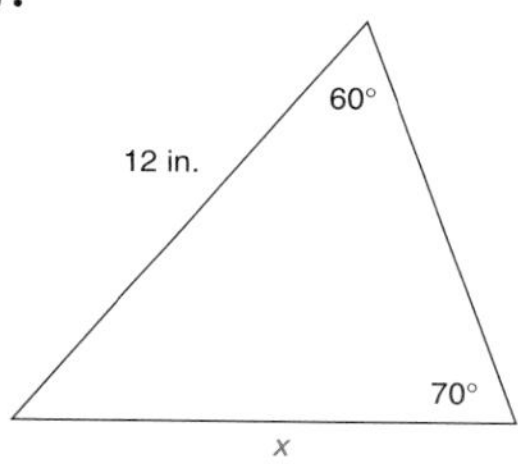

18.

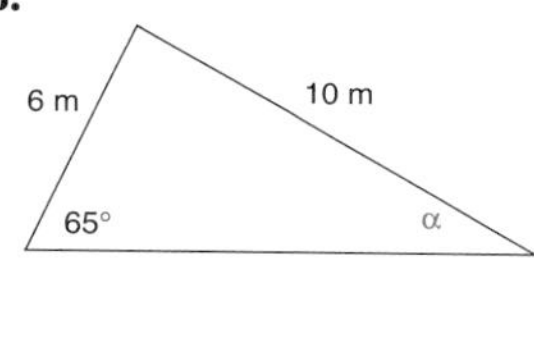

19.

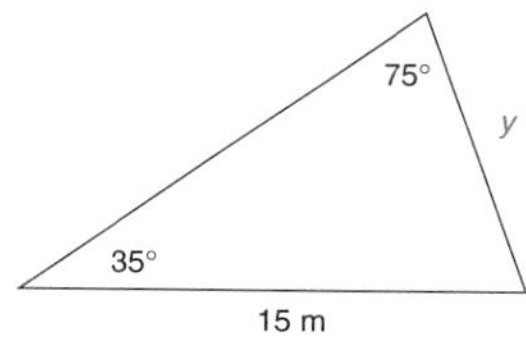

20.

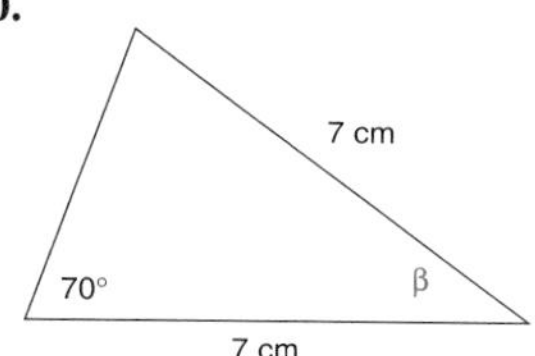

21.

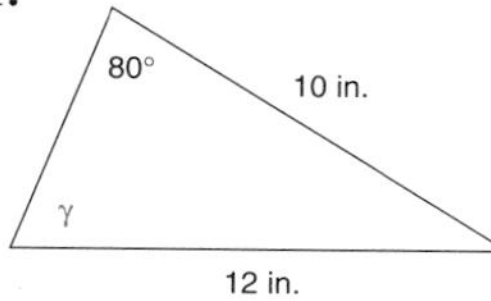

22.

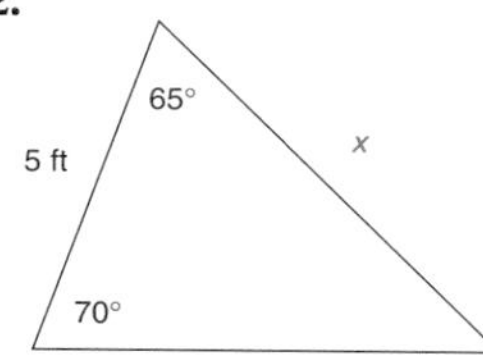

In Exercises 23 to 28, use a form of the Law of Cosines to find the measure of the indicated side or angle. Angle measures should be found to the nearest degree and lengths to the nearest tenth of a unit.

23.

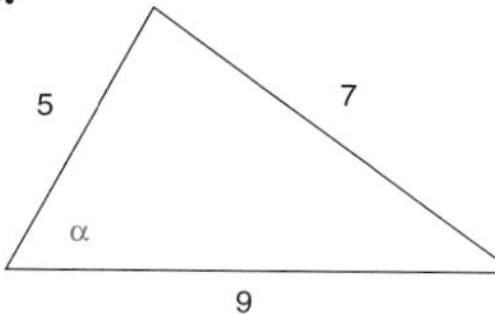

24.

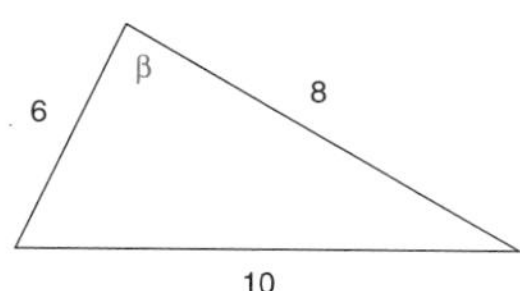

25.

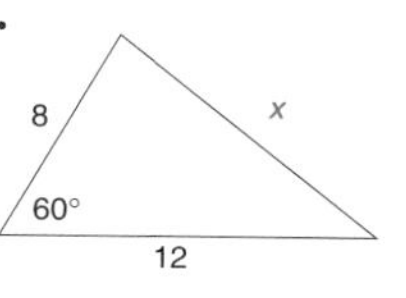

26.

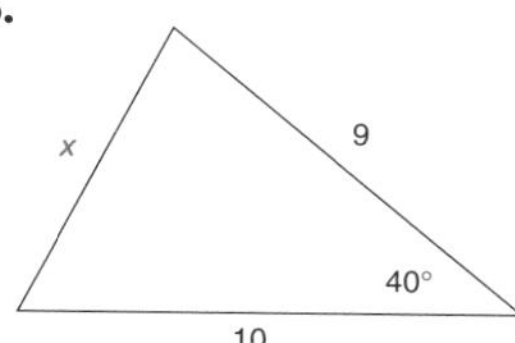

27.

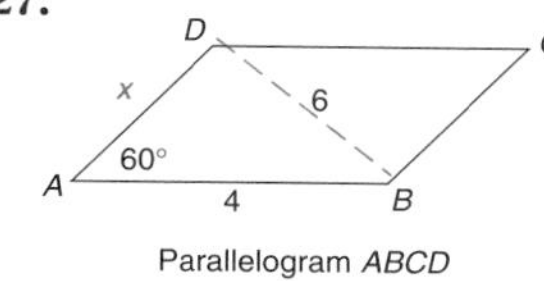

Parallelogram *ABCD*

28.

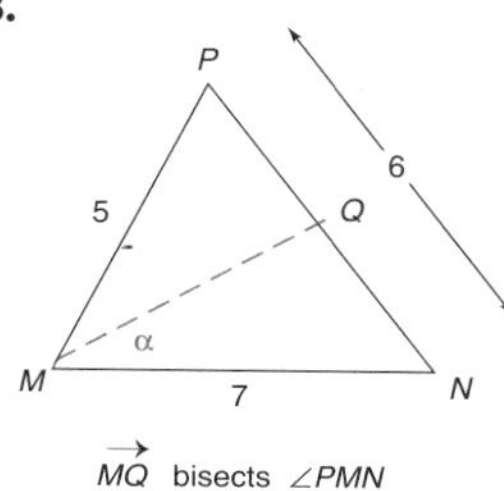

$\overrightarrow{MQ}$ bisects $\angle PMN$

In Exercises 29 to 34, use the Law of Sines or the Law of Cosines to solve each problem. Angle measures should be found to the nearest degree and areas and distances to the nearest tenth of a unit.

29. A triangular lot has street dimensions of 150 ft and 180 ft, and an included angle of 80° for these two sides.

a) Find the length of the remaining side of the lot.
b) Find the area of the lot in square feet.

30. Two people observe a balloon. They are 500 ft apart, and their angles of observation are 47° and 65° as shown. Find the distance x from the second observer to the balloon.

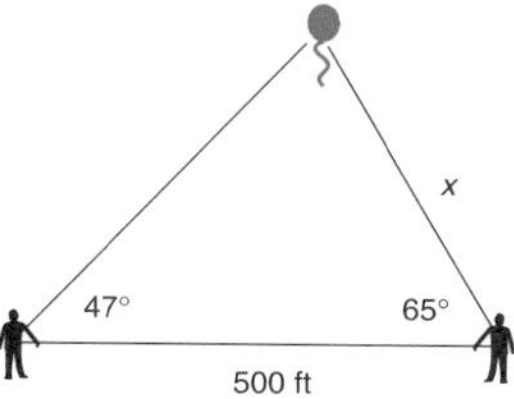

31. A surveillance aircraft at point C sights an ammunition warehouse at A and enemy headquarters at B through the angles indicated. If points A and B are 10,000 m apart, what is the distance from the aircraft to enemy headquarters?

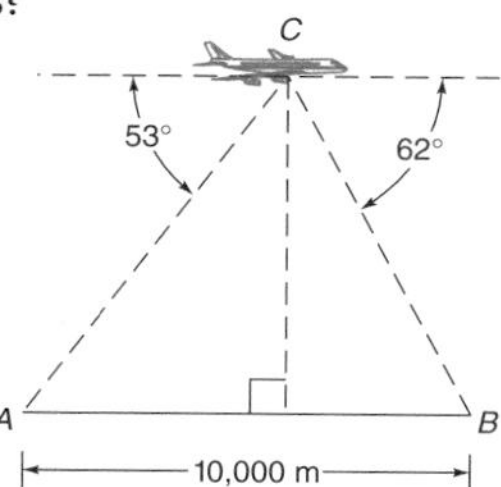

32. Above one room of a house, the rafters meet as shown. What is the measure of the angle α at which they meet?

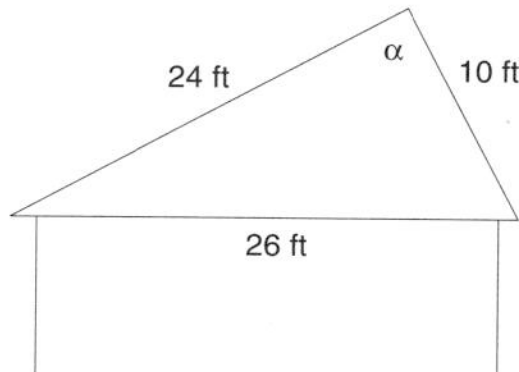

33. In an A-frame house, a bullet is found embedded at a point 8 ft up the sloped wall. If it was fired at a 30° angle with the horizontal, how far from the base of the wall was the gun fired?

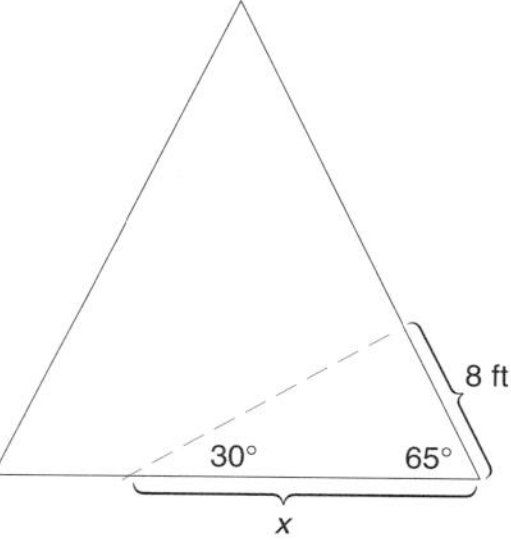

34. Clay pigeons are released at an angle of 30° with the horizontal. A sharpshooter hits one of the clay pigeons when shooting through an angle of elevation of 70°. If the point of release is 120 m from the sharpshooter, how far (x) is the sharpshooter from the target when it is hit?

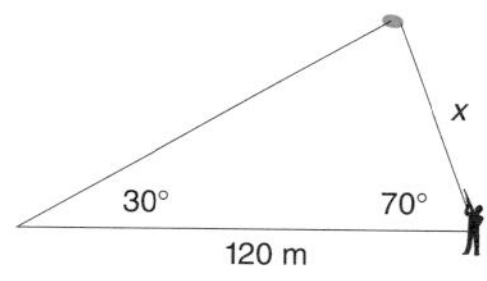

35. For the triangle shown, the area is exactly $18\sqrt{3}$ units². Determine the value of x.

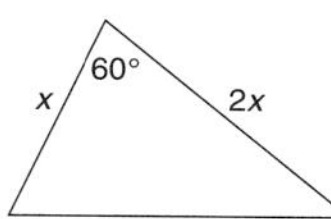

36. For the triangle shown, use the Law of Cosines to determine b.

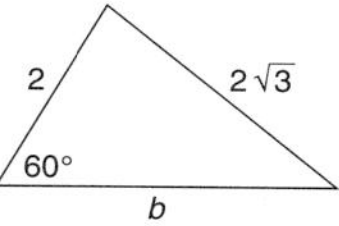

37. In the support structure for the Ferris wheel, $m\angle CAB = 30°$. If $AB = AC = 27$ ft, find BC.

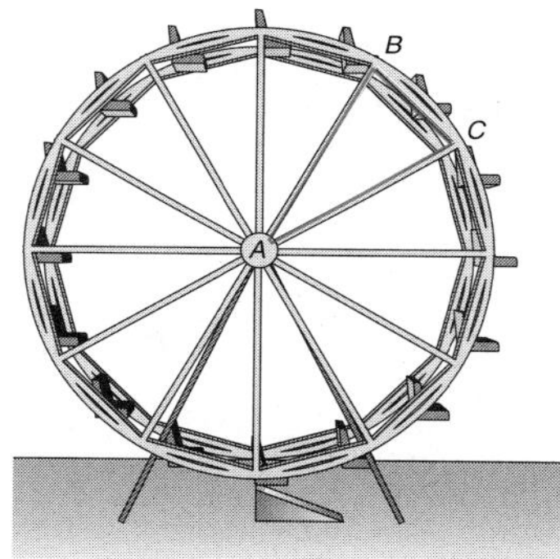

38. Show that the form of the Law of Cosines written $c^2 = a^2 + b^2 - 2ab \cos \gamma$ reduces to the Pythagorean Theorem when $\gamma = 90°$.

A Look Beyond: Radian Measure of Angles

In much of this textbook, we have considered angle measures from 0° to 180°. As you apply geometry, you will find that two things are true:

1. Angle measures do not have to be limited to degree measures from 0° to 180°.
2. The degree is not the only unit used in measuring angles.

We will address the first of these issues in Examples 1, 2, and 3.

EXAMPLE 1

As the time changes from 1 P.M. to 1:45 P.M., through what angle does the minute hand rotate? See Figure 10.47.

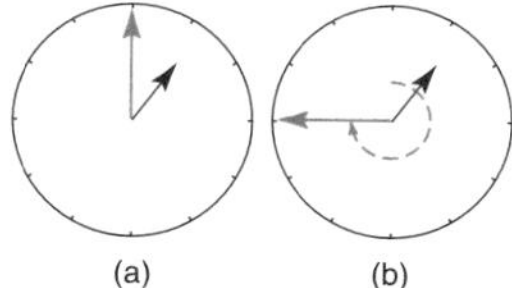

FIGURE 10.47

Solution

Since the rotation is $\frac{3}{4}$ of a complete circle (360°), the result is $\left(\frac{3}{4}\right)$ 360° or 270°.

EXAMPLE 2

An airplane pilot is instructed to circle the control tower twice during a holding pattern before receiving clearance to land. Through what angle does the airplane move? See Figure 10.48.

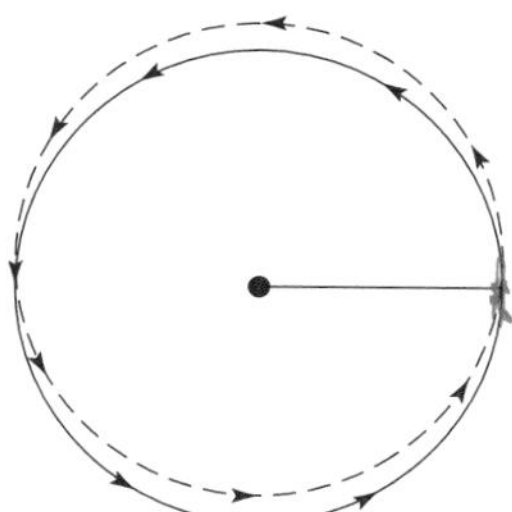

FIGURE 10.48

Solution

Two circular rotations give 2(360°) or 720°.

In trigonometry, negative measures for angles are used to distinguish the direction of rotation. A counterclockwise rotation is measured as positive, while a clockwise rotation is measured as negative. The arrowed arcs in Figure 10.49 are used to indicate the direction of rotation.

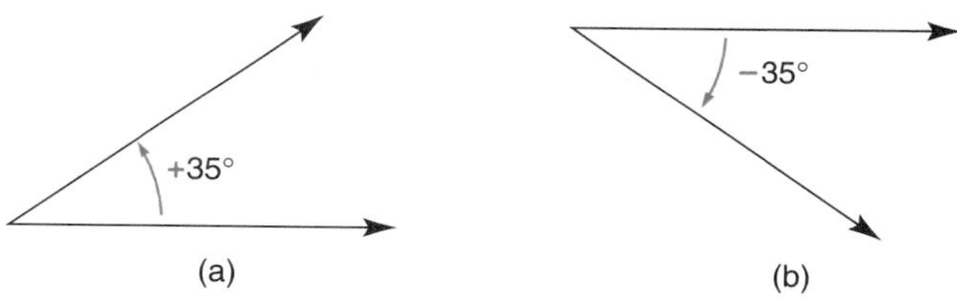

FIGURE 10.49

EXAMPLE 3

To tighten a hex bolt, a mechanic applies rotations of 45° several times. What is the measure of each rotation? See Figure 10.50.

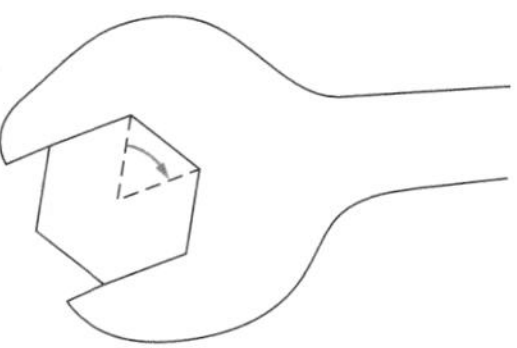

FIGURE 10.50

Solution

Tightening occurs if the angle is −45°.

NOTE: If the angle of rotation is 45° (that is, +45°), the bolt is loosened.

Our second concern is with an alternative unit of measuring angles, a unit often used in the study of trigonometry and calculus.

DEFINITION: In a circle, a **radian** (rad) is the measure of a central angle that intercepts an arc whose length is equal to the radius of the circle.

In Figure 10.51, the length of each radius and the intercepted arc are all equal to r. Thus, the central angle shown measures 1 rad. A complete rotation about the circle corresponds to 360° and to $2\pi r$. Thus, the arc length of 1 radius corresponds to the central angle measure of 1 rad, and the circumference of 2π radii corresponds to the complete rotation of 2π rad.

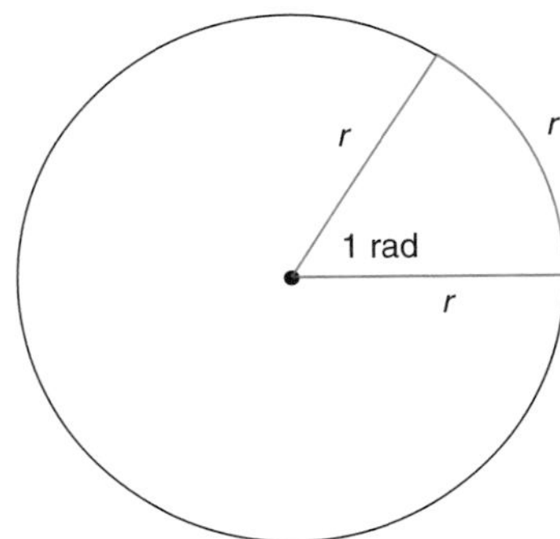

FIGURE 10.51

This relationship for the complete rotation allows us to equate 360° and 2π radians. As suggested by Figure 10.52, there are approximately 6.28 rad (or exactly 2π radians) about the circle. The exact result leads to an important relationship.

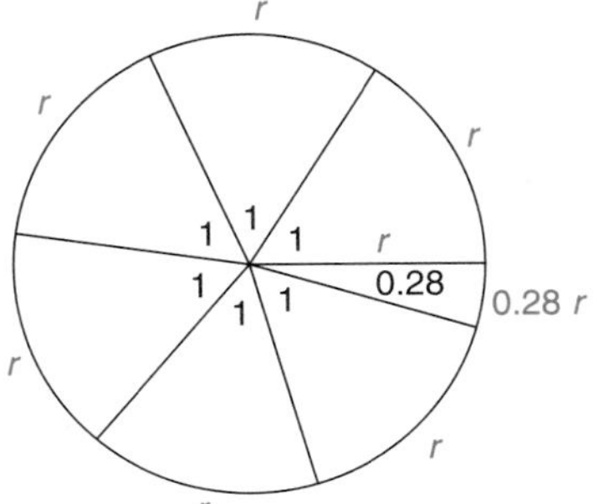

FIGURE 10.52

$$2\pi \text{ rad} = 360^\circ$$
or
$$360^\circ = 2\pi \text{ rad}$$

Through division by 2, the relationship is often restated as follows:

$$\pi \text{ rad} = 180^\circ$$
or
$$180^\circ = \pi \text{ rad}$$

With π rad $= 180°$, we divide each side of this equation by π to obtain the following relationship:

$$1 \text{ rad} = \frac{180^\circ}{\pi} \approx 57.3^\circ$$

To compare angle measures, we may also wish to divide each side of the equation $180° = \pi$ rad by 180 to get the following relationship.

$$1^\circ = \frac{\pi}{180} \text{ rad}$$

EXAMPLE 4

Using the fact that $1° = \frac{\pi}{180}$ rad, find the radian equivalencies for:

a) 30° **b)** 45° **c)** 60° **d)** −90°

Solution

a) $30° = 30(1°) = 30\left(\frac{\pi}{180}\right) \text{ rad} = \frac{\pi}{6} \text{ rad}$

b) $45° = 45(1°) = 45\left(\frac{\pi}{180}\right) \text{ rad} = \frac{\pi}{4} \text{ rad}$

c) $60° = 60(1°) = 60\left(\frac{\pi}{180}\right) \text{ rad} = \frac{\pi}{3} \text{ rad}$

d) $-90° = -90(1°) = -90\left(\frac{\pi}{180}\right) \text{ rad} = -\frac{\pi}{2} \text{ rad}$

EXAMPLE 5

Using the fact that π rad = 180°, find the degree equivalencies for the following angles measured in radians:

a) $\frac{\pi}{6}$ **b)** $\frac{2\pi}{5}$ **c)** $\frac{-3\pi}{4}$ **d)** $\frac{\pi}{2}$

Solution

a) $\frac{\pi}{6} = \frac{180°}{6} = 30°$

b) $\frac{2\pi}{5} = \frac{2}{5} \cdot \pi = \frac{2}{5} \cdot 180° = 72°$

c) $\frac{-3\pi}{4} = \frac{-3}{4} \cdot \pi = \frac{-3}{4} \cdot 180° = -135°$

d) $\frac{\pi}{2} = \frac{180°}{2} = 90°$

While we did not use this method of measuring angles in this textbook, you may need to use this method of angle measurement in a more advanced course.

Summary

A Look Back at Chapter 10

One goal of this chapter was to define the sine, cosine, and tangent ratios in terms of the sides of a right triangle. We derived a formula for finding the area of a triangle, given two sides and the included angle. We also proved the Law of Sines and the Law of Cosines for acute triangles. Another unit, called the "radian," was introduced for the purpose of measuring angles.

Important Terms and Concepts of Chapter 10

10.1 Greek Letters to Name Angles: $\alpha, \beta, \gamma, \theta$
Opposite Side
Hypotenuse
Sine Ratio: $\sin\theta = \dfrac{\text{opposite}}{\text{hypotenuse}}$
Angle of Elevation
Angle of Depression

10.2 Adjacent Side
Cosine Ratio: $\cos\theta = \dfrac{\text{adjacent}}{\text{hypotenuse}}$
Identity

10.3 Tangent Ratio: $\tan\theta = \dfrac{\text{opposite}}{\text{adjacent}}$
Cotangent
Secant
Cosecant

10.4 Trigonometric Identities

Reciprocal: $\tan\theta \cdot \cot\theta = 1$
$\cos\theta \cdot \sec\theta = 1$
$\sin\theta \cdot \csc\theta = 1$

Quotient: $\tan\theta = \dfrac{\sin\theta}{\cos\theta}$; $\cot\theta = \dfrac{\cos\theta}{\sin\theta}$

Pythagorean: $\sin^2\theta + \cos^2\theta = 1$
$\tan^2\theta + 1 = \sec^2\theta$
$\cot^2\theta + 1 = \csc^2\theta$

Area of a Triangle: $A = \frac{1}{2}bc\sin\alpha$
$A = \frac{1}{2}ac\sin\beta$
$A = \frac{1}{2}ab\sin\gamma$

Law of Sines: $\dfrac{\sin\alpha}{a} = \dfrac{\sin\beta}{b} = \dfrac{\sin\gamma}{c}$

Law of Cosines: $c^2 = a^2 + b^2 - 2ab\cos\gamma$
$b^2 = a^2 + c^2 - 2ac\cos\beta$
$a^2 = b^2 + c^2 - 2bc\cos\alpha$

A Look Beyond: *Radian Measure of Angles*

Review Exercises

In Exercises 1 to 4, state the ratio needed, and use it to find the measure of the indicated segment to the nearest tenth of a unit.

1.

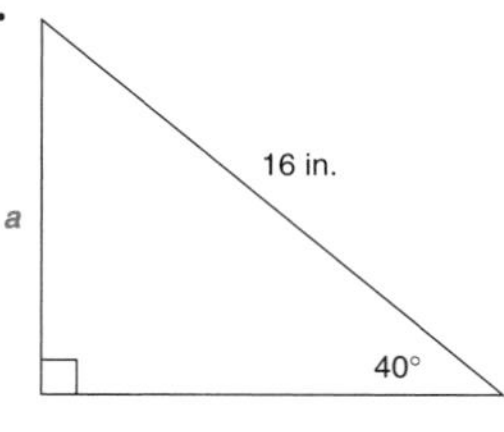

2.

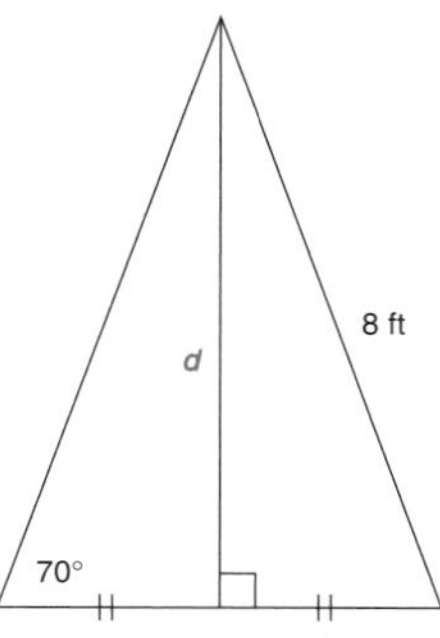

3.

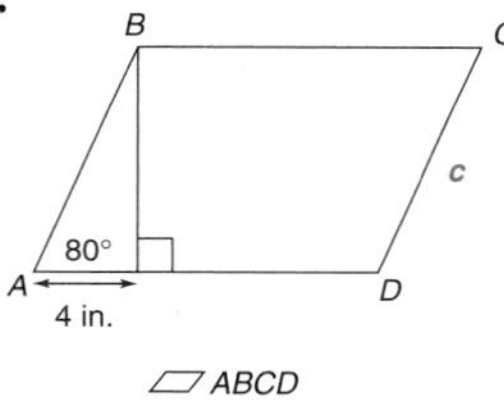

▱ABCD

4.

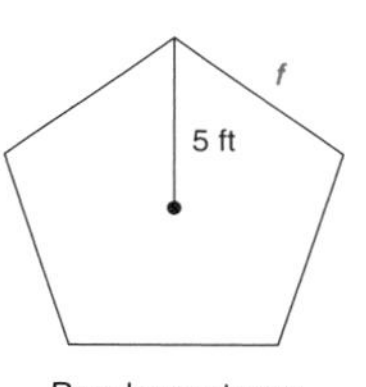

Regular pentagon with radius = 5 ft

In Exercises 5 to 8, state the ratio needed, and use it to find the measure of the angle to the nearest degree. Use a calculator.

5.

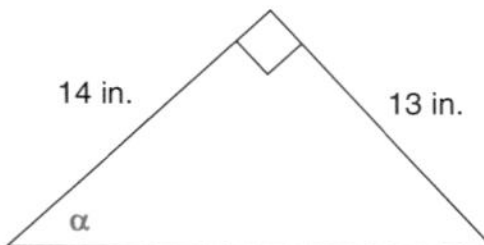

6.

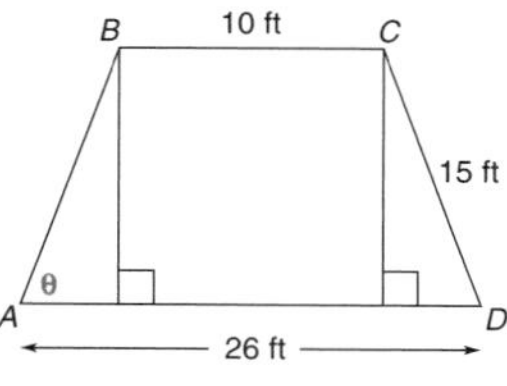

Isosceles trapezoid *ABCD*

7.

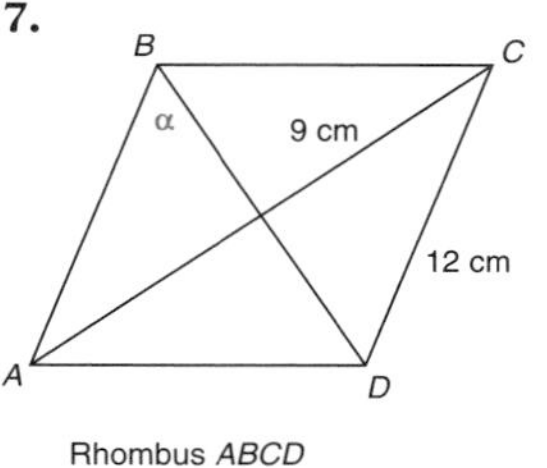

Rhombus *ABCD*

8.

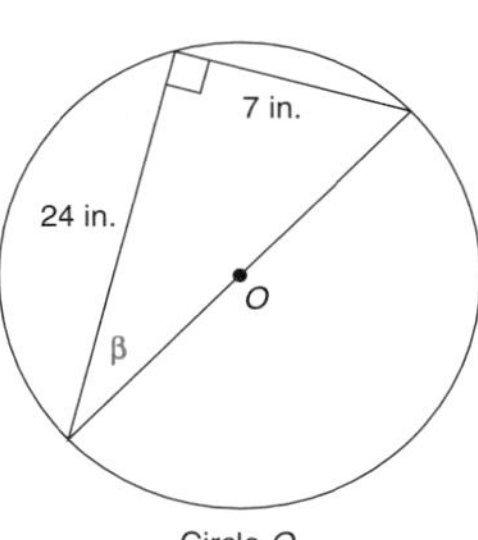

Circle *O*

In Exercises 9 to 12, use the Law of Sines or the Law of Cosines to solve each triangle for the indicated side or angle. Angle measures should be found to the nearest degree; distances should be found to the nearest tenth of a unit.

9.

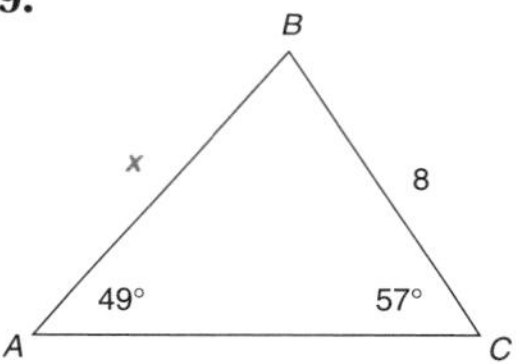

10.

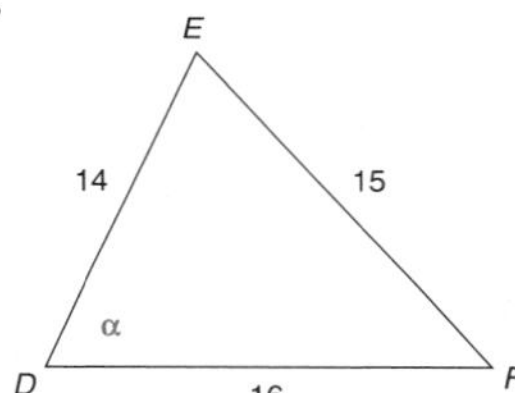

11.

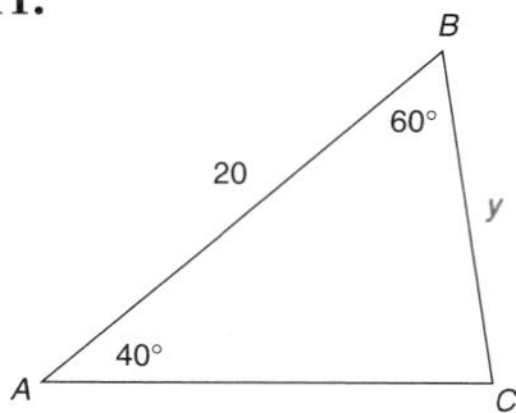

12.

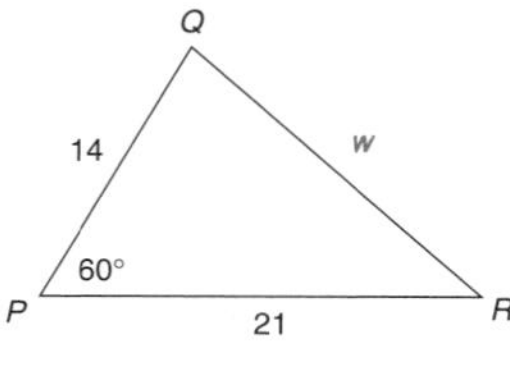

In Exercises 13 to 17, use the Law of Sines or the Law of Cosines to solve each problem. Angle measures should be found to the nearest degree; distances should be found to the nearest tenth of a unit.

13. A building 50 ft tall is on a hillside. A surveyor at a point on the hill observes that the angle of elevation to the top of the building measures 43° and the angle of depression to the base of the building measures 16°. How far is the surveyor from the base of the building?

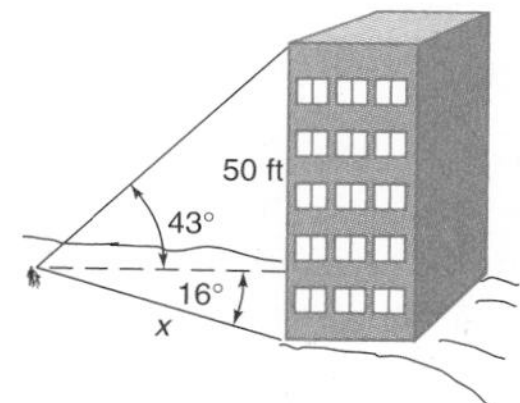

14. Two sides of a parallelogram are 50 cm and 70 cm long. Find the length of the shorter diagonal if a larger angle of the parallelogram measures 105°.

15. The sides of a rhombus are 6 in. each and the longer diagonal is 11 in. Find the measure of the acute angle in the rhombus.

16. The area of $\triangle ABC$ is 9.7 in.^{2} If $a = 6$ in. and $c = 4$ in., find the measure of angle B.

17. Find the area of the rhombus in Exercise 15.

In Exercises 18 to 20, prove each statement without using a table or a calculator. Draw an appropriate right triangle.

18. If $m\angle R = 45°$, then $\tan R = 1$.

19. If $m\angle S = 30°$, then $\sin S = \frac{1}{2}$.

20. If $m\angle T = 60°$, then $\sin T = \frac{\sqrt{3}}{2}$.

In Exercises 21 to 30, use the drawings where provided to solve each problem. Angle measures should be found to the nearest degree; lengths should be found to the nearest tenth of a unit.

21. In the evening, a tree that stands 12 ft tall casts a long shadow. If the angle of depression from the top of the tree to the top of the shadow is 55°, what is the length of the shadow?

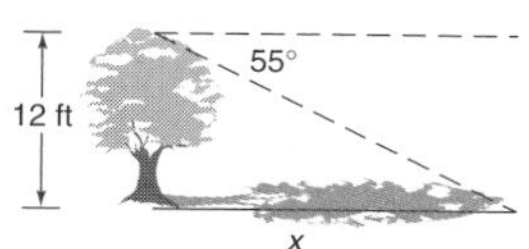

22. A rocket is shot into the air at an angle of 60°. If it is traveling at 200 feet per second, how high in the air is it after 5 seconds? (Ignoring gravity, assume that the path of the rocket is a straight line.)

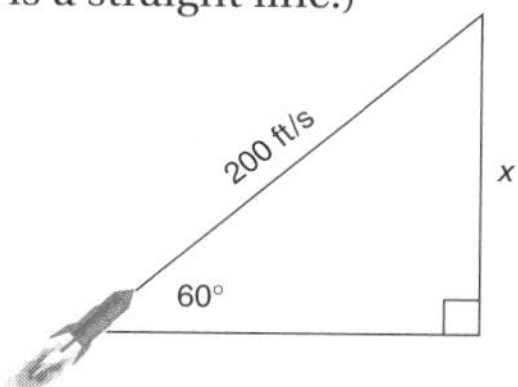

23. A 4-m beam is used to brace a wall. If the bottom of the beam is 3 m from the base of the wall, what is the angle of elevation to the top of the wall?

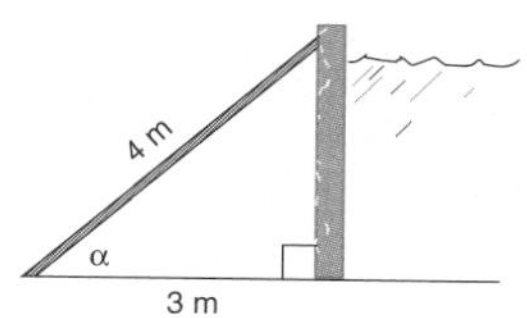

24. A hot air balloon is 300 ft high. The pilot of the balloon sights a stadium 2200 ft away. What is the angle of depression?

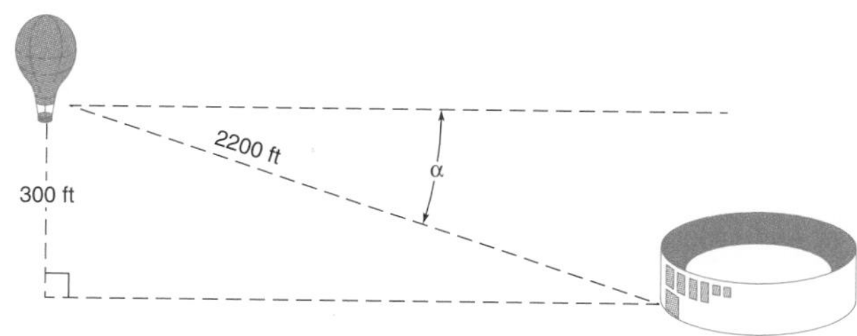

25. The apothem of a regular pentagon is approximately 3.44 cm. What is the approximate length of each side of the pentagon?

26. What is the approximate length of the radius of the pentagon in Exercise 25?

27. The legs of an isosceles triangle are each 40 cm in length. The base is 30 cm in length. Find the measure of a base angle.

28. The diagonals of a rhombus measure 12 in. and 16 in. Find the measure of the obtuse angle in the rhombus.

29. The unit used for measuring the steepness of a hill is the **grade.** A grade of a to b means the hill rises a vertical units for every b horizontal units. If, at some point, the hill is 3 ft above the horizontal and the angle of elevation to that point is 23°, what is the grade of this hill?

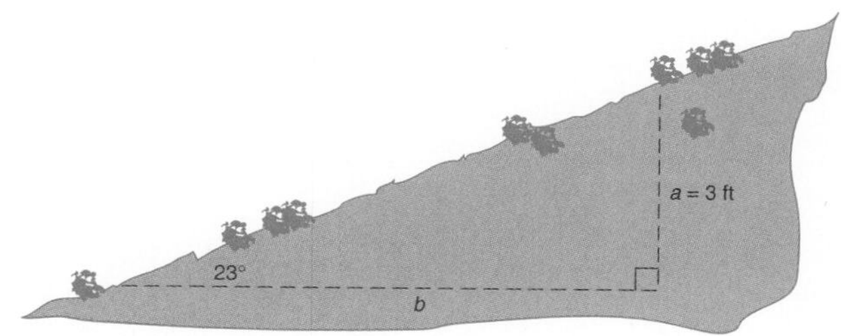

30. An observer in a plane 2500 m high sights two ships below. The angle of depression to one ship is 32°, and the angle of depression to the other ship is 44°. How far apart are the ships?

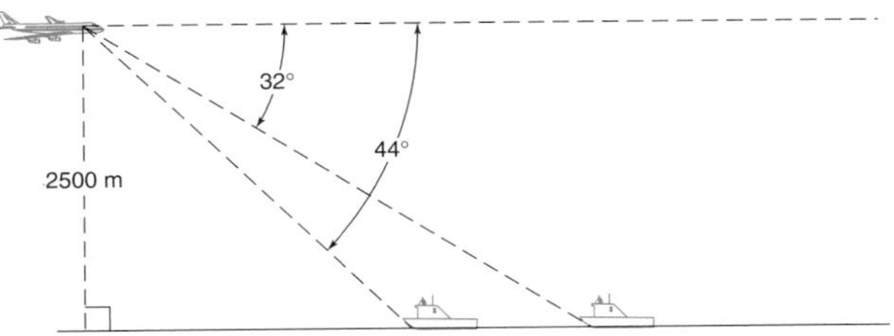

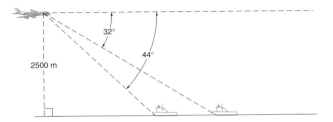

31. If $\sin \theta = \frac{7}{25}$, find $\cos \theta$ and $\sec \theta$.

32. If $\tan \theta = \frac{11}{60}$, find $\sec \theta$ and $\cot \theta$.

33. If $\cot \theta = \frac{21}{20}$, find $\csc \theta$ and $\sin \theta$.

34. In a right circular cone, the radius of the base is 3.2 ft in length and the angle formed by the radius and slant height measures $\theta = 65°$. To the nearest tenth of a foot, find the length of the altitude of the cone. Then use this length of altitude to find the volume of the cone to the nearest tenth of a cubic foot.

Appendices

Appendix A: Quadratic Equations

An equation that can be written in the form $ax^2 + bx + c = 0$ $(a \neq 0)$ is a **quadratic equation.** For example, $x^2 - 7x + 12 = 0$ and $6x^2 = 7x + 3$ are quadratic. Many quadratic equations can be solved by a factoring method that depends on the Zero Product Property.

> **ZERO PRODUCT PROPERTY**
>
> If $a \cdot b = 0$, then $a = 0$ or $b = 0$.

When this property is stated in words, it reads: "If the product of two expressions equals 0, then at least one of the factors must equal 0."

EXAMPLE 1

Solve $x^2 - 7x + 12 = 0$.

Solution

First you must factor the polynomial; then check the factors by using the FOIL method of multiplication.

$$(x - 3)(x - 4) = 0 \qquad \text{factoring}$$
$$x - 3 = 0 \quad \text{or} \quad x - 4 = 0 \qquad \text{Zero Product Property}$$
$$x = 3 \quad \text{or} \quad x = 4 \qquad \text{Addition Property}$$

To check $x = 3$, substitute into the given equation:

$$3^2 - (7 \cdot 3) + 12 = 9 - 21 + 12 = 0$$

Similarly, to check $x = 4$, substitute again:

$$4^2 - (7 \cdot 4) + 12 = 16 - 28 + 12 = 0$$

Checks for later problems will not be provided. The solutions are usually expressed as a set, {3, 4}.

If you were asked to solve the quadratic equation

$$6x^2 = 7x + 3$$

it would be necessary to change the equation so that one side would be equal to 0. The form $ax^2 + bx + c = 0$ is the **standard form** of a quadratic equation.

Solving a Quadratic Equation by the Factoring Method

1. Be sure the equation is in standard form (one side = 0).
2. Factor the polynomial side of the equation.
3. Set each factor containing the variable equal to 0.
4. Solve each equation found in step 3.
5. Check solutions by substituting into the original equation.

EXAMPLE 2

Solve $6x^2 = 7x + 3$.

Solution

First changing to standard form, we have

$$\begin{array}{rcrl} & 6x^2 - 7x - 3 = 0 & & \text{standard form} \\ & (2x - 3)(3x + 1) = 0 & & \text{factoring} \\ 2x - 3 = 0 & \text{or} & 3x + 1 = 0 & \text{Zero Product Property} \\ 2x = 3 & \text{or} & 3x = -1 & \text{Addition-Subtraction Property} \\ x = \dfrac{3}{2} & \text{or} & x = \dfrac{-1}{3} & \text{division} \end{array}$$

Therefore $\left\{\frac{3}{2}, -\frac{1}{3}\right\}$ is the solution set.

In some instances, a common factor can be extracted from each term in the factoring step. In the equation $2x^2 + 10x - 48 = 0$, the left side of the equation has the common factor 2. Factoring leads to $2(x^2 + 5x - 24) = 0$ and then to $2(x + 8)(x - 3) = 0$. Of course, only the factors containing variables can equal 0, so the solutions to this equation are -8 and 3.

Equations such as $4x^2 = 9$ and $4x^2 - 12x = 0$ are **incomplete quadratic equations** because one term is missing from the standard form; the linear term (having exponent 1) is missing from the first equation, and the constant term is omitted in the second. Either equation can, however, be solved by factoring; in particular, the factoring is given by

$$4x^2 - 9 = (2x + 3)(2x - 3)$$

and

$$4x^2 - 12x = 4x(x - 3)$$

When solutions to $ax^2 + bx + c = 0$ cannot be found by factoring, they may be determined by the following formula, in which a is the number multiplied by x^2,

b is the number multiplied by x, and c is the constant term. The $\pm$ symbol tells us that there are generally two solutions, one found by adding and one by subtracting. The symbol $\sqrt{a}$ is read "square root of a."

QUADRATIC FORMULA

$x = \dfrac{-b \pm \sqrt{b^2 - 4ac}}{2a}$ are solutions for $ax^2 + bx + c = 0$, where $a \neq 0$.

Although the formula may provide two solutions for the equation, an application problem in geometry may have a single positive solution representing a segment (or angle) measure. Recall that for $a > 0$, $\sqrt{a}$ represents the principal square root of a.

DEFINITION: Where $a > 0$, the number $\sqrt{a}$ is the positive number for which $(\sqrt{a})^2 = a$.

EXAMPLE 3

a) Explain why $\sqrt{25}$ is equal to 5.
b) Without a calculator, find the value of $\sqrt{3} \cdot \sqrt{3}$.
c) Use a calculator to show that $\sqrt{5} \approx 2.236$.

Solution

a) We see that $\sqrt{25}$ must equal 5 because $5^2 = 25$.
b) By definition, $\sqrt{3}$ is the number for which $(\sqrt{3})^2 = 3$.
c) By using a calculator, we see that $2.236^2 \approx 5$.

EXAMPLE 4

Simplify each expression, if possible.

a) $\sqrt{16}$ **b)** $\sqrt{0}$ **c)** $\sqrt{7}$ **d)** $\sqrt{400}$ **e)** $\sqrt{-4}$

Solution

a) $\sqrt{16} = 4$ because $4^2 = 16$.
b) $\sqrt{0} = 0$ because $0^2 = 0$.
c) $\sqrt{7}$ cannot be simplified; however, $\sqrt{7} \approx 2.646$.
d) $\sqrt{400} = 20$ because $20^2 = 400$; calculator can be used.
e) $\sqrt{-4}$ is not a real number; calculator gives an "ERROR" message.

Whereas $\sqrt{25}$ represents the principal square root of 25 (namely 5), the expression $-\sqrt{25}$ can be interpreted as "the negative number whose square is 25"; thus, $-\sqrt{25} = -5$ because $(-5)^2 = 25$. In expressions such as $\sqrt{9 + 16}$ and

$\sqrt{4 + 9}$, we first simplify the radicand (expression under the bar of the square root); thus, $\sqrt{9 + 16} = \sqrt{25} = 5$ and $\sqrt{4 + 9} = \sqrt{13} \approx 3.606$.

Just as fractions are reduced to lower terms $\left(\frac{6}{8}\right.$ is replaced by $\left.\frac{3}{4}\right)$, it is also common to reduce the size of the radicand when possible. To accomplish this, we use the Product Property of Square Roots.

PRODUCT PROPERTY OF SQUARE ROOTS

For $a \geq 0$ and $b \geq 0$, $\sqrt{a \cdot b} = \sqrt{a} \cdot \sqrt{b}$.

When simplifying, the radicand is replaced by a product in which the largest possible number from the list of perfect squares below is selected as one of the factors:

$$4, 9, 16, 25, 36, 49, 64, 81, 100, 121, \ldots$$

For example,

$$\begin{aligned} \sqrt{45} &= \sqrt{9 \cdot 5} \\ &= \sqrt{9} \cdot \sqrt{5} \\ &= 3\sqrt{5} \end{aligned}$$

The radicand has now been reduced from 45 to 5. Using a calculator, we see that $\sqrt{45} \approx 6.708$. Also, $3\sqrt{5}$ means 3 times $\sqrt{5}$ and with the calculator we see that $3\sqrt{5} \approx 6.708$.

Leave the smallest possible integer under the square root symbol.

EXAMPLE 5

Simplify the following radicals:

a) $\sqrt{27}$ **b)** $\sqrt{125}$

Solution

a) 9 is the largest perfect square factor of 27. Therefore

$$\sqrt{27} = \sqrt{9 \cdot 3} = \sqrt{9} \cdot \sqrt{3} = 3\sqrt{3}$$

b) 25 is a perfect square factor of 125. Therefore

$$\sqrt{125} = \sqrt{25 \cdot 5} = \sqrt{25} \cdot \sqrt{5} = 5\sqrt{5}$$

The Product Property of Square Roots has a symmetric form that reads $\sqrt{a} \cdot \sqrt{b} = \sqrt{ab}$; for example, $\sqrt{2} \cdot \sqrt{3} = \sqrt{6}$ and $\sqrt{5} \cdot \sqrt{5} = \sqrt{25} = 5$.

> When calculator answers are requested or provided, the answers in this textbook will generally be rounded to two decimal places. For instance, $\sqrt{125} \approx 11.18$ (rounded from 11.1803).

The expression $ax^2 + bx + c$ may be **prime** (meaning "not factorable"). Because $x^2 - 5x + 3$ is prime, we solve the equation $x^2 - 5x + 3 = 0$ by using the Quadratic Formula $x = \frac{-b \pm \sqrt{b^2 - 4ac}}{2a}$; see Example 6.

NOTE: When square root radicals are left in an answer, the answer is considered to be exact. Once we use the calculator, the solutions are only approximate.

EXAMPLE 6

Find exact solutions for $x^2 - 5x + 3 = 0$. Then use a calculator to approximate these solutions correct to two decimal places.

Solution

With the equation in standard form, we see that $a = 1$, $b = -5$, and $c = 3$. So

$$x = \frac{-(-5) \pm \sqrt{(-5)^2 - 4(1)(3)}}{2(1)}$$

$$x = \frac{5 \pm \sqrt{25 - 12}}{2} \quad \text{or} \quad x = \frac{5 \pm \sqrt{13}}{2}$$

The exact solutions are $\frac{5 + \sqrt{13}}{2}$ and $\frac{5 - \sqrt{13}}{2}$. Using a calculator, the approximate solutions are 4.30 and 0.70.

In solving the equation $x^2 - 6x + 7 = 0$, the Quadratic Formula leads to $x = \frac{6 \pm \sqrt{8}}{2}$. In Example 7, we focus on the simplification of such an expression.

EXAMPLE 7

Simplify $\frac{6 \pm \sqrt{8}}{2}$.

Solution

Because $\sqrt{8} = \sqrt{4} \cdot \sqrt{2}$ or $2\sqrt{2}$, we simplify the expression as follows.

$$\frac{6 \pm \sqrt{8}}{2} = \frac{6 \pm 2\sqrt{2}}{2} = \frac{\cancel{2}(3 \pm \sqrt{2})}{\cancel{2}} = 3 \pm \sqrt{2}$$

NOTE 1: The number 2 was a common factor for the numerator's terms. We then reduced the fraction to lowest terms.

NOTE 2: The approximate values of $3 \pm \sqrt{2}$ are 4.41 and 1.59. Use your calculator to show that these values are the approximate solutions of the equation $x^2 - 6x + 7 = 0$.

Our final method for solving quadratic equations is used if the equation has the form $ax^2 + c = 0$.

SQUARE ROOTS PROPERTY

If $x^2 = p$ where $p \geq 0$, then $x = \pm\sqrt{p}$.

According to the Square Roots Property, the equation $x^2 = 6$ has the solutions $\pm\sqrt{6}$.

EXAMPLE 8

Use the Square Roots Property to solve the equation $2x^2 - 56 = 0$.

Solution

$$2x^2 - 56 = 0 \rightarrow 2x^2 = 56 \rightarrow x^2 = 28$$

Then

$$x = \pm\sqrt{28} = \pm\sqrt{4} \cdot \sqrt{7} = \pm 2\sqrt{7}$$

The exact solutions are $2\sqrt{7}$ and $-2\sqrt{7}$; the approximate solutions are 5.29 and −5.29.

In Example 10, the solutions for the quadratic equation will involve fractions. For this reason, we consider the Quotient Property of Square Roots. The Quotient Property allows us to replace the square root of a fraction by the square root of the numerator divided by the square root of the denominator.

QUOTIENT PROPERTY OF SQUARE ROOTS

For $a \geq 0$ and $b > 0$, $\sqrt{\frac{a}{b}} = \frac{\sqrt{a}}{\sqrt{b}}$.

EXAMPLE 9

Simplify the following square root expressions:

a) $\sqrt{\frac{16}{9}}$

b) $\sqrt{\frac{3}{4}}$

Solution

a) $\sqrt{\frac{16}{9}} = \frac{\sqrt{16}}{\sqrt{9}} = \frac{4}{3}$

b) $\sqrt{\frac{3}{4}} = \frac{\sqrt{3}}{\sqrt{4}} = \frac{\sqrt{3}}{2}$

EXAMPLE 10

Solve the equation $4x^2 - 9 = 0$.

Solution

$$4x^2 - 9 = 0 \rightarrow 4x^2 = 9 \rightarrow x^2 = \frac{9}{4}$$

Then

$$x = \pm\sqrt{\frac{9}{4}} = \pm\frac{\sqrt{9}}{\sqrt{4}} = \pm\frac{3}{2}$$

In summary, quadratic equations have the form $ax^2 + bx + c = 0$ and are solved by one of the following methods:

1. Factoring, when $ax^2 + bx + c$ is easily factored.
2. The Quadratic Formula

$$x = \frac{-b \pm \sqrt{b^2 - 4ac}}{2a}$$

when $ax^2 + bx + c$ is not easily factored or cannot be factored.

3. The Square Roots Property, when $b = 0$.

Exercises

1. Use your calculator to find the approximate value of each number, correct to two decimal places:

a) $\sqrt{13}$ **b)** $\sqrt{8}$ **c)** $-\sqrt{29}$ **d)** $\sqrt{\frac{3}{5}}$

2. Use your calculator to find the approximate value of each number, correct to two decimal places:

a) $\sqrt{17}$ **b)** $\sqrt{400}$ **c)** $-\sqrt{7}$ **d)** $\sqrt{1.6}$

3. Which equations are quadratic?

a) $2x^2 - 5x + 3 = 0$ **d)** $\frac{1}{2}x^2 - \frac{1}{4}x - \frac{1}{8} = 0$
b) $x^2 = x^2 + 4$ **e)** $\sqrt{2x - 1} = 3$
c) $x^2 = 4$ **f)** $(x + 1)(x - 1) = 15$

4. Which equations are incomplete quadratic equations?

a) $x^2 - 4 = 0$ **d)** $2x^2 - 4 = 2x^2 + 8x$
b) $x^2 - 4x = 0$ **e)** $x^2 = \frac{9}{4}$
c) $3x^2 = 2x$ **f)** $x^2 - 2x - 3 = 0$

5. Simplify each expression by using the Product Property of Square Roots:

a) $\sqrt{8}$ **c)** $\sqrt{900}$
b) $\sqrt{45}$ **d)** $(\sqrt{3})^2$

6. Simplify each expression by using the Product Property of Square Roots:

a) $\sqrt{28}$ **c)** $\sqrt{54}$
b) $\sqrt{32}$ **d)** $\sqrt{200}$

7. Simplify each expression by using the Quotient Property of Square Roots:

a) $\sqrt{\frac{9}{16}}$ **c)** $\sqrt{\frac{7}{16}}$
b) $\sqrt{\frac{25}{49}}$ **d)** $\sqrt{\frac{6}{9}}$

8. Simplify each square root expression by using the Quotient Property of Square Roots:

a) $\sqrt{\frac{1}{4}}$ **c)** $\sqrt{\frac{5}{36}}$
b) $\sqrt{\frac{16}{9}}$ **d)** $\sqrt{\frac{3}{16}}$

9. Use your calculator to verify that the following expressions are equivalent:

a) $\sqrt{54}$ and $3\sqrt{6}$ **b)** $\sqrt{\frac{5}{16}}$ and $\frac{\sqrt{5}}{4}$

10. Use your calculator to verify that the following expressions are equivalent:

a) $\sqrt{48}$ and $4\sqrt{3}$ **b)** $\sqrt{\frac{7}{9}}$ and $\frac{\sqrt{7}}{3}$

In Exercises 11 to 18, solve each quadratic equation by factoring.

11. $x^2 - 6x + 8 = 0$

12. $x^2 + 4x = 21$

13. $3x^2 - 51x + 180 = 0$
(**HINT:** There is a common factor.)

14. $2x^2 + x - 6 = 0$

15. $3x^2 = 10x + 8$

16. $8x^2 + 40x - 112 = 0$

17. $6x^2 = 5x - 1$

18. $12x^2 + 10x = 12$

In Exercises 19 to 26, solve each equation using the Quadratic Formula. Give exact solutions in simplified form. When answers contain square roots, approximate the solutions rounded to two decimal places.

19. $x^2 - 7x + 10 = 0$

20. $x^2 + 7x + 12 = 0$

21. $x^2 + 9 = 7x$

22. $2x^2 + 3x = 6$

23. $x^2 - 4x - 8 = 0$

24. $x^2 - 6x - 2 = 0$

25. $5x^2 = 3x + 7$

26. $2x^2 = 8x - 1$

In Exercises 27 to 32, solve each incomplete quadratic equation. Use the Square Roots Property as needed.

27. $2x^2 = 14$

28. $2x^2 = 14x$

29. $4x^2 - 25 = 0$

30. $4x^2 - 25x = 0$

31. $ax^2 - bx = 0$

32. $ax^2 - b = 0$

33. The length of a rectangle is 3 more than its width. If the area of the rectangle is 40, the dimensions x and $x + 3$ can be found by solving the equation $x(x + 3) = 40$. Find these dimensions.

34. To find the lengths of $\overline{CP}$ (which is x), $\overline{PD}$, $\overline{AP}$, and $\overline{PB}$, one must solve the equation

$$x \cdot (x + 5) = (x + 1) \cdot 4$$

Find the length of $\overline{CP}$.

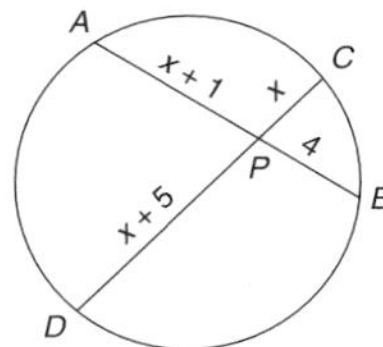

In Exercises 35 and 36, use Theorem 2.5.1 to solve the problem. According to this theorem, the number of diagonals in a polygon of n sides is given by $D = \frac{n(n - 3)}{2}$.

35. Find the number of sides in a polygon that has 9 diagonals.

36. Find the number of sides in a polygon that has the same number of diagonals as it has sides.

37. In the right triangle, find c if $a = 3$ and $b = 4$.
(**HINT:** $c^2 = a^2 + b^2$)

38. In the right triangle, find b if $a = 6$ and $c = 10$.
(**HINT:** $c^2 = a^2 + b^2$)

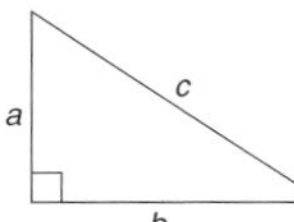

Exercises 37, 38

Appendix B: Truth Tables

P	$\sim P$
T	F
F	T

Recall that a *statement* is a group of words or symbols that can be classified collectively as true or false. The claim "$5 + 7 = 12$" is a true statement whereas "A pentagon has six sides" is a false statement. However, "Man overboard!" is not a statement at all. Symbolically, we represent statements by letters such as P, Q, and R. The negation of a given statement P is written $\sim P$; when P is true, $\sim P$ is false, and vice versa. This relationship between a statement and its negation can be expressed by a *truth table.* The first line of the truth table shown indicates "If P is true, then $\sim P$ is false." The second line in the table indicates "When P is false, $\sim P$ is true." Suppose that P is the true statement "Abraham Lincoln lived in Illinois." Then $\sim P$ = "Abraham Lincoln did not live in Illinois" is false.

EXAMPLE 1

Determine which of the following are statements. If a statement, is it true or false?

a) $4 + 3 < 5$
b) Babe Ruth played baseball.
c) Are you Mike?
d) $\sim P$, if P is true.

Solution

a) False statement
b) True statement
c) Not a statement
d) False statement

A truth table is a valuable tool for examining the truth of a statement that is more complex. What exactly is a truth table?

DEFINITION: A **truth table** is a table that provides the truth values of a statement by considering all possible true/false combinations of the statement's components.

P	Q	$P \wedge Q$
T	T	T
T	F	F
F	T	F
F	F	F

Conjunction

Statements can be combined to form **compound statements.** For example, a statement of the form "P and Q" is called the **conjunction** of P and Q. In symbols, the conjunction is written $P \wedge Q$. For the conjunction to be true, it is necessary for P to be true *and* Q to be true. If either statement is false, the conjunction is false. To allow for all possible true/false combinations, four rows are needed in the truth table of the conjunction. When P is true, Q may be true or false (two rows of the table). When P is false, Q may be true or false (two additional rows).

EXAMPLE 2

Let $P =$ "Babe Ruth played baseball" and $Q =$ "$4 + 3 < 5$." Classify as true or false:

a) $P \wedge Q$ **b)** $P \wedge \sim Q$

Solution

a) The conjunction is false because Q is false.
b) The conjunction is true because P is true and $\sim Q$ is also true.

P	Q	$P \vee Q$
T	T	T
T	F	T
F	T	T
F	F	F

Disjunction

A compound statement of the form "P or Q" is called the **disjunction** of P and Q. In symbols, the disjunction is written $P \vee Q$. For the disjunction to be true, either P is true, Q is true, or both P and Q are true. A disjunction is false only if P and Q are both false. To understand the first line in the truth table, consider this situation: You can join the Math Club if you have an A average *or* you are enrolled in a mathematics class. If you satisfy both requirements, you may still join the club.

EXAMPLE 3

Let $P =$ "Babe Ruth played baseball" and $Q =$ "$4 + 3 < 5$." Classify as true or false:

a) $P \vee Q$ **b)** $P \vee \sim Q$

Solution

a) The disjunction is true because P is true even though Q is false.
b) The disjunction is true because P is true and $\sim Q$ is also true.

In some cases, we use parentheses to clarify the meaning of a compound statement. In this context, parentheses are given priority just as they are in numerical expressions. See Example 4.

EXAMPLE 4

Where P, Q, and R are statements, suppose that P is true, Q is false, and R is false. Classify the statement $P \wedge (\sim Q \vee R)$ as true or false.

Solution

The statement in parentheses is a disjunction of the form "T or F" and is true. Then $P \wedge (\sim Q \vee R)$ is a conjunction of the form "T and T," so the given statement is true.

NOTE: The steps used to determine the truth of the statement in Example 4 can be written as follows:

$$P \wedge (\sim Q \vee R)$$
$$T \wedge (T \vee F)$$
$$T \wedge (T)$$
$$T$$

Implication

P	Q	$P \rightarrow Q$
T	T	T
T	F	F
F	T	T
F	F	T

The final compound statement that we consider is of the form "If P, then Q." This statement is called an **implication** or a **conditional statement.** In symbols, we write $P \rightarrow Q$. The conditional statement makes a *promise* and fails to satisfy the conditions of this promise only when P is true and Q is false (see the truth table). Consider the claim, "If you are good, then I'll give you a dollar." The only way the claim is false is when "you are good, but I don't give you the dollar."

In $P \rightarrow Q$, P is called the *antecedent* and Q is called the *consequent.*

Truth tables have several applications. They are used to show that:

1. Two statements are *logically equivalent,* meaning their truth values are the same.
2. The negation of a compound statement has a particular form. For instance, we can show that the negation of $(P \wedge Q)$ is $(\sim P \vee \sim Q)$.
3. Some statements are always true. Such statements are called *tautologies.*
4. An argument is *valid.* Valid arguments are discussed in Appendix C.

Recall that the conditional statement "If P, then Q" is expressed symbolically by $P \rightarrow Q$. Related to the conditional are its:

Converse — If Q, then P. $(Q \rightarrow P)$

Inverse — If not P, then not Q. $(\sim P \rightarrow \sim Q)$

Contrapositive — If not Q, then not P. $(\sim Q \rightarrow \sim P)$

Logical Equivalence of Statements

DEFINITION: Two statements are **logically equivalent** if their truth values are the same for all possible true/false combinations of their components.

In Example 5, we will show that an implication and its contrapositive are logically equivalent. Because the example involves two simple statements P and Q, the truth table shown has four horizontal rows. The columns at the top of the truth table show P and Q as well as certain statements that involve statement P, statement Q, or both. Because our goal is to compare the truth values of $P \rightarrow Q$ and $\sim Q \rightarrow \sim P$, the column headings of the truth table are chosen accordingly.

EXAMPLE 5

Use a truth table to show that the statement $P \to Q$ and its contrapositive $\sim Q \to \sim P$ are logically equivalent.

Solution

In the following truth table, the final columns show that the implication and its contrapositive have the same truth values for all combinations of P and Q.

P	Q	$\sim P$	$\sim Q$	$P \to Q$	$\sim Q \to \sim P$
T	T	F	F	T	T
T	F	F	T	F	F
F	T	T	F	T	T
F	F	T	T	T	T

A double arrow is used to show that two statements are logically equivalent. In Example 5, $(P \to Q) \leftrightarrow (\sim Q \to \sim P)$.

EXAMPLE 6

Based on Example 5, write a statement that is logically equivalent to, "If a person lives in London, then the person lives in England."

Solution

The given statement is true. Its contrapositive (also true) is, "If a person does not live in England, then the person does not live in London."

DeMorgan's Laws

In the study of logic, DeMorgan's Laws are used to describe the negations of the conjunction and disjunction. Augustin DeMorgan was a nineteenth-century English mathematician and logician.

DEMORGAN'S LAWS

1. $[\sim(P \wedge Q)] \leftrightarrow [\sim P \vee \sim Q]$
The negation of a conjunction is the disjunction of negations.

2. $[\sim(P \vee Q)] \leftrightarrow [\sim P \wedge \sim Q]$
The negation of a disjunction is the conjunction of negations.

EXAMPLE 7

Use DeMorgan's Laws to write the negation of:

a) $2 + 3 = 5$ and 13 is prime.

b) Clint is cool or Tim is handsome.

Solution

a) $2 + 3 \neq 5$ or 13 is composite (opposite of prime).
b) Clint is not cool and Tim is not handsome.

EXAMPLE 8

Use a truth table to establish DeMorgan's first law $[\sim(P \wedge Q)] \leftrightarrow [\sim P \vee \sim Q]$.

Solution

We need to show that $[\sim(P \wedge Q)]$ and $[\sim P \vee \sim Q]$ have identical truth values. The results are shown in the fourth and seventh columns.

P	Q	$P \wedge Q$	$\sim(P \wedge Q)$	$\sim P$	$\sim Q$	$\sim P \vee \sim Q$
T	T	T	F	F	F	F
T	F	F	T	F	T	T
F	T	F	T	T	F	T
F	F	F	T	T	T	T

NOTE: The final column is completed by looking at the columns headed $\sim P$ and $\sim Q$.

The Tautology

A truth table can be used to show that some statements are always true. For instance, $P \vee \sim P$ is always true.

DEFINITION: A **tautology** is a statement that is true for all possible truth values of its components.

EXAMPLE 9

Show that the statement $(P \wedge Q) \rightarrow Q$ is a tautology.

Solution

Here we look at the truth value of $P \wedge Q$ and then back to the column headed Q to determine the truth or falsity of $(P \wedge Q) \rightarrow Q$.

P	Q	$P \wedge Q$	$(P \wedge Q) \rightarrow Q$
T	T	T	T
T	F	F	T
F	T	F	T
F	F	F	T

NOTE: A tautology must have a final column consisting only of T's (true).

In the following example, there are three component statements: P, Q, and R. In order to consider all true/false possibilities of these statements, we need eight horizontal rows in the table.

EXAMPLE 10

Is the statement $P \vee (Q \vee R)$ a tautology?

Solution

To list all possible truth value combinations for P, Q, and R, we use the eight horizontal rows shown. It will also be convenient to have one column headed $Q \vee R$. We will then decide on the truth/falsity of $P \vee (Q \vee R)$ by considering the columns headed P and $(Q \vee R)$.

P	Q	R	$(Q \vee R)$	$P \vee (Q \vee R)$
T	T	T	T	T
T	T	F	T	T
T	F	T	T	T
T	F	F	F	T
F	T	T	T	T
F	T	F	T	T
F	F	T	T	T
F	F	F	F	F

The statement $P \vee (Q \vee R)$ is *not* a tautology because the final column contains an F.

In Appendix C, we will show that certain forms of deductive reasoning lead to conclusions that cannot be refuted. Such arguments are known as *valid* arguments and are used throughout the study of mathematics. To establish that an argument is valid, we must show that its premises and conclusion form a compound statement that is a tautology.

Exercises

In Exercises 1 to 8, statement P is true, Q is true, and R is false. Classify each statement as true or false.

1. $\sim Q \vee R$
2. $P \vee \sim R$
3. $P \wedge Q$
4. $Q \wedge R$
5. $P \rightarrow R$
6. $P \rightarrow Q$
7. $P \wedge (Q \vee R)$
8. $(P \wedge Q) \vee R$

In Exercises 9 to 12, let P = "Mary is an accountant" and Q = "Hamburgers are health food." Write each symbolic statement in words.

9. $\sim Q$

10. $P \vee Q$

11. $P \rightarrow Q$

12. $P \wedge \sim Q$

In Exercises 13 to 18, form a truth table and determine all possible truth values for the given statement. Is the given statement a tautology?

13. $P \vee \sim P$

14. $P \wedge \sim P$

15. $(P \vee Q) \rightarrow P$

16. $(P \wedge Q) \rightarrow Q$

17. $[(P \rightarrow Q) \wedge Q] \rightarrow P$

18. $[(P \rightarrow Q) \wedge P] \rightarrow Q$

In Exercises 19 to 24, use DeMorgan's Laws to write the negation of the given statement.

19. $P \wedge Q$

20. $P \vee R$

21. Mary is an accountant or hamburgers are health food.

22. Mary is an accountant and hamburgers are health food.

23. It is cold and snowing.

24. We will go to dinner or to the movie.

25. Use a truth table to prove DeMorgan's second law: $[\sim(P \vee Q)] \leftrightarrow [\sim P \wedge \sim Q]$.
(**HINT:** See Example 8.)

26. Use a truth table to prove $[Q \rightarrow P] \leftrightarrow [\sim P \rightarrow \sim Q]$.
(**NOTE:** This proof establishes that the converse and inverse of an implication are logically equivalent.)

27. Use a truth table to show that $[(P \rightarrow Q) \wedge (Q \rightarrow R)] \rightarrow (P \rightarrow R)$ is a tautology.

28. Use a truth table to show that $[P \vee (Q \wedge R)]$ and $[(P \vee Q) \wedge (P \vee R)]$ are logically equivalent.

29. Use a truth table to show that $[P \wedge \sim Q]$ is the negation of $P \rightarrow Q$.
(**HINT:** The truth values of these statements must be opposites.)

In Exercises 30 to 33, use the result of Exercise 29 to write the negation of the given statement.

30. If it is medicine, then it tastes bad.

31. If I am good, then I can go to the movie.

32. If I am 18 or older, then I can vote.

33. If I study hard and make an A, then I can be a member of Phi Theta Kappa.

Appendix C: Valid Arguments

What is an argument? By definition, an **argument** is a set of statements called **premises,** followed by a statement called the **conclusion.** In a **valid argument,** the truth of the premises forces a conclusion that must also be true.

Law of Detachment

One form of valid argument used often in this textbook is called the Law of Detachment. This type of deductive reasoning takes the following form:

LAW OF DETACHMENT

1. $P \rightarrow Q$	Premise 1
2. P	Premise 2
C. Q	Conclusion

EXAMPLE 1

Use the Law of Detachment to determine the conclusion in the following argument. Assume that premises 1 and 2 are true.

1. If a person lives in London, then he lives in England.
2. Simon lives in London.

C. ?

Solution Simon lives in England.

The symbolic form of an argument is (*conjunction of premises*) *implies* (*conclusion*).

To prove that the Law of Detachment is valid, we need to establish that $[(P \to Q) \wedge P] \to Q$ is a tautology. In the truth table, study the five columns from left to right for each possible true/false combination of P and Q. Consider the first line of the table:

If P *and* Q are true, then $P \to Q$ is true.

Because $P \to Q$ is true *and* P is true, the conjunction $[(P \to Q) \wedge P]$ must be true.

If $[(P \to Q) \wedge P]$ is true *and* Q is true, then $[(P \to Q) \wedge P] \to Q$ is true.

The student should verify the entries in each remaining row of the table.

Proof of the Law of Detachment

P	Q	$P \to Q$	$(P \to Q) \wedge P$	$[(P \to Q) \wedge P] \to Q$
T	T	T	T	T
T	F	F	F	T
F	T	T	F	T
F	F	T	F	T

EXAMPLE 2

Use the Law of Detachment to find the conclusion of the geometry argument.

1. If a triangle is isosceles, then it has two congruent sides.
2. $\triangle ABC$ is isosceles.

C. ?

Solution

$\triangle ABC$ has two congruent sides.

An Invalid Argument

A common error in reasoning occurs when one *asserts the conclusion.* This form of argument looks similar to the Law of Detachment. The mistake lies in the fact that the second premise is statement Q (and not P). Because the first premise of this argument states that P *implies* Q, the argument is **invalid** (not valid).

INVALID ARGUMENT

1. $P \to Q$	Premise 1
2. Q	Premise 2
C. P	Conclusion

EXAMPLE 3

If possible, draw a conclusion in the following argument.

1. If Morgan works a lot of hours this week, Morgan can put money into savings.
2. Morgan put \$50 in savings this week.

C. ?

Solution No conclusion! Morgan may have money for savings due to his grandparents' generosity. We cannot conclude that Morgan worked a lot.

To establish that an argument is not valid, we must show that its symbolic form is not a tautology. Again the form of the argument is (conjunction of premises) implies (conclusion). For the invalid argument preceding Example 3, the symbolic form is $[(P \to Q) \wedge Q] \to P$. To understand why this argument is not valid, see the last column in the following table.

P	Q	$P \to Q$	$(P \to Q) \wedge Q$	$[(P \to Q) \wedge Q] \to P$
T	T	T	T	T
T	F	F	F	T
F	T	T	T	F
F	F	T	F	T

Law of Negative Inference

A second form of valid argument, the Law of Negative Inference, is shown next and then illustrated in Examples 4 and 5. This form of deductive reasoning was used to complete an indirect proof.

LAW OF NEGATIVE INFERENCE

1. $P \to Q$ Premise 1
2. $\sim Q$ Premise 2

C. $\sim P$ Conclusion

EXAMPLE 4

Give the symbolic form of the Law of Negative Inference.

Solution (Conjunction of premises) implies (conclusion), so the form is

$$[(P \to Q) \wedge \sim Q] \to \sim P$$

EXAMPLE 5

Use the Law of Negative Inference to determine the conclusion in the following argument. Assume that premises 1 and 2 are true.

1. If a person plays on a major league baseball team, then he earns a good salary.
2. Bill McAllen does not earn a good salary.

C. ?

Solution Bill McAllen does not play on a major league baseball team.

To prove that the Law of Negative Inference is valid requires that we establish that $[(P \to Q) \wedge \sim Q] \to \sim P$ is a tautology. Again, the columns of the following truth table were developed from left to right, with the rightmost column showing that the symbolic form of the argument is a tautology.

Proof of the Law of Negative Inference

P	Q	$\sim P$	$\sim Q$	$P \to Q$	$(P \to Q) \wedge \sim Q$	$[(P \to Q) \wedge \sim Q] \to \sim P$
T	T	F	F	T	F	T
T	F	F	T	F	F	T
F	T	T	F	T	F	T
F	F	T	T	T	T	T

EXAMPLE 6

Use the Law of Negative Inference to find the conclusion in the geometry argument.

1. If a quadrilateral is a parallelogram, then its opposite sides are parallel.
2. The opposite sides of quadrilateral $MNPQ$ are not parallel.

C. ?

Solution Quadrilateral $MNPQ$ is not a parallelogram.

Law of Syllogism

In every direct proof in geometry, there is a "chain" of conclusions that depend on a form of argument called the Law of Syllogism. This form of argument builds upon three or more simple statements. For simplicity, we will illustrate this principle of logic with three statements P, Q, and R.

LAW OF SYLLOGISM

1. $P \rightarrow Q$	Premise 1
2. $Q \rightarrow R$	Premise 2
C. $P \rightarrow R$	Conclusion

EXAMPLE 7

Use the Law of Syllogism to find the conclusion in the argument.

1. If an integer is even, then it has a factor of 2.
2. If a number has a factor of 2, then it can be divided exactly by 2.

C. ?

Solution If an integer is even, then it can be divided exactly by 2.

In the proof of the Law of Syllogism, the truth table requires eight lines because there are eight different true/false combinations for three simple statements *P*, *Q*, and *R*. A generalization follows:

If a statement is composed of n simple statements, then there are 2^n lines in the truth table.

Proof of the Law of Syllogism

P	Q	R	$P \rightarrow Q$	$Q \rightarrow R$	$P \rightarrow R$	$(P \rightarrow Q) \wedge (Q \rightarrow R)$	$[(P \rightarrow Q) \wedge (Q \rightarrow R)] \rightarrow (P \rightarrow R)$
T	T	T	T	T	T	T	T
T	T	F	T	F	F	F	T
T	F	T	F	T	T	F	T
T	F	F	F	T	F	F	T
F	T	T	T	T	T	T	T
F	T	F	T	F	T	F	T
F	F	T	T	T	T	T	T
F	F	F	T	T	T	T	T

EXAMPLE 8

Use the Law of Syllogism to find the conclusion in the geometric argument.

1. If a triangle is isosceles, then it has two congruent sides.
2. If a triangle has two congruent sides, then it has two congruent angles.

C. ?

Solution If a triangle is isosceles, then it has two congruent angles.

Many forms of deductive reasoning are used to reach conclusions. In each form, the truth table corresponding to the argument reveals a tautology.

Exercises

In Exercises 1 to 4, use the Law of Detachment to draw a conclusion.

1. If two angles are complementary, their sum of measures is 90 degrees. $\angle 1$ and $\angle 2$ are complementary.

2. If it gets hot this morning, we will have to turn on the air conditioner. It is hot this morning.

3. If Tina goes ice skating, she will have a good time. Tina goes ice skating.

4. If Gloria is scheduled to appear at the concert, then we will go to the concert. Gloria is scheduled to appear at the concert.

In Exercises 5 to 8, use the Law of Negative Inference to draw a conclusion.

5. If two angles are complementary, their sum of measures is 90 degrees. $m\angle 1 + m\angle 2 \neq 90°$.

6. If an animal lives in the zoo, it should have a companion of the same species. Fido does not have a companion of the same species.

7. If Tom doesn't finish the job, then I will not pay him. I did pay Tom for the job.

8. If the traffic light changes, then you can travel through the intersection. You cannot travel through the intersection.

In Exercises 9 to 12, use the Law of Syllogism to draw a conclusion.

9. If Izzi lives in Chicago, then she lives in Illinois. If a person lives in Illinois, then she lives in the Midwest.

10. If you pay your tuition, then you will need to pay additional fees. If you need to pay additional fees, then you will need to write a check.

11. If Ken Travis gets a hit, then my favorite baseball team will win the game. If my favorite baseball team wins the game, then I will be happy.

12. If Tom speaks at the rally, the union members will listen. If the union members listen, the union will decide not to strike.

In Exercises 13 to 16, determine which arguments are valid.

13. **1.** If I go to the football game, I'll cheer for the Cowboys.
2. I went to the football game.

C. I cheered for the Cowboys.

14. **1.** If I go to the football game, I'll cheer for the Cowboys.
2. I cheered for the Cowboys.

C. I went to the football game.

15. **1.** If Bill and Mary stop to visit, I'll prepare a meal.
2. Bill stopped to visit at 5 P.M.

C. I prepared a meal.

16. **1.** If it turns cold and snows, I'll build a fire in the fireplace.
2. The temperature began to fall around 3 P.M.
3. It began snowing before 5 P.M.

C. I built a fire in the fireplace.

In Exercises 17 to 20, which law of reasoning was used to reach the conclusion?

17. **1.** If it is cloudy, then it will rain.
2. If it rains, the garden will grow.
C. If it is cloudy, the garden will grow.

18. **1.** If you leave the apartment unlocked, someone will steal your CD player.
2. Your CD player was not stolen.
C. You did not leave the apartment unlocked.

19. **1.** If it snows more than 3 in. today, then we will go skiing.
2. It snowed 6 in. today.
C. We will go skiing.

20. **1.** If the mosquitoes are bad, I won't get out of the car.
2. If I don't get out of the car, we cannot have a picnic.
C. If the mosquitoes are bad, we cannot have a picnic.

21. A form of deductive reasoning known as the Law of Denial (or Denial of Alternative) is shown below.

LAW OF DENIAL

1. $P \vee Q$
2. $\sim Q$

C. P

a) Write the symbolic form of this argument.
b) Complete a truth table to establish the validity of this argument.

In Exercises 22 to 24, use the Law of Denial (see Exercise 21) to draw a conclusion.

22. Terry is sick or hurt.

Terry is not hurt.

23. Mary's family will visit us at Thanksgiving or at Christmas.

Mary's family did not visit us at Thanksgiving.

24. Wendell will have to study geometry or he will fail that course.

Wendell did not fail the geometry course.

In Exercises 25 to 27, complete a truth table to validate each form of reasoning. Do not refer back to the truth tables in this section.

25. Law of Detachment.

26. Law of Negative Inference.

27. Law of Syllogism.

Appendix D:

Summary of Constructions, Postulates, and Theorems and Corollaries

Constructions

Section 1.2

1. To construct a segment congruent to a given segment.
2. To construct the midpoint M of a given line segment AB.

Section 1.4

3. To construct an angle congruent to a given angle.
4. To construct the angle bisector of a given angle.

Section 1.5

5. To construct the line perpendicular to a given line at a specified point on the given line.

Section 2.1

6. To construct the line that is perpendicular to a given line from a point not on the given line.

Section 2.3

7. To construct the line parallel to a given line from a point not on that line.

Section 6.4

8. To construct a tangent to a circle at a point on the circle.
9. To construct a tangent to a circle from an external point.

Postulates

Section 1.3

1. Through two distinct points, there is exactly one line.
2. (Ruler Postulate) The measure of any line segment is a unique positive number.
3. (Segment-Addition Postulate) If X is a point on $\overline{AB}$ and A-X-B, then $AX + XB = AB$.
4. If two lines intersect, they intersect at a point.
5. Through three noncollinear points, there is exactly one plane.
6. If two distinct planes intersect, then their intersection is a line.
7. Given two distinct points in a plane, the line containing these points also lies in the plane.

Section 1.4

8. (Protractor Postulate) The measure of an angle is a unique positive number.
9. (Angle-Addition Postulate) If a point D lies in the interior of angle ABC, then $m\angle ABD + m\angle DBC = m\angle ABC$.

Section 2.1

10. (Parallel Postulate) Through a point not on a line, exactly one line is parallel to the given line.
11. If two parallel lines are cut by a transversal, then the corresponding angles are congruent.

Section 3.1

12. If the three sides of one triangle are congruent to the three sides of a second triangle, then the triangles are congruent (SSS).
13. If two sides and the included angle of one triangle are congruent to two sides and the included angle of a second triangle, then the triangles are congruent (SAS).

14. If two angles and the included side of one triangle are congruent to two angles and the included side of a second triangle, then the triangles are congruent (ASA).

Section 5.2

15. If the three angles of one triangle are congruent to the three angles of a second triangle, then the triangles are similar (AAA).

Section 6.1

16. (Central Angle Postulate) In a circle, the degree measure of a central angle is equal to the degree measure of its intercepted arc.

17. (Arc-Addition Postulate) If B lies between A and C on a circle, then $m\overset{\frown}{AB} + m\overset{\frown}{BC} = m\overset{\frown}{ABC}$.

Section 7.1

18. (Area Postulate) Corresponding to every bounded region is a unique positive number A, known as the area of that region.

19. If two closed plane figures are congruent, then their areas are equal.

20. (Area-Addition Postulate) Let R and S be two enclosed regions that do not intersect. Then $A_{R \cup S} = A_R + A_S$.

21. The area A of a rectangle whose base has length b and whose altitude has length h is given by $A = bh$.

Section 7.4

22. The ratio of the circumference of a circle to the length of its diameter is a unique positive constant.

Section 7.5

23. The ratio of the degree measure m of the central angle of a sector to 360° is the same as the ratio of the area of the sector to the area of the circle; that is, $\frac{\text{area of sector}}{\text{area of circle}} = \frac{m}{360°}$.

Section 8.1

24. (Volume Postulate) Corresponding to every solid is a unique positive number V known as the volume of that solid.

25. The volume of a right rectangular prism is given by

$$V = \ell wh$$

where ℓ measures the length, w the width, and h the altitude of the prism.

26. The volume of a right prism is given by

$$V = Bh$$

where B is the area of a base and h is the altitude of the prism.

Theorems and Corollaries

1.3.1 The midpoint of a line segment is unique.

1.4.1 There is one and only one angle bisector for a given angle.

1.5.1 There is exactly one line perpendicular to a given line at any point on the line.

1.5.2 The perpendicular bisector of a line segment is unique.

1.6.1 If two lines are perpendicular, then they meet to form right angles.

1.6.2 If two lines meet to form a right angle, then these lines are perpendicular.

1.6.3 If two angles are complementary to the same angle (or to congruent angles), then these angles are congruent.

1.6.4 If two angles are supplementary to the same angle (or to congruent angles), then these angles are congruent.

1.6.5 If two lines intersect, then the vertical angles formed are congruent.

1.6.6 Any two right angles are congruent.

1.6.7 If the exterior sides of two adjacent acute angles form perpendicular rays, then these angles are complementary.

1.6.8 If the exterior sides of two adjacent angles form a straight line, then these angles are supplementary.

1.6.9 If two segments are congruent, then their midpoints separate these segments into four congruent segments.

1.6.10 If two angles are congruent, then their bisectors separate these angles into four congruent angles.

2.1.1 From a point not on a given line, there is exactly one line perpendicular to the given line.

2.1.2 If two parallel lines are cut by a transversal, then the alternate interior angles are congruent.

2.1.3 If two parallel lines are cut by a transversal, then the alternate exterior angles are congruent.

2.1.4 If two parallel lines are cut by a transversal, then the interior angles on the same side of the transversal are supplementary.

2.1.5 If two parallel lines are cut by a transversal, then the exterior angles on the same side of the transversal are supplementary.

2.3.1 If two lines are cut by a transversal so that the corresponding angles are congruent, then these lines are parallel.

2.3.2 If two lines are cut by a transversal so that the alternate interior angles are congruent, then these lines are parallel.

2.3.3 If two lines are cut by a transversal so that the alternate exterior angles are congruent, then these lines are parallel.

2.3.4 If two lines are cut by a transversal so that the interior angles on the same side of the transversal are supplementary, then these lines are parallel.

2.3.5 If two lines are cut by a transversal so that the exterior angles on the same side of the transversal are supplementary, then these lines are parallel.

2.3.6 If two lines are each parallel to a third line, then these lines are parallel to each other.

2.3.7 If two coplanar lines are each perpendicular to a third line, then these lines are parallel to each other.

2.4.1 In a triangle, the sum of the measures of the interior angles is 180°.

2.4.2 Each angle of an equiangular triangle measures 60°.

2.4.3 The acute angles of a right triangle are complementary.

2.4.4 If two angles of one triangle are congruent to two angles of another triangle, then the third angles are also congruent.

2.4.5 The measure of an exterior angle of a triangle equals the sum of the measures of the two nonadjacent interior angles.

2.5.1 The total number of diagonals D in a polygon of n sides is given by the formula $D = \frac{n(n-3)}{2}$.

2.5.2 The sum S of the measures of the interior angles of a polygon with n sides is given by $S = (n-2) \cdot 180°$. Note that $n > 2$ for any polygon.

2.5.3 The measure I of each interior angle of a regular polygon of n sides is $I = \frac{(n-2) \cdot 180°}{n}$.

2.5.4 The sum of the measures of the four interior angles of a quadrilateral is 360°.

2.5.5 The sum of the measures of the exterior angles, one at each vertex, of a polygon is 360°.

2.5.6 The measure E of each exterior angle of a regular polygon of n sides is $E = \frac{360°}{n}$.

3.1.1 If two angles and the nonincluded side of one triangle are congruent to two angles and the nonincluded side of a second triangle, then the triangles are congruent (AAS).

3.2.1 If the hypotenuse and a leg of one right triangle are congruent to the hypotenuse and a leg of a second right triangle, then the triangles are congruent (HL).

3.3.1 Corresponding altitudes of congruent triangles are congruent.

3.3.2 The bisector of the vertex angle of an isosceles triangle separates the triangle into two congruent triangles.

3.3.3 If two sides of a triangle are congruent, then the angles opposite these sides are also congruent.

3.3.4 If two angles of a triangle are congruent, then the sides opposite these angles are also congruent.

3.3.5 An equilateral triangle is also equiangular.

3.3.6 An equiangular triangle is also equilateral.

3.5.1 The measure of a line segment is greater than the measure of any of its parts.

3.5.2 The measure of an angle is greater than the measure of any of its parts.

3.5.3 The measure of an exterior angle of a triangle is greater than the measure of either nonadjacent interior angle.

3.5.4 If a triangle contains a right or an obtuse angle, then the measure of this angle is greater than the measure of either of the remaining angles.

3.5.5 (Addition Property of Inequality): If $a > b$ and $c > d$, then $a + c > b + d$.

3.5.6 If one side of a triangle is longer than a second side, then the measure of the angle opposite the first side is greater than the measure of the angle opposite the second side.

3.5.7 If the measure of one angle of a triangle is greater than the measure of a second angle, then the side opposite the larger angle is longer than the side opposite the smaller angle.

3.5.8 The perpendicular segment from a point to a line is the shortest segment that can be drawn from the point to the line.

3.5.9 The perpendicular segment from a point to a plane is the shortest segment that can be drawn from the point to the plane.

3.5.10 (Triangle Inequality) The sum of the lengths of any two sides of a triangle is greater than the length of the third side.

3.5.10 (*Alternate*) The length of one side of a triangle must be between the sum and difference of the lengths of the other two sides.

4.1.1 A diagonal of a parallelogram separates it into two congruent triangles.

4.1.2 Opposite angles of a parallelogram are congruent.

4.1.3 Opposite sides of a parallelogram are congruent.

4.1.4 Diagonals of a parallelogram bisect each other.

4.1.5 Consecutive angles of a parallelogram are supplementary.

4.1.6 If two sides of one triangle are congruent to two sides of a second triangle and the included angle of the first triangle is greater than the included angle of the second, then the length of the side opposite the included angle of the first triangle is greater than the length of the side opposite the included angle of the second.

4.1.7 In a parallelogram with unequal pairs of consecutive angles, the longer diagonal lies opposite the obtuse angle.

4.2.1 If two sides of a quadrilateral are both congruent and parallel, then the quadrilateral is a parallelogram.

4.2.2 If both pairs of opposite sides of a quadrilateral are congruent, then it is a parallelogram.

4.2.3 If the diagonals of a quadrilateral bisect each other, then the quadrilateral is a parallelogram.

4.2.4 In a kite, one pair of opposite angles is congruent.

4.2.5 The segment that joins the midpoints of two sides of a triangle is parallel to the third side and has a length equal to one-half the length of the third side.

4.3.1 All angles of a rectangle are right angles.

4.3.2 The diagonals of a rectangle are congruent.

4.3.3 All sides of a square are congruent.

4.3.4 All sides of a rhombus are congruent.

4.3.5 The diagonals of a rhombus are perpendicular.

4.4.1 The base angles of an isosceles trapezoid are congruent.

4.4.2 The diagonals of an isosceles trapezoid are congruent.

4.4.3 The length of the median of a trapezoid equals one-half the sum of the lengths of the two bases.

4.4.4 The median of a trapezoid is parallel to each base.

4.4.5 If two consecutive angles of a quadrilateral are supplementary, the quadrilateral is a trapezoid.

4.4.6 If two base angles of a trapezoid are congruent, the trapezoid is an isosceles trapezoid.

4.4.7 If the diagonals of a trapezoid are congruent, the trapezoid is an isosceles trapezoid.

4.4.8 If three (or more) parallel lines intercept congruent segments on one transversal, then they intercept congruent segments on any transversal.

5.2.1 If two angles of one triangle are congruent to two angles of another triangle, then the triangles are similar (AA).

5.2.2 The lengths of the corresponding altitudes of similar triangles have the same ratio as the lengths of any pair of corresponding sides.

5.3.1 The altitude drawn to the hypotenuse of a right triangle separates the right triangle into two right triangles that are similar to each other and to the original right triangle.

5.3.2 The length of the altitude to the hypotenuse of a right triangle is the geometric mean of the lengths of the segments of the hypotenuse.

5.3.3 The length of each leg of a right triangle is the geometric mean of the length of the hypotenuse and the length of the segment of the hypotenuse adjacent to that leg.

5.3.4 (Pythagorean Theorem) The square of the length of the hypotenuse of a right triangle is equal to the sum of the squares of the lengths of the legs.

5.3.5 (Converse of Pythagorean Theorem) If a, b, and c are the lengths of the three sides of a triangle, with c the length of the longest side, and if $c^2 = a^2 + b^2$, then the triangle is a right triangle with the right angle opposite the side of length c.

5.3.6 If the hypotenuse and a leg of one right triangle are congruent to the hypotenuse and a leg of a second right triangle, then the triangles are congruent (HL).

5.3.7 Let a, b, and c represent the lengths of the three sides of a triangle with c the length of the longest side.

1. If $c^2 > a^2 + b^2$, then the triangle is obtuse and the obtuse angle lies opposite the side of length c.
2. If $c^2 < a^2 + b^2$, then the triangle is acute.

5.4.1 (45-45-90 Theorem) In a triangle whose angles measure 45°, 45°, and 90°, the hypotenuse has a length equal to the product of $\sqrt{2}$ and the length of either leg.

5.4.2 (30-60-90 Theorem) In a triangle whose angles measure 30°, 60°, and 90°, the hypotenuse has a length equal to twice the length of the shorter leg, while the length of the longer leg is the product of $\sqrt{3}$ and the length of the shorter leg.

5.4.3 If the length of the hypotenuse of a right triangle equals the product of $\sqrt{2}$ and the length of either leg, then the angles of the triangle measure 45°, 45°, and 90°.

5.4.4 If the length of the hypotenuse of a right triangle is twice the length of one leg of the triangle, then the angle of the triangle opposite that leg measures 30°.

5.5.1 If a line is parallel to one side of a triangle and intersects the other two sides, then it divides these sides proportionally.

5.5.2 When three (or more) parallel lines are cut by a pair of transversals, the transversals are divided proportionally by the parallel lines.

5.5.3 If a ray bisects one angle of a triangle, then it divides the opposite side into segments whose lengths are proportional to the two sides which form that angle.

6.1.1 A radius that is perpendicular to a chord bisects the chord.

6.1.2 The measure of an inscribed angle of a circle is one-half the measure of its intercepted arc.

6.1.3 In a circle or in congruent circles, congruent minor arcs have congruent central angles.

6.1.4 In a circle (or in congruent circles), congruent central angles have congruent arcs.

6.1.5 In a circle (or in congruent circles), congruent chords have congruent minor (major) arcs.

6.1.6 In a circle (or in congruent circles), congruent arcs have congruent chords.

6.1.7 Chords that are at the same distance from the center of a circle are congruent.

6.1.8 Congruent chords are located at the same distance from the center of a circle.

6.1.9 An angle inscribed in a semicircle is a right angle.

6.1.10 If two inscribed angles intercept the same arc, then these angles are congruent.

6.2.1 If a quadrilateral is inscribed in a circle, the opposite angles are supplementary.

6.2.2 The measure of an angle formed by two chords intersecting within a circle is one-half the sum of the measures of the arcs intercepted by the angle and its vertical angle.

6.2.3 The radius (or any other line through the center of a circle) drawn to a tangent at the point of tangency is perpendicular to the tangent at that point.

6.2.4 The measure of an angle formed by a tangent and a chord drawn to the point of tangency is one-half the measure of the intercepted arc.

6.2.5 The measure of an angle formed when two secants intersect at a point outside the circle is one-half the difference of the measures of the two intercepted arcs.

6.2.6 If an angle is formed by a secant and tangent that intersect in the exterior of a circle, then the measure of the angle is one-half the difference of the measures of its intercepted arcs.

6.2.7 If an angle is formed by two intersecting tangents, then the measure of the angle is one-half the difference of the measures of the intercepted arcs.

6.2.8 If two parallel lines intersect a circle, the intercepted arcs between these lines are congruent.

6.3.1 If a line is drawn through the center of a circle perpendicular to a chord, then it bisects the chord and its arc.

6.3.2 If a line through the center of a circle bisects a chord other than a diameter, then it is perpendicular to the chord.

6.3.3 The perpendicular bisector of a chord contains the center of the circle.

6.3.4 The tangent segments to a circle from an external point are congruent.

6.3.5 If two chords intersect within a circle, then the product of the lengths of the segments (parts) of one chord is equal to the product of the lengths of the segments of the other.

6.3.6 If two secant segments are drawn to a circle from an external point, then the products of the lengths of each secant with its external segment are equal.

6.3.7 If a tangent segment and secant segment are drawn to a circle from an external point, then the square of the length of the tangent equals the product of the lengths of the secant with its external segment.

6.4.1 The line that is perpendicular to the radius of a circle at its endpoint on the circle is a tangent to the circle.

6.4.2 In a circle (or in congruent circles) containing two unequal central angles, the larger angle corresponds to the larger intercepted arc.

6.4.3 In a circle (or in congruent circles) containing two unequal arcs, the larger arc corresponds to the larger central angle.

6.4.4 In a circle (or in congruent circles) containing two unequal chords, the shorter chord is at the greater distance from the center of the circle.

6.4.5 In a circle (or in congruent circles) containing two unequal chords, the chord nearer the center of the circle has the greater length.

6.4.6 In a circle (or in congruent circles) containing two unequal chords, the longer chord corresponds to the greater minor arc.

6.4.7 In a circle (or in congruent circles) containing two unequal minor arcs, the greater minor arc corresponds to the longer of the chords related to these arcs.

6.5.1 The locus of points in a plane and equidistant from the sides of an angle is the angle bisector.

6.5.2 The locus of points in a plane that are equidistant from the endpoints of a line segment is the perpendicular bisector of that line segment.

6.6.1 The three angle bisectors of the angles of a triangle are concurrent.

6.6.2 The three perpendicular bisectors of the sides of a triangle are concurrent.

6.6.3 The three altitudes of a triangle are concurrent.

6.6.4 The three medians of a triangle are concurrent at a point that is two-thirds the distance from any vertex to the midpoint of the opposite side.

7.1.1 The area A of a square whose sides are each of length s is given by $A = s^2$.

7.1.2 The area A of a parallelogram with a base of length b and with corresponding altitude of length h is given by

$$A = bh$$

7.1.3 The area A of a triangle whose base has length b and whose corresponding altitude has length h is given by

$$A = \frac{1}{2}bh$$

7.2.1 (Heron's Formula) If the three sides of a triangle have lengths a, b, and c, then the area A of the triangle is given by

$$A = \sqrt{s(s - a)(s - b)(s - c)}$$

where the semiperimeter of the triangle is

$$s = \frac{1}{2}(a + b + c)$$

7.2.2 The area A of a trapezoid whose bases have lengths b_1 and b_2 and whose altitude has length h is given by

$$A = \frac{1}{2}h(b_1 + b_2)$$

7.2.3 The area of any quadrilateral with perpendicular diagonals of lengths d_1 and d_2 is given by

$$A = \frac{1}{2}d_1d_2$$

7.2.4 The area A of a rhombus whose diagonals have lengths d_1 and d_2 is given by

$$A = \frac{1}{2}d_1d_2$$

7.2.5 The area A of a kite whose diagonals have lengths d_1 and d_2 is given by

$$A = \frac{1}{2}d_1d_2$$

7.2.6 The ratio of the areas of two similar triangles equals the square of the ratio of the lengths of any two corresponding sides; that is,

$$\frac{A_1}{A_2} = \left(\frac{a_1}{a_2}\right)^2$$

7.3.1 A circle can be circumscribed about (or inscribed in) any regular polygon.

7.3.2 The measure of the central angle of a regular polygon of n sides is given by $c = \frac{360}{n}$.

7.3.3 Any radius of a regular polygon bisects the angle at the vertex to which it is drawn.

7.3.4 Any apothem to a side of a regular polygon bisects the side of the polygon to which it is drawn.

7.3.5 The area A of a regular polygon whose apothem has length a and whose perimeter is P is given by

$$A = \frac{1}{2}aP$$

7.4.1 The circumference of a circle is given by the formula

$$C = \pi d \text{ or } C = 2\pi r$$

7.4.2 In a circle whose circumference is C, the length ℓ of an arc whose degree measure is m is given by

$$\ell = \frac{m}{360} \cdot C$$

7.4.3 The area A of a circle whose radius is of length r is given by $A = \pi r^2$.

7.5.1 In a circle of radius r, the area A of a sector whose arc has degree measure m is given by

$$A = \frac{m}{360}\pi r^2$$

7.5.2 The area of a semicirclular region of radius r is $A = \frac{1}{2}\pi r^2$.

7.5.3 Where P represents the perimeter of a triangle and r represents the length of the radius of its inscribed circle, the area of the triangle is given by

$$A = \frac{1}{2}rP$$

8.1.1 The lateral area L of a right prism whose altitude has measure h and whose base has perimeter P is given by $L = hP$.

8.1.2 The total area T of any prism with lateral area L and base area B is given by $T = L + 2B$.

8.2.1 In a regular pyramid, the length a of the apothem of the base, the altitude h, and the slant height ℓ satisfy the Pythagorean Theorem; that is, $\ell^2 = a^2 + h^2$ in every regular pyramid.

8.2.2 The lateral area L of a regular pyramid with slant height of length ℓ and perimeter P of the base is given by

$$L = \frac{1}{2}\ell P$$

8.2.3 The total area (surface area) T of a pyramid with lateral area L and base area B is given by $T = L + B$.

8.2.4 The volume V of a pyramid having a base area B and an altitude of length h is given by

$$V = \frac{1}{3}Bh$$

8.3.1 The lateral area L of a right circular cylinder with altitude of length h and circumference C of the base is given by $L = hC$.

Alternate: Where r is the length of the radius of the base, $L = 2\pi rh$.

8.3.2 The total area T of a right circular cylinder with base area B and lateral area L is given by $T = L + 2B$.

Alternate: Where r is the length of the radius of the base and h is the length of the altitude, $T = 2\pi rh + 2\pi r^2$.

8.3.3 The volume V of a right circular cylinder with base area B and altitude of length h is given by $V = Bh$.

Alternate: Where r is the length of the radius of the base, $V = \pi r^2 h$.

8.3.4 The lateral area L of a right circular cone with slant height of length ℓ and circumference C of the base is given by $L = \frac{1}{2}\ell C$.

Alternate: Where r is the length of the radius of the base, $L = \pi r\ell$.

8.3.5 The total area T of a right circular cone with base area B and lateral area L is given by $T = B + L$.

Alternate: Where r is the length of the radius of the base and ℓ is the length of the slant height, $T = \pi r^2 + \pi r\ell$.

8.3.6 In a right circular cone, the lengths of the radius r (of the base), the altitude h, and the slant height ℓ satisfy the Pythagorean Theorem; that is, $\ell^2 = r^2 + h^2$ in every right circular cone.

8.3.7 The volume V of a right circular cone with base area B and altitude of length h is given by $V = \frac{1}{3}Bh$.

Alternate: Where r is the length of the radius of the base, $V = \frac{1}{3}\pi r^2 h$.

8.4.1 (Euler's Equation) The numbers of vertices V, the number of edges E, and the number of faces F of a polyhedron are related by the equation $V + F = E + 2$.

8.4.2 The surface area S of a sphere whose radius has length r is given by $S = 4\pi r^2$.

8.4.3 The volume V of a sphere with radius of length r is given by $V = \frac{4}{3}\pi r^3$.

9.1.1 (Distance Formula) The distance between two points (x_1, y_1) and (x_2, y_2) is given by the formula

$$d = \sqrt{(x_2 - x_1)^2 + (y_2 - y_1)^2}$$

9.1.2 (Midpoint Formula) The midpoint M of the line segment joining (x_1, y_1) and (x_2, y_2) has coordinates x_M and y_M, where

$$(x_M, y_M) = \left(\frac{x_1 + x_2}{2}, \frac{y_1 + y_2}{2}\right)$$

that is, $$M = \left(\frac{x_1 + x_2}{2}, \frac{y_1 + y_2}{2}\right)$$

9.2.1 If two nonvertical lines are parallel, then their slopes are equal.

9.2.2 If two lines (neither horizontal nor vertical) are perpendicular, then the product of their slopes is -1.

9.3.1 (Slope-Intercept Form of a Line) The line whose slope is m and whose y intercept is b has the equation $y = mx + b$.

9.3.2 (Point-Slope Form of a Line) The line with slope m and containing the point (x_1, y_1) has the equation

$$y - y_1 = m(x - x_1)$$

9.5.1 The line segment determined by the midpoints of two sides of a triangle is parallel to the third side.

9.5.2 The diagonals of a parallelogram bisect each other.

9.5.3 The diagonals of a rhombus are perpendicular.

9.5.4 If the diagonals of a parallelogram are equal in length, then the parallelogram is a rectangle.

9.5.5 The three medians of a triangle are concurrent at a point that is two-thirds the distance from any vertex to the midpoint of the opposite side.

9.6.1 The circle whose center is (h, k) and whose radius has length r, where $r > 0$, has the equation

$$(x - h)^2 + (y - k)^2 = r^2$$

9.6.2 The equation of a circle with center $(0, 0)$ and radius r is $x^2 + y^2 = r^2$.

10.2.1 In any right triangle in which α is the measure of an acute angle,

$$\sin^2 \alpha + \cos^2 \alpha = 1$$

10.4.1 The area of a triangle equals one-half the product of the lengths of two sides and the sine of the included angle.

10.4.2 (Law of Sines) In any triangle, the three ratios between the sines of the angles and the lengths of the opposite sides are equal. That is,

$$\frac{\sin \alpha}{a} = \frac{\sin \beta}{b} = \frac{\sin \gamma}{c}$$

10.4.3 (Law of Cosines) In acute triangle *ABC*,

$$c^2 = a^2 + b^2 - 2ab \cos \gamma$$
$$b^2 = a^2 + c^2 - 2ac \cos \beta$$
$$a^2 = b^2 + c^2 - 2bc \cos \alpha$$

Answers

Selected Odd-Numbered Exercises and Proofs

CHAPTER 1

1.1 EXERCISES

1. (a) Not a statement (b) Statement; true (c) Statement; true (d) Statement; false
3. (a) Christopher Columbus did not cross the Atlantic Ocean. (b) Some jokes are not funny.
5. Conditional **7.** Simple **9.** Simple
11. H: You go to the game. C: You will have a great time.
13. H: The diagonals of a parallelogram are perpendicular. C: The parallelogram is a rhombus.
15. H: Two parallel lines are cut by a transversal. C: Corresponding angles are congruent.
17. First write the statement in "If, then" form: If a figure is a square, then it is a rectangle. H: A figure is a square. C: It is a rectangle.
19. True **21.** True **23.** False **25.** Induction
27. Deduction **29.** Intuition **31.** None
33. Angle 1 looks equal in measure to angle 2.
35. Three angles in one triangle are equal in measure to the three angles in the other triangle.
37. *A Prisoner of Society* might be nominated for an Academy Award. **39.** The instructor is a math teacher.
41. Angles 1 and 2 are complementary.
43. Alex has a strange sense of humor. **45.** None
47. June Jesse will be in the public eye.
49. Marilyn is a happy person.

1.2 EXERCISES

1. $AB < CD$ **3.** Two; one **5.** One; none
7. $\angle ABC, \angle ABD, \angle DBC$ **9.** Yes; no; yes
11. $\angle ABC, \angle CBA$ **13.** Yes; no
15. (a) 3 (b) $2\frac{1}{2}$ **17.** (a) 40° (b) 50°
19. Congruent; congruent **21.** Equal **23.** No
25. Yes **27.** Congruent **29.** $\overline{MN}$ and $\overline{QP}$
31. $\overline{AB}$ **33.** 22 **35.** $x = 9$ **37.** 124 **39.** 71
41. $x = 23$ **43.** 10.9 **45.** $x = 102$; $y = 78$
47. N 22° E

1.3 EXERCISES

1. (a) A-C-D (b) A, B, C or B, C, D or A, B, D
3. $\overleftrightarrow{CD}$ means line CD; $\overline{CD}$ means segment CD; CD means the measure or length of $\overline{CD}$; $\overrightarrow{CD}$ means ray CD with endpoint C.
5. (a) m and t (b) m and $\overleftrightarrow{AD}$ or $\overleftrightarrow{AD}$ and t
7. $x = 3$; $AM = 7$ **9.** $x = 7$; $AB = 38$
11. (a) $\overrightarrow{OA}$ and $\overrightarrow{OD}$ (b) $\overrightarrow{OA}$ and $\overrightarrow{OB}$ (There are other possible answers.)
15. Planes M and N intersect at $\overleftrightarrow{AB}$. **17.** A
19. (a) C (b) C (c) H
25. (a) No (b) Yes (c) No (d) Yes
27. Six **29.** Nothing

1.4 EXERCISES

1. (a) Yes (b) No
3. (a) Obtuse (b) Straight (c) Acute (d) Obtuse
5. $m\angle FAC + m\angle CAD = 180$; $\angle FAC$ and $\angle CAD$ are supplementary.
7. 42° **9.** $x = 20$; $m\angle RSV = 56°$
11. $y = 8$; $x = 24$ **13.** $\angle CAB \cong \angle DAB$
15. $\angle$s measure 128° and 52°
17. (a) $180 - x$ (b) $192 - 3x$ (c) $180 - 2x - 5y$
19. $x = 143$
25. It appears that the angle bisectors meet at one point.
27. It appears that the two sides opposite $\angle$s A and B are congruent.

1.5 Exercises

1. 1. Given 2. If two ∠s are ≅ , then they are equal in measure. 3. Angle-Addition Postulate 4. Addition Property of Equality 5. Substitution 6. If two ∠s are equal in measure, then they are ≅ .
3. 1. ∠1 ≅ ∠2 and ∠2 ≅ ∠3 2. ∠1 ≅ ∠3
11. 1. Given 3. Substitution 4. m∠1 = m∠2 5. ∠1 ≅ ∠2 **13.** No; yes; no **15.** No; yes; yes
17. (a) perpendicular (b) angles (c) supplementary (d) right (e) measure of angle (f) adjacent (g) complementary (h) ray *AB* (i) is congruent to (j) vertical
19. In space, there are an infinite number of lines that perpendicularly bisect a given line segment at its midpoint.

1.6 Exercises

1. H: A line segment is bisected. C: Each of the equal segments has half the length of the original segment.
3. First write the statement in "If, then" form. If a figure is a square, then it is a quadrilateral. H: A figure is a square. C: It is a quadrilateral.
5. H: Each is a right angle. C: Two angles are congruent. **7.** Statement, Drawing, Given, Prove, Proof
9. 1. Given 2. If two ∠s are complementary, the sum of their measures is 90 3. Substitution 4. Subtraction Property of Equality 5. If two ∠s are equal in measure, then they are ≅
13. 1. Given 2. ∠*ABC* is a right ∠ 3. The measure of a rt. ∠ = 90 4. Angle-Addition Postulate 6. ∠1 is comp. to ∠2

1.6 Selected Proof

15.

PROOF

Statements	Reasons
1. ∠*ABC* ≅ ∠*EFG*	**1.** Given
2. m∠*ABC* = m∠*EFG*	**2.** If two ∠s are ≅ , their measures are =
3. m∠*ABC* = m∠1 + m∠2 m∠*EFG* = m∠3 + m∠4	**3.** Angle-Addition Postulate
4. m∠1 + m∠2 = m∠3 + m∠4	**4.** Substitution
5. $\overrightarrow{BD}$ bisects ∠*ABC* $\overrightarrow{FH}$ bisects ∠*EFG*	**5.** Given
6. m∠1 = m∠2 and m∠3 = m∠4	**6.** If a ray bisects an ∠, then two ∠s of equal measure are formed

(continued)

(continued)

7. m∠1 + m∠1 = m∠3 + m∠3 or 2 · m∠1 = 2 · m∠3	**7.** Substitution
8. m∠1 = m∠3	**8.** Division Prop. of Equality
9. m∠1 = m∠2 = m∠3 = m∠4	**9.** Substitution
10. ∠1 ≅ ∠2 ≅ ∠3 ≅ ∠4	**10.** If ∠s are = in measure, then they are ≅

Review Exercises

1. Undefined terms, defined terms, axioms or postulates, theorems **2.** Induction, deduction, intuition
3. 1. Names the term being defined 2. Places the term into a set or category 3. Distinguishes the term from other terms in the same category 4. Reversible
4. Intuition **5.** Induction **6.** Deduction
7. H: The diagonals of a trapezoid are equal in length. C: The trapezoid is isosceles
8. H: The parallelogram is a rectangle. C: The diagonals of a parallelogram are congruent.
9. No conclusion **10.** Jody Smithers has a college degree.
11. The measure of angle *A* is not 90° **12.** C
13. ∠*RST*; ∠*S*; greater than 90° **14.** Perpendicular
18. 98 **19.** 47 **20.** 22 **21.** 17 **22.** 34
23. 152 **24.** 39 **25.** $67\frac{1}{2}$ **26.** 28 and 152
27. (a) $6x + 8$ (b) $x = 4$ (c) 11; 10; 11
28. The measure of angle 3 is less than 50. **29.** 10 pegs
30. S **31.** S **32.** A **33.** S **34.** N
35. 2. ∠4 ≅ ∠*P* 3. ∠1 ≅ ∠4 4. If two ∠s are ≅ , then their measures are = 5. Given 6. m∠2 = m∠3 7. m∠1 + m∠2 = m∠4 + m∠3 8. Angle-Addition Postulate 9. Substitution 10. ∠*TVP* ≅ ∠*MVP*

Review Exercises Selected Proofs

36.

PROOF

Statements	Reasons
1. $\overline{KF} \perp \overline{FH}$	**1.** Given
2. ∠*KFH* is a rt. ∠	**2.** If two segments are ⊥ , then they form a rt. ∠
3. ∠*JHF* is a rt. ∠	**3.** Given
4. ∠*KFH* ≅ ∠*JHF*	**4.** Any two rt. ∠s are ≅

37.

PROOF

Statements	Reasons
1. $\overline{KH} \cong \overline{FJ}$; G is the midpoint of both $\overline{KH}$ and $\overline{FJ}$	**1.** Given
2. $\overline{KG} \cong \overline{GJ}$	**2.** If two segments are ≅, then their midpoints separate these segments into four ≅ segments

38.

PROOF

Statements	Reasons
1. $\overline{KF} \perp \overline{FH}$	**1.** Given
2. $\angle KFH$ is a rt. ∠	**2.** If two lines are ⊥, then they form a rt. ∠
3. $\angle KFJ$ is comp. to $\angle JFH$	**3.** If the exterior sides of two adjacent ∠s form ⊥ rays, then these ∠s are comp.

39.

PROOF

Statements	Reasons
1. ∠1 is comp. to $\angle M$	**1.** Given
2. ∠2 is comp. to $\angle M$	**2.** Given
3. ∠1 ≅ ∠2	**3.** If two ∠s are comp. to the same ∠, then these angles are ≅

40.

PROOF

Statements	Reasons
1. $\angle MOP \cong \angle MPO$	**1.** Given
2. $\overrightarrow{OR}$ bisects $\angle MOP$; $\overrightarrow{PR}$ bisects $\angle MPO$	**2.** Given
3. ∠1 ≅ ∠2	**3.** If two ∠s are ≅, then their bisectors separate these ∠s into four ≅ ∠s

41.

PROOF

Statements	Reasons
1. ∠4 ≅ ∠6	**1.** Given
2. ∠4 ≅ ∠5	**2.** If two angles are vertical ∠s, then they are ≅
3. ∠5 ≅ ∠6	**3.** Transitive Property

42.

PROOF

Statements	Reasons
1. Figure as shown	**1.** Given
2. ∠4 is supp. to ∠2	**2.** If the exterior sides of two adjacent ∠s form a line, then the ∠s are supp.

43.

PROOF

Statements	Reasons
1. ∠3 is supp. to ∠5; ∠4 is supp. to ∠6	**1.** Given
2. ∠4 ≅ ∠5	**2.** If two lines intersect, the vertical angles formed are ≅
3. ∠3 ≅ ∠6	**3.** If two ∠s are supp. to congruent angles, then these angles are ≅

CHAPTER 2

2.1 EXERCISES

1. (a) No (b) Yes (c) Yes
3. Angle 9 appears to be a right angle.
5. (a) m∠3 = 87 (b) m∠6 = 87 (c) m∠1 = 93 (d) m∠7 = 87 **7.** (a) ∠5 (b) ∠5 (c) ∠8 (d) ∠5
9. (a) m∠2 = 68 (b) m∠4 = 112 (c) m∠5 = 112 (d) m∠MOQ = 34
11. $x = 10$; m∠4 = 110 **13.** $x = 12$; $y = 4$; m∠7 = 76
15. 1. Given 2. If two parallel lines are cut by a transversal, then the corresponding angles are ≅
3. If two lines intersect, the vertical angles are ≅
4. ∠3 ≅ ∠4 5. ∠1 ≅ ∠4

21. (a) $\angle 4 \cong \angle 2$ and $\angle 5 \cong \angle 3$ (b) 180 (c) 180
25. No

2.1 Selected Proof

17. PROOF

Statements	Reasons
1. $\overleftrightarrow{CE} \parallel \overleftrightarrow{DF}$; transversal $\overleftrightarrow{AB}$	1. Given
2. $\angle ACE \cong \angle ADF$	2. If two $\parallel$ lines are cut by a transversal, then the corresponding $\angle$s are $\cong$
3. $\overrightarrow{CX}$ bisects $\angle ACE$ $\overrightarrow{DE}$ bisects $\angle CDF$	3. Given
4. $\angle 1 \cong \angle 3$	4. If two $\angle$s are $\cong$, then their bisectors separate these $\angle$s into four $\cong$ $\angle$s

2.2 Exercises

1. *Converse:* If Juan is rich, then he won the state lottery. FALSE.
Inverse: If Juan does not win the state lottery, then he will not be rich. FALSE.
Contrapositive: If Juan is not rich, then he did not win the state lottery. TRUE.
3. *Converse:* If two angles are complementary, then the sum of their measures is 90. TRUE.
Inverse: If the sum of the measures of two angles is not 90, then the two angles are not complementary. TRUE
Contrapositive: If two angles are not complementary, then the sum of their measures is not 90. TRUE.
5. No conclusion **7.** $x = 5$ **9.** (a), (b), and (e)
11. Parallel

2.2 Selected Proofs

13. Assume that $\overleftrightarrow{DC} \parallel \overleftrightarrow{EG}$. Then $\angle AOD \cong \angle AFE$ because they are corresponding angles. But this contradicts the Given information. Therefore, our assumption is false and $\overleftrightarrow{DC} \nparallel \overleftrightarrow{EG}$.
17. Assume that the angles are vertical angles. If they are vertical angles, then they are congruent. But this contradicts the hypothesis that the two angles are not congruent. Hence, our assumption must be false and the angles are not vertical angles.
21. If M is a midpoint of $\overline{AB}$, then $AM = \frac{1}{2}AB$. Assume that N is also a midpoint of $\overline{AB}$ so that $AN = \frac{1}{2}AB$. By substitution, $AM = AN$. By the Segment-Addition Postulate, $AM = AN + NM$. Using substitution again, $AN + NM = AN$. Subtracting gives $NM = 0$. But this contradicts the Ruler Postulate, which states that the measure of a line segment is a positive number. Therefore, our assumption is wrong and M is the only midpoint for $\overline{AB}$.

2.3 Exercises

1. $p \parallel q$ **3.** None **5.** $\ell \parallel n$ **7.** None **9.** $\ell \parallel n$
11. 1. Given 2. If two $\angle$s are comp. to the same $\angle$, then they are $\cong$ 3. $\overline{BC} \parallel \overline{DE}$ **17.** $x = 9$ **19.** $x = 6$

2.3 Selected Proof

13. PROOF

Statements	Reasons
1. $\overline{AD} \perp \overline{DC}$ and $\overline{BC} \perp \overline{DC}$	1. Given
2. $\overline{AD} \parallel \overline{BC}$	2. If two lines are each $\perp$ to a third line, then these lines are $\parallel$ to each other

2.4 Exercises

1. If two $\angle$s of one $\triangle$ are $\cong$ to two $\angle$s of another $\triangle$, then the third $\angle$s of the triangle are $\cong$.
3. $m\angle 1 = 122$; $m\angle 2 = 58$; $m\angle 5 = 72$
5. $m\angle 2 = 57.7$; $m\angle 3 = 80.8$; $m\angle 4 = 41.5$ **7.** 35
9. 40 **11.** 360 **13.** $x = 45$; $y = 45$
15. $y = 20$; $x = 100$; $m\angle 5 = 60$ **21.** 44
23. $m\angle N = 49$; $m\angle P = 98$ **25.** 75 **33.** $m\angle M = 84$

2.5 Exercises

1. Increase **3.** $x = 113$; $y = 67$; $z = 36$
5. (a) 5 (b) 35 **7.** (a) 540 (b) 1440
9. (a) 90 (b) 150 **11.** (a) 90 (b) 30
13. (a) 7 (b) 9 **15.** (a) $n = 5$ (b) $n = 10$
17. (a) 15 (b) 20 **19.** 135
25. Figure (a): 90, 90, 120, 120, 120 Figure (b): 90, 90, 90, 135, 135
27. 36
29. The resulting polygon is also a regular polygon.
31. 150
33. (a) $n - 3$ (b) $\frac{n(n-3)}{2}$

2.5 Selected Proof

23. **PROOF**

Statements	Reasons
1. Quad. $RSTV$ with diagonals $\overline{RT}$ and $\overline{SV}$ intersecting at W	1. Given
2. $m\angle RWS = m\angle 1 + m\angle 2$	2. The measure of an exterior $\angle$ of a $\triangle$ equals the sum of the measures of the nonadjacent interior $\angle$s of the $\triangle$
3. $m\angle RWS = m\angle 3 + m\angle 4$	3. Same as (2)
4. $m\angle 1 + m\angle 2 = m\angle 3 + m\angle 4$	4. Substitution

Review Exercises

1. (a): $\overline{BC} \parallel \overline{AD}$ (b): $\overline{AB} \parallel \overline{CD}$ **2.** 110 **3.** $x = 37$
4. $m\angle D = 75$; $m\angle DEF = 125$ **5.** $x = 20$; $y = 10$
6. $x = 30$; $y = 35$ **7.** $\overline{AE} \parallel \overline{BF}$
8. None **9.** $\overline{BE} \parallel \overline{CF}$ **10.** $\overline{BE} \parallel \overline{CF}$
11. $\overline{AC} \parallel \overline{DF}$ and $\overline{AE} \parallel \overline{BF}$ **12.** $x = 120$; $y = 70$
13. $x = 32$; $y = 30$ **14.** $y = -8$; $x = 24$ **15.** $x = 140$
16. $x = 6$ **17.** $m\angle 3 = 69$; $m\angle 4 = 67$; $m\angle 5 = 44$
18. 110 **19.** S **20.** N **21.** N **22.** S **23.** S
24. A
25.

Number of sides	8	12	20	15	10	16	180
Measure of each ext. $\angle$	45	30	18	24	36	22.5	2
Measure of each int. $\angle$	135	150	162	156	144	157.5	178
Number of diagonals	20	54	170	90	35	104	15,930

28. Not possible
30. *Statement:* If two angles are right angles, then the angles are congruent.
Converse: If two angles are congruent, then the angles are right angles.
Inverse: If two angles are not right angles, then the angles are not congruent.
Contrapositive: If two angles are not congruent, then the angles are not right angles.
31. *Statement:* If it is not raining, then I am happy.
Converse: If I am happy, then it is not raining.
Inverse: If it is raining, then I am not happy.
Contrapositive: If I am not happy, then it is raining.
32. Contrapositive **37.** Assume $x = -3$.
38. Assume the sides opposite these angles are $\cong$.
39. Assume that $\angle 1 \cong \angle 2$. Then $m \parallel n$ since congruent corresponding angles are formed. But this contradicts our hypothesis. Therefore, our assumption must be false and $\angle 1 \not\cong \angle 2$.
40. Assume that $m \parallel n$. Then $\angle 1 \cong \angle 3$ since alternate exterior angles are congruent when parallel lines are cut by a transversal. But this contradicts the given fact that $\angle 1 \not\cong \angle 3$. Therefore, our assumption must be false and it follows that $m \nparallel n$.

Review Exercises Selected Proofs

33. **PROOF**

Statements	Reasons
1. $\overline{AB} \parallel \overline{CF}$	1. Given
2. $\angle 1 \cong \angle 2$	2. If two $\parallel$ lines are cut by a transversal, then corresponding $\angle$s are $\cong$
3. $\angle 2 \cong \angle 3$	3. Given
4. $\angle 1 \cong \angle 3$	4. Transitive Prop. of Congruence

34. **PROOF**

Statements	Reasons
1. $\angle 1$ is comp. to $\angle 2$; $\angle 2$ is comp. to $\angle 3$	1. Given
2. $\angle 1 \cong \angle 3$	2. If two $\angle$s are comp. to the same $\angle$, then these $\angle$s are $\cong$
3. $\overline{BD} \parallel \overline{AE}$	3. If two lines are cut by a transversal so that corresponding $\angle$s are $\cong$, then the lines are $\parallel$

35. PROOF

Statements	Reasons
1. $\overline{BE} \perp \overline{DA}$ $\overline{CD} \perp \overline{DA}$	1. Given
2. $\overline{BE} \parallel \overline{CD}$	2. If two lines are each $\perp$ to a third line, then these lines are parallel to each other
3. $\angle 1 \cong \angle 2$	3. If two $\parallel$ lines are cut by a transversal, then the alternate interior $\angle$s are $\cong$

36. PROOF

Statements	Reasons
1. $\angle A \cong \angle C$	1. Given
2. $\overrightarrow{DC} \parallel \overrightarrow{AB}$	2. Given
3. $\angle C \cong \angle 1$	3. If two $\parallel$ lines are cut by a transversal, the alt. int. $\angle$s are $\cong$
4. $\angle A \cong \angle 1$	4. Transitive Prop. of Congruence
5. $\overline{DA} \parallel \overline{CD}$	5. If two lines are cut by a transversal so that corr. $\angle$s are $\cong$, then these lines are $\parallel$

Chapter 3

3.1 Exercises

1. $\angle A$; $\overline{AB}$; No; No **3.** SAS **5.** $\triangle AED \cong \triangle FDE$
7. SSS **9.** AAS **11.** ASA **13.** ASA **15.** SSS
17. (a) $\angle A \cong \angle A$ (b) ASA **19.** $\overline{AD} \cong \overline{EC}$
21. $\overline{MO} \cong \overline{MO}$ **23.** 1. Given 2. $\overline{AC} \cong \overline{AC}$ 3. SSS
31. Yes; SAS or SSS **33.** No
35. (a) $\triangle CBE, \triangle ADE, \triangle CDE$ (b) $\triangle ADC$ (c) $\triangle CBD$

3.1 Selected Proofs

25. PROOF

Statements	Reasons
1. $\overrightarrow{PQ}$ bisects $\angle MPN$	1. Given
2. $\angle MPQ \cong \angle NPQ$	2. If a ray bisects an $\angle$, it forms two $\cong$ $\angle$s
3. $\overline{MP} \cong \overline{NP}$	3. Given
4. $\overline{PQ} \cong \overline{PQ}$	4. Identity
5. $\triangle MQP \cong \triangle NQP$	5. SAS

29. PROOF

Statements	Reasons
1. $\angle VRS \cong \angle TSR$ and $\overline{RV} \cong \overline{TS}$	1. Given
2. $\overline{RS} \cong \overline{RS}$	2. Identity
3. $\triangle RST \cong \triangle SRV$	3. SAS

3.2 Exercises

9. $m\angle 2 = 48$; $m\angle 3 = 48$; $m\angle 5 = 42$; $m\angle 6 = 42$
11. 1. Given
2. If two lines are $\perp$, then they form right $\angle$s
3. Identity 4. $\triangle HJK \cong \triangle HJL$ 5. $\overline{KJ} \cong \overline{JL}$
25. (a) 8 (b) 37° (c) 53°

3.2 Selected Proofs

1. PROOF

Statements	Reasons
1. $\angle 1$ and $\angle 2$ are right $\angle$s $\overline{CA} \cong \overline{DA}$	1. Given
2. $\overline{AB} \cong \overline{AB}$	2. Identity
3. $\triangle ABC \cong \triangle ABD$	3. HL

5. PROOF

Statements	Reasons
1. $\angle R$ and $\angle V$ are right $\angle$s $\angle 1 \cong \angle 2$	1. Given
2. $\angle R \cong \angle V$	2. All right $\angle$s are $\cong$.
3. $\overline{ST} \cong \overline{ST}$	3. Identity
4. $\triangle RST \cong \triangle VST$	4. AAS

13. PROOF

Statements	Reasons
1. $\angle$s P and R are right $\angle$s	**1.** Given
2. $\angle P \cong \angle R$	**2.** All right $\angle$s are $\cong$
3. M is the midpoint of $\overline{PR}$	**3.** Given
4. $\overline{PM} \cong \overline{MR}$	**4.** The midpoint of a segment forms two $\cong$ segments
5. $\angle NMP \cong \angle QMR$	**5.** If two lines intersect, the vertical angles formed are $\cong$
6. $\triangle NPM \cong \triangle QRM$	**6.** ASA
7. $\angle N \cong \angle Q$	**7.** CPCTC

17. PROOF

Statements	Reasons
1. $\overline{DF} \cong \overline{DG}$ and $\overline{FE} \cong \overline{EG}$	**1.** Given
2. $\overline{DE} \cong \overline{DE}$	**2.** Identity
3. $\triangle FDE \cong \triangle GDE$	**3.** SSS
4. $\angle FDE \cong \angle GDE$	**4.** CPCTC
5. $\overrightarrow{DE}$ bisects $\angle FDG$	**5.** If a ray divides an $\angle$ into two $\cong$ $\angle$s, then the ray bisects the angle

21. PROOF

Statements	Reasons
1. $\angle 1 \cong \angle 2$ and $\overline{MN} \cong \overline{QP}$	**1.** Given
2. $\overline{MP} \cong \overline{MP}$	**2.** Identity
3. $\triangle NMP \cong \triangle QPM$	**3.** SAS
4. $\angle 3 \cong \angle 4$	**4.** CPCTC
5. $\overline{MQ} \parallel \overline{NP}$	**5.** If two lines are cut by a transversal so that the alt. int. $\angle$s are $\cong$, then the lines are $\parallel$

3.3 Exercises

1. Underdetermined **3.** Overdetermined
5. Determined **7.** 55 **9.** $m\angle 2 = 68$; $m\angle 1 = 44$
11. $m\angle 5 = 124$ **13.** $m\angle A = 52$; $m\angle B = 64$; $m\angle C = 64$
15. 26 **17.** 12 **19.** Yes
21. 1. Given 2. $\angle 3 \cong \angle 2$ 3. $\angle 1 \cong \angle 2$ 4. If two $\angle$s of a $\triangle$ are $\cong$, then the opposite sides are $\cong$
27. $m\angle A = 40$ **29.** 75

3.3 Selected Proof

23. PROOF

Statements	Reasons
1. $\angle 1 \cong \angle 3$	**1.** Given
2. $\overline{RU} \cong \overline{VU}$	**2.** Given
3. $\angle R \cong \angle V$	**3.** If two sides of a $\triangle$ are $\cong$, then the $\angle$s opposite these sides are also $\cong$
4. $\triangle RUS \cong \triangle VUT$	**4.** ASA
5. $\overline{SU} \cong \overline{TU}$	**5.** CPCTC
6. $\triangle STU$ is isosceles	**6.** If a $\triangle$ has two $\cong$ sides, it is an isosceles $\triangle$

3.4 Exercises

19. Construct a 90° angle; bisect it to form two 45° $\angle$s. Bisect one of the 45° angles to get a 22.5° $\angle$.
29. 120° **37.** D is on the bisector of $\angle A$.

3.5 Exercises

1. False **3.** True **5.** True **7.** False **9.** True
13. Nashville
15. 1. $m\angle ABC > m\angle DBE$ and $m\angle CBD > m\angle EBF$
3. Angle-Addition Postulate 4. $m\angle ABD > m\angle DBF$
19. $BC < EF$ **21.** $2 < x < 10$
23. $x + 2 < y < 5x + 12$
25. *Proof:* Assume $PM = PN$. Then $\triangle MPN$ is isosceles. But that contradicts the hypothesis; thus, our assumption must be wrong and $PM \neq PN$.

3.5 Selected Proof

17. PROOF

Statements	Reasons
1. Quad. $RSTU$ with diagonal $\overline{US}$; $\angle R$ and $\angle TUS$ are right $\angle$s	**1.** Given
2. $TS > US$	**2.** The shortest distance from a point to a line is the $\perp$ distance
3. $US > UR$	**3.** Same as (2)
4. $TS > UR$	**4.** Transitive Prop. of Inequality

Review Exercises

15. (a) $\overline{PR}$ (b) $\overline{PQ}$ **17.** $\angle R, \angle Q, \angle P$ **19.** (b)
21. 20 **23.** $m\angle C = 64$
25. The triangle is also equilateral.

Review Exercises Selected Proofs

1.

PROOF

Statements	Reasons
1. $\angle AEB \cong \angle DEC$	1. Given
2. $\overline{AE} \cong \overline{ED}$	2. Given
3. $\angle A \cong \angle D$	3. If two sides of a $\triangle$ are $\cong$, then the $\angle$s opposite these sides are also $\cong$
4. $\triangle AEB \cong \triangle DEC$	4. ASA

5.

PROOF

Statements	Reasons
1. $\overline{AB} \cong \overline{DE}$ and $\overline{AB} \parallel \overline{DE}$	1. Given
2. $\angle A \cong \angle D$	2. If two $\parallel$ lines are cut by a transversal, then the alt. int. $\angle$s are $\cong$
3. $\overline{AC} \cong \overline{DF}$	3. Given
4. $\triangle BAC \cong \triangle EDF$	4. SAS
5. $\angle BCA \cong \angle EFD$	5. CPCTC
6. $\overline{BC} \parallel \overline{FE}$	6. If two lines are cut by a transversal so that alt. int. $\angle$s are $\cong$, then the lines are $\parallel$

9.

PROOF

Statements	Reasons
1. $\overline{YZ}$ is the base of an isosceles triangle	1. Given
2. $\angle Y \cong \angle Z$	2. Base $\angle$s of an isosceles $\triangle$ are $\cong$
3. $\overrightarrow{XA} \parallel \overline{YZ}$	3. Given
4. $\angle 1 \cong \angle Y$	4. If two $\parallel$ lines are cut by a transversal, then the corresponding $\angle$s are $\cong$
5. $\angle 2 \cong \angle Z$	5. If two $\parallel$ lines are cut by a transversal, then the alt. int. $\angle$s are $\cong$
6. $\angle 1 \cong \angle 2$	6. Transitive Prop. for Congruence

13.

PROOF

Statements	Reasons
1. $\overline{AB} \cong \overline{CD}$	1. Given
2. $\angle BAD \cong \angle CDA$	2. Given
3. $\overline{AD} \cong \overline{AD}$	3. Identity
4. $\triangle BAD \cong \triangle CDA$	4. SAS
5. $\angle CAD \cong \angle BDA$	5. CPCTC
6. $\overline{AE} \cong \overline{ED}$	6. If two $\angle$s of a $\triangle$ are $\cong$, then the sides opposite these $\angle$s are also $\cong$
7. $\triangle AED$ is isosceles	7. If a $\triangle$ has two $\cong$ sides, it is an isosceles $\triangle$

Chapter 4

4.1 Exercises

1. (a) $AB = DC$ (b) $AD = BC$
3. (a) 8 (b) 5 (c) 70 (d) 110
5. $AB = DC = 8$; $BC = AD = 9$
7. $m\angle A = m\angle C = 83$; $m\angle B = m\angle D = 97$
9. Parallelogram **11.** Parallelogram
13. 1. Given 2. $\overline{RV} \perp \overline{VT}$ and $\overline{ST} \perp \overline{VT}$ 3. $\overline{RV} \parallel \overline{ST}$
4. $RSTV$ is a parallelogram
21. $\angle P$ is a right angle **23.** $\overline{RT}$ **25.** 255 mph
27. $\overline{AC}$

4.1 Selected Proof

15.

PROOF

Statements	Reasons
1. Parallelogram $RSTV$	1. Given
2. $\overline{RS} \parallel \overline{VT}$	2. Opposite sides of a parallelogram are $\parallel$
3. $\overline{XY} \parallel \overline{VT}$	3. Given
4. $\overline{RS} \parallel \overline{XY}$	4. If two lines are each $\parallel$ to a third line, then the lines are $\parallel$
5. $RSYX$ is a parallelogram	5. If a quadrilateral has opposite sides $\parallel$, then the quadrilateral is a parallelogram
6. $\angle 1 \cong \angle S$	6. Opposite angles of a parallelogram are $\cong$

4.2 Exercises

1. (a) Yes (b) No
3. Parallelogram **5.** (a) Kite (b) Parallelogram
7. $\overline{AC}$ **9.** 6.18 **11.** (a) 8 (b) 7 (c) 6
13. 10 **15.** Parallel and congruent
17. 1. Given 2. Identity 3. $\triangle NMQ \cong \triangle NPQ$
4. CPCTC 5. *MNPQ* is a kite.
27. $y = 6$; $MN = 9$; $ST = 18$
29. $x = 5$; $RM = 11$; $ST = 22$

4.2 Selected Proofs

19. PROOF

Statements	Reasons
1. *M-Q-T* and *P-Q-R* so that *MNPQ* and *QRST* are parallelograms	**1.** Given
2. $\angle N \cong \angle MQP$	**2.** Opposite $\angle$s in a parallelogram are $\cong$
3. $\angle MQP \cong \angle RQT$	**3.** If two lines intersect, the vertical $\angle$s formed are $\cong$
4. $\angle RQT \cong \angle S$	**4.** Same as (2).
5. $\angle N \cong \angle S$	**5.** Transitive Prop. for Congruence

21. PROOF

Statements	Reasons
1. Kite *HJKL* with diagonal $\overline{HK}$	**1.** Given
2. $\overline{LH} \cong \overline{HJ}$ and $\overline{LK} \cong \overline{JK}$	**2.** A kite is a quadrilateral with two distinct pairs of $\cong$ adjacent sides
3. $\overline{HK} \cong \overline{HK}$	**3.** Identity
4. $\triangle LHK \cong \triangle JHK$	**4.** SSS
5. $\angle LHK \cong \angle JHK$	**5.** CPCTC
6. $\overrightarrow{HK}$ bisects $\angle LHJ$	**6.** If a ray divides an $\angle$ into two $\cong$ $\angle$s, then the ray bisects the $\angle$

4.3 Exercises

1. $m\angle A = 60$; $m\angle ABC = 120$
3. The parallelogram is a rectangle.
5. The quadrilateral is a rhombus.
7. $\overline{MN} \parallel$ to both $\overline{AB}$ and $\overline{DC}$; $MN = AB = DC$
9. $x = 5$; $AD = 19$ **11.** 5
13. 1. Given 4. The line joining the midpoints of two sides of a $\triangle$ is $\parallel$ to the third side 5. If two lines are each $\parallel$ to a third line, then the two lines are $\parallel$ 6. Same as (2) 7. Same as (3) 8. Same as (4) 9. Same as (5) 10. *ABCD* is a parallelogram
25. 20.4

4.4 Exercises

1. $m\angle D = 122$; $m\angle B = 55$
3. The trapezoid is an isosceles trapezoid.
5. The quadrilateral is a rhombus. **7.** 9.7 **9.** 10.8
11. $7x + 2$ **15.** 12 **17.** 22 **19.** 14

Review Exercises

1. A **2.** S **3.** N **4.** S **5.** S **6.** A **7.** A
8. A **9.** A **10.** N **11.** S **12.** N
13. $AB = DC = 17$; $AD = BC = 31$
14. 106 **15.** 52 **16.** $m\angle M = 100$; $m\angle P = 80$
17. $\overline{PN}$ **18.** Kite
19. $m\angle G = m\angle F = 72$; $m\angle E = 108$ **20.** 14.9
21. $MN = 23$; $PO = 7$ **22.** 26
23. $MN = 6$; $m\angle FMN = 80$; $m\angle FNM = 40$
24. $x = 3$; $MN = 15$; $JH = 30$

Review Exercises Selected Proofs

25. PROOF

Statements	Reasons
1. *ABCD* is a parallelogram	**1.** Given
2. $\overline{AD} \cong \overline{CB}$	**2.** Opposite sides of a parallelogram are $\cong$
3. $\overline{AD} \parallel \overline{CB}$	**3.** Opposite sides of a parallelogram are $\parallel$
4. $\angle 1 \cong \angle 2$	**4.** If two $\parallel$ lines are cut by a transversal, then the alt. int. $\angle$s are $\cong$
5. $\overline{AF} \cong \overline{CE}$	**5.** Given
6. $\triangle DAF \cong \triangle BCE$	**6.** SAS
7. $\angle DFA \cong \angle BEC$	**7.** CPCTC
8. $\overline{DF} \parallel \overline{EB}$	**8.** If two lines are cut by a transversal so that alt. ext. $\angle$s are $\cong$, then the lines are $\parallel$

26.

PROOF

Statements	Reasons
1. *ABEF* is a rectangle	1. Given
2. *ABEF* is a parallelogram	2. A rectangle is a parallelogram with a rt. ∠
3. $\overline{AF} \cong \overline{BE}$	3. Opposite sides of a parallelogram are ≅
4. *BCDE* is a rectangle	4. Given
5. ∠*F* and ∠*BED* are rt. ∠s	5. Same as (2)
6. $\angle F \cong \angle BED$	6. Any two rt. ∠s are ≅
7. $\overline{FE} \cong \overline{ED}$	7. Given
8. $\triangle AFE \cong \triangle BED$	8. SAS
9. $\overline{AE} \cong \overline{BD}$	9. CPCTC
10. $\angle AEF \cong \angle BDE$	10. CPCTC
11. $\overline{AE} \parallel \overline{BD}$	11. If lines are cut by a transversal so that the corresponding ∠s are ≅, then the lines are ∥

27.

PROOF

Statements	Reasons
1. $\overline{DE}$ is a median in △*ADC*	1. Given
2. *E* is the midpoint of $\overline{AC}$	2. A median of a △ is a line segment drawn from a vertex to the midpoint of the opposite side
3. $\overline{AE} \cong \overline{EC}$	3. Midpoint of a segment forms two ≅ segments
4. $\overline{BE} \cong \overline{FD}$ and $\overline{EF} \cong \overline{FD}$	4. Given
5. $\overline{BE} \cong \overline{EF}$	5. Transitive Prop. for Congruence
6. *ABCF* is a parallelogram	6. If the diagonals of a quadrilateral bisect each other, then the quad. is a parallelogram

28.

PROOF

Statements	Reasons
1. $\triangle FAB \cong \triangle HCD$	1. Given
2. $\overline{AB} \cong \overline{DC}$	2. CPCTC
3. $\triangle EAD \cong \triangle GCB$	3. Given
4. $\overline{AD} \cong \overline{BC}$	4. CPCTC

(continued)

(continued)

Statements	Reasons
5. *ABCD* is a parallelogram	5. If a quadrilateral has both pairs of opposite sides ≅, then the quad. is a parallelogram

29.

PROOF

Statements	Reasons
1. *ABCD* is a parallelogram	1. Given
2. $\overline{DC} \cong \overline{BN}$	2. Given
3. $\angle 3 \cong \angle 4$	3. Given
4. $\overline{BN} \cong \overline{BC}$	4. If two ∠s of a △ are ≅, then the sides opposite these ∠s are also ≅
5. $\overline{DC} \cong \overline{BC}$	5. Transitive Prop. for Congruence
6. *ABCD* is a rhombus	6. If a parallelogram has two ≅ adjacent sides, then the parallelogram is a rhombus

30.

PROOF

Statements	Reasons
1. △*TWX* is isosceles with base $\overline{WX}$	1. Given
2. $\angle W \cong \angle X$	2. Base ∠s of an isosceles △ are ≅
3. $\overline{RY} \parallel \overline{WX}$	3. Given
4. $\angle TRY \cong \angle W$ and $\angle TYR \cong \angle X$	4. If two ∥ lines are cut by a transversal, then the corresp. ∠s are ≅
5. $\angle TRY \cong \angle TYR$	5. Transitive Prop. for Congruence
6. $\overline{TR} \cong \overline{TY}$	6. If two ∠s of a △ are ≅, then the sides opposite these ∠s are also ≅
7. $\overline{TW} \cong \overline{TX}$	7. An isosceles △ has two ≅ sides
8. $TR = TY$ and $TW = TX$	8. If two segments are ≅, then they are equal in length
9. $TW = TR + RW$ and $TX = TY + YX$	9. Segment-Addition Postulate
10. $TR + RW = TY + YX$	10. Substitution

(continued)

(continued)

11. $RW = YX$	11. Subtraction Prop. of Equality
12. $\overline{RW} \cong \overline{YX}$	12. If segments are = in length, then they are ≅
13. *RWXY* is an isosceles trapezoid	13. If a quadrilateral has one pair of ∥ sides and the non-parallel sides are ≅, then the quad. is an isosceles trapezoid

CHAPTER 5

5.1 EXERCISES

1. (a) $\frac{4}{5}$ (b) $\frac{4}{5}$ (c) $\frac{2}{3}$ (d) incommensurable
3. (a) $\frac{5}{8}$ (b) $\frac{1}{3}$ (c) $\frac{4}{3}$ (d) incommensurable
5. (a) 3 (b) 8 **7.** (a) 6 (b) 4
9. (a) $\pm 2\sqrt{7} \approx \pm 5.29$ (b) $\pm 3\sqrt{2} \approx \pm 4.24$
11. (a) 4 (b) $-\frac{5}{6}$ or 3
13. (a) $\dfrac{3 \pm \sqrt{33}}{4} \approx 2.19$ or -0.69
(b) $\dfrac{7 \pm \sqrt{89}}{4} \approx 4.11$ or -0.61
15. 6.3 m/sec **17.** $10\frac{1}{2}$ **19.** ≈24 outlets
21. (a) $4\sqrt{3} \approx 6.93$ (b) $4\frac{1}{2}$
23. Secretary's salary is \$18,500; salesperson's salary is \$27,750; vice-president's salary is \$46,250
25. 40 and 50 **27.** 30.48 cm **29.** $2\frac{4}{7} \approx 2.57$
31. $a = 12$; $b = 16$ **33.** 4 in. by $4\frac{2}{3}$ in.

5.2 EXERCISES

1. (a) $\triangle ABC \sim \triangle XTN$ (b) $\triangle ACB \sim \triangle NXT$
3. Yes; yes. Spheres have the same shape; one is generally an enlargement of the other.
5. $\triangle RST \sim \triangle UVW$; $\frac{WU}{TR} = \frac{WV}{TS} = \frac{UV}{RS} = \frac{3}{2}$
7. (a) 82 (b) 42 (c) $10\frac{1}{2}$ (d) 8
9. (a) Yes (b) Yes (c) Yes
11. 1. $\overline{MN} \perp \overline{NP}$ and $\overline{QR} \perp \overline{RP}$
2. If two lines are ⊥, then they form a rt. ∠
3. $\angle N \cong \angle QRP$ 4. Identity 5. $\triangle MNP \sim \triangle QRP$; AA
13. 1. $\angle H \cong \angle F$ 2. If two ∠s are vertical ∠s, then they are ≅ 3. $\triangle HJK \sim \triangle FGK$; AA
23. $4\frac{1}{2}$ **25.** 16 **27.** 12
29. $10 + 2\sqrt{5}$ or $10 - 2\sqrt{5} \approx 14.47$ or 5.53 **31.** 3 ft 9 in.
33. 74 ft

5.2 SELECTED PROOFS

15. PROOF

Statements	Reasons
1. $\overline{AB} \parallel \overline{DF}$ and $\overline{BD} \parallel \overline{FG}$	1. Given
2. $\angle A \cong \angle FEG$ and $\angle BCA \cong \angle G$	2. If two ∥ lines are cut by a transversal, then the corresponding ∠s are ≅
3. $\triangle ABC \sim \triangle EFG$	3. AA

19. *Proof:* By hypothesis, $\overline{RS} \parallel \overline{UV}$. $\angle R \cong \angle V$ and $\angle S \cong \angle U$ because they are alternate interior angles. $\triangle RTS \sim \triangle VTU$ by AA. It follows that $\frac{RT}{VT} = \frac{RS}{VU}$, since corresponding sides of similar △s are proportional.

21. PROOF

Statements	Reasons
1. $\triangle DEF \sim \triangle MNP$; $\overline{DG}$ and $\overline{MQ}$ are altitudes	1. Given
2. $\overline{DG} \perp \overline{EF}$ and $\overline{MQ} \perp \overline{NP}$	2. An altitude is a segment drawn from a vertex ⊥ to the opposite side
3. $\angle DGE$ and $\angle MQN$ are rt. ∠s	3. ⊥ lines form a rt. ∠
4. $\angle DGE \cong \angle MQN$	4. Right ∠s are ≅
5. $\angle E \cong \angle N$	5. If two △s are ~, then the corresponding ∠s are ≅
6. $\triangle DGE \sim \triangle MQN$	6. AA
7. $\frac{DG}{MQ} = \frac{DE}{MN}$	7. Corresponding sides of ~ △s are proportional

5.3 EXERCISES

1. $\triangle RST \sim \triangle RVS \sim \triangle SVT$ **3.** 4.5
5. (a) 10 (b) $\sqrt{34} \approx 5.83$ **7.** (a) 8 (b) 4
9. (a) Right (b) Acute (c) Right (d) No △
11. 15 ft **13.** $6\sqrt{5} \approx 13.4$m **15.** 12
17. The base is 8; the altitude is 6; the diagonal is 10.
19. $6\sqrt{7} \approx 15.87$ **21.** 12 **23.** 4 **25.** $9\frac{3}{13}$
27. $5\sqrt{5} \approx 11.18$ **33.** 60
35. $TS = 13$; $RT = 13\sqrt{2} \approx 18.38$

5.4 Exercises

1. $YZ = 8$ and $XY = 8\sqrt{2} \approx 11.31$
3. $DF = 5\sqrt{3} \approx 8.66$ and $FE = 10$
5. $HL = 6$; $HK = 12$; $MK = 6$
7. $AC = 6$ and $AB = 6\sqrt{2} \approx 8.49$
9. $RS = 6$ and $RT = 6\sqrt{3} \approx 10.39$
11. $DB = 5\sqrt{6} \approx 12.25$ **13.** $6\sqrt{3} + 6 \approx 16.39$
15. 45° **17.** 60°; 146 ft further
19. $DC = 2\sqrt{3} \approx 3.46$
$DB = 4\sqrt{3} \approx 6.93$
21. $6\sqrt{3} \approx 10.39$ **23.** $4\sqrt{3} \approx 6.93$
25. $6 + 6\sqrt{3} \approx 16.39$

5.5 Exercises

1. 30 oz of ingredient A; 24 oz of ingredient B; 36 oz of ingredient C
3. (a) Yes (b) Yes **5.** $EF = 4\frac{1}{6}$, $FG = 3\frac{1}{3}$, $GH = 2\frac{1}{2}$
7. $x = 5\frac{1}{3}$, $DE = 5\frac{1}{3}$, $EF = 6\frac{2}{3}$ **9.** $EC = 16\frac{4}{5}$
11. $a = \frac{1}{2}$ or $a = 5$; $AD = 4$ **13.** (a) No (b) Yes
15. 9 **17.** $4\sqrt{6} \approx 9.80$ **19.** $\frac{AC}{CE} = \frac{AD}{DE}$; $\frac{DC}{CB} = \frac{DE}{EB}$
21. $SV = 2\sqrt{3} \approx 3.46$; $VT = 4\sqrt{3} \approx 6.93$
23. $x = \frac{1 + \sqrt{73}}{2}$ or $x = \frac{1 - \sqrt{73}}{2}$; reject both because each will give a negative number for the length of a side.
25. 1. Given 2. Means-Extremes Property
3. Addition Property of Equality 4. Distributive Property
5. Means-Extremes Property 6. Substitution
31. $\frac{-1 + \sqrt{5}}{2} \approx 0.62$

5.5 Selected Proof

27. PROOF

Statements	Reasons
1. $\triangle RST$ with M the midpoint of $\overline{RS}$; $\overleftrightarrow{MN} \parallel \overline{ST}$	1. Given
2. $RM = MS$	2. The midpoint of a segment divides the segment into two segments of equal measure
3. $\frac{RM}{MS} = \frac{RN}{NT}$	3. If a line is $\parallel$ to one side of a $\triangle$ and intersects the other two sides, then it divides these sides proportionally
4. $\frac{MS}{MS} = 1 = \frac{RN}{NT}$	4. Substitution
5. $RN = NT$	5. Means-Extremes Property
(continued)	
6. N is the midpoint of $\overline{RT}$	6. If a point divides a segment into two segments of equal measure, then the point is a midpoint

Review Exercises

1. False **2.** True **3.** False **4.** True **5.** True
6. False **7.** True
8. (a) $\pm 3\sqrt{2} \approx \pm 4.24$ (b) 26 (c) -1 (d) 2
(e) 7 or -1 (f) $\frac{-9}{5}$ or 4 (g) 6 or -1 (h) -6 or 3
9. \$3.78 **10.** 6 packages **11.** \$79.20
12. The length of the sides are 8, 12, 20 and 28. **13.** 18
14. 20 and $22\frac{1}{2}$ **15.** 150 **18.** $x = 5$; $m\angle F = 97$
19. $AB = 6$ and $BC = 12$ **20.** 3 **21.** $4\frac{1}{2}$ **22.** $6\frac{1}{4}$
23. $5\frac{3}{5}$ **24.** 10 **25.** 6 **26.** $EO = 1\frac{1}{5}$; $EK = 9$
29. (a) $8\frac{1}{3}$ (b) 21 (c) $2\sqrt{3} \approx 3.46$ (d) 3
30. (a) 16 (b) 40 (c) $2\sqrt{5} \approx 4.47$ (d) 4
31. (a) 30 (b) 24 (c) 20 (d) 16
32. $AE = 20$; $EF = 15$; $AF = 25$ **33.** $4\sqrt{2} \approx 5.66$ in.
34. $3\sqrt{2} \approx 4.24$ cm **35.** 25 cm **36.** $5\sqrt{3} \approx 8.66$ in.
37. $4\sqrt{3} \approx 6.93$ in. **38.** 12 cm
39. (a) $y = 9$; $x = 9\sqrt{2} \approx 12.73$ (b) $y = 6$; $x = 4\frac{1}{2}$
(c) $y = 3$; $x = 12$ (d) $x = 2\sqrt{14} \approx 7.48$; $y = 13$
40. 11 km
41. (a) Acute (b) No $\triangle$ (c) Obtuse (d) Right
(e) No $\triangle$ (f) Acute (g) Obtuse (h) Obtuse

Review Exercises Selected Proofs

16. PROOF

Statements	Reasons
1. $ABCD$ is a parallelogram; $\overline{DB}$ intersects $\overline{AE}$ at point F	1. Given
2. $\overline{DC} \parallel \overline{AB}$	2. Opposite sides of a parallelogam are $\parallel$
3. $\angle CDB \cong \angle ABD$	3. If two $\parallel$ lines are cut by a transversal, then the alt. int. $\angle$s are $\cong$
4. $\angle DEF \cong \angle BAF$	4. Same as (3)
5. $\triangle DFE \sim \triangle BFA$	5. AA
6. $\frac{AF}{EF} = \frac{AB}{DE}$	6. Corresponding sides of $\sim$ $\triangle$s are proportional

17. **PROOF**

Statements	Reasons
1. $\angle ADC \cong \angle 2$	1. If two lines intersect, then the vertical $\angle$s formed are $\cong$
2. $\angle 1 \cong \angle 2$	2. Given
3. $\angle ADC \cong \angle 1$	3. Transitive Prop. for Congruence
4. $\angle A \cong \angle A$	4. Identity
5. $\triangle BAE \sim \triangle CAD$	5. AA
6. $\frac{AB}{AC} = \frac{BE}{CD}$	6. Corresponding sides of $\sim \triangle$s are proportional

CHAPTER 6

6.1 EXERCISES

1. (a) 90 (b) 270 (c) 135 (d) 135
3. (a) 80 (b) 120 (c) 160 (d) 80 (e) 120 (f) 160 (g) 10 (h) 50 (i) 30
5. (a) 72 (b) 144 (c) 36 (d) 72 (e) 18
7. (a) 12 (b) $6\sqrt{2}$ **9.** 3 **11.** $\sqrt{7} + 3\sqrt{3}$
13. 90°; square
15. (a) The measure of an arc equals the measure of its corresponding central angle. Therefore, congruent arcs would have to have congruent central angles. (b) The measure of a central angle equals the measure of its intercepted arc. Therefore, congruent central angles have congruent arcs. (c) Draw the radii to the endpoints of the congruent chords. The two triangles formed are congruent by SSS. The central angles of each triangle are congruent by CPCTC. Therefore, the arcs corresponding to the central angles are also congruent. Hence, congruent chords will have congruent arcs. (d) Draw the four radii to the endpoints of the congruent arcs. Also draw the chords corresponding to the congruent arcs. The central angles corresponding to the congruent arcs are also congruent. Therefore, the triangles are congruent by SAS. The chords are congruent by CPCTC. Hence, congruent arcs will have congruent chords.
(e) Congruent central angles will have congruent arcs (from b). Congruent arcs will have congruent chords (from d). Hence, congruent central angles will have congruent chords.
(f) Congruent chords have congruent arcs (from c). Congruent arcs have congruent central angles (from a). Therefore, congruent chords will have congruent central angles.
17. (a) 15° (b) 70° **19.** 72° **21.** 45°
23. 1. $\overline{MN} \parallel \overline{OP}$ in $\odot O$ 2. If two $\parallel$ lines are cut by a transversal, then the alt. int. $\angle$s are $\cong$ 3. If two $\angle$s are $\cong$, then their measures are $=$ 4. The measure of an inscribed $\angle$ equals $\frac{1}{2}$ the measure of its intercepted arc 5. The measure of a central $\angle$ equals the measure of its arc 6. Substitution
31. If $\overparen{ST} \cong \overparen{TV}$, $\overline{ST} \cong \overline{TV}$ since $\cong$ arcs in a circle have $\cong$ chords. $\triangle STV$ is an isosceles $\triangle$ because it has two $\cong$ sides.

6.1 SELECTED PROOF

25. *Proof:* Using the chords $\overline{AB}$, $\overline{BC}$, $\overline{CD}$, and $\overline{AD}$ in $\odot O$ as sides of inscribed angles, $\angle B \cong \angle D$ and $\angle A \cong \angle C$ because they are inscribed angles intercepting the same arc. $\triangle ABE \sim \triangle CDE$ by AA.

6.2 EXERCISES

1. (a) 8 (b) 46 (c) 38 (d) 54 (e) 126
3. No. If $\overline{RS}$ is a diameter and $\overrightarrow{SW}$ is a tangent, then $\overline{RS} \perp \overrightarrow{SW}$. $\overline{TS}$ cannot be $\perp$ to $\overrightarrow{SW}$ because there is already a segment $\perp$ to $\overrightarrow{SW}$ at S. **5.** (a) 22 (b) 7 (c) 15
7. (a) 136 (b) 224 (c) 68 (d) 44
9. (a) 96 (b) 60 **11.** (a) 120 (b) 240 (c) 60
13. 28 **15.** $m\overparen{CE} = 88$; $m\overparen{BD} = 36$
17. 1. $\overline{AB}$ and $\overline{AC}$ are tangents to $\odot O$ from A 2. The measure of an $\angle$ formed by a tangent and a chord equals $\frac{1}{2}$ the arc measure 3. Substitution 4. If two $\angle$s are $=$ in measure, they are $\cong$ 5. $\overline{AB} \cong \overline{AC}$ 6. $\triangle ABC$ is isosceles
23. ≈ 154.95 mi **25.** $m\angle 1 = 36$; $m\angle 2 = 108$ **27.** 10
37. With O-D-X, $OX = OD + DX$. Since $DX > 0$, $OX > OD$ or $OD < OX$. (If $a > b$, then $a = b + k$, where k is a positive number.)

6.2 SELECTED PROOF

19. *Given:* Tangent $\overline{AB}$ to $\odot O$ at point B; $m\angle A = m\angle B$
Prove: $m\overparen{BD} = 2 \cdot m\overparen{BC}$
Proof: $m\angle BCD = m\angle A + m\angle B$; but since $m\angle A = m\angle B$, $m\angle BCD = m\angle B + m\angle B$ or $m\angle BCD = 2 \cdot m\angle B$. $m\angle BCD$ also equals $\frac{1}{2}m\overparen{BD}$ since it is an inscribed $\angle$. Therefore, $\frac{1}{2}m\overparen{BD} = 2 \cdot m\angle B$ or $m\overparen{BD} = 4 \cdot m\angle B$. But if $\overline{AB}$ is a tangent to $\odot O$ at B, $m\angle B = \frac{1}{2}m\overparen{BC}$. By substitution, $m\overparen{BD} = 4(\frac{1}{2}m\overparen{BC})$ or $m\overparen{BD} = 2 \cdot m\overparen{BC}$.

6.3 EXERCISES

1. 30 **3.** $6\sqrt{5}$ **7.** 3
9. $DE = 4$ and $EC = 12$ or $DE = 12$ and $EC = 4$ **11.** 4
13. $9\frac{2}{5}$ **15.** 9 **17.** $5\frac{1}{3}$ **19.** $3 + 3\sqrt{5}$
21. (a) None (b) One (c) 4
27. Yes; $\overline{AE} \cong \overline{CE}$; $\overline{DE} \cong \overline{EB}$ **29.** $AM = 5$; $PC = 7$; $BN = 9$
31. 12 **33.** 45

6.3 Selected Proofs

23. If $\overline{AF}$ is a tangent to $\odot O$ and $\overline{AC}$ is a secant to $\odot O$, then $(AF)^2 = AC \cdot AB$. If $\overline{AF}$ is a tangent to $\odot Q$ and $\overline{AE}$ is a secant to $\odot Q$, then $(AF)^2 = AE \cdot AD$. By substitution, $AC \cdot AB = AE \cdot AD$.

25. In $\odot Q$, if tangents $\overline{MN}$ and $\overline{MP}$ are $\perp$, then $\angle M$ is a right $\angle$. $\angle$s N and P are right $\angle$s since a radius drawn to the point of tangency is $\perp$ to the tangent. Therefore $\overline{QN} \parallel \overline{PM}$ and $\overline{NM} \parallel \overline{QP}$. Hence, $QNMP$ is a parallelogram. But since $MNQP$ has a right angle and two adjacent congruent sides, $(\overline{QN} \cong \overline{QP})$ it is also a square.

35. *Given:* $\overleftrightarrow{AB}$ contains O, the center of the circle and $\overleftrightarrow{AB}$ contains M, the midpoint of $\overline{RS}$

Prove: $\overleftrightarrow{AB} \perp \overline{RS}$

Proof: If M is the midpoint of $\overline{RS}$ in $\odot O$, then $\overline{RM} \cong \overline{MS}$. Draw $\overline{RO}$ and $\overline{OS}$, which are $\cong$ since they are radii in the same circle. Using $\overline{OM} \cong \overline{OM}$, $\triangle ROM \cong \triangle SOM$ by SSS. By CPCTC, $\angle OMS \cong \angle OMR$ and hence $\overleftrightarrow{AB} \perp \overline{RS}$.

6.4 Exercises

7. $\overline{AB}$; $\overline{GH}$; for a circle containing unequal chords, the chord nearest the center has the greatest length and the chord at the greatest distance from the center has the least length.
9. (a) $\overline{OT}$ (b) $\overline{OD}$
11. (a) $m\widehat{MN} > m\widehat{QP}$ (b) $m\widehat{MPN} < m\widehat{PMQ}$
13. Obtuse
15. (a) $m\angle AOB > m\angle BOC$ (b) $AB > BC$
17. (a) $m\widehat{AB} > m\widehat{BC}$ (b) $AB > BC$
19. (a) $\angle C$ (b) $\overline{AC}$
21. $\overline{AB}$ is $(4\sqrt{3} - 4\sqrt{2})$ closer than $\overline{CD}$. **27.** 7

6.5 Exercises

9. The locus of points at a given distance from a fixed line is two parallel lines on either side of the fixed line at the same (given) distance from the fixed line.
11. The locus of points at a distance of 3 inches from point O is a circle with center O and radius 3.
13. The locus of points equidistant from the three noncollinear points D, E, and F is the circumcenter of $\triangle DEF$.
15. The locus of the midpoints of the chords in $\odot Q$ parallel to diameter $\overline{PR}$ is the perpendicular bisector of $\overline{PR}$.
17. The locus of points equidistant from two given intersecting lines are two perpendicular lines that bisect the angles formed by the two intersecting lines.
23. The locus of points at a distance of 2 cm from a sphere whose radius is 5 cm are two concentric spheres with the same center. The radius of one sphere is 3 cm and the radius of the other sphere is 7 cm.
25. The locus is another sphere with the same center and a radius of length 2.5 m.
27. The locus of points equidistant from an 8-ft ceiling and the floor is a parallel plane in the middle of the ceiling and floor.

6.6 Exercises

1. (a) Angle bisectors (b) Perpendicular bisectors of sides (c) Altitudes (d) Medians **3.** No
5. Equilateral triangle **7.** Midpoint of the hypotenuse
17. No **19.** $\frac{10\sqrt{3}}{3}$ **21.** $RQ = 10$; $SQ = \sqrt{89}$
23. (a) 4 (b) 6 (c) 10.5 **27.** Equilateral
29. (a) Yes (b) Yes **31.** (a) Yes (b) No

Review Exercises

1. 9 mm **2.** 30 cm **3.** $\sqrt{41}$ in. **4.** $6\sqrt{2}$ cm
5. 130
6. 35 **7.** 80 **8.** 35
9. $m\widehat{AC} = m\widehat{DC} = 93\frac{1}{3}$; $m\widehat{AD} = 173\frac{1}{3}$
10. $m\widehat{AC} = 110$ and $m\widehat{AD} = 180$
11. $m\angle 2 = 44$; $m\angle 3 = 90$; $m\angle 4 = 46$; $m\angle 5 = 44$
12. $m\angle 1 = 50$; $m\angle 2 = 40$; $m\angle 3 = 90$; $m\angle 4 = 50$
13. 24 **14.** 10 **15.** A **16.** S **17.** N **18.** S
19. A **20.** N **21.** A **22.** N
23. (a) 70 (b) 28 (c) 64 (d) $m\angle P = 21$ (e) $m\widehat{AB} = 90$; $m\widehat{CD} = 40$ (f) 260 **24.** (a) 3 (b) 8 (c) 16 (d) 4 (e) 4 (f) 8 or 1 (g) $3\sqrt{5}$ (h) 3 (i) $4\sqrt{3}$ (j) 3
25. 29
26. If $x = 7$, then $AC = 35$; $DE = 17\frac{1}{2}$. If $x = -4$, then $AC = 24$; $DE = 12$.
30. $m\angle 1 = 93$; $m\angle 2 = 25$; $m\angle 3 = 43$; $m\angle 4 = 68$; $m\angle 5 = 90$; $m\angle 6 = 22$; $m\angle 7 = 68$; $m\angle 8 = 22$; $m\angle 9 = 50$; $m\angle 10 = 112$ **31.** $24\sqrt{2}$ cm **32.** $15 + 5\sqrt{3}$cm
33. 14 cm and 15 cm **34.** $AD = 3$; $BE = 6$; $FC = 7$
35. (a) $AB > CD$ (b) $QP < QR$ (c) $m\angle A < m\angle C$
39. The locus of the midpoints of the radii of a circle is a concentric circle with radius half the length of the given radius.
40. The locus of the centers of all circles passing through two given points is the perpendicular bisector of the segment joining the two given points.
41. The locus of the centers of a penny that rolls around a half-dollar is a circle.

42. The locus of points in space less than three units from a given point is the interior of a sphere.
43. The locus of points equidistant from two parallel planes is a parallel plane midway between the two planes.
50. (a) 12 (b) 2 (c) $2\sqrt{3}$ **51.** $BF = 6$; $AE = 9$

Review Exercises Selected Proofs

27. *Proof:* If $\overline{DC}$ is tangent to circles B and A at points D and C, then $\overline{BD} \perp \overline{DC}$ and $\overline{AC} \perp \overline{DC}$. $\angle$s D and C are congruent since they are right angles. $\angle DEB \cong \angle CEA$ because of vertical angles. $\triangle BDE \sim \triangle ACE$ by AA. It follows that $\frac{AC}{CE} = \frac{BD}{ED}$ since corresponding sides are proportional. Hence, $AC \cdot ED = CE \cdot BD$.

28. *Proof:* In $\odot O$, if $\overline{EO} \perp \overline{BC}$, $\overline{DO} \perp \overline{BA}$, and $\overline{EO} \cong \overline{OD}$, then $\overline{BC} \cong \overline{BA}$. (Chords equidistant from the center of the circle are congruent.) It follows then that $\overline{BC} \cong \overline{BA}$.

29. *Proof:* If $\overline{AP}$ and $\overline{BP}$ are tangent to $\odot Q$ at A and B, then $\overline{AP} \cong \overline{BP}$. $\overline{AC} \cong \overline{BC}$ since C is the midpoint of $\overline{AB}$. It follows that $\overline{AC} \cong \overline{BC}$ and, using $\overline{CP} \cong \overline{CP}$, we have $\triangle APC \cong \triangle BCP$ by SSS. $\angle APC \cong \angle BPC$ by CPCTC and hence $\overrightarrow{PC}$ bisects $\angle APB$.

Chapter 7

7.1 Exercises

1. Two triangles with equal areas are not necessarily congruent. Two squares with equal areas must be congruent because the sides are congruent. **3.** 37 units^2 **5.** The altitudes to $\overline{PN}$ and to $\overline{MN}$ are congruent. This is because $\triangle$s QMN and QPN are congruent; corresponding altitudes of $\cong \triangle$s are $\cong$ **7.** 54 cm^2 **9.** 9 m^2 **11.** 72 in.^{2}
13. 100 in.^{2} **15.** 126 in.^{2} **17.** 264 units^2
19. 144 units^2 **21.** $\frac{27}{4}\sqrt{3}$ units^2 **23.** 192 ft^2
25. (a) 300 ft^2 (b) 3 gallons (c) \$46.50
27. $156 + 24\sqrt{10}$ ft^2 **29.** (a) 9 sq ft = 1 sq yd (b) 1296 sq in. = 1 sq yd **31.** $\overline{MN}$ joins the midpoints of $\overline{CA}$ and $\overline{CB}$ so $MN = \frac{1}{2}(AB)$. Therefore $\overline{AP} \cong \overline{PB} \cong \overline{MN}$. $\overline{PN}$ joins the midpoints of $\overline{CB}$ and $\overline{AB}$ so $PN = \frac{1}{2}(AC)$. Therefore $\overline{AM} \cong \overline{MC} \cong \overline{PN}$. $\overline{MP}$ joins the midpoints of $\overline{AB}$ and $\overline{AC}$ so $MP = \frac{1}{2}(BC)$. Therefore $\overline{CN} \cong \overline{NB} \cong \overline{MP}$. The four triangles are all $\cong$ by SSS. Therefore, the area of each triangle is the same. Hence, the area of the big triangle is equal to four times the area of one of the smaller triangles.
35. 8 in. **37.** (a) 12 in. (b) 84 in.^{2} **39.** 56%
41. By the Area-Addition Postulate, $A_{R \cup S} = A_R + A_S$. Now $A_{R \cup S}$, A_R, and A_S are all positive numbers. Let p represent the area of region S, so that $A_{R \cup S} = A_R + p$. By the definition of inequality, $A_R < A_{R \cup S}$ or $A_{R \cup S} > A_R$.
43. $(a + b)(c + d) = ac + ad + bc + bd$ **45.** $4\frac{8}{13}$ in.
47. 8 **49.** (a) 10 (b) 26 (c) 18 (d) No

7.1 Selected Proof

33. *Proof:* $A = (LH)(HJ) = s^2$. By the Pythagorean Theorem, $s^2 + s^2 = d^2$.

$$2s^2 = d^2$$
$$s^2 = \frac{d^2}{2}$$
Thus $$A = \frac{d^2}{2}.$$

7.2 Exercises

1. 30 in. **3.** $4\sqrt{29}$ m **5.** $7\sqrt{6} + 7\sqrt{3} + 41$ **7.** 38
9. 84 in.^{2} **11.** 40 ft^2 **13.** 80 units^2
15. $36 + 36\sqrt{3}$ units^2 **17.** 16 in., 32 in., and 28 in.
19. 15 cm **21.** (a) $\frac{9}{4}$ (b) $\frac{4}{1}$ **25.** $24 + 4\sqrt{21}$ units^2
27. 96 units^2 **29.** 6 yd by 8 yd **31.** (a) 770 ft (b) \$454.30 **33.** 624 ft^2 **35.** Square with sides of length 10 in. **37.** (a) 52 units (b) 169 units^2
39. (a) No (b) Yes

7.2 Selected Proofs

23. Using Heron's Formula, the semiperimeter is $\frac{1}{2}(3s)$ or $\frac{3s}{2}$. Then

$$A = \sqrt{\frac{3s}{2}\left(\frac{3s}{2} - s\right)\left(\frac{3s}{2} - s\right)\left(\frac{3s}{2} - s\right)}$$
$$A = \sqrt{\frac{3s}{2}\left(\frac{s}{2}\right)\left(\frac{s}{2}\right)\left(\frac{s}{2}\right)}$$
$$A = \sqrt{\frac{3s^4}{16}} = \frac{\sqrt{3} \cdot \sqrt{s^4}}{\sqrt{16}}$$
$$A = \frac{s^2\sqrt{3}}{4}$$

41. The area of a trapezoid $= \frac{1}{2}h(b_1 + b_2) = h \cdot \frac{1}{2}(b_1 + b_2)$. The length of the median of a trapezoid is $m = \frac{1}{2}(b_1 + b_2)$. By substitution, the area of a trapezoid is $A = hm$.

7.3 Exercises

1. First, construct the angle bisectors of two consecutive angles, say A and B. The point of intersection, O, is the center of the inscribed circle.

Second, construct the line segment $\overline{OM}$ perpendicular to $\overline{AB}$. Then, using the radius $r = OM$, construct the inscribed circle with center O.

3. Draw the diagonals (angle bisectors) $\overline{JL}$ and $\overline{MK}$. These determine center O of the inscribed circle. Now construct the line segment $\overline{OR} \perp \overline{MJ}$. Use OR as the length of the radius of the inscribed circle.

9. $a = 5$ in.; $r = 5\sqrt{2}$ in. **11.** $16\sqrt{3}$ ft; 16 ft.
13. (a) 120° (b) 90° (c) 72° (d) 60°
15. $54\sqrt{3}$ cm^2 **17.** $75\sqrt{3}$ in.^{2} **19.** $(24 + 12\sqrt{3})$ in.^{2}
21. $\frac{2}{1}$ **23.** 168°

7.3 Selected Proof

25. *Proof:* Let $M, N, P,$ and Q be the points of tangency for $\overline{DC}, \overline{DA}, \overline{AB},$ and $\overline{BC}$, respectively. Because the tangent segments from an external point are congruent, $AP = AN$, $PB = BQ$, $CM = CQ$, and $MD = DN$. Thus $AP + PB + CM + MD = AN + BQ + CQ + DN$.

Reordering and associating,
$(AP + PB) + (CM + MD) = (AN + DN) + (BQ + CQ)$ or
$AB + CD = DA + BC$.

7.4 Exercises

1. $C = 16\pi$ cm; $A = 64\pi$ cm^2 **3.** $C = 66$ in.; $A = 346\frac{1}{2}$ in.^{2}
5. (a) $r = 22$ in.; $d = 44$ in. (b) $r = 30$ ft; $d = 60$ ft
7. (a) $r = 5$ in.; $d = 10$ in. (b) $r = 1.5$ cm; $d = 3.0$ cm
9. $\frac{8}{3}\pi$ in. **11.** $C \approx 77.79$ in. **13.** $r \approx 6.7$ cm
15. $\ell \approx 7.33$ in. **17.** 16 in.^{2} **19.** $5 < AN < 13$
21. $(32\pi - 64)$ in.^{2} **23.** $(600 - 144\pi)$ ft^2 **25.** ≈ 7 cm
27. 8 in.
29. $A = A_{\text{LARGER CIRCLE}} - A_{\text{SMALLER CIRCLE}}$

$A = \pi R^2 - \pi r^2$

$A = \pi(R^2 - r^2)$

But $R^2 - r^2$ is a difference of two squares, so $A = \pi(R + r)(R - r)$.

31. 3 in. and 4 in. **33.** (a) ≈ 201.06 ft^2
(b) 2.87 pints. Thus, 3 pints need to be purchased.
(c) \$8.85
35. (a) ≈ 1256 ft^2 (b) 20.93. Thus, 21 pounds of seed are needed. (c.) \$34.65 **37.** ≈ 43.98 cm
39. ≈ 14.43 in. **41.** $\approx 27{,}488.94$ mi

7.5 Exercises

1. 34 in. **3.** 150 cm^2 **5.** $\frac{3}{2}rs$ **7.** 54 mm **9.** 24 in.^{2}
11. 1 in. **13.** $P = 16 + \frac{8}{3}\pi$ in. and $A = \frac{32}{3}\pi$ in.^{2}
15. ≈ 30.57 in. **17.** $P = (12 + 4\pi)$ in.; $A = (24\pi - 36\sqrt{3})$ in.^{2} **19.** $\left(25\sqrt{3} - \frac{25}{2}\pi\right)$ cm^2
21. $\frac{9}{2}$ cm **23.** 36π **25.** 90°
27. Cut the pizza into 8 slices. **29.** $A = \left(\frac{\pi}{2}\right)s^2 - s^2$
31. $r = 3\frac{1}{3}$ ft or 3 ft 4 in. **35.** (a) 3 (b) 2
37. $\frac{308\pi}{3} \approx 322.54$ in.^{2}

Review Exercises

1. 480 **2.** (a) 40 (b) $40\sqrt{3}$ (c) $40\sqrt{2}$ **3.** 50
4. 204 **5.** 336 **6.** 36 **7.** (a) $24\sqrt{2} + 18$
(b) $24 + 9\sqrt{3}$ (c) $33\sqrt{3}$ **8.** $A = 216$ in.^{2}; $P = 60$ in.
9. (a) 19,000 ft^2 (b) 4 bags (c) \$72
10. (a) 3 double rolls (b) 3 rolls
11. a. $\frac{289}{4}\sqrt{3} + 8\sqrt{33}$ (b) $50 + \sqrt{33}$ **12.** 168
13. 5 cm by 7 cm **14.** (a) 15 cm, 25 cm, and 20 cm
(b) 150 cm^2 **15.** 36 **16.** $36\sqrt{3}$ cm^2 **17.** 20
18. (a) 72° (b) 108° (c) 72° **19.** $96\sqrt{3}$ ft^2
20. 6 in. **21.** $162\sqrt{3}$ in.^{2} **22.** (a) 8 (b) ≈ 120 cm^2
23. (a) No. $\perp$ bisectors of sides of a parallelogram are not necessarily concurrent. (b) No. $\perp$ bisectors of sides of a rhombus are not necessarily concurrent.
(c) Yes. $\perp$ bisectors of sides of a rectangle are concurrent.
(d) Yes. $\perp$ bisectors of sides of a square are concurrent.
24. (a) No. $\angle$ bisectors of a parallelogram are not necessarily concurrent. (b) Yes. $\angle$ bisectors of a rhombus are concurrent. (c) No. $\angle$ bisectors of a rectangle are not necessarily concurrent.
(d) Yes. $\angle$ bisectors of a square are concurrent.
25. $147\sqrt{3} \approx 254.61$ in.^{2} **26.** (a) 312 ft^2 (b) 35 yd^2
(c) \$348.95 **27.** $64 - 16\pi$ **28.** $\frac{49}{2}\pi - \frac{49}{2}\sqrt{3}$
29. $\frac{8}{3}\pi - 4\sqrt{3}$ **30.** $288 - 72\pi$ **31.** $25\sqrt{3} - \frac{25}{3}\pi$
32. $\ell = \frac{2\pi\sqrt{5}}{3}$ cm; $A = 5\pi$ cm^2 **33.** (a) 21 ft
(b) $\approx 346\frac{1}{2}$ ft^2 **34.** (a) 6π ft^2 (b) $\left(6\sqrt{3} + \frac{4\pi}{3}\sqrt{3}\right)$ ft
35. $(9\pi - 18)$ in.^{2} **39.** (a) ≈ 28 (b) ≈ 21.2 ft^2
40. (a) ≈ 905 ft^2 (b) \$162.90
(c) Approximately 151 flowers

Review Exercises Selected Proof

36. *Proof:* By an earlier theorem,

$$\begin{aligned} A_{\text{RING}} &= \pi R^2 - \pi r^2 \\ &= \pi(OC)^2 - \pi(OB)^2 \\ &= \pi[(OC)^2 - (OB)^2] \end{aligned}$$

In rt. $\triangle OBC$,

$$(OB)^2 + (BC)^2 = (OC)^2$$

Thus $(OC)^2 - (OB)^2 = (BC)^2$.

In turn, $A_{\text{RING}} = \pi(BC)^2$.

CHAPTER 8

8.1 EXERCISES

1. (a) Yes (b) Oblique (c) Hexagon (d) Oblique hexagonal prism (e) Parallelogram **3.** (a) 12 (b) 18 (c) 8 **5.** (a) cm^2 (b) cm^3 **7.** 132 cm^2 **9.** 120 cm^3 **11.** (a) 16 (b) 8 (c) 16 **13.** (a) $2n$ (b) n (c) $2n$ (d) $3n$ (e) n (f) 2 (g) $n + 2$ **15.** (a) 671.6 cm^2 (b) 961.4 cm^2 (c) 2115.54 cm^3 **17.** (a) 72 ft^2 (b) 84 ft^2 (c) 36 ft^3 **19.** 6 in. by 6 in. by 3 in. **21.** $x = 3$ **23.** $4.44 **25.** 640 ft^3 **27.** (a) $T = L + 2B$, $T = hP + 2(e \cdot e)$, $T = e(4e) + 2e^2$, $T = 4e^2 + 2e^2$, $T = 6e^2$ (b) 96 cm^2 (c) $V = Bh$, $V = e^2 \cdot e$, $V = e^3$ (d) 64 cm^3 **31.** $128 **33.** 720 cm^2 **35.** 2952 cm^3

8.2 EXERCISES

1. (a) Right pentagonal prism (b) Oblique pentagonal prism **3.** (a) Regular square pyramid (b) Oblique square pyramid **5.** (a) Pyramid (b) E (c) $\overline{EA}, \overline{EB}, \overline{EC}, \overline{ED}$ (d) $\triangle EAB, \triangle EBC, \triangle ECD, \triangle EAD$ (e) No **7.** (a) 5 (b) 8 (c) 5 **9.** 66 $in.^2$ **11.** 32 cm^3 **13.** (a) $n + 1$ (b) n (c) n (d) $2n$ (e) n (f) $n + 1$ **15.** 4 in. **17.** (a) 144.9 cm^2 (b) 705.18 cm^3 **19.** (a) 60 ft^2 (b) 96 ft^2 (c) 48 ft^3 **21.** $36\sqrt{5} + 36 \approx 116.5$ $in.^2$ **23.** 480 ft^2 **25.** 900 ft^3 **27.** ≈ 24 ft **29.** 336 $in.^3$

8.3 EXERCISES

1. (a) $60\pi \approx 188.5$ $in.^2$ (b) $110\pi \approx 345.58$ $in.^2$ (c) $150\pi \approx 471.24$ $in.^3$ **3.** ≈ 54.19 $in.^2$ **5.** 5 cm **7.** The radius has a length of 2 in. and the altitude has a length of 3 in. **9.** $32\pi \approx 100.53$ $in.^3$ **11.** $2\sqrt{13} \approx 7.21$ cm **13.** 2 m **15.** $4\sqrt{3} \approx 6.93$ in. **17.** $3\sqrt{5} \approx 6.71$ cm **19.** (a) $6\pi\sqrt{85} \approx 173.78$ $in.^2$ (b) $6\pi\sqrt{85} + 36\pi \approx 286.88$ $in.^2$ (c) $84\pi \approx 263.89$ $in.^3$ **21.** 54π $in.^3$ **23.** 2000π cm^3 **25** 1200π cm^3 **27.** $65\pi \approx 204.2$ cm^2 **29.** $192\pi \approx 603.19$ $in.^3$ **33.** $60\pi \approx 188.5$ $in.^2$ **35.** $\frac{4}{1}$ or 4:1 **37.** ≈ 471.24 gal **41.** ≈ 290.60 cm^3

8.4 EXERCISES

1. Polyhedron *EFGHIJK* is concave.
3. Polyhedron *EFGHIJK* has nine faces (F), seven vertices (V), and 14 edges (E). $V + F = E + 2$
$7 + 9 = 14 + 2$
5. A regular hexahedron has six faces (F), eight vertices (V), and 12 edges (E). $V + F = E + 2$
$8 + 6 = 12 + 2$
7. (a) $\frac{1}{2}$ (b) $\frac{5}{12}$ (c) $\frac{5}{6}$ **9.** (a) $6\sqrt{2} \approx 8.49$ in. (b) $6\sqrt{3} \approx 10.39$ in. **11.** (a) $\frac{3}{2}$ or 3:2 (b) $\frac{3}{2}$ or 3:2 **13.** $r = 3\sqrt{2} \approx 4.24$ in.; $h = 6\sqrt{2} \approx 8.49$ in. **15.** (a) $3\sqrt{3} \approx 5.20$; (b) 9 in. **17.** (a) $36\pi \approx 113.1$ m^2 (b) $36\pi \approx 113.1$ m^3 **19.** 1.5 in. **21.** 113.1 ft^2; ≈ 3 pints **23.** $7.4\pi \approx 23.24$ $in.^3$ **25.** $S = 36\pi$ $units^2$; $V = 36\pi$ $units^3$

REVIEW EXERCISES

1. 672 $in.^2$ **2.** 297 cm^2 **3.** Dimensions are 6 in. by 6 in. by 20 in.; $V = 720$ $in.^3$ **4.** $T = 468$ cm^2; $V = 648$ cm^3 **5.** (a) 360 $in.^2$ (b) 468 $in.^2$ (c) 540 $in.^3$ **6.** (a) 624 cm^2 (b) $624 + 192\sqrt{3} \approx 956.55$ cm^2 (c) $1248\sqrt{3} \approx 2161.6$ cm^3 **7.** $\sqrt{89} \approx 9.43$ cm **8.** $3\sqrt{7} \approx 7.94$ in. **9.** $\sqrt{74} \approx 8.60$ in. **10.** $2\sqrt{3} \approx 3.46$ cm **11.** (a) 540 $in.^2$ (b) 864 $in.^2$ (c) 1296 $in.^3$ **12.** (a) $36\sqrt{19} \approx 156.92$ cm^2 (b) $36\sqrt{19} + 36\sqrt{3} \approx 219.27$ cm^2 (c) $96\sqrt{3} \approx 166.28$ cm^3 **13.** (a) 120π $in.^2$ (b) 192π $in.^2$ (c) 360π $in.^3$ **14.** (a) ≈ 351.68 ft^3 (b) ≈ 452.16 ft^2 **15.** (a) $72\pi \approx 226.19$ cm^2 (b) $108\pi \approx 339.29$ cm^2 (c) $72\pi\sqrt{3} \approx 391.78$ cm^3 **16.** $\ell = 10$ **17.** ≈ 616 $in.^2$ **18.** ≈ 904.32 cm^3 **19.** 120π $units^3$ **20.** $\frac{\text{surface area of smaller}}{\text{surface area of larger}} = \frac{1}{9}$; $\frac{\text{volume of smaller}}{\text{volume of larger}} = \frac{1}{27}$ **21.** $\approx 183\frac{1}{3}$ $in.^3$ **22.** 288π cm^3 **23.** $\frac{32\pi}{3}$ $in.^3$ **24.** ≈ 1017.36 $in.^3$ **25.** $(2744 - \frac{1372}{3}\pi)$ $in.^3$ **26.** (a) An octahedron has eight faces that are equilateral triangles. (b) A tetrahedron has four faces that are equilateral triangles. (c) A dodecahedron has twelve faces that are regular pentagons. **27.** 40π mm^3
28. (a) $V = 16, E = 24, F = 10$ $V + F = E + 2$
$16 + 10 = 24 + 2$
(b) $V = 4, E = 6, F = 4$ $V + F = E + 2$
$4 + 4 = 6 + 2$
(c) $V = 6, E = 12, F = 8$ $V + F = E + 2$
$6 + 8 = 12 + 2$
29. 114 $in.^3$

CHAPTER 9

9.1 EXERCISES

3. (a) 4 (b) 8 (c) 5 (d) 9
5. $b = 3.5$ or $b = 10.5$
7. (a) 5 (b) 10 (c) $2\sqrt{5}$ (d) $\sqrt{a^2 + b^2}$
9. (a) $\left(2, \frac{-3}{2}\right)$ (b) (1, 1) (c) (4, 0) (d) $\left(\frac{a}{2}, \frac{b}{2}\right)$
11. (a) $(-3, 4)$ (b) $(0, -2)$ (c) $(-a, 0)$
(d) $(-b, -c)$ **13.** (a) $(-3, -4)$ (b) $(-2, 0)$
(c) $(-a, 0)$ (d) $(-b, c)$ **15.** $(2.5, -13.7)$
17. (2, 3); 16 **19.** (a) Isosceles (b) Equilateral
(c) Isosceles right triangle
21. $x + y = 6$ **23.** $(a, a\sqrt{3})$ or $(a, -a\sqrt{3})$
25. $(0, 1 + 3\sqrt{3})$ and $(0, 1 - 3\sqrt{3})$ **27.** 17 **29.** 9
31. (a) 135π units^3 (b) 75π units^3
33. (a) 96π units^3 (b) 144π units^3
35. (a) 90π units^2 (b) 90π units^2

9.2 EXERCISES

1. (4, 0) and (0, 3) **3.** (5, 0) and $\left(0, -\frac{5}{2}\right)$ **5.** $(-3, 0)$
7. (6, 0) and (0, 3) **9.** (a) 4 (b) Undefined (c) -1
(d) 0 (e) $\frac{d - b}{c - a}$ (f) $-\frac{b}{a}$ **11.** (a) 10
(b) 15 **13.** (a) Collinear (b) Noncollinear
15. (a) $\frac{3}{4}$ (b) $-\frac{5}{3}$ (c) -2 (d) $\frac{a - b}{c}$
17. None of these **19.** Perpendicular **21.** $\frac{3}{2}$
23. 23 **31.** Right triangle **33.** (4, 7); (0, -1); (10, -3)
37. $m_{\overline{EH}} = \frac{2c - 0}{2b - 0} = \frac{2c}{2b} = \frac{c}{b}$
$m_{\overline{FG}} = \frac{c - 0}{(a + b) - a} = \frac{c}{b}$
Due to equal slopes, $\overline{EH} \parallel \overline{FG}$. Thus, $EFGH$ is a trapezoid.

9.2 SELECTED PROOF

35. $m_{\overline{VT}} = \dfrac{e - e}{(c - d) - (a + d)} = \dfrac{0}{c - a - 2d} = 0$

$m_{\overline{RS}} = \dfrac{b - b}{c - a} = \dfrac{0}{c - a} = 0$

$\therefore \overline{VT} \parallel \overline{RS}$

$RV = \sqrt{[(a + d) - a]^2 + (e - b)^2}$
$= \sqrt{d^2 + (e - b)^2}$
$= \sqrt{d^2 + e^2 - 2be + b^2}$

$ST = \sqrt{[c - (c - d)]^2 + (b - e)^2}$
$= \sqrt{(d)^2 + (b - e)^2}$
$= \sqrt{d^2 + b^2 - 2be + e^2}$

$\therefore RV = ST$

$RSTV$ is an isosceles trapezoid.

9.3 EXERCISES

1. $x + 2y = 6$; $y = -\frac{1}{2}x + 3$
3. $-x + 3y = -40$; $y = \frac{1}{3}x - \frac{40}{3}$ **11.** $2x + 3y = 15$
13. $x + y = 6$ **15.** $-2x + 3y = -3$ **17.** $-x + y = -2$
19. $5x + 2y = 5$ **21.** $4x + 3y = -12$
23. $-x + 3y = 2$ **25.** (6, 0) **27.** (5, -4)
29. (6, -2) **31.** (3, 2) **33.** (6, -1) **35.** (5, -2)
37. $a = 2$; $b = 3$ **39.** (3, -2); $k = 5$

9.3 SELECTED PROOF

41. For $B \neq 0$, the equation $Ax + By = C$ is equivalent to

$$By = -Ax + C \text{ or } y = -\frac{A}{B}x + \frac{C}{B}$$

For $B \neq 0$, the equation $Ax + By = D$ is equivalent to $y = -\frac{A}{B}x + \frac{D}{B}$. Since $m_1 = m_2 = -\frac{A}{B}$, the graphs (lines) are parallel.

NOTE: If $B = 0$, both graphs (lines) are vertical and parallel.

9.4 EXERCISES

1. (a) $a\sqrt{2}$ if $a > 0$ (b) $\frac{d - b}{c - a}$
3. (a) $y = -1x + a$ (b) $y = -\frac{b}{a}x + b$
5. $\overline{AB}$ is horizontal and $\overline{BC}$ is vertical; $\therefore \overline{AB} \perp \overline{BC}$. Hence, $\angle B$ is a right $\angle$ and $\triangle ABC$ is a right triangle.
7. $m_{\overline{QM}} = \frac{c - 0}{b - 0} = \frac{c}{b}$
$m_{\overline{PN}} = \frac{c - 0}{(a + b) - a} = \frac{c}{b}$
$\therefore \overline{QM} \parallel \overline{PN}$
$m_{\overline{QP}} = \frac{c - c}{(a + b) - b} = \frac{0}{a} = 0$
$m_{\overline{MN}} = \frac{0 - 0}{a - 0} = \frac{0}{a} = 0$
$\therefore \overline{QP} \parallel \overline{MN}$
Since both pairs of opposite sides are parallel, $MNPQ$ is a parallelogram.
9. $m_{\overline{MN}} = 0$ and $m_{\overline{QP}} = 0$; $\therefore \overline{MN} \parallel \overline{QP}$. $\overline{QM}$ and $\overline{PN}$ are both vertical; $\therefore \overline{QM} \parallel \overline{PN}$. Hence, $MQPN$ is a parallelogram. Since $\overline{QM}$ is vertical and $\overline{MN}$ is horizontal, $\angle QMN$ is a right angle. Because parallelogram $MQPN$ has a right $\angle$, it is also a rectangle.
11. $A = (0, 0)$; $B = (a, 0)$; $C = (a, b)$
13. $M = (0, 0)$; $N = (r, 0)$; $P = (r + s, t)$
15. $A = (0, 0)$; $B = (a, 0)$; $C = (a - c, d)$
17. (a) Square
$A = (0, 0)$; $B = (a, 0)$; $C = (a, a)$; $D = (0, a)$
(b) Square (with midpoints of sides)
$A = (0, 0)$; $B = (2a, 0)$; $C = (2a, 2a)$; $D = (0, 2a)$
19. (a) Parallelogram
$A = (0, 0)$; $B = (a, 0)$; $C = (a + b, c)$; $D = (b, c)$
NOTE: D chosen before C

(b) Parallelogram (with midpoints of sides)
$A = (0, 0)$; $B = (2a, 0)$; $C = (2a + 2b, 2c)$; $D = (2b, 2c)$

21. (a) Isosceles triangle
$R = (0, 0)$; $S = (2a, 0)$; $T = (a, b)$
(b) Isosceles triangle (with midpoints)
$R = (0, 0)$; $S = (4a, 0)$; $T = (2a, 2b)$

23. $r^2 = s^2 + t^2$ **25.** $c^2 = a^2 - b^2$ **27.** $b^2 = 3a^2$

29. (a) Positive (b) Negative (c) $2a$

31. (a) Slope Formula (b) Distance Formula (c) Midpoint Formula (d) Slope Formula **37.** $\left(c, \frac{ac}{b}\right)$

9.5 Exercises

23. True. The quadrilateral which results is a parallelogram.

9.5 Selected Proofs

3. The diagonals of a square are perpendicular bisectors of each other.

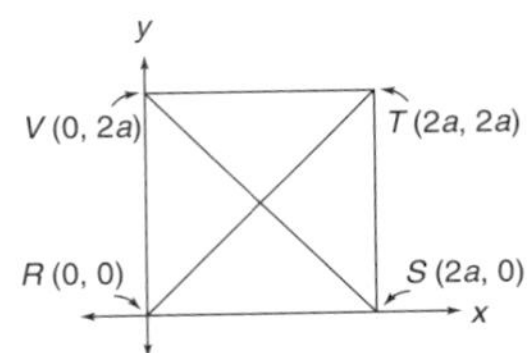

Proof: Let square $RSTV$ have the vertices shown. Then the midpoints of the diagonals are $M_{\overline{RT}} = (a, a)$ and $M_{\overline{VS}} = (a, a)$. Also, $m_{\overline{RT}} = 1$ and $m_{\overline{VS}} = -1$. Because the two diagonals share the midpoint (a, a) and the product of their slopes is -1, they are perpendicular bisectors of each other.

7. The segments that join the midpoints of the consecutive sides of a quadrilateral form a parallelogram.

Proof: The midpoints, as shown, of the sides of quadrilateral $ABCD$ are

$$R = \left(\frac{0 + 2a}{2}, \frac{0 + 0}{2}\right) = (a, 0)$$

$$S = \left(\frac{2a + 2b}{2}, \frac{0 + 2c}{2}\right) = (a + b, c)$$

$$T = \left(\frac{2d + 2b}{2}, \frac{2e + 2c}{2}\right) = (d + b, e + c)$$

$$V = \left(\frac{0 + 2d}{2}, \frac{0 + 2e}{2}\right) = (d, e)$$

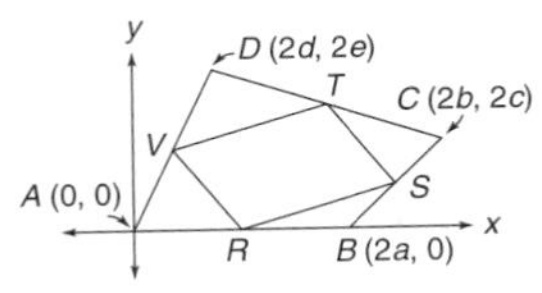

Now we determine slopes as follows.

$$m_{\overline{RS}} = \frac{c - 0}{(a + b) - a} = \frac{c}{b}$$

$$m_{\overline{ST}} = \frac{(e + c) - c}{(d + b) - (a + b)} = \frac{e}{d - a}$$

$$m_{\overline{TV}} = \frac{(e + c) - e}{(d + b) - d} = \frac{c}{b}$$

$$m_{\overline{VR}} = \frac{e - 0}{d - a} = \frac{e}{d - a}$$

Because $m_{\overline{RS}} = m_{\overline{TV}}$, $\overline{RS} \parallel \overline{TV}$. Also $m_{\overline{ST}} = m_{\overline{VR}}$ so $\overline{ST} \parallel \overline{VR}$. Then $RSTV$ is a parallelogram.

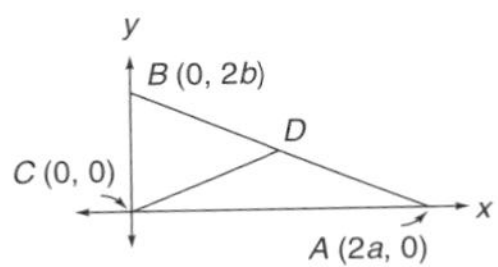

11. The midpoint of the hypotenuse of a right triangle is equidistant from the three vertices of the triangle.

Proof: Let right $\triangle ABC$ have vertices as shown. Then D, the midpoint of the hypotenuse, is given by

$$D = \left(\frac{0 + 2a}{2}, \frac{2b + 0}{2}\right) = (a, b)$$

Now $$BD = DA = \sqrt{(2a - a)^2 + (0 - b)^2} = \sqrt{a^2 + (-b)^2} = \sqrt{a^2 + b^2}$$

Also, $$CD = \sqrt{(a - 0)^2 + (b - 0)^2} = \sqrt{a^2 + b^2}$$

Then D is equidistant from A, B, and C.

15. If the midpoint of one side of a rectangle is joined to the endpoints of the opposite sides, an isosceles triangle is formed.

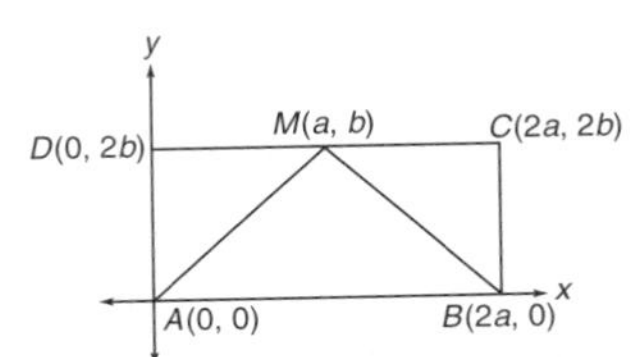

Proof: Let rectangle $ABCD$ have endpoints as shown on page A49. With M the midpoint of $\overline{DC}$,

$$M = \left(\frac{0 + 2a}{2}, \frac{2b + 2b}{2}\right) = (a, b)$$
$$MA = \sqrt{(a - 0)^2 + (b - 0)^2}$$
$$MA = \sqrt{a^2 + b^2}$$
$$MB = \sqrt{(a - 2a)^2 + (b - 0)^2}$$
$$MB = \sqrt{a^2 + b^2}$$

Since $MA = MB$, $\triangle AMB$ is isosceles.

19. The altitudes of a triangle are concurrent.

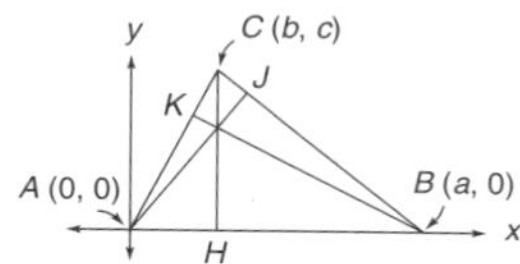

Proof: For $\triangle ABC$, let $\overline{CH}$, $\overline{AJ}$, and $\overline{BK}$ name the altitudes. Because $\overline{AB}$ is horizontal ($m_{\overline{AB}} = 0$), $\overline{CH}$ is vertical and has the equation $x = b$.

Because $m_{\overline{BC}} = \frac{c - 0}{b - a} = \frac{c}{b - a}$, the slope of altitude $\overline{AJ}$ is $m_{\overline{AJ}} = -\frac{b - a}{c} = \frac{a - b}{c}$. Since $\overline{AJ}$ contains (0, 0), its equation is $y = \frac{a - b}{c}x$.

The intersection of altitudes $\overline{CH}$ ($x = b$) and $\overline{AJ}$ $\left(y = \frac{a - b}{c}x\right)$ is at $x = b$ so that $y = \frac{a - b}{c} \cdot b = \frac{b(a - b)}{c} = \frac{ab - b^2}{c}$. That is, $\overline{CH}$ and $\overline{AJ}$ intersect at $\left(b, \frac{ab - b^2}{c}\right)$. The remaining altitude is $\overline{BK}$. Since $m_{\overline{AC}} = \frac{c - 0}{b - 0} = \frac{c}{b}$, $m_{\overline{BK}} = -\frac{b}{c}$. Because $\overline{BK}$ contains $(a, 0)$, its equation is $y - 0 = -\frac{b}{c}(x - a)$ or $y = \frac{-b}{c}(x - a)$.

For the three altitudes to be concurrent, $\left(b, \frac{ab - b^2}{c}\right)$ must lie on the line $y = \frac{-b}{c}(x - a)$. Substitution leads to

$$\frac{ab - b^2}{c} = \frac{-b}{c}(b - a)$$
$$\frac{ab - b^2}{c} = \frac{-b(b - a)}{c}$$
$$\frac{ab - b^2}{c} = \frac{-b^2 + ab}{c}, \text{ which is true.}$$

Then the three altitudes are concurrent.

9.6 Exercises

1. $(x - 0)^2 + (y - 0)^2 = 5^2$
Center (0, 0); $r = 5$
3. $(x - 2)^2 + [y - (-3)]^2 = 3^2$
Center $(2, -3)$; $r = 3$
5. $(x - 1)^2 + [y - (-2)]^2 = 2^2$
Center $(1, -2)$; $r = 2$
7. The radius equals 0; thus $x^2 + y^2 = 0$ represents the point (0, 0). **9.** The radius would be $\sqrt{-4}$; no graph possible. **11.** (5, 0) and $(-4, 3)$
13. There is no point of intersection.
15. (5, 0) and $(-4, 3)$ **17.** $(-2, 4)$ and (1, 1)
19. $x^2 + y^2 - 4x + 10y + 13 = 0$
21. $x^2 + y^2 + 6x + 2y - 15 = 0$
23. $x^2 + y^2 - 2x + 2y - 23 = 0$
25. $x^2 + y^2 + 10x - 10y + 25 = 0$ **27.** $3x + 4y = 25$
29. Because the tangent lines have slopes $-\frac{3}{4}$ and $\frac{4}{3}$, they are perpendicular. **31.** 2 **33.** (2, 3)

Review Exercises

1. (a) 7 (b) 6 (c) 13 (d) 5
2. (a) 8 (b) 10 (c) $4\sqrt{5}$ (d) 10
3. (a) $\left(6, \frac{1}{2}\right)$ (b) $(-2, 4)$ (c) $\left(1, -\frac{1}{2}\right)$ (d) $\left(\frac{2x - 3}{2}, y\right)$
4. (a) (2, 1) (b) $(-2, -2)$ (c) (0, 3) (d) $(x + 1, y + 1)$
5. (a) Undefined (b) 0 (c) $-\frac{5}{12}$ (d) $-\frac{4}{3}$
6. (a) Undefined (b) 0 (c) $\frac{1}{2}$ (d) $\frac{4}{3}$
7. $(-4, -8)$ **8.** (3, 7) **9.** $x = \frac{4}{3}$ **10.** $y = -4$
11. (a) Perpendicular (b) Parallel (c) Neither (d) Perpendicular **12.** Not collinear
13. $x = 4$ **14.** (7, 0) and (0, 3)
17. (a) $3x + 5y = 21$ (b) $3x + y = -7$ (c) $-2x + y = -8$ (d) $y = 5$ (e) $x^2 + y^2 - 6x + 2y - 6 = 0$ (f) $x^2 + y^2 - 4x + 6y + 9 = 0$
18. $m_{\overline{AB}} = \frac{4}{3}$, $m_{\overline{BC}} = \frac{1}{2}$; $m_{\overline{AC}} = -2$. Because $m_{\overline{AC}} \cdot m_{\overline{BC}} = -1$, $\overline{AC} \perp \overline{BC}$ and $\angle C$ is a rt. $\angle$.
19. $AB = \sqrt{85}$; $BC = \sqrt{85}$. Because $AB = BC$, the triangle is isosceles.
20. $m_{\overline{RS}} = \frac{-4}{3}$; $m_{\overline{ST}} = \frac{5}{6}$; $m_{\overline{TV}} = \frac{-4}{3}$; $m_{\overline{RV}} = \frac{5}{6}$. Therefore $\overline{RS} \parallel \overline{VT}$ and $\overline{RV} \parallel \overline{ST}$ and $RSTV$ is a parallelogram.
21. (a) Circle: $C = (0, 0)$; $r = 7$ (b) Circle: $C = (1, -3)$; $r = 5$ (c) Circle: $C = (2, -5)$; $r = 3$
22. (a) Circle: $C = (0, 0)$; $r = 3\sqrt{2}$ (b) Circle: $C = (-2, 1)$; $r = 4$ (c) Circle: $C = (-5, -1)$; $r = 7$
23. (3, 5) **24.** $(6, -2)$ **25.** (1, 4) and (3, 12)
26. $(0, -6)$ **27.** (3, 5) **28.** $(6, -2)$
29. (1, 4), (3, 12) **30.** $(0, -6)$
31. $x^2 + y^2 - 10x - 6y + 9 = 0$ **32.** $4x + 3y = 54$
33. (16, 11), $(4, -9)$, $(-4, 5)$ **34.** $C = (4, -7)$; $r = 10$
35. $C = (-5, 6)$; $r = \sqrt{51}$ **36.** a. $\sqrt{53}$ b. -4 c. $\frac{1}{4}$

37. $A = (-a, 0)$; $B = (0, b)$; $C = (a, 0)$
38. $D = (0, 0)$; $E = (a, 0)$; $F = (a, 2a)$; $G = (0, 2a)$
39. $R = (0, 0)$; $U = (0, a)$; $T = (a, a + b)$
40. $M = (0, 0)$; $N = (a, 0)$; $Q = (a + b, c)$; $P = (b, c)$
41. (a) $\sqrt{(a + c)^2 + (b + d - 2e)^2}$ (b) $-\frac{a}{b - e}$ or $\frac{a}{e - b}$
(c) $y - 2d = \frac{a}{e - b}(x - 2c)$

CHAPTER 10

10.1 EXERCISES

1. $\sin \alpha = \frac{5}{13}$; $\sin \beta = \frac{12}{13}$ **3.** $\sin \alpha = \frac{8}{17}$; $\sin \beta = \frac{15}{17}$
5. $\sin \alpha = \frac{\sqrt{15}}{5}$; $\sin \beta = \frac{\sqrt{10}}{5}$ **7.** 1 **9.** 0.2924
11. 0.9903 **13.** 0.9511 **15.** $a \approx 6.9$ in.; $b \approx 9.8$ in.
17. $a \approx 10.9$ ft; $b \approx 11.7$ ft **19.** $c \approx 8.8$ cm; $d \approx 28.7$ cm
21. $\alpha \approx 29°$; $\beta \approx 61°$ **23.** $\alpha \approx 17°$; $\beta \approx 73°$
25. $\alpha \approx 19°$; $\beta \approx 71°$ **27.** $\alpha \approx 23°$ **29.** $d \approx 103.5$ ft
31. $d \approx 128.0$ ft **33.** $\alpha \approx 24°$
35. (a) ≈ 5.4 ft (b) ≈ 54 ft^2 **37.** $\theta \approx 50°$

10.2 EXERCISES

1. $\cos \alpha = \frac{12}{13}$; $\cos \beta = \frac{5}{13}$ **3.** $\cos \alpha = \frac{3}{5}$; $\cos \beta = \frac{4}{5}$
5. $\cos \alpha = \frac{\sqrt{10}}{5}$; $\cos \beta = \frac{\sqrt{15}}{5}$
7. (a) $\sin \alpha = \frac{a}{c}$; $\cos \beta = \frac{a}{c}$. Thus $\sin \alpha = \cos \beta$.
(b) $\cos \alpha = \frac{b}{c}$; $\sin \beta = \frac{b}{c}$. Thus $\cos \alpha = \sin \beta$. **9.** 0.9205
11. 0.9563 **13.** 0 **15.** 0.1392
17. $a \approx 84.8$ ft; $b \approx 53.0$ ft **19.** $a = b = 5$ cm
21. $c \approx 19.1$ in.; $d \approx 14.8$ in. **23.** $\alpha = 60°$; $\beta = 30°$
25. $\alpha \approx 51°$; $\beta \approx 39°$ **27.** $\alpha \approx 65°$; $\beta \approx 25°$
29. $\theta \approx 34°$ **31.** $x \approx 1147.4$ ft **33.** ≈ 8.1 in.
35. ≈ 13.1 cm **37.** $\alpha \approx 55°$
39. (a) $m\angle A = 68°$; (b) $m\angle B = 112°$

10.3 EXERCISES

1. $\tan \alpha = \frac{3}{4}$; $\tan \beta = \frac{4}{3}$ **3.** $\tan \alpha = \frac{\sqrt{5}}{2}$; $\tan \beta = \frac{2\sqrt{5}}{5}$
5. $\sin \alpha = \frac{5}{13}$; $\cos \alpha = \frac{12}{13}$; $\tan \alpha = \frac{5}{12}$; $\cot \alpha = \frac{12}{5}$; $\sec \alpha = \frac{13}{12}$;
$\csc \alpha = \frac{13}{5}$
7. $\sin \alpha = \frac{a}{c}$; $\cos \alpha = \frac{b}{c}$; $\tan \alpha = \frac{a}{b}$; $\cot \alpha = \frac{b}{a}$; $\sec \alpha = \frac{c}{b}$;
$\csc \alpha = \frac{c}{a}$
9. $\sin \alpha = \frac{x\sqrt{x^2 + 1}}{x^2 + 1}$; $\cos \alpha = \frac{\sqrt{x^2 + 1}}{x^2 + 1}$; $\tan \alpha = \frac{x}{1}$; $\cot \alpha = \frac{1}{x}$;
$\sec \alpha = \sqrt{x^2 + 1}$; $\csc \alpha = \frac{\sqrt{x^2 + 1}}{x}$ **11.** 0.2679
13. 1.5399 **15.** $x \approx 7.5$; $z \approx 14.2$
17. $y \approx 5.3$; $z \approx 8.5$ **19.** $d \approx 8.1$
21. $\alpha \approx 37°$; $\beta \approx 53°$ **23.** $\theta \approx 56°$; $\gamma \approx 34°$
25. $\alpha \approx 29°$; $\beta \approx 61°$ **27.** 1.4826 **29.** 2.0000
31. 1.3456 **33.** ≈ 1376.8 ft **35.** ≈ 4.1 in.

37. $\approx 72°$
39. $\alpha \approx 47°$. The heading may be described as N 47° W.
41. $\approx 26{,}730$ ft
43. (a) $h \approx 9.2$ units (b) $V \approx 110.4$ units^3

10.4 EXERCISES

1. $\frac{\cos \theta}{\sin \theta} = \frac{\frac{b}{c}}{\frac{a}{c}} = \frac{b}{c} \cdot \frac{c}{a} = \frac{b}{a} = \cot \theta$
3. $\tan \theta = 0.75$ or $\frac{3}{4}$; $\cot \theta \approx 1.3333$ or $\frac{4}{3}$ **5.** $\cos \theta = \frac{4}{5}$
7. $\sec \theta = \frac{13}{12}$ **11.** 8 in.^{2} **13.** ≈ 11.6 ft^2
15. ≈ 15.2 ft^2 **17.** ≈ 11.1 in. **19.** ≈ 8.9 m
21. $\approx 55°$ **23.** $\approx 51°$ **25.** ≈ 10.6 **27.** ≈ 6.9
29. (a) ≈ 213.4 ft (b) $\approx 13{,}294.9$ ft^2 **31.** ≈ 8812 m
33. ≈ 15.9 ft **35.** 6 **37.** ≈ 14.0 ft

REVIEW EXERCISES

1. sine; ≈ 10.3 in. **2.** sine; ≈ 7.5 ft
3. cosine; ≈ 23.0 in. **4.** sine; ≈ 5.9 ft
5. tangent; $\approx 43°$ **6.** cosine; $\approx 58°$ **7.** sine; $\approx 49°$
8. tangent; $\approx 16°$ **9.** ≈ 8.9 units **10.** $\approx 60°$
11. ≈ 13.1 units **12.** ≈ 18.5 units **13.** ≈ 42.7 ft
14. ≈ 74.8 cm **15.** $\approx 47°$ **16.** $\approx 54°$
17. ≈ 26.3 in.^{2}
19. If $m\angle S = 30$ and $m\angle Q = 90$, then the sides of $\triangle RQS$ can be represented by $RQ = x$, $RS = 2x$, and $SQ = x\sqrt{3}$. $\sin S = \sin 30° = \frac{x}{2x} = \frac{1}{2}$. **21.** ≈ 8.4 ft **22.** ≈ 866 ft
23. $\approx 41°$ **24.** $\approx 8°$ **25.** ≈ 5.0 cm
26. ≈ 4.3 cm **27.** $\approx 68°$ **28.** $\approx 106°$
29. 3 to 7 (or 3:7) **30.** ≈ 1412.0 m
31. $\cos \theta = \frac{24}{25}$; $\sec \theta = \frac{25}{24}$ **32.** $\sec \theta = \frac{61}{60}$; $\cot \theta = \frac{60}{11}$
33. $\csc \theta = \frac{29}{20}$; $\sin \theta = \frac{20}{29}$ **34.** $h \approx 6.9$ ft; $V \approx 74.0$ ft^3

APPENDIX A:

QUADRATIC EQUATIONS

1. (a) 3.61 (b) 2.83 (c) −5.39 (d) 0.77
3. a, c, d, f **5.** (a) $2\sqrt{2}$ (b) $3\sqrt{5}$ (c) 30 (d) 3
7. (a) $\frac{3}{4}$ (b) $\frac{5}{7}$ (c) $\frac{\sqrt{7}}{4}$ (d) $\frac{\sqrt{6}}{3}$
9. (a) $\sqrt{54} \approx 7.35$ and $3\sqrt{6} \approx 7.35$
(b) $\sqrt{\frac{5}{16}} \approx 0.56$ and $\frac{\sqrt{5}}{4} \approx 0.56$
11. $x = 4$ or $x = 2$ **13.** $x = 12$ or $x = 5$
15. $x = -\frac{2}{3}$ or $x = 4$ **17.** $x = \frac{1}{3}$ or $x = \frac{1}{2}$
19. $x = 5$ or $x = 2$ **21.** $x = \frac{7 \pm \sqrt{13}}{2} \approx 5.30$ or 1.70
23. $x = 2 \pm 2\sqrt{3} \approx 5.46$ or -1.46

25. $x = \frac{3 \pm \sqrt{149}}{10} \approx 1.52$ or -0.92
27. $x = \pm\sqrt{7} \approx \pm 2.65$
29. $x = \pm\frac{5}{2}$ **31.** $x = 0$ or $x = \frac{b}{a}$ **33.** 5 by 8
35. $n = 6$ **37.** $c = 5$

APPENDIX B:

TRUTH TABLES

1. F **3.** T **5.** F **7.** T
9. Hamburgers are not health food.
11. If Mary is an accountant, then hamburgers are health food.

13.

P	$\sim P$	$P \vee \sim P$
T	F	T
F	T	T

Yes, a tautology.

15.

P	Q	$P \vee Q$	$(P \vee Q) \rightarrow P$
T	T	T	T
T	F	T	T
F	T	T	F
F	F	F	T

No, not a tautology.

17.

P	Q	$P \rightarrow Q$	$(P \rightarrow Q) \wedge Q$	$[(P \rightarrow Q) \wedge Q] \rightarrow P$
T	T	T	T	T
T	F	F	F	T
F	T	T	T	F
F	F	T	F	T

No, not a tautology.

19. $\sim P \vee \sim Q$

21. Mary is not an accountant and hamburgers are not health food.
23. It is not cold or it is not snowing.
25. *Prove:* $[\sim(P \vee Q)] \leftrightarrow [\sim P \wedge \sim Q]$

P	Q	$P \vee Q$	$\sim(P \vee Q)$	$\sim P$	$\sim Q$	$\sim P \wedge \sim Q$
T	T	T	F	F	F	F
T	F	T	F	F	T	F
F	T	T	F	T	F	F
F	F	F	T	T	T	T

Because the fourth and seventh columns match, the equivalence is established.

27. Show that $[(P \to Q)] \wedge (Q \to R)] \to (P \to R)$ is a tautology.

P	Q	R	$P \to Q$	$Q \to R$	$[(P \to Q) \wedge (Q \to R)]$	$P \to R$	$[(P \to Q) \wedge (Q \to R)] \to (P \to R)$
T	T	T	T	T	T	T	T
T	T	F	T	F	F	F	T
T	F	T	F	T	F	T	T
T	F	F	F	T	F	F	T
F	T	T	T	T	T	T	T
F	T	F	T	F	F	T	T
F	F	T	T	T	T	T	T
F	F	F	T	T	T	T	T

Because the final column consists only of T's, the given statement is a tautology.

29. Show $[P \wedge \sim Q)]$ is the negative of $P \to Q$.

P	Q	$\sim Q$	$P \wedge \sim Q$	$P \to Q$
T	T	F	F	T
T	F	T	T	F
F	T	F	F	T
F	F	T	F	T

The fourth and fifth columns have opposite truth values.

31. I am good and I cannot go to the movie.
33. I studied hard and made an A and I cannot be a member of Phi Theta Kappa.

Appendix C:

Valid Arguments

1. The sum of the measures of ∠s 1 and 2 is 90 degrees.
3. Tina will have a good time.
5. ∠1 and ∠2 are not complementary.
7. Tom finished the job.
9. If Izzi lives in Chicago, then she lives in the Midwest.
11. If Ken Travis gets a hit, then I will be happy.
13. Valid **15.** Not valid **17.** Law of Syllogism
19. Law of Detachment
21. (a) $[(P \vee Q) \wedge \sim Q] \rightarrow P$

(b)

P	Q	$P \vee Q$	$\sim Q$	$(P \vee Q) \wedge \sim Q$	$[(P \vee Q) \wedge \sim Q] \rightarrow P$
T	T	T	F	F	T
T	F	T	T	T	T
F	T	T	F	F	T
F	F	F	T	F	T

23. Mary's family will visit at Christmas.
25. Law of Detachment: $[(P \rightarrow Q) \wedge P] \rightarrow Q$

P	Q	$P \rightarrow Q$	$(P \rightarrow Q) \wedge P$	$[(P \rightarrow Q) \wedge P] \rightarrow Q$
T	T	T	T	T
T	F	F	F	T
F	T	T	F	T
F	F	T	F	T

27. Law of Syllogism: $[(P \rightarrow Q) \wedge (Q \rightarrow R)] \rightarrow (P \rightarrow R)$

P	Q	R	$P \rightarrow Q$	$Q \rightarrow R$	$(P \rightarrow Q) \wedge (Q \rightarrow R)$	$P \rightarrow R$	$[(P \rightarrow Q) \wedge (Q \rightarrow R)] \rightarrow (P \rightarrow R)$
T	T	T	T	T	T	T	T
T	T	F	T	F	F	F	T
T	F	T	F	T	F	T	T
T	F	F	F	T	F	F	T
F	T	T	T	T	T	T	T
F	T	F	T	F	F	T	T
F	F	T	T	T	T	T	T
F	F	F	T	T	T	T	T

Glossary

acute angle. an angle whose measure is between 0° and 90°

acute triangle. a triangle whose three interior angles are all acute

adjacent angles. two angles that have a common vertex and a common side between them

altitude of plane figure. a line segment drawn perpendicularly from a vertex or side of a parallelogram, rhombus, or trapezoid to the opposite side; the length of the altitude is the height of the plane figure

altitude of triangle. a line segment drawn perpendicularly from a vertex of the triangle to the opposite side of the triangle; the length of the altitude is the height of the triangle

angle. the plane figure formed by two rays that share a common endpoint

angle bisector. *see* bisector of angle

angle of depression (elevation). acute angle formed by a horizontal ray and a ray determined by a downward (an upward) rotation

apothem of regular polygon. any line segment drawn from the center of the regular polygon perpendicular to one of its sides

arc. the segment (part) of a circle determined by two points on the circle and all points between them

area. the measurement in square units of the amount of region within an enclosed plane figure

auxiliary line. a line (or part of a line) added to a drawing to help complete a proof or solve a problem

axiom. *see* postulate

base. a side (of plane figure) or face (of solid figure) to which an altitude is drawn

base angles of isosceles triangle. the two congruent angles of the isosceles triangle

base of isosceles triangle. the side of the triangle whose length is unique

bases of trapezoid. the two parallel sides of the trapezoid

bisector of angle. a ray that separates the given angle into two smaller congruent angles

center of circle. the interior point of the circle whose distance from all points on the circle is the same

center of regular polygon. the common center of the inscribed and circumscribed circles of the regular polygon

center of sphere. the interior point of the sphere whose distance from all points on the sphere is the same

central angle of circle. an angle whose vertex is at the center of the circle and whose sides are radii of the circle

central angle of regular polygon. an angle whose vertex is at the center of the regular polygon and whose sides are two consecutive radii of the polygon

centroid of triangle. the point determined by the intersection of the three medians of the triangle

chord of circle. any line segment that joins two points on the circle

circle. the set of points in a plane that are at a fixed distance from a point (the center of the circle) in the plane

circumcenter of triangle. the center of the circumscribed circle of a triangle; the point determined by the intersection of the perpendicular bisectors of the three sides of the triangle

circumference. the linear measure of the distance around a circle

circumscribed circle. a circle that contains all vertices of a polygon such that the sides of the polygon are chords of the circle

circumscribed polygon. a polygon whose sides are all tangent to a circle in the interior of the polygon

collinear points. points that lie on the same line

common tangent. a line (or segment) that is tangent to more than one circle; can be a common external tangent or a common internal tangent

complementary angles. two angles whose sum of measures is 90°

concave polygon. a polygon in which at least one diagonal lies in the exterior of the polygon

concentric circles (spheres). used to describe two circles (spheres) having the same center

conclusion. the "then" clause of an "If, then" statement; the part of a theorem indicating the claim to be proved

concurrent lines. three or more lines that contain the same point

congruent. used to describe figures (such as angles) that can be made to coincide

converse. relative to the statement "If *P*, then *Q*," this statement has the form "If *Q*, then *P*"

convex polygon. a polygon in which all diagonals lie in the interior of the polygon

coplanar points. points that lie in the same plane

corollary. a theorem that follows from another theorem as a "by-product;" a theorem that is easily proved as the consequence of another theorem

cosecant. in a right triangle, the ratio $\frac{\text{hypotenuse}}{\text{opposite}}$

cosine. in a right triangle, the ratio $\frac{\text{adjacent}}{\text{hypotenuse}}$

cotangent. in a right triangle, the ratio $\frac{\text{adjacent}}{\text{opposite}}$

cylinder (circular). the solid generated by using parallel line segments to join each point of one circle to each point of a second circle that lies in a plane parallel to that of the first circle

decagon. a polygon with exactly ten sides

deduction. a form of reasoning in which conclusions are reached through the use of established principles

degree. the unit of measure that corresponds to $\frac{1}{360}$ of a complete revolution

diagonal of polygon. a line segment that joins two nonconsecutive vertices of a polygon

diameter. any line segment that joins two points on a circle (or sphere) and contains the center of the circle (or sphere)

dodecagon. a polygon that has exactly twelve sides

dodecahedron (regular). a polyhedron that has exactly twelve faces that are congruent regular pentagons

edge of polyhedron. any line segment that joins two consecutive vertices of the polyhedron (includes prisms and pyramids)

equiangular polygon. a type of polygon whose angles are congruent (equal)

equilateral polygon. a type of polygon whose sides are congruent (equal)

extended proportion. a proportion that has three or more members, such as $\frac{a}{b} = \frac{c}{d} = \frac{e}{f}$.

extended ratio. a ratio that compares three or more numbers, such as $a:b:c$

exterior. refers to all points that lie outside an enclosed (bounded) plane or solid figure

exterior angle of polygon. an angle formed by one side of a polygon and an extension of a second side having a common endpoint with the first side

extremes of a proportion. the first and last terms of a proportion; in $\frac{a}{b} = \frac{c}{d}$, a and d are the extremes

face of polyhedron. any one of the polygons that lies in a plane determined by the vertices of the polyhedron; includes base(s) and lateral faces of prisms and pyramids

geometric mean. the repeated second and third terms of certain proportions; in $\frac{a}{b} = \frac{b}{c}$, b is the geometric mean of a and c

height. *see* altitude

heptagon. a polygon that has exactly seven sides

hexagon. a polygon that has exactly six sides

hexahedron (regular). a polyhedron that has six congruent square faces; also called a cube

hypotenuse of a right triangle. the side of a right triangle that lies opposite the right angle

hypothesis. the "if" clause of an "If, then" statement; the part of a theorem providing the given information

icosahedron (regular). a polyhedron with twenty faces that are equilateral triangles

incenter of triangle. the center of the inscribed circle of a triangle; the point determined by the intersection of the three angle bisectors of the angles of the triangle

induction. a form of reasoning in which a number of specific observations are used to draw a general conclusion

inscribed angle of circle. an angle whose vertex is on a circle and whose sides are chords of the circle

inscribed circle. a circle that lies inside a polygon such that the sides of the polygon are tangents of the circle

inscribed polygon. a polygon whose vertices all lie on a circle such that the sides of the polygon are chords of the circle

intercepted arc. the arc (an arc) of a circle that is cut off in the interior of an angle

intercepts. the points at which the graph of an equation intersects the axes

interior. refers to all points that lie inside an enclosed (bounded) plane or solid figure

interior angle of polygon. any angle formed by two sides of the polygon such that the angle lies in the interior of the polygon

intersection. the points shared in common by two geometric figures

intuition. drawing a conclusion through insight

inverse. relative to the statement "If *P*, then *Q*," this statement has the form "If not *P*, then not *Q*"

isosceles trapezoid. a trapezoid that has two congruent legs (its nonparallel sides)

isosceles triangle. a triangle that has two congruent sides

kite. a quadrilateral which has two distinct pairs of congruent adjacent sides

lateral area. the sum of areas of the faces of a solid excluding the base area(s) (as in prisms, pyramids, cylinders, and cones)

legs of an isosceles triangle. the two congruent sides of the triangle

legs of a right triangle. the two sides that form the right angle of the triangle

legs of a trapezoid. the two nonparallel sides of the trapezoid

lemma. a theorem that is introduced and proved so that a later theorem can be proved

line of centers. the line (or line segment) that joins the centers of two circles

line segment. the part of a line determined by two points and all points on the line that lie between those two points

locus. the set of all points that satisfy a given condition or conditions

major arc. an arc whose measure is between 180° and 360°

mean proportional. *see* geometric mean

means of a proportion. the second and third terms of a proportion; in $\frac{a}{b} = \frac{c}{d}$, b and c are the means

median of trapezoid. the line segment that joins the midpoints of the two legs (nonparallel sides) of the trapezoid

median of triangle. the line segment joining a vertex of the triangle to the midpoint of the opposite side

midpoint. the point on a line segment (or arc) that separates the line segment (arc) into two congruent parts

minor arc. an arc whose measure is between 0° and 180°

nonagon. a polygon that has exactly nine sides

noncollinear points. three or more points that do not lie on the same line

noncoplanar points. four or more points that do not lie in the same plane

obtuse angle. an angle whose measure is between 90° and 180°

obtuse triangle. a triangle that has exactly one interior obtuse angle

octagon. a polygon that has exactly eight sides

octahedron (regular). a polyhedron with eight congruent faces that are equilateral triangles

opposite rays. two rays having a common endpoint that together form a line

orthocenter of triangle. the point determined by the intersection of the three altitudes of the triangle

parallel lines (planes). two lines in a plane (or two planes) that do not intersect

parallelogram. a quadrilateral that has two pairs of parallel sides

pentagon. a polygon that has exactly five sides

perimeter of polygon. the sum of lengths of the sides of the polygon

perpendicular bisector of a line segment. a line (or part of a line) that is both perpendicular to and bisects a given line segment

perpendicular lines. two lines that intersect to form congruent adjacent angles

pi (π). the constant ratio of the circumference of a circle to the length of its diameter; this ratio is commonly approximated by the fraction $\frac{22}{7}$ or the decimal 3.1416

point of tangency (contact). the point at which a tangent to a circle touches the circle

polygon. a plane figure whose sides are line segments that intersect only at their endpoints

polyhedron. a solid figure whose faces are polygons that intersect other faces along common sides of the polygons

postulate. a statement that is assumed true but is not proved

Quadratic Formula. the formula $x = \frac{-b \pm \sqrt{b^2 - 4ac}}{2a}$, which provides solutions for the equation $ax^2 + bx + c = 0$, where a, b, and c are real numbers and $a \neq 0$

quadrilateral. a polygon that has exactly four sides

radian. the measure of a central angle of a circle whose intercepted arc has length equal to the radius of the circle

radius. the line segment that joins the center of a circle (or sphere) to any point on the circle (or sphere)

ratio. a comparison between two quantities a and b, generally written $\frac{a}{b}$ or $a{:}b$

ray. the part of a line that begins at a point and extends infinitely far in one direction

rectangle. a parallelogram that contains a right angle

regular polygon. a polygon whose sides are congruent and whose interior angles are congruent

regular polyhedron. a polyhedron whose edges are congruent and whose faces are congruent

regular prism. a prism whose bases are regular polygons

regular pyramid. a pyramid whose base is a regular polygon and whose lateral faces are congruent isosceles triangles

rhombus. a parallelogram with two congruent adjacent sides

right angle. an angle whose measure is exactly 90°

right circular cone. a cone in which the line segment joining the vertex to the center of the circular base is perpendicular to the base

right circular cylinder. a cylinder in which the line segment joining the centers of the circular bases is perpendicular to the plane of each base

right prism. a prism in which lateral edges are perpendicular to the base edges they intersect

right triangle. a triangle in which exactly one of the interior angles is a right angle

scalene triangle. a triangle in which no two sides are congruent

secant. in a right triangle, the ratio $\frac{\text{hypotenuse}}{\text{adjacent}}$

secant of circle. a line (or part of a line) that intersects a circle at two points

sector of circle. the plane region bounded by two radii of the circle and the arc that is intercepted by the central angle formed by those radii

segment of circle. the plane region bounded by a chord and a minor arc (major arc) that has the same endpoints as that chord

semicircle. the arc of a circle determined by a diameter; an arc of a circle whose measure is exactly 180°

similar polygons. polygons that have the same shape

sine. in a right triangle, the ratio $\frac{\text{opposite}}{\text{hypotenuse}}$

skew quadrilateral. a quadrilateral whose sides do not all lie in one plane

slant height of cone. any line segment joining the vertex of the cone to a point on the circular base

slant height of regular pyramid. a line segment joining the vertex of the pyramid to the midpoint of a base edge of the pyramid

slope. a measure of the steepness of a line; in the rectangular coordinate system, the slope m of the line through (x_1, y_1) and (x_2, y_2) is $m = \frac{y_2 - y_1}{x_2 - x_1}$

sphere. the set of points in space that are at a fixed distance from a point (the center of the sphere)

straight angle. an angle whose measure is exactly 180°; an angle whose sides are opposite rays

straightedge. an idealized instrument used to construct parts of lines

supplementary angles. two angles whose sum of measures is 180°

surface area. the measure of the total area (lateral area plus base area) of any solid figure

tangent. in a right triangle, the ratio $\frac{\text{opposite}}{\text{adjacent}}$

tangent circles. two circles that have one point in common; the circles may be externally tangent or internally tangent

tangent of circle. a line (or part of a line) that touches a circle at only one point

tetrahedron (regular). a four-faced solid in which the faces are congruent equilateral triangles

theorem. a statement that follows logically from previous definitions and principles; a statement that can be proved

torus. a three dimensional solid that has a "doughnut" shape

transversal. a line that intersects two or more lines, intersecting each at one point

trapezoid. a quadrilateral having exactly two parallel sides

triangle. a polygon that has exactly three sides

triangle inequality. a statement that the sum of the lengths of two sides of a triangle cannot be greater than the length of the third side

union. the joining together of geometric figures

valid argument. an argument in which the conclusion follows logically from previously stated (and accepted) premises or assumptions

vertex angle of isosceles triangle. the angle formed by the two congruent sides of the triangle

vertex of angle. the point at which the two sides of the angle meet

vertex of isosceles triangle. the point at which the two congruent sides of the triangle meet

vertex of polygon. any point at which two sides of the polygon meet

vertex of polyhedron. any point at which two edges of the polyhedron meet

vertical angles. a pair of angles that lie in opposite positions when formed by two intersecting lines

volume. the measurement in cubic units of the amount of space within a bounded region of space

Index

Abbreviations

AA	angle-angle (proves △s similar)
ASA	angle-side-angle (proves △s congruent)
AAS	angle-angle-side (proves △s congruent)
add.	addition
adj.	adjacent
alt.	altitude, alternate
ax.	axiom
cm	centimeters
cm^2	square centimeters
cm^3	cubic centimeters
comp.	complementary
corr.	corresponding
cos	cosine
cot	cotangent
CPCTC	Corresponding parts of congruent triangles are congruent.
csc	cosecant
CSSTP	Corresponding sides of similar triangles are proportional.
diag.	diagonal
exs.	exercises
ext.	exterior
eq.	equality
ft	foot (or feet)
gal	gallon
HL	hypotenuse-leg (proves △s congruent)
hr	hour
in.	inch (or inches)
ineq.	inequality
int.	interior
isos.	isosceles
km	kilometers
m	meters
mi	miles
mm	millimeters
n-gon	polygon of n sides
opp.	opposite
pent.	pentagon
post.	postulate
prop.	property
pt.	point
quad.	quadrilateral
rect.	rectangle
rt.	right
SAS	side-angle-side (proves △s congruent)
sec, sec.	secant, section
sin	sine
SSS	side-side-side (proves △s congruent)
st.	straight
supp.	supplementary
tan	tangent
trans.	transversal
trap.	trapezoid
vert.	vertical (angles)
yd	yards